Advances in Modal Logic

Volume 9

Advances in Modal Logic
Volume 9

Edited by

Thomas Bolander,
Torben Braüner,
Silvio Ghilardi
and
Lawrence Moss

ISBN 978-1-84890-068-4

College Publications
Scientific Director: Dov Gabbay
Managing Director: Jane Spurr
Department of Computer Science
King's College London, Strand, London WC2R 2LS, UK

http://www.collegepublications.co.uk

Original cover design by Richard Fraser, Avalon Arts, UK
Printed by Lightning Source, Milton Keynes, UK

Contents

Preface vii

Steve Awodey and Kohei Kishida
Topological Completeness of First-Order Modal Logic 1

Franz Baader, Stefan Borgwardt and Barbara Morawska
Computing Minimal $\mathcal{EL}$-unifiers is Hard 18

Philippe Balbiani, Hans van Ditmarsch, Andreas Herzig and Tiago de Lima
Some Truths Are Best Left Unsaid 36

Philippe Balbiani and Stanislav Kikot
Sahlqvist Theorems for Precontact Logics 55

Philippe Balbiani and Levan Uridia
Completeness and Definability of a Modal Logic Interpreted over Iterated Strict Partial Orders 71

Lev Beklemishev
Calibrating Provability Logic: From Modal Logic to Reflection Calculus 89

Johan van Benthem
Foundational Issues in Logical Dynamics 95

Johan van Benthem, David Fernández-Duque and Eric Pacuit
Evidence Logic: A New Look at Neighborhood Structures 97

Marta Bílková, Rostislav Horčík and Jiří Velebil
Distributive Substructural Logics as Coalgebraic Logics over Posets .. 119

Lars Birkedal
First Steps in Synthetic Guarded Domain Theory 143

Patrick Blackburn and Klaus Frovin Jørgensen
Indexical Hybrid Tense Logic 144

Facundo Carreiro and Stéphane Demri
Beyond Regularity for Presburger Modal Logics 161

BALDER TEN CATE
Guarded Negation 183

DAVID FERNÁNDEZ-DUQUE AND JOOST J. JOOSTEN
Kripke Models of Transfinite Provability Logic 185

DAVID FERNÁNDEZ-DUQUE
Non-finite Axiomatizability of Dynamic Topological Logic 200

TIM FRENCH, JOHN M^{C}CABE-DANSTED AND MARK REYNOLDS
Synthesis for Temporal Logic over the Reals 217

PATRICK GIRARD, JEREMY SELIGMAN AND FENRONG LIU
General Dynamic Dynamic Logic 239

STEFAN GÖLLER, ARNE MEIER, MARTIN MUNDHENK, THOMAS SCHNEIDER, MICHAEL THOMAS AND FELIX WEISS
The Complexity of Monotone Hybrid Logics over Linear Frames and the Natural Numbers 261

RAJEEV GORÉ AND REVANTHA RAMANAYAKE
Labelled Tree Sequents, Tree Hypersequents and Nested (Deep) Sequents 279

DANIEL GORÍN AND LUTZ SCHRÖDER
Extending $\mathcal{ALCQ}$ with Bounded Self-Reference 300

JAMES HALES, TIM FRENCH AND ROWAN DAVIES
Refinement Quantified Logics of Knowledge and Belief for Multiple Agents 317

CHRISTOPHER HAMPSON AND AGI KURUCZ
On Modal Products with the Logic of 'Elsewhere' 339

WESLEY H. HOLLIDAY, TOMOHIRO HOSHI AND THOMAS F. ICARD, III
A Uniform Logic of Information Dynamics 348

KRZYSZTOF KAPULKIN, ALEXANDER KURZ AND JIŘÍ VELEBIL
Expressiveness of Positive Coalgebraic Logic 368

ANDREY KUDINOV
Modal Logic of Some Products of Neighborhood Frames 386

ANDREY KUDINOV, ILYA SHAPIROVSKY AND VALENTIN SHEHTMAN
On Modal Logics of Hamming Spaces 395

2.1 Topological-Sheaf Interpretation

Let us first lay out topological-sheaf semantics for first-order modal logic introduced in [3]; see [3] and also [13] for a detailed exposition.

Topological semantics interprets propositional modal logic by using a set X and the Boolean structure on the powerset for the classical part of the logic, and a topology $\mathcal{O}(X)$ on X for the modal part. We present topological-sheaf semantics for first-order modal logic in a similar, two-part fashion. In the classical part of the semantics, instead of a set X, we take structures from the slice category **Sets**/X of sets over X to interpret the first-order vocabulary.

It is helpful to consider structures of **Sets**/X from the *bundle* point of view as follows. Take an object of **Sets**/X, that is, any map $\pi : D \to X$. Each $w \in X$ has its inverse image $D_w = \pi^{-1}\{w\}$, called the *fiber* over w. We may regard $w \in X$ as (an index for) a model of first-order logic and D_w as the domain of individuals for w. D is then the bundle of all the fibers taken *over* X, that is, the disjoint union of all D_w, written $D = \sum_{w \in X} D_w$; it is the domain of all individuals from some model or other—for $a \in D$, $\pi(a) \in X$ is the model it is from. Each model w interprets a unary formula φ with its extension $\llbracket x \mid \varphi \rrbracket_w \subseteq D_w$; then the entire bundle D interprets φ by the bundle $\llbracket x \mid \varphi \rrbracket = \sum_{w \in X} \llbracket x \mid \varphi \rrbracket_w \subseteq D$ of the extensions.

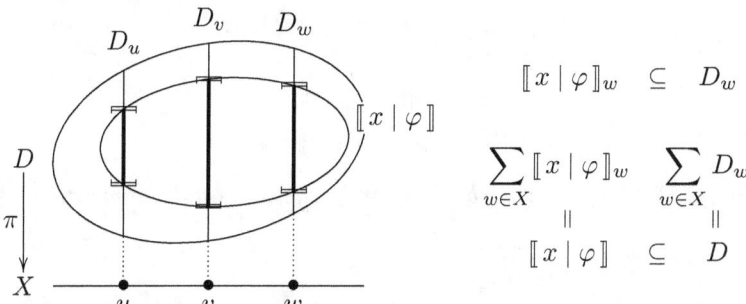

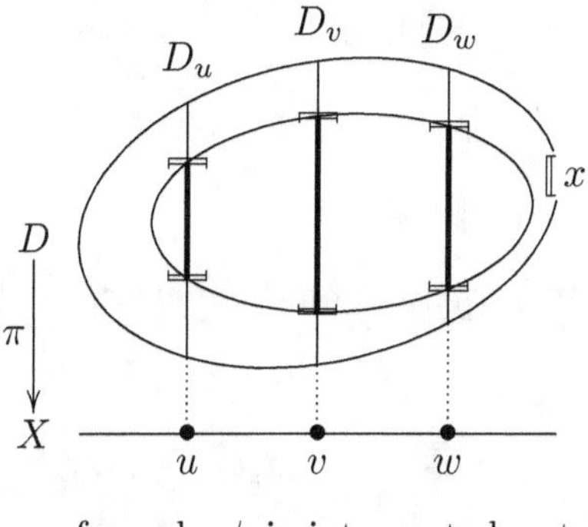

$$\begin{array}{ccc} \llbracket x \mid \varphi \rrbracket_w & \subseteq & D_w \\ & & \\ \sum_{w \in X} \llbracket x \mid \varphi \rrbracket_w & & \sum_{w \in X} D_w \\ \| & & \| \\ \llbracket x \mid \varphi \rrbracket & \subseteq & D \end{array}$$

An n-ary formula ψ is interpreted not in the cartesian product of D, but in the *fibered* product of D *over* X (that is, the product in **Sets**/X). It is $D^n = \sum_{w \in X} D^n_w$, the bundle of n-fold cartesian products of D_w, together with the map $\pi^n : D^n \to X$ that sends $\bar{a} = (a_1, \ldots, a_n) \in D^n_w$ to w.[1] In other words, D^n is the set of n-tuples from the same model. In particular, $D^0 = \sum_{w \in X} D^0_w = \sum_{w \in X} \{w\} = X$. Each model w has an extension $\llbracket \bar{x} \mid \psi \rrbracket_w$ of ψ, and the bundle D^n interprets ψ with the bundle $\llbracket \bar{x} \mid \psi \rrbracket = \sum_{w \in X} \llbracket \bar{x} \mid \psi \rrbracket_w \subseteq D^n$. An n-ary term t is interpreted not just by a map from D^n to D, but by a map *over* X (that is, an arrow in **Sets**/X). Given two maps $\pi_D : D \to X$ and $\pi_E : E \to X$, a map $f : D \to E$ is said to be over X if $\pi_E \circ f = \pi_D$, or in other words, it is a bundle $\sum_{w \in X} f_w$ of maps $f_w : D_w \to E_w$. Each model w interprets t with $\llbracket \bar{x} \mid t \rrbracket_w : D^n_w \to D_w$, and the entire interpretation is these interpretations

[1] Throughout this article, we write $\bar{x}$, $\bar{c}$, $\bar{t}$, or $\bar{a}$ for a finite sequence $x_1, \ldots, x_n$, and so on, and assume that such tuples have the appropriate arity (which we often denote by n).

bundled up, that is, the "map of bundles" $[\![\, \bar{x} \mid t \,]\!] = \sum_{w \in X} [\![\, \bar{x} \mid t \,]\!]_w : D^n \to D$. In short, we can think of the classical part of the semantics as interpreting the first-order vocabulary with standard interpretations bundled up over X. (This idea will prove crucial later in our completeness proof.)

The formal definition of the semantics goes directly, without $\sum$. First let us note that, in the formulation of first-order modal logic in [3], modal languages are assumed to have a unary modal operator $\Box$ which respects the substitution of terms, in the sense that $[\bar{t}/\bar{x}]\Box\varphi = \Box[\bar{t}/\bar{x}]\varphi$.[2]

Definition 2.1 *Given a first-order (modal) language $\mathcal{L}$, by a* bundle interpretation *for $\mathcal{L}$ we mean a pair $(\pi, [\![\cdot]\!])$ of*

- *a surjection $\pi : D \to X$ of some domain (set) D and codomain (set) X;*[3]
- *a map $[\![\cdot]\!]$ that assigns a set $[\![\, \bar{x} \mid \varphi \,]\!] \subseteq D^n$ to each formula φ of $\mathcal{L}$ in the context of variables $\bar{x}$, that is, in which no variables occur freely except $\bar{x}$, and a map $[\![\, \bar{x} \mid t \,]\!] : D^n \to D$ over X to each term t of $\mathcal{L}$ in which no variables occur (freely) except $\bar{x}$, and that satisfies suitable conditions such as*

$$[\![\, x, y \mid x = y \,]\!] = \{\, (a, a) \in D^2 \mid a \in D \,\} \subseteq D^2; \tag{1}$$
$$[\![\, \bar{x} \mid \neg\varphi \,]\!] = D^n \setminus [\![\, \bar{x} \mid \varphi \,]\!]; \tag{2}$$
$$[\![\, \bar{x} \mid \varphi \wedge \psi \,]\!] = [\![\, \bar{x} \mid \varphi \,]\!] \cap [\![\, \bar{x} \mid \psi \,]\!]; \tag{3}$$
$$[\![\, \bar{x} \mid \varphi \to \psi \,]\!] = (W \setminus [\![\, \bar{x} \mid \varphi \,]\!]) \cup [\![\, \bar{x} \mid \psi \,]\!]; \tag{4}$$
$$[\![\, \bar{x} \mid \top \,]\!] = D^n; \tag{5}$$
$$[\![\, \bar{x} \mid \exists y\, \varphi \,]\!] = p[[\![\, \bar{x}, y \mid \varphi \,]\!]]; \tag{6}$$
$$[\![\, \bar{x}, y \mid \varphi \,]\!] = p^{-1}[[\![\, \bar{x} \mid \varphi \,]\!]], \tag{7}$$

where $p : D^{n+1} \to D^n :: (\bar{a}, b) \mapsto \bar{a}$; and, when z is not among $\bar{x}$,

$$[\![\, \bar{x}, \bar{y} \mid [t/z]\varphi \,]\!] = \langle p_1, \ldots, p_n, [\![\, \bar{x}, \bar{y} \mid t \,]\!] \rangle^{-1}[[\![\, \bar{x}, z \mid \varphi \,]\!]], \tag{8}$$

where we write $\langle p_1, \ldots, p_n, f \rangle : D^{n+m} \to D^{n+1} :: (\bar{a}, \bar{b}) \mapsto (\bar{a}, f(\bar{a}, \bar{b}))$ for $f : D^{n+m} \to D$.

We say that, in such $(\pi, [\![\cdot]\!])$, φ is valid iff $[\![\, \bar{x} \mid \varphi \,]\!] = D^n$, and an inference is valid iff it preserves validity.

Topological semantics for *propositional* modal logic interprets $\Box$ by adding a topology $\mathcal{O}(X)$ to a set X and using $[\![\Box]\!] = \mathbf{int}$, the interior operation of $\mathcal{O}(X)$. Extending this to *first-order logic*, topological-sheaf semantics interprets $\Box$ by adding topologies to structures in $\mathbf{Sets}/X$; specifically, it takes structures in $\mathbf{LH}/X$, the category of "local homeomorphisms," or sheaves, over a topological space X.

Recall that, given topological spaces X and Y, a map $f : Y \to X$ is called a *homeomorphism* if it is a continuous bijection with a continuous inverse (that

[2] $[\bar{t}/\bar{x}]\psi$ is the formula obtained by substituting terms $\bar{t}$ for free variables $\bar{x}$ in a formula ψ (with each t_i for x_i), which is defined when (and only when) $\bar{t}$ are free for $\bar{x}$ in ψ.

[3] We require π to be surjective, so that $D_w \neq \varnothing$ for every $w \in X$.

is, if X and Y share the same topological structure, along the relabeling of points via f). Then the topological notion of a sheaf is defined as follows.[4]

Definition 2.2 *A continuous map $\pi : D \to X$ is called a* local homeomorphism *if every $a \in D$ has some $U \in \mathcal{O}(D)$ such that $a \in U$, $\pi[U] \in \mathcal{O}(X)$, and the restriction $\pi \restriction U : U \to \pi[U]$ of π to U is a homeomorphism. We say that such a pair (D, π) is a* sheaf over the space X, *and call π its* projection*; X and D are respectively called the* base space *and* total space *of π.*

Given sheaves (D, π_D) and (E, π_E) over a space X, we say that a map $f : D \to E$ is a *map of sheaves over X* (from (D, π_D) to (E, π_E)) if it is over X and is continuous. Since maps of sheaves are themselves local homeomorphisms,[5] the category of sheaves and maps of sheaves is just $\mathbf{LH}/X$, the category $\mathbf{LH}$ of topological spaces and local homeomorphisms over X.

We should emphasize that topological-sheaf semantics interprets $\Box$ by not just $\mathbf{int}_X$ but by the family of $\mathbf{int}_{D^n}$, corresponding to the arity of $[\![\, \bar{x} \mid \varphi \,]\!]$ to which $[\![\Box]\!]$ is applied. Hence the semantics requires topologies on all D^n. It uses the n-fold product of D in $\mathbf{LH}/X$, that is, the coarsest topology on D^n that makes every projection $p_i^n : D^n \to D :: \bar{a} \mapsto a_i$ continuous, together with the projection $\pi^n : D^n \to X :: \bar{a} \mapsto \pi(a_1)$. (D^n, π^n) is in fact a sheaf over X, and all projections $p : D^n \to D^m$ are maps of sheaves.

Now, topological-sheaf semantics consists in equipping the bundle semantics above with topologies, by using structures from $\mathbf{LH}/X$ rather than from $\mathbf{Sets}/|X|$.[6] So here is the topological part of topological-sheaf semantics:

Definition 2.3 *Given any first-order modal language $\mathcal{L}$, by a* topological-sheaf interpretation *for $\mathcal{L}$ we mean a bundle interpretation $(\pi, [\![\cdot]\!])$ for $\mathcal{L}$ such that $\pi : D \to X$ is a local homeomorphism, $[\![\, \bar{x} \mid f\bar{x} \,]\!]$ is continuous (and hence is a map of sheaves) for each n-ary function symbol f of $\mathcal{L}$,[7] and, for each $n \in \mathbb{N}$, $[\![\Box]\!] : \mathcal{P}(D^n) \to \mathcal{P}(D^n) :: [\![\, \bar{x} \mid \varphi \,]\!] \mapsto [\![\, \bar{x} \mid \Box\varphi \,]\!]$ is $\mathbf{int}_{D^n}$, the interior operation on $\mathfrak{D}^n$, that is,*

$$[\![\, \bar{x} \mid \Box\varphi \,]\!] = \mathbf{int}_{D^n}([\![\, \bar{x} \mid \varphi \,]\!]). \tag{9}$$

We emphasize that such sheaf semantics for non-modal systems of intuitionistic first-order logic are quite standard; see [17].

2.2 First-Order Modal Logic FOS4

Topological-sheaf semantics unifies the semantics in $\mathbf{Sets}/X$ (for first-order logic) and topological semantics (for propositional **S4**) naturally, in the sense that its logic is a simple union of classical first-order logic and **S4**.

Let us say that a theory in a first-order modal language is FOS4 if it has

---

4 See e.g. [17] for the relation between this and the "functorial" or "variable set" notion of a sheaf.

5 Exercise II.10(b) in [17], 105.

6 Given a space X, we write $|X|$ for its underlying set.

7 This implies that $[\![\, \bar{x} \mid t \,]\!]$ is a map of sheaves for any term t in any suitable context $\bar{x}$.

1) all the rules and axioms of classical first-order logic, and

2) the rules and axioms of propositional modal logic **S4**, that is,

$$\frac{\varphi}{\Box\varphi}, \tag{N}$$

$$\frac{\varphi \to \psi}{\Box\varphi \to \Box\psi}, \tag{M}$$

$$\Box\varphi \wedge \Box\psi \to \Box(\varphi \wedge \psi), \tag{C}$$

$$\Box\varphi \to \varphi, \tag{T}$$

$$\Box\varphi \to \Box\Box\varphi. \tag{S4}$$

In a FOS4 theory, schemes of first-order rules and axioms do *not* distinguish formulas containing $\Box$ from ones not. In the axiom $x = y \to ([x/z]\varphi \to [y/z]\varphi)$ of identity, for instance, φ may contain $\Box$. Also, modal rules and axioms are insensitive to the first-order structure of formulas. Hence, letting **FOS4** be the smallest FOS4 theory, we regard it as a simple union of first-order logic and **S4**. The soundness of **FOS4** with respect to topological-sheaf semantics can be checked straightforwardly [3]; it is the goal of this article to show the completeness.

To give examples of theorems of **FOS4**, any FOS4 theory $\mathbb{T}$ proves

$$x = y \to \Box(x = y), \tag{10}$$

because $x = y \to (\Box(x = x) \to \Box(x = y))$ is an instance of the above-mentioned axiom of identity (with $\Box(x = z)$ for φ), while $\mathbb{T} \vdash x = x$ implies $\mathbb{T} \vdash \Box(x = x)$ by N. Also, from $\mathbb{T} \vdash \varphi \to \exists x\, \varphi$, M implies $\mathbb{T} \vdash \Box\varphi \to \Box\exists x\, \varphi$; this then implies, since x is not free in $\Box\varphi$, that $\mathbb{T} \vdash \exists x\, \Box\varphi \to \Box\exists x\, \varphi$. Similarly, $\mathbb{T} \vdash \Box\forall x\, \varphi \to \forall x\, \Box\varphi$. In contrast, **FOS4** proves neither $x \neq y \to \Box(x \neq y)$, $\Box\exists x\, \varphi \to \exists x\, \Box\varphi$, nor $\forall x\, \Box\varphi \to \Box\forall x\, \varphi$.

Writing $\mathbb{T} \vdash \varphi \equiv \psi$ for the conjunction of $\mathbb{T} \vdash \varphi \to \psi$ and $\mathbb{T} \vdash \psi \to \varphi$, let us observe the following (11)–(15), which will be useful in our completeness proof.

$$\mathbb{T} \vdash \Box(\varphi \wedge \psi) \equiv \Box\varphi \wedge \Box\psi \tag{11}$$

by M and C. Given M, S4 is equivalent to (and hence $\mathbb{T}$ has) the rule

$$\frac{\Box\psi \to \varphi}{\Box\psi \to \Box\varphi}. \tag{12}$$

(10) implies the following for terms t, t', $\bar{t}$, $\bar{t}'$ (we write $\bar{t} = \bar{t}'$ for $t_1 = t'_1 \wedge \cdots \wedge t_n = t'_n$); (13) also uses T, (14) uses C and M, and (15) uses (11) and (13).

$$\mathbb{T} \vdash \Box(t = t') \equiv t = t'; \tag{13}$$

$$\mathbb{T} \vdash \Box([\bar{t}/\bar{x}]\varphi) \wedge \bar{x} = \bar{t} \to \Box\varphi; \tag{14}$$

$$\mathbb{T} \vdash \Box\varphi \wedge \bar{t} = \bar{t}' \equiv \Box(\varphi \wedge t_1 = t'_1) \wedge \cdots \wedge \Box(\varphi \wedge t_n = t'_n). \tag{15}$$

3 Preliminary Constructions

We will prove the completeness of **FOS4** with respect to the topological-sheaf semantics of Section 2. In this section, we introduce two general constructions that we will employ in Section 4.

3.1 De-Modalization

The first construction we introduce is what may be called "de-modalization." Given a first-order modal language, the construction gives a first-order non-modal language and a surjective translation from the former to the latter; along this translation, we can have a non-modal version of a given modal theory.

Fix a first-order modal language $\mathcal{L}$. Then write $\approx_\alpha$ for the α-equivalence among formulas of $\mathcal{L}$; that is, $\varphi \approx_\alpha \psi$ iff φ and ψ share the same variable structure possibly with relabeling of bound variables. Also write $\approx_\mathrm{f}$ for sharing the same variable structure possibly with relabeling of *free* variables. More precisely, $\varphi \approx_\mathrm{f} \psi$ iff $\varphi \precsim \psi$ and $\psi \precsim \varphi$ for the transitive closure $\precsim$ of the (reflexive) relation $\precsim_0$ such that $\varphi \precsim_0 \psi$ iff $\psi = [t/x]\varphi$ for some term t that is free for x in φ. Moreover, write $\approx$ for the equivalence relation generated by the union of $\approx_\alpha$ and $\approx_\mathrm{f}$; that is, $\varphi \approx \psi$ iff φ and ψ share the same variable structure possibly with relabeling of (bound or free) variables.

Let $\Box_{\mathrm{min}}(\mathcal{L})$ be the set of $\precsim$-minimal formulas of $\mathcal{L}$ of the form $\Box\varphi$. (Note that, in a $\precsim$-minimal formula, each free variable has exactly one free occurrence.) For each $\varphi \in \Box_{\mathrm{min}}(\mathcal{L})$, write $[\varphi]$ for the $\approx$-equivalence class of φ.

Definition 3.1 *Given a first-order modal language $\mathcal{L}$, we write $\mathcal{L}^\natural$ for the first-order non-modal language obtained by adding to $\mathcal{L}$ new primitive predicates $[\varphi]$ for all $\varphi \in \Box_{\mathrm{min}}(\mathcal{L})$. We let $[\varphi]$ be n-ary if φ has exactly n free variables.*[8]

Let us call a formula of $\mathcal{L}$ *classically atomic* if it is either atomic or of the form $\Box\varphi$. Then all the formulas of $\mathcal{L}$ can be constructed from classically atomic formulas with classical connectives. Hence we can define a map $\natural$ as follows, by "induction on the classical construction" of formulas of $\mathcal{L}$.

Definition 3.2 *Given a first-order modal language $\mathcal{L}$, we recursively define a map $\natural$ from the formulas of $\mathcal{L}$ to those of $\mathcal{L}^\natural$, as follows. Base clauses are*

(i) *If $\varphi = F\bar{t}$ for a primitive predicate F of $\mathcal{L}$, then simply $\varphi^\natural = F\bar{t}$.*

(ii) *If $\varphi = [\bar{t}/\bar{x}]\psi$ for $\psi \in \Box_{\mathrm{min}}(\mathcal{L})$, $\bar{x}$ are exactly the free variables of ψ, and $\bar{x}$ occur in ψ in the order of $x_1, \ldots, x_n$, then $\varphi^\natural = [\psi]\bar{t}$.*

And then we have obvious inductive clauses for classical *connectives, namely, $(\neg\varphi)^\natural = \neg\varphi^\natural$, $(\varphi \wedge \psi)^\natural = \varphi^\natural \wedge \psi^\natural$, $(\forall x\, \varphi)^\natural = \forall x.\varphi^\natural$, and so on. Moreover, given any theory $\mathbb{T}$ in $\mathcal{L}$, we write $\mathbb{T}^\natural$ for the theory $\mathbb{T}^\natural = \{\varphi^\natural \mid \mathbb{T} \vdash \varphi\}$ in $\mathcal{L}^\natural$.*

Clause (ii) above is well defined because, if $\varphi = [\bar{t}/\bar{x}]\psi_0 = [\bar{t'}/\bar{x'}]\psi_1$ and if $\bar{x}$ and $\bar{x'}$ occur in ψ in the order of $x_1, \ldots, x_n$ and $x'_1, \ldots, x'_n$ respectively, then

[8] Even though we assume in this article that $\mathcal{L}$ is first-order, that $\mathcal{L}$ has a unary operator $\Box$ and that $[\bar{t}/\bar{x}]\Box\varphi = \Box[\bar{t}/\bar{x}]\varphi$, the de-modalization construction works for languages without these assumptions, with any number of modal operators of any arities.

$\bar{t} = \bar{t}'$ and also $\psi_0 \approx_{\mathrm{f}} \psi_1$, which implies $[\psi_0] = [\psi_1]$ and hence $[\psi_0]\bar{t} = [\psi_1]\bar{t}'$.

Observe that, if an atomic formula φ of $\mathcal{L}^\natural$ has the form $[\psi]\bar{t}$ for $\psi \in \Box_{\min}(\mathcal{L})$, then there is $\psi_0 \in [\psi]$ such that $\bar{x}$ are exactly the free variables of ψ_0 and $\bar{x}$ occur in ψ_0 in the order of $x_1, \ldots, x_n$, so that $\varphi = ([\bar{t}/\bar{x}]\psi_0)^\natural$. Therefore

Fact 3.3 $\natural$ *is surjective.*

Proof. By induction on the construction of formulas φ of $\mathcal{L}^\natural$. □

Also, induction on the classical construction of formulas of $\mathcal{L}$ shows

Fact 3.4 *For any formula φ of $\mathcal{L}$, $([t/x]\varphi)^\natural = [t/x](\varphi^\natural)$.*

Fact 3.5 *For any formula φ of $\mathcal{L}$, φ and $\varphi^\natural$ have the same set of free variables.*

Fact 3.6 *For any formulas φ, ψ of $\mathcal{L}$, $\varphi^\natural = \psi^\natural$ only if $\varphi \approx_\alpha \psi$.*

Let us say that $\mathbb{T}$ *respects* $\approx_\alpha$ (or $\approx_{\mathrm{f}}$, respectively) if $\mathbb{T} \vdash \varphi$ and $\varphi \approx_\alpha \psi$ (or $\varphi \approx_{\mathrm{f}} \psi$) imply $\mathbb{T} \vdash \psi$.

Fact 3.7 *If a theory $\mathbb{T}$ in a language $\mathcal{L}$ respects $\approx_\alpha$, then*

$$\mathbb{T}^\natural \vdash \varphi^\natural \iff \mathbb{T} \vdash \varphi \tag{16}$$

for any formula φ of $\mathcal{L}$.

Proof. The "$\Leftarrow$" direction is by the definition of $\mathbb{T}^\natural$. For the other direction, suppose $\mathbb{T}$ respects $\approx_\alpha$ and that $\mathbb{T}^\natural \vdash \varphi^\natural$. Then the definition of $\mathbb{T}^\natural$ means that there is a formula ψ of $\mathcal{L}$ such that $\psi^\natural = \varphi^\natural$ and $\mathbb{T} \vdash \psi$. By Fact 3.6, $\psi \approx_\alpha \varphi$. Therefore $\mathbb{T} \vdash \varphi$ since $\mathbb{T}$ respects $\approx_\alpha$. □

Note that (16) determines $\mathbb{T}^\natural$ uniquely, since $\natural$ is surjective (Fact 3.3). Moreover observe the following consequence of Fact 3.7.

Fact 3.8 *If a theory $\mathbb{T}$ in a language $\mathcal{L}$ respects $\approx_\alpha$ and contains classical logic, then $\mathbb{T}^\natural$ contains classical logic as well.*

Proof. Suppose the antecedent. Then, for instance, $\mathbb{T}^\natural$ has the rule

$$\frac{\varphi \to \psi}{\varphi \to \forall x\, \psi} \ (x \text{ does not occur freely in } \varphi)$$

for the following reason. Given any formulas φ, ψ of $\mathcal{L}^\natural$ such that x does not occur freely in φ, there are formulas φ_0, ψ_0 of $\mathcal{L}$ such that ${\varphi_0}^\natural = \varphi$ (and so x does not occur freely in φ_0 by Fact 3.5) and ${\psi_0}^\natural = \psi$, and then (16) implies

$$\begin{array}{ccc} \mathbb{T}^\natural \vdash \varphi \to \psi \ \ (= (\varphi_0 \to \psi_0)^\natural) & \xRightarrow{(16)} & \mathbb{T} \vdash \varphi_0 \to \psi_0 \\ & & \Downarrow \\ \mathbb{T}^\natural \vdash \varphi \to \forall x\, \psi \ \ (= (\varphi_0 \to \forall x\, \psi_0)^\natural) & \xLeftarrow[(16)]{} & \mathbb{T} \vdash \varphi_0 \to \forall x\, \psi_0. \end{array}$$

Similarly for other rules and axioms. □

3.2 Lazy Henkinization

In this subsection we introduce a construction that may be called "lazy Henkinization." Given a set of first-order structures, the construction introduces new constant symbols and a new set of structures in which every individual is referred to, but which retains essentially the same theory, with no new axioms added such as Henkin axioms.[9]

Let us first add new constant symbols.

Definition 3.9 *Given a first-order language $\mathcal{L}$ (which may be modal or not) and a cardinal κ, we write $\mathcal{L}_\kappa$ for the first-order language (modal or not, depending on whether $\mathcal{L}$ is modal or not) obtained by adding to $\mathcal{L}$ a set of new constant symbols $C_\kappa = \{\, c_\alpha \mid \alpha < \kappa \,\}$. Given a theory $\mathbb{T}$ in $\mathcal{L}$, we write $\mathbb{T}_\kappa$ for the theory $\mathbb{T}_\kappa = \{\, [\bar{c}/\bar{x}]\varphi \mid \mathbb{T} \vdash \varphi$ and $\bar{c} \in C_\kappa \,\}$ in $\mathcal{L}_\kappa$. (Here it is not assumed that $\bar{x}$ are the only free variables of φ or that all of $\bar{x}$ occur freely in φ.)*

In a manner similar to Fact 3.7, we have

Fact 3.10 *If a theory $\mathbb{T}$ in a language $\mathcal{L}$ respects $\approx_{\mathrm{f}}$, then*

$$\mathbb{T}_\kappa \vdash [\bar{c}/\bar{x}]\varphi \iff \mathbb{T} \vdash \varphi \tag{17}$$

for any formula φ of $\mathcal{L}$ and any $\bar{c} \in C_\kappa$.

Note that (17) determines $\mathbb{T}_\kappa$ uniquely, since any formula of $\mathcal{L}_\kappa$ has the form $[\bar{c}/\bar{x}]\varphi$ for a formula φ of $\mathcal{L}$ and $\bar{c} \in C_\kappa$. Note moreover that, when Fact 3.10 applies, $\mathbb{T}_\kappa$ is not just a conservative extension of $\mathbb{T}$, but moreover is essentially the same theory as $\mathbb{T}$, since $\mathbb{T}_\kappa$ and $\mathbb{T}$ share the same schemes of rules and axioms. For instance, $\mathbb{T}_\kappa$ has M iff $\mathbb{T}$ does:

$$\begin{array}{ccc} \mathbb{T}_\kappa \vdash [\bar{c}/\bar{x}]\varphi & \overset{(17)}{\Longleftrightarrow} & \mathbb{T} \vdash \varphi \\ \Downarrow & & \Downarrow \\ \mathbb{T}_\kappa \vdash \Box[\bar{c}/\bar{x}]\varphi \;\; (= [\bar{c}/\bar{x}]\Box\varphi) & \underset{(17)}{\Longleftrightarrow} & \mathbb{T} \vdash \Box\varphi \end{array}$$

Hence, in particular, because $\mathbb{T}$ respects $\approx_{\mathrm{f}}$ if it is FOS4, we have

Fact 3.11 *If a theory $\mathbb{T}$ in a language $\mathcal{L}$ is FOS4, then so is $\mathbb{T}_\kappa$.*

Fact 3.10 extends to the semantic level, too, by the following construction. (We only lay out a version for a classical language here, but it can be extended to modal and other languages as well.)

Definition 3.12 *Given a set $\mathfrak{M}$ of structures for a first-order* classical *language $\mathcal{L}$, let κ be a cardinal such that $||M|| \leqslant \kappa$ for every $M \in \mathfrak{M}$. Given any*

[9] If we were to add a constant symbol c_φ to a modal language $\mathcal{L}$ that has $\Box[t/x]\varphi = [t/x]\Box\varphi$ and add a Henkin axiom $\exists x\, \varphi \to [c_\varphi/x]\varphi$ to a theory $\mathbb{T}$ in $\mathcal{L}$ that has M and the classical rule on $\exists$ (for instance, **FOS4** has them), the new theory $\mathbb{T}^+$ may fail to extend $\mathbb{T}$ conservatively, as follows. $\mathbb{T}^+ \vdash \exists x\, \varphi \to [c_\varphi/x]\varphi$ implies $\mathbb{T}^+ \vdash \Box\exists x\, \varphi \to \Box[c_\varphi/x]\varphi$ by M, where $\Box[c_\varphi/x]\varphi = [c_\varphi/x]\Box\varphi$; therefore, by the rule on $\exists$, $\mathbb{T}^+$ proves $\Box\exists x\, \varphi \to \exists x\, \Box\varphi$. This is a formula of $\mathcal{L}$; it may, however, not be provable in $\mathbb{T}$ (for instance, it is not in **FOS4**).

$M \in \mathfrak{M}$ and any surjection $e : \kappa \twoheadrightarrow |M|$, let M_e be the expansion of M to $\mathcal{L}_\kappa$ with ${c_\alpha}^{M_e} = e(\alpha)$ for each $\alpha < \kappa$. Then we write $\mathfrak{M}_\kappa = \{ M_e \mid M \in \mathfrak{M}$ and $e : \kappa \twoheadrightarrow |M| \}$.

Lemma 3.13 *If $\mathbb{T}$ is the theory of $\mathfrak{M}$ (and if $||M|| \leqslant \kappa$ for every $M \in \mathfrak{M}$), then $\mathbb{T}_\kappa$ is the theory of $\mathfrak{M}_\kappa$.*[10]

Proof. Suppose $\mathbb{T}$ is the theory of $\mathfrak{M}$; then it respects $\approx_{\mathrm{f}}$ and hence Fact 3.10 applies. Hence (because (17) determines $\mathbb{T}_\kappa$) it is enough to show that, for any formula φ of $\mathcal{L}$ and $\bar{c} = (c_{\alpha_1}, \ldots, c_{\alpha_n}) \in C_\kappa$, the following are equivalent:

(i) $M_e \models_{[\bar{b}/\bar{y}]} [\bar{c}/\bar{x}]\varphi$ for all $M_e \in \mathfrak{M}_\kappa$ and $\bar{b} \in |M_e|$.[11]

(ii) $M \models_{[\bar{a},\bar{b}/\bar{x},\bar{y}]} \varphi$ for all $M \in \mathfrak{M}$ and $\bar{a}, \bar{b} \in |M|$.

Observe the equivalence below (where we write $e(\bar{\alpha}) = (e(\alpha_1), \ldots, e(\alpha_n))$) for every $M \in \mathfrak{M}$, $e : \kappa \twoheadrightarrow |M|$, and $\bar{b} \in |M| = |M_e|$. $(*)$ holds because $e(\alpha_i) = {c_{\alpha_i}}^{M_e}$ for every $i \leqslant n$ and $(\dagger)$ holds because M_e expands M.

$$M_e \models_{[\bar{b}/\bar{y}]} [\bar{c}/\bar{x}]\varphi \overset{(*)}{\Longleftrightarrow} M_e \models_{[e(\bar{\alpha}),\bar{b}/\bar{x},\bar{y}]} \varphi \overset{(\dagger)}{\Longleftrightarrow} M \models_{[e(\bar{\alpha}),\bar{b}/\bar{x},\bar{y}]} \varphi$$

Hence (ii) entails (i). Assume (i) and fix any $M \in \mathfrak{M}$ and $\bar{a}, \bar{b} \in |M|$. There is $e : \kappa \twoheadrightarrow |M|$ such that $e(\bar{\alpha}) = \bar{a}$; then $M \models_{[e(\bar{\alpha}),\bar{b}/\bar{x},\bar{y}]} \varphi$ for this e. Thus (ii). □

3.3 The Two Constructions Commute

The two constructions we just introduced can be combined in a natural way.

Fact 3.14 *$(\mathcal{L}^\natural)_\kappa = (\mathcal{L}_\kappa)^\natural$ for any first-order modal language $\mathcal{L}$.*

Proof. Observe that the equivalence relation $\approx$ on $\mathcal{L}_\kappa$ is an extension of $\approx$ on $\mathcal{L}$. Moreover, $\Box_{\min}(\mathcal{L}_\kappa) = \Box_{\min}(\mathcal{L})$, since any new formula of $\mathcal{L}_\kappa$ has the form $[\bar{c}/\bar{x}]\varphi$ and hence is not $\precsim$-minimal. Therefore the same set of new primitive predicates is added to $\mathcal{L}^\natural$ and $(\mathcal{L}_\kappa)^\natural$. ($[\varphi]$ for $\varphi \in \Box_{\min}(\mathcal{L})$ has the same arity as a predicate of $\mathcal{L}^\natural$ and as a predicate of $(\mathcal{L}_\kappa)^\natural$, since φ has the same number of free variables as a formula of $\mathcal{L}$ and as a formula of $\mathcal{L}_\kappa$.) Thus $(\mathcal{L}^\natural)_\kappa$ and $(\mathcal{L}_\kappa)^\natural$ have the same sets of new primitive predicates and new constant symbols. □

Fact 3.15 *The map $\natural$ from formulas of $\mathcal{L}_\kappa$ to those of $\mathcal{L}^\natural_\kappa$ is an extension of the map $\natural$ from formulas of $\mathcal{L}$ to those of $\mathcal{L}^\natural$.*

Proof. By induction on the classical construction of formulas of $\mathcal{L}_\kappa$. □

Fact 3.16 *$(\mathbb{T}^\natural)_\kappa = (\mathbb{T}_\kappa)^\natural$ for any theory $\mathbb{T}$ in a first-order modal language $\mathcal{L}$.*

Proof. $(\mathbb{T}^\natural)_\kappa = \{ [\bar{c}/\bar{x}](\varphi^\natural) \mid \mathbb{T} \vdash \varphi$ and $\bar{c} \in C_\kappa \}$ equals $(\mathbb{T}_\kappa)^\natural = \{ ([\bar{c}/\bar{x}]\varphi)^\natural \mid \mathbb{T} \vdash \varphi$ and $\bar{c} \in C_\kappa \}$ by Facts 3.4 and 3.15. □

---

[10] In this article, we use the notion of a theory as a set of formulas rather than sequents, so that $\mathbb{T}$ is the theory of $\mathfrak{M}$ if $\mathbb{T}$ and $\mathfrak{M}$ agree on every formula ($\mathbb{T} \vdash \varphi$ iff all $M \in \mathfrak{M}$ validate φ). Yet all the results in this section extend to the sequent formulation as well.

[11] $M \models_{[\bar{a}/\bar{x}]} \psi$ means that, in the model M, the formula ψ is true of the individuals $\bar{a} \in |M|$ (with each a_i in place of the free variable x_i). We assume here that the arities of $\bar{b}$ and $\bar{y}$ are the same, but not that they are n.

Fact 3.17 *If a theory $\mathbb{T}$ in a first-order modal language $\mathcal{L}$ respects $\approx_{\mathrm{f}}$, so does $\mathbb{T}^{\natural}$.*

Proof. Given formulas φ, ψ of $\mathcal{L}^{\natural}$ such that $\varphi \approx_{\mathrm{f}} \psi$, suppose $\mathbb{T}^{\natural} \vdash \varphi$, which means that $\mathbb{T} \vdash \varphi_0$ for some formula φ_0 of $\mathcal{L}$ such that $\varphi = {\varphi_0}^{\natural}$. On the other hand, $\varphi \approx_{\mathrm{f}} \psi$ means $\psi = [\bar{y}/\bar{x}]\varphi$ for some variables $\bar{x}$, $\bar{y}$. Let $\psi_0 = [\bar{y}/\bar{x}]\varphi_0$. Then $\varphi_0 \approx_{\mathrm{f}} \psi_0$; so, if $\mathbb{T}$ respects $\approx_{\mathrm{f}}$, $\mathbb{T} \vdash \psi_0$ and hence $\mathbb{T}^{\natural} \vdash {\psi_0}^{\natural}$. This means $\mathbb{T}^{\natural} \vdash \psi$ since $(\psi_0)^{\natural} = ([\bar{y}/\bar{x}]\varphi_0)^{\natural} = [\bar{y}/\bar{x}]({\varphi_0}^{\natural}) = [\bar{y}/\bar{x}]\varphi = \psi$ by Fact 3.4. □

Similar relabeling of variables (this time, bound ones) also shows

Fact 3.18 *If a theory $\mathbb{T}$ in a first-order language $\mathcal{L}$ respects $\approx_{\alpha}$, so does $\mathbb{T}_{\kappa}$.*

Combining these with Facts 3.4, 3.7 and 3.10, we have

Lemma 3.19 *If a theory $\mathbb{T}$ in a first-order modal language $\mathcal{L}$ respects both $\approx_{\alpha}$ and $\approx_{\mathrm{f}}$, then, for any formulas φ of $\mathcal{L}$, φ_0 of $\mathcal{L}^{\natural}$, and φ_1 of $\mathcal{L}_{\kappa}$, and for any $\bar{c} \in C_{\kappa}$, the following equivalences hold:*

$$\begin{array}{ccccc}
\mathbb{T} \vdash \varphi & \xLeftrightarrow{(17)} & \mathbb{T}_{\kappa} \vdash [\bar{c}/\bar{x}]\varphi & & \mathbb{T}_{\kappa} \vdash \varphi_1 \\
(16)\Updownarrow & & \Updownarrow(16) & & \Updownarrow(16) \\
\mathbb{T}^{\natural} \vdash \varphi^{\natural} & \xLeftrightarrow{(17)} & \mathbb{T}^{\natural}_{\kappa} \vdash [\bar{c}/\bar{x}]\varphi^{\natural} & & \mathbb{T}^{\natural}_{\kappa} \vdash {\varphi_1}^{\natural}. \\
\mathbb{T}^{\natural} \vdash \varphi_0 & \xLeftrightarrow{(17)} & \mathbb{T}^{\natural}_{\kappa} \vdash [\bar{c}/\bar{x}]\varphi_0 & &
\end{array}$$

4 Topological Completeness

In this section, we prove the completeness of **FOS4** with respect to the topological-sheaf semantics of Section 2. More precisely, we prove

Theorem 4.1 *For any consistent FOS4 theory $\mathbb{T}$ in a first-order modal language $\mathcal{L}$, there exist a topological space X and a topological-sheaf interpretation $(\pi : D \to X, [\![\cdot]\!])$ for $\mathcal{L}$ such that, for every formula φ of $\mathcal{L}$,*

$$\mathbb{T} \vdash \varphi \iff [\![\bar{x} \mid \varphi]\!] = D^n. \tag{18}$$

To prove this, let us fix any such language $\mathcal{L}$ and theory $\mathbb{T}$.

4.1 Constructing an Interpretation

Applying the constructions we introduced in Section 3, we first construct a bundle interpretation $(\pi, [\![\cdot]\!])$ that satisfies (18) of Theorem 4.1. It will be shown in Subsection 4.2 to be a topological-sheaf interpretation as desired.

First de-modalize $\mathcal{L}$ and obtain $\mathcal{L}^{\natural}$ along with $\natural$ and $\mathbb{T}^{\natural}$ as in Definitions 3.1 and 3.2. Observe that, because $\mathbb{T}$ contains classical logic and has E, $\mathbb{T} \vdash \varphi \equiv \psi$ whenever $\varphi \approx_{\alpha} \psi$ (which can be shown by induction on the construction of φ, ψ). Hence $\mathbb{T}$ respects $\approx_{\alpha}$ and Fact 3.8 applies, so that $\mathbb{T}^{\natural}$ contains classical logic. Thus, the completeness theorem for classical first-order logic applies to the theory $\mathbb{T}^{\natural}$ in the classical language $\mathcal{L}^{\natural}$, providing a class $\mathbf{M}$ of $\mathcal{L}^{\natural}$-structures whose theory is $\mathbb{T}^{\natural}$. Moreover, although $\mathbf{M}$ may well be a proper class, the downward Löwenheim-Skolem theorem enables us to cut $\mathbf{M}$ down to a set $\mathfrak{M}_0 = \{ M \in \mathbf{M} \mid ||M|| \leqslant \kappa \}$, for some cardinal κ, whose theory is $\mathbb{T}^\natural$.

Next, for lazy Henkinization, add a set of new constant symbols $C_\kappa = \{ c_\alpha \mid \alpha < \kappa \}$ to $\mathcal{L}$ and $\mathcal{L}^\natural$ and obtain $\mathcal{L}_\kappa$ and $\mathcal{L}^\natural_\kappa$, respectively. Since $\mathbb{T}$ contains classical first-order logic, it respects $\approx_{\mathrm{f}}$ (in addition to $\approx_\alpha$ as seen above); therefore Lemma 3.19 implies (17) for any formula φ of $\mathcal{L}$ and (16) for any formula φ of $\mathcal{L}_\kappa$. Applying Lemma 3.13 to $\mathfrak{M}_0$ yields a set $\mathfrak{M} = (\mathfrak{M}_0)_\kappa$ of $\mathcal{L}^\natural_\kappa$-structures whose theory is $\mathbb{T}^\natural_\kappa$, that is, for any formula φ of $\mathcal{L}^\natural_\kappa$,

$$\mathbb{T}^\natural_\kappa \vdash \varphi \iff M \models_{[\bar{a}/\bar{x}]} \varphi \text{ for all } M \in \mathfrak{M} \text{ and } \bar{a} \in |M|, \tag{19}$$

and such that,

$$\text{for every } M \in \mathfrak{M} \text{ and } a \in |M|,\ a = c^M \text{ for some } c \in C_\kappa. \tag{20}$$

To construct our interpretation from $\mathfrak{M}$, it is crucial to recall the bundle idea we laid out in Subsection 2.1: given a bundle interpretation $(\pi, [\![\cdot]\!])$ (Definition 2.1), the codomain (set) X of $\pi : D \to X$ can be regarded as a set of models, the domain (set) D as a bundle of domains, and $[\![\cdot]\!]$ as a bundle of interpretations. Accordingly, we use $\mathfrak{M}$ as (the underlying set of) the base space of our interpretation. We obtain the total space $\mathfrak{D}$ and its fibered products $\mathfrak{D}^n$ by bundling up domains $|M|$ of $M \in \mathfrak{M}$ and their cartesian products $|M|^n$; more precisely, for each $n \in \mathbb{N}$, let $\mathfrak{D}^n$ be the disjoint union

$$\mathfrak{D}^n = \sum_{M \in \mathfrak{M}} |M|^n = \{ (M, a_1, \ldots, a_n) \mid M \in \mathfrak{M} \text{ and } a_1, \ldots, a_n \in |M| \}$$

along with the projection $\pi^n : \mathfrak{D}^n \to \mathfrak{M} :: (M, \bar{a}) \mapsto M$. In particular, π will be the projection of the interpretation $(\pi, [\![\cdot]\!])$ we construct.

By the same token, we obtain $[\![\cdot]\!]$ by bundling up interpretations in $M \in \mathfrak{M}$. More precisely, let us first write

$$[\![\bar{x} \mid \varphi]\!]^\natural_\kappa = \{ (M, \bar{a}) \in \mathfrak{D}^n \mid M \models_{[\bar{a}/\bar{x}]} \varphi^\natural \} \subseteq \mathfrak{D}^n,$$
$$[\![\bar{x} \mid t]\!]^\natural_\kappa : \mathfrak{D}^n \to \mathfrak{D} :: (M, \bar{a}) \mapsto (M, t^M(\bar{a}))$$

for each formula φ and term t of $\mathcal{L}^\natural_\kappa$. Note that then (19) means that

$$\mathbb{T}^\natural_\kappa \vdash \varphi \iff [\![\bar{x} \mid \varphi]\!]^\natural_\kappa = \mathfrak{D}^n \tag{21}$$

for any formula φ of $\mathcal{L}^\natural_\kappa$. Then, given any formula φ and term t of $\mathcal{L}_\kappa$ (note that $\varphi^\natural$ and t are a formula and term of $\mathcal{L}^\natural_\kappa$), define

$$[\![\bar{x} \mid \varphi]\!]_\kappa = [\![\bar{x} \mid \varphi^\natural]\!]^\natural_\kappa, \qquad [\![\bar{x} \mid t]\!]_\kappa = [\![\bar{x} \mid t]\!]^\natural_\kappa.$$

This map $[\![\cdot]\!]_\kappa$ from $\mathcal{L}_\kappa$, of the type as in an interpretation $(\pi, [\![\cdot]\!]_\kappa)$, can in fact be restricted to a map $[\![\cdot]\!]$ from $\mathcal{L}$; that is, for φ and t of $\mathcal{L}$, define

$$[\![\bar{x} \mid \varphi]\!] = [\![\bar{x} \mid \varphi]\!]_\kappa = [\![\bar{x} \mid \varphi^\natural]\!]^\natural_\kappa, \qquad [\![\bar{x} \mid t]\!] = [\![\bar{x} \mid t]\!]_\kappa = [\![\bar{x} \mid t]\!]^\natural_\kappa.$$

Since $\mathbb{T}$ is FOS4, Fact 3.11 implies $\mathbb{T}_\kappa$ is FOS4 as well. In particular, $\mathbb{T}_\kappa$ contains classical first-order logic, which implies $(\pi, [\![\cdot]\!]_\kappa)$ satisfies the classical part of the definition of topological-sheaf interpretation; that is,

Fact 4.2 $(\pi, [\![\cdot]\!]_\kappa)$ *is a bundle interpretation.*

Proof. First, $[\![\bar{x} \mid \varphi]\!]_\kappa \subseteq \mathfrak{D}^n$ by definition. Every $[\![\bar{x} \mid t]\!]_\kappa : \mathfrak{D}^n \to \mathfrak{D}$ is over $\mathfrak{M}$ since $\pi \circ [\![\bar{x} \mid t]\!]_\kappa(M, \bar{a}) = M = \pi^n(M, \bar{a})$. Moreover, $(\pi, [\![\cdot]\!]_\kappa)$ satisfies (1)–(8) because each $M \in \mathfrak{M}$ is a classical structure. For instance, (6) holds since

$$\begin{aligned}
(M, \bar{a}) \in [\![\bar{x} \mid \exists y\, \varphi]\!]_\kappa = [\![\bar{x} \mid (\exists y\, \varphi)^\natural]\!]_\kappa^\natural = [\![\bar{x} \mid \exists y . \varphi^\natural]\!]_\kappa^\natural \\
\iff M \models_{[\bar{a}/\bar{x}]} \exists y . \varphi^\natural \\
\iff M \models_{[\bar{a}, b/\bar{x}, y]} \varphi^\natural \text{ for some } b \in |M| \\
\iff (M, \bar{a}, b) \in [\![\bar{x}, y \mid \varphi^\natural]\!]_\kappa^\natural = [\![\bar{x}, y \mid \varphi]\!]_\kappa \text{ for some } b \in |M| \\
\iff (M, \bar{a}) \in p_n[[\![\bar{x}, y \mid \varphi]\!]_\kappa]
\end{aligned}$$

for every formula φ of $\mathcal{L}$. □

This $(\pi, [\![\cdot]\!])$ forms an interpretation of the desired kind, namely,

Fact 4.3 $(\pi, [\![\cdot]\!])$ *satisfies* (18) *of Theorem 4.1.*

Proof. First observe that, for any formula φ of $\mathcal{L}_\kappa$,

$$\mathbb{T}_\kappa \vdash \varphi \overset{(16)}{\iff} \mathbb{T}_\kappa^\natural \vdash \varphi^\natural \overset{(21)}{\iff} [\![\bar{x} \mid \varphi]\!]_\kappa = [\![\bar{x} \mid \varphi^\natural]\!]_\kappa^\natural = \mathfrak{D}^n. \tag{22}$$

Therefore (18), that is, for any formula φ of $\mathcal{L}$,

$$\mathbb{T} \vdash \varphi \overset{(17)}{\iff} \mathbb{T}_\kappa \vdash \varphi \overset{(22)}{\iff} [\![\bar{x} \mid \varphi]\!] = [\![\bar{x} \mid \varphi]\!]_\kappa = \mathfrak{D}^n.$$

□

It is useful to observe that (4) enables us to rewrite (22) as follows.

$$\mathbb{T}_\kappa \vdash \varphi \to \psi \iff [\![\bar{x} \mid \varphi]\!]_\kappa \subseteq [\![\bar{x} \mid \psi]\!]_\kappa. \tag{23}$$

4.2 McKinsey-Tarski Topologies

Now that we have constructed a bundle interpretation $(\pi, [\![\cdot]\!])$ that satisfies (18) of Theorem 4.1, we finish our completeness proof by showing that, equipped with suitable topologies, it is in fact a topological-sheaf interpretation.

We shall define suitable topologies on $\mathfrak{M}$ and $\mathfrak{D}$. For this purpose, it is useful to observe the following consequences (24) and (25) of $\mathbb{T}_\kappa$ being FOS4 and (23); (24) is by N and (5), and (25) by (11) and (3). (26)–(30) follow similarly and will be useful later; they are by T, (12), (13), (14), and (15), respectively ((30) uses (3), too).

$$[\![\bar{x} \mid \Box\top]\!]_\kappa = [\![\bar{x} \mid \top]\!]_\kappa = \mathfrak{D}^n, \tag{24}$$

$$[\![\bar{x} \mid \Box(\varphi \wedge \psi)]\!]_\kappa = [\![\bar{x} \mid \Box\varphi \wedge \Box\psi]\!]_\kappa = [\![\bar{x} \mid \Box\varphi]\!]_\kappa \cap [\![\bar{x} \mid \Box\psi]\!]_\kappa, \tag{25}$$

$$[\![\bar{x} \mid \Box\varphi]\!]_\kappa \subseteq [\![\bar{x} \mid \varphi]\!]_\kappa, \tag{26}$$

$$\text{if } [\![\bar{x} \mid \Box\psi]\!]_\kappa \subseteq [\![\bar{x} \mid \varphi]\!]_\kappa \text{ then } [\![\bar{x} \mid \Box\psi]\!]_\kappa \subseteq [\![\bar{x} \mid \Box\varphi]\!]_\kappa, \tag{27}$$

$$[\![\bar{x} \mid \Box(t_0 = t_1)]\!]_\kappa = [\![\bar{x} \mid t_0 = t_1]\!]_\kappa, \tag{28}$$

$$[\![\, \bar{x} \mid \Box([\bar{t}/\bar{x}]\varphi) \wedge \bar{x} = \bar{t} \,]\!]_\kappa \subseteq [\![\, \bar{x} \mid \Box\varphi \,]\!]_\kappa, \tag{29}$$

$$[\![\, \bar{x} \mid \Box\varphi \wedge \bar{t} = \bar{t'} \,]\!]_\kappa = \bigcap_{i \leqslant n} [\![\, \bar{x} \mid \Box(\varphi \wedge t_i = t'_i) \,]\!]_\kappa. \tag{30}$$

Now we define topologies on $\mathfrak{M}$ and $\mathfrak{D}$. We do so by extending the idea that McKinsey and Tarski [20] used for the propositional case—using the family of (the interpretations of) formulas of the form $\Box\varphi$ as a basis for a topology—to the first-order case. That is, for each $n \in \mathbb{N}$, writing

$$B^n_\varphi = [\![\, \bar{x} \mid \Box\varphi \,]\!]_\kappa$$

for each formula φ of $\mathcal{L}_\kappa$ (that has no free variables except possibly $\bar{x}$), and $\mathcal{B}^n$ for the family of all B^n_φ, we define $\mathcal{O}(\mathfrak{D}^n)$ to be the topology on $\mathfrak{D}^n$ generated by $\mathcal{B}^n$; so, $U \in \mathcal{O}(\mathfrak{D}^n)$ iff U is a union of sets of the form B^n_φ. Each $\mathcal{B}^n$ in fact forms a basis for a topology, because $\mathfrak{D}^n = B^n_\top \in \mathcal{B}^n$ by (24) and $B^n_\varphi \cap B^n_\psi = B^n_{\varphi\wedge\psi} \in \mathcal{B}^n$ by (25).

We then show that, equipped with the topologies $\mathcal{O}(\mathfrak{M}) = \mathcal{O}(\mathfrak{D}^0)$ and $\mathcal{O}(\mathfrak{D}) = \mathcal{O}(\mathfrak{D}^1)$, $(\pi, [\![\cdot]\!]_\kappa)$ is a topological-sheaf interpretation for $\mathcal{L}_\kappa$—that is, it satisfies Definition 2.3. We only need to show: (a) that π is a sheaf; (b) that $[\![\, \bar{x} \mid f\bar{x} \,]\!]_\kappa : \mathfrak{D}^n \to \mathfrak{D}$ is continuous; and (c) that $[\![\, \bar{x} \mid \Box\varphi \,]\!]_\kappa = \mathbf{int}_{\mathfrak{D}^n}([\![\, \bar{x} \mid \varphi \,]\!]_\kappa)$. We should note that the topology on $\mathfrak{D}^n$ in (b) and (c) that is relevant to Definition 2.3 is the n-fold fibered product topology of $\mathcal{O}(\mathfrak{D})$ over $\mathcal{O}(\mathfrak{M})$; it is, however, enough to show (b) and (c) with respect to the topology $\mathcal{O}(\mathfrak{D}^n)$ generated by $\mathcal{B}^n$ as above, due to

Fact 4.4 *$\mathcal{O}(\mathfrak{D}^n)$ is the n-fold fibered product topology of $\mathcal{O}(\mathfrak{D})$ over $\mathcal{O}(\mathfrak{M})$.*

Proof. $\mathcal{O}(\mathfrak{D}^0) = \mathcal{O}(\mathfrak{M})$ is the 0-fold fibered product topology over $\mathcal{O}(\mathfrak{M})$ by definition. $\mathcal{O}(\mathfrak{D}^1) = \mathcal{O}(\mathfrak{D})$ is the 1-fold fibered product topology of $\mathcal{O}(\mathfrak{D})$ by definition. So let us fix any $n > 1$ and write $\mathcal{O}^n$ for the n-fold fibered product topology of $\mathcal{O}(\mathfrak{D})$ over $\mathcal{O}(\mathfrak{M})$. Observe that, for each $i \leqslant n$, the projection $p_i : \mathfrak{D}^n \to \mathfrak{D} :: (M, \bar{a}) \mapsto (M, a_i)$ satisfies

$$p_i{}^{-1}[[\![\, x_i \mid \varphi \,]\!]_\kappa] = [\![\, \bar{x} \mid \varphi \,]\!]_\kappa, \tag{31}$$

by applying (7) $n - 1$ times.

Note that, to show a map $f : X \to Y$ continuous, it is enough to show that $f^{-1}[B]$ is open in X for every B in a basis for Y (since f^{-1} commutes with union). Therefore p_i is continuous, because, for every $B^1_\varphi \in \mathcal{B}^1$, (31) implies

$$p_i{}^{-1}[B^1_\varphi] = p_i{}^{-1}[[\![\, x_i \mid \Box\varphi \,]\!]_\kappa] = [\![\, \bar{x} \mid \Box\varphi \,]\!]_\kappa = B^n_\varphi \in \mathcal{O}(\mathfrak{D}^n).$$

Thus $\mathcal{O}^n \subseteq \mathcal{O}(\mathfrak{D}^n)$, since $\mathcal{O}^n$ is the coarsest topology making all p_i continuous.

On the other hand, given any $B^n_\varphi \in \mathcal{B}^n$, fix any $(M, \bar{a}) \in B^n_\varphi$; this means $(M, \bar{a}) \in [\![\, \bar{x} \mid \Box\varphi \,]\!]_\kappa$ and hence $M \models_{[\bar{a}/\bar{x}]} (\Box\varphi)^\natural$. For each $i \leqslant n$, (20) implies $a_i = c_i{}^M$ for some $c_i \in C_\kappa$. It follows that $M \models_{[\bar{a}/\bar{x}]} x_i = c_i$, and moreover that $M \models_{[\bar{a}/\bar{x}]} (\Box([\bar{c}/\bar{x}]\varphi))^\natural$ (since $(\Box([\bar{c}/\bar{x}]\varphi))^\natural = [\bar{c}/\bar{x}](\Box\varphi)^\natural$ by Fact 3.4); therefore

$M \models_{[\bar{a}/\bar{x}]} (\Box([\bar{c}/\bar{x}]\varphi) \wedge \bar{x} = \bar{c})^{\natural}$ (note that $(x_i = c_i)^{\natural}$ is $x_i = c_i$) and hence $(M, \bar{a}) \in [\![\, \bar{x} \mid \Box([\bar{c}/\bar{x}]\varphi) \wedge \bar{x} = \bar{c} \,]\!]_{\kappa}$. Then, while

$$[\![\, \bar{x} \mid \Box([\bar{c}/\bar{x}]\varphi) \wedge \bar{x} = \bar{c} \,]\!]_{\kappa} \subseteq [\![\, \bar{x} \mid \Box\varphi \,]\!]_{\kappa} = B^n_{\varphi}$$

by (29), (30) implies

$$\begin{aligned} [\![\, \bar{x} \mid \Box([\bar{c}/\bar{x}]\varphi) \wedge \bar{x} = \bar{c} \,]\!]_{\kappa} &= \bigcap_{i \leqslant n} [\![\, \bar{x} \mid \Box([\bar{c}/\bar{x}]\varphi \wedge x_i = c_i) \,]\!]_{\kappa} \\ &= \bigcap_{i \leqslant n} p_i{}^{-1}[B^1_{[\bar{c}/\bar{x}]\varphi \wedge x_i = c_i}] \end{aligned}$$

because, for each $i \leqslant n$, (31) implies

$$\begin{aligned} [\![\, \bar{x} \mid \Box([\bar{c}/\bar{x}]\varphi \wedge x_i = c_i) \,]\!]_{\kappa} &= p_i{}^{-1}[[\![\, x_i \mid \Box([\bar{c}/\bar{x}]\varphi \wedge x_i = c_i) \,]\!]_{\kappa}] \\ &= p_i{}^{-1}[B^1_{[\bar{c}/\bar{x}]\varphi \wedge x_i = c_i}]. \end{aligned}$$

Note that $\bigcap_{i \leqslant n} p_i{}^{-1}[B^1_{[\bar{c}/\bar{x}]\varphi \wedge x_i = c_i}] \in \mathcal{O}^n$ since each p_i is continuous from $\mathcal{O}^n$. Thus every $(M, \bar{a}) \in B^n_{\varphi}$ has $(M, \bar{a}) \in U \subseteq B^n_{\varphi}$ for some $U \in \mathcal{O}^n$; this means $B^n_{\varphi} \in \mathcal{O}^n$. Hence $\mathcal{O}(\mathfrak{D}^n) \subseteq \mathcal{O}^n$. □

Note the use of (20) in showing $\mathcal{O}(\mathfrak{D}^n) \subseteq \mathcal{O}^n$; we emphasize that we introduced lazy Henkinization to make sure that Fact 4.4 holds.

Now let us show (a)–(c) (in the order of (b), (a), (c)) to complete our proof.

Fact 4.5 *For each n-ary function symbol f of $\mathcal{L}_{\kappa}$, $[\![\, \bar{x} \mid f\bar{x} \,]\!]_{\kappa} : \mathfrak{D}^n \to \mathfrak{D}$ is continuous from $\mathcal{O}(\mathfrak{D}^n)$ to $\mathcal{O}(\mathfrak{D})$.*

Proof. (8) immediately implies $[\![\, \bar{x} \mid f\bar{x} \,]\!]_{\kappa}{}^{-1}[B^1_{\varphi}] = B^n_{[f\bar{x}/x]\varphi} \in \mathcal{O}(\mathfrak{D}^n)$. □

Fact 4.6 *$\pi : \mathfrak{D} \to \mathfrak{M}$ is a local homeomorphism from $\mathcal{O}(\mathfrak{D})$ to $\mathcal{O}(\mathfrak{M})$.*

Proof. π is continuous since, for every $B^0_{\varphi} \in \mathcal{B}^0$,

$$\pi^{-1}[B^0_{\varphi}] = \pi^{-1}[[\![\Box\varphi]\!]_{\kappa}] \overset{(7)}{=} [\![\, x \mid \Box\varphi \,]\!]_{\kappa} = B^1_{\varphi} \in \mathcal{O}(\mathfrak{D}).$$

Fix any $(M, a) \in \mathfrak{D}$. By (20), there is $c \in C_{\kappa}$ such that $a = c^M$, that is, $(M, a) = [\![c]\!]_{\kappa}(M)$ for $[\![c]\!]_{\kappa} = [\![c]\!]^{\natural}_{\kappa} : \mathfrak{M} \to \mathfrak{D} :: N \mapsto (N, c^N)$. Note that the image of $[\![c]\!]_{\kappa}$ is $[\![\, x \mid x = c \,]\!]_{\kappa} = [\![\, x \mid \Box(x = c) \,]\!]_{\kappa} = B^1_{x=c}$ by (28); so $(M, a) \in B^1_{x=c} \in \mathcal{O}(\mathfrak{D})$. Clearly, $\pi \restriction B^1_{x=c}$ and $[\![c]\!]_{\kappa}$ are inverse to each other. $[\![c]\!]_{\kappa}$ is continuous by Fact 4.5. $\pi \restriction B^1_{x=c}$ is continuous (from the subspace of $\mathcal{O}(\mathfrak{D})$ on $B^1_{x=c}$) since π is continuous and $B^1_{x=c} \in \mathcal{O}(\mathfrak{D})$. □

It is worth noting that, even without lazy Henkinization and (20), another (longer) proof would show that **FOS4** *forces* (D, π) to be a sheaf with respect to $\mathcal{O}(\mathfrak{D})$ and $\mathcal{O}(\mathfrak{M})$ (though Fact 4.4 fails without (20)).

Fact 4.7 $[\![\, \bar{x} \mid \Box\varphi \,]\!]_{\kappa} = \mathbf{int}_{\mathfrak{D}^n}([\![\, \bar{x} \mid \varphi \,]\!]_{\kappa})$ *for the interior operation* $\mathbf{int}_{\mathfrak{D}^n}$ *of* $\mathcal{O}(\mathfrak{D}^n)$.

Proof. $[\![\bar{x} \mid \Box\varphi]\!]_\kappa = B^n_\varphi \in \mathcal{O}(\mathfrak{D}^n)$ and, by (26), $[\![\bar{x} \mid \Box\varphi]\!]_\kappa \subseteq [\![\bar{x} \mid \varphi]\!]_\kappa$. Moreover, for every $B^n_\psi \in \mathcal{B}^n$ such that $B^n_\psi \subseteq [\![\bar{x} \mid \varphi]\!]$, (27) implies $B^n_\psi \subseteq [\![\bar{x} \mid \Box\varphi]\!]_\kappa$. Thus $[\![\bar{x} \mid \Box\varphi]\!]_\kappa$ is the largest open subset of $[\![\bar{x} \mid \varphi]\!]_\kappa$, that is, $[\![\bar{x} \mid \Box\varphi]\!]_\kappa = \mathbf{int}_{\mathfrak{D}^n}([\![\bar{x} \mid \varphi]\!]_\kappa)$. □

By Fact 4.4, Facts 4.5–4.7 along with Fact 4.2 mean that $(\pi, [\![\cdot]\!]_\kappa)$ is indeed a topological interpretation for $\mathcal{L}_\kappa$, and therefore that $(\pi, [\![\cdot]\!])$, the reduct of $(\pi, [\![\cdot]\!]_\kappa)$ to $\mathcal{L}$, is a topological-sheaf interpretation for $\mathcal{L}$. Thus Fact 4.3 completes our proof of Theorem 4.1.

5 Conclusion

In this article, we introduced two constructions—de-modalization of a first-order modal language and lazy Henkinization of a set of models—and applied them to the particular case of first-order **FOS4** to prove its completeness with respect to its topological-sheaf semantics. We emphasize that these constructions can in fact be applied to a wider range of logics. For instance, they can be applied to first-order modal logics weaker than **FOS4** and their neighborhood-sheaf semantics [13] (in which T, S4, and even N may fail), yielding a completeness result by replacing McKinsey's and Tarski's [20] construction of open sets by Segerberg's [24] of neighborhoods. Moreover, even though **FOS4** contains classical first-order logic and we applied the completeness theorem for classical first-order logic, de-modalization also works for theories in modal logic that do not contain classical first-order logic: As long as there is a completeness result for the de-modalized version of a given theory, de-modalization extends that result to the theory. For example, a similar approach can be applied to *intuitionistic* **FOS4**, in virtue of [21], or classical *higher-order* **S4**, in virtue of [2], with respect to sheaf models.

Acknowledgments

The research of the second author is currently funded by the VIDI research program 639.072.904, which is financed by the Netherlands Organization for Scientific Research (NWO).

We are grateful to Guram Bezhanishvili, Henrik Forssell, and Dana Scott for helpful discussions of these results, which originally grew from related work [1] of the late Horacio Arló-Costa. We benefited greatly from his influence and input, and take this opportunity to express our gratitude for his generous collaboration and deep sense of loss at his untimely passing.

References

[1] Arló-Costa, H. and E. Pacuit, *First-order classical modal logic*, Studia Logica **84** (2006), pp. 171–210.

[2] Awodey, S. and C. Butz, *Topological completeness for higher-order logic*, Journal of Symbolic Logic **65** (2000), pp. 1168–1182.

[3] Awodey, S. and K. Kishida, *Topology and modality: The topological interpretation of first-order modal logic*, Review of Symbolic Logic **1** (2008), pp. 146–166.

[4] Braüner, T. and S. Ghilardi, *First-order modal logic*, in: P. Blackburn, J. van Benthem and F. Wolter, editors, *Handbook of Modal Logic*, Elsevier, 2007 pp. 549–620.
[5] Butz, C. and I. Moerdijk, *Topological representation of sheaf cohomology of sites*, Compositio Mathematica **118** (1999), pp. 217–233.
[6] Drágalin, A. G., "Mathematical Intuitionism: Introduction to Proof Theory," Translations of Mathematical Monographs **67**, American Mathematical Society, 1988, translated by E. Mendelson from the original in Russian, Nauka, 1979.
[7] Fourman, M. P. and D. S. Scott, *Sheaves and logic*, in: M. P. Fourman, C. J. Mulvey and D. S. Scott, editors, *Applications of Sheaves: Proceedings of the Research Symposium on Applications of Sheaf Theory to Logic, Algebra, and Analysis, Durham, July 9–21, 1977* (1979), pp. 302–401.
[8] Gabbay, D. M., V. Shehtman and D. Skvortsov, "Quantification in Nonclassical Logic, Volume 1," Elsevier, 2009.
[9] Ghilardi, S. and G. Meloni, *Modal and tense predicate logic: Models in presheaves and categorical conceptualization*, in: F. Borceux, editor, *Categorical Algebra and Its Applications: Proceedings of a Conference, Held in Louvain-La-Neuve, Belgium, July 26–August 1, 1987* (1988), pp. 130–142.
[10] Goldblatt, R., "Topoi: The Categorial Analysis of Logic," North-Holland, 1979.
[11] Hilken, B. and D. Rydeheard, *A first order modal logic and its sheaf models*, in: M. Fairtlough, M. Mendler and E. Moggi, editors, *FLoC Satellite Workshop on Intuitionistic Modal Logics and Applications (IMLA'99)*, Trento, Italy, 1999.
[12] Kaplan, D., *Review of [14]*, Journal of Symbolic Logic **31** (1966), pp. 120–122.
[13] Kishida, K., *Neighborhood-sheaf semantics for first-order modal logic*, Electronic Notes in Theoretical Computer Science **278** (2011), pp. 129–143.
[14] Kripke, S., *Semantical analysis of modal logic I: Normal modal propositional calculi*, Zeitschrift für mathematische Logik und Grundlagen der Mathematik **9** (1963), pp. 67–96.
[15] Lawvere, F. W., *Adjointness in foundations*, Dialectica **23** (1969), pp. 281–296.
[16] Lawvere, F. W., *Quantifiers and sheaves*, in: *Actes du Congrès International des Mathématiciens, Nice, 1970*, 1970, pp. 329–334.
[17] Mac Lane, S. and I. Moerdijk, "Sheaves in Geometry and Logic: A First Introduction to Topos Theory," Springer-Verlag, 1992.
[18] Makinson, D., *On some completeness theorems in modal logic*, Zeitschrift für mathematische Logik und Grundlagen der Mathematik **12** (1966), pp. 379–384.
[19] Makkai, M. and G. E. Reyes, *Completeness results for intuitionistic and modal logic in a categorical setting*, Annals of Pure and Applied Logic **72** (1995), pp. 25–101.
[20] McKinsey, J. C. C. and A. Tarski, *The algebra of topology*, Annals of Mathematics **45** (1944), pp. 141–191.
[21] Moerdijk, I., *Some topological spaces which are universal for intuitionistic predicate logic*, Indagationes Mathematicae **85** (1982), pp. 227–235.
[22] Rasiowa, H. and R. Sikorski, "The Mathematics of Metamathematics," Polish Scientific Publishers, 1963.
[23] Reyes, G. E., *A topos-theoretic approach to reference and modality*, Notre Dame Journal of Formal Logic **32** (1991), pp. 359–391.
[24] Segerberg, K., "An Essay in Classical Modal Logic," Number 13 in Filosofiska Studier, Uppsala Universitet, 1971.

Computing Minimal $\mathcal{EL}$-unifiers is Hard

Franz Baader Stefan Borgwardt Barbara Morawska [1]

Theoretical Computer Science, TU Dresden, Germany

Abstract

Unification has been investigated both in modal logics and in description logics, albeit with different motivations. In description logics, unification can be used to detect redundancies in ontologies. In this context, it is not sufficient to decide unifiability, one must also compute appropriate unifiers and present them to the user. For the description logic $\mathcal{EL}$, which is used to define several large biomedical ontologies, deciding unifiability is an NP-complete problem. It is known that every solvable $\mathcal{EL}$-unification problem has a *minimal unifier*, and that every minimal unifier is a *local unifier*. Existing unification algorithms for $\mathcal{EL}$ compute all minimal unifiers, but additionally (all or some) non-minimal local unifiers. Computing only the minimal unifiers would be better since there are considerably less minimal unifiers than local ones, and their size is usually also quite small.
In this paper we investigate the question whether the known algorithms for $\mathcal{EL}$-unification can be modified such that they compute exactly the minimal unifiers without changing the complexity and the basic nature of the algorithms. Basically, the answer we give to this question is negative.

Keywords: Unification, Description Logics, Complexity

1 Introduction

It is well-known that there is a close connection between modal logics (MLs) and description logics (DLs). In fact, many DLs are syntactic variants of classical MLs. Unification has been introduced in both areas [5], with the same formal meaning, but with different applications in mind. In ML, unification [16,17,23] was mainly investigated in the context of the admissibility problem for inference rules [22,18,12]. Unification is simpler than the admissibility problem in the sense that it can easily be reduced to it, but in some cases (e.g., if the unification problem is effectively finitary, i.e., finite complete sets of unifiers can be computed) there is also a reduction in the other direction (see, e.g., [20]). An important open problem in the area is the question whether unification in the basic modal logic K, which corresponds to the DL $\mathcal{ALC}$, is decidable. It is only know that relatively minor extensions of K have an undecidable unification problem [24].

[1] Supported by DFG under grant BA 1122/14-1

Unification in DLs has been introduced as a novel inference service that can be used to detect redundancies in ontologies [10]. For example, assume that one developer of a medical ontology defines the concept of a *patient with severe head injury* as

$$\mathsf{Patient} \sqcap \exists \mathsf{finding}.(\mathsf{Head\_injury} \sqcap \exists \mathsf{severity}.\mathsf{Severe}), \tag{1}$$

whereas another one represents it as

$$\mathsf{Patient} \sqcap \exists \mathsf{finding}.(\mathsf{Severe\_finding} \sqcap \mathsf{Injury} \sqcap \exists \mathsf{finding\_site}.\mathsf{Head}). \tag{2}$$

Formally, these two concept descriptions are not equivalent, but they are nevertheless meant to represent the same concept. They can obviously be made equivalent by treating the concept names $\mathsf{Head\_injury}$ and $\mathsf{Severe\_finding}$ as variables, and substituting the first one by $\mathsf{Injury} \sqcap \exists \mathsf{finding\_site}.\mathsf{Head}$ and the second one by $\exists \mathsf{severity}.\mathsf{Severe}$. In this case, we say that the descriptions are unifiable, and call the substitution that makes them equivalent a *unifier*. Intuitively, such a unifier proposes definitions for the concept names that are used as variables: in our example, we know that, if we define $\mathsf{Head\_injury}$ as $\mathsf{Injury} \sqcap \exists \mathsf{finding\_site}.\mathsf{Head}$ and $\mathsf{Severe\_finding}$ as $\exists \mathsf{severity}.\mathsf{Severe}$, then the two concept descriptions (1) and (2) are equivalent w.r.t. these definitions.

Of course, this example was constructed such that the unifier actually provides sensible definitions for the concept names used as variables. In general, the existence of a unifier only says that there is a structural similarity between the two concepts. The developer that uses unification as a tool for finding redundancies in an ontology or between two different ontologies needs to inspect the unifier(s) to see whether the definitions it suggests really make sense. Thus, a decision procedure for unifiability is not sufficient in this context. One needs a procedure that also produces appropriate unifiers.

Due to the fact that the decidability status of unification in the DL $\mathcal{ALC}$ is a long-standing open problem (at least in its ML variant of unification in K), the work on unification in DLs has mostly concentrated on sub-Boolean fragments of K. Originally, unification in DLs has been investigated in [10] for the DL $\mathcal{FL}_0$, which offers the constructors conjunction ($\sqcap$), value restrictions ($\forall r.C$), and the top-concept ($\top$). However, the usability of unification in this DL is impaired by the facts that, on the one hand, there are almost no ontologies that use only $\mathcal{FL}_0$, and on the other hand, the complexity of the unification problem is quite high (ExpTime-complete).

In this paper, we consider unification in the DL $\mathcal{EL}$, which differs from $\mathcal{FL}_0$ by offering existential restrictions ($\exists r.C$) in place of value restrictions, and thus corresponds to the fragment of K that uses only diamond, conjunction, and the truth constant "true." $\mathcal{EL}$ has recently drawn considerable attention since, on the one hand, important inference problems such as the subsumption problem are polynomial in $\mathcal{EL}$ [1,13]. On the other hand, though quite inexpressive, $\mathcal{EL}$ can be used to define biomedical ontologies. For example, both the large medical ontology SNOMED CT and the Gene Ontology[2] can be expressed in

[2] see http://www.ihtsdo.org/snomed-ct/ and http://www.geneontology.org/

$\mathcal{EL}$. In [7], we were able to show that unification in $\mathcal{EL}$ is of considerably lower complexity than unification in $\mathcal{FL}_0$: the decision problem for $\mathcal{EL}$ is NP-complete. The main steps in the proof of this statement given in [7] were the following. First, the inverse subsumption order on concept descriptions was used to define an order on substitutions:

$$\sigma \succeq \theta \text{ iff } \sigma(X) \sqsubseteq \theta(X) \text{ holds for all variables } X,$$

and it was shown that this order is well-founded. As an immediate consequence of the well-foundedness of $\succeq$, every solvable unification problem has a minimal unifier. Second, it was shown that every minimal unifier is a local substitution, where local substitutions are built from a polynomial number of so-called atoms determined by the unification problem. Finally, a brute-force "guess and then test" NP-algorithm was described, which guesses a local substitution and then checks (in polynomial time) whether it is a unifier.

An obvious disadvantage of this brute-force algorithm is that it blindly guesses a local substitution and only afterwards checks whether the guessed substitution is a unifier. Thus, in general many substitutions will be generated that only in the subsequent check turn out not to be unifiers. In contrast, the SAT reduction presented in [8] is such that only unifiers are generated. To be more precise, it was shown in [8] how a given unification problem Γ can be translated in polynomial time into a propositional formula ϕ_Γ such that the satisfying valuations of ϕ_Γ correspond to the local unifiers of Γ. The translation into SAT allows us to employ existing highly optimized state-of-the-art SAT solvers for implementing the unification algorithm. While this yields a quite efficient decision procedure for unifiability, the fact that all local unifiers, rather than only minimal ones, are generated turned out to be problematic if one wants to show the unifiers to the user. In fact, even very small unification problems can have hundreds of local unifiers, many of which do not make sense in the application. The set of all minimal unifiers is a subset of the set of all local unifiers, whose cardinality is usually much smaller.[3] Another advantage of minimal unifiers is that they are usually of smaller size (where the size of a substitution is the sum of the sizes of the concept terms substituted for the variables), and are thus easier to read and comprehend.

In [9] we describe a goal-oriented unification algorithm for $\mathcal{EL}$, in which nondeterministic decisions are only made if they are triggered by "unsolved parts" of the unification problem. By construction, this algorithm can only compute local unifiers, and it is shown in [9] that all minimal ones are among the ones computed by it. Though in our initial tests the number of unifiers computed by the goal-oriented algorithm turned out to be usually much smaller than of the ones computed by the SAT reduction, the goal-oriented algorithm is not guaranteed to compute only minimal unifiers.

[3] In the above example, the unifier we have described is the only minimal unifier, but the SAT-translation computes 64 local unifiers, albeit first the minimal one.

Name	Syntax	Semantics
concept name	A	$A^{\mathcal{I}} \subseteq \Delta^{\mathcal{I}}$
role name	r	$r^{\mathcal{I}} \subseteq \Delta^{\mathcal{I}} \times \Delta^{\mathcal{I}}$
top-concept	$\top$	$\top^{\mathcal{I}} = \Delta^{\mathcal{I}}$
conjunction	$C \sqcap D$	$(C \sqcap D)^{\mathcal{I}} = C^{\mathcal{I}} \cap D^{\mathcal{I}}$
existential restriction	$\exists r.C$	$(\exists r.C)^{\mathcal{I}} = \{x \mid \exists y : (x, y) \in r^{\mathcal{I}} \wedge y \in C^{\mathcal{I}}\}$

Table 1
Syntax and semantics of $\mathcal{EL}$.

Following the assumption that it is desirable to compute only minimal unifiers rather than all (or some additional non-minimal) local ones, this paper asks the question whether the NP decision procedures for unification in $\mathcal{EL}$ presented in [8] and [9] can be appropriately modified such that the successful runs of the procedure produce exactly the minimal unifiers of the given $\mathcal{EL}$-unification problem. We show in Section 4 that the answer to this question is negative if we use a slightly more general definition of the order $\succeq$, where the subsumption test $\sigma(X) \sqsubseteq \theta(X)$ can be restricted to a subset of all variables. This restriction is justified by the fact that the user may be interested only in the substitution-images of some of the variables. In fact, the algorithms in [8] and [9] first flatten the input problem, which introduces auxiliary variables. These auxiliary variables are internal to the unification procedure and are not shown to the user.

All three $\mathcal{EL}$-unification algorithms mentioned above (the brute-force algorithm, the goal-oriented algorithm, and the one based on a reduction to SAT) actually do not directly compute local unifiers, but so-called acyclic assignments, which can be seen as compact representations of local unifiers. In Section 3 we ask what properties of the acyclic assignment make the induced unifiers small. To this purpose, we introduce a natural order on acyclic assignments and compare it with the order $\succeq$ on the induced unifiers.

2 Unification in $\mathcal{EL}$

Starting with a finite set N_C of *concept names* and a finite set N_R of *role names*, *$\mathcal{EL}$-concept descriptions* are built using the concept constructors *top-concept* ($\top$), *conjunction* ($C \sqcap D$), and *existential restriction* ($\exists r.C$ for every $r \in N_R$).

An *interpretation* $\mathcal{I} = (\Delta^{\mathcal{I}}, \cdot^{\mathcal{I}})$ consists of a nonempty domain $\Delta^{\mathcal{I}}$ and an interpretation function $\cdot^{\mathcal{I}}$ that assigns binary relations on $\Delta^{\mathcal{I}}$ to role names and subsets of $\Delta^{\mathcal{I}}$ to concept descriptions, as shown in the semantics column of Table 1.

The concept description C *is subsumed by* the concept description D (written $C \sqsubseteq D$) iff $C^{\mathcal{I}} \subseteq D^{\mathcal{I}}$ holds for all interpretations $\mathcal{I}$. We say that C *is equivalent to* D (written $C \equiv D$) iff $C \sqsubseteq D$ and $D \sqsubseteq C$, i.e., iff $C^{\mathcal{I}} = D^{\mathcal{I}}$ holds for all interpretations $\mathcal{I}$.

We will also need the notion of an acyclic TBox $\mathcal{T}$, which is a finite set of concept definitions of the form $A \equiv C$, where A is a concept name and C a concept description, that is unambiguous and acyclic (see [4] for details). The interpretation $\mathcal{I}$ is a model of $\mathcal{T}$ iff it satisfies all concept definitions in $\mathcal{T}$, i.e., $A^{\mathcal{I}} = C^{\mathcal{I}}$ holds for all $A \equiv C$ in $\mathcal{T}$. The concept description C *is subsumed by* the concept description D w.r.t. the acyclic TBox $\mathcal{T}$ (written $C \sqsubseteq_{\mathcal{T}} D$) iff $C^{\mathcal{I}} \subseteq D^{\mathcal{I}}$ holds for all models $\mathcal{I}$ of $\mathcal{T}$.

An $\mathcal{EL}$-concept description is an *atom* if it is an existential restriction or a concept name. The atoms of an $\mathcal{EL}$-concept description C are the subdescriptions of C that are atoms, and the top-level atoms of C are the atoms occurring in the top-level conjunction of C. Obviously, any $\mathcal{EL}$-concept description is the conjunction of its top-level atoms, where the empty conjunction corresponds to the top-concept $\top$.

When defining unification in $\mathcal{EL}$, we assume that the set of concept names is partitioned into a set N_v of concept variables (which may be replaced by substitutions) and a set N_c of concept constants (which must not be replaced by substitutions). A *substitution* σ is a mapping from N_v into the set of all $\mathcal{EL}$-concept descriptions. This mapping is extended to concept descriptions in the usual way, i.e., by replacing all occurrences of variables in the description by their σ-images. Unification tries to make concept descriptions equivalent by applying a substitution.

Definition 2.1 *An* $\mathcal{EL}$-unification problem *is of the form* $\Gamma = \{C_1 \equiv^? D_1, \ldots, C_n \equiv^? D_n\}$, *where* $C_1, D_1, \ldots C_n, D_n$ *are* $\mathcal{EL}$*-concept descriptions. The substitution* σ *is a* unifier *(or* solution*) of* Γ *iff* $\sigma(C_i) \equiv \sigma(D_i)$ *for* $i = 1, \ldots, n$. *In this case,* Γ *is called* solvable *or* unifiable.

We will sometimes use the subsumption $C \sqsubseteq^? D$ as abbreviation for the equivalence $C \sqcap D \equiv^? C$. Obviously, the substitution σ solves this subsumption iff $\sigma(C) \sqsubseteq \sigma(D)$.

Flattening

As mentioned before, the algorithms in [8] and [9] first flatten the unification problem. An atom is called *flat* if it is a concept name or an existential restriction of the form $\exists r.A$ for a concept name A. The unification problem Γ is called *flat* if it contains only flat subsumptions of the form $C_1 \sqcap \cdots \sqcap C_n \sqsubseteq^? D$, where $n \geq 0$ and $C_1, \ldots, C_n, D$ are flat atoms. [4]

Let Γ be a unification problem. By introducing auxiliary variables, Γ can be transformed in polynomial time into a flat unification problem Γ' such that the unifiability status remains unchanged, i.e., Γ has a unifier iff Γ' has a unifier. More precisely, it can be shown that, restricted to the variables of Γ, every unifier of Γ' is also a unifier of Γ. Conversely, every unifier of Γ can be extended to a unifier of Γ' by defining appropriate images for the auxiliary variables. Thus, we may assume without loss of generality that our input $\mathcal{EL}$-unification problems are flat.

[4] If $n = 0$, then we have an empty conjunction on the left-hand side, which as usual stands for $\top$.

Local unifiers

Let Γ be a flat unification problem. The atoms of Γ are the atoms of all the concept descriptions occurring in Γ. We define

$$\mathrm{At} := \{C \mid C \text{ is an atom of } \Gamma\} \text{ and}$$

$$\mathrm{At}_{\mathsf{nv}} := \mathrm{At} \setminus N_v \quad \text{(non-variable atoms)}.$$

Every assignment S of subsets S_X of $\mathrm{At}_{\mathsf{nv}}$ to the variables X in N_v induces the following relation $>_S$ on N_v: $>_S$ is the transitive closure of

$$O_S := \{(X, Y) \in N_v \times N_v \mid Y \text{ occurs in an element of } S_X\}.$$

We call the assignment S *acyclic* if $>_S$ is irreflexive (and thus a strict partial order). Any acyclic assignment S induces a unique substitution σ_S, which can be defined by induction along $>_S$:

- If X is a minimal element of N_v w.r.t. $>_S$, then we define $\sigma_S(X) := \bigsqcap_{D \in S_X} D$.
- Assume that $\sigma_S(Y)$ is already defined for all Y such that $X >_S Y$. Then we define $\sigma_S(X) := \bigsqcap_{D \in S_X} \sigma_S(D)$.

We call a substitution σ *local* if it is of this form, i.e., if there is an acyclic assignment S such that $\sigma = \sigma_S$. If the unifier σ of Γ is a local substitution, then we call it a *local unifier* of Γ.

Theorem 2.2 ([7]) *Let Γ be a flat unification problem. If Γ has a unifier, then it also has a local unifier.*

The theorem shows that, in order to decide unifiability of Γ, it is sufficient to guess an acyclic assignment and then check whether the induced substitution is a unifier. The remaining problem is that the induced unifier may be of exponential size. However, in order to check whether a given acyclic assignment S induces a unifier of Γ, one does not need to construct the unifier σ_S explicitly. In fact, S can be turned into an acyclic TBox

$$\mathcal{T}_S := \{X \equiv \bigsqcap_{D \in S_X} D \mid X \in N_v\},$$

and it is easy to see that the following holds for arbitrary concept descriptions E, F: $\sigma_S(E) \sqsubseteq \sigma_S(F)$ iff $E \sqsubseteq_{\mathcal{T}_S} F$. Since subsumption in $\mathcal{EL}$ w.r.t. acyclic TBoxes can be decided in polynomial time [1], this obviously yields a way for checking, in polynomial time, whether σ_S solves all equations of Γ.

The original proof of Theorem 2.2 in [7] was based on the notion of a minimal unifier, though subsequent simpler proofs [8,3] no longer need this notion.

For readers familiar with unification theory [11], it should be pointed out that the order we use to define minimality of local unifiers (see below) is *not* the instantiation pre-order on substitutions. In fact, it is an easy consequence of the definition of local substitutions that they are ground (i.e., the images of variables under these substitutions do not contain variables), and thus there is no further instantiation possible.

Minimal unifiers

Given a set of variables $\mathcal{X} \subseteq N_v$, we define

$$\sigma \succeq_{\mathcal{X}} \theta \text{ iff } \sigma(X) \sqsubseteq \theta(X) \text{ holds for all variables } X \in \mathcal{X},$$

$$\sigma \succ_{\mathcal{X}} \theta \text{ iff } \sigma \succeq_{\mathcal{X}} \theta \text{ and } \theta \not\succeq_{\mathcal{X}} \sigma.$$

We say that the unifier σ of Γ is *$\mathcal{X}$-minimal* iff there is no unifier θ of Γ such that $\sigma \succ_{\mathcal{X}} \theta$. We say that two substitutions σ, θ are *equivalent* ($\sigma \equiv \theta$) iff $\sigma(X) \equiv \theta(X)$ holds for all $X \in N_v$. Note that we have $\sigma \equiv \theta$ iff $\sigma \succeq_{N_v} \theta$ and $\theta \succeq_{N_v} \sigma$.

Lemma 2.3 (**[9]**) *Let Γ be a flat unification problem.*

(i) *If Γ is solvable, then it also has an N_v-minimal unifier.*

(ii) *Every N_v-minimal unifier is equivalent to a local unifier.*

The first part of the lemma is an immediate consequence of the fact [9] that $\succ_{N_v}$ is well-founded, whereas the proof of the second part in [9] is rather long and intricate. Theorem 2.2 is an immediate consequence of Lemma 2.3.

3 Minimal unifiers versus minimal assignments

As mentioned before, we are interested in computing only the N_v-minimal unifiers rather than all local unifiers of a given unification problem. All three $\mathcal{EL}$-unification algorithms mentioned in the introduction (the brute-force algorithm, the goal-oriented algorithm, and the one based on a reduction to SAT) actually compute acyclic assignments rather than directly local unifiers. Thus, one can ask what properties of the assignment make unifiers small w.r.t. $\succeq_{N_v}$. To answer this question, we define an order similar to $\succeq_{N_v}$ on acyclic assignments. Let S, T be acyclic assignments of subsets S_X, T_X of $\mathsf{At}_{\mathsf{nv}}$ to the variables X in N_v. We define

$$S \geq T \text{ iff } S_X \supseteq T_X \text{ holds for all } X \in N_v.$$

As usual, we write $S > T$ if $S \geq T$ and $S \neq T$. First, we show that smaller assignments indeed yield smaller unifiers.

Lemma 3.1 *If $S \geq T$, then $\sigma_S \succeq_{N_v} \sigma_T$.*

Proof. Obviously, $S \geq T$ implies $O_S \supseteq O_T$, and thus $>_S \supseteq >_T$. We show $\sigma_S(X) \sqsubseteq \sigma_T(X)$ for all $X \in N_v$ by induction along $>_S$.

If X is a minimal element of N_v w.r.t. $>_S$, then it is also a minimal element of N_v w.r.t. $>_T$ since $>_S \supseteq >_T$. Thus, $\sigma_S(X) = \bigsqcap_{D \in S_X} D$ and $\sigma_T(X) = \bigsqcap_{E \in T_X} E$. Consequently, $S_X \supseteq T_X$ implies that $\sigma_S(X) \sqsubseteq \sigma_T(X)$.

Assume that $\sigma_S(Y) \sqsubseteq \sigma_T(Y)$ holds for all Y such that $X >_S Y$. Since $>_S \supseteq >_T$, this implies that $\sigma_S(Z) \sqsubseteq \sigma_T(Z)$ holds for all Z such that $X >_T Z$. Since the concept constructors of $\mathcal{EL}$ are monotonic w.r.t. subsumption and $S_X \supseteq T_X$, this implies

$$\sigma_S(X) = \bigsqcap_{D \in S_X} \sigma_S(D) \sqsubseteq \bigsqcap_{D \in T_X} \sigma_S(D) \sqsubseteq \bigsqcap_{D \in T_X} \sigma_T(D) = \sigma_T(X). \qquad \square$$

As an easy consequence of this lemma we obtain that minimal unifiers are induced by minimal acyclic assignments.

Theorem 3.2 *Let Γ be a flat unification problem. Then the set*

$$\{\sigma_S \mid \sigma_S \text{ is a unifier of } \Gamma \text{ and there is no acyclic assignment } T \text{ for } \Gamma \text{ such that } \sigma_T \text{ is a unifier of } \Gamma \text{ and } S > T\}$$

contains all N_v-minimal unifiers of Γ up to equivalence.

Proof. Let θ be an N_v-minimal unifier of Γ. By Lemma 2.3, θ is (equivalent to) a local unifier, and thus there exists an acyclic assignment T such that $\theta \equiv \sigma_T$. Let S be minimal among all assignments that induce a substitution equivalent to θ, i.e., $\theta \equiv \sigma_S$ and there is no acyclic assignment T for Γ such that $\sigma_T \equiv \theta$ and $S > T$.

We claim that this implies that there is no acyclic assignment T for Γ such that σ_T is a unifier of Γ and $S > T$. Assume that such an assignment T exists. Then Lemma 3.1 implies that $\sigma_S \succeq_{N_v} \sigma_T$. Minimality of S among all assignments that induce a unifier equivalent to θ implies that $\sigma_S \not\equiv \sigma_T$, and thus $\sigma_S \succ_{N_v} \sigma_T$. This contradicts the assumption that $\theta \equiv \sigma_S$ is N_v-minimal.□

Thus, if one wants to generate *all* minimal unifiers, it is enough to generate only the minimal acyclic assignments yielding unifiers. If the converse of Lemma 3.1 were true, we could also show that these assignments yield *only* minimal unifiers. Unfortunately, the converse of Lemma 3.1 is not true, as demonstrated by the following example.

Example 3.3 Let

$$\Gamma := \{X \equiv^? \exists r.Y, X \equiv^? \exists r.Z, Y \equiv^? A, Z \equiv^? A\}.$$

Consider the acyclic assignments S, T with

$$S_X := \{\exists r.Y\},\ S_Y := \{A\},\ S_Z := \{A\};$$
$$T_X := \{\exists r.Z\},\ T_Y := \{A\},\ T_Z := \{A\}.$$

Then $\sigma_S(Y) = A = \sigma_T(Y)$, $\sigma_S(Z) = A = \sigma_T(Z)$, and $\sigma_S(X) = \exists r.A = \sigma_T(X)$, i.e., $\sigma_S = \sigma_T$ and this substitution is a unifier of Γ. In particular, this implies $\sigma_S \succeq_{N_v} \sigma_T$. However, $S \geq T$ obviously does *not* hold since $S_X \not\supseteq T_X$.

It is easy to see that S and T are minimal among the acyclic assignments generating unifiers of Γ. This shows that the same N_v-minimal unifier can be generated by different minimal assignments.

We can also use a unifier σ of Γ to define an acyclic assignment S^σ:

$$S^\sigma_X := \{D \in \mathsf{At}_{\mathsf{nv}} \mid \sigma(X) \sqsubseteq \sigma(D)\}.$$

As shown in [3], this assignment is indeed acyclic.

Surprisingly, the analog of Lemma 3.1 does not hold: going from unifiers to the induced acyclic assignments is neither monotone nor antitone.

Example 3.4 We introduce an $\mathcal{EL}$-unification problem that demonstrates that $\theta \succeq_{N_v} \sigma$ implies neither $S^\theta \geq S^\sigma$ nor $S^\sigma \geq S^\theta$, even if σ and θ are equivalent to local unifiers of the given unification problem Γ. For this purpose, consider the unification problem

$$\Gamma := \{A \sqsubseteq^? X,\ \exists r.X \sqsubseteq^? Y,\ Y \sqsubseteq^? \exists r.X'\}.$$

The following substitutions are obviously unifiers of Γ:

$$\sigma := \{X \mapsto \top, X' \mapsto \top, Y \mapsto \exists r.\top\},$$
$$\theta := \{X \mapsto A, X' \mapsto \top, Y \mapsto \exists r.\top\}.$$

and they satisfy $\theta \succeq_{N_v} \sigma$.

The non-variable atoms of Γ are A, $\exists r.X$, and $\exists r.X'$, and thus

$$S^\sigma_X = \emptyset, \quad S^\sigma_{X'} = \emptyset, \quad S^\sigma_Y = \{\exists r.X, \exists r.X'\},$$
$$S^\theta_X = \{A\}, \quad S^\theta_{X'} = \emptyset, \quad S^\theta_Y = \{\exists r.X'\}.$$

Since $S^\sigma_Y \not\subseteq S^\theta_Y$, we do *not* have $S^\theta \geq S^\sigma$; and since $S^\theta_X \not\subseteq S^\sigma_X$ we do *not* have $S^\sigma \geq S^\theta$.

Finally, note that (up to equivalence) σ and θ are local since they are the substitutions respectively induced by S^σ and S^θ, i.e., $\sigma \equiv \sigma_{S^\sigma}$ and $\theta \equiv \sigma_{S^\theta}$.

This example actually strengthens Example 3.3 in the sense that it shows that the converse of Lemma 3.1 is not even true if we assume $\sigma_S \succ_{N_v} \sigma_T$ rather than $\sigma_S \succeq_{N_v} \sigma_T$. Just take $S = S^\theta$ and $T = S^\sigma$. Indeed, we have $\sigma_{S^\theta} \succ_{N_v} \sigma_{S^\sigma}$, but $S^\theta \not\geq S^\sigma$.

A similar example can be used to show that the set

$$\{\sigma_S \mid \sigma_S \text{ is a unifier of } \Gamma \text{ and there is no acyclic assignment } T \text{ for } \Gamma \text{ such that } \sigma_T \text{ is a unifier of } \Gamma \text{ and } S > T\}$$

may in general contain unifiers that are not N_v-minimal.

Example 3.5 Let

$$\Gamma := \{A \sqsubseteq^? X,\ Y \equiv^? \exists r.X,\ \exists r.A \sqsubseteq^? Y\}.$$

The non-variable atoms of Γ are A, $\exists r.X$, and $\exists r.A$. The acyclic assignments

$$S_X = \emptyset, \quad S_Y = \{\exists r.X\},$$
$$T_X = \{A\}, \quad T_Y = \{\exists r.A\}$$

generate the unifiers $\sigma_S = \{X \mapsto \top, Y \mapsto \exists r.\top\}$ and $\sigma_T = \{X \mapsto A, Y \mapsto \exists r.A\}$ of Γ. We have $\sigma_T \succ_{N_v} \sigma_S$, and thus σ_T is not N_v-minimal. However, it is easy to see that there is no acyclic assignment $U < T$ such that σ_U is a unifier of Γ.

In fact, assume that $U < T$. If $U_Y = \emptyset$, then $\sigma_U(Y) = \top$, and thus σ_U does not solve the equivalence $Y \equiv^? \exists r.X$ independent of whether $U_X = \{A\}$ or $U_X = \emptyset$. Consequently, we must have $U_Y = \{\exists r.A\} = T_Y$, and thus $\sigma_U(Y) = \exists r.A$. However, then $U < T$ implies $U_X = \emptyset$, i.e., $\sigma_U(X) = \top$. But then σ_U again does not solve the equivalence $Y \equiv^? \exists r.X$.

This example shows that, even if we only generate minimal acyclic assignments that induce unifiers, this may yield additional local unifiers that are not N_v-minimal.

We finish this section by investigating what happens if we compose the two transformations $S \mapsto \sigma_S$ and $\sigma \mapsto S^\sigma$.

Lemma 3.6 *Let S be an assignment and σ a substitution. Then $S^{\sigma_S} \geq S$ and $\sigma \succeq_{N_v} \sigma_{S^\sigma}$. If σ is a local substitution, then we even have $\sigma \equiv \sigma_{S^\sigma}$.*

Proof. If $D \in S_X$, then $\sigma_S(D)$ is a top-level conjunct of $\sigma_S(X)$, and thus $\sigma_S(X) \sqsubseteq \sigma_S(D)$, which shows $D \in S_X^{\sigma_S}$.

We show the other inequality by induction along $>_{S^\sigma}$. If X is minimal, then no variables occur in S_X^σ, and thus $\sigma(D) = D$ for all $D \in S_X^\sigma$. This yields $\sigma_{S^\sigma}(X) = \bigsqcap_{D \in S_X^\sigma} D \sqsupseteq \sigma(X)$ since all $D \in S_X^\sigma$ satisfy $\sigma(X) \sqsubseteq \sigma(D) = D$.

Assume that for all $Y \in N_v$ with $X >_{S^\sigma} Y$ we have $\sigma_{S^\sigma}(Y) \sqsupseteq \sigma(Y)$. Consider $\sigma_{S^\sigma}(X) = \bigsqcap_{D \in S_X^\sigma} \sigma_S^\sigma(D)$. Since $D \in S_X^\sigma$ contains only variables that are smaller than X w.r.t. $>_{S^\sigma}$ and the concept constructors of $\mathcal{EL}$ are monotone w.r.t. subsumption, the induction assumption yields

$$\bigsqcap_{D \in S_X^\sigma} \sigma_S^\sigma(D) \sqsupseteq \bigsqcap_{D \in S_X^\sigma} \sigma(D).$$

Finally, since all $D \in S_X^\sigma$ satisfy $\sigma(X) \sqsubseteq \sigma(D)$, we have

$$\bigsqcap_{D \in S_X^\sigma} \sigma(D) \sqsupseteq \sigma(X).$$

Since the subsumption relation is transitive, this completes the proof that $\sigma \succeq_{N_v} \sigma_{S^\sigma}$.

Finally, assume that σ is local, i.e., there is an acyclic assignment T such that $\sigma = \sigma_T$. We must show $\sigma_{S^\sigma} \succeq_{N_v} \sigma$. Because of the first statement of the lemma, we have $S^{\sigma_T} \geq T$, and thus $\sigma_{S^\sigma} = \sigma_{S^{\sigma_T}} \succeq_{N_v} \sigma_T = \sigma$ by Lemma 3.1.□

4 The complexity of computing exactly the minimal unifiers

The three $\mathcal{EL}$-unification algorithms mentioned in the introduction (the brute-force algorithm, the goal-oriented algorithm, and the one based on a reduction to SAT) are NP-decision procedures for unifiability that additionally compute local unifiers in the following sense: each successful run of the nondeterministic algorithm generates an acyclic assignment that induces a unifier. The brute-force algorithm and the SAT-based algorithm generate all local unifiers, whereas the goal-oriented algorithm generates all N_v-minimal unifiers, but may also generate some additional, non-minimal local unifiers. In [2] we sketch a variant of the SAT reduction of [8] that generates exactly the minimal assignments that induce unifiers. It is based on a reduction into a special case of the partial MAX-SAT problem [21]. By Lemma 3.2, this procedure generates all N_v-minimal unifiers, but Example 3.5 shows that—like the goal-oriented algorithm—it may also produce additional, non-minimal local unifiers. Unlike the other three algorithms, this is no longer an NP-algorithm.

In this section we investigate the question whether there can exist an NP-algorithm that produces exactly the minimal unifiers in the sense that the successful runs of this algorithm yield a set of acyclic assignments that induces exactly the set of minimal unifiers. For the general case of $\mathcal{X}$-minimality for an arbitrary subset $\mathcal{X}$ of N_v, we give a negative answer to this question. We also show an analogous result for the problem of computing $\mathcal{X}$-minimal assignments that induce unifiers, where $\mathcal{X}$-minimality for assignments is defined in the obvious way.

To show these negative results, we consider the following decision problem, which we call the *minimal unifier containment problem*:

Given: A flat $\mathcal{EL}$-unification problem Γ, a set $\mathcal{X} \subseteq N_v$, a concept constant $A \in N_c$, and a concept variable $X \in \mathcal{X}$.

Question: Is there an $\mathcal{X}$-minimal unifier σ of Γ such that $\sigma(X) \sqsubseteq A$?

Theorem 4.1 *The minimal unifier containment problem is Σ_2^p-complete.*

Proof. Containment in Σ_2^p is easy to see. Guess an acyclic assignment S of Γ and check (in polynomial time, using $\mathcal{T}_S$) whether it induces a unifier σ_S of Γ that satisfies $\sigma_S(X) \sqsubseteq A$. If this check succeeds, then use an NP-oracle to check whether σ_S is $\mathcal{X}$-minimal. In fact, an NP procedure for testing whether σ_S is *not* $\mathcal{X}$-minimal guesses an acyclic assignment T, and then uses subsumption tests w.r.t. $\mathcal{T}_S$ to check whether $\sigma_S \succ_{\mathcal{X}} \sigma_T$.

To show Σ_2^p-hardness, we use a reduction from the *minimal model deduction problem*:

Given: A propositional formula ϕ in conjunctive normal form and a propositional variable x.

Question: Is there a minimal model M of ϕ such that $M \models x$?

Here, minimality of propositional models is defined w.r.t. the following order on propositional valuations: $V' \geq V$ iff for all propositional variables x, $V \models x$ implies $V' \models x$. Σ_2^p-completeness of the minimal model deduction problem is an immediate consequence of Lemma 3.1 in [15].

In order to reduce the minimal model deduction problem (as specified above) to the minimal unifier containment problem, we adapt the proof of NP-hardness of $\mathcal{EL}$-matching given in [6]. Let $\phi = \phi_1 \wedge \cdots \wedge \phi_m$ be a propositional formula in conjunctive normal form and let $\{x_1, \ldots, x_n\}$ be the propositional variables of this problem. Assume without loss of generality that $x = x_1$.

For the propositional variables, we introduce the concept variables

$$\{X_1, \ldots, X_n, \overline{X}_1, \ldots, \overline{X}_n\},$$

which encode x_i and $\neg x_i$, respectively. In addition, we introduce concept variables $\{Y_1, \ldots, Y_n\}$, which are used for minimization, i.e.,

$$\mathcal{X} = \{Y_1, \ldots, Y_n\}.$$

Furthermore, we need concept constants A and B (encoding the truth values) and a role name r. The unification problem $\Gamma_{\phi,x}$ constructed from the given minimal model deduction problem consists of the equations introduced below.

First, we specify equations that ensure that A, B encode the truth values. For all $i, 1 \leq i \leq n$, we add the equation

$$\exists r.X_i \sqcap \exists r.\overline{X}_i \equiv^? \exists r.A \sqcap \exists r.B.$$

Obviously, any solution of this equation replaces

- either X_i by a concept description equivalent to A and $\overline{X}_i$ by a concept description equivalent to B (corresponding to $x_i = \mathit{true}$),
- or X_i by a concept description equivalent to B and $\overline{X}_i$ by a concept description equivalent to A (corresponding to $x_i = \mathit{false}$).

In order to encode ϕ, we introduce an equation for every conjunct ϕ_j of ϕ, where we view ϕ_j as the set of its disjuncts:

$$B \sqcap \bigsqcap_{x_i \in \phi_j} X_i \sqcap \bigsqcap_{\neg x_i \in \phi_j} \overline{X}_i \equiv^? A \sqcap B$$

For example, if $\phi_j = x_1 \vee \overline{x}_2 \vee x_3 \vee \overline{x}_4$, then the corresponding equation is $B \sqcap X_1 \sqcap \overline{X}_2 \sqcap X_3 \sqcap \overline{X}_4 \equiv^? A \sqcap B$. The above equation ensures that, among all the concept variables occurring on the left-hand side, at least one must be replaced by a concept description equivalent to A. This corresponds to the fact that, in the conjunct ϕ_j, there must be at least one literal that evaluates to *true*. Note that we need the concept name B on the left-hand side to cover the case where all the variables occurring in it are substituted with A.

It is easy to see that (up to equivalence) the unifiers of the equations introduced until now (which do not contain the variables in $\mathcal{X}$) correspond exactly to the propositional models of ϕ. Given a propositional valuation V of ϕ, we define the corresponding substitution σ_V as follows:

- if $V \models x_i$, then $\sigma_V(X_i) := A$ and $\sigma_V(\overline{X}_i) := B$;
- if $V \models \neg x_i$, then $\sigma_V(X_i) := B$ and $\sigma_V(\overline{X}_i) := A$.

According to our observations above, V is a model of ϕ iff σ_V is a unifier of the equations introduced above. In addition, any unifier of these equations is equivalent to a unifier of the form σ_M for a model M of ϕ.

It remains to express minimal models as $\mathcal{X}$-minimal unifiers. For this purpose, we add the equations

$$A \sqcap B \equiv^? B \sqcap \overline{X}_i \sqcap Y_i \tag{3}$$

for all $i, 1 \leq i \leq n$. This completes the description of the unification problem $\Gamma_{\phi,x}$.

The effect of equation (3) is the following:

- If X_i is substituted with a concept description equivalent to A (corresponding to x_i being evaluated to *true*), then $\overline{X}_i$ is substituted with a concept description equivalent to B, and thus Y_i must be substituted by a concept description equivalent to A or $A \sqcap B$. In an $\mathcal{X}$-minimal unifier, it is thus substituted with a concept description equivalent to A.

- If X_i is substituted with a concept description equivalent to B (corresponding to x_i being evaluated to *false*), then $\overline{X}_i$ is substituted with a concept description equivalent to A, and thus Y_i can be substituted by a concept description equivalent to $\top$, A, B, or $A \sqcap B$. In an $\mathcal{X}$-minimal unifier, it is thus substituted with a concept description equivalent to $\top$.

We extend the definition of the substitution σ_V induced by a propositional valuation V by setting:

- if $\sigma_V(X_i) = A$, then $\sigma_V(Y_i) := A$;
- if $\sigma_V(X_i) = B$, then $\sigma_V(Y_i) := \top$.

We claim that the minimal models of ϕ correspond to the $\mathcal{X}$-minimal unifiers of $\Gamma_{\phi,x}$.

Let M be a minimal model of ϕ, and σ_M the corresponding unifier of $\Gamma_{\phi,x}$, as defined above. Assume that σ_M is not $\mathcal{X}$-minimal. Then there is an $\mathcal{X}$-minimal unifier θ of $\Gamma_{\phi,x}$ such that $\sigma_M \succ_{\mathcal{X}} \theta$. Define the propositional valuation U by setting $U(x_i) := \mathit{true}$ iff $\theta(X_i) \equiv A$. We claim that $\theta \equiv \sigma_U$. For $X \in \{X_1, \ldots, X_n, \overline{X}_1, \ldots, \overline{X}_n\}$, we clearly have $\theta(X) \equiv \sigma_U(X)$. For $X \in \{Y_1, \ldots, Y_n\}$, $\theta(X) \equiv \sigma_U(X)$ is a consequence of the fact that, for a $\mathcal{X}$-minimal unifier θ, $\theta(Y_i) \equiv A$ iff $\theta(X_i) \equiv A$ and $\theta(Y_i) \equiv \top$ iff $\theta(X_i) \equiv B$. Since θ is a unifier of $\Gamma_{\phi,x}$, the same is true for σ_U, and thus U is a model of ϕ. However, it is easy to see that $\sigma_M \succ_{\mathcal{X}} \theta \equiv \sigma_U$ implies that $M > U$, which contradicts minimality of M. In fact, assume that $U \models x_i$, i.e., $\sigma_U(X_i) = A$. Then $\sigma_U(Y_i) = A$, which implies $\sigma_M(Y_i) = A$ (due to $\sigma_M \succeq_{\mathcal{X}} \sigma_U$), and thus $M \models x_i$. This shows $M \geq U$. Since $\sigma_M \succ_{\mathcal{X}} \sigma_U$, there is an index i such that $\sigma_M(Y_i) \sqsubset \sigma_U(Y_i)$. This is only possible if $\sigma_M(Y_i) = A$ and $\sigma_U(Y_i) = \top$. But then $\sigma_M(X_i) = A$ and $\sigma_U(X_i) = B$, and thus $M \models x_i$ and $U \not\models x_i$. This yields $M > U$. To sum up, we have shown:

If M is a minimal model of ϕ, then σ_M is an $\mathcal{X}$-minimal unifier of $\Gamma_{\phi,x}$.

Conversely, assume that θ is a minimal unifier of $\Gamma_{\phi,x}$. As shown above, the propositional valuation U defined as $U(x_i) := \mathit{true}$ iff $\theta(X_i) \equiv A$ is such that U is a model of ϕ and $\theta \equiv \sigma_U$. We claim that U is a *minimal* model of ϕ. Assume that M is a model of ϕ such that $U > M$. First, note that $U \geq M$ implies $\sigma_U \succeq_{\mathcal{X}} \sigma_M$, i.e., $\sigma_U(Y_i) \sqsubseteq \sigma_M(Y_i)$ for all $i, 1 \leq i \leq n$. To see this, it is enough to show that $\sigma_M(Y_i) = A$ implies $\sigma_U(Y_i) = A$. However, $\sigma_M(Y_i) = A$ implies $\sigma_M(X_i) = A$, which in turn implies $M \models x_i$. But then $U \geq M$ yields $U \models x_i$, and thus $\sigma_U(X_i) = A$, which finally implies $\sigma_U(Y_i) = A$. Since $U > M$, there is an index i such that $U \models x_i$, but $M \not\models x_i$. But then $\sigma_U(Y_i) = A$ and $\sigma_M(Y_i) = \top$, and thus $\sigma_U(Y_i) \sqsubset \sigma_M(Y_i)$. This shows $\sigma_U \succ_{\mathcal{X}} \sigma_M$, which contradicts the $\mathcal{X}$-minimality of $\theta \equiv \sigma_U$. To sum up, we have shown:

If θ is an $\mathcal{X}$-minimal unifier of $\Gamma_{\phi,x}$, then there is a minimal model M of ϕ

such that $\theta \equiv \sigma_M$.

To finish the proof of the theorem, first assume that there is a minimal model M of ϕ such that $M \models x_1$. Then the $\mathcal{X}$-minimal unifier σ_M of $\Gamma_{\phi,x}$ satisfies $\sigma_M(Y_1) = A$, and thus $\sigma_M(Y_1) \sqsubseteq A$. Conversely, assume that there is an $\mathcal{X}$-minimal unifier θ of $\Gamma_{\phi,x}$ such that $\theta(Y_1) \sqsubseteq A$. Then there is a minimal model M of ϕ such that $\theta \equiv \sigma_U$. But then $\sigma_U(Y_1) \equiv \theta(Y_1) \sqsubseteq A$ yields $\sigma_U(Y_1) = A$, which implies $M \models x_1$.

To sum up, we have described a polynomial-time reduction of the minimal model deduction problem to the minimal unifier containment problem. Since the former problem is known to be Σ_2^p-hard, this shows Σ_2^p-hardness of the latter problem. □

As an immediate consequence of this theorem, we can show that there cannot be an NP-algorithm that generates exactly the minimal unifiers of the given $\mathcal{EL}$-unification problem.

Corollary 4.2 *Unless the polynomial hierarchy collapses, there cannot exist an NP-decision procedure for unifiability in $\mathcal{EL}$ that, given a flat $\mathcal{EL}$-unification problem Γ and a subset $\mathcal{X}$ of the concept variables occurring in Γ, not only decides unifiability of Γ, but additionally computes exactly the $\mathcal{X}$-minimal unifiers of Γ in the following sense:*

- *each successful run of the nondeterministic procedure generates an acyclic assignment S such that the induced local unifier σ_S is an $\mathcal{X}$-minimal unifier of Γ, and*
- *for every $\mathcal{X}$-minimal unifier θ of Γ there is a successful run of the nondeterministic procedure that generates an acyclic assignment S such that $\sigma_S \equiv \theta$.*

Proof. Assume that there exists an NP-decision procedure for unifiability in $\mathcal{EL}$ that computes exactly the $\mathcal{X}$-minimal unifiers of Γ in the sense introduced above. Then we could decide the minimal unifier containment problem within NP. In fact, the NP-procedure for deciding this problem is obtained by using the one that computes exactly the $\mathcal{X}$-minimal unifiers, but for every successful path of that procedure checks whether the generated acyclic assignment S satisfies $X \sqsubseteq_{\mathcal{T}_S} A$. This test can be performed in polynomial time, and it yields the same result as testing whether $\sigma_S(X) \sqsubseteq A$. Since the acyclic assignments generated by the original NP-procedure correspond exactly to the $\mathcal{X}$-minimal unifiers, there is a successful path of the extended NP-procedure iff there is an $\mathcal{X}$-minimal unifier θ satisfying $\theta(X) \sqsubseteq A$. Thus, this extended procedure decides the minimal unifier containment problem within NP. Obviously, membership of a Σ_2^p-complete problem in NP would imply $\Sigma_2^p = \text{NP}$. It is well-known that this would imply that the whole polynomial hierarchy collapses. □

Although in general minimal assignments need not induce minimal unifiers, this is not the case for the unification problem constructed in the proof of Theorem 4.1. For this reason, $\mathcal{X}$-minimal assignments that yield unifiers can also not be computed with an NP-procedure. Here, $\mathcal{X}$-minimality of assignments is defined with the following variant of the order on assignments introduced in

Section 3. Let S, T be acyclic assignments of subsets S_X, T_X of $\mathsf{At}_{\mathsf{nv}}$ to the variables X in N_v, and $\mathcal{X} \subseteq N_v$ a set of variables. We define

$S \geq_{\mathcal{X}} T$ iff $S_X \supseteq T_X$ holds for all variables $X \in \mathcal{X}$, and

$S >_{\mathcal{X}} T$ iff $S \geq_{\mathcal{X}} T$ and $T \not\geq_{\mathcal{X}} S$.

If we consider the unification problem Γ constructed in the proof of Theorem 4.1, then its non-variable atoms are $\exists r.X_i$, $\exists r.\overline{X}_i$, $\exists r.A$, $\exists r.B$, A, and B. However, assignments that induce unifiers can obviously only assign subsets of $\{A, B\}$ to variables. As an easy consequence, we have

$$S \geq_{\mathcal{X}} T \text{ iff } \sigma_S \succeq_{\mathcal{X}} \sigma_T \quad \text{and} \quad S >_{\mathcal{X}} T \text{ iff } \sigma_S \succ_{\mathcal{X}} \sigma_T$$

for all acyclic assignments S, T such that σ_S, σ_T are unifiers. This actually holds for all subsets $\mathcal{X}$ of the set of variables of Γ, and thus in particular for the one employed in the proof of Theorem 4.1. In addition, since unifiers can only substitute the variables with $\top$, A, B, or $A \sqcap B$, all unifiers are local.

Lemma 4.3 *Let Γ be the unification problem constructed in the proof of Theorem 4.1 and let $\mathcal{X} := \{Y_1, \ldots, Y_n\}$. Then the set*

$$\{\sigma_S \mid \sigma_S \text{ is a unifier of } \Gamma \text{ and there is no acyclic assignment } T \text{ for } \Gamma \text{ such that } \sigma_T \text{ is a unifier of } \Gamma \text{ and } S >_{\mathcal{X}} T\}$$

consists of exactly the $\mathcal{X}$-minimal unifiers of Γ up to equivalence.

Proof. First, assume that S is an acyclic assignment such that σ_S is a unifier of Γ and there is no acyclic assignment T for Γ such that σ_T is a unifier of Γ and $S >_{\mathcal{X}} T$. If σ_S is not $\mathcal{X}$-minimal, then there is a unifier τ of Γ such that $\sigma_S \succ_{\mathcal{X}} \tau$. Since all unifiers are local, there is an acyclic assignment T such that $\tau \equiv \sigma_T$. But then σ_T is a unifier of Γ, and $\sigma_S \succ_{\mathcal{X}} \tau \equiv \sigma_T$ implies $S >_{\mathcal{X}} T$, which contradicts our assumption on S.

Conversely, let us assume that σ is an $\mathcal{X}$-minimal unifier of Γ. Since all unifiers are local, there is an acyclic assignment S such that $\sigma_S \equiv \sigma$. Assume that there is an acyclic assignment T for Γ such that σ_T is a unifier of Γ and $S >_{\mathcal{X}} T$. But then $\sigma_S \succ_{\mathcal{X}} \sigma_T$, which contradicts our assumption that σ is $\mathcal{X}$-minimal. □

We say that S is an *$\mathcal{X}$-minimal assignment that induces a unifier of* Γ if σ_S is a unifier of Γ and there is no acyclic assignment T for Γ such that σ_T is a unifier of Γ and $S >_{\mathcal{X}} T$. Obviously, the lemma implies that an NP-algorithm that computes all $\mathcal{X}$-minimal assignments that induce unifiers also computes all $\mathcal{X}$-minimal unifiers in the sense defined in Corollary 4.2. Thus, Corollary 4.2 implies that such an NP-algorithm cannot exist.

Corollary 4.4 *Unless the polynomial hierarchy collapses, there cannot exist an NP-decision procedure for unifiability in $\mathcal{EL}$ that, given a flat $\mathcal{EL}$-unification problem Γ and a subset $\mathcal{X}$ of the concept variables occurring in Γ, not only decides unifiability of Γ, but additionally computes exactly the $\mathcal{X}$-minimal assignments for Γ that induces unifiers in the following sense:*

- *each successful run of the nondeterministic procedure generates an $\mathcal{X}$-minimal assignment that induces a unifier of* Γ, *and*
- *for every $\mathcal{X}$-minimal assignment that induces a unifier of* Γ, *there is a successful run of the nondeterministic procedure that generates it.*

5 Conclusion

The results of this paper indicate that it is not easy to compute all and only the minimal unifiers of a given $\mathcal{EL}$-unification problem. On the one hand, while it is sufficient to compute only minimal acyclic assignment to obtain all minimal unifiers, this restriction does not guarantee that only minimal unifiers are generated. In addition, one cannot even compute with an NP-algorithm all $\mathcal{X}$-minimal assignments for subsets $\mathcal{X}$ of the set of all variables. On the other hand, NP-procedures cannot generate exactly the $\mathcal{X}$-minimal unifiers for subsets $\mathcal{X}$ of the set of all variables. It is an open problem whether this last fact is also true if $\mathcal{X}$ is required to be the set of all variables. As mentioned in the introduction, the restriction to a subset of all variables can be justified by the fact that the user may be interested only in the substitution-images of some of the variables. In fact, the algorithms in [8] and [9] first flatten the input problem, which introduces auxiliary variables. These auxiliary variables are internal to the unification procedure and their substitution-images are not shown to the user [2]. Thus, the unifiers need not be minimized w.r.t. these auxiliary variables. However, since the auxiliary variables are introduced in very specific way, leaving them out of the minimization process can probably not produce the effects responsible for the hardness result in Theorem 4.1.

The problem considered here is reminiscent of, but not identical to, some problems considered in complexity theory that are concerned with enumerating (minimal or maximal) [5] solutions. *Maximality problems (MAXPs)* [14] are concerned with computing a maximal solution or some maximal solutions (under set inclusion) for a given decision problem. The main difference to our setting is that algorithms solving MAXPs are not required to compute all maximal solutions. Nevertheless, it is an interesting topic for future work to consider MAXP-variants of our problem (e.g., computation of one minimal unifier) and determine the exact complexity of these problems.

For problems that can potentially have an exponential number of solutions, like ours, one can also investigate their so-called *enumeration complexity* [19]. Even though computing all solutions in the worst-case takes exponential time (in the size of the input), one can distinguish between different kinds of algorithms: for polynomial-delay algorithms, the time needed to compute the next solution is polynomial in the size of the input; for incrementally polynomial algorithms, the time needed to compute the next solution is polynomial in the size of the input and the already computed solutions; and for output-polynomial algorithms, the time needed to compute all solutions is polynomial

[5] Note that, with an appropriate change of the representation of solutions, minimal solutions can be seen as maximal solutions and vice versa.

in the size of the input and output. Since existence of a minimal unifier is an NP-complete problem, there cannot be polynomial-delay algorithms or incrementally polynomial algorithms for enumerating all minimal unifiers (unless P = NP) since otherwise the first minimal unifier would be computable in polynomial time, and thus existence of a solution could be decided in polynomial time. However, there could still exist an output-polynomial algorithm, though we conjecture that this is not the case.

References

[1] Baader, F., *Terminological cycles in a description logic with existential restrictions*, in: G. Gottlob and T. Walsh, editors, *Proc. of the 18th Int. Joint Conf. on Artificial Intelligence (IJCAI 2003)* (2003), pp. 325–330.

[2] Baader, F., S. Borgwardt, J. A. Mendez and B. Morawska, *UEL: Unification solver for $\mathcal{EL}$*, in: *Proceedings of the 25th International Workshop on Description Logics (DL'12)*, CEUR Workshop Proceedings **846**, Rome, Italy, 2012.

[3] Baader, F., S. Borgwardt and B. Morawska, *Extending unification in $\mathcal{EL}$ towards general TBoxes*, in: *Proc. of the 13th Int. Conf. on Principles of Knowledge Representation and Reasoning (KR 2012)* (2012), pp. 568–572, short paper.

[4] Baader, F., D. Calvanese, D. McGuinness, D. Nardi and P. F. Patel-Schneider, editors, "The Description Logic Handbook: Theory, Implementation, and Applications," Cambridge University Press, 2003.

[5] Baader, F. and S. Ghilardi, *Unification in modal and description logics*, Logic Journal of the IGPL **19** (2011), pp. 705–730.

[6] Baader, F. and R. Küsters, *Matching in description logics with existential restrictions*, in: *Proc. of the 7th Int. Conf. on Principles of Knowledge Representation and Reasoning (KR 2000)*, 2000, pp. 261–272.

[7] Baader, F. and B. Morawska, *Unification in the description logic $\mathcal{EL}$*, in: R. Treinen, editor, *Proc. of the 20th Int. Conf. on Rewriting Techniques and Applications (RTA 2009)*, Lecture Notes in Computer Science **5595** (2009), pp. 350–364.

[8] Baader, F. and B. Morawska, *SAT encoding of unification in $\mathcal{EL}$*, in: C. Fermüller and A. Voronkov, editors, *Proc. of the 17th Int. Conf. on Logic for Programming, Artifical Intelligence, and Reasoning (LPAR-17)*, Lecture Notes in Computer Science **6397** (2010), pp. 97–111.

[9] Baader, F. and B. Morawska, *Unification in the description logic $\mathcal{EL}$*, Logical Methods in Computer Science **6** (2010).

[10] Baader, F. and P. Narendran, *Unification of concept terms in description logics*, J. of Symbolic Computation **31** (2001), pp. 277–305.

[11] Baader, F. and W. Snyder, *Unification theory*, in: J. Robinson and A. Voronkov, editors, *Handbook of Automated Reasoning, Vol. I*, Elsevier Science Publishers, 2001 pp. 447–533.

[12] Babenyshev, S., V. V. Rybakov, R. Schmidt and D. Tishkovsky, *A tableau method for checking rule admissibility in S4*, in: *Proc. of the 6th Workshop on Methods for Modalities (M4M-6)*, Copenhagen, 2009.

[13] Brandt, S., *Polynomial time reasoning in a description logic with existential restrictions, GCI axioms, and—what else?*, in: R. L. de Mántaras and L. Saitta, editors, *Proc. of the 16th Eur. Conf. on Artificial Intelligence (ECAI 2004)*, 2004, pp. 298–302.

[14] Chen, Z.-Z. and S. Toda, *The complexity of selecting maximal solutions*, Inf. Comput. **119** (1995), pp. 231–239.

[15] Eiter, T. and G. Gottlob, *Propositional circumscription and extended closed world reasoning are Π_2^P-complete*, Theoretical Computer Science **114** (1993), pp. 231–245.

[16] Ghilardi, S., *Unification through projectivity*, J. of Logic and Computation **7** (1997), pp. 733–752.

[17] Ghilardi, S., *Unification in intuitionistic logic*, J. of Symbolic Logic **64** (1999), pp. 859–880.

[18] Iemhoff, R. and G. Metcalfe, *Proof theory for admissible rules*, Ann. Pure Appl. Logic **159** (2009), pp. 171–186.

[19] Johnson, D. S., C. H. Papadimitriou and M. Yannakakis, *On generating all maximal independent sets*, Inf. Process. Lett. **27** (1988), pp. 119–123.

[20] Kracht, M., *Modal consequence relations*, in: P. Blackburn, J. van Benthem and F. Wolter, editors, *The Handbook of Modal Logic*, Elsevier, 2006 pp. 491–545.

[21] Li, C. M. and F. Manyà, *MaxSAT, hard and soft constraints*, in: *Handbook of Satisfiability*, IOS Press, 2009 pp. 613–631.

[22] Rybakov, V. V., "Admissibility of logical inference rules," Studies in Logic and the Foundations of Mathematics **136**, North-Holland Publishing Co., Amsterdam, 1997.

[23] Rybakov, V. V., *Multi-modal and temporal logics with universal formula - reduction of admissibility to validity and unification*, J. of Logic and Computation **18** (2008), pp. 509–519.

[24] Wolter, F. and M. Zakharyaschev, *Undecidability of the unification and admissibility problems for modal and description logics*, ACM Trans. Comput. Log. **9** (2008).

Some Truths Are Best Left Unsaid

Philippe Balbiani [1]

University of Toulouse and CNRS

Hans van Ditmarsch [2]

University of Sevilla

Andreas Herzig [3]

University of Toulouse and CNRS

Tiago de Lima [4]

University of Artois

Abstract

We study the formal properties of extensions of the basic public announcement logic by standard modal axioms such as D, T, 2, 4, and 5. We show that some of them fail to be conservative extensions of the underlying modal logic. This leads us to propose new truth conditions for announcements that better suit these extensions. The corresponding reduction axioms postulate the suitability of the updated model to the underlying logic. We show that if the fact can be expressed that the frame of an updated model is in the class of frames of the underlying modal logic, then the public announcement extension is axiomatisable. This is the case for, for example, K, KT, S4, and S4.3. We also show that such a formula does not exist for several logics whose frame condition involves existential quantification. This is the case for, for example, S4.2.

Keywords: public announcement logic, dynamic epistemic logic, reduction axioms.

[1] Address: Institut de recherche en informatique de Toulouse, CNRS — Université de Toulouse, 118 route de Narbonne, 31062 Toulouse Cedex 9, FRANCE; Philippe.Balbiani@irit.fr.

[2] Address: Department of Logic, University of Sevilla, Calle Camilo José Cela s/n, 41018 Sevilla, SPAIN; hvd@us.es.

[3] Address: Institut de recherche en informatique de Toulouse, CNRS — Université de Toulouse, 118 route de Narbonne, 31062 Toulouse Cedex 9, FRANCE; Andreas.Herzig@irit.fr.

[4] Address: CRIL, Université d'Artois, Rue Jean Souvraz, SP 18, 62307 Lens Cedex, FRANCE; delima@cril.fr.

1 Introduction

In public announcement logics, one can write formulas like $\langle\varphi!\rangle\psi$ standing for "the announcement of φ can be made and, after that, ψ holds". Such logics have been extensively studied over the last years, starting with Plaza's paper [13] (reprinted in [14]); see [16] for an overview. In the literature, the term "Public Announcement Logic" is often used in the singular. But there is more than one such logic because the underlying modal logic of knowledge or belief may vary. In the literature, it is mainly the basic modal logic K and the modal logic S5 that are investigated, or rather their multimodal versions. The reasons that are given for these choices are that K is an appropriate basis for a logic of belief, while S5 is *the* logic of knowledge.

One may argue against S5 as a logic of knowledge. It is typically taken for granted in artificial intelligence and game theory that it is S5; however, this choice has been criticised in the philosophical literature, most prominently so by Lenzen [11,12]. The latter argues for S4.2 and S4.3 as the appropriate logics of knowledge. Let us remark that we only consider knowledge of a perfect reasoner, i.e. we leave the omniscience problem aside.

The base modal logic K is not a suitable logic of belief: philosophers insisted that such a logic should contain the D axiom since the seminal work of Hintikka [7]; Hintikka took KD, while later authors rather took KD45 [11,12].

These considerations motivate a more systematic study of extensions of modal logics by public announcements. It turns out that many of these extensions are problematic. Semantically speaking, the problem is that the update of a model by an announcement may no longer be in the intended class of models: typically existential properties such as seriality may be lost after an update. Axiomatically speaking, as soon as we take it for granted that the public announcement operator is a normal modal operator, i.e. it obeys the K axiom and the necessitation rule, then the extensions of different modal logics may collapse.

In order to illustrate this let us show that the public announcement extension of KD coincides with the public announcement extension of KT if the announcement operator is normal. First, from the KD theorem $\neg\Box_i\bot$ we can infer

$$[\neg\varphi!]\neg\Box_i\bot$$

by the necessitation rule for announcements. Second, we have

$$\Box_i\varphi \to [\neg\varphi!]\Box_i\bot$$

by the usual reduction axioms for public announcements $[\neg\varphi!]\Box_i\bot \leftrightarrow (\neg\varphi \to \Box_i[\neg\varphi!]\bot)$ and $[\neg\varphi!]\bot \leftrightarrow \varphi$. From the above two and the fact that $[\neg\varphi!]$ is normal it follows that $\Box_i\varphi \to [\neg\varphi!]\bot$ is a theorem. The application of the reduction axioms uses the inference rule

$$\frac{\varphi \leftrightarrow \varphi'}{[\psi!]\varphi \leftrightarrow [\psi!]\varphi'}$$

which is derivable for normal modal operators. As $[\neg\varphi!]\bot$ reduces to φ we obtain that $\Box_i\varphi \to \varphi$ is a theorem. Another way of formulating this result is the following: if we extend KD by Plaza's reduction axioms (plus rules of replacement of equivalents for announcements) then one can prove the T axiom.

Given the above negative result we propose a new truth condition for public announcements that has not been studied before. It differs from the standard interpretation where the update is conditioned by the truth of the announcement: it requires moreover that the updated model is a *legal frame of the underlying logic.* We denote that interpretation by the superscript '$\mathcal{C}$' where $\mathcal{C}$ is a class of frames validating the underlying logic. We investigate the axiomatisability of the resulting public announcement logics. We give axiomatisations for all those of our logics where the fact that a frame is a $\mathcal{C}$ frame can be characterised in the language (more precisely, in its extension by the universal modal operator). On the negative side, the $\mathcal{C}$-semantics still does not allow to axiomatise a public announcement extension of S4.2; the reason is that it cannot be characterised in the language of S4.2 that a frame is a legal frame of the underlying logic $\mathcal{L}$.

The paper is organised as follows: In Section 2 we recall the standard presentation of public announcement logics: a semantics in terms of updates that are conditioned by the truth of the announcement and an axiomatics in terms of reduction axioms. In Section 3 we present our version of public announcement logics in terms of an enhanced truth condition. In Section 4 we characterise the validities of different classes of frames axiomatically by means of reduction axioms. In Section 5 we present several examples of classes of frames that still cannot be axiomatised. In Section 6 we discuss various other semantical options allowing to define variants of public announcement logics. In Section 7 we discuss some related work and in Section 8 we conclude.

2 Public announcement logics: standard version

Let $\mathbb{P}$ be a countable set of propositional letters and let $\mathbb{J}$ be a finite set of agent names. The public announcement language $\mathsf{L}_{\mathsf{PAL}}$ is defined by the following BNF:

$$\varphi ::= p \mid \bot \mid \neg\varphi \mid (\varphi \vee \varphi) \mid \Diamond_i\varphi \mid \langle\varphi!\rangle\varphi$$

where p ranges over $\mathbb{P}$ and i ranges over $\mathbb{J}$. The length of φ, i.e. the number of occurrences of symbols in φ, will be denoted $\mathrm{l}(\varphi)$.

The epistemic language L_{EL} is the fragment of $\mathsf{L}_{\mathsf{PAL}}$ without announcement operators $\langle\varphi!\rangle$. As usual, $\top$ abbreviates $\neg\bot$, $\Box_i\varphi$ abbreviates $\neg\Diamond_i\neg\varphi$, and $[\varphi!]\psi$ abbreviates $\neg\langle\varphi!\rangle\neg\psi$. We adopt the standard rules for omission of the parentheses.

2.1 Models and their updates

A *Kripke frame* is a tuple $\langle W, R\rangle$ such that:

- W is a nonempty set of possible worlds;

- $R : \mathbb{J} \to \wp(W \times W)$ associates to every agent $i \in \mathbb{J}$ a binary relation R_i on W.

The class of all Kripke frames is denoted by $\mathcal{C}_{\mathsf{all}}$.

A *Kripke model* is a tuple $M = \langle W, R, V\rangle$ such that $\langle W, R\rangle$ is a Kripke frame and $V : \mathbb{P} \to \wp(W)$ associates an interpretation $V(p) \subseteq W$ to each $p \in \mathbb{P}$. For every $x \in W$, the pair (M, x) is a *pointed model.* For convenience, we define $R_i(x) = \{x' \mid (x, x') \in R_i\}$. In an epistemic or doxastic interpretation, the elements of $R_i(x)$ are the worlds agent i considers possible at x: the worlds that are compatible with i's knowledge or respectively i's belief.

Let $M = \langle W, R, V\rangle$ be a Kripke model and let U be some subset of W. The *world update* (alias relativisation) of M by U is defined as $M \circ U = \langle W', R', V'\rangle$, with:

$$\begin{aligned} W' &= U \\ R'_i &= R_i \cap (U \times U), \quad \text{for every } i \in \mathbb{J} \\ V'(p) &= V(p) \cap U, \quad \text{for every } p \in \mathbb{P} \end{aligned}$$

If U is empty then $\langle W', R'\rangle$ is not a Kripke frame. We shall see that the standard truth condition avoids this by conditioning the update, preventing thus the semantics from being ill-defined.

If U is non-empty then $\langle W', R'\rangle$ is a Kripke frame. Things are less straightforward if we want to preserve membership in some subclass $\mathcal{C}$ of the class of all Kripke frames: it may for example happen that existential properties such as seriality, density, or confluence of the relations R_i are not always preserved. This will be the *raison d'être* of our enhanced semantics.

2.2 Standard truth conditions and validity

We recall the interpretation of $\mathsf{L_{PAL}}$ formulas in a given model $M = \langle W, R, V\rangle$:

$$\begin{aligned} \|p\|_M &= V(p), \text{ for } p \in \mathbb{P} \\ \|\neg\varphi\|_M &= W \setminus \|\varphi\|_M \\ \|\varphi \vee \psi\|_M &= \|\varphi\|_M \cup \|\psi\|_M \\ \|\Diamond_i\varphi\|_M &= R_i^{-1}(\|\varphi\|_M) \\ &= \{x \in W \mid R_i(x) \cap \|\varphi\|_M \neq \emptyset\} \\ \|\langle\varphi!\rangle\psi\|_M &= \|\varphi\|_M \cap \|\psi\|_{M \circ \|\varphi\|_M} \end{aligned}$$

If we write $M, x \Vdash \varphi$ instead of $x \in \|\varphi\|_M$, then we can restate the last condition in a form that is perhaps more customary:

$$M, x \Vdash \langle\varphi!\rangle\psi \quad \text{iff} \quad M, x \Vdash \varphi \text{ and } M \circ \|\varphi\|_M, x \Vdash \psi$$

An $\mathsf{L_{PAL}}$ formula φ is *globally true in the Kripke model* $M = \langle W, R, V\rangle$ if and only if $\|\varphi\|_M = W$. (This is sometimes noted $M \Vdash \varphi$.) We say that φ is *valid in the Kripke frame* $\langle W, R\rangle$ if and only if φ is globally true in every Kripke model over $\langle W, R\rangle$.

We are interested in several particular classes of Kripke frames: $\mathcal{C}_{\mathsf{all}}$ is the class of all Kripke frames; $\mathcal{C}_{\mathsf{serial}}$ is the class of frames where each accessibility relation is serial; $\mathcal{C}_{\mathsf{refl}}$ is the class of frames where each accessibility relation is reflexive; $\mathcal{C}_{\mathsf{confl}}$ is the class of frames where each accessibility relation is confluent; $\mathcal{C}_{\mathsf{refl,trans,eucl}}$ is the class of frames where each accessibility relation is an equivalence relation (reflexive, transitive and Euclidean); and so on.

Given a class of frames $\mathcal{C}$, let $\Lambda(\mathcal{C})$ be the set of $\mathsf{L_{PAL}}$ formulas that are valid in every frame of $\mathcal{C}$. An example of a formula that is valid in the class of all Kripke frames $\mathcal{C}_{\mathsf{all}}$ is $[p!]\Box_i p$, for atomic p. In contrast, the schema $[\varphi!]\Box_i\varphi$ is not valid in every Kripke frame. Another schema that is valid in every Kripke frame is $\Box_i\varphi \to [\neg\varphi!]\Box_i\bot$. It plays an important role in this paper: remember that we have already used it in the introduction when proving the T axiom.

Let us immediately say that the definition of $\Lambda(\mathcal{C})$ is problematic. Consider a Kripke model $M = \langle W, R, V\rangle$ such that R is serial (i.e., $\langle W, R\rangle \in \mathcal{C}_{\mathsf{serial}}$) and such that $M, x \Vdash \varphi \wedge \Box_i\neg\varphi$, and suppose we want to check whether $M, x \Vdash \langle\varphi!\rangle\Box_i\bot$. This involves checking whether $M \circ \|\varphi\|_M, x \Vdash \Box_i\bot$; which is the case, and therefore $M, x \Vdash \langle\varphi!\rangle\Box_i\bot$. This means that it may happen that after an announcement, agent i gets crazy and starts to believe everything. Formally speaking, while $\neg\Box_i\bot$ is valid in serial frames, its necessitation $[\varphi!]\neg\Box_i\bot$ is not: necessitation by announcements does not preserve validity! This is clearly undesirable. The key observation is that in $M \circ \|\varphi\|_M$ we have $R'_i(x) = \emptyset$: the accessibility relation R'_i is not serial any more.

The above discussion was about the preservation of seriality, but the same problem arises for example for confluence. More generally, it arises for classes of frames that are defined by an *existential condition*.

2.3 Axiomatisation of K-PAL

Let K-PAL be the least set of formulas in our language $\mathsf{L_{PAL}}$ that contains all instances of axiom schemas of the basic modal logic K for every $\Diamond_i$ and of the reduction axiom schemas

$$\langle\psi!\rangle p \leftrightarrow \psi \wedge p, \quad \text{for } p \in \mathbb{P} \qquad \mathsf{Red}_{\langle!\rangle,\mathbb{P}}$$
$$\langle\psi!\rangle\bot \leftrightarrow \bot \qquad \mathsf{Red}_{\langle!\rangle,\bot}$$
$$\langle\psi!\rangle\neg\varphi \leftrightarrow \psi \wedge \neg\langle\psi!\rangle\varphi \qquad \mathsf{Red}_{\langle!\rangle,\neg}$$
$$\langle\psi!\rangle(\varphi_1 \vee \varphi_2) \leftrightarrow \langle\psi!\rangle\varphi_1 \vee \langle\psi!\rangle\varphi_2 \qquad \mathsf{Red}_{\langle!\rangle,\vee}$$
$$\langle\psi!\rangle\Diamond_i\varphi \leftrightarrow \psi \wedge \Diamond_i\langle\psi!\rangle\varphi \qquad \mathsf{Red}_{\langle!\rangle,\Diamond_i}$$

and that is closed with respect to the inference rules of Modus Ponens, necessitation by announcements, and the two rules of equivalents for $\langle\psi!\rangle$:

$$\frac{\varphi \leftrightarrow \varphi'}{\langle\psi!\rangle\varphi \leftrightarrow \langle\psi!\rangle\varphi'} \qquad \mathsf{RE}^{\mathsf{r}}_{\langle!\rangle}$$
$$\frac{\psi \leftrightarrow \psi'}{\langle\psi!\rangle\varphi \leftrightarrow \langle\psi'!\rangle\varphi} \qquad \mathsf{RE}^{\mathsf{l}}_{\langle!\rangle}$$

Completeness of K-PAL w.r.t. $\mathcal{C}_{\mathsf{all}}$ follows from Wang's result that the axiom schemas of K for every $\Diamond_i$ plus the above reduction axioms plus the K axiom and the necessitation rule for announcements make up a complete axiomatisation of K-PAL [17, Corollary 1]: indeed, the K axiom and the necessitation rule for announcements may equivalently be replaced by $\mathsf{Red}_{\langle!\rangle,\bot}$, $\mathsf{Red}_{\langle!\rangle,\vee}$, and $\mathsf{RE}^{\mathsf{l}}_{\langle!\rangle}$ (cf. [4, Theorem 4.3]).

As customary in dynamic epistemic logics, our axiomatic system allows for a proof procedure in terms of reduction axioms. As there is no axiom schema for two consecutive announcements, reduction has to be performed 'bottom-up' (or 'inside-out' as Wang calls it [17]), starting by some innermost dynamic operator. The sound performance of such 'deep replacements' requires the rule of replacement of proved equivalents RRE; and indeed, our two rules of equivalents for announcements $\mathsf{RE}^{\mathsf{r}}_{\langle!\rangle}$ and $\mathsf{RE}^{\mathsf{l}}_{\langle!\rangle}$ enable the derivation of the rule

$$\frac{\chi \leftrightarrow \chi'}{(\varphi)^p_\chi \leftrightarrow (\varphi)^p_{\chi'}} \qquad \mathsf{RRE}$$

where $(\varphi)^p_\chi$ denotes the result of replacing all occurrences of p in φ by χ.

Proposition 2.1 *The rule* RRE *is derivable from the axiom schemas of* K *for* $\Diamond_i$ *by Modus Ponens, necessitation by announcements,* $\mathsf{RE}^{\mathsf{r}}_{\langle!\rangle}$*, and* $\mathsf{RE}^{\mathsf{l}}_{\langle!\rangle}$.

Proof. This follows from the fact that rules of equivalents can be derived for every (Boolean and modal) operator; remember that for the dynamic operators $\langle\psi!\rangle$ we directly have the two rules $\mathsf{RE}^{\mathsf{r}}_{\langle!\rangle}$ and $\mathsf{RE}^{\mathsf{l}}_{\langle!\rangle}$). For a proof see e.g. [4]. □

Here is an example of proof by means of reduction axioms:

$$\begin{array}{rll} \langle p!\rangle\langle\neg q!\rangle r & \leftrightarrow \langle p!\rangle(\neg q \wedge r) & \mathsf{Red}_{\langle!\rangle,\mathbb{P}} \\ & \leftrightarrow \langle p!\rangle\neg q \wedge \langle p!\rangle r & \mathsf{Red}_{\langle!\rangle,\wedge} \\ & \leftrightarrow p \wedge \neg\langle p!\rangle q \wedge \langle p!\rangle r & \mathsf{Red}_{\langle!\rangle,\neg} \\ & \leftrightarrow p \wedge \neg(p \wedge q) \wedge p \wedge r & \mathsf{Red}_{\langle!\rangle,\mathbb{P}} \text{ (twice)} \\ & \leftrightarrow p \wedge \neg q \wedge r & \end{array}$$

In the second step, $\mathsf{Red}_{\langle!\rangle,\wedge}$ stands for the equivalence $\langle\psi!\rangle(\varphi_1 \wedge \varphi_2) \leftrightarrow \langle\psi!\rangle\varphi_1 \wedge \langle\psi!\rangle\varphi_2$ that can be proved from $\mathsf{Red}_{\langle!\rangle,\vee}$ and $\mathsf{Red}_{\langle!\rangle,\neg}$.

Observe that RRE is used in each of the steps. Observe also that we have to start by reducing the innermost dynamic operator $\langle\neg q!\rangle$ by means of the $\mathsf{Red}_{\langle!\rangle,\neg}$ rule —which requires the application of $\mathsf{RE}^{\mathsf{r}}_{\langle!\rangle}$— because our axiomatisation does not provide for a reduction axiom for the case of two consecutive dynamic operators.

2.4 Alternative axiomatisation

The axiomatisations in the literature typically lack $\mathsf{RE}^{\mathsf{r}}_{\langle!\rangle}$ and $\mathsf{RE}^{\mathsf{l}}_{\langle!\rangle}$. They instead have the following axiom schema for the composition of announcements:

$$\langle\psi_1!\rangle\langle\psi_2!\rangle\varphi \leftrightarrow \langle\langle\psi_1!\rangle\psi_2!\rangle\varphi \qquad \mathsf{Red}_{\langle!\rangle,\langle.!\rangle}$$

(see e.g. [13] or [16, Proposition 4.22]). To illustrate the difference we give a proof of the above example formula via $\mathsf{Red}_{\langle !\rangle,\langle .!\rangle}$.

$$\begin{aligned}
\langle p!\rangle\langle \neg q!\rangle r &\leftrightarrow \langle\langle p!\rangle \neg q!\rangle r && \mathsf{Red}_{\langle !\rangle,\langle .!\rangle}\\
&\leftrightarrow \langle p!\rangle \neg q \wedge r && \mathsf{Red}_{\langle !\rangle,\mathbb{P}}\\
&\leftrightarrow p \wedge \neg\langle p!\rangle q \wedge r && \mathsf{Red}_{\langle !\rangle,\neg}\\
&\leftrightarrow p \wedge \neg(p \wedge q) \wedge r && \mathsf{Red}_{\langle !\rangle,\mathbb{P}}\\
&\leftrightarrow p \wedge \neg q \wedge r
\end{aligned}$$

This is an 'outside-in' reduction, as opposed to reductions without $\mathsf{Red}_{\langle !\rangle,\langle .!\rangle}$ which have to proceed 'inside-out' and require our above rule of replacement of proved equivalents RRE [17].

RRE is stated in Proposition 4.46 in [16], and its proof (Exercise 4.48) says that it follows from the fact that the necessitation rule for announcements is an admissible inference rule (cf. also Exercise 4.52 there). We note in passing that Wang proved that neither RRE nor $\mathsf{RE}^{\mathsf{r}}_{\langle !\rangle}$ can be derived from the axiom schemas of K plus the above reduction axioms alone, i.e., without $\mathsf{RE}^{\mathsf{r}}_{\langle !\rangle}$, $\mathsf{RE}^{\mathsf{l}}_{\langle !\rangle}$, or $\mathsf{Red}_{\langle !\rangle,\langle .!\rangle}$ [17].

Remark 2.2 It follows from the completeness theorem (and from the fact that RRE preserves validity) that RRE is admissible, i.e., it preserves theoremhood: for every *formula instance* $\chi \leftrightarrow \chi'$ and every *formula instance* φ, if there is a proof of $\chi \leftrightarrow \chi'$ then there is a proof of $(\varphi)^p_\chi \leftrightarrow (\varphi)^p_{\chi'}$. However, this does not mean that RRE is derivable, i.e., that there is a derivation of the inference rule RRE itself. This situation can be compared to the completeness theorem for K-PAL which only says that every valid formula is provable, but does not guarantee that there are proofs of all valid formula schemas. These two versions of completeness —w.r.t. schemas and w.r.t. instances— coincide for logics where the rule of uniform substitution preserves validity, but K-PAL does not have that property. We note in passing that it is only recently that a complete axiomatisation of the schematic validities of K-PAL was given [8].

2.5 When things get wrong: public announcement extensions of KD, KD45, S4.2, etc.

We have just seen how K-PAL completely axiomatises the set of formulas that are valid in the class of all Kripke frames. Let KT-PAL denote the least set of formulas in our language that contains K-PAL and all instances of the T axiom. Then KT-PAL completely axiomatises the set of formulas $\Lambda(\mathcal{C}_{\mathsf{refl}})$, i.e., the set of formulas that are valid in the class of all reflexive Kripke frames. In the same way, S5-PAL —the extension of K-PAL by the axiom schemas T, 4, and 5— axiomatises $\Lambda(\mathcal{C}_{\mathsf{refl,trans,eucl}})$.

More generally, one might naively expect that if the validities of a class of frames $\mathcal{C}$ in the non-dynamic language $\mathsf{L_{EL}}$ can be axiomatised by some set of schemas and rules $\mathcal{AX}_\mathcal{C}$, then the validities of $\mathcal{C}$ in $\mathsf{L_{PAL}}$ can be axiomatised by the set of schemas and rules for K-PAL plus $\mathcal{AX}_\mathcal{C}$. The following argument shows what happens when we do this for the logic KD.

By $\mathsf{Red}_{\langle ! \rangle,\neg}$, $\mathsf{Red}_{\langle ! \rangle,\Diamond_i}$, and $\mathsf{Red}_{\langle ! \rangle,\mathbb{P}}$, one can derive $p \wedge \neg\Diamond_i p \to \langle p! \rangle \neg\Diamond_i \top$. We have seen in Section 2.3 that the rule of necessitation for announcements is derivable in our axiomatics of K-PAL and therefore also in the —hypothetical— axiomatisation of KD-PAL. Thus, in KD, by the axiom $\Diamond_i \top$ and the necessitation by announcement, one can derive $[p!]\Diamond_i \top$. From $p \wedge \neg\Diamond_i p \to \langle p! \rangle \neg\Diamond_i \top$ and $[p!]\Diamond_i \top$, one can obviously derive $\Box_i p \to p$.

This makes that φ is a theorem of the latter system if and only if φ is a theorem of KT-PAL, which is clearly undesirable.

It seems that in the case of serial frames the only way to 'save' the standard truth condition is to abandon the rule of necessitation by announcements. While this is technically possible, the price to pay is that in many cases, the extension of an underlying modal logic by public announcements cannot be a conservative extension of that underlying logic. We think that this should better be avoided.

3 A new semantics

The preceding observation has motivated us to design a new semantics for the logics of public announcements.

Our results typically require a master modality, such as the universal modality. We therefore add the latter to our language: we define the language $\mathsf{L}_{\mathsf{PAL},\forall}$ by the following BNF:

$$\varphi ::= p \mid \bot \mid \neg\varphi \mid (\varphi \vee \varphi) \mid \Diamond_i \varphi \mid \langle \varphi! \rangle \varphi \mid \forall \varphi$$

The formula $\exists\varphi$ abbreviates $\neg\forall\neg\varphi$. The language $\mathsf{L}_{\mathsf{EL},\forall}$ is the fragment of $\mathsf{L}_{\mathsf{PAL},\forall}$ without dynamic operators.

The universal modality $\forall$ is interpreted as follows:

$$\|\forall\varphi\|_M = \begin{cases} W & \text{if } \|\varphi\|_M = W \\ \emptyset & \text{else} \end{cases}$$

3.1 A parametrised truth condition

Let $\mathcal{C}$ be a given class of Kripke frames. Let M be some model over some frame of $\mathcal{C}$. In order to distinguish our semantics from the standard semantics we write $\|\varphi\|_M^{\mathcal{C}}$ instead of $\|\varphi\|_M$.

The truth conditions for all but the dynamic operator take the same form as before. For the latter we define:

$$\|\langle \varphi! \rangle \psi\|_M^{\mathcal{C}} = \begin{cases} \emptyset & \text{if the frame of } M \circ \|\varphi\|_M^{\mathcal{C}} \text{ is not in } \mathcal{C} \\ \|\varphi\|_M^{\mathcal{C}} \cap \|\psi\|_{M \circ \|\varphi\|_M^{\mathcal{C}}}^{\mathcal{C}} & \text{otherwise} \end{cases}$$

If we write $M, x \Vdash^{\mathcal{C}} \varphi$ instead of $x \in \|\varphi\|_M^{\mathcal{C}}$, then we can restate this condition in a form that is perhaps more customary:

$$\begin{aligned} M, x \Vdash^{\mathcal{C}} \langle \varphi! \rangle \psi \quad \text{iff} \quad & \text{the frame of } M \circ \|\varphi\|_M^{\mathcal{C}} \text{ is in } \mathcal{C}, \\ & M, x \Vdash^{\mathcal{C}} \varphi \text{ and } M \circ \|\varphi\|_M^{\mathcal{C}}, x \Vdash^{\mathcal{C}} \psi \end{aligned}$$

Remark that the above truth condition for the dynamic operator makes it a normal modal operator, i.e. the dynamic operator obeys the K axiom and the necessitation rule.

Obviously, given a class $\mathcal{C}$ of Kripke frames, what interests us is the set $\Lambda(\mathcal{C})^{\mathcal{C}}$ of $\mathsf{L_{PAL}}$ formulas that are valid in every Kripke frame of the class $\mathcal{C}$ under our $\mathcal{C}$ conditioned interpretation. Traditionally, if a class $\mathcal{C}'$ of frames is included in a class $\mathcal{C}''$, then every formula valid in $\mathcal{C}''$ is also valid in $\mathcal{C}'$. In our new semantics, seeing that the truth conditions are conditioned by classes of frames, validity with respect to $\mathcal{C}''$ and validity with respect to $\mathcal{C}'$ are not equivalent notions. For some classes $\mathcal{C}'$ and $\mathcal{C}''$, it might appear that $\mathcal{C}' \subseteq \mathcal{C}''$ and $\Lambda(\mathcal{C}'')^{\mathcal{C}''} \not\subseteq \Lambda(\mathcal{C}')^{\mathcal{C}'}$. To see an example, let $\mathcal{C}'$ be the set of all strict total orders with at least 3 points and $\mathcal{C}''$ be the set of all strict total orders with at least 2 points. Obviously, $\mathcal{C}' \subseteq \mathcal{C}''$. In order to show that $\Lambda(\mathcal{C}'')^{\mathcal{C}''} \not\subseteq \Lambda(\mathcal{C}')^{\mathcal{C}'}$, let us consider the formula $\varphi = p \wedge [p!]\bot \rightarrow \Box\neg p$. We claim the following:

(i) $\varphi \in \Lambda(\mathcal{C}'')^{\mathcal{C}''}$;

(ii) $\varphi \notin \Lambda(\mathcal{C}')^{\mathcal{C}'}$.

To demonstrate **(i)**, let $M = \langle W, R, V\rangle$ be some model based on a linear order $\langle W, R\rangle$ in $\mathcal{C}''$. Hence, W contains at least 2 points. Let $x \in W$ be such that $M, x \Vdash^{\mathcal{C}''} p \wedge [p!]\bot$. Thus, $x \in V(p)$ but $V(p)$ does not contain at least 2 points. Therefore, $V(p) = \{x\}$ and $M, x \Vdash^{\mathcal{C}''} \Box\neg p$. To demonstrate **(ii)**, let $M = \langle W, R, V\rangle$ be a model based on the linear order $\{0, 1, 2\}$ in $\mathcal{C}'$ such that $V(p) = \{0, 1\}$. Obviously, $M, 0 \Vdash^{\mathcal{C}'} p \wedge [p!]\bot$ and $M, 0 \nVdash^{\mathcal{C}'} \Box\neg p$.

In sections 4 and 5 we will explore this $\mathcal{C}$ conditioned interpretation. We focus on axiomatisability in terms of reduction axioms. In that perspective, the crucial point is whether we are able to characterise the condition "the frame of $M \circ \|\varphi\|^{\mathcal{C}}_M$ belongs to class $\mathcal{C}$" in the logical language. For the cases where this is possible, our characterisations require in general a 'master modality' such as the common knowledge operator or the universal modality [1] (i.e., they require the language $\mathsf{L_{EL,\forall}}$); the only exception is the class $\mathcal{C}^1_{\mathsf{tr}}$ where $\mathbb{J}$ is a singleton and where the frames have transitive accessibility relations: no master modality is needed in that case. An example of such a frame class characterising condition, relative to a given announced formula ψ, is $\forall(\psi \rightarrow \bigwedge_{i\in\mathbb{J}} \Diamond_i\psi)$. Not surprisingly, this is the characterising formula for the class of serial frames. Before we formally introduce that, we have to further prepare the theoretical ground.

3.2 Reduction axioms

We define the *announcement degree* of a $\mathsf{L}_{\mathsf{PAL},\forall}$ formula φ as follows:

$$\begin{aligned}
\mathrm{d}(p) &= 0 \\
\mathrm{d}(\bot) &= 0 \\
\mathrm{d}(\neg\varphi) &= \mathrm{d}(\varphi) \\
\mathrm{d}(\varphi \vee \psi) &= \max(\mathrm{d}(\varphi), \mathrm{d}(\psi)) \\
\mathrm{d}(\Diamond_i\varphi) &= \mathrm{d}(\varphi) \\
\mathrm{d}(\langle\varphi!\rangle\psi) &= \max(\mathrm{d}(\varphi), \mathrm{d}(\psi)) + 1 \\
\mathrm{d}(\forall\varphi) &= \mathrm{d}(\varphi)
\end{aligned}$$

For example, the announcement degree of both $\langle p!\rangle\Diamond_i\langle\neg q!\rangle\Diamond_j r$ and $\langle\langle p!\rangle\neg q!\rangle(\Diamond_j r \vee \langle p!\rangle\top)$ is 2.

Consider some class of Kripke frames $\mathcal{C}$. Suppose the fact that the frame of the updated model $M \circ \|\varphi\|_M^{\mathcal{C}}$ belongs to $\mathcal{C}$ can be characterised by an $\mathsf{L}_{\mathsf{PAL},\forall}$ formula $\mathrm{f}(\varphi)$ whose announcement degree is at most that of φ. We then obtain the following reduction axioms:

$$\begin{aligned}
\langle\psi!\rangle p &\leftrightarrow \psi \wedge \mathrm{f}(\psi) \wedge p \quad \text{for } p \in \mathbb{P} && \mathsf{Red}^{\mathcal{C}}_{\langle!\rangle,\mathbb{P}} \\
\langle\psi!\rangle\bot &\leftrightarrow \bot && \mathsf{Red}^{\mathcal{C}}_{\langle!\rangle,\bot} \\
\langle\psi!\rangle\neg\varphi &\leftrightarrow \psi \wedge \mathrm{f}(\psi) \wedge \neg\langle\psi!\rangle\varphi && \mathsf{Red}^{\mathcal{C}}_{\langle!\rangle,\neg} \\
\langle\psi!\rangle(\varphi_1 \vee \varphi_2) &\leftrightarrow \langle\psi!\rangle\varphi_1 \vee \langle\psi!\rangle\varphi_2 && \mathsf{Red}^{\mathcal{C}}_{\langle!\rangle,\vee} \\
\langle\psi!\rangle\Diamond_i\varphi &\leftrightarrow \psi \wedge \mathrm{f}(\psi) \wedge \Diamond_i\langle\psi!\rangle\varphi && \mathsf{Red}^{\mathcal{C}}_{\langle!\rangle,\Diamond_i} \\
\langle\psi!\rangle\exists\varphi &\leftrightarrow \psi \wedge \mathrm{f}(\psi) \wedge \exists\langle\psi!\rangle\varphi && \mathsf{Red}^{\mathcal{C}}_{\langle!\rangle,\exists}
\end{aligned}$$

Observe that the announcement degree of formulas does not increase from the left to the right due to our hypothesis that the announcement degree of $\mathrm{f}(\varphi)$ is at most that of φ; this would be violated e.g. if $\mathrm{f}(\varphi)$ was $\langle\varphi!\rangle\top$.

Let us associate to every formula ψ in $\mathsf{L}_{\mathsf{PAL},\forall}$, its measure $\mathrm{m}(\psi) = (\mathrm{d}(\psi), \mathrm{l}(\psi))$ in $\mathbb{N}_0 \times \mathbb{N}_0$, $\mathbb{N}_0$ denoting the set of all the non-negative integers, and $\mathrm{d}(\psi)$ and $\mathrm{l}(\psi)$ respectively denoting the announcement degree and the length of ψ. Let $\ll$ be the well-founded ordering on $\mathbb{N}_0 \times \mathbb{N}_0$ defined by $(m_1, n_1) \ll (m_2, n_2)$ iff either $m_1 < m_2$, or $m_1 = m_2$ and $n_1 < n_2$.

The above reduction axioms suggest us to consider a function $\tau : \mathsf{L}_{\mathsf{PAL},\forall} \longrightarrow$

$\mathsf{L}_{\mathsf{EL},\forall}$ defined by the following equations:

$$\begin{aligned}
\tau(p) &= p\\
\tau(\bot) &= \bot\\
\tau(\neg\varphi) &= \neg\tau(\varphi)\\
\tau(\varphi\vee\psi) &= \tau(\varphi)\vee\tau(\psi)\\
\tau(\Diamond_i\varphi) &= \Diamond_i\tau(\varphi)\\
\tau(\langle\psi!\rangle p) &= \tau(\psi)\wedge\tau(f(\psi))\wedge p\\
\tau(\langle\psi!\rangle\bot) &= \bot\\
\tau(\langle\psi!\rangle\neg\varphi) &= \tau(\psi)\wedge\tau(f(\psi))\wedge\neg\tau(\langle\psi!\rangle\varphi)\\
\tau(\langle\psi!\rangle(\varphi_1\vee\varphi_2)) &= \tau(\langle\psi!\rangle\varphi_1)\vee\tau(\langle\psi!\rangle\varphi_2)\\
\tau(\langle\psi!\rangle\Diamond_i\varphi) &= \tau(\psi)\wedge\tau(f(\psi))\wedge\Diamond_i\tau(\langle\psi!\rangle\varphi)\\
\tau(\langle\psi!\rangle\exists\varphi) &= \tau(\psi)\wedge\tau(f(\psi))\wedge\exists\tau(\langle\psi!\rangle\varphi)\\
\tau(\exists\varphi) &= \exists\tau(\varphi)
\end{aligned}$$

These equations really define a function by $\ll$-induction from $\mathsf{L}_{\mathsf{PAL},\forall}$ to $\mathsf{L}_{\mathsf{EL},\forall}$, seeing that if $\tau(\psi)$ occurs on the right side of the equation defining $\tau(\varphi)$ then $\mathrm{m}(\psi)\ll\mathrm{m}(\varphi)$.

In other respect, remark that for every ψ in $\mathsf{L}_{\mathsf{PAL},\forall}$, the formula

$$\psi\leftrightarrow\tau(\psi)$$

is valid in $\mathcal{C}$. It follows that when applying τ, we can eliminate step by step every occurrence of a dynamic operator. Therefore we can prove completeness of the axiomatisation obtained by replacing the standard reduction axioms by the above ones in the very same way as the completeness of the standard axiomatisation for K-PAL. Moreover, the validity problem of $\mathsf{L}_{\mathsf{PAL},\forall}$ formulas in the class $\mathcal{C}$ is reducible to the validity problem of $\mathsf{L}_{\mathsf{EL},\forall}$ formulas in $\mathcal{C}$.

So it remains to find out for which classes $\mathcal{C}$ such a function exists. This is the same as looking for a function $\mathrm{f}:\mathsf{L}_{\mathsf{PAL},\forall}\longrightarrow\mathsf{L}_{\mathsf{PAL},\forall}$ such that for every ψ in $\mathsf{L}_{\mathsf{PAL},\forall}$, the formula

$$\langle\psi!\rangle\top\leftrightarrow\psi\wedge\mathrm{f}(\psi)$$

is valid in $\mathcal{C}$ and the announcement degree of $\mathrm{f}(\psi)$ is at most that of ψ (the latter ensuring that reduction terminates). We do so in the next two sections.

4 Positive results

For which classes of frames $\mathcal{C}$ can we express that the frame of the updated model belongs to $\mathcal{C}$ by means of a formula $\mathrm{f}(\psi)$ of the language $\mathsf{L}_{\mathsf{PAL},\forall}$? Clearly, when $\mathcal{C}$ is the class of all Kripke frames $\mathcal{C}_{\mathsf{all}}$ then $\mathrm{f}(\psi)=\top$. The same is the case when $\mathcal{C}$ is a class of frames that is defined by universal first-order conditions[5], such as reflexivity, transitivity, symmetry, Euclideanity, linearity,

[5] A first-order formula is universal if it is of the form $\forall x_1\ldots\forall x_n\varphi$ where φ is quantifier-free and where the variables of φ are among $x_1,\ldots,x_n$.

or combinations thereof: these conditions are preserved under any update. This accounts for the public announcement extensions of modal logics such as KT, K4, KT4 = S4, KB, KTB, KB4, KT45 = S5 and S4.3.

Things are less straightforward for frames defined by seriality and combinations of seriality with other conditions such as transitivity and Euclideanity. In this section we exhibit formulas f(.) for some of these cases, thus accounting in particular for the public announcement logics KD-PAL and KD45-PAL.

4.1 Universal frame conditions

Let Φ be a universal first-order sentence over $\{R, =\}$. Let $\mathcal{C}_\Phi$ be the class of frames satisfying Φ. Then

$$\langle\psi!\rangle\top \leftrightarrow \psi$$

is valid in $\mathcal{C}_\Phi$. Therefore, we can set $\mathrm{f}_\Phi(\psi) = \top$.

This covers in particular the case where $\mathcal{C}$ is the class of frames for S4.3, i.e., the class of reflexive, transitive and linear frames.

4.2 Seriality

The equivalence

$$\langle\psi!\rangle\top \leftrightarrow \psi \wedge \forall(\psi \to \bigwedge_{i\in\mathbb{J}} \Diamond_i\psi)$$

is valid in the class of all serial frames $\mathcal{C}_{\mathsf{serial}}$. We can therefore set $\mathrm{f}(\psi) = \forall(\psi \to \bigwedge_{i\in\mathbb{J}} \Diamond_i\psi)$.

We observe that there is no $\mathsf{L_{EL}}$ formula φ such that $\langle p!\rangle\top \leftrightarrow \varphi$ is valid in serial frames. Suppose such a formula exists. Let n be its modal degree. Consider the frame $\langle \mathbb{N}_0, R\rangle$ where $\mathbb{N}_0$ is the set of all the non-negative integers and $\langle x, y\rangle \in R$ iff $y = x + 1$. This is clearly a serial frame. We define two valuations V and V' on that frame by stipulating $V(p) = \mathbb{N}_0$, $V'(p) = \{0, \cdots, n\}$, and $V(q) = \emptyset$ for every $q \neq p$. We have $\langle \mathbb{N}_0, R, V\rangle, 0 \Vdash^{\mathcal{C}_{\mathsf{serial}}} \varphi$ iff $\langle \mathbb{N}_0, R, V'\rangle, 0 \Vdash^{\mathcal{C}_{\mathsf{serial}}} \varphi$ because n is the modal degree of φ. However, $\langle \mathbb{N}_0, R, V\rangle, 0 \Vdash^{\mathcal{C}_{\mathsf{serial}}} \langle p!\rangle\top$ while $\langle \mathbb{N}_0, R, V'\rangle, 0 \nVdash^{\mathcal{C}_{\mathsf{serial}}} \langle p!\rangle\top$; the latter is the case because the frame of the updated model $M \circ \|p\|_M^{\mathcal{C}_{\mathsf{serial}}}$ is not serial. We therefore have a contradiction.

4.3 Seriality and transitivity

As the reader can check, the equivalence

$$\langle\psi!\rangle\top \leftrightarrow \psi \wedge \forall(\psi \to \textstyle\bigwedge_{i\in\mathbb{J}} \Diamond_i\psi)$$

is also valid in the class of all serial and transitive frames $\mathcal{C}_{\mathsf{serial,trans}}$. We can simplify that equivalence when the set of agents $\mathbb{J}$ is a singleton, say $\{i\}$.

Consider the class $\mathcal{C}^g_{\mathsf{serial,trans}}$ of frames for KD4 that are point-generated, i.e. the class of serial and transitive frames $\langle W, R\rangle$ with a world $x \in W$ such that $W = \{x\} \cup R_i(x)$. Then

$$\langle\psi!\rangle\top \leftrightarrow \psi \wedge \Diamond_i\psi \wedge \Box_i(\psi \to \Diamond_i\psi)$$

is valid in $\mathcal{C}^g_{\mathsf{serial,trans}}$. We may therefore set $\mathrm{f}(\psi) = \Diamond_i\psi \wedge \Box_i(\psi \to \Diamond_i\psi)$.

Example 4.1 Let us check that for the logic KD4-PAL the formula

$$(p \wedge \neg\Diamond_i p) \rightarrow \langle p!\rangle\neg\Diamond_i\top$$

is not valid. For $\mathcal{C}_{\mathsf{serial,trans}}$ we have the condition $\mathrm{f}(p) = \forall(p \rightarrow \bigwedge_{i\in\mathbb{J}} \Diamond_i p)$. By the reduction axioms of Section 3.2 we obtain:

$$\begin{aligned}
\langle p!\rangle\neg\Diamond_i\top \leftrightarrow\ & p \wedge \forall\left(p \rightarrow \bigwedge_{i\in\mathbb{J}} \Diamond_i p\right) \wedge \neg\langle p!\rangle\Diamond_i\top && \mathsf{Red}^{\mathcal{C}}_{\langle!\rangle,\neg}\\
\leftrightarrow\ & p \wedge \forall\left(p \rightarrow \bigwedge_{i\in\mathbb{J}} \Diamond_i p\right) \wedge\\
& \neg\left(p \wedge \forall\left(p \rightarrow \bigwedge_{i\in\mathbb{J}} \Diamond_i p\right) \wedge \Diamond_i\langle p!\rangle\top\right) && \mathsf{Red}^{\mathcal{C}}_{\langle!\rangle,\Diamond_i}\\
\leftrightarrow\ & p \wedge \forall\left(p \rightarrow \bigwedge_{i\in\mathbb{J}} \Diamond_i p\right) \wedge \neg\Diamond_i\langle p!\rangle\top && \text{(propos. simplif.)}\\
\leftrightarrow\ & p \wedge \forall\left(p \rightarrow \bigwedge_{i\in\mathbb{J}} \Diamond_i p\right) \wedge\\
& \neg\Diamond_i\left(p \wedge \forall\left(p \rightarrow \bigwedge_{i\in\mathbb{J}} \Diamond_i p\right)\right) && \mathsf{Red}^{\mathcal{C}}_{\langle!\rangle,\top}
\end{aligned}$$

where the reduction axiom used in the last step can be obtained from $\mathsf{Red}^{\mathcal{C}}_{\langle!\rangle,\neg}$ and $\mathsf{Red}^{\mathcal{C}}_{\langle!\rangle,\bot}$. The $\mathsf{L}_{\mathsf{PAL},\forall}$ formula

$$(p \wedge \neg\Diamond_i p) \rightarrow \left(p \wedge \forall(p \rightarrow \bigwedge_{i\in\mathbb{J}} \Diamond_i p) \wedge \neg\Diamond_i(p \wedge \forall(p \rightarrow \bigwedge_{i\in\mathbb{J}} \Diamond_i p))\right)$$

is not valid in serial and transitive frames. (It is actually even invalid in $\mathcal{C}_{\mathsf{all}}$.) Therefore

$$(p \wedge \neg\Diamond_i p) \rightarrow \langle p!\rangle\neg\Diamond_i\top$$

is not valid either. This contrasts with the hypothetical proof system that we have discussed in Section 2.5.

4.4 Seriality, transitivity, and Euclideanity: single agent case

Let us suppose that the set of agents $\mathbb{J}$ is the singleton $\{i\}$. Consider the class $\mathcal{C}^g_{\mathsf{serial,trans,eucl}}$ of frames for KD45 that are point-generated, i.e. the class of serial, transitive and Euclidean frames $\langle W, R\rangle$ with a world $x \in W$ such that $W = \{x\} \cup R_i(x)$.

Just as for serial frames the equivalence

$$\langle \psi ! \rangle \top \leftrightarrow \psi \wedge \forall(\psi \rightarrow \bigwedge_{i \in \mathbb{J}} \Diamond_i \psi)$$

is valid in $\mathcal{C}^g_{\mathsf{serial,trans,eucl}}$. However, we can do better because there is only one agent: the schema

$$\langle \psi ! \rangle \top \leftrightarrow \psi \wedge \Diamond_i \psi$$

is valid in $\mathcal{C}^g_{\mathsf{serial,trans,eucl}}$, i.e., $\mathrm{f}(\psi) = \Diamond_i \psi$.

We observe that the restriction to point-generated frames cannot be avoided if we want a characterisation by a $\mathsf{L_{EL}}$ formula: there is no $\mathsf{L_{EL}}$ formula φ such that

$$\langle p! \rangle \top \leftrightarrow \varphi$$

is valid in the class of serial, transitive and Euclidean frames. Indeed, suppose such a formula φ exists. Consider the frame $\langle W, R \rangle$ where $W = \{0, 1, 2\}$ and $R = \{\langle 0, 0 \rangle, \langle 1, 2 \rangle, \langle 2, 2 \rangle\}$. That frame is serial, transitive and Euclidean. Let $V_1(p) = \{0, 1\}$ and let $V_2(p) = \{0\}$. However, we have $\langle W, R, V_1 \rangle, 0 \not\Vdash^{\mathcal{C}_{\mathsf{serial,trans,eucl}}} \langle p! \rangle \top$, while $\langle W, R, V_2 \rangle, 0 \Vdash^{\mathcal{C}_{\mathsf{serial,trans,eucl}}} \langle p! \rangle \top$. Then we have $\langle W, R, V_1 \rangle, 0 \not\Vdash \varphi$ and $\langle W, R, V_2 \rangle, 0 \Vdash \varphi$. Since φ is without $\forall$ this leads us to a contradiction.

5 Negative results

We now show that there is no such formula as $\mathrm{f}(\psi)$ for the class of confluent frames, for the class of reflexive, transitive and confluent frames —i.e., for the basic logic $\mathsf{S4.2}$—, and for the class of dense frames.

5.1 Confluence

Let $\mathcal{C}_{\mathsf{confl}}$ be the class of all confluent frames. Is there a $\mathsf{L_{EL,\forall}}$ formula φ such that $\langle p! \rangle \top \leftrightarrow \varphi$ is valid in $\mathcal{C}_{\mathsf{confl}}$?

Suppose such a formula exists.

Let $W = \{x, y, z\}$ and let R_i be the reflexive and transitive closure of the relation $\{\langle x, y \rangle, \langle y, z \rangle\}$, for every i. The frame $\langle W, R \rangle$ is in $\mathcal{C}_{\mathsf{confl}}$. Let V be a valuation on $\langle W, R \rangle$ such that $V(p) = \{x, y\}$. The model $M = \langle W, R, V \rangle$ and its update by $\|p\|_M$ are depicted (without the reflexive edges) in Figure 1.

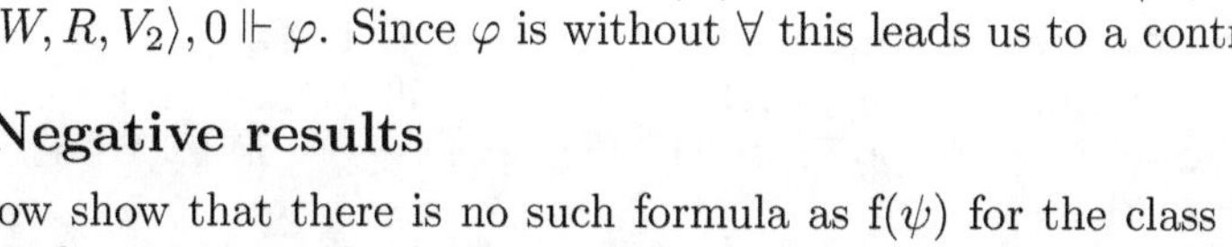

Fig. 1. M and its update by $\|p\|^{\mathcal{C}}_M$ (reflexive edges omitted)

Now let $W' = \{x, y, y', z\}$ and let R'_i be the reflexive and transitive closure of the relation $\{\langle x, y \rangle, \langle y, z \rangle, \langle x, y' \rangle, \langle y', z \rangle\}$, for every i. The frame $\langle W', R' \rangle$ is in $\mathcal{C}_{\mathsf{confl}}$, too. Let V' be a valuation on $\langle W', R' \rangle$ such that $V'(p) = \{x, y, y'\}$. The model $M' = \langle W', R', V' \rangle$ and its update by $\|p\|_{M'}$ are depicted (without the reflexive edges) in Figure 2.

The models M and M' are bisimilar [6], and we therefore have $M, x \Vdash \varphi$ iff

[6] The definition of bisimilarity has to take the universal modality into account. So $M =$

Fig. 2. M' and its update by $\|p\|^{\mathcal{C}}_{M'}$ (reflexive edges omitted)

$M', x \Vdash \varphi$ for every $\mathsf{L}_{\mathsf{EL},\forall}$ formula φ. However, we have $M, x \Vdash^{\mathcal{C}_{\mathsf{confl}}} \langle p! \rangle \top$, while $M', x \nVdash^{\mathcal{C}_{\mathsf{confl}}} \langle p! \rangle \top$; the former is the case because the frame of the updated model $M \circ \|p\|^{\mathcal{C}}_M$ is confluent, and the latter is the case because the frame of the updated model $M' \circ \|p\|^{\mathcal{C}}_M$ is not. We therefore have a contradiction.

5.2 Reflexivity, transitivity, and confluence

Let $\mathcal{C}_{\mathsf{refl,trans,confl}}$ be the class of all reflexive, transitive, and confluent frames. There is no $\mathsf{L}_{\mathsf{EL},\forall}$ formula φ such that $\langle p! \rangle \top \leftrightarrow \varphi$ is valid in that class. Indeed, suppose such a formula exists. We take over the two above counterexample models for confluence. Observe that both frames are in $\mathcal{C}_{\mathsf{refl,trans,confl}}$. Again, M and M' are bisimilar, and therefore $M, x \Vdash \varphi$ iff $M', x \Vdash \varphi$ for every φ. However, $M, x \Vdash^{\mathcal{C}_{\mathsf{refl,trans,confl}}} \langle p! \rangle \top$, while $M', x \nVdash^{\mathcal{C}_{\mathsf{refl,trans,confl}}} \langle p! \rangle \top$. We therefore have a contradiction.

5.3 Density

Let $\mathcal{C}_{\mathsf{dense}}$ be the class of all dense frames. There is no $\mathsf{L}_{\mathsf{EL},\forall}$ formula φ such that $\langle p! \rangle \top \leftrightarrow \varphi$ is valid in $\mathcal{C}_{\mathsf{dense}}$. Indeed, suppose such a formula exists. Let $\langle W, R \rangle$ be the frame defined by $W = \{\alpha, \omega, 1, 2, 3, 4, 5\}$ and

$$\begin{aligned} R_i = \; & \{\langle \alpha, \omega \rangle\} \cup \\ & \{\langle \alpha, y \rangle \mid 1 \leq y \leq 5\} \cup \\ & \{\langle x, \omega \rangle \mid 1 \leq x \leq 5\} \cup \\ & \{\langle x, y \rangle \mid 1 \leq x, y \leq 5, x \neq y\} \end{aligned}$$

for every $i \in \mathbb{J}$. The reader may check that $\langle W, R \rangle$ is indeed dense. Let $V_1(p) = \{\alpha, \omega, 1, 5\}$ and let $V_2(p) = \{\alpha, \omega, 1, 3, 5\}$. The models $M_1 = \langle W, R, V_1 \rangle$ and $M_2 = \langle W, R, V_2 \rangle$ are bisimilar, and therefore $M_1, \alpha \Vdash \varphi$ iff $M_2, \alpha \Vdash \varphi$ for every $\mathsf{L}_{\mathsf{EL},\forall}$ formula φ. However, $M_1, \alpha \nVdash^{\mathcal{C}_{\mathsf{dense}}} \langle p! \rangle \top$, while $M_2, \alpha \Vdash^{\mathcal{C}_{\mathsf{dense}}} \langle p! \rangle \top$. We therefore have a contradiction.

$\langle W, R, V \rangle$ and $M' = \langle W', R', V' \rangle$ are bisimilar if there is a relation $Z \subseteq W \times W$ such that:

(i) if $(x, x') \in Z$ then $x \in V(p)$ iff $x' \in V'(p)$, for every $p \in \mathbb{P}$

(ii) if $(x, x') \in Z$ and $(x, y) \in R_i$ then there is $y' \in W'$ such that $(x', y') \in R'_i$ and $(x, y) \in Z$

(iii) if $(x, x') \in Z$ and $(x', y') \in R'_i$ then there is $y \in W$ such that $(x, y) \in R_i$ and $(x, y) \in Z$

(iv) for every $x \in W$ there is $x' \in W'$ such that $(x, x') \in Z$

(v) for every $x' \in W'$ there is $x \in W$ such that $(x, x') \in Z$

The last two conditions say that both Z and its converse Z^{-1} are serial.

6 Discussion: semantic alternatives

We now explore some alternative semantics for public announcements that one can find in the literature: first, a proposal to update models in a different way, and second, a different formulation of the truth condition for public announcements.

6.1 Relation updates

Beyond the standard way of updating a Kripke model by eliminating *worlds* of Section 2, there are proposals in the literature where instead of worlds it is *edges* that are eliminated [5,6]. Let $M = \langle W, R, V \rangle$ be a model, and let U be some subset of W. The *relation update* of M by U is defined as $M \overset{\mathrm{r}}{\circ} U = \langle W, R', V \rangle$, with:

$$R'_i = R_i \cap (W \times U)$$

Let $\Lambda(\mathcal{C})^{\mathrm{r}}$ be the set of $\mathsf{L}_{\mathsf{PAL}}$ formulas valid in $\mathcal{C}$ under relation update (under our truth condition for announcements of Section 2.2 requiring truth of the announcement).

We argue that if one wants the underlying modal logic to be a customary logic of knowledge or belief, then this way of extending a modal logic by announcements is not very interesting, for two reasons. First, as far as the language $\mathsf{L}_{\mathsf{PAL}}$ is concerned we have $\Lambda(\mathcal{C}) = \Lambda(\mathcal{C})^{\mathrm{r}}$ (because the generated submodels of $M \overset{\mathrm{r}}{\circ} U$ and $M \circ U$ are equal); the logics differ only when the universal modal operator comes into play. Second, while membership in the class of all models $\mathcal{C}_{\mathsf{all}}$ is preserved under relation update, membership in a particular class of models is preserved in fewer cases: not only do existential first-order conditions such as seriality, density and confluence fail, but also universal conditions such as reflexivity and symmetry.

6.2 An unconditioned truth condition

Remember that the standard formulation of the truth condition requires announcements to be *truthful*. This means that the agents acquire knowledge. In the literature one can find not only another definition of model update, but also another formulation of the truth condition for public announcements. To the contrary, in Gerbrandy's formulation announcements may be false [5,6]. The latter formulation is therefore often claimed to be more appropriate for agents acquiring beliefs, see e.g. [9]. We call this the *unconditioned* truthcondition and highlight it by "u":

$$M, x \Vdash^{\mathrm{u}} \langle \varphi ! \rangle \psi \quad \text{iff} \quad M \circ \|\varphi\|^{\mathrm{u}}_M, x \Vdash \psi$$

Then we have two options, according to whether we use world update or relation update. Call $\Lambda(\mathcal{C})^{\mathrm{u}}$ and $\Lambda(\mathcal{C})^{\mathrm{ru}}$ the resulting logics of the class of frames $\mathcal{C}$. For example, Kooi's basic public announcement logic is $\Lambda(\mathcal{C}_{\mathsf{all}})^{\mathrm{ru}}$ [9].

Observe that none of the logics $\Lambda(\mathcal{C})^{\mathrm{u}}$ makes sense. Indeed, consider the case where a model M is updated by a formula that is false at every point of

M; then the set of possible worlds of the updated model is empty, and therefore the update of the model is not a legal Kripke model: the unconditioned interpretation is ill-defined. Let us finally notice that when we 'repair' the logics $\Lambda(\mathcal{C})^{\mathrm{u}}$ by replacing the unconditioned truth condition by our enhanced version then we obtain $\Lambda(\mathcal{C})^{\mathcal{C}}$.

7 Related work

Yanjing Wang has recently investigated axiomatisations of public announcement logics [17]. He focusses on public announcement extensions of the basic modal logic K and subtleties with different versions of the axiomatization. In particular, he highlights the role of the rule of replacement of equivalents $\mathsf{RE}^{\mathsf{r}}_{\langle ! \rangle}$. His work certainly provide a stimulating background to our own. Updates that preserve $\mathsf{KD45}$ have been investigated in [15,2,10]. Guillaume Aucher [2] defines a language fragment that makes you go mad ('crazy formulas'). The formula characterising the cases where this can be avoided is the same as ours in Section 4. David Steiner [15] proposes that the agent does not incorporate the new information if he already believes to the contrary. In that case, nothing happens. Otherwise, access to states where the information is not believed is eliminated, just as for believed public announcements. This solution to model unbelievable information is similarly proposed in the elegant [10], where it is called 'cautious update' — a suitable term. The difference between these approaches and ours is that the agent simply keeps his old beliefs in case the new information is unbelievable (i.e., if there is no accessible state where the announced formula is true). In our KD45 preserving updates the update cannot be executed if it is unbelievable.

8 Conclusion

In this paper we had a closer look at the axiomatization and the semantics of various public announcement logics. We highlighted problems that arise for epistemic or doxastic logics with existential frame conditions such as seriality or confluence and proposed an enhanced truth condition avoiding these problems in some cases. Our new truth condition amounts to the original condition if the basic logic is K or $\mathsf{S5}$. We have studied the limitations of our solution; in particular the case of confluence remains without a satisfactory solution, and with it the extension of the logic of knowledge $\mathsf{S4.2}$ by public announcements.

Our results required to extend the language by a master modality. We opted for the universal modality; however, the common knowledge modality would do, too.

Everything said here transfers to other kinds of updates such as assignments. More precisely, in dynamic epistemic logics with assignments, we can model announcements that stay within a certain frame class in the same way, but for the dynamics involving assignments there are no additional complications: an assignment is a total function that can always be executed, and that never changes the frame properties of the transformed model. It would be interesting to study whether (and how) it transfers to dynamic epistemic logics with event models and product update [3]. In [3], Baltag and Moss showed that reflexivity, transitivity and Euclideanity are preserved under standard product updates. In [2], Aucher provided a characterisation of the condition $\mathrm{f}(\psi)$ under which product update preserves seriality.

Here is a proposal for a way to overcome the expressive limitations that we have highlighted in Section 5. The idea is to enrich the language by a modal constant $\delta_\mathcal{C}$ whose interpretation is that it is true exactly when the frame it is evaluated in is part of the class $\mathcal{C}$. Let us call that language $\mathsf{L}_{\mathsf{PAL},\delta_\mathcal{C}}$. Its truth condition is:

$$\langle W, R, V\rangle, x \Vdash \delta_\mathcal{C} \text{ iff } \langle W, R\rangle \in \mathcal{C}$$

Let us define a translation from $\mathsf{L}_{\mathsf{PAL}}$ to $\mathsf{L}_{\mathsf{PAL},\delta_\mathcal{C}}$ whose main clauses are:

$$p^{\mathsf{t}} = p$$
$$(\langle\psi!\rangle\varphi)^{\mathsf{t}} = \langle\psi^{\mathsf{t}}!\rangle(\delta_\mathcal{C} \wedge \varphi^{\mathsf{t}})$$

and homomorphic for the other cases. We then have that for every frame $\langle W, R\rangle$, every valuation V over that frame, and every world $x \in W$, $\langle W, R, V\rangle, x \Vdash^\mathcal{C} \varphi$ iff $\langle W, R, V\rangle, x \Vdash \varphi^{\mathsf{t}}$. It remains however to axiomatise the $\mathcal{C}$ validities in the augmented language.

Acknowledgements

Thanks are due to the three AiML reviewers for their thorough comments on the submitted version. We wish to express our special appreciation for the second AiML reviewer, who went far beyond his duties in giving us very precise and detailed comments that helped to improve the paper and make it more readable. Hans van Ditmarsch also wants to thank the IMSc (Institute of Mathematical Sciences), Chennai, India, where he is affiliated as a research associate.

References

[1] Carlos Areces and Balder ten Cate. Hybrid logics. In Patrick Blackburn, Johan van Benthem, and Frank Wolter, editors, *Handbook of Modal Logic*, volume 3. Elsevier Science, 2006.

[2] Guillaume Aucher. Consistency preservation and crazy formulas in BMS. In S. Hölldobler, C. Lutz, and H. Wansing, editors, *Logics in Artificial Intelligence, 11th European Conference, JELIA 2008. Proceedings*, pages 21–33. Springer, 2008. LNCS 5293.

[3] Alexandru Baltag and Lawrence Moss. Logics for epistemic programs. *Synthese*, 139(2):165–224, 2004.

[4] Brian Chellas. *Modal logic: An introduction.* Cambridge University Press, 1980.

[5] Jelle Gerbrandy. *Bisimulations on Planet Kripke.* PhD thesis, ILLC, University of Amsterdam, 1999.

[6] Jelle Gerbrandy and Willem Groeneveld. Reasoning about information change. *J. of Logic, Language and Information*, 6(2), 1997.

[7] Jaakko Hintikka. *Knowledge and Belief.* Cornell University Press, 1962.

[8] Wesley H. Holliday, Tomohiro Hoshi, and Thomas F. Icard III. Schematic validity in dynamic epistemic logic: decidability. In *Proceedings of the Third international conference on Logic, rationality, and interaction*, LORI'11, pages 87–96, Berlin, Heidelberg, 2011. Springer-Verlag.

[9] Barteld Kooi. Expressivity and completeness for public update logic via reduction axioms. *Journal of Applied Non-Classical Logics*, 17(2):231–253, 2007.

[10] Barteld Kooi and Bryan Renne. Arrow update logic. *Review of Symbolic Logic*, 4:536–559, 2011.

[11] Wolfgang Lenzen. *Recent work in epistemic logic.* North Holland Publishing Company, Amsterdam, 1978.

[12] Wolfgang Lenzen. On the semantics and pragmatics of epistemic attitudes. In Armin Laux and Heinrich Wansing, editors, *Knowledge and belief in philosophy and AI*, pages 181–197. Akademie Verlag, Berlin, 1995.

[13] Jan Plaza. Logics of public communications. In M. L. Emrich, M. Hadzikadic, M. S. Pfeifer, and Z. W. Ras, editors, *Proceedings of the Fourth International Symposium on Methodologies for Intelligent Systems (ISMIS)*, pages 201–216, 1989.

[14] Jan Plaza. Logics of public communications. *Synthese*, 158(2):165–179, 2007.

[15] D. Steiner. A system for consistency preserving belief change. In *Proceedings of the ESSLLI Workshop on Rationality and Knowledge*, pages 133–144, 2006.

[16] Hans van Ditmarsch, Wiebe van der Hoek, and Barteld Kooi. *Dynamic Epistemic Logic*, volume 337 of *Synthese Library.* Springer, 2007.

[17] Yanjing Wang. On axiomatizations of PAL. In *Proc. LORI'11*, pages 314–327, 2011.

Sahlqvist Theorems for Precontact Logics

Philippe Balbiani [1]

Institut de recherche en informatique de Toulouse
CNRS — Université de Toulouse

Stanislav Kikot [2]

Department of Computer Science and Information Systems
Birbeck — University of London

Abstract

Precontact logics are propositional modal logics that have been recently considered in order to obtain decidable fragments of the region-based theories of space introduced by De Laguna and Whitehead. We give the definition of Sahlqvist formulas to this region-based setting and we prove correspondence and canonicity results. Together, these results give rise to a completeness result for precontact logics that are axiomatized by Sahlqvist axioms.

Keywords: Sahlqvist theory; Precontact logics; Correspondence; Canonicity.

1 Introduction

In modal logic, Sahlqvist formulas are modal formulas with remarkable properties [30,31]: the Sahlqvist correspondence theorem says that every Sahlqvist formula corresponds to a first-order definable class of frames; the Sahlqvist completeness theorem says that when Sahlqvist formulas are used as axioms in a normal logic, the logic is complete with respect to the elementary class of frames the axioms define. Roughly speaking, modal formulas in the Sahlqvist fragment are implications the antecedents of which do not contain occurrences of boxes taking scope over diamonds. As a result, their main characteristic consists in this: second-order quantifier elimination is complete for their standard translation in a second-order setting [10,17].

The Sahlqvist fragment does not contain all modal formulas corresponding to a first-order definable class of frames: there exists non-Sahlqvist formulas that correspond to first-order conditions. Moreover, it is undecidable, given a modal formula, to determine whether it has a first-order correspondent. As well, it is

[1] Address: Institut de recherche en informatique de Toulouse, CNRS — Université de Toulouse, 118 route de Narbonne, 31062 Toulouse Cedex 9, FRANCE; balbiani@irit.fr.

[2] Address: Department of Computer Science and Information Systems, Birkbeck — University of London, Malet Street, London WC1E 7HX, UK; kikot@dcs.bbk.ac.uk.

undecidable, given a first-order sentence, to determine whether it has a modal correspondent. For more on this, see [8,9].
There is quite a lot of literature around Sahlqvist theorem, which roughly can be divided into the following groups. A first group concerns the study of algorithms performing second-order quantifier elimination, see [32] for a recent account of this area. A second group deals with generalizations of Sahlqvist theorem within the classical syntax and semantics of modal logic [21,34,35]. A third group deals with generalization of Sahlqvist theorem to stronger or weaker variants of modal language: hybrid logics [7], distributive modal logic [16], polyadic modal languages [19,20], relevant modal logics [33], modal fixed point logics [5]. This paper certainly belongs to this group.
Recently, in order to obtain decidable fragments of the region-based theories of space introduced by De Laguna [28] and Whitehead [38], propositional languages with topological semantics have been considered [15,25,36,37]. The main tools in the completeness proofs of the associated logics are the representation theorems for precontact algebras and adjacency spaces presented in [11,12,13,14]. At first sight, the modal nature of the logics in question is not patently visible. Nevertheless, almost all known tools and techniques in modal logic — e.g. the method of canonical models and the filtration method — can be transferred to them with slight modifications for obtaining the abovementioned completeness proofs [1,2,3,4].
Hence, a natural question is to ask whether a Sahlqvist-like theory — i.e. a theory that identifies a set of formulas that correspond to first-order definable classes of frames and that define logics complete with respect to the elementary classes of frames they correspond to — can be elaborated on the setting of the region-based propositional modal logics of space (RBPMLS). With the object of answering this question, we give the definition of Sahlqvist formulas to this RBPMLS setting and we prove correspondence and canonicity results. Together, these results give rise to a completeness result for RBPMLS that are axiomatized by Sahlqvist axioms. Note that the translation $C(a,b) = \Diamond_u(b \wedge \Diamond a)$ conservatively embeds RBPMLS into the basic modal language extended with universal modality, hence the correspondence part for RBPMLS follows from the classical Sahlqvist theorem. However, the completeness part for RBPMLS is genuinely new, since it provides an inference in a relatively weak calculus for every RBPMLS formula, which follows semantically from the initial 'Sahlqvist' RBPMLS formula. We assume the reader is at home with tools and techniques in modal logic. For more on this, see [6,27].

2 Syntax

The language is defined using a countable set BV of Boolean variables (with typical members denoted by p, q, etc). We inductively define the set $t(BV)$ of terms (with typical members denoted by a, b, etc) as follows:

- $a ::= p \mid 0 \mid -a \mid (a \cup b)$.

The other Boolean constructs for RBPMLS terms are defined as usual: 1 for -0 and $(a \cap b)$ for $-(-a \cup -b)$. A term a is positive iff a is built up from Boolean variables using only 1, $\cup$ and $\cap$. We inductively define the set $f(BV)$ of formulas (with typical members denoted by ϕ, ψ, etc) as follows:

- $\phi ::= a \equiv b \mid C(a,b) \mid \bot \mid \neg\phi \mid (\phi \vee \psi)$.

The other Boolean constructs for RBPMLS formulas are defined as usual: $\top$ for $\neg\bot$, $(\phi \wedge \psi)$ for $\neg(\neg\phi \vee \neg\psi)$, $(\phi \to \psi)$ for $(\neg\phi \vee \psi)$ and $(\phi \leftrightarrow \psi)$ for $(\neg(\phi \vee \psi) \vee \neg(\neg\phi \vee \neg\psi))$. We obtain the formulas $a \not\equiv b$ and $\bar{C}(a,b)$ as abbreviations:

- $a \not\equiv b ::= \neg a \equiv b$,
- $\bar{C}(a,b) ::= \neg C(a,b)$.

If a formula ϕ is built up from $a \not\equiv 0$ and $C(a,b)$ (where a and b are positive terms) using only $\top$, $\vee$ and $\wedge$ then we say that ϕ is negation-free. A formula ϕ is positive iff ϕ is built up from $a \not\equiv 0$, $-a \equiv 0$, $C(a,b)$ and $\bar{C}(-a,-b)$ (where a and b are positive terms) using only $\top$, $\vee$ and $\wedge$. The notion of subterm and the notion of subformula are standard. We adopt the standard rules for omission of the parentheses. If a formula ϕ is an implication $\psi \to \chi$ in which ψ is negation-free and χ is positive then we say that ϕ is a Sahlqvist formula. Let us consider the following 8 formulas:

(i) $\top \to C(1,1)$,

(ii) $p \not\equiv 0 \to C(p,1)$,

(iii) $p \not\equiv 0 \to C(p,p)$,

(iv) $C(p,q) \to C(q,p)$,

(v) $C(p,q) \to C(p,r) \vee C(-r,q)$,

(vi) $p \not\equiv 0 \wedge -p \not\equiv 0 \to C(p,-p)$,

(vii) $(p \cup q) \equiv 1 \wedge (p \cap q) \equiv 0 \to C(p,p) \vee C(q,q)$,

(viii) $(p \cap -q) \not\equiv 0 \to C(p,-q) \vee C(q,-q)$.

Obviously, the first 4 formulas are Sahlqvist formulas whereas the last 4 formulas are not Sahlqvist formulas.

3 Kripke-type semantics

RBPMLS have 3 kinds of semantics:

- an algebraic semantics based on some classes of abstract contact algebras of regions,
- a topological semantics based on concrete contact algebras of regions over some classes of topological spaces,
- a Kripke-type semantics based on some classes of Kripke frames regarded as adjacency spaces.

The main tools in the equivalence of these 3 kinds of semantics are the representation theorems for precontact algebras and adjacency spaces presented in [11,12,13,14]. In this paper, seeing that we want to elaborate a Sahlqvist-like theory on the setting of RBPMLS, we concentrate attention to the Kripke-type semantics. A Kripke frame is an ordered pair $\mathcal{F} = (W, R)$ where W is a non-empty set of possible worlds and R is a binary relation on W. A valuation based on $\mathcal{F}$ is a function V assigning to each Boolean variable p a subset $V(p)$ of W. As usual, V induces a homomorphism $(\cdot)^V$ assigning to each term a a subset $(a)^V$ of W as follows:

- $(p)^V = V(p)$,
- $(0)^V = \emptyset$,
- $(-a)^V = W \setminus (a)^V$,
- $(a \cup b)^V = (a)^V \cup (b)^V$.

We shall say that V is smaller than a valuation V' based on $\mathcal{F}$, in symbols $V \leq V'$, iff for all Boolean variables p, $V(p) \subseteq V'(p)$.

Lemma 3.1 *Let V, V' be valuations based on $\mathcal{F}$ such that $V \leq V'$. For all positive terms a, $(a)^V \subseteq (a)^{V'}$.*

Proof. The proof is done by induction on a. □

A Kripke model is an ordered triple $\mathcal{M} = (W, R, V)$ where $\mathcal{F} = (W, R)$ is a frame and V is a valuation based on $\mathcal{F}$. The satisfiability of a formula ϕ in $\mathcal{M}$, in symbols $\mathcal{M} \models \phi$, is defined as follows:

- $\mathcal{M} \models a \equiv b$ iff $(a)^V = (b)^V$,
- $\mathcal{M} \models C(a, b)$ iff there exists $x, y \in W$ such that xRy, $x \in (a)^V$ and $y \in (b)^V$,
- $\mathcal{M} \not\models \bot$,
- $\mathcal{M} \models \neg\phi$ iff $\mathcal{M} \not\models \phi$,
- $\mathcal{M} \models \phi \vee \psi$ iff $\mathcal{M} \models \phi$ or $\mathcal{M} \models \psi$.

As a result, $\mathcal{M} \models a \not\equiv b$ iff $(a)^V \neq (b)^V$ and $\mathcal{M} \models \bar{C}(a, b)$ iff for all $x, y \in W$, if xRy then $x \notin (a)^V$ or $y \notin (b)^V$.

Lemma 3.2 *Let V, V' be valuations based on $\mathcal{F}$ such that $V \leq V'$.*

(i) *For all negation-free formulas ϕ, if $(\mathcal{F}, V) \models \phi$ then $(\mathcal{F}, V') \models \phi$.*

(ii) *For all positive formulas ϕ, if $(\mathcal{F}, V) \models \phi$ then $(\mathcal{F}, V') \models \phi$.*

Proof. Both items follow by induction on ϕ, using Lemma 3.1. □

Let $\mathcal{F}$ be a frame. A formula ϕ is valid in $\mathcal{F}$, in symbols $\mathcal{F} \models \phi$, iff for all models $\mathcal{M}$ based on $\mathcal{F}$, $\mathcal{M} \models \phi$.

4 Standard translation into a first-order language

In the above Kripke-type semantics, satisfaction is a binary relation between models and formulas whereas in the semantics for the basic modal language, satisfaction is a ternary relation between models, possible worlds and formulas. Such a difference is illustrated by the following translation of our language into a first-order language. Let $\mathcal{L}^1(BV)$ be the first-order language with equality which has the unary predicates P_0, P_1, etc corresponding to the Boolean variables p_0, p_1, etc in BV and the binary predicate R_C corresponding to the modal operator C. If u is a first-order variable and a is a term then the first-order formula $ST(u, a)$ is inductively defined as follows:

- $ST(u, p_n) = P_n(u)$,
- $ST(u, 0) = \bot$,
- $ST(u, -a) = \neg ST(u, a)$,
- $ST(u, a \cup b) = ST(u, a) \vee ST(u, b)$.

If ϕ is a formula then the first-order sentence $ST(\phi)$ is inductively defined as follows:

- $ST(a \equiv b) = \forall u\ (ST(u, a) \leftrightarrow ST(u, b))$,
- $ST(C(a, b)) = \exists u\ \exists v\ (R_C(u, v) \wedge ST(u, a) \wedge ST(v, b))$,
- $ST(\bot) = \bot$,
- $ST(\neg\phi) = \neg ST(\phi)$,
- $ST(\phi \vee \psi) = ST(\phi) \vee ST(\psi)$.

Proposition 4.1 *Let $\mathcal{M} = (W, R, V)$ be a model.*

(i) *For all terms a and for all $x \in W$, $x \in (a)^V$ iff $\mathcal{M} \models ST(u, a)[x]$.*

(ii) *For all formulas ϕ, $\mathcal{M} \models \phi$ iff $\mathcal{M} \models ST(\phi)$.*

Proof. The first item follows by induction on a and the second one follows by induction on ϕ, using the first item. □

Proposition 4.2 *Let $\mathcal{F} = (W, R)$ be a frame. For all formulas ϕ, $\mathcal{F} \models \phi$ iff $\mathcal{F} \models ST(\phi)$.*

Proof. By Proposition 4.1. □

Obviously, $ST(\phi)$ belongs to the 2-variable fragment of $\mathcal{L}^1(BV)$ for each formula ϕ. Since the satisfiability problem for the 2-variable fragment of any first-order language with equality is decidable in nondeterministic exponential time [22,29], then the embedding of our language into $\mathcal{L}^1(BV)$ considered in Proposition 4.2 implies that if $\mathcal{C}$ is a class of frames definable by a first-order sentence with at most 2 variables then the satisfiability problem in models based on $\mathcal{C}$-frames for RBPMLS formulas is decidable in nondeterministic exponential time.

5 Correspondence theorem

We shall say that a formula ϕ and a first-order sentence α of the first-order language $\mathcal{L}^1(\emptyset)$ with equality which has the binary predicate R_C corresponding to the modal operator C are frame correspondents iff for all frames $\mathcal{F} = (W, R)$, $\mathcal{F} \models \phi$ iff $\mathcal{F} \models \alpha$.

Theorem 5.1 *Let ϕ be a Sahlqvist formula. There exists a first-order sentence α of the first-order language $\mathcal{L}^1(\emptyset)$ such that ϕ and α are frame correspondents.*

Moreover, α is effectively computable from ϕ.

Proof. Since ϕ is a Sahlqvist formula, then ϕ is an implication $\psi \to \chi$ in which ψ is negation-free and χ is positive. Without loss of generality, we may assume that ψ is equal either to $\top$ or to a disjunction $\psi_1 \vee \ldots \vee \psi_n$ of $\top$-free $\vee$-free negation-free formulas. Consider a frame $\mathcal{F} = (W, R)$. Let $p_1, \ldots, p_N$ be an enumeration of the Boolean variables occuring in χ and $P_1, \ldots, P_N$ be the corresponding unary predicates. We need to consider the following 2 cases.
Case "ψ is equal to $\top$". The following properties are equivalent:

(i) $\mathcal{F} \models \phi$,

(ii) for all valuations V based on $\mathcal{F}$, $(\mathcal{F}, V) \models \phi$,

(iii) for all valuations V based on $\mathcal{F}$, $(\mathcal{F}, V) \models \chi$.

Let V_{min} be the empty valuation. Since χ is positive, then by the second item of Lemma 3.2, **(iii)** is equivalent to the following property:

(iv) $(\mathcal{F}, V_{min}) \models \chi$.

By Proposition 4.1, **(iv)** is equivalent to the following property:

(v) $(\mathcal{F}, V_{min}) \models ST(\chi)$.

Since V_{min} is definable in $\mathcal{L}^1(\emptyset)$ by $P_m(\cdot) ::= \bot$ for each $m \in \{1, \ldots, N\}$, then **(v)** is equivalent to the following property:

(vi) $\mathcal{F} \models \theta(ST(\chi))$, θ being a substitution that replaces $P_m(\cdot)$ by $\bot$ for each $m \in \{1, \ldots, N\}$.

As a result, one may take α to be $\theta(ST(\chi))$.
Case "ψ is equal to a disjunction $\psi_1 \vee \ldots \vee \psi_n$ of $\top$-free $\vee$-free negation-free formulas". Let $i \in \{1, \ldots, n\}$. Without loss of generality, we may assume that ψ_i is equal to a conjunction of the form $a_{i,1} \not\equiv 0 \wedge \ldots \wedge a_{i,k_i} \not\equiv 0 \wedge C(b_{i,1}, c_{i,1}) \wedge \ldots \wedge C(b_{i,l_i}, c_{i,l_i})$ where all $a_{i,\star}$ are equal to an intersection of Boolean variables and all $b_{i,\star}, c_{i,\star}$ are equal either to 1 or to an intersection of Boolean variables. The following properties are equivalent:

(i) $\mathcal{F} \models \psi_i \to \chi$,

(ii) for all valuations V based on $\mathcal{F}$, $(\mathcal{F}, V) \models \psi_i \to \chi$,

(iii) for all valuations V based on $\mathcal{F}$, if $(\mathcal{F}, V) \models \psi_i$ then $(\mathcal{F}, V) \models \chi$.

For all $y_1, z_1, \ldots, y_{l_i}, z_{l_i} \in W$, if $y_1 R z_1$, ..., $y_{l_i} R z_{l_i}$ then for all $x_1, \ldots, x_{k_i} \in W$, let $V_{i,min}$ be a valuation such that for all $m \in \{1, \ldots, N\}$, $V_{i,min}(p_m) = \{x_j\colon 1 \leq j \leq k_i \text{ and } p_m \text{ occurs in } a_{i,j}\} \cup \{y_j\colon 1 \leq j \leq l_i \text{ and } p_m \text{ occurs in } b_{i,j}\} \cup \{z_j\colon 1 \leq j \leq l_i \text{ and } p_m \text{ occurs in } c_{i,j}\}$. By Proposition 4.1, the following properties are equivalent:

(iv) $(\mathcal{F}, V) \models \psi_i$,

(v) $(\mathcal{F}, V) \models ST(\psi_i)$,

(vi) $(\mathcal{F}, V) \models \exists u_1\, ST(u_1, a_{i,1}) \wedge \ldots \wedge \exists u_{k_i}\, ST(u_{k_i}, a_{i,k_i}) \wedge \exists v_1\, \exists w_1\, (R_C(v_1, w_1) \wedge ST(v_1, b_{i,1}) \wedge ST(w_1, c_{i,1})) \wedge \ldots \wedge \exists v_{l_i}\, \exists w_{l_i}\, (R_C(v_{l_i}, w_{l_i}) \wedge ST(v_{l_i}, b_{i,l_i}) \wedge$

$ST(w_{l_i}, c_{i,l_i}))$,

(vii) there exists $y_1, z_1, \ldots, y_{l_i}, z_{l_i} \in W$ such that $y_1 R z_1, \ldots, y_{l_i} R z_{l_i}$ and there exists $x_1, \ldots, x_{k_i} \in W$ such that $(\mathcal{F}, V) \models ST(u_1, a_{i,1}) \wedge \ldots \wedge ST(u_{k_i}, a_{i,k_i}) \wedge (ST(v_1, b_{i,1}) \wedge ST(w_1, c_{i,1})) \wedge \ldots \wedge (ST(v_{l_i}, b_{i,l_i}) \wedge ST(w_{l_i}, c_{i,l_i}))[u_1 := x_1, \ldots, u_{k_i} := x_{k_i}, v_1 := y_1, w_1 := z_1, \ldots, v_{l_i} := y_{l_i}, w_{l_i} := z_{l_i}]$,

(viii) there exists $y_1, z_1, \ldots, y_{l_i}, z_{l_i} \in W$ such that $y_1 R z_1, \ldots, y_{l_i} R z_{l_i}$ and there exists $x_1, \ldots, x_{k_i} \in W$ such that $V_{i,min} \leq V$.

Since χ is positive, then by the second item of Lemma 3.2, **(iii)** is equivalent to the following property:

(ix) for all $y_1, z_1, \ldots, y_{l_i}, z_{l_i} \in W$, if $y_1 R z_1, \ldots, y_{l_i} R z_{l_i}$ then for all $x_1, \ldots, x_{k_i} \in W$, $(\mathcal{F}, V_{i,min}) \models \chi$.

By Proposition 4.1, **(ix)** is equivalent to the following property:

(x) for all $y_1, z_1, \ldots, y_{l_i}, z_{l_i} \in W$, if $y_1 R z_1, \ldots, y_{l_i} R z_{l_i}$ then for all $x_1, \ldots, x_{k_i} \in W$, $(\mathcal{F}, V_{i,min}) \models ST(\chi)$.

Since $V_{i,min}$ is definable in $\mathcal{L}^1(\emptyset)$ by $P_m(\cdot) ::= \bigvee\{\cdot = u_j\colon 1 \leq j \leq k_i$ and p_m occurs in $a_{i,j}\} \vee \bigvee\{\cdot = v_j\colon 1 \leq j \leq l_i$ and p_m occurs in $b_{i,j}\} \vee \bigvee\{\cdot = w_j\colon 1 \leq j \leq l_i$ and p_m occurs in $c_{i,j}\}$ for each $m \in \{1, \ldots, N\}$, then **(x)** is equivalent to the following properties:

(xi) for all $y_1, z_1, \ldots, y_{l_i}, z_{l_i} \in W$, if $y_1 R z_1, \ldots, y_{l_i} R z_{l_i}$ then for all $x_1, \ldots, x_{k_i} \in W$, $\mathcal{F} \models \theta_i(ST(\chi))$, θ_i being a substitution that replaces $P_m(\cdot)$ by $\bigvee\{\cdot = u_j\colon 1 \leq j \leq k_i$ and p_m occurs in $a_{i,j}\} \vee \bigvee\{\cdot = v_j\colon 1 \leq j \leq l_i$ and p_m occurs in $b_{i,j}\} \vee \bigvee\{\cdot = w_j\colon 1 \leq j \leq l_i$ and p_m occurs in $c_{i,j}\}$ for each $m \in \{1, \ldots, N\}$,

(xii) $\mathcal{F} \models \forall u_1 \ldots \forall u_{k_i} \forall v_1 \forall w_1 \ldots \forall v_{l_i} \forall w_{l_i} (R_C(v_1, w_1) \wedge \ldots \wedge R_C(v_{l_i}, w_{l_i}) \to \theta_i(ST(\chi)))$.

As a result, one may take α to be the conjunction of all $\forall u_1 \ldots \forall u_{k_i} \forall v_1 \forall w_1 \ldots \forall v_{l_i} \forall w_{l_i} (R_C(v_1, w_1) \wedge \ldots \wedge R_C(v_{l_i}, w_{l_i}) \to \theta_i(ST(\chi)))$ for each $1 \leq i \leq n$. □

By way of examples, we determine the first-order sentences corresponding to the 4 Sahlqvist formulas considered at the end of Section 2.

(i) Concerning the formula $\top \to C(1, 1)$, its frame correspondent is the first-order sentence $\exists u' \exists v' (R_C(u', v') \wedge ST(u', 1) \wedge ST(v', 1))$. It is equivalent to $\exists u' \exists v' R_C(u', v')$.

(ii) As for the formula $p \not\equiv 0 \to C(p, 1)$, its frame correspondent is the first-order sentence $\forall u\, \theta(\exists u' \exists v' (R_C(u', v') \wedge ST(u', p) \wedge ST(v', 1)))$ where $\theta(P(\cdot))$ is $\cdot = u$. It is equivalent to $\forall u \exists v' R_C(u, v')$.

(iii) Concerning the formula $p \not\equiv 0 \to C(p, p)$, its frame correspondent is the first-order sentence $\forall u\, \theta(\exists u' \exists v' (R_C(u', v') \wedge ST(u', p) \wedge ST(v', p)))$ where $\theta(P(\cdot))$ is $\cdot = u$. It is equivalent to $\forall u R_C(u, u)$.

(iv) As for the formula $C(p, q) \to C(q, p)$, its frame correspondent is the first-order sentence $\forall v \forall w (R_C(v, w) \to \theta(\exists u' \exists v' (R_C(u', v') \wedge ST(u', q) \wedge$

$ST(v', p))))$ where $\theta(P(\cdot))$ is $\cdot = v$ and $\theta(Q(\cdot))$ is $\cdot = w$. It is equivalent to $\forall v \, \forall w \, (R_C(v, w) \rightarrow R_C(w, v))$.

6 Logics

We shall say that a set L of formulas is a logic iff

- L is closed under the rule of modus ponens,
- L is closed under the rule of uniform substitution,
- L contains all instances of tautologies of the classical propositional logic,
- L contains all instances of axioms for non-degenerate Boolean algebras in terms of $\equiv$,
- L contains all instances of the following 3 formulas:
  · $C(a, b) \rightarrow a \not\equiv 0 \wedge b \not\equiv 0$,
  · $C(a \cup b, c) \leftrightarrow C(a, c) \vee C(b, c)$,
  · $C(a, b \cup c) \leftrightarrow C(a, b) \vee C(a, c)$.

We will use L, M, etc, for logics. Obviously, the set of all logics is a partially ordered set with respect to set inclusion. Since the intersection of any collection of logics is again a logic, then there exists a least logic, denoted L_{min}. Note that the greatest logic is the set of all formulas. Of course, a logic L is the set of all formulas iff $\bot \in L$. A logic L will be defined to be consistent iff $\bot \notin L$. We now come to an important convention of notation:

until the end of this paper, L will denote a consistent logic.

For all formulas ϕ, let $L + \phi$ be the least logic containing L and ϕ.

7 Theories

We shall say that a set Γ of formulas is an L-theory iff

- Γ is closed under the rule of modus ponens,
- Γ contains L.

We will use Γ, Δ, etc, for L-theories. Let us be clear that the set of all L-theories is a partially ordered set with respect to set inclusion. The least L-theory is L and the greatest L-theory is the set of all formulas. Of course, an L-theory Γ is the set of all formulas iff $\bot \in \Gamma$. An L-theory Γ will be defined to be consistent iff $\bot \notin \Gamma$. Since each intersection of L-theories is an L-theory, then there exists a least L-theory, denoted $\Gamma \oplus \phi$, containing a given L-theory Γ and a given formula ϕ: namely, $\Gamma \oplus \phi = \{\psi\colon \phi \rightarrow \psi \in \Gamma\}$. Obviously, if $\neg\phi \notin \Gamma$ then $\Gamma \oplus \phi$ is consistent. We shall say that an L-theory Γ is maximal iff for all formulas ϕ, $\phi \in \Gamma$ or $\neg\phi \in \Gamma$. In Lemma 7.2 below, the expression "maximal consistent set of terms" refers to the notions of maximality and consistency in Boolean logic which can be found in most elementary logic texts.

Lemma 7.1 *Let Γ be a consistent L-theory. There exists a maximal consistent L-theory Δ such that $\Gamma \subseteq \Delta$.*

Proof. This is the Lindenbaum's lemma, a standard result. □

Lemma 7.2 *Let Γ be a maximal consistent L-theory.*

(i) *For all terms a, if $a \not\equiv 0 \in \Gamma$ then there exists a maximal consistent set x of terms such that $a \in x$ and for all terms a', if $a' \in x$ then $a' \not\equiv 0 \in \Gamma$.*

(ii) *For all terms a, b, if $C(a, b) \in \Gamma$ then there exists maximal consistent sets x, y of terms such that $a \in x$, $b \in y$ and for all terms a', b', if $a' \in x$ and $b' \in y$ then $C(a', b') \in \Gamma$.*

Proof. See [4]. □

8 Canonical model

Let Γ be a maximal consistent L-theory. The canonical model for Γ is the ordered triple $\mathcal{M}_\Gamma = (W_\Gamma, R_\Gamma, V_\Gamma)$ where:

- W_Γ is the set of all maximal consistent sets x of terms such that for all terms a, if $a \in x$ then $a \not\equiv 0 \in \Gamma$,
- R_Γ is the binary relation on W_Γ such that $x R_\Gamma y$ iff for all terms a, b, if $a \in x$ and $b \in y$ then $C(a, b) \in \Gamma$,
- V_Γ is the function assigning to each Boolean variable p the subset $V_\Gamma(p)$ of W_Γ such that $x \in V_\Gamma(p)$ iff $p \in x$.

Lemma 8.1 below plays for our Kripke-type semantics the role usually played by the truth lemma in the semantics for the basic modal language.

Lemma 8.1 (i) *For all terms a, $x \in (a)^{V_\Gamma}$ iff $a \in x$.*

(ii) *For all formulas ϕ, $(W_\Gamma, R_\Gamma, V_\Gamma) \models \phi$ iff $\phi \in \Gamma$.*

Proof. The first item follows by induction on a and the second one follows by induction on ϕ, using Lemma 7.2 and the first item. □

9 Finite valuations and admissible valuations

Let Γ be a maximal consistent L-theory. The pair $\mathcal{F}_\Gamma = (W_\Gamma, R_\Gamma)$ is called the canonical frame for Γ. V_Γ is called the canonical valuation for Γ. We shall say that a valuation V based on $\mathcal{F}_\Gamma$ is finite iff for all Boolean variables p, $V(p)$ is a finite subset of W_Γ. A valuation V based on $\mathcal{F}_\Gamma$ is said to be admissible iff for all Boolean variables p, there exists a term a such that $V(p) = (a)^{V_\Gamma}$. For all valuations V based on $\mathcal{F}_\Gamma$, let $adm(V)$ be the set of all admissible valuations V' based on $\mathcal{F}_\Gamma$ such that $V \leq V'$.

Lemma 9.1 *Let V be an admissible valuation based on $\mathcal{F}_\Gamma$. For all $\phi \in L$, $(\mathcal{F}_\Gamma, V) \models \phi$.*

Proof. Let $\phi \in L$. Let $p_1, \ldots, p_n$ be an enumeration of the Boolean variables occurring in ϕ. Since V is admissible, then there exists terms $a_1, \ldots, a_n$ such that $V(p_1) = (a_1)^{V_\Gamma}, \ldots, V(p_n) = (a_n)^{V_\Gamma}$. Obviously, for all terms $b(p_1, \ldots, p_n)$ and for all formulas $\psi(p_1, \ldots, p_n)$:

- $(b(p_1, \ldots, p_n))^V = (b(a_1, \ldots, a_n))^{V_\Gamma}$,
- $(\mathcal{F}_\Gamma, V) \models \psi(p_1, \ldots, p_n)$ iff $(\mathcal{F}_\Gamma, V_\Gamma) \models \psi(a_1, \ldots, a_n)$.

The first item follows by induction on b and the second one follows by induction on ψ, using the first item. Since $\phi(p_1,\ldots,p_n) \in L$, then $\phi(a_1,\ldots,a_n) \in L$. Hence, $\phi(a_1,\ldots,a_n) \in \Gamma$. By the second item of Lemma 8.1, $(\mathcal{F}_\Gamma, V_\Gamma) \models \phi(a_1,\ldots,a_n)$. By the second item above, $(\mathcal{F}_\Gamma, V) \models \phi(p_1,\ldots,p_n)$. □

Lemma 9.2 *Let $A \subseteq W_\Gamma$. If A is finite then $A \supseteq \bigcap\{(a)^{V_\Gamma}$: a is a term such that $A \subseteq (a)^{V_\Gamma}\}$.*

Proof. Suppose A is finite. Hence, there exists a nonnegative integer n such that $Card(A) = n$. We need to consider the following 3 cases.
Case "$n = 0$". Hence, A is empty. Thus, 0 is a term such that $A \subseteq (0)^{V_\Gamma}$. Since $(0)^{V_\Gamma} = \emptyset$, then $\bigcap\{(a)^{V_\Gamma}$: a is a term such that $A \subseteq (a)^{V_\Gamma}\} = \emptyset$. Therefore, $A \supseteq \bigcap\{(a)^{V_\Gamma}$: a is a term such that $A \subseteq (a)^{V_\Gamma}\}$.
Case "$n = 1$". Hence, there exists $x \in W_\Gamma$ such that $A = \{x\}$. By the first item of Lemma 8.1, for all terms a, $x \in (a)^{V_\Gamma}$ iff $a \in x$. Thus, the following sets are equal:

- $\bigcap\{(a)^{V_\Gamma}$: a is a term such that $A \subseteq (a)^{V_\Gamma}\}$,
- $\bigcap\{(a)^{V_\Gamma}$: a is a term such that $x \in (a)^{V_\Gamma}\}$,
- $\bigcap\{(a)^{V_\Gamma}$: a is a term such that $a \in x\}$.

Obviously, x is the only element in $\bigcap\{(a)^{V_\Gamma}$: a is a term such that $a \in x\}$. Therefore, $A \supseteq \bigcap\{(a)^{V_\Gamma}$: a is a term such that $A \subseteq (a)^{V_\Gamma}\}$.
Case "$n \geq 2$". Hence, there exists $x_1,\ldots,x_n \in W_\Gamma$ such that $A = \{x_1,\ldots,x_n\}$. For all $i = 1\ldots n$, by the second case, $\{x_i\} \supseteq \bigcap\{(a)^{V_\Gamma}$: a is a term such that $\{x_i\} \subseteq (a)^{V_\Gamma}\}$. If $A \not\supseteq \bigcap\{(a)^{V_\Gamma}$: a is a term such that $A \subseteq (a)^{V_\Gamma}\}$ then there exists $x \in W_\Gamma$ such that $x \notin A$ and $x \in \bigcap\{(a)^{V_\Gamma}$: a is a term such that $A \subseteq (a)^{V_\Gamma}\}$. Since $x \notin A$, then for all $i = 1\ldots n$, $x \neq x_i$ and there exists a term a_i such that $x_i \in (a_i)^{V_\Gamma}$ and $x \notin (a_i)^{V_\Gamma}$. Thus, $x \notin (a_1 \cup \ldots \cup a_n)^{V_\Gamma}$. Since for all $i = 1\ldots n$, $x_i \in (a_i)^{V_\Gamma}$, then $A \subseteq (a_1 \cup \ldots \cup a_n)^{V_\Gamma}$. Since $x \in \bigcap\{(a)^{V_\Gamma}$: a is a term such that $A \subseteq (a)^{V_\Gamma}\}$, then $x \in (a_1 \cup \ldots \cup a_n)^{V_\Gamma}$: a contradiction. □

Lemma 9.3 *Let V be a valuation based on $\mathcal{F}_\Gamma$. If V is finite then $V \geq \bigcap\{V'$: $V' \in adm(V)\}$.*

Proof. Suppose V is finite. If $V \not\geq \bigcap\{V'$: $V' \in adm(V)\}$ then there exists a Boolean variable p such that $V(p) \not\supseteq \bigcap\{V'(p)$: $V' \in adm(V)\}$. Hence, there exists $x \in W_\Gamma$ such that $x \notin V(p)$ and $x \in \bigcap\{V'(p)$: $V' \in adm(V)\}$. Since V is finite, then by Lemma 9.2, $V(p) \supseteq \bigcap\{(a)^{V_\Gamma}$: a is a term such that $V(p) \subseteq (a)^{V_\Gamma}\}$. Since $x \notin V(p)$, then there exists a term a such that $V(p) \subseteq (a)^{V_\Gamma}$ and $x \notin (a)^{V_\Gamma}$. Let V' be the valuation based on $\mathcal{F}_\Gamma$ such that $V'(p) = (a)^{V_\Gamma}$ and $V'(q) = W_\Gamma$ for every Boolean variable q distinct from p. Obviously, $V' \in adm(V)$. Since $x \in \bigcap\{V'(p)$: $V' \in adm(V)\}$, then $x \in V'(p)$. Since $V'(p) = (a)^{V_\Gamma}$, then $x \in (a)^{V_\Gamma}$: a contradiction. □

Lemma 9.4 *Let V be a valuation based on $\mathcal{F}_\Gamma$. If V is finite then for all positive terms a, $(a)^V \supseteq \bigcap\{(a)^{V'}$: $V' \in adm(V)\}$.*

Proof. Suppose V is finite and let a be a positive term. The proof is done by induction on a.

Case "$a = p$". Since V is finite, then by Lemma 9.3, $V(p) \supseteq \bigcap\{V'(p)\colon V' \in adm(V)\}$. Hence, $(a)^V \supseteq \bigcap\{(a)^{V'}\colon V' \in adm(V)\}$.
Case "$a = 1$". Left to the reader.
Case "$a = b \cap c$" where b and c are positive terms. Left to the reader.
Case "$a = b \cup c$" where b and c are positive terms. By induction hypothesis, $(b)^V \supseteq \bigcap\{(b)^{V'}\colon V' \in adm(V)\}$ and $(c)^V \supseteq \bigcap\{(c)^{V'}\colon V' \in adm(V)\}$. Hence, it suffices to demonstrate that $\bigcap\{(b)^{V'}\colon V' \in adm(V)\} \cup \bigcap\{(c)^{V'}\colon V' \in adm(V)\} \supseteq \bigcap\{(b)^{V'} \cup (c)^{V'}\colon V' \in adm(V)\}$. Let $x \in \bigcap\{(b)^{V'} \cup (c)^{V'}\colon V' \in adm(V)\}$. If $x \notin \bigcap\{(b)^{V'}\colon V' \in adm(V)\} \cup \bigcap\{(c)^{V'}\colon V' \in adm(V)\}$ then $x \notin \bigcap\{(b)^{V'}\colon V' \in adm(V)\}$ and $x \notin \bigcap\{(c)^{V'}\colon V' \in adm(V)\}$. Thus, there exists $V'_b \in adm(V)$ such that $x \notin (b)^{V'_b}$ and there exists $V'_c \in adm(V)$ such that $x \notin (c)^{V'_c}$. Let $V' = V'_b \cap V'_c$. Obviously, $V' \in adm(V)$, $V' \leq V'_b$ and $V' \leq V'_c$. Since b and c are positive terms, then by Lemma 3.1, $(b)^{V'} \subseteq (b)^{V'_b}$ and $(c)^{V'} \subseteq (c)^{V'_c}$. Since $V' \in adm(V)$ and $x \in \bigcap\{(b)^{V'} \cup (c)^{V'}\colon V' \in adm(V)\}$, then $x \in (b)^{V'}$ or $x \in (c)^{V'}$. Since $(b)^{V'} \subseteq (b)^{V'_b}$ and $(c)^{V'} \subseteq (c)^{V'_c}$, then $x \in (b)^{V'_b}$ or $x \in (c)^{V'_c}$: a contradiction. □

Lemma 9.5 *Let V be a valuation based on $\mathcal{F}_\Gamma$. If V is finite then for all positive terms a, b, $((a)^V \times W_\Gamma) \cup (W_\Gamma \times (b)^V) \supseteq \bigcap\{((a)^{V'} \times W_\Gamma) \cup (W_\Gamma \times (b)^{V'})\colon V' \in adm(V)\}$.*

Proof. Suppose V is finite and let a and b be positive terms. If $((a)^V \times W_\Gamma) \cup (W_\Gamma \times (b)^V) \not\supseteq \bigcap\{((a)^{V'} \times W_\Gamma) \cup (W_\Gamma \times (b)^{V'})\colon V' \in adm(V)\}$ then there exists $x, y \in W_\Gamma$ such that $x \notin (a)^V$, $y \notin (b)^V$ and $x \in (a)^{V'}$ or $y \in (b)^{V'}$ for each $V' \in adm(V)$. Let $\cong$ be the binary relation on $adm(V)$ defined as follows: $V' \cong V''$ iff for all Boolean variables p occurring in a, $V'(p) = V''(p)$. Obviously, $\cong$ is an equivalence relation on $adm(V)$. Moreover, $adm(V)_{|\cong}$, the quotient set of $adm(V)$ modulo $\cong$, is countable. Hence, there exists an ω-sequence $(\mid V'_n \mid)_{n \in \mathbb{N}}$ of equivalence classes modulo $\cong$ enumerating $adm(V)_{|\cong}$. Let $(\mid V''_n \mid)_{n \in \mathbb{N}}$ be the ω-sequence of equivalence classes modulo $\cong$ defined as follows: if $n = 0$ then $\mid V''_n \mid = \mid V'_0 \mid$ else $\mid V''_n \mid = \mid V''_{n-1} \cap V'_n \mid$. Since $x \in (a)^{V'}$ or $y \in (b)^{V'}$ for each $V' \in adm(V)$, then $x \in (a)^{V''_n}$ or $y \in (b)^{V''_n}$ for each $n \in \mathbb{N}$. Since V is finite and a and b are positive terms, then by construction of $(\mid V''_n \mid)_{n \in \mathbb{N}}$ and by Lemma 9.4, $(a)^V \supseteq \bigcap\{(a)^{V''_n}\colon n \in \mathbb{N}\}$ and $(b)^V \supseteq \bigcap\{(b)^{V''_n}\colon n \in \mathbb{N}\}$. Since $x \in (a)^{V''_n}$ or $y \in (b)^{V''_n}$ for each $n \in \mathbb{N}$, then by construction of $(\mid V''_n \mid)_{n \in \mathbb{N}}$ and by Lemma 3.1, $x \in (a)^{V''_n}$ for each $n \in \mathbb{N}$ or $y \in (b)^{V''_n}$ for each $n \in \mathbb{N}$. Since $(a)^V \supseteq \bigcap\{(a)^{V''_n}\colon n \in \mathbb{N}\}$ and $(b)^V \supseteq \bigcap\{(b)^{V''_n}\colon n \in \mathbb{N}\}$, then $x \in (a)^V$ or $y \in (b)^V$: a contradiction. □

Lemma 9.6 *Let V be a finite valuation based on $\mathcal{F}_\Gamma$. Let a be a positive term such that for all $V' \in adm(V)$, $(a)^{V'} \neq \emptyset$. Then $(a)^V \neq \emptyset$.*

Proof. Let $\cong$, $(\mid V'_n \mid)_{n \in \mathbb{N}}$ and $(\mid V''_n \mid)_{n \in \mathbb{N}}$ be defined as in the proof of Lemma 9.5. Since $(a)^{V'} \neq \emptyset$ for each $V' \in adm(V)$, then $(a)^{V''_n} \neq \emptyset$ for each $n \in \mathbb{N}$. Since V is finite and a is a positive term, then by construction of $(\mid V''_n \mid)_{n \in \mathbb{N}}$ and by Lemma 9.4, $(a)^V \supseteq \bigcap\{(a)^{V''_n}\colon n \in \mathbb{N}\}$. Since V''_n is admissible for each $n \in \mathbb{N}$, then there exists a term a_n such that $(a)^{V''_n} = (a_n)^{V_\Gamma}$

for each $n \in \mathbb{N}$. Remark that $(a_0)^{V_\Gamma} \supseteq (a_1)^{V_\Gamma} \supseteq \ldots$. Since $(a)^{V''_n} \neq \emptyset$ for each $n \in \mathbb{N}$, then $(a_n)^{V_\Gamma} \neq \emptyset$ for each $n \in \mathbb{N}$. Since $(a_0)^{V_\Gamma} \supseteq (a_1)^{V_\Gamma} \supseteq \ldots$, then there exists $x \in W_\Gamma$ such that $a_n \in x$ for each $n \in \mathbb{N}$. Thus, $\bigcap\{(a_n)^{V_\Gamma} : n \in \mathbb{N}\} \neq \emptyset$. Therefore, $\bigcap\{(a)^{V''_n} : n \in \mathbb{N}\} \neq \emptyset$. Since $(a)^V \supseteq \bigcap\{(a)^{V''_n} : n \in \mathbb{N}\}$, then $(a)^V \neq \emptyset$. □

Lemma 9.7 *Let V be a finite valuation based on $\mathcal{F}_\Gamma$. Let a and b be positive terms such that for all $V' \in adm(V)$, there exists $x, y \in W_\Gamma$ such that $xR_\Gamma y$, $x \in (a)^{V'}$ and $y \in (b)^{V'}$. Then there exists $x, y \in W_\Gamma$ such that $xR_\Gamma y$, $x \in (a)^V$ and $y \in (b)^V$.*

Proof. Let $\cong$, $(\mid V'_n \mid)_{n \in \mathbb{N}}$ and $(\mid V''_n \mid)_{n \in \mathbb{N}}$ be defined as in the proof of Lemma 9.5. Since there exists $x, y \in W_\Gamma$ such that $xR_\Gamma y$, $x \in (a)^{V'}$ and $y \in (b)^{V'}$ for each $V' \in adm(V)$, then there exists $x, y \in W_\Gamma$ such that $xR_\Gamma y$, $x \in (a)^{V''_n}$ and $y \in (b)^{V''_n}$ for each $n \in \mathbb{N}$. Since V is finite and a and b are positive terms, then by construction of $(\mid V''_n \mid)_{n \in \mathbb{N}}$ and by Lemma 9.4, $(a)^V \supseteq \bigcap\{(a)^{V''_n} : n \in \mathbb{N}\}$ and $(b)^V \supseteq \bigcap\{(b)^{V''_n} : n \in \mathbb{N}\}$. Since V''_n is admissible for each $n \in \mathbb{N}$, then there exists terms a_n, b_n such that $(a)^{V''_n} = (a_n)^{V_\Gamma}$ and $(b)^{V''_n} = (b_n)^{V_\Gamma}$ for each $n \in \mathbb{N}$. Remark that $(a_0)^{V_\Gamma} \supseteq (a_1)^{V_\Gamma} \supseteq \ldots$ and $(b_0)^{V_\Gamma} \supseteq (b_1)^{V_\Gamma} \supseteq \ldots$. Since there exists $x, y \in W_\Gamma$ such that $xR_\Gamma y$, $x \in (a)^{V''_n}$ and $y \in (b)^{V''_n}$ for each $n \in \mathbb{N}$, then there exists $x, y \in W_\Gamma$ such that $xR_\Gamma y$, $x \in (a_n)^{V_\Gamma}$ and $y \in (b_n)^{V_\Gamma}$ for each $n \in \mathbb{N}$. Since $(a_0)^{V_\Gamma} \supseteq (a_1)^{V_\Gamma} \supseteq \ldots$ and $(b_0)^{V_\Gamma} \supseteq (b_1)^{V_\Gamma} \supseteq \ldots$, then there exists $x, y \in W_\Gamma$ such that $xR_\Gamma y$, $a_n \in x$ and $b_n \in y$ for each $n \in \mathbb{N}$. Thus, $x \in \bigcap\{(a_n)^{V_\Gamma} : n \in \mathbb{N}\}$ and $y \in \bigcap\{(b_n)^{V_\Gamma} : n \in \mathbb{N}\}$. Therefore, $x \in \bigcap\{(a)^{V''_n} : n \in \mathbb{N}\}$ and $y \in \bigcap\{(b)^{V''_n} : n \in \mathbb{N}\}$. Since $(a)^V \supseteq \bigcap\{(a)^{V''_n} : n \in \mathbb{N}\}$ and $(b)^V \supseteq \bigcap\{(b)^{V''_n} : n \in \mathbb{N}\}$, then $x \in (a)^V$ and $y \in (b)^V$. □

Lemma 9.8 *Let V be a valuation based on $\mathcal{F}_\Gamma$. Let ϕ be a negation-free formula such that $(\mathcal{F}_\Gamma, V) \models \phi$. Then there exists a finite valuation V_0 based on $\mathcal{F}_\Gamma$ such that $V_0 \leq V$ and $(\mathcal{F}_\Gamma, V_0) \models \phi$.*

Proof. Without loss of generality, we may assume that ϕ is equal either to $\top$ or to a disjunction $\phi_1 \vee \ldots \vee \phi_n$ of $\top$-free $\vee$-free negation-free formulas. In the former case, let V_0 be the empty valuation. In the latter case, there exists $i \in \{1, \ldots, n\}$ such that $(\mathcal{F}_\Gamma, V) \models \phi_i$. Since we may also assume that ϕ_i is equal to a conjunction of the form $a_1 \not\equiv 0 \wedge \ldots \wedge a_k \not\equiv 0 \wedge C(b_1, c_1) \wedge \ldots \wedge C(b_l, c_l)$ where all $a_\star$ are equal to an intersection of Boolean variables and all $b_\star, c_\star$ are equal either to 1 or to an intersection of Boolean variables, then there exists $x_1, \ldots, x_k, y_1, z_1, \ldots, y_l, z_l \in W_\Gamma$ such that $x_1 \in (a_1)^V$, ..., $x_1 \in (a_k)^V$, $y_1 \in (b_1)^V$, $z_1 \in (c_1)^V$ and $y_1 R_\Gamma z_1$, ..., $y_l \in (b_l)^V$, $z_l \in (c_l)^V$ and $y_l R_\Gamma z_l$. Let V_0 be the finite valuation based on $\mathcal{F}_\Gamma$ such that for all Boolean variables p, $V_0(p) = V(p) \cap (\{x_1, \ldots, x_k\} \cup \{y_1, z_1, \ldots, y_l, z_l\})$. Obviously, $V_0 \leq V$ and $(\mathcal{F}_\Gamma, V_0) \models \phi$. □

Lemma 9.9 *Let V be a finite valuation based on $\mathcal{F}_\Gamma$. Let ϕ be a positive formula such that for all $V' \in adm(V)$, $(\mathcal{F}_\Gamma, V') \models \phi$. Then $(\mathcal{F}_\Gamma, V) \models \phi$.*

Proof. The proof is done by induction on ϕ.

Case "$\phi = a \not\equiv 0$". Since $(\mathcal{F}_\Gamma, V') \models \phi$ for each $V' \in adm(V)$, then $(a)^{V'} \neq \emptyset$ for each $V' \in adm(V)$. Since a is a positive term, then by Lemma 9.6, $(a)^V \neq \emptyset$. Hence, $(\mathcal{F}_\Gamma, V) \models \phi$.
Case "$\phi = -a \equiv 0$". Since $(\mathcal{F}_\Gamma, V') \models \phi$ for each $V' \in adm(V)$, then $(a)^{V'} = W_\Gamma$ for each $V' \in adm(V)$. Since V is finite and a is a positive term, then by Lemma 9.4, $(a)^V = W_\Gamma$. Hence, $(\mathcal{F}_\Gamma, V) \models \phi$.
Case "$\phi = C(a, b)$". Since $(\mathcal{F}_\Gamma, V') \models \phi$ for each $V' \in adm(V)$, then there exists $x, y \in W_\Gamma$ such that $xR_\Gamma y$, $x \in (a)^{V'}$ and $y \in (b)^{V'}$ for each $V' \in adm(V)$. Since a and b are positive terms, then by Lemma 9.7, there exists $x, y \in W_\Gamma$ such that $xR_\Gamma y$, $x \in (a)^V$ and $y \in (b)^V$. Hence, $(\mathcal{F}_\Gamma, V) \models \phi$.
Case "$\phi = \bar{C}(-a, -b)$". Since $(\mathcal{F}_\Gamma, V') \models \phi$ for each $V' \in adm(V)$, then for all $x, y \in W_\Gamma$, if $xR_\Gamma y$ then $x \in (a)^{V'}$ or $y \in (b)^{V'}$ for each $V' \in adm(V)$. Since V is finite and a and b are positive terms, then by Lemma 9.5, for all $x, y \in W_\Gamma$, if $xR_\Gamma y$ then $x \in (a)^V$ or $y \in (b)^V$. Hence, $(\mathcal{F}_\Gamma, V) \models \phi$.
Case "$\phi = \top$". Left to the reader.
Case "$\phi = \psi \vee \chi$" where ψ and χ are positive formulas. Let $\cong$, $(\mid V'_n \mid)_{n \in \mathbb{N}}$ and $(\mid V''_n \mid)_{n \in \mathbb{N}}$ be defined as in the proof of Lemma 9.5. Since $(\mathcal{F}_\Gamma, V') \models \phi$ for each $V' \in adm(V)$, then $(\mathcal{F}_\Gamma, V''_n) \models \phi$ for each $n \in \mathbb{N}$. Since ψ and χ are positive, then by construction of $(\mid V''_n \mid)_{n \in \mathbb{N}}$ and by the second item of Lemma 3.2, $(\mathcal{F}_\Gamma, V''_n) \models \psi$ for each $n \in \mathbb{N}$ or $(\mathcal{F}_\Gamma, V''_n) \models \chi$ for each $n \in \mathbb{N}$. Since ψ and χ are positive, then by construction of $(\mid V''_n \mid)_{n \in \mathbb{N}}$ and by the second item of Lemma 3.2, $(\mathcal{F}_\Gamma, V') \models \psi$ for each $V' \in adm(V)$ or $(\mathcal{F}_\Gamma, V') \models \chi$ for each $V' \in adm(V)$. Hence, by induction hypothesis, $(\mathcal{F}_\Gamma, V) \models \phi$.
Case "$\phi = \psi \wedge \chi$" where ψ and χ are positive formulas. Left to the reader. □

10 Completeness theorem

We shall say that L is canonical iff for all maximal consistent L-theories Γ, $\mathcal{F}_\Gamma \models L$.

Theorem 10.1 *Let ϕ be a Sahlqvist formula. If L is canonical then $L + \phi$ is canonical.*

Proof. Suppose L is canonical. If $L + \phi$ is not canonical then there exists a maximal consistent $L+\phi$-theory Γ such that $\mathcal{F}_\Gamma \not\models L+\phi$. Hence, Γ is a maximal consistent L-theory such that $\mathcal{F}_\Gamma \not\models L$ or $\mathcal{F}_\Gamma \not\models \phi$. Since L is canonical, then $\mathcal{F}_\Gamma \models L$. Since $\mathcal{F}_\Gamma \not\models L$ or $\mathcal{F}_\Gamma \not\models \phi$, then $\mathcal{F}_\Gamma \not\models \phi$. Thus, there exists a valuation V based on $\mathcal{F}_\Gamma$ such that $(\mathcal{F}_\Gamma, V) \not\models \phi$. Since ϕ is a Sahlqvist formula, then ϕ is an implication $\psi \to \chi$ in which ψ is negation-free and χ is positive. Since $(\mathcal{F}_\Gamma, V) \not\models \phi$, then $(\mathcal{F}_\Gamma, V) \models \psi$ and $(\mathcal{F}_\Gamma, V) \not\models \chi$. Since ψ is negation-free, then by Lemma 9.8, there exists a finite valuation V_0 based on $\mathcal{F}_\Gamma$ such that $V_0 \leq V$ and $(\mathcal{F}_\Gamma, V_0) \models \psi$. Since χ is positive and $(\mathcal{F}_\Gamma, V) \not\models \chi$, then by the second item of Lemma 3.2, $(\mathcal{F}_\Gamma, V_0) \not\models \chi$. Since V_0 is finite and χ is positive, then by Lemma 9.9, $(\mathcal{F}_\Gamma, V') \not\models \chi$ for some $V' \in adm(V_0)$. Since ψ is negation-free and $(\mathcal{F}_\Gamma, V_0) \models \psi$, then by Lemma 3.2, $(\mathcal{F}_\Gamma, V') \models \psi$. Since V' is admissible and $\phi \in L + \phi$, then by Lemma 9.1, $(\mathcal{F}_\Gamma, V') \models \phi$. Since $(\mathcal{F}_\Gamma, V') \models \psi$, then

$(\mathcal{F}_\Gamma, V') \models \chi$: a contradiction. □

As a result,

Theorem 10.2 *Let ϕ be a Sahlqvist formula and α be the first-order sentence of the first-order language $\mathcal{L}^1(\emptyset)$ that corresponds to it by Theorem 5.1. For all formulas ψ, $\psi \in L_{min} + \phi$ iff for all frames $\mathcal{F} = (W, R)$, if $\mathcal{F} \models \alpha$ then $\mathcal{F} \models \psi$.*

Proof. Firstly, let us prove the direction from left to right. Let ψ be a formula. If $\psi \in L_{min} + \phi$ then let $\mathcal{F} = (W, R)$ be a frame such that $\mathcal{F} \models \alpha$. Hence, by Theorem 5.1, $\mathcal{F} \models \phi$. Since $\psi \in L_{min} + \phi$, then there exists a proof of ψ in the axiomatic system based on the rules and the instances of formulas considered at the beginning of Section 6 and ϕ. By induction on the length of this proof, one can show that $\mathcal{F} \models \psi$.
Secondly, let us prove the direction from right to left. Let ψ be a formula. If $\psi \notin L_{min} + \phi$ then $(L_{min} + \phi) \oplus \neg\psi$ is a consistent $(L_{min} + \phi)$-theory. Hence, by Lemma 7.1, there exists a maximal consistent $(L_{min} + \phi)$-theory Γ such that $(L_{min} + \phi) \oplus \neg\psi \subseteq \Gamma$. Now, it suffices to demonstrate that $\mathcal{F}_\Gamma \models \alpha$ and $\mathcal{F}_\Gamma \not\models \psi$. Since ϕ is a Sahlqvist formula, then by Theorem 10.1, $L_{min} + \phi$ is canonical. Thus, $\mathcal{F}_\Gamma \models \phi$. Therefore, by Theorem 5.1, $\mathcal{F}_\Gamma \models \alpha$. Since $\neg\psi \in \Gamma$, then $\psi \notin \Gamma$. Consequently, by Lemma 8.1, $\mathcal{M}_\Gamma \not\models \psi$. Hence, $\mathcal{F}_\Gamma \not\models \psi$. □

11 Conclusion

As we already said, the last 4 formulas considered at the end of Section 2 are not Sahlqvist formulas. However, there could be the possibility of finding 4 Sahlqvist formulas corresponding to them. At this point, it might be useful to remark that the first of these 4 formulas, namely $C(p, q) \to C(p, r) \vee C(-r, q)$, corresponds to a first-order property whereas the last 3 of them, namely $p \not\equiv 0 \wedge -p \not\equiv 0 \to C(p, -p)$, $(p \cup q) \equiv 1 \wedge (p \cap q) \equiv 0 \to C(p, p) \vee C(q, q)$ and $(p \cap -q) \not\equiv 0 \to C(p, -q) \vee C(q, -q)$, corresponds to second-order properties. For more on this, see [3,4]. Hence, a first question presents itself: the decidability of determining whether a given RBPMLS formula is equivalent to a Sahlqvist RBPMLS formula.
Important problems are the so-called algorithmic problems in correspondence theory: given a RBPMLS formula, determine whether it has a first-order correspondent; given a first-order sentence, determine whether it has a RBPMLS correspondent. In modal logic, such problems have been proved to be undecidable by Chagrova in 1989. For more on this, see [8,9]. Chagrova's proof of the undecidability of modal definability of first-order sentences can be almost reproduced word for word in the setting of RBPMLS (Tinko Tinchev, personal communication, Sofia (Bulgaria), February 24, 2012). Hence, a second question presents itself: the decidability of determining whether a given RBPMLS formula corresponds to a first-order sentence.
The undecidability of RBPMLS definability of first-order sentences shows that any sufficient condition for RBPMLS definability is very interesting by itself. In modal logic, Kracht formulas are the first-order counterparts of Sahlqvist formulas [26]. By means of an algorithm constructing a Sahlqvist formula from

a given Kracht formula, one can axiomatize validity in such and such elementary class of frames determined by Kracht formulas. Recently, Kracht theorem has been extended to the class of generalized Sahlqvist formulas introduced by Goranko and Vakarelov [20]. For more on this, see [23,24]. Hence, a third question presents itself: the definition of Kracht formulas in the setting of RBPMLS.

Acknowledgements

The research of the first author has been partly supported by the project DID02/32/2009 of Bulgarian Science Fund. Both authors want to express their appreciation to Tinko Tinchev for his kind help in the course of this research.

References

[1] Balbiani, P., Tinchev, T.: *Definability over the class of all partitions.* Journal of Logic and Computation **16** (2006) 541–557.

[2] Balbiani, P., Tinchev, T.: *Boolean logics with relations.* Journal of Logic and Algebraic Programming **79** (2010) 707–721.

[3] Balbiani, P., Tinchev, T., Vakarelov, D.: *Dynamic logics of the region-based theory of discrete spaces.* Journal of Applied Non-Classical Logics **17** (2007) 39–61.

[4] Balbiani, P., Tinchev, T., Vakarelov, D.: *Modal logics for region-based theories of space.* Fundamenta Informaticæ**81** (2007) 29–82.

[5] Bezhanishvili, N., Hodkinson, I.: *Sahlqvist theorem for modal fixed point logic.* Theoretical Computer Science **424** (2012) 1–19.

[6] Blackburn, P., de Rijke, M., Venema, Y.: *Modal Logic.* Cambridge University Press (2001).

[7] Ten Cate, B., Marx, M., Viana, P.: *Hybrid logics with Sahlqvist axioms.* Logic Journal of the IGPL **13** (2005) 293–300.

[8] Chagrov, A., Chagrova, L.: *The truth about algorithmic problems in correspondence theory.* In: *Advances in Modal Logic. Volume 6.* College Publications (2006) 121–138.

[9] Chagrova, L.: *An undecidable problem in correspondence theory.* Journal of Symbolic Logic **56** (1991) 1261–1272.

[10] Conradie, W., Goranko, V., Vakarelov, D.: *Algorithmic correspondence and completeness in modal logic. I. The core algorithm SQEMA.* Logical Methods in Computer Science **2** (2006) 1–26.

[11] Dimov, G., Vakarelov, D.: *Contact algebras and region-based theory of space: a proximity approach – I.* Fundamenta Informaticæ**74** (2006) 209–249.

[12] Dimov, G., Vakarelov, D.: *Contact algebras and region-based theory of space: proximity approach – II.* Fundamenta Informaticæ**74** (2006) 251–282.

[13] Düntsch, I., Vakarelov, D.: *Region-based theory of discrete spaces: a proximity approach.* Annals of Mathematics and Artificial Intelligence **49** (2007) 5–14.

[14] Düntsch, I., Winter, M.: *A representation theorem for Boolean contact algebras.* Theoretical Computer Science **347** (2005) 498–512.

[15] Gabelaia, D., Kontchakov, R., Kurucz, A., Wolter, F., Zakharyaschev, M.: *Combining spatial and temporal logics: expressiveness vs. complexity.* Journal of Artificial Intelligence Research **23** (2005) 167–243.

[16] Gehrke, M., Nagahashi, H., Venema, Y.: *A Sahlqvist theorem for distributive modal logic.* Annals of Pure and Applied Logic **131** (2005) 65–102.

[17] Goranko, V., Hustadt, U., Schmidt, R., Vakarelov, D.: *SCAN Is complete for all Sahlqvist formulae.* In: *Relational and Kleene-Algebraic Methods.* Springer (2004) 149–162.

[18] Goranko, V., Otto, M.: *Model theory of modal logic.* In: *Handbook of Modal Logic.* Elsevier (2007) 249–329.

[19] Goranko, V. and D. Vakarelov, *Sahlqvist formulas in hybrid polyadic modal logics*, Journal of Logic and Computation **11** (2001), pp. 737–54.

[20] Goranko, V., Vakarelov, D.: *Sahlqvist formulas unleashed in polyadic modal languages.* In: *Advances in Modal Logic, Volume 3.* World Scientific Publishing (2002) 221–240.

[21] Goranko, V., Vakarelov, D.: *Elementary canonical formulae: extending Sahlqvist's theorem.* Annals of Pure and Applied Logic **141** (2006) 180–217.

[22] Grädel, E., Kolaitis, P., Vardi, M.: *On the decision problem for two-variable first-order logic.* Bulletin of Symbolic Logic **3** (1997) 53–69.

[23] Kikot, S.: *An extension of Kracht's theorem to generalized Sahlqvist formulas.* Journal of Applied Non-Classical Logics **19** (2009) 227–251.

[24] Kikot, S.: *Semantic characterization of Kracht formulas.* In: *Advances in Modal Logic. Volume 8.* College Publications (2010) 218–234.

[25] Kontchakov, R., Pratt-Hartmann, I., Wolter, F., Zakharyaschev, M.: *Spatial logics with connectedness predicates.* Logical Methods in Computer Science **6** (2010) 1–43.

[26] Kracht, M.: *How completeness and correspondence theory got married.* In: *Diamonds and Defaults.* Kluwer (1993) 175–214.

[27] Kracht, M.: *Tools and Techniques in Modal Logic.* Elsevier (1999).

[28] De Laguna, T.: *Point, line and surface, as sets of solids.* The Journal of Philosophy **19** (1922) 449–461.

[29] Mortimer, M.: *On languages with two variables.* Zeitschrift für mathematische Logik und Grundlagen der Mathematik **21** (1975) 135–140.

[30] Sahlqvist, H.: *Completeness and correspondence in the first and second order semantics for modal logic.* In: *Proceedings of the Third Scandinavian Logic Symposium.* North-Holland (1975) 110–143.

[31] Sambin, G., Vaccaro, V.: *A new proof of Sahlqvist theorem on modal definability and completeness.* The Journal of Symbolic Logic **54** (1989) 992–999.

[32] Schmidt, R.: *The Ackermann approach for modal logic, correspondence theory and second-order reduction.* Journal of Applied Logic **10** (2012) 52–74.

[33] Seki, T.: *A Sahlqvist theorem for relevant modal logics.* Studia Logica **73** (2003) 383–411.

[34] Vakarelov, D.: *Modal definability in languages with a finite number of propositional variables and a new extension of the Sahlqvist's class.* In: *Advances in Modal Logic, Volume 4.* King's College Publications (2003) 499–518.

[35] Vakarelov, D.: *Extended Sahlqvist formulas and solving equations in modal algebras.* In: *12th International Congress of Logic, Methodology and Philosophy of Science.* Oviedo (Spain), August 7–13, 2003.

[36] Vakarelov, D.: *Region-based theory of space: algebras of regions, representation theory, and logics.* In: *Mathematical Problems from Applied Logic. Logics for the XXIst Century. II.* Springer (2007) 267–348.

[37] Wolter, F., Zakharyaschev, M.: *Spatial representation and reasoning in RCC-8 with Boolean region terms.* In: *Proceedings of the 14th European Conference on Artificial Intelligence.* IOS Press (2000) 244–248.

[38] Whitehead, A.: *Process and Reality.* MacMillan (1929).

Completeness and Definability of a Modal Logic Interpreted over Iterated Strict Partial Orders

Philippe Balbiani [1]

Institut de recherche en informatique de Toulouse
CNRS — Université de Toulouse

Levan Uridia [2]

Escuela Técnica Superior de Ingeniería Informática
Universidad Rey Juan Carlos

Abstract

Any strict partial order R on a nonempty set X defines a function θ_R which associates to each strict partial order $S \subseteq R$ on X the strict partial order $\theta_R(S) = R \circ S$ on X. Owing to the strong relationships between Alexandroff T_D derivative operators and strict partial orders, this paper firstly calls forth the links between the Cantor-Bendixson ranks of Alexandroff T_D topological spaces and the greatest fixpoints of the θ-like functions defined by strict partial orders. It secondly considers a modal logic with modal operators $\square$ and $\square^\star$ respectively interpreted by strict partial orders and the greatest fixpoints of the θ-like functions they define. It thirdly addresses the question of the complete axiomatization of this modal logic.

Keywords: Topologies; Derivative operators; Strict partial orders; Cantor-Bendixson rank; Modal logic; Completeness and definability.

1 Introduction

The τ-derived set $d_\tau(A)$ of a set $A \subseteq X$ of points is the set of all limit points of A with respect to a given topology τ on a nonempty set X. Introduced by Cantor, the derivative operator d_τ possesses interesting properties. In particular, a set $A \subseteq X$ of points is τ-closed iff $d_\tau(A) \subseteq A$. A consequence of the entire description of τ in terms of derived sets is the possibility to use derivative operators d as the primitive notion in topology. What happens if we iterate the derivative operator d_τ, considering the sequence d_τ, $d_\tau \circ d_\tau$, ... of operators? If τ is T_D then each element d_τ^α of this sequence is a derivative operator. Now,

[1] Address: Institut de recherche en informatique de Toulouse, CNRS — Université de Toulouse, 118 route de Narbonne, 31062 Toulouse Cedex 9, FRANCE; Philippe.Balbiani@irit.fr.

[2] Address: Escuela Técnica Superior de Ingeniería Informática, Universidad Rey Juan Carlos, C Tulipán, s/n 28933 Móstoles, SPAIN; uridia@ia.urjc.es.

a question arises: what is the link between the topologies τ_α corresponding to the elements d_τ^α of the sequence? The answer is simple: the topologies τ_α are getting finer when α increases. Since the lattice of all T_D topologies on a given nonempty set X is complete, this iteration process should stop. The Cantor-Bendixson rank of (X, τ) is then defined as the least ordinal α such that $d_\tau(d_\tau^\alpha(X)) = d_\tau^\alpha(X)$. A consequence of Tarski's fixpoint theorem [21] is that there exists an ordinal $\alpha^\star$ such that $\alpha \leq \alpha^\star$ and $d_\tau \circ d_\tau^{\alpha^\star} = d_\tau^{\alpha^\star}$, the greatest fixpoint of d_τ.
Owing to the strong relationships between Alexandroff T_D derivative operators and strict partial orders, the notion of rank of a strict partial order can also be defined. More precisely, any strict partial order R on a given nonempty set X defines a function θ_R which associates to each strict partial order $S \subseteq R$ on X the strict partial order $\theta_R(S) = R \circ S$ on X. What happens if we iterate the function θ_R, considering the sequence R, $\theta_R(R)$, ... of partial orders? Simply, the partial orders $\theta_R^\alpha(R)$ are getting smaller when α increases. Since the lattice of all strict partial orders on X is complete, this iteration process should stop. And again, there exists an ordinal $\alpha^\star$ — called the rank of R — such that $\theta_R(\theta_R^{\alpha^\star}(R)) = \theta_R^{\alpha^\star}(R)$, the greatest fixpoint of θ_R. Moreover, if R is the strict partial order on X corresponding to a given Alexandroff T_D derivative operator d, then $\theta_R^{\alpha^\star}(R)$ is a strict partial order on X corresponding to the derivative operator $d_\tau^{\alpha^\star}$ considered above. Hence, it is natural to consider a modal logic with modal operators $\Box$ and $\Box^\star$ respectively interpreted by strict partial orders and the greatest fixpoints of the θ-like functions they define. The goal of this paper is to address the question of its complete axiomatization.
Sections 2, 3 and 4 consider, on one hand, the strong relationships between topologies and derivative operators and, on the other hand, the strong relationships between Alexandroff T_D derivative operators and strict partial orders. Most of the results they contain are well-known. See [8,9,11] for more on these. Sections 5, 6 and 7 present the above-mentioned modal logic and axiomatize it. The proof of its completeness is based on the step-by-step method.

2 Topologies and derivative operators

In this section, we present topologies and derivative operators. We also call forth the fact that topologies and derivative operators are the two sides of the same medal. See [8,9,11] for more on these.

2.1 Topologies

A *topology* on X is a set τ of subsets of X such that: (i) $\emptyset \in \tau$, (ii) $X \in \tau$, (iii) each union of members of τ is in τ, (iv) each finite intersection of members of τ is in τ. We shall say that $A \subseteq X$ is τ-*closed* iff $X \setminus A \in \tau$. τ is said to be T_D iff for all $x \in X$, there exists $A, B \in \tau$ such that $A \setminus B = \{x\}$. We shall say that τ is *Alexandroff* iff each intersection of members of τ is in τ. Let $\leq$ be the binary relation between topologies on X defined by $\tau \leq \tau'$ iff $\tau \subseteq \tau'$. It follows immediately from the definition that for all topologies τ, τ' on X, if τ is T_D and $\tau \leq \tau'$ then τ' is T_D.

Example 2.1 If $X = \{x, y\}$ then let $\tau = \{\emptyset, \{x\}, X\}$, the Sierpiński space. Obviously, τ is a topology on X such that the τ-closed subsets of X are $\emptyset$, $\{y\}$ and X. Moreover, since $\{x\} \setminus \emptyset = \{x\}$ and $X \setminus \{x\} = \{y\}$, τ is T_D. Finally, since X is finite, τ is Alexandroff.

Given a topology τ on X, let L_τ be the set of all topologies τ' on X such that $\tau \leq \tau'$. Remark that the least element of L_τ is τ and the greatest element of L_τ is the topology $\mathcal{P}(X)$. Moreover, the least upper bound of a family $\{\tau_i'\colon i \in I\}$ in L_τ is the intersection of all $\tau' \in L_\tau$ such that $\bigcup\{\tau_i'\colon i \in I\} \subseteq \tau'$ (note that the collection of all such τ' is nonempty, seeing that the topology $\mathcal{P}(X)$ belongs to it) and the greatest lower bound of a family $\{\tau_i'\colon i \in I\}$ in L_τ is $\bigcap\{\tau_i'\colon i \in I\}$. Hence, $(L_\tau, \leq)$ is a complete lattice.

2.2 Derivative operators

A *derivative operator* on X is a function $d\colon \mathcal{P}(X) \to \mathcal{P}(X)$ such that: (i) $d(\emptyset) = \emptyset$, (ii) for all $A, B \subseteq X$, $d(A \cup B) = d(A) \cup d(B)$, (iii) for all $A \subseteq X$, $d(d(A)) \subseteq d(A) \cup A$, (iv) for all $x \in X$, $x \not\in d(\{x\})$. $A \subseteq X$ is said to be d-closed iff $d(A) \subseteq A$. We shall say that d is T_D iff for all $A \subseteq X$, $d(d(A)) \subseteq d(A)$. d is said to be *Alexandroff* iff for all $x \in X$, there exists a greatest $A \subseteq X$ such that A is d-closed and $x \not\in A$. Let $\leq$ be the binary relation between derivative operators on X defined by $d \leq d'$ iff for all $A \subseteq X$, $d(A) \subseteq d'(A)$. It follows immediately from the definition and from the results stated in Section 2.3 that for all derivative operators d, d' on X, if $d \leq d'$ and d' is T_D then d is T_D.

Example 2.2 If $X = \{x, y\}$ then let $d(\emptyset) = \emptyset$, $d(\{x\}) = \{y\}$, $d(\{y\}) = \emptyset$ and $d(X) = \{y\}$. Obviously, d is a derivative operator on X such that the d-closed subsets of X are $\emptyset$, $\{y\}$ and X. Moreover, since $d(d(\emptyset)) \subseteq d(\emptyset)$, $d(d(\{x\})) \subseteq d(\{x\})$, $d(d(\{y\})) \subseteq d(\{y\})$ and $d(d(X)) \subseteq d(X)$, d is T_D. Finally, since X is finite, d is Alexandroff.

Given a derivative operator d on X, let L_d be the set of all derivative operators d' on X such that $d' \leq d$. Remark that the least element of L_d is the derivative operator $d_\emptyset\colon \mathcal{P}(X) \to \mathcal{P}(X)$ such that for all $A \subseteq X$, $d_\emptyset(A) = \emptyset$ and the greatest element of L_d is d. What about the least upper bound of a family $\{d_i'\colon i \in I\}$ in L_d and the greatest lower bound of a family $\{d_i'\colon i \in I\}$ in L_d? We do not know any representation of them using set-theoretic operations of the complete Boolean algebra of all subsets of X. Nevertheless, by the results stated in Sections 2.1 and 2.3, $(L_d, \leq)$ is a complete lattice.

2.3 Topologies v. derivative operators

Given a topology τ on X, let d_τ be the function $d_\tau\colon \mathcal{P}(X) \to \mathcal{P}(X)$ such that for all $A \subseteq X$, $d_\tau(A) = \{x\colon x \text{ is a } \tau\text{-limit point of } A\}$ where $x \in X$ is a τ-*limit point* of $A \subseteq X$ iff for all $B \in \tau$, if $x \in B$ then $(B \setminus \{x\}) \cap A \neq \emptyset$. Remark that d_τ is a derivative operator on X such that for all $A \subseteq X$, A is d_τ-closed iff A is τ-closed. Moreover, (i) d_τ is T_D iff τ is T_D, (ii) d_τ is Alexandroff iff τ is Alexandroff, (iii) $d_{\tau'} \leq d_\tau$ iff $\tau \leq \tau'$.

Example 2.3 If $X = \{x, y\}$ and τ is the topology on X considered in Example 2.1 then d_τ is the derivative operator on X considered in Example 2.2.

Given a derivative operator d on X, let τ_d be the set of subsets of X such that for all $A \subseteq X$, $A \in \tau_d$ iff $X \setminus A$ is d-closed. Remark that τ_d is a topology on X such that for all $A \subseteq X$, A is τ_d-closed iff A is d-closed. Moreover, (i) τ_d is T_D iff d is T_D, (ii) τ_d is Alexandroff iff d is Alexandroff, (iii) $\tau_{d'} \leq \tau_d$ iff $d \leq d'$.

Example 2.4 If $X = \{x, y\}$ and d is the derivative operator on X considered in Example 2.2 then τ_d is the topology on X considered in Example 2.1.

To continue, let us further remark that $\tau_{d_\tau} = \tau$ and $d_{\tau_d} = d$. Given a topology τ on X, let f be the function $f\colon L_\tau \to L_{d_\tau}$ such that $f(\tau') = d_{\tau'}$. By the results stated above, f is an anti-isomorphism between $(L_{d_\tau}, \leq)$ and $(L_\tau, \leq)$. Given a derivative operator d on X, let f be the function $f\colon L_d \to L_{\tau_d}$ such that $f(d') = \tau_{d'}$. By the results stated above, f is an anti-isomorphism between $(L_{\tau_d}, \leq)$ and $(L_d, \leq)$.

3 Alexandroff T_D derivative operators and strict partial orders

In this section, we present Alexandroff T_D derivative operators and strict partial orders. We also call forth the fact that Alexandroff T_D derivative operators and strict partial orders are the two sides of the same medal. See [8,9,11] for more on these. In the sequel, if R is a binary relation on a nonempty set X then for all $x \in X$, $R(x)$ and $R^{-1}(x)$ will respectively denote the set of all $y \in X$ such that xRy and the set of all $y \in X$ such that yRx. Moreover, for all $A \subseteq X$, $R(A)$ and $R^{-1}(A)$ will respectively denote the set $\bigcup\{R(x)\colon x \in A\}$ and the set $\bigcup\{R^{-1}(x)\colon x \in A\}$.

3.1 Alexandroff T_D derivative operators

Given an Alexandroff T_D derivative operator d on X, let L_d^A be the set of all Alexandroff T_D derivative operators d' on X such that $d' \leq d$. Remark that the least element of L_d^A is the derivative operator $d_\emptyset$ considered in Section 2.2 and the greatest element of L_d^A is d. What about the least upper bound of a family $\{d_i'\colon i \in I\}$ in L_d^A and the greatest lower bound of a family $\{d_i'\colon i \in I\}$ in L_d^A? We do not know any representation of them using set-theoretic operations of the complete Boolean algebra of all subsets of X. Nevertheless, by the results stated in Sections 3.2 and 3.3, $(L_d^A, \leq)$ is a complete lattice.

3.2 Strict partial orders

A *strict partial order* on X is a binary relation R on X such that: (i) for all $x \in X$, $x \not\in R(x)$, (ii) for all $x \in X$, $R(R(x)) \subseteq R(x)$. We shall say that $A \subseteq X$ is *R-closed* iff $R^{-1}(A) \subseteq A$. Let $\leq$ be the binary relation between strict partial orders on X defined by $R \leq R'$ iff $R \subseteq R'$. Given a strict partial order R on X, let L_R be the set of all strict partial orders R' on X such that $R' \leq R$. Remark that the least element of L_R is the strict partial order $\emptyset$ and the greatest element of L_R is R. Moreover, the least upper bound of a family $\{R'_i\colon i \in I\}$ in L_R is the transitive closure of $\bigcup\{R'_i\colon i \in I\}$ and the greatest lower bound of a family $\{R'_i\colon i \in I\}$ in L_R is $\bigcap\{R'_i\colon i \in I\}$. Hence, $(L_R, \leq)$ is a complete lattice.

3.3 Alexandroff T_D derivative operators v. strict partial orders

Given an Alexandroff T_D derivative operator d on X, let R_d be the binary relation on X such that for all $x, y \in X$, xR_dy iff $x \in d(\{y\})$. Remark that R_d is a strict partial order on X such that for all $A \subseteq X$, A is R_d-closed iff A is d-closed. Moreover, $R_d \leq R_{d'}$ iff $d \leq d'$. Given a strict partial order R on X, let d_R be the function $d_R\colon \mathcal{P}(X) \to \mathcal{P}(X)$ such that for all $A \subseteq X$, $d_R(A) = R^{-1}(A)$. Remark that d_R is an Alexandroff T_D derivative operator on X such that for all $A \subseteq X$, A is d_R-closed iff A is R-closed. Moreover, $d_R \leq d_{R'}$ iff $R \leq R'$. To continue, let us further remark that $d_{R_d} = d$ and $R_{d_R} = R$. Given an Alexandroff T_D derivative operator d on X, let f be the function $f\colon L_d^A \to L_{R_d}$ such that $f(d') = R_{d'}$. By the results stated above, f is an isomorphism between $(L_{R_d}, \leq)$ and $(L_d^A, \leq)$. Given a strict partial order R on X, let $f\colon L_R \to L_{d_R}^A$ such that $f(R') = d_{R'}$. By the results stated above, f is an isomorphism between $(L_{d_R}^A, \leq)$ and $(L_R, \leq)$.

4 Cantor-Bendixson ranks

In this section, we present Cantor-Bendixson ranks of Alexandroff T_D derivative operators and strict partial orders.

4.1 Cantor-Bendixson ranks of Alexandroff T_D derivative operators

Given an Alexandroff T_D derivative operator d on X, let θ_d be the function $\theta_d\colon L_d \to L_d$ such that for all $d' \in L_d$, $\theta_d(d') = d \circ d'$, i.e. $\theta_d(d')$ is the function $\theta_d(d')\colon \mathcal{P}(X) \to \mathcal{P}(X)$ such that for all $A \subseteq X$, $\theta_d(d')(A) = d(d'(A))$. Clearly, the function θ_d is monotonic. Since $(L_d, \leq)$ is a complete lattice, the function θ_d has a least fixpoint $\mathrm{lfp}(\theta_d)$ and a greatest fixpoint $\mathrm{gfp}(\theta_d)$. Obviously, $\mathrm{lfp}(\theta_d)$ is the derivative operator $d_\emptyset$ considered in Section 2.2. So, let us concentrate on $\mathrm{gfp}(\theta_d)$. A consequence of Tarski's fixpoint theorem [21] is that $\mathrm{gfp}(\theta_d)$ is the least upper bound of the family $\{d'\colon d' \leq \theta_d(d')\}$ in L_d. Next, we give the well-known characterization of $\mathrm{gfp}(\theta_d)$ in terms of ordinal powers of θ_d. For all ordinals α, we inductively define $\theta_d{\downarrow}\alpha$ as follows:

- $\theta_d{\downarrow}0$ is d,
- for all successor ordinals α, $\theta_d{\downarrow}\alpha$ is $\theta_d(\theta_d{\downarrow}(\alpha - 1))$,
- for all limit ordinals α, $\theta_d{\downarrow}\alpha$ is the greatest lower bound of the family $\{\theta_d{\downarrow}\beta\colon \beta \in \alpha\}$ in L_d.

The next result follows from the definition of $\theta_d{\downarrow}\alpha$ as being the greatest lower bound of the family $\{\theta_d{\downarrow}\beta\colon \beta \in \alpha\}$ in L_d for each limit ordinal α: (i) for all $x, y \in X$, $x \in \theta_d{\downarrow}\alpha(\{y\})$ iff for all ordinals β, if $\beta \in \alpha$ then $x \in \theta_d{\downarrow}\beta(\{y\})$, (ii) for all $A \subseteq X$, $\theta_d{\downarrow}\alpha(A) = \bigcap\{\theta_d{\downarrow}\beta(A)\colon \beta \in \alpha\}$. The next result is, again, a consequence of Tarski's fixpoint theorem [21]: (i) for all ordinals α, $\mathrm{gfp}(\theta_d) \leq \theta_d{\downarrow}\alpha$, (ii) there exists an ordinal α such that $\mathrm{gfp}(\theta_d) = \theta_d{\downarrow}\alpha$. The least ordinal α such that $\theta_d{\downarrow}\alpha = \mathrm{gfp}(\theta_d)$ is called the *Cantor-Bendixson rank* of d.

Example 4.1 If $X = \mathbb{Z}$ then let $d_{\mathbb{Z}}$ be the derivative operator on X defined by $d_{\mathbb{Z}}(A) = \{x\colon$ there exists $y \in X$ such that $x <_{\mathbb{Z}} y$ and $y \in A\}$ for each $A \subseteq X$. Obviously, $\theta_{d_{\mathbb{Z}}}(\theta_{d_{\mathbb{Z}}}{\downarrow}\omega) = \theta_{d_{\mathbb{Z}}}{\downarrow}\omega$. Moreover, no finite iteration of $\theta_{d_{\mathbb{Z}}}$ gives the greatest fixpoint. Hence, the Cantor-Bendixson rank of $d_{\mathbb{Z}}$ is ω.

Remark that the Cantor-Bendixson rank of d does not always coincide with the usual Cantor-Bendixson rank of the space X. Actually, it is the supremum of the Cantor-Bendixson ranks of all subspaces of X.

4.2 Cantor-Bendixson ranks of strict partial orders

Given a strict partial order R on X, let θ_R be the function $\theta_R\colon L_R \to L_R$ such that for all $R' \in L_R$, $\theta_R(R') = R \circ R'$, i.e. $\theta_R(R')$ is the binary relation on X such that for all $x, y \in X$, $x\theta_R(R')y$ iff there exists $z \in X$ such that xRz and $zR'y$. Clearly, the function θ_R is monotonic. Since $(L_R, \leq)$ is a complete lattice, the function θ_R has a least fixpoint $\mathrm{lfp}(\theta_R)$ and a greatest fixpoint $\mathrm{gfp}(\theta_R)$. Obviously, $\mathrm{lfp}(\theta_R)$ is the strict partial order $\emptyset$. So, let us concentrate on $\mathrm{gfp}(\theta_R)$. A consequence of Tarski's fixpoint theorem [21] is that $\mathrm{gfp}(\theta_R)$ is the least upper bound of the family $\{R'\colon R' \leq \theta_R(R')\}$ in L_R. Next, we give the well-known characterization of $\mathrm{gfp}(\theta_R)$ in terms of ordinal powers of θ_R. For all ordinals α, we inductively define $\theta_R{\downarrow}\alpha$ as follows:

- $\theta_R{\downarrow}0$ is R,
- for all successor ordinals α, $\theta_R{\downarrow}\alpha$ is $\theta_R(\theta_R{\downarrow}(\alpha - 1))$,
- for all limit ordinals α, $\theta_R{\downarrow}\alpha$ is the greatest lower bound of the family $\{\theta_R{\downarrow}\beta\colon \beta \in \alpha\}$ in L_R.

The next result follows from the definition of $\theta_R{\downarrow}\alpha$ as being the greatest lower bound of the family $\{\theta_R{\downarrow}\beta\colon \beta \in \alpha\}$ in L_R for each limit ordinal α: (i) for all $x, y \in X$, $x\theta_R{\downarrow}\alpha y$ iff for all ordinals β, if $\beta \in \alpha$ then $x\theta_R{\downarrow}\beta y$, (ii) for all $A \subseteq X$, $\theta_R{\downarrow}\alpha^{-1}(A) \subseteq \bigcap\{\theta_R{\downarrow}\beta^{-1}(A)\colon \beta \in \alpha\}$. The next result is, again, a consequence of Tarski's fixpoint theorem [21]: (i) for all ordinals α, $\mathrm{gfp}(\theta_R) \leq \theta_R{\downarrow}\alpha$, (ii) there exists an ordinal α such that $\mathrm{gfp}(\theta_R) = \theta_R{\downarrow}\alpha$. The least ordinal α such that $\theta_R{\downarrow}\alpha = \mathrm{gfp}(\theta_R)$ is called the *Cantor-Bendixson rank of* R.

Example 4.2 If $X = \mathbb{Q}$ then let $R_{\mathbb{Q}}$ be the strict partial order on X defined by $xR_{\mathbb{Q}}y$ iff $x <_{\mathbb{Q}} y$ for each $x, y \in X$. Obviously, $\theta_{R_{\mathbb{Q}}}(\theta_{R_{\mathbb{Q}}}{\downarrow}0) = \theta_{R_{\mathbb{Q}}}{\downarrow}0$. Hence, the Cantor-Bendixson rank of $R_{\mathbb{Q}}$ is 0.

4.3 Alexandroff T_D derivative operators v. strict partial orders

Let d be an Alexandroff T_D derivative operator on X and R be a strict partial order on X such that for all $x, y \in X$, xRy iff $x \in d(\{y\})$ and for all $A \subseteq X$, $d(A) = R^{-1}(A)$. By the results stated in Sections 4.1 and 4.2, one can prove by induction on the ordinal α that (i) for all $x, y \in X$, $x\theta_R{\downarrow}\alpha y$ iff $x \in \theta_d{\downarrow}\alpha(\{y\})$,

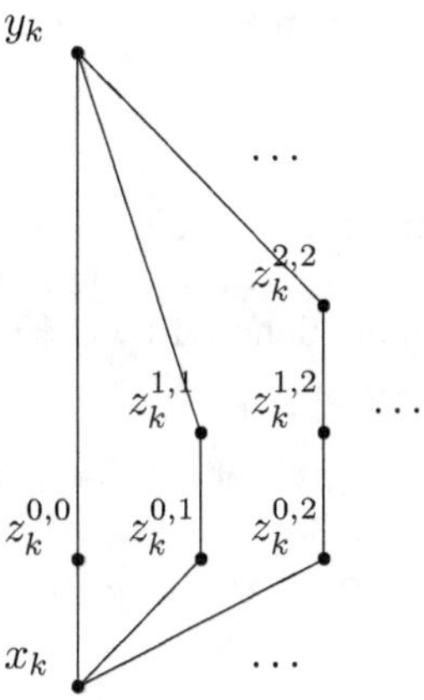

Fig. 1. The relational structure $(X_k, <_k)$.

(ii) for all $A \subseteq X$, $\theta_d{\downarrow}\alpha(A) \supseteq \theta_R{\downarrow}\alpha^{-1}(A)$. Let α_d be the Cantor-Bendixson rank of d and α_R be the Cantor-Bendixson rank of R. The above considerations prove that (i) for all $x, y \in X$, $x\theta_R{\downarrow}\alpha_R y$ iff $x \in \theta_d{\downarrow}\alpha_d(\{y\})$, (ii) for all $A \subseteq X$, $\theta_d{\downarrow}\alpha_d(A) \supseteq \theta_R{\downarrow}{\alpha_R}^{-1}(A)$. Example 4.3 shows that the last inclusion can be strict.

Example 4.3 For all $k \in \mathbb{N}$, let $X_k = \{x_k, y_k\} \cup \{z_k^{i,j}\colon i, j \in \mathbb{N}$ are such that $0 \leq i \leq j\}$ and $<_k$ be the least transitive relation on X_k such that: (i) for all $i, j \in \mathbb{N}$ such that $0 \leq i \leq j$, $x_k <_k z_k^{i,j}$, (ii) for all $i_1, j_1, i_2, j_2 \in \mathbb{N}$ such that $0 \leq i_1 \leq j_1$ and $0 \leq i_2 \leq j_2$, $z_k^{i_1,j_1} <_k z_k^{i_2,j_2}$ iff $i_1 < i_2$ and $j_1 = j_2$, (iii) for all $i, j \in \mathbb{N}$ such that $0 \leq i \leq j$, $z_k^{i,j} <_k y_k$. See Figure 1. Take $X = \bigcup\{X_k\colon k \in \mathbb{N}\}$. Let d be the function $d\colon \mathcal{P}(X) \to \mathcal{P}(X)$ such that for all $A \subseteq X$, $d(A) = \{x\colon$ there exists $y \in X$ such that $x < y$ and $y \in A\}$ and R be the least transitive relation on X such that: (i) for all $k \in \mathbb{N}$, $<_k \subseteq R$, (ii) for all $k, l \in \mathbb{N}$, if $k < l$ then $x_k R x_l$. Obviously, d is a derivative operator on X and R is a strict partial order on X. Moreover, the Cantor-Bendixson ranks of d and R are both equal to $\omega + \omega$. Finally, $\theta_d{\downarrow}(\omega + \omega)$ is not the derivative operator $d_\emptyset$ considered in Section 2.2 and $\theta_R{\downarrow}(\omega + \omega)$ is the strict partial order $\emptyset$.

5 A modal logic

In this section, we present a modal logic with modal operators $\Box$ and $\Box^\star$. Section 5.2 presents the relational semantics where $\Box$ and $\Box^\star$ are respectively interpreted by strict partial orders and the greatest fixpoints of the θ-like functions they define whereas Section 5.3 presents the topological semantics where $\Box$ and $\Box^\star$ are respectively interpreted by Alexandroff T_D derivative operators and the greatest fixpoints of the θ-like functions they define. Note that by 1944, McKinsey and Tarski [17] had already given an interpretation of $\Box$ in terms of derivative operators. For more on this, see also [3,11,19]. We assume the reader is at home with tools and techniques in modal logic; see [4,6,14] for more on these.

5.1 Syntax

The language is defined using a countable set BV of Boolean variables (with typical members denoted by p, q, ...). We inductively define the set $f(BV)$ of *formulas* (with typical members denoted by ϕ, ψ, ...) as follows:

- $\phi ::= p \mid \bot \mid \neg\phi \mid (\phi \vee \psi) \mid \Box\phi \mid \Box^{\star}\phi$.

The other Boolean constructs are defined as usual. We obtain the formulas $\Diamond\phi$ and $\Diamond^{\star}\phi$ as abbreviations: $\Diamond\phi ::= \neg\Box\neg\phi$, $\Diamond^{\star}\phi ::= \neg\Box^{\star}\neg\phi$. The notion of subformula is standard. We adopt the standard rules for omission of the parentheses.

5.2 Relational semantics

A *relational frame* is a structure of the form $\mathcal{F} = (X, R, S)$ such that (i) X is a nonempty set, (ii) R is a strict partial order on X, (iii) S is the greatest fixpoint of the function θ_R in L_R. The following lemma is basic.

Lemma 5.1 *Let $\mathcal{F} = (X, R, S)$ be a relational frame. (i) $R \circ R \leq R$, (ii) $S \circ S \leq S$, (iii) $S \leq R$, (iv) $R \circ S \leq S$, (v) $S \circ R \leq S$, (vi) $S \leq R \circ S$.*

Proof. (i), (ii) and (iii) follow from the fact that R is a strict partial order on X, S is a strict partial order on X and $S \in L_R$. (iv), (v) and (vi) follow from the fact that S is the greatest fixpoint of the function θ_R in L_R. □

A *relational model* is a structure of the form $\mathcal{M} = (X, R, S, V)$ where (i) (X, R, S) is a relational frame, (ii) V is a *valuation* on X, i.e. a function $V\colon BV \rightarrow \mathcal{P}(X)$. The *satisfiability* of $\phi \in f(BV)$ in a relational model $\mathcal{M} = (X, R, S, V)$ at $x \in X$, in symbols $\mathcal{M}, x \models \phi$, is inductively defined as follows:

- $\mathcal{M}, x \models p$ iff $x \in V(p)$,
- $\mathcal{M}, x \not\models \bot$,
- $\mathcal{M}, x \models \neg\phi$ iff $\mathcal{M}, x \not\models \phi$,
- $\mathcal{M}, x \models \phi \vee \psi$ iff either $\mathcal{M}, x \models \phi$ or $\mathcal{M}, x \models \psi$,
- $\mathcal{M}, x \models \Box\phi$ iff for all $y \in X$, if xRy then $\mathcal{M}, y \models \phi$,
- $\mathcal{M}, x \models \Box^{\star}\phi$ iff for all $y \in X$, if xSy then $\mathcal{M}, y \models \phi$.

As a result: $\mathcal{M}, x \models \Diamond\phi$ iff there exists $y \in X$ such that xRy and $\mathcal{M}, y \models \phi$, $\mathcal{M}, x \models \Diamond^{\star}\phi$ iff there exists $y \in X$ such that xSy and $\mathcal{M}, y \models \phi$. $\phi \in f(BV)$ is said to be *true* in a relational model $\mathcal{M} = (X, R, S, V)$, in symbols $\mathcal{M} \models \phi$, iff for all $x \in X$, $\mathcal{M}, x \models \phi$. We shall say that $\phi \in f(BV)$ is *valid* in a relational frame $\mathcal{F} = (X, R, S)$, in symbols $\mathcal{F} \models \phi$, iff for all valuations V on X, $(X, R, S, V) \models \phi$. It is worth noting at this point the following:

Lemma 5.2 *Let $\mathcal{F} = (X, R, S)$ be a relational frame. The following formulas are valid in $\mathcal{F}$: $\Box\phi \rightarrow \Box\Box\phi$, $\Box^{\star}\phi \rightarrow \Box^{\star}\Box^{\star}\phi$, $\Box\phi \rightarrow \Box^{\star}\phi$, $\Box^{\star}\phi \rightarrow \Box\Box^{\star}\phi$,*

$\Box^\star\phi \to \Box^\star\Box\phi$, $\Box\Box^\star\phi \to \Box^\star\phi$.

Proof. The above formulas are Sahlqvist formulas. By Sahlqvist Correspondence Theorem [4, Theorem 3.54], they correspond to the first-order conditions considered in Lemma 5.1. Hence, they are valid in $\mathcal{F}$. □

Let Λ_{rf} be the set of all formulas that are valid in the class of all relational frames.

5.3 Topological semantics

A *topological frame* is a structure of the form $\mathcal{F} = (X, d, e)$ such that (i) X is a nonempty set, (ii) d is an Alexandroff T_D derivative operator on X, (iii) e is the greatest fixpoint of the function θ_d in L_d. The following lemma is basic.

Lemma 5.3 *Let $\mathcal{F} = (X, d, e)$ be a topological frame. (i) $d \circ d \leq d$, (ii) $e \circ e \leq e$, (iii) $e \leq d$, (iv) $d \circ e \leq e$, (v) $e \circ d \leq e$, (vi) $e \leq d \circ e$.*

Proof. (i), (ii) and (iii) follow from the fact that d is an Alexandroff T_D derivative operator on X, e is an Alexandroff T_D derivative operator on X and $e \in L_d$. (iv), (v) and (vi) follow from the fact that e is the greatest fixpoint of the function θ_d in L_d. □

A *topological model* is a structure of the form $\mathcal{M} = (X, d, e, V)$ where (i) (X, d, e) is a topological frame, (ii) V is a *valuation* on X, i.e. a function V: $BV \to \mathcal{P}(X)$. The interpretation of $\phi \in f(BV)$ in a topological model $\mathcal{M} = (X, d, e, V)$, in symbols $\parallel \phi \parallel_{\mathcal{M}}$, is inductively defined as follows:

- $\parallel p \parallel_{\mathcal{M}} = V(p)$,
- $\parallel \bot \parallel_{\mathcal{M}} = \emptyset$,
- $\parallel \neg\phi \parallel_{\mathcal{M}} = X \setminus \parallel \phi \parallel_{\mathcal{M}}$,
- $\parallel \phi \vee \psi \parallel_{\mathcal{M}} = \parallel \phi \parallel_{\mathcal{M}} \cup \parallel \psi \parallel_{\mathcal{M}}$,
- $\parallel \Box\phi \parallel_{\mathcal{M}} = X \setminus d(X \setminus \parallel \phi \parallel_{\mathcal{M}})$,
- $\parallel \Box^\star\phi \parallel_{\mathcal{M}} = X \setminus e(X \setminus \parallel \phi \parallel_{\mathcal{M}})$.

As a result: $\parallel \Diamond\phi \parallel_{\mathcal{M}} = d(\parallel \phi \parallel_{\mathcal{M}})$, $\parallel \Diamond^\star\phi \parallel_{\mathcal{M}} = e(\parallel \phi \parallel_{\mathcal{M}})$. $\phi \in f(BV)$ is said to be *true* in a topological model $\mathcal{M} = (X, d, e, V)$, in symbols $\mathcal{M} \models \phi$, iff $\parallel \phi \parallel_{\mathcal{M}} = X$. We shall say that $\phi \in f(BV)$ is *valid* in a topological frame $\mathcal{F} = (X, d, e)$, in symbols $\mathcal{F} \models \phi$, iff for all valuations V on X, $(X, d, e, V) \models \phi$. It is worth noting at this point the following:

Lemma 5.4 *Let $\mathcal{F} = (X, d, e)$ be a topological frame. The following formulas are valid in $\mathcal{F}$: $\Box\phi \to \Box\Box\phi$, $\Box^\star\phi \to \Box^\star\Box^\star\phi$, $\Box\phi \to \Box^\star\phi$, $\Box^\star\phi \to \Box\Box^\star\phi$, $\Box^\star\phi \to \Box^\star\Box\phi$, $\Box\Box^\star\phi \to \Box^\star\phi$.*

Proof. The above formulas are Sahlqvist formulas. By Sahlqvist Correspondence Theorem [18], they correspond to the conditions considered in Lemma 5.3. Hence, they are valid in $\mathcal{F}$. □

Let Λ_{tf} be the set of all formulas that are valid in the class of all topological frames.

6 Axiomatization and completeness

In this section, we present a complete axiomatization of Λ_{rf}.

6.1 Axiomatization

Let L be the least normal modal logic in our language containing the formulas considered in Lemmas 5.2 and 5.4:

- $\Box\phi \to \Box\Box\phi$,
- $\Box^{\star}\phi \to \Box^{\star}\Box^{\star}\phi$,
- $\Box\phi \to \Box^{\star}\phi$,
- $\Box^{\star}\phi \to \Box\Box^{\star}\phi$,
- $\Box^{\star}\phi \to \Box^{\star}\Box\phi$,
- $\Box\Box^{\star}\phi \to \Box^{\star}\phi$.

Since these formulas are valid in the class of all relational frames and in the class of all topological frames,

Proposition 6.1 *Let $\phi \in f(BV)$. If $\phi \in L$ then $\phi \in \Lambda_{rf}$ and $\phi \in \Lambda_{tf}$.*

It follows that L is sound with respect to the class of all relational frames and with respect to the class of all topological frames. In spite of the connection between Alexandroff T_D derivative operators and strict partial orders studied in Section 3, the class of all relational frames and the class of all topological frames do not validate the same formulas. By Proposition 6.1 and Theorem 6.17, $\Lambda_{rf} \subseteq \Lambda_{tf}$. Example 6.2 shows that the inclusion is strict (David Gabelaia, personal communication, Tbilisi (Georgia), March 24, 2012).

Example 6.2 Let $\phi = \Box(p \to \Diamond p) \to (\Diamond p \to \Diamond^{\star} p)$, we demonstrate $\phi \notin \Lambda_{rf}$ and $\phi \in \Lambda_{tf}$. Intuitively, ϕ says that, in a relational frame $\mathcal{F} = (X, R, S)$, if we have an infinite sequence $y_0 R y_1 R \ldots$ then there exists $i, j \in \mathbb{N}$ such that $0 \leq i \leq j$ and $y_i S y_j$. Firstly, let $\mathcal{M} = (\mathbb{Z}, <_{\mathbb{Z}}, \emptyset, V)$ be the model defined over the integers and such that for all $q \in BV$, $V(q) = \mathbb{Z}$ and $x \in \mathbb{Z}$, we demonstrate $\mathcal{M}, x \not\models \phi$. Obviously, $\mathcal{M}, x \models \Box(p \to \Diamond p)$, $\mathcal{M}, x \models \Diamond p$ and $\mathcal{M}, x \not\models \Diamond^{\star} p$. Hence, $\mathcal{M}, x \not\models \phi$. Secondly, let $\mathcal{M} = (X, d, e, V)$ be a topological model, we demonstrate $\mathcal{M} \models \phi$. It suffices to demonstrate that $\parallel \phi \parallel_{\mathcal{M}} = X$, i.e. $d(V(p)) \setminus d(V(p) \setminus d(V(p))) \subseteq e(V(p))$. Let $A = d(V(p)) \setminus d(V(p) \setminus d(V(p)))$. Obviously, $A \subseteq d(V(p))$. Moreover, by [9, Section 8.5], $A \subseteq d(A)$. Thus, $d(A) \subseteq d(d(A))$. Since d is a T_D derivative operator on X, $d(d(A)) \subseteq d(A)$. Since $d(A) \subseteq d(d(A))$, $d(A) = d(d(A))$. Since e is the greatest fixpoint of the function θ_d in L_d, $e(A) = d(A)$. Since $A \subseteq d(A)$, $A \subseteq e(A)$. Since $A \subseteq d(V(p))$, $e(A) \subseteq e(d(V(p)))$. Since e is the greatest fixpoint of the function θ_d in L_d, $e(d(V(p)) \subseteq e(V(p))$. Since $e(A) \subseteq e(d(V(p)))$, $e(A) \subseteq e(V(p))$. Since $A \subseteq e(A)$, $A \subseteq e(V(p))$.

In the sequel, all frames and all models will be relational. The completeness of L with respect to the class of all frames is more difficult to establish that its soundness and we defer proving it till the end of this section. $\Gamma \subseteq f(BV)$ is said to be an$L$-*theory* iff Γ contains L and Γ is closed under the rule of modus ponens. Let us be clear that the set of all L-theories is a partially ordered set with respect to set inclusion. The least L-theory is L and the greatest L-theory is $f(BV)$. Of course, an L-theory Γ is equal to $f(BV)$ iff $\bot \in \Gamma$. We shall say that an L-theory Γ is *consistent* iff $\bot \notin \Gamma$. $\phi \in f(BV)$ is said to be L-*consistent* iff there exists a consistent L-theory Γ such that $\phi \in \Gamma$. Of course, $\phi \in f(BV)$ is L-consistent iff $\neg\phi \notin L$. We shall say that an L-theory Γ is *maximal* iff for all $\phi \in f(BV)$, either $\phi \in \Gamma$, or $\neg\phi \in \Gamma$. The set of all maximal consistent L-theories will be denoted MCT_L. For all L-theories Γ and for all $\phi \in f(BV)$, let $\Gamma + \phi = \{\psi\colon \phi \to \psi \in \Gamma\}$. For all L-theories Γ, let $\Box\Gamma = \{\phi\colon \Box\phi \in \Gamma\}$ and $\Box^\star\Gamma = \{\phi\colon \Box^\star\phi \in \Gamma\}$. One can easily establish the following results.

Lemma 6.3 *Let Γ be an L-theory. (i) For all $\phi \in f(BV)$, $\Gamma + \phi$ is the least L-theory containing Γ and ϕ, (ii) for all $\phi \in f(BV)$, $\Gamma + \phi$ is consistent iff $\neg\phi \notin \Gamma$, (iii) $\Box\Gamma$ is an L-theory, (iv) $\Box^\star\Gamma$ is an L-theory.*

Our next results are variants of Lindenbaum's Lemma [4, Lemma 4.17] and the Existence Lemma [4, Lemma 4.20].

Lemma 6.4 *Let Γ be a consistent L-theory. There exists $\Delta \in MCT_L$ such that $\Gamma \subseteq \Delta$.*

Lemma 6.5 *Let $\Gamma \in MCT_L$ and $\phi \in f(BV)$. (i) If $\Box\phi \notin \Gamma$ then there exists $\Delta \in MCT_L$ such that $\Box\Gamma \subseteq \Delta$ and $\phi \notin \Delta$, (ii) if $\Box^\star\phi \notin \Gamma$ then there exists $\Delta \in MCT_L$ such that $\Box^\star\Gamma \subseteq \Delta$ and $\phi \notin \Delta$.*

Moreover,

Lemma 6.6 *Let $\Gamma, \Delta \in MCT_L$. If $\Box^\star\Gamma \subseteq \Delta$ then there exists $\Lambda \in MCT_L$ such that $\Box\Gamma \subseteq \Lambda$ and $\Box^\star\Lambda \subseteq \Delta$.*

Proof. The proof is very similar to the one considered, for example, in [12, Theorem 3.6] to derive density conditions. □

What we have in mind is to demonstrate that if $\phi \in f(BV)$ is valid in the class of all frames then $\phi \in L$. In this respect, the concept of subordination structure will be needed. A *subordination structure* is a structure of the form $\mathcal{S} = (X, R, S, \mu)$ where (i) X is a finite nonempty subset of $\mathbb{Z}$, (ii) R is a strict partial order on X, (iii) S is a strict partial order on X, (iv) $S \subseteq R$, (v) $R \circ S \subseteq S$, (vi) $S \circ R \subseteq S$, (vii) μ is an *interpretation* on X, i.e. a function $\mu\colon X \to MCT_L$ such that (vii-a) for all $x, y \in X$, if xRy then $\Box\mu(x) \subseteq \mu(y)$, (vii-b) for all $x, y \in X$, if xSy then $\Box^\star\mu(x) \subseteq \mu(y)$. $\phi \in f(BV)$ is said to be *true* in a subordination structure $\mathcal{S} = (X, R, S, \mu)$, in symbols $\mathcal{S} \models \phi$, iff for all $x \in X$, $\phi \in \mu(x)$. Given two subordination structures $\mathcal{S} = (X, R, S, \mu)$ and $\mathcal{S}' = (X', R', S', \mu')$, we shall say that $\mathcal{S}'$ *contains* $\mathcal{S}$, in symbols $\mathcal{S} \ll \mathcal{S}'$, iff $X \subseteq X'$, $R \subseteq R'$, $S \subseteq S'$ and for all $x \in X$, $\mu(x) = \mu'(x)$. In a subordination structure $\mathcal{S} = (X, R, S, \mu)$, for all $x, y \in X$, if xRy then let $\Pi_{\mathcal{S}}(x, y)$ be the set of all sequences $z_0, \ldots, z_n \in X$ such that $xRz_0 \ldots z_nRy$. Why are subordination structures so interesting? The following proposition contains a fact which helps to prove the starting point of our enterprise: L is complete with respect to the class of all subordination structures of cardinal 1.

Proposition 6.7 *Let $\phi \in f(BV)$. If ϕ is true in the class of all subordination structures of cardinal 1 then $\phi \in L$.*

Proof. Suppose $\phi \notin L$. Hence, by Lemma 6.3, $L + \neg\phi$ is a consistent L-theory. Thus, by Lemma 6.4, there exists $\Gamma \in MCT_L$ such that $L + \neg\phi \subseteq \Gamma$. Thus, $\neg\phi \in \Gamma$. Since Γ is consistent, $\phi \notin \Gamma$. Let $\mathcal{S} = (X, R, S, \mu)$ be the structure such that $X = \{0\}$, $R = \emptyset$, $S = \emptyset$ and μ is the function μ: $X \to MCT_L$ such that $\mu(0) = \Gamma$. Obviously, $\mathcal{S}$ is a subordination structure of cardinal 1 such that $\phi \notin \mu(0)$. Therefore, ϕ is not true in the class of all subordination structures of cardinal 1. □

It follows from Proposition 6.7 that we have reduced the task of proving the completeness of L with respect to the class of all frames to the task of showing how to transform any subordination structure of cardinal 1 into a model satisfying the same formulas. One remark is in order here. Given a subordination structure $\mathcal{S} = (X, R, S, \mu)$, it may contain imperfections:

- $\Box$-*imperfections*, i.e. triples of the form $(x, \Box, \phi)$ where $x \in X$ and $\phi \in f(BV)$ are such that $\Box\phi \notin \mu(x)$ and for all $y \in X$, if xRy then $\phi \in \mu(y)$,
- $\Box^\star$-*imperfections*, i.e. triples of the form $(x, \Box^\star, \phi)$ where $x \in X$ and $\phi \in f(BV)$ are such that $\Box^\star\phi \notin \mu(x)$ and for all $y \in X$, if xSy then $\phi \in \mu(y)$,
- *imperfections of density*, i.e. pairs of the form (x, y) where $x, y \in X$ are such that xSy and for all $z \in X$, not xRz or not zSy.

Remark that for all subordination structures $\mathcal{S} = (X, R, S, \mu)$, the imperfections of $\mathcal{S}$ are elements of $(\mathbb{Z} \times \{\Box, \Box^\star\} \times f(BV)) \cup (\mathbb{Z} \times \mathbb{Z})$.

6.2 Repairing imperfections

Lemmas 6.8, 6.10 and 6.12 state that every imperfection can be repaired.

Lemma 6.8 *Let $\mathcal{S} = (X, R, S, \mu)$ be a subordination structure and $(x, \Box, \phi)$ be a $\Box$-imperfection in $\mathcal{S}$. There exists a subordination structure $\mathcal{S}' = (X', R', S', \mu')$ such that $\mathcal{S} \ll \mathcal{S}'$ and $(x, \Box, \phi)$ is not a $\Box$-imperfection in $\mathcal{S}'$.*

Proof. Since $(x, \Box, \phi)$ is a $\Box$-imperfection in $\mathcal{S}$, $x \in X$ and $\phi \in f(BV)$ are such that $\Box\phi \notin \mu(x)$ and for all $y \in X$, if xRy then $\phi \in \mu(y)$. Since $\Box\phi \notin \mu(x)$, by Lemma 6.5, there exists $\Gamma \in MCT_L$ such that $\Box\mu(x) \subseteq \Gamma$ and $\phi \notin \Gamma$. Let $y \in \mathbb{Z} \setminus X$. We define the structure $\mathcal{S}' = (X', R', S', \mu')$ as follows:

- $X' = X \cup \{y\}$,
- R' is the binary relation on X' such that for all $x', y' \in X'$, $x'R'y'$ iff one of the following conditions holds:
  · $x', y' \in X$ and $x'Ry'$,
  · $x' \in X$, $y' = y$ and $x'Rx$,
  · $x' \in X$, $y' = y$ and $x' = x$,
- S' is the binary relation on X' such that for all $x', y' \in X'$, $x'S'y'$ iff one of the following conditions holds:
  · $x', y' \in X$ and $x'Sy'$,
  · $x' \in X$, $y' = y$ and $x'Sx$,

- μ' is the function μ': $X' \to MCT_L$ such that for all $x' \in X'$,
  - if $x' \in X$ then $\mu'(x') = \mu(x')$,
  - if $x' = y$ then $\mu'(x') = \Gamma$.

Obviously, R' is a strict partial order on X', S' is a strict partial order on X', $S' \subseteq R'$, $R' \circ S' \subseteq S'$ and $S' \circ R' \subseteq S'$. Moreover, for all $x', y' \in X'$, if $x'R'y'$ then $\Box\mu'(x') \subseteq \mu'(y')$ and for all $x', y' \in X'$, if $x'S'y'$ then $\Box^\star\mu'(x') \subseteq \mu'(y')$. Hence, $\mathcal{S}'$ is a subordination structure. In other respect, as the reader can check, $\mathcal{S} \ll \mathcal{S}'$ and $(x, \Box, \phi)$ is not a $\Box$-imperfection in $\mathcal{S}'$. □

Remark 6.9 Note that for all $x', y' \in X$, $\Pi_{\mathcal{S}'}(x', y') = \Pi_{\mathcal{S}}(x', y')$.

Lemma 6.10 *Let $\mathcal{S} = (X, R, S, \mu)$ be a subordination structure and $(x, \Box^\star, \phi)$ be a $\Box^\star$-imperfection in $\mathcal{S}$. There exists a subordination structure $\mathcal{S}' = (X', R', S', \mu')$ such that $\mathcal{S} \ll \mathcal{S}'$ and $(x, \Box^\star, \phi)$ is not a $\Box^\star$-imperfection in $\mathcal{S}'$.*

Proof. Since (x, ϕ) is a $\Box^\star$-imperfection in $\mathcal{S}$, $x \in X$ and $\phi \in f(BV)$ are such that $\Box^\star\phi \notin \mu(x)$ and for all $y \in X$, if xSy then $\phi \in \mu(y)$. Since $\Box^\star\phi \notin \mu(x)$, by Lemma 6.5, there exists $\Gamma \in MCT_L$ such that $\Box^\star\mu(x) \subseteq \Gamma$ and $\phi \notin \Gamma$. Let $y \in \mathbb{Z} \setminus X$. We define the structure $\mathcal{S}' = (X', R', S', \mu')$ as follows:

- $X' = X \cup \{y\}$,
- R' is the binary relation on X' such that for all $x', y' \in X'$, $x'R'y'$ iff one of the following conditions holds:
  - $x', y' \in X$ and $x'Ry'$,
  - $x' \in X$, $y' = y$ and $x'Rx$,
  - $x' \in X$, $y' = y$ and $x' = x$,
- S' is the binary relation on X' such that for all $x', y' \in X'$, $x'S'y'$ iff one of the following conditions holds:
  - $x', y' \in X$ and $x'Sy'$,
  - $x' \in X$, $y' = y$ and $x'Rx$,
  - $x' \in X$, $y' = y$ and $x' = x$,
- μ' is the function μ': $X' \to MCT_L$ such that for all $x' \in X'$,
  - if $x' \in X$ then $\mu'(x') = \mu(x')$,
  - if $x' = y$ then $\mu'(x') = \Gamma$.

Obviously, R' is a strict partial order on X', S' is a strict partial order on X', $S' \subseteq R'$, $R' \circ S' \subseteq S'$ and $S' \circ R' \subseteq S'$. Moreover, for all $x', y' \in X'$, if $x'R'y'$ then $\Box\mu'(x') \subseteq \mu'(y')$ and for all $x', y' \in X'$, if $x'S'y'$ then $\Box^\star\mu'(x') \subseteq \mu'(y')$. Hence, $\mathcal{S}'$ is a subordination structure. In other respect, as the reader can check, $\mathcal{S} \ll \mathcal{S}'$ and $(x, \Box^\star, \phi)$ is not a $\Box^\star$-imperfection in $\mathcal{S}'$. □

Remark 6.11 Note that for all $x', y' \in X$, $\Pi_{\mathcal{S}'}(x', y') = \Pi_{\mathcal{S}}(x', y')$.

Lemma 6.12 *Let $\mathcal{S} = (X, R, S, \mu)$ be a subordination structure and (x, y) be an imperfection of density in $\mathcal{S}$. There exists a subordination structure $\mathcal{S}' = (X', R', S', \mu')$ such that $\mathcal{S} \ll \mathcal{S}'$ and (x, y) is not an imperfection of density in $\mathcal{S}'$.*

Proof. Since (x, y) is an imperfection of density in $\mathcal{S}$, $x, y \in X$ are such that xSy and for all $z \in X$, not xRz or not zSy. Since xSy, $\square^\star\mu(x) \subseteq \mu(y)$. Hence, by Lemma 6.6, there exists $\Gamma \in MCT_L$ such that $\square\mu(x) \subseteq \Gamma$ and $\square^\star\Gamma \subseteq \mu(y)$. Let $z \in \mathbb{Z} \setminus X$. We define the structure $\mathcal{S}' = (X', R', S', \mu')$ as follows:

- $X' = X \cup \{z\}$,
- R' is the binary relation on X' such that for all $x', y' \in X'$, $x'R'y'$ iff one of the following conditions holds:
  · $x', y' \in X$ and $x'Ry'$,
  · $x' \in X$, $y' = z$ and $x'Rx$,
  · $x' \in X$, $y' = z$ and $x' = x$,
  · $x' = z$, $y' \in X$ and yRy',
  · $x' = z$, $y' \in X$ and $y' = y$,
- S' is the binary relation on X' such that for all $x', y' \in X'$, $x'S'y'$ iff one of the following conditions holds:
  · $x', y' \in X$ and $x'Sy'$,
  · $x' \in X$, $y' = z$ and $x'Sx$,
  · $x' = z$, $y' \in X$ and yRy',
  · $x' = z$, $y' \in X$ and $y' = y$,
- μ' is the function μ': $X' \to MCT_L$ such that for all $x' \in X'$,
  · if $x' \in X$ then $\mu'(x') = \mu(x')$,
  · if $x' = z$ then $\mu'(x') = \Gamma$.

Obviously, R' is a strict partial order on X', S' is a strict partial order on X', $S' \subseteq R'$, $R' \circ S' \subseteq S'$ and $S' \circ R' \subseteq S'$. Moreover, for all $x', y' \in X'$, if $x'R'y'$ then $\square\mu'(x') \subseteq \mu'(y')$ and for all $x', y' \in X'$, if $x'S'y'$ then $\square^\star\mu'(x') \subseteq \mu'(y')$. Thus, $\mathcal{S}'$ is a subordination structure. In other respect, as the reader can check, $\mathcal{S} \ll \mathcal{S}'$ and (x, y) is not an imperfection of density in $\mathcal{S}'$. □

Remark 6.13 Note that for all $x', y' \in X$, $x'S'y'$ or $\Pi_{\mathcal{S}'}(x', y') = \Pi_{\mathcal{S}}(x', y')$.

Let the structures defined in the proofs of Lemmas 6.8, 6.10 and 6.12 be respectively called *completion* of $\mathcal{S}$ with respect to $(x, \square, \phi)$, *completion* of $\mathcal{S}$ with respect to $(x, \square^\star, \phi)$ and *completion* of $\mathcal{S}$ with respect to (x, y).

6.3 Completeness

The following proposition constitutes the heart of our method.

Proposition 6.14 *Let $\phi \in f(BV)$. If ϕ is valid in the class of all frames then ϕ is true in the class of all subordination structures of cardinal* 1.

Proof. Suppose ϕ is not true in the class of all subordination structures of cardinal 1. Hence, there exists a subordination structure $\mathcal{S} = (X, R, S, \mu)$ of cardinal 1 such that $\mathcal{S} \not\models \phi$. Let i_0, i_1, ... be an enumeration of $(\mathbb{Z} \times \{\square, \square^\star\} \times f(BV)) \cup (\mathbb{Z} \times \mathbb{Z})$ where each item is repeated infinitely often. We inductively define the sequence $\mathcal{S}_0 = (X_0, R_0, S_0, \mu_0)$, $\mathcal{S}_1 = (X_1, R_1, S_1, \mu_1)$, ... of subordination structures as follows:

- let $\mathcal{S}_0$ be $\mathcal{S}$,

- for all nonnegative integers n, if i_n is an imperfection in $\mathcal{S}_n$ then let $\mathcal{S}_{n+1}$ be the completion of $\mathcal{S}_n$ with respect to i_n else let $\mathcal{S}_{n+1}$ be $\mathcal{S}_n$.

Let $\mathcal{M}' = (X', R', S', V')$ be the structure defined as follows: $X' = \bigcup\{X_n$: n is a nonnegative integer$\}$, $R' = \bigcup\{R_n$: n is a nonnegative integer$\}$, $S' = \bigcup\{S_n$: n is a nonnegative integer$\}$ and V' is the function V': $BV \to \mathcal{P}(X)$ such that for all $p \in BV$, $V'(p) = \{x'$: there exists a nonnegative integer n such that $x' \in X_n$ and $p \in \mu_n(x')\}$. Obviously, R' is a strict partial order on X', S' is a strict partial order on X', $S' \subseteq R'$, $R' \circ S' = S'$ and $S' \circ R' \subseteq S'$. Hence, S' is a fixpoint of the θ-like function defined by R'. Now, let S'' be a fixpoint of the θ-like function defined by R', we demonstrate $S'' \leq S'$. Let $x', y' \in X'$ be such that $s'S''y'$, we demonstrate $x'S'y'$. Since S'' is a fixpoint of the θ-like function defined by R', $R' \circ S'' = S''$. Since $x'S''y'$, we can inductively construct an infinite sequence $z'_0, z'_1, \ldots \in X'$ such that $x'R'z'_0R'z'_1\ldots$ and for all nonnegative integers n, $z'_nS''y'$. By Remarks 6.9, 6.11 and 6.13, there exists a nonnegative integer n such that $x', y' \in X_n$ and $x'S_ny'$. Thus, $x'S'y'$. In conclusion, we have proved that

Claim 6.15 *S' is the greatest fixpoint of the θ-like function defined by R'.*

Moreover, for all $x', y' \in X'$, if $x'R'y'$ then $\Box\mu'(x') \subseteq \mu'(y')$ and for all $x', y' \in X'$, if $x'S'y'$ then $\Box^\star\mu'(x') \subseteq \mu'(y')$. Now, let $\psi \in f(BV)$, we prove for all $x' \in X'$, $\mathcal{M}', x' \models \psi$ iff there exists a nonnegative integer n such that $x' \in X_n$ and $\psi \in \mu_n(x')$. The proof is done by induction on ψ.
Induction hypothesis. Let $\psi \in f(BV)$ be such that for all $\chi \in f(BV)$, if χ is a subformula of ψ then for all $x' \in X'$, $\mathcal{M}', x' \models \chi$ iff there exists a nonnegative integer n such that $x' \in X_n$ and $\chi \in \mu_n(x')$.
Induction step. We have to consider the six following cases.
Case $\psi = p$. By definition of V'.
Cases $\psi = \bot$, $\psi = \neg\chi$, $\psi = \chi' \vee \chi''$. By the induction hypothesis.
Cases $\psi = \Box\chi$, $\psi = \Box^\star\chi$. By the induction hypothesis, by the fact that for all $x', y' \in X'$, if $x'R'y'$ then $\Box\mu'(x') \subseteq \mu'(y')$, by the fact that for all $x', y' \in X'$, if $x'S'y'$ then $\Box^\star\mu'(x') \subseteq \mu'(y')$, by the fact that for all $x' \in X'$ if $\Box\chi \notin \mu'(x')$ then there exists $y' \in X'$ such that $x'R'y'$ and $\chi \notin \mu'(y')$ and by the fact that for all $x' \in X'$ if $\Box^\star\chi \notin \mu'(x')$ then there exists $y' \in X'$ such that $x'S'y'$ and $\chi \notin \mu'(y')$.
In conclusion, we have proved that

Claim 6.16 *Let $\psi \in f(BV)$. For all $x' \in X'$, $\mathcal{M}', x' \models \psi$ iff there exists a nonnegative integer n such that $x' \in X_n$ and $\psi \in \mu_n(x')$.*

Since $\mathcal{S} \not\models \phi$, $\phi \notin \mu_0(0)$. By the above claim, $\mathcal{M}', 0 \not\models \phi$. Therefore, ϕ is not valid in the class of all frames. □

The result that emerges from the above discussion is the following theorem.

Theorem 6.17 *Let $\phi \in f(BV)$. The following conditions are equivalent: (i) $\phi \in L$, (ii) ϕ is valid in the class of all frames, (iii) ϕ is true in the class of all subordination structures of cardinal 1.*

Proof. (i)→(ii): By Proposition 6.1.
(ii)→(iii): By Proposition 6.14.
(iii)→(i): By Proposition 6.7. □

7 Definability

In this section, we show that $\Box^\star$ is not definable in the ordinary language of modal logic and that the class of all frames is not first-order definable.

7.1 Modal definability

Suppose there exists a $\Box^\star$-free formula ϕ such that $\Box^\star p \leftrightarrow \phi \in L$. Let $\mathcal{M} = (\mathbb{Z}, <_{\mathbb{Z}}, \emptyset, V)$ be the model defined over the integers and such that for all $q \in BV$, $V(q) = \emptyset$ and $\mathcal{M}' = (\mathbb{Q}, <_{\mathbb{Q}}, <_{\mathbb{Q}}, V')$ be the model defined over the rationals and such that for all $q \in BV$, $V'(q) = \emptyset$. Obviously, for all $\Box^\star$-free formulas ψ, for all $x \in \mathbb{Z}$ and for all $x' \in \mathbb{Q}$, $\mathcal{M}, x \models \psi$ iff $\mathcal{M}', x' \models \psi$. Hence, $\mathcal{M}, 0 \models \phi$ iff $\mathcal{M}', 0 \models \phi$ Since $S_{\mathcal{M}} = \emptyset$, $\mathcal{M}, 0 \models \Box^\star p$. Since $\Box^\star p \leftrightarrow \phi \in L$, by Proposition 6.1, $\mathcal{M}, 0 \models \phi$. Since $S_{\mathcal{M}'} = <_{\mathbb{Q}}$, $\mathcal{M}', 0 \not\models \Box^\star p$. Since $\Box^\star p \leftrightarrow \phi \in L$, by Proposition 6.1, $\mathcal{M}', 0 \not\models \phi$: a contradiction. These considerations prove

Proposition 7.1 *There exists no $\Box^\star$-free formula ϕ such that $\Box^\star p \leftrightarrow \phi \in L$.*

That is to say, $\Box^\star$ is not definable in the ordinary language of modal logic.

7.2 First-order definability

Suppose there exists a first-order sentence ϕ in $\tilde{R}$, $\tilde{S}$ and $\equiv$ (interpreted in a relational structure $\mathcal{F} = (X, R, S)$ by R, S and equality) such that for all relational structures $\mathcal{F} = (X, R, S)$, $\mathcal{F}$ is a frame iff $\mathcal{F} \models \phi$. For all $n \in \mathbb{N}$, let $\mathcal{F}_n = (X_n, R_n, S_n)$ be the relational structure defined as follows: $X_n = \{0, \ldots, n\}$, $R_n = \{(i,j)\colon\ 0 \leq i < j \leq n\}$ and $S_n = \emptyset$, Obviously, for all $n \in \mathbb{N}$, $\mathcal{F}_n \models \phi \wedge \exists y \forall x (x\tilde{R}y \vee x \equiv y) \wedge \forall x \forall y \neg x\tilde{S}y$. Let U be an ultrafilter over $\mathbb{N}$ and $\mathcal{F}_U = (X_U, R_U, S_U)$ be the ultraproduct of the family $\{\mathcal{F}_n\colon n \in \mathbb{N}\}$ modulo U. Since for all $n \in \mathbb{N}$, $\mathcal{F}_n \models \phi \wedge \exists y \forall x (x\tilde{R}y \vee x \equiv y) \wedge \forall x \forall y \neg x\tilde{S}y$, by the Fundamental Theorem of Ultraproducts [7, Theorem 4.1.9], $\mathcal{F}_U \models \phi \wedge \exists y \forall x (x\tilde{R}y \vee x \equiv y) \wedge \forall x \forall y \neg x\tilde{S}y$. Since $\mathcal{F}_U \models \phi$, $\mathcal{F}_U$ is a frame. For all $i \in \mathbb{N}$, let $[i]$ be the class of $(i, i, \ldots)$ modulo U. Remark that for all $i, j \in \mathbb{N}$, $[i]R_U[j]$ iff $i < j$. Since $\mathcal{F}_U \models \exists y \forall x (x\tilde{R}y \vee x \equiv y)$, there exists $M_U \in X_U$ such that for all $i \in \mathbb{N}$, $[i]R_U M_U$ or $[i] = M_U$. Since for all $i, j \in \mathbb{N}$, $[i]R_U[j]$ iff $i < j$, for all $i \in \mathbb{N}$, $[i]R_U M_U$. Let R'_U be the binary relation on X_U such that for all $x, y \in X_U$, $xR'_U y$ iff there exists $i \in \mathbb{N}$ such that $x = [i]$ and $y = M_U$, we demonstrate $R'_U \leq \theta_{R_U}(R'_U)$, i.e. $R'_U \leq R_U \circ R'_U$. Remark that R'_U is a strict partial order on X_U and $R'_U \subseteq R_U$. Moreover, $R'_U \neq \emptyset$. Let $x, y \in X_U$ be such that $xR'_U y$, we demonstrate there exists $z \in X_U$ such that $xR_U z$ and $zR'_U y$. Since $xR'_U y$, there exists $i \in \mathbb{N}$ such that $x = [i]$ and $y = M_U$. Hence, it suffices to take $z = [i+1]$ and we have $xR_U z$ and $zR'_U y$. In conclusion, we have proved that

Claim 7.2 $R'_U \leq \theta_{R_U}(R'_U)$.

By the results stated in Section 4.2, $R'_U \leq \text{gfp}(\theta_{R_U})$. Since $\mathcal{F}_U \models \forall x \forall y \neg x\tilde{S}y$, $\text{gfp}(\theta_{R_U}) = \emptyset$. Since $R'_U \leq \text{gfp}(\theta_{R_U})$, $R'_U = \emptyset$: a contradiction. The conclusion can be summarized as follows.

Proposition 7.3 *There exists no first-order sentence ϕ in $\tilde{R}$, $\tilde{S}$ and $\equiv$ (interpreted in a relational structure $\mathcal{F} = (X, R, S)$ by R, S and equality) such that for all relational structures $\mathcal{F} = (X, R, S)$, $\mathcal{F}$ is a frame iff $\mathcal{F} \models \phi$.*

That is to say, the class of all frames is not first-order definable.

8 Conclusion

In this article, we considered a modal logic with modal operators $\Box$ and $\Box^\star$ respectively interpreted by strict partial orders and the greatest fixpoints of the θ-like functions they define. Much remains to be done. Firstly, there is the issue of the complete axiomatization of the set of all formulas in the $\Box$-free fragment of our language that are valid in the class of all frames. Are the axioms of the form $\Box^\star\phi \to \Box^\star\Box^\star\phi$ sufficient in this respect? Secondly, there is the question of the computability and complexity of the membership problem in L. Obviously, L is a conservative extension of $K4$. Hence, by Ladner's Theorem [4, Theorem 6.50], the membership problem in L is $PSPACE$-hard. Is it possible to demonstrate that it is in $PSPACE$? Thirdly, there is the issue of the finite model property (fmp) of L. There are possibly two ways to ask whether L has the fmp, depending on the class of relational structures one considers. One possibility is to consider the fmp with respect to the class of all frames. Another possibility is to consider the fmp with respect to the class of all relational structures satisfying the conditions considered in Lemma 5.1. Fourthly, there is the question of the modal definability of the class of all frames. More precisely, is there $\Gamma \subseteq f(BV)$ such that for all relational structures $\mathcal{F} = (X, R, S)$, $\mathcal{F}$ is a frame iff for all $\phi \in f(BV)$, if $\phi \in \Gamma$ then $\mathcal{F} \models \phi$? If such $\Gamma \subset f(BV)$ exists, can it be finite? Fifthly, there is the issue of the addition to our language of the global operator $[U]$ and the difference operator $[\neq]$ respectively interpreted by the universal relation and the inequality relation. As is well-known, see [4, Chapter 7] or [15], these modal operators greatly increase the expressive power of a modal language whether it is interpreted in relational structures as in Section 5.2 or in topological structures as in Section 5.3. Sixthly, there is the issue of the complete axiomatization of the set of all formulas that are valid in the class of all topological frames. Are the axioms of L together with the axioms of the form $\Box(p \to \Diamond p) \to (\Diamond p \to \Diamond^\star p)$ sufficient in this respect? Seventhly, there is the question of the possible readings of $\Box^\star$ in terms of knowledge and belief. See [2,22] for more details.

Acknowledgements

The research of the first author has been partly supported by the project DID02/32/2009 of Bulgarian Science Fund. The research of the second author has been partly supported by the MICINN projects TIN2009-14562-C05 and CSD2007-00022 and by the Rustaveli Science Foundation grant ♯ FR/489/5-

105/11. Both authors want to express their appreciation to David Gabelaia for his kind help in the course of this research. Thanks are due as well to our reviewers for their thorough comments on the submitted version of our paper that helped us to make its final version more readable.

References

[1] Van Benthem, J., Bezhanishvili, G.: *Modal logics of space.* In Aiello, M., Pratt-Hartmann, I., van Benthem, J. (editors): *Handbook of Spatial Logics.* Springer (2007) 217–298.
[2] Van Benthem, J., Sarenac, D.: *The geometry of knowledge.* In Béziau, J.-Y., Costa Leite, A., Facchini, A. (editors): *Aspects of Universal Logic.* Centre de recherches sémiologiques de l'université de Neuchatel (2004) 1–31.
[3] Bezhanishvili, G., Esakia, L., Gabelaia, D.: *Some results on modal axiomatization and definability for topological spaces.* Studia Logica **81** (2005) 325–355.
[4] Blackburn, P., de Rijke, M., Venema, Y.: *Modal Logic.* Cambridge University Press (2001).
[5] Ten Cate, B., Gabelaia, D., Sustretov, D.: *Modal languages for topology: expressivity and definability.* Annals of Pure and Applied Logic **159** (2009) 146–170.
[6] Chagrov, A., Zakharyaschev, M.: *Modal Logic.* Oxford University Press (1997).
[7] Chang, C., Keisler, H.: Model Theory. Elsevier Science (1990).
[8] Esakia, L.: *Weak transitivity — restitution.* Logical Studies **6** (2001).
[9] Esakia, L.: *Intuitionistic logic and modality via topology.* Annals of Pure and Applied Logic **127** (2004) 155–170.
[10] Gabelaia, D.: *Modal Definability in Modal Logic.* University of Amsterdam (2001) Master Thesis.
[11] Gabelaia, D.: *Topological, Algebraic and Spatio-Temporal Semantics for Multi-Dimensional Modal Logics.* King's College London (2004) Doctoral Thesis.
[12] Goldblatt, R.: *Logics of Time and Computation.* CSLI (1992).
[13] Goranko, V., Otto, M.: *Model theory of modal logic.* In: *Handbook of Modal Logic.* Elsevier (2007) 249–329.
[14] Kracht, M.: *Tools and Techniques in Modal Logic.* Elsevier (1999).
[15] Kudinov, A.: *Topological modal logics with difference modality.* In Governatori, G., Hodkinson, I., Venema, Y. (editors): *Advances in Modal Logic.* College Publications (2006) 319–332.
[16] Lucero-Bryan, J.: *The d-logic of the rational numbers: a fruitful construction.* Studia Logica **97** (2011) 265–295.
[17] McKinsey, J., Tarski, A.: *The algebra of topology.* Annals of Mathematics **45** (1944) 141–191.
[18] De Rijke, M., Venema, Y.: *Sahlqvist's theorem for Boolean algebras with operators with an application to cylindric algebras.* Studia Logica **54** (1995) 61–78.
[19] Shehtman, V.: *Derived sets in Euclidean spaces and modal logic.* University of Amsterdam (1990) ITLI Prepublication Series X-90-05.
[20] Shehtman, V.: *"Everywhere" and "here".* Journal of Applied Non-Classical Logics **9** (1999) 369–380.
[21] Tarski, A.: *A lattice-theoretical fixpoint theorem and its applications.* Pacific Journal of Mathematics **5** (1955) 285–309.
[22] Uridia, L.: *Boolean modal logic wK4Dyn — doxastic interpretation.* In Bezhanishvili, N., Löbner, S., Schwabe, K., Spada, L. (editors): *Logic, Language, and Computation.* Springer (2011) 158–169.

Calibrating Provability Logic: From Modal Logic to Reflection Calculus

Lev Beklemishev [1]

Steklov Institute of Mathematics, Moscow

Several interesting applications of provability logic in proof theory made use of a polymodal logic **GLP** due to Giorgi Japaridze. This system, although decidable, is not very easy to handle. In particular, it is not Kripke complete. It is complete w.r.t. neighborhood semantics, however this could only be established recently by rather complicated techniques [1].

In this talk we will advocate the use of a weaker system, called *Reflection Calculus*, which is much simpler than **GLP**, yet expressive enough to regain its main proof-theoretic applications, and more. From the point of view of modal logic, **RC** can be seen as a fragment of polymodal logic consisting of implications of the form $A \to B$, where A and B are formulas built-up from $\top$ and the variables using just $\wedge$ and the diamond modalities. In this paper we formulate it in a somewhat more succinct self-contained format.

Further, we state its arithmetical interpretation, and provide some evidence that **RC** is much simpler than **GLP**. We then outline a consistency proof for Peano arithmetic based on **RC** and state a simple combinatorial statement, the so-called Worm principle, that was suggested by the use of **GLP** but is even more directly related to the Reflection Calculus.

1 Reflection calculus RC

Basic symbols of **RC** are propositional variables $p, q, \dots$, constant $\top$, conjunction $\wedge$, the symbols n, for each $n \in \omega$, and the brackets. Informally, n corresponds to the n-th modality $\langle n \rangle$.

The formulas α of **RC** are generated by the following grammar:

$$\alpha ::= \top \mid p \mid (\alpha \wedge \alpha) \mid n\alpha \qquad n \in \omega.$$

Example: $\alpha = 3(2\top \wedge 32\top)$. The symbols $\top$ occurring after a number symbol can be omitted without impairing the readability of the formula, e.g., the previous formula can be shortened to $3(2 \wedge 32)$.

Derivable objects of **RC** are *sequents*, that is, expressions of the form $\alpha \vdash \beta$ with α, β formulas.

[1] Supported by the Russian Foundation for Basic Research (RFBR), Russian Presidential Council for Support of Leading Scientific Schools, and the Swiss–Russian cooperation project STCP–CH–RU "Computational proof theory."

RC rules:

(i) $\alpha \vdash \alpha$; $\alpha \vdash \top$; if $\alpha \vdash \beta$ and $\beta \vdash \gamma$ then $\alpha \vdash \gamma$;

(ii) $\alpha \wedge \beta \vdash \alpha, \beta$; if $\alpha \vdash \beta$ and $\alpha \vdash \gamma$ then $\alpha \vdash \beta \wedge \gamma$;

(iii) $nn\alpha \vdash n\alpha$; if $\alpha \vdash \beta$ then $n\alpha \vdash n\beta$;

(iv) $n\alpha \vdash m\alpha$ for $n > m$;

(v) $n\alpha \wedge m\beta \vdash n(\alpha \wedge m\beta)$ for $n > m$.

For example, the following is derivable in **RC**:

$$3 \wedge 23 \vdash 3(\top \wedge 23) \vdash 323.$$

Notice that Axioms 1 and 2 express that $\vdash$ induces a Tarskian consequence relation and that $\wedge$ has the usual properties of conjunction. Axioms 3 correspond to the modal axioms of **K4**. (Notably, any principle related to Löb's axiom is absent.) Axioms 4 and 5 relate different modalities to each other.

In the following, the variable-free fragment of **RC** will, in a sense, be more important than **RC** itself. We denote it $\mathbf{RC}^0$.

2 Arithmetical interpretation of RC

Let S be a first order r.e. theory containing enough arithmetic to satisfy the assumptions of Gödel's second incompleteness theorem. For each $n \in \omega$, *reflection principles* $R_n(S)$ are the formulas in the language of S naturally expressing that *each arithmetical* Σ^0_n*-sentence provable in* S *is true.* Reflection principles are well-known in proof theory; their use is going back to Rosser, Turing, Kreisel and Feferman. They are best to be seen as generalizations of Gödel's consistency assertion to higher levels of arithmetical complexity.

Having fixed the formulas $R_n(S)$, we now define an interpretation of the language of **RC** in the style of provability logic.

Let f be a substitution mapping propositional variables to sentences in the language of S. *Arithmetical translation* $f_S(\alpha)$ of a formula α is defined inductively as follows:

- $f_S(\top) = \top$; $f_S(p) = f(p)$; $f_S(\alpha \wedge \beta) = (f_S(\alpha) \wedge f_S(\beta))$;
- $f_S(n\alpha) = R_n(S + f_S(\alpha))$.

Suppose $\mathbb{N} \models S$ and S contains Peano arithmetic PA.

Theorem 2.1 *$\alpha \vdash \beta$ in* **RC** *iff $S \vdash f_S(\alpha) \to f_S(\beta)$, for all f.*

We note that if α is variable-free, then $f_S(\alpha)$ does not depend on f. We abbreviate $f_S(\alpha)$ by α_S.

3 Interpretation of RC in GLP

As we mentioned before, **RC** can be seen as a fragment of polymodal provability logic **GLP**. We translate **RC**-formulas α to **GLP**-formulas α^* as follows: $\top^* = \top$, $p^* = p$, $(\alpha \wedge \beta)^* = (\alpha^* \wedge \beta^*)$, and $(n\alpha)^* = \langle n \rangle \alpha^*$. Thus, $3(2 \wedge 32)$ translates to $\langle 3 \rangle(\langle 2 \rangle \top \wedge \langle 3 \rangle \langle 2 \rangle \top)$.

The following theorem is an adaptation of the results of Dashkov [6].

Theorem 3.1 (i) **GLP** *is a conservative extension of* **RC**, *that is, for each* α, β, **RC** *proves* $\alpha \vdash \beta$ *iff* **GLP** $\vdash \alpha^* \to \beta^*$;

(ii) **RC** *is polytime decidable;*

(iii) **RC** *enjoys the finite model property.*

We note that by the results of Shapirovsky, **GLP** is PSPACE-complete. We also note that Theorem 2.1 follows from part 1 and Japaridze's arithmetical completeness theorem for **GLP**.

From now on we shall mainly work in the variable-free fragment of **RC**.

4 $\mathbf{RC}^0$ as an ordinal notation system

Let W denote the set of all $\mathbf{RC}^0$-formulas. Using derivability in **RC** we define the following relations on W:

- $\alpha \sim \beta$ if ($\alpha \vdash \beta$ and $\beta \vdash \alpha$);
- $\alpha <_n \beta$ if $\beta \vdash n\alpha$.

Obviously, $\sim$ is an equivalence relation and $<_n$ is correctly defined on the equivalence classes. We note that by the results of the previous section both of these relations are polynomially decidable.

A formula without variables and $\wedge$ is called a *word.* In fact, any such formula syntactically is a sequence of numbers (followed by $\top$).

Theorem 4.1 (i) *Every* $\alpha \in W$ *is equivalent to a word;*

(ii) $(W/{\sim}, <_0)$ *is isomorphic to* $(\varepsilon_0, <)$.

Here, ε_0 is the first ordinal α such that $\omega^\alpha = \alpha$. The isomorphism can be established by the following function $o : W/{\sim} \to \varepsilon_0$.

First, define $o(0^k) = k$, for each $k \in \omega$. Any other word can be written in the form $\alpha = \alpha_1 0 \alpha_2 0 \cdots 0 \alpha_n$, where each α_i does not contain 0 and not all of them are empty. Then we define

$$o(\alpha) = \omega^{o(\alpha_n^-)} + \cdots + \omega^{o(\alpha_1^-)},$$

where β^- means subtracting 1 from each letter of a word β.

Example 4.2 $o(1012) = \omega^{o(01)} + \omega^{o(0)} = \omega^{\omega^1 + \omega^0} + \omega = \omega^{\omega+1} + \omega$

Thus, calculating the ordinal $o(\alpha)$ gives a criterium for the equivalence and comparison of words. It is useful, however, to regard the set of words as a specific notation system for ordinals alternative to Cantor normal forms. In fact, in what follows we can completely disregard Cantor normal forms.

5 Reduction property

For the proof-theoretic applications of **RC** we need to state a basic property of reflection principles called the *reduction property.* Finitely iterated reflection principles are defined as follows:

$$R_n^1(S) = R_n(S), \quad R_n^{k+1}(S) = R_n(S + R_n^k(S)).$$

Let A and B be two sets of formulas over a given arithmetical theory S. We write $A \equiv_n B$ modulo S, if $S + A$ and $S + B$ prove the same arithmetical Π^0_{n+1}-sentences. The following theorem is proved in [2].

Theorem 5.1 (reduction) *Suppose* $S \subseteq \Pi^0_{n+2}$ *and* $V \vdash S$. *Then*

$$R_{n+1}(V) \equiv_n \{R_n^k(V) : k < \omega\} \text{ modulo } S.$$

Let us now apply this theorem to the situation when $V = S + \beta_S$, for some $\beta \in W$.

Denote $\alpha = (n+1)\beta$ and $\alpha[\![0]\!] := n\beta$, $\alpha[\![k+1]\!] := n(\beta \wedge \alpha[\![k]\!])$.

It is easy to check that $\alpha[\![0]\!] <_0 \alpha[\![1]\!] <_0 \alpha[\![2]\!] <_0 \cdots \to \alpha$. Moreover, the formulas $\alpha[\![k]\!]$ correspond to k-fold iterated reflection principles $R_n^k(V)$. Thus, from the reduction property we infer

Corollary 5.2 $\alpha_S \equiv_n \{\alpha[\![k]\!]_S : k < \omega\}$, *whenever* $S \subseteq \Pi^0_{n+2}$.

6 Consistency proof for PA

Theorem 6.1 *Primitive recursive arithmetic together with transfinite induction over* $(W, <_0)$ *proves the consistency of* PA.

We sketch a proof of this version of Gentzen's theorem. As our basic system we take $S =$ EA, the Elementary Arithmetic, aka $I\Delta_0 + \exp$. We have that EA $\subseteq \Pi^0_2$, so Corollary 5.2 applies for each n. We will also use the fact that PRA proves $R_1(\mathsf{EA})$.

Let $\Diamond\varphi$ denote a standard arithmetical formula expressing the consistency of a sentence (with the Gödel number) φ over S. In fact, $\Diamond\varphi$ is equivalent to $R_0(S + \varphi)$.

First, we prove $\forall\alpha\, \Diamond\alpha_S$ within PRA together with $(W, <_0)$-induction. Here and below, quantifiers $\forall\alpha$ are understood as ranging over Gödel numbers of words (under some natural Gödel numbering in S). Binary relation $<_0$ on W is arithmetized in a similar way.

It is sufficient to prove:

$$\mathsf{PRA} \vdash \forall\alpha\, (\forall\beta <_0 \alpha\, \Diamond\beta_S \to \Diamond\alpha_S).$$

The following argument can be formalized in PRA.

Assume $\forall\beta <_0 \alpha\, \Diamond\beta_S$.

- If $\alpha = 0\beta$, then $\Diamond\beta_S$. Since PRA $\vdash R_1(S)$, every Π^0_1-sentence π implies $\Diamond\pi$. Taking $\Diamond\beta_S$ for π we infer $\Diamond\Diamond\beta_S$ and $\Diamond\alpha_S$.
- If $\alpha = (n+1)\beta$, then $\forall k\, \Diamond\alpha[\![k]\!]_S$, because $\alpha[\![k]\!] <_0 \alpha$.

  By Corollary 5.2 (formalizable in PRA),

$$\alpha_S \equiv_n \{\alpha[\![k]\!]_S : k < \omega\}.$$

Therefore $\forall k\, \Diamond \alpha[\![k]\!]_S$ yields $\Diamond \alpha_S$.

Thus, we have proved $\forall \alpha\, \Diamond \alpha_S$. What remains to be seen is that $\forall \alpha\, \Diamond \alpha_S$ implies the consistency of PA. This follows from a well known observation (originally due to Kreisel) that any instance of arithmetical induction follows from EA together with $R_n(\mathsf{EA})$, for an appropriate n. If, for each $n \in \omega$, the theory $S + n_S$ is consistent, then so is PA. In other words, $\forall n\, \Diamond n_S$ implies the consistency of PA.

7 The Worm principle

For any word α, we say that α *is higher than* n if each letter of α exceeds n. Given a word α, consider the following sequence $(\alpha_n)_{n \in \omega}$ of words.

Set $\alpha_0 := \alpha$ and suppose α_k is given. Define α_{k+1} by the following two rules:

- If $\alpha_k = 0\beta$ then $\alpha_{k+1} := \beta$.
- If $\alpha_k = (n+1)\beta$, find the longest (possibly empty) prefix β_0 of β such that β_0 is higher than n. Assume $\beta = \beta_0\gamma$. Then let $\alpha_{k+1} := (n\beta_0)^{k+2}\gamma$.

The *Worm principle* states that, for each α, the sequence α_k terminates in an empty word. A proof of the following theorem is based on the observation that the words α_k are equivalent to the formulas $\alpha[\![k]\!]$ in $\mathbf{RC}^0$ (see [4,3]).

Theorem 7.1 *The Worm principle is true but unprovable in Peano arithmetic. In fact, it is equivalent in* PRA *to the* Σ_1*-reflection* $R_1(\mathsf{PA})$ *for* PA.

The Worm principle can be seen as an analog of the well-known *Hydra battle* principle due to Paris and Kirby. However, it deals with words rather than finite trees. It can also be viewed, modulo some minor details, as a linear version of the so-called *Buchholz hydra battle* which deals with labeled trees. A version of the Worm principle deriving from Buchholz hydra battle has been analyzed by Hamano and Okada [7]. Independently but later, the Worm principle has been found (and baptized in the current form) in [4]. This paper was based on different, provability logical, considerations. A detailed correspondence between the Worm principle and the Hydra battle has been established by Carlucci [5], see also Lee [8].

References

[1] L. Beklemishev and D. Gabelaia. Topological completeness of the provability logic GLP. Preprint arXiv:1106.5693v1 [math.LO], 2011.

[2] L.D. Beklemishev. Provability algebras and proof-theoretic ordinals, I. *Annals of Pure and Applied Logic*, 128:103–123, 2004.

[3] L.D. Beklemishev. Reflection principles and provability algebras in formal arithmetic. *Uspekhi Matematicheskikh Nauk*, 60(2):3–78, 2005. In Russian. English translation in: *Russian Mathematical Surveys*, 60(2): 197–268, 2005.

[4] L.D. Beklemishev. The Worm Principle. In Z. Chatzidakis, P. Koepke, and W. Pohlers, editors, *Lecture Notes in Logic 27. Logic Colloquium '02*, pages 75–95. AK Peters, 2006. Preprint: Logic Group Preprint Series 219, Utrecht University, March 2003.

[5] L. Carlucci. Worms, gaps and hydras. *Mathematical Logic Quarterly*, 51(4):342–350, 2005.
[6] E.V. Dashkov. On a positive fragment of polymodal provability logic GLP. *Matematicheskie Zametki*, 91(3):331–336, 2012. English translation in: *Mathematical Notes* 91(3):318–333, 2012.
[7] M. Hamano and M. Okada. A relationship among Gentzen's proof-reduction, Kirbi-Paris' Hydra game, and Buchholz's Hydra game. *Mathematical Logic Quarterly*, 43(1):103–120, 1997.
[8] Gyesik Lee. A comparison of well-known ordinal notation systems for ε_0. *Annals of Pure and Applied Logic*, 147(1-2):48–70, 2007.

Foundational Issues in Logical Dynamics

Johan van Benthem [1]

Amsterdam and Stanford

A healthy field combines two driving forces, extension of coverage and theoretical reflection – and modal logic is no exception. In this talk, we look at some broader theoretical issues raised by recent developments in dynamic logics of information flow, whose main technical feature is the modal study of definable model change. Our pilot system will be public announcement logic PAL, the base calculus of update with hard information. We identify several general themes behind the particulars of its design. These include model-theoretic issues like the balance of expressive power for the static and dynamic components of the language, and learnability results in the form of preservation theorems. But we also discuss PAL's calculus of reasoning, via correspondence analysis of its axioms as constraints on update functions in domains of inquiry, showing how this leads to two recent extensions: protocol versions where no reduction to the static base language occurs, and substitution-closed core versions of the system. Both moves exemplify more general operations on modal logics.

Next, we consider other dynamic-epistemic logics, and amplify on the above themes, while adding some new ones. First, we discuss event models and product update in DEL, a calculus of model construction that shows analogies with process algebra, but also with the epistemic μ–calculus. Then we turn to dynamic logics for belief change under incoming hard and soft information that upgrade plausibility relations without eliminating worlds. We discuss what logical format best captures the generality needed to avoid a jungle of revision policies. In this setting also, we look at the general phenomenon of defining new static modalities by dynamic considerations.

While all this broadens our view of single-step model transformations, another conspicuous dimension in current research is the global temporal horizon of a process of inquiry, a conversation, or a game, adding procedural information to local updates. We discuss representation results linking local dynamic and global temporal logics, while also pointing at specific model-theoretic issues arising with concrete processes involving iterated hard or soft announcements, and their limit behavior. Here we link up with modal fixed-point logics, though the general situation is far from clear. As a very concrete instance of merging local and global perspectives, we discuss some recent links between dynamic epistemic logics and formal learning theory, where modal models acquire a new meaning as priors pre-encoding learning methods.

---

[1] `http://staff.science.uva.nl/~johan`

Drawing these threads together, we will discuss what it means to be a dynamic logic, looking for postulates on model transformations at the level of abstract model theory.

This survey by no means exhausts the foundational themes behind current dynamic logics of information and agency. We refer to the cited literature for further issues of logic combination and entanglement of modal operators in areas of agency such as games, with new questions about expressiveness and computational complexity for modal languages capturing reasoning about best action in interactive social settings. Likewise, recent work has questioned the use of standard relational models, using more finely-grained neighborhood models generalizing many of the above concerns. A final general theme are interfaces between modal and probabilistic approaches, which are emerging naturally across many of the areas discussed in this lecture.

Our presentation will refer to earlier literature plus talks at AiML 2012, but its main sources are J. van Benthem, 1996, *Exploring Logical Dynamics*, CSLI Publications, Stanford, 2011, *Logical Dynamics of Information and Interaction*, Cambridge University Press, Cambridge, 2012, *Logic in Games*, The MIT Press, Cambridge Mass.

Evidence Logic: A New Look at Neighborhood Structures

Johan van Benthem [1]

Institute for Logic, Language and Computation
University of Amsterdam
Department of Philosophy
Stanford University

David Fernández-Duque [2]

Department of Computer Science and Artificial Intelligence
University of Seville

Eric Pacuit [3]

Department of Philosophy
University of Maryland, College Park

Abstract

Two of the authors (van Benthem and Pacuit) recently introduced *evidence logic* as a way to model epistemic agents faced with possibly contradictory evidence from different sources. For this the authors used neighborhood semantics, where a neighborhood N indicates that the agent has reason to believe that the true state of the world lies in N. A normal belief modality is defined in terms of the neighborhood structure. In this paper we consider four variants of evidence logic which hold for different classes of evidence models. For each of these logics we give a representation theorem using *extended evidence models*, where the belief operator is replaced by a standard relational modality. With this, we axiomatize all four logics, and determine whether each has the finite model property.

Keywords: Neighborhood Models, Logics of Belief, Combining Logics

1 Introduction

Neighborhood models are a generalization of the usual relational semantics for modal logic. In a neighborhood model, each state is assigned a collection of subsets of the set of states. Such structures provide a semantics for both normal

[1] johan.vanbenthem@uva.nl

[2] dfduque@us.es

[3] epacuit@umd.ed

and non-normal modal logics. See [17] for an early discussion of neighborhood semantics for modal logic, and [6,12,7] for modern motivations and mathematical details. Concrete uses of neighborhood models include logics for knowledge [22], players' powers in games [13,14], concurrent PDL [15], and beliefs in the epistemic foundations of game theory [24,10].

In this paper, we study yet another concrete interpretation of neighborhood models. The idea is to interpret the neighborhood functions as describing the evidence that an agent has accepted (in general, we assume the agent accepts different evidence at different states). The agent then uses this evidence to form her beliefs. A dynamic extension of this logic of evidence-based belief was introduced in [21]. The main technical contribution of that paper is a series of of *relative* completeness results in the style familiar to much of the literature on dynamic epistemic logic: validity in a language with dynamic modalities is reduced to a modal language without the dynamic modalities via the validity of so-called *recursion axioms* (see [19] for a general discussion of this technique). In this paper, we continue the project started in [21] focusing on the underlying static logics of belief and evidence.

To do this, we shall consider four variants of evidence logic, which depend on the fundamental assumptions one may make about evidence models. For each of the resulting logics, we prove two main results.

The first is a characterization theorem in terms of *extended evidence models.* These differ from evidence models only in that belief is interpreted via an explicit accessibility relation rather than in terms of the neigbhorhood structure. We show that, up to *p*-morphism, the class of extended evidence models and the class of evidence models are equivalent for each logic considered.

The second is to give a complete deductive calculus for each logic. Here our representation using extended evidence models is crucial, since it permits us to employ familiar techniques from modal logic.

2 A Logic of Evidence and Belief

We start by presenting our formal framework for evidence logics, leaving a more detailed discussion of its motivation to the end of this section. Given a set W of possible worlds or states, one of which represents the "actual" situation, an agent gathers evidence about this situation from a variety of sources. To simplify things, we assume these sources provide *binary* evidence, i.e., subsets of W which (may) contain the actual world. The agent uses this evidence (i.e., collection of subsets of W) to form her beliefs.

The following modal language can be used to describe what the agent believes given her available evidence (cf. [21]).

Definition 2.1 *Let* At *be a fixed set of atomic propositions. Let* $\mathcal{L}$ *be the smallest set of formulas generated by the following grammar*

$$p \mid \neg\varphi \mid \varphi \wedge \psi \mid B\varphi \mid \Box\varphi \mid A\varphi$$

where $p \in \mathsf{At}$. *Additional propositional connectives* $(\wedge, \to, \leftrightarrow)$ *are defined as usual and the duals*[4] *of* $\Box, B$ *and* A *are* $\Diamond, \widehat{B}$ *and* $\widehat{A}$*, respectively.*

The intended interpretation of $\Box\varphi$ is "the agent has evidence for φ" and $B\varphi$ says that "the agents believes that φ is true." We also include the universal modality ($A\varphi$: "φ is true in all states") for technical convenience.[5]

Since we do not assume that the sources of the evidence are jointly consistent (or even that a single source is guaranteed to be consistent and provide *all* the available evidence), the "evidence for" operator ($\Box\varphi$) is not a normal modal operator. That is, the agent may have evidence for φ and evidence for ψ ($\Box\varphi \wedge \Box\psi$) without having evidence for their conjunction ($\neg\Box(\varphi \wedge \psi)$). Of course, both the belief and universal operators are normal modal operators. So, the logical system we study in this paper *combines* a non-normal modal logic with a normal one.

2.1 Neighborhood Models for $\mathcal{L}$

In the intended interpretation of evidence logic, there are many possible states of the world, and the agent possesses evidence for these states in the form of neighborhoods; all other epistemic operators are derived from this neighborhood structure.

Thus we define *evidence models* as follows:

Definition 2.2 *An* **evidence model** *is a tuple* $\mathcal{M} = \langle W, E, V \rangle$*, where* W *is a non-empty set of worlds,* $E \subseteq W \times \wp(W)$ *is an evidence relation,* $V : \mathsf{At} \to \wp(W)$ *is a valuation function. We write* $E(w)$ *for the set* $\{X \mid wEX\}$*. Two constraints are imposted on the evidence sets: For each* $w \in W$*,* $\emptyset \notin E(w)$ *and* $W \in E(w)$*. A* **uniform evidence model** *is an evidence model where* $E(\cdot)$ *is a constant function (each state has the same set of evidence).*

We do not assume that the collection of evidence sets $E(w)$ is closed under supersets. Also, even though evidence pieces are non-empty, their combination through the obvious operations of taking *intersections* need not yield consistent evidence: we allow for disjoint evidence sets, whose combination may lead (and should lead) to trouble. But importantly, even though an agent may not be able to consistently combine *all* of her evidence, there will be maximal collections of admissible evidence that she can safely put together to form *scenarios*:

Definition 2.3 *A* **w-scenario** *is a maximal collection* $\mathcal{X} \subseteq E(w)$ *that has the fip (i.e., the finite intersection property: for each finite subfamily* $\{X_1, \ldots, X_n\} \subseteq \mathcal{X}$*,* $\bigcap_{1 \le i \le n} X_i \neq \emptyset$*). A collection is called a* **scenario** *if it is a w-scenario for some state w.*

Truth of formulas in $\mathcal{L}$ is defined as follows:

Definition 2.4 *Let* $\mathcal{M} = \langle W, E, V \rangle$ *be an evidence model. Truth of a formula* $\varphi \in \mathcal{L}$ *is defined inductively as follows:*

- $\mathcal{M}, w \models p$ *iff* $w \in V(p)$ $\quad$ ($p \in \mathsf{At}$)

---

[4] In other words, $\widehat{B} = \neg B \neg$, and similarly for other operators.

[5] A natural interpretation of $A\varphi$ in the context of this paper is "the agent knows that φ".

- $\mathcal{M}, w \models \neg\varphi$ *iff* $\mathcal{M}, w \not\models \varphi$
- $\mathcal{M}, w \models \varphi \wedge \psi$ *iff* $\mathcal{M}, w \models \varphi$ *and* $\mathcal{M}, w \models \psi$
- $\mathcal{M}, w \models \Box\varphi$ *iff there exists* X *such that* wEX *and for all* $v \in X$, $\mathcal{M}, v \models \varphi$
- $\mathcal{M}, w \models B\varphi$ *for each* w*-scenario* $\mathcal{X}$ *and for all* $v \in \bigcap\mathcal{X}$, $\mathcal{M}, v \models \varphi$
- $\mathcal{M}, w \models A\varphi$ *iff for all* $v \in W$, $\mathcal{M}, v \models \varphi$

The truth set of φ *is the set* $[\![\varphi]\!]_{\mathcal{M}} = \{w \mid \mathcal{M}, w \models \varphi\}$. *The standard logical notions of* **satisfiability** *and* **validity** *are defined as usual.*

Our notion of having evidence for φ need not imply that the agent *believes* φ. In order to believe a proposition φ, the agent must consider *all* her evidence for or against φ. The idea is that each w-scenario represents a maximally consistent theory based on (some of) the evidence collected at w.[6] Note that the definition of truth of the "evidence for" operator builds in monotonicity. That is, the agent has evidence for φ at w provided there is some evidence available at w that implies φ.

The class of evidence models we have described gives the most general setting such an agent may face. However, there are natural additional assumptions one may consider:

Definition 2.5 *An evidence model* $\mathcal{M}$ *is* **flat** *if every scenario on* $\mathcal{M}$ *has non-empty intersection. In the interest of brevity, we may write* **$\flat$-evidence model** *instead of flat evidence model.*

An evidence model $\mathcal{M} = \langle W, E, V\rangle$ *is* **uniform** *if* E *is a constant. In this case, we shall treat* E *as a set (of neighborhoods) rather than a function.*

Flatness and uniformity are natural assumptions which, as we will see, may be captured by adding different axioms to our logic.

2.2 The Logics

We now turn to logics for reasoning about distinct classes of evidence models. Our first observation is that the language $\mathcal{L}$ is sensitive to flatness:

Lemma 2.6 *If* $\mathcal{M}$ *is a* $\flat$*-evidence model, then* $\mathcal{M} \models \Box\varphi \to \widehat{B}\varphi$.

Proof. If $X \in E(w)$ is an evidence set witnessing φ (i.e., $X \subseteq [\![\varphi]\!]_{\mathcal{M}}$), then the singleton $\{X\}$ can be extended to a w-scenario using Zorn's Lemma, which, in flat structures, has non-empty intersection. □

Meanwhile, the formula $\Box\varphi \to \widehat{B}\varphi$ is not valid in general: Consider a uniform evidence model $\mathcal{M}_\infty = \langle W, E, V\rangle$ with domain $W = \mathbb{N}$ and evidence sets $E(w) = \{[N, \infty) \mid N \in \mathbb{N}\}$ for each $w \in W$. The valuation is unimportant, so we may let $V = \varnothing$. Clearly, the only scenario on $\mathcal{M}_\infty$ is all of E, but $\bigcap E = \varnothing$. Hence $\mathcal{M}_\infty \models B\bot$, i.e., $\mathcal{M}_\infty \not\models \widehat{B}\top$; yet $\mathcal{M}_\infty \models \Box\top$ (this formula is universally valid), and we conclude

$$\mathcal{M}_\infty \not\models \Box\top \to \widehat{B}\top.$$

[6] Analogous ideas occur in semantics of conditionals [9,23] and belief revision [3,16].

From this we get the following corollary:

Corollary 2.7 *The logic of evidence models does not have the finite model property, nor does the logic of uniform evidence models.*

Proof. Every finite model is flat, and hence validates $\Box\top \to \widehat{B}\top$; but as we have just shown, this formula is not valid over all uniform evidence models. □

With this in mind, we state a list of axioms and rules for evidence logics:

taut	all propositional tautologies
$\mathsf{S5}_A$	$\mathsf{S5}$ axioms for A
K_B	K axioms for B
$\top$**-evidence**	$\Box\top$
pullout	$\Box\varphi \wedge A\psi \leftrightarrow \Box(\varphi \wedge A\psi)$
universal belief	$A\varphi \to BA\varphi$
$\Box$**-monotonicity**	$\dfrac{\varphi \to \psi}{\Box\varphi \to \Box\psi}$
$\flat$	$\Box\varphi \to \widehat{B}\varphi$
$\bigcirc$**-uniformity**	$\bigcirc\varphi \to A\bigcirc\varphi$ for $\bigcirc = B, \widehat{B}, \Box, \Diamond$
MP	Modus Ponens
$\mathbf{N}_{\bigcirc}$	Necessitation for $\bigcirc = A, B$

We let Log denote the logic which uses all axioms and rules *except* for $\flat$ or the uniformity axioms. The subscripts $\flat, u$ denote the addition of the respective axioms. We will denote derivability in the logic λ by $\vdash_\lambda$, where λ is any one of the four combinations that we may form, that is, $\lambda \in \{\mathsf{Log}, \mathsf{Log}_\flat, \mathsf{Log}_u, \mathsf{Log}_{\flat u}\}$.

The weakest logic Log will be called *general* evidence logic, while $\mathsf{Log}_\flat, \mathsf{Log}_{\flat u}$ will be called *flat logics* and $\mathsf{Log}_u, \mathsf{Log}_{\flat u}$ will be called *uniform logics.* We will write λ*-consistency* for consistency over the logic λ.

2.3 Extended Neighborhood Models for $\mathcal{L}$

The finite model property fails for evidence logic, a fact which may be rather inconvenient. Fortunately, there is a way to sidestep this problem.

In evidence models, the belief operator is interpreted using the neighborhoods by taking intersections of scenarios. There is another natural class of models for the language $\mathcal{L}$ which avoids the use of scenarios. The key idea is to extend a neighborhood model with a relation R that will be used to interpret the belief modality. These models will allow us, later, to employ standard techniques from modal logic to prove completeness.

Definition 2.8 *An* **extended evidence model** *is a structure* $\mathfrak{M} = \langle W, R, E, V\rangle$ *where such that* W *is a set,* R *an (arbitrary) binary relation on* W*,* $E \subseteq W \times \wp(W)$ *is a relation such that* $w\ E\ W$ *and* $w \not\!E\ \varnothing$ *for all* $w \in W$*, and* $V : \mathsf{At} \to \wp(W)$*.*

Truth in an extended evidence model is defined in the standard way: Boolean connectives are as usual, $\mathfrak{M}, w \models A\varphi$ iff for all $v \in W$, $\mathcal{M}, v \models \varphi$; $\mathfrak{M}, w \models B\varphi$ iff for all $v \in W$, if wRv then $\mathcal{M}, v \models \varphi$; and $\mathfrak{M}, w \models \Box\varphi$ iff there exists a $X \subseteq W$ such that $w\ E\ X$ and for all $v \in X$, $\mathcal{M}, v \models \varphi$. We write $[\![\varphi]\!]_{\mathfrak{M}} = \{w \mid \mathfrak{M}, w \models \varphi\}$ for the truth set of φ in $\mathfrak{M}$.

As we shall see below, for every evidence model $\mathcal{M}$ there is an extended evidence model which is naturally associated to it and satisfies the same $\mathcal{L}$-formulas. But first, let us also define a notion of "flatness" for extended evidence models:

Definition 2.9 *Given* $w\ E\ X$*, we will say* X *is* **flat for** w *if there is* $v \in X$ *such that* $w\ R\ v$*. The set* X *is* **flat** *if it is flat for all* $w \in W$ *such that* $w\ E\ X$*. The extended evidence model* $\mathfrak{M}$ *is* **flat** *if every evidence set is flat.*

For uniform extended evidence models, we will also demand that the accessibility relation be independent of the current state:

Definition 2.10 *An extended evidence model* $\mathfrak{M} = \langle W, R, E, V\rangle$ *is* uniform *if both* R *and* E *are constant; that is, given* $w, v, u \in W$ *and* $X \subseteq W$*,* $w\ R\ u$ *if and only if* $v\ R\ u$ *and, likewise,* $w\ E\ X$ *if and only if* $v\ E\ X$*.*

2.4 Motivating the Logics

Having stated our formal framework, we digress from the main technical goal of this paper to discuss the intended interpretation of evidence models in some more detail. The reader who is interested in technical aspects only can safely skip ahead to the next section.

In a number of areas, the need has long been recognized for models that keep track of the reasons, or the *evidence* for beliefs and other informational attitudes (cf. [11,8]). At one extreme, the evidence is encoded as the current range of worlds the agent considers possible. However, this ignores how the agent arrived at this epistemic state. At the other extreme, models record the complete syntactic details of what the agent has learned so far (including the precise formulation and sources for each piece of evidence). In this paper, we explore an intermediate level, viz. neighborhood structures, where evidence is recorded as a family of sets of worlds. In particular, we want to mention three general issues:

We start by making explicit the underlying assumptions motivating the logical framework defined in Section 2.1. Let W be a set of states (or possible worlds) one of which represents the "actual" state. We are interested in a situation where an agent gathers evidence about this state from a variety of sources. To simplify things, we assume these sources provide *binary* evidence, i.e., subsets of W which (may) contain the actual world. The following basic assumptions are implicit in the above definitions:

(i) Sources may or may not be *reliable*: a subset recording a piece of evidence need not contain the actual world. Also, agents need not know which evidence is reliable.

(ii) The evidence gathered from different sources (or even the same source) may be jointly inconsistent. And so, the intersection of all the gathered evidence may be empty.

(iii) Despite the fact that sources may not be reliable or jointly inconsistent, they are all the agent has for forming beliefs. [7]

The *evidential state* of the agent is the set of all propositions (i.e., subsets of W) identified by the agent's sources. In general, this could be any collection of subsets of W; but we do impose some constraints:

- No evidence set is empty (evidence per se is never contradictory),
- The whole universe W is an evidence set (agents know their 'space').

In addition, one might expect a 'monotonicity' assumption:

If the agent has evidence X and $X \subseteq Y$ then the agent has evidence Y.

To us, however, this is a property of propositions supported by evidence, not of the evidence itself. Therefore, we model this feature differently through the definition of our "evidence for" modality ($\Box$).

This brings us to a second point of discussion. The evidence models discussed in this paper do not *directly* represent the agent's sources of evidence. However, the neighborhood models can be used to distinguish between a wide range of evidential situations. Consider the following three evidential states:

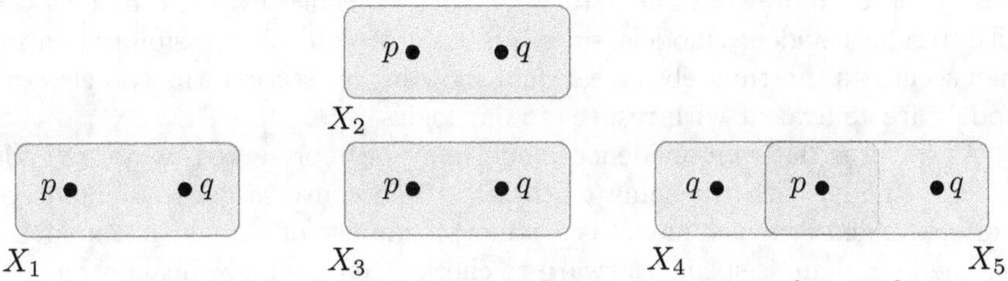

$\mathcal{E}_1 = \{X_1\}$ $\qquad$ $\mathcal{E}_2 = \{X_2, X_3\}$ $\qquad$ $\mathcal{E}_3 = \{X_4, X_5\}$

In each state, the agent has evidence that $p \vee q$ is true (but not both). However, the evidential situation underlying each state is very different. In the first situation, the agent has received the information from a single source (which the agent "trusts") that the actual state satisfies either $p \wedge \neg q$ or $\neg p \wedge q$. In the second situation, the agent has received the same information from two

[7] Modeling sources and agents' *trust* in these is possible – but we will not pursue this here.

different sources (or perhaps the same source reinforced its evidence). The sources agree that either $p \wedge \neg q$ or $\neg p \wedge q$ is true, but disagree about the conditions under which these formulas are true. Finally, the most interesting case is the third one where the agent received information from two sources that agree on the conditions under which $p \wedge \neg q$ is true, but disagree about the conditions needed to make $\neg p \wedge q$ true. In this case, the agent believes p, whereas in the first two situations the agent only believes the weaker proposition that $p \vee q$ is true. Note that our language (see Definition 2.1) can distinguish between the third situation and the first two, but cannot distinguish between the first two evidential situations.

Our final point of discussion concerns the relationship between evidence models and existing modal logics of knowledge and belief based on so-called "plausibility models". [8] The models proposed here are not intended to *replace* plausibility models, but rather to complement them. So, what exactly is the relationship between these two frameworks for modeling beliefs? This question is explored in detail in [21, Section 5], but we only mention the highlights here. Every plausibility model can be transformed into a *uniform* evidence model where the set of evidence are all the downward closed subsets (according to the plausibility ordering). This, in part, motivates our interest in the logic of uniform evidence models. There is also a translation from evidence models to plausibility models (using the well-known definition of a *specialization (pre)-order*). However, it is easy to see that not every evidence model, not even a uniform one, comes from a plausibility model. So, evidence models generalize the standard plausibility models which have been successfully used to represent an agent's knowledge and different flavors of belief. Once again, this shows the additional level of generality provided by neighborhood models. [9]

3 Representation Theorems

It is generally more convenient to work with extended evidence models than with standard evidence models, since it is easier to control accessibility relations than scenarios. Fortunately, as we shall show in this section, the two classes of models are equivalent with respect to our logics.

As it turns out, any evidence model may be represented as an extended evidence model with the same truth sets. The converse does not hold; yet, every extended evidence model is a p-morphic image of an evidence model.

The first claim is straightforward to check given the following construction.

Definition 3.1 *Given an evidence model $\mathcal{M} = \langle W, E, V\rangle$, define an extended evidence model $\mathfrak{M}^* = \langle W, R_E, E, V\rangle$ where $w\ R_E\ v$ if and only if v lies in the intersection of some w-scenario.*

---

[8] A *plausibility* model is a tuple $\langle W, \preceq, V\rangle$ where W is a nonempty set, V is a valuation function and $\preceq$ is a reflexive, transitive and well-founded order on W. We assume the reader is familiar with these well-studied models and the modal languages used to reason about them (see [19] for details and pointers to the relevant literature).

[9] Both transformations extend to ternary world-dependent plausibility relations.

The following result is obvious by the definition of R_E and we present it without proof:

Theorem 3.2 *Given any evidence model $\mathcal{M}$ and any formula φ,*

$$[\![\varphi]\!]_{\mathcal{M}^*} = [\![\varphi]\!]_{\mathcal{M}}.$$

Further, it is immediate that if $\mathcal{M}$ is uniform then so is $\mathcal{M}^*$. Flatness is also preserved by this operation; if $w\ E\ X$, then use Zorn's lemma to extend $\{X\}$ to a w-scenario $\mathcal{X}$. Then, if $\mathcal{M}$ is flat, $\bigcap \mathcal{X} \neq \varnothing$, and hence for any $v \in \bigcap \mathcal{X}$ we have that $w\ R_E\ v \in X$.

Thus every (flat, uniform) evidence model can be represented as a (flat, uniform) extended evidence model. The opposite is true as well, but a bit more subtle. For this, we first recall the definition of a p-morphism:

Definition 3.3 *Given extended evidence models $\mathfrak{M}_1 = \langle W_1, R_1, E_1, V_1\rangle$ and $\mathfrak{M}_2 = \langle W_2, R_2, E_2, V_2\rangle$, we say a function $\pi : W_1 \to W_2$ is a* p-morphism *if the following conditions hold:*

atoms $V_1 = \pi^{-1} V_2$

forth$_R$ *if $w\ R_1\ v$ then $\pi(w)\ R_2\ \pi(v)$*

back$_R$ *if $\pi(w)\ R_2\ u$ then there is $v \in \pi^{-1}(u)$ such that $w\ R_1\ v$*

forth$_E$ *if $w\ E_1\ X$ then there is $Y \subseteq W_2$ such that $\pi(w)\ E_2\ Y$ and $\pi[X] = Y$*

back$_E$ *if $\pi(w)\ E_2\ Y$ then there is X such that $\pi[X] = Y$ and $w\ E_1\ X$*

Then we obtain the following familiar result which we present without proof:

Theorem 3.4 *If π is a p-morphism between extended evidence models $\mathfrak{M}_1$ and $\mathfrak{M}_2$ and φ is any formula, then*

$$[\![\varphi]\!]_{\mathfrak{M}_1} = \pi^{-1}[[\![\varphi]\!]_{\mathfrak{M}_2}].$$

If a surjective p-morphism exists from $\mathfrak{M}_1$ to $\mathfrak{M}_2$, we will write $\mathfrak{M}_1 \gg \mathfrak{M}_2$. Our goal is to show now that, given an extended evidence model $\mathfrak{M}$, there is an evidence model $\mathfrak{M}^+$ such that $\mathfrak{M}^+ \gg \mathfrak{M}$; or, more precisely, $(\mathfrak{M}^+)^* \gg \mathfrak{M}$. Thus an extended evidence model may also be represented as a p-morphic image of an evidence model. The latter evidence model, however, is often much larger.

Definition 3.5 *Given a extended evidence model $\mathfrak{M} = \langle W, R, E, V\rangle$, we define an evidence model $\mathfrak{M}^+ = \langle W^+, E^+, V^+\rangle$ and a map $\pi : W^+ \to W$ as follows:*

(i) *W^+ is the set of all triples $\langle w, X, n\rangle$ such that $X \subseteq W$ and either*
  (a) *X is flat and $n \in \{0, 1\}$ or*
  (b) *X is not flat and $n \in \mathbb{N}$.*

(ii) $\pi(\langle w, X, n\rangle) = w$

(iii) $V^+(p) = \pi^{-1}[V(p)]$

(iv) $E^+(\langle w, X, n\rangle) = \bigcup_{wEY} \mathcal{B}^Y(w) \cup \{W^+\}$, *where*

(a) *if* Y *is flat for* w*, then* $\mathcal{B}^Y(w) = \{U_0, U_1\}$*, with*

$$U_i = \{\langle v, Y, 0\rangle \,|\, \text{there is a } v \text{ with } w\ R\ v \text{ and } v \in Y\} \cup \{\langle v, Y, i\rangle \,|\, v \in Y\};$$

(b) *if* Y *is not flat for* w*, then* $\mathcal{B}^Y(w) = \{V_n\}_{n\in\mathbb{N}}$*, where*

$$V_N = \{\langle v, Y, n\rangle \mid v \in Y, n \geq N\}.$$

As a simple example, consider two extended evidence models, $\mathfrak{M}_1 = \langle W, R, E, V\rangle$ and $\mathfrak{M}_2 = \langle W, R', E, V\rangle$, so that the two are very similar, differing only on the accessibility relation. Suppose that $W = \{w\}$, $w\ R\ w$ but $R' = \varnothing$, and $E(w) = \{W\}$. The valuations are not too important for this example, so we may assume $V \equiv \varnothing$.

Then, $\mathfrak{M}_1$ is flat, but $\mathfrak{M}_2$ is not. Thus we have that $\mathfrak{M}_1^+ = \langle W_1^+, E_1^+, V_1^+\rangle$, where W_1^+ consists of two copies of w ($\langle w, W, 0\rangle$ and $\langle w, W, 1\rangle$, but let us call them w_0, w_1 for simplicity). Both points have the same evidence sets, namely $\{w_0\}$ and $\{w_0, w_1\}$; thus the only scenario is $\big\{\{w_0\}, \{w_0, w_1\}\big\}$ which has intersection $\{w_0\}$.

Meanwhile, $\mathfrak{M}_2^+$ consists of countably many copies of w (of the form $\langle w, W, n\rangle$, but let us call them w_n) and is isomorphic to the model $\mathcal{M}_\infty$ used in the proof of Corollary 2.7. As we saw then, the only scenario on $\mathfrak{M}_2^+$ has empty intersection.

More generally, it is always the case that flat, finite extended evidence models give rise to finite models, and uniform extended evidence models give rise to uniform evidence models:

Lemma 3.6 *If* $\mathfrak{M}$ *is flat and finite, then* $\mathfrak{M}^+$ *is finite. Further, if* $\mathfrak{M}$ *is uniform, then so is* $\mathfrak{M}^+$

Proof. If all evidence sets are flat, we have $W^+ = W \times \wp(W) \times \{0, 1\}$, which is a finite set (provided W is finite).

The second claim is easy to check using the definition of E^+. □

The following key lemma shows a close relationship between the accessibility relations on $\mathfrak{M}$ and $(\mathfrak{M}^+)^*$.

Lemma 3.7 *Let* $\mathfrak{M} = \langle W, R, E, V\rangle$ *be a extended evidence model with associated evidence model* $\mathfrak{M}^+$*. Let* $\alpha \in W^+$ *and* $v \in W$ *be arbitrary.*

Then, $\pi(\alpha)\ R\ v$ *if and only if there is* $\beta \in \pi^{-1}(v)$ *such that* $\alpha\ R_{E^+}\ \beta$*.*

The proof can be found in Appendix A. With this one can check that π is a p-morphism and $(\mathfrak{M}^+)^* \gg \mathfrak{M}$:

Theorem 3.8 *If* $\mathfrak{M}$ *is an extended evidence model and* π *is the associated map for* $\mathfrak{M}^+$*, then* π *is a surjective p-morphism between* $(\mathfrak{M}^+)^*$ *and* $\mathfrak{M}$*.*

Proof. The atoms clause holds by the definition of V^+ and the clauses for R hold by Lemma 3.7.

It remains to check that forth_E and back_E hold as well.

To check that forth_E holds, note that neighborhoods of $\alpha = \langle w, X, n\rangle$ are all of the form

$$N = \{\langle v, Y, m\rangle : v \in Y \text{ and } w\ E\ Y\};$$

but then, $\pi[N] = Y$ and hence $\pi(\alpha) = w\ E\ Y$.

Meanwhile, back_E holds because if $w\ E\ Y$ and $\pi(\alpha) = w$, we note that $\alpha\ E\ N$, where

$$N = Y \times \{Y\} \times \{0\}$$

and $\pi[N] = Y$. □

Corollary 3.9 *Given a extended evidence model $\mathfrak{M}$ with map $\pi : W^+ \to W$ and a formula $\varphi \in \mathcal{L}_0$, $[\![\varphi]\!]_{\mathfrak{M}^+} = \pi^{-1}[[\![\varphi]\!]_{\mathfrak{M}}]$.*

Proof. Immediate from Theorems 3.8 and 3.4. □

From this we obtain the following, very useful result:

Theorem 3.10 *A set of formulas Φ is satisfiable on a (flat, uniform) extended evidence model if and only if it is satisfiable on a (flat, uniform) evidence model.*

4 Completeness

In view of the previous section, it suffices to build extended evidence models for consistent sets of formulas in order to prove completeness. Extended evidence models are much closer to standard semantics of modal logic than evidence models and hence we can apply familiar techniques.

We assume that all formulas are in 'negation-normal form' in which negations are only applied to propositional variables. Of course, in order to do this we must allow for dual operators $(\Diamond, \widehat{B}, \widehat{A})$ to appear. Henceforth, we assume all formulas are in this form unless we explicitly indicate otherwise. Note that this convention is only for the sake of exposition, as we allow negation in our calculus and dual operators are really abbreviations.

We denote the normal-form negation of φ by $\sim\varphi$. The **closure of** φ, denoted $cl(\varphi)$ contains all subformulas of φ, is closed under normal-from negation (if $\psi \in cl(\varphi)$ then $\sim \psi \in cl(\varphi)$), and we stipulate that $\top, \Box\top, \Diamond\top$ and $B\top$ are all in $cl(\varphi)$. If Ω is a set of formulas, then $cl(\Omega) = \bigcup\{cl(\varphi) \mid \varphi \in \Omega\}$. As usual, the states in our canonical extended evidence model are maximally consistent sets of formulas (which we call types).

In the remainder of this paper, by an *evidence logic* we will mean exclusively an element of $\{\mathsf{Log}, \mathsf{Log}_\flat, \mathsf{Log}_u, \mathsf{Log}_{\flat u}\}$.

Definition 4.1 *Let Ω be a set of formulas and λ an evidence logic. An (Ω, λ)-type is a maximal λ-consistent subset of $cl(\Omega)$.*

A set of formulas Γ is a **λ-type** *if it is a (Γ, λ)-type, i.e., Γ is λ-consistent and for each $\psi \in cl(\Gamma)$, either $\psi \in \Gamma$ or $\sim\psi \in \Gamma$.*

Note that (Ω, λ)-types may be finite or infinite, depending on whether Ω is. We denote the set of (Ω, λ)-types by $\mathrm{type}_\lambda(\Omega)$. Given an (Ω, λ)-type Φ, define Φ^A as $\{\psi \mid A\psi \in \Phi\}$, and similarly for the other modalities.

Of course, in a given model all points must satisfy the same universal formulas, so it is convenient to consider the collection of all such types.

Given a set of formulas Γ, if $\lambda \in \{\mathsf{Log}, \mathsf{Log}_\flat\}$, define $\Gamma^\lambda = A\Gamma^A \cup \widehat{A}\Gamma^{\widehat{A}}$; if $\lambda \in \{\mathsf{Log}_u, \mathsf{Log}_{\flat u}\}$, define

$$\Gamma^\lambda = \bigcup\{\bigcirc\Gamma^\bigcirc : \bigcirc = A, \widehat{A}, B, \widehat{B}, \Box, \Diamond\}.$$

Definition 4.2 *Given a λ-type Φ, we define*

(i) $\text{type}^A_\lambda(\Phi)$ *to be the set of all (Φ, λ)-types Ψ such that $\Psi^\lambda = \Phi^\lambda$;*

(ii) $\text{type}^B_\lambda(\Phi) = \{\Psi \in \text{type}^A_\lambda(\Phi) \mid \Phi^B \subseteq \Psi\}$.

Now we need to define evidence sets on our extended evidence models.

Definition 4.3 *Given a λ-type Φ with $\Box\alpha \in \Phi$, we define the* α-neighborhood of Φ *as*

$$\mathcal{N}^\alpha_\lambda(\Phi) = \{\Psi \in \text{type}^A_\lambda(\Phi) \mid \alpha \in \Psi\}.$$

We are now ready to define a canonical extended evidence model :

Definition 4.4 *Let Φ be a λ-type. The* **λ-canonical extended evidence model for** Φ *is the extended evidence model $\mathfrak{M}_\lambda(\Phi) = \langle W^\Phi_\lambda, E^\Phi_\lambda, R^\Phi_\lambda, V^\Phi_\lambda\rangle$ where*

(i) $W^\Phi_\lambda = \text{type}^A_\lambda(\Phi)$

(ii) *For each $p \in \mathsf{At} \cap \Gamma$, $\Gamma \in V^\Phi_\lambda(p)$ iff $p \in \Gamma$,*

(iii) $E^\Phi_\lambda(\Gamma) = \{\mathcal{N}^\alpha_\lambda(\Gamma) \mid \Box\alpha \in \Gamma\}$.

(iv) $\Gamma\, R^\Phi_\lambda\, \Delta$ *iff* $\Delta \in \text{type}^B_\lambda(\Gamma)$ *(i.e., $\Gamma^B \subseteq \Delta$).*

Note that our 'canonical' extended evidence models are not unique as they depend on the type Φ we wish to satisfy and the logic λ we wish to work in; in this sense they could be considered 'semicanonical'. Considering different extended evidence models is unavoidable, given that our language includes a universal modality and it is impossible to satisfy all types on a single model. However, we are also taking advantage of this in order to obtain the finite (extended) model property directly, since canonical extended evidence models for finite types are themselves finite.

Let us observe that canonical models for uniform types are uniform:

Lemma 4.5 *If $\lambda \in \{\mathsf{Log}_u, \mathsf{Log}_{\flat u}\}$ and Φ is λ-consistent, then the model $\mathfrak{M}_\lambda(\Phi)$ is uniform.*

Proof. Given $\Psi \in W^\Phi_\lambda$, we have that both the accessible states and the neighborhoods of Ψ depend only on Ψ^λ, which is constant amongst all of W^Φ_λ. □

Our goal will now be to prove a version of the Truth Lemma which will imply that the canonical extended evidence model s we have defined satisfy the right formulas. For this, we need some preliminaries. We start by gathering the main results about the axiom system that we use to prove the Truth Lemma. We defer proofs to the Appendix.

Lemma 4.6 *Let Γ, Δ be sets of formulas and λ an evidence logic.*

(i) *Suppose that $\varphi, A\Gamma, \widehat{A}\Delta \vdash_\lambda \psi$. Then, $\Box\varphi, A\Gamma, \widehat{A}\Delta \vdash_\lambda \Box\psi$.*

(ii) *Suppose that $\Phi, A\Gamma, \widehat{A}\Delta \vdash_\lambda \psi$. Then, $B\Phi, A\Gamma, \widehat{A}\Delta \vdash_\lambda B\psi$.*

It will be very useful to draw some dual conclusions from the above result.

Lemma 4.7 *Let Γ be a set of formulas and λ an evidence logic.*

(i) *If Γ is λ-consistent, $\alpha \in \Gamma^\Box$ and $\delta \in \Gamma^\Diamond$, then $\{\alpha, \delta\} \cup \Gamma^\lambda$ is λ-consistent.*

(ii) *If Γ is λ-consistent and $\psi \in \Gamma^{\widehat{B}}$, then $\{\psi\} \cup \Gamma^B \cup \Gamma^\lambda$ is λ-consistent.*

(iii) *If $\lambda \in \{\mathsf{Log}_\flat, \mathsf{Log}_{\flat u}\}$, Γ is λ-consistent and $\psi \in \Gamma^\Box$, then $\{\psi\} \cup \Gamma^B \cup \Gamma^\lambda$ is λ-consistent.*

The next lemmas show the key steps in the proof of the Truth Lemma.

Lemma 4.8 *Let Γ be a (Φ, λ)-type.*

(i) *If $\Box\alpha, \Diamond\beta \in \Gamma$, there is $\Delta \in \mathcal{N}^\alpha_\lambda(\Gamma)$ with $\beta \in \Delta$.*

(ii) *If $\lambda \in \{\mathsf{Log}_\flat, \mathsf{Log}_{\flat u}\}$ and $\alpha \in \Gamma^\Box$, then $\mathcal{N}^\alpha_\lambda(\Gamma) \cap \mathrm{type}^B_\lambda(\Gamma)$ is non-empty.*

(iii) *Suppose that $\widehat{B}\alpha \in \Gamma$. Then, there is $\Delta \in \mathrm{type}^B_\lambda(\Gamma)$ with $\beta \in \Delta$.*

The proofs of the above three lemmas can be found in Appendix B. For what follows, it will be convenient to flesh out some of the model-theoretic consequences of Lemma 4.8.

Lemma 4.9 *Given a λ-type Φ with a λ-canonical extended evidence model $\mathfrak{M}_\lambda(\Phi) = \langle W^\Phi_\lambda, R^\Phi_\lambda, E^\Phi_\lambda, V^\Phi_\lambda \rangle$,*

(i) *for each $\Gamma \in W^\Phi_\lambda$, $\Gamma\ E^\Phi_\lambda\ W^\Phi_\lambda$;*

(ii) *if $\Box\alpha \in \Phi$, $\mathcal{N}^\alpha_\lambda(\Phi)$ is non-empty; and*

(iii) *if Φ is flat and $\Gamma\ E^\Phi_\lambda\ X$, there is $\Delta \in X$ such that $\Gamma\ R^\Phi_\lambda\ \Delta$.*

Proof. Suppose Γ is a (Φ, λ)-type and let $\mathfrak{M}_\lambda(\Phi)$ be as above.

(i) Recall that we stipulated that $\Box\top, \top \in cl(\Phi)$, and these formulas are valid so it follows that $\top \in \Delta$ for all $\Delta \in W^\Phi_\lambda$; thus $\mathcal{N}^\top_\lambda(\Phi) = W^\Phi_\lambda$. Similarly, $\Box\top \in \Gamma$, so that $\Gamma\ E^\Phi_\lambda\ W^\Phi_\lambda$, as claimed.

(ii) Put $\beta = \top$ in Lemma 4.8(i).

(iii) Immediate from the definition of the accessibility relation and neighborhoods on canonical extended evidence models and Lemma 4.8(ii).

□

Putting everything together, we have:

Corollary 4.10 *If λ is an evidence logic and Φ is λ-consistent, then $\mathfrak{M}_\lambda(\Phi)$ is a extended λ-evidence model.*

Proof. From Lemma 4.9(i) we see that, given $\Gamma \in W^\Phi_\lambda$, $\Gamma\ E\ W^\Phi_\lambda$, while from Lemma 4.9(ii) we obtain $\Gamma\ \not{E}\ \varnothing$. It follows that $\mathfrak{M}_\lambda(\Phi)$ is a extended evidence model.

If further we have that $\lambda \in \{\mathsf{Log}_\flat, \mathsf{Log}_{\flat u}\}$, by Lemma 4.9(iii), every neighborhood on $\mathfrak{M}_\lambda(\Phi)$ is flat, i.e., $\mathfrak{M}_\lambda(\Phi)$ is flat.

Finally, if $\lambda \in \{\mathsf{Log}_u, \mathsf{Log}_{\flat u}\}$, we use Lemma 4.5 to see that the model $\mathfrak{M}_\lambda(\Phi)$ is uniform. □

We are now ready to prove our own version of the 'Truth Lemma':

Proposition 4.11 (Truth Lemma) *For every formula $\psi \in cl(\Phi)$ and every set $\Gamma \in \mathrm{type}^A_\lambda(\Phi)$,*

$$\psi \in \Gamma \Rightarrow \Gamma \in [\![\psi]\!]_{\mathfrak{M}_\lambda(\Phi)}.$$

Proof. The proof is by induction on the structure of ψ. The only interesting cases are the modalities $\Box$ and B (and their duals).

Suppose first that $\Box\psi \in \Gamma$, and let $X = \mathcal{N}^\psi_\lambda(\Gamma)$. By definition, we have $\psi \in \Theta$ for every $\Theta \in X$, and by the induction hypothesis this implies that for each $\Theta \in X$, $\Theta \in [\![\psi]\!]_{\mathfrak{M}_\lambda(\Phi)}$. Thus, X is a neighborhood of Γ ($X \in E^\Phi(\Gamma)$) with $X \subseteq [\![\psi]\!]_{\mathfrak{M}_\lambda(\Phi)}$, and so, $\Gamma \in [\![\Box\psi]\!]_{\mathfrak{M}_\lambda(\Phi)}$, as desired.

Now, assume that $\Diamond\psi \in \Gamma$. Let X be any neighborhood of Γ. Then $X = \mathcal{N}^\alpha(\Gamma)$ for some α. Hence, $\Box\alpha \in \Gamma$ and $\Diamond\psi \in \Gamma$. By Lemma 4.8.1, there is $\Theta \in \mathcal{N}^\alpha(\Gamma)$ with $\psi \in \Theta$. By the induction hypothesis, $\Theta \in [\![\psi]\!]_{\mathfrak{M}_\lambda(\Phi)}$, and, since X was arbitrary, we conclude that $\Gamma \in [\![\Diamond\psi]\!]_{\mathfrak{M}_\lambda(\Phi)}$.

Assume that $B\psi \in \Gamma$. Now, suppose that $\Gamma R^\Phi_\lambda \Delta$. Then, by definition $\psi \in \Delta$, which by the induction hypothesis implies $\Delta \in [\![\psi]\!]_{\mathfrak{M}_\lambda(\Phi)}$. Thus, $\Gamma \in [\![B\psi]\!]_{\mathfrak{M}_\lambda(\Phi)}$

Assume that $\widehat{B}\psi \in \Gamma$. By Lemma 4.8.3, $\psi \in \Theta$ for some $\Theta \in \mathrm{type}^B_\lambda(\Gamma)$. But then $\Gamma\ R^\Phi_\lambda\ \Theta$, and thus $\Gamma \in [\![\widehat{B}\psi]\!]_{\mathfrak{M}_\lambda(\Phi)}$. □

4.1 Proof of the Main Theorem

With the work developed in previous sections, we are ready to state and prove our main results.

Theorem 4.12 *Each evidence logic λ is sound and strongly complete both for the class of λ-evidence models and the class of extended λ-evidence models.*

Proof. Let Φ be λ-consistent and let $\mathfrak{M}_\lambda(\Phi)$ be the canonical extended evidence model for Φ. By Corollary 4.10, $\mathfrak{M}_\lambda(\Phi)$ is a λ-extended evidence model, and Φ can be extended to a (Φ, λ)-type Γ. By Proposition 4.11, given $\varphi \in \Gamma$, $\Gamma \in [\![\varphi]\!]_{\mathfrak{M}_\lambda(\Phi)}$, so that Φ is satisfiable in $\mathfrak{M}_\lambda(\Phi)$.

By Lemma 3.9, $(\mathfrak{M}(\Phi))^+$ is a λ-model also satisfying Φ. □

Theorem 4.13 *The flat evidence logics $\mathsf{Log}_\flat$, $\mathsf{Log}_{\flat u}$ have the finite model property. In fact, if φ has length ℓ, then it has a λ-model of size at most*[10] $2^\ell \cdot 2^{2^\ell} \cdot 2$.

Non-flat evidence logics do not have the finite model property, but they do have the finite extended evidence model property.

Proof. If φ is λ-consistent, we can extend it to a $(\{\varphi\}, \lambda)$-type Φ and let $\mathfrak{M}$ be the canonical extended λ-evidence model for Φ. Clearly, $\mathfrak{M}$ has at most 2^ℓ states, where ℓ is the length of φ, and thus it is finite.

[10] This bound could be improved to $2^\ell \cdot \ell \cdot 2$.

If φ is further $\flat$-consistent, then the model $\mathfrak{M}^+$ is finite as well, and it has at most $2^\ell \cdot 2^{2^\ell} \cdot 2$ points. □

Corollary 4.14 *Both flat and general evidence logic are decidable.*

Our analysis does not yield the exact computational complexity of deciding validity or satisfiability. Note however that the general evidence logic is a conservative extension of K_u (i.e., K with a universal modality) and hence the validity problem is EXPTIME-hard; the same should be true of $\mathsf{Log}_\flat$. Meanwhile, we expect the uniform logics to be much simpler, as they extend $\mathsf{S5}$ rather than K, but do not have a specific conjecture on their complexity.

5 Language Extensions

The logical systems studied in the previous sections are interesting in their own right as they combine features of both normal and non-normal modal logics. An additional appealing feature of the logical systems studied here and in [21] is the fresh new interpretation of neighborhood structures. In particular, interpreting neighborhoods as bodies of evidence suggests a number of new and interesting modalities beyond the usual repertoire studied in modal neighborhood logics. In this section, we briefly explore this rich landscape of extensions of our basic language, and point out some resulting problems of axiomatization.

Two operators that immediately suggest themselves in anticipation of the dynamic extensions found in [21] are conditional versions of our evidence and belief operators. The conditional belief operator ($B^\varphi\psi$: "the agent believes ψ conditional on φ") is well-known, but is given a new twist in our setting. Some of the agent's current evidence may be inconsistent with φ (i.e., disjoint with $[\![\varphi]\!]_\mathcal{M}$). If one is restricting attention to situations where φ is true, then such inconsistent evidence must be "ignored". Here is how we do this:

Definition 5.1 (Relativized maximal overlapping evidence.) *Suppose that $X \subseteq W$. Given a collection $\mathcal{X}$ of subsets of W (i.e., $\mathcal{X} \subseteq \wp(W)$), the relativization of $\mathcal{X}$ to X is the set $\mathcal{X}^X = \{Y \cap X \mid Y \in \mathcal{X}\}$. We say that a collection $\mathcal{X}$ of subsets of W has the* **finite intersection property relative to** *X (X-f.i.p.) if, for each $\{X_1, \ldots, X_n\} \subseteq \mathcal{X}^X$, $\bigcap_{1\leq i\leq n} X_i \neq \emptyset$. We say that $\mathcal{X}$ has the* **maximal** *X***-f.i.p.** *if $\mathcal{X}$ has X-f.i.p. and no proper extension $\mathcal{X}'$ of $\mathcal{X}$ has the X-f.i.p.*

To simplify notation, when X is the truth set of formula φ, we write "maximal φ-f.i.p." for "maximal $[\![\varphi]\!]_\mathcal{M}$-f.i.p." and "$\mathcal{X}^\varphi$" for "$\mathcal{X}^{[\![\varphi]\!]_\mathcal{M}}$". Now we define a natural notion of conditional belief:

- $\mathcal{M}, w \models B^\varphi\psi$ iff for each maximal φ-f.i.p. $\mathcal{X} \subseteq E(w)$, for each $v \in \bigcap \mathcal{X}^\varphi$, $\mathcal{M}, v \models \psi$

This notion suggests a logical investigation beyond what we have provided so far. First of all, strikingly, $B\varphi \to B^\psi\varphi$ is not valid. One can compare this to the failure of monotonicity for antecedents in conditional logic. In our more general setting which allows inconsistencies among accepted evidence, we also see that even the following variant is not valid: $B\varphi \to (B^\psi\varphi \vee B^{\neg\psi}\varphi)$. To see this, consider an evidence model with $E(w) = \{X_1, Y_1, X_2, Y_2\}$ where the sets are defined as follows:

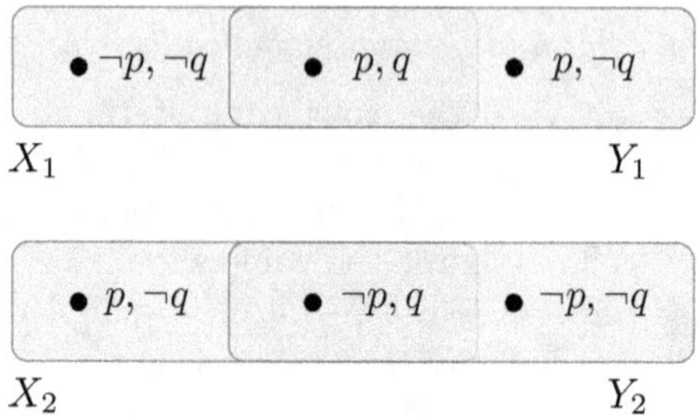

Then, $\mathcal{M}, w \models Bq$; however, $\mathcal{M}, w \not\models B^{p}q \vee B^{\neg p}q$. This is interesting as it is valid on *connected* plausibility models for conditional belief (cf. [2] for a complete modal logic of conditional belief on such models). Extending our completeness proof from the previous section requires new ideas as there is no obvious way to interpret conditional belief operators on our auxiliary "extended evidence models".

The conditional evidence operator $\Box^{\varphi}\psi$ ("the agent has evidence that ψ is true conditional on φ being true") is a new modality on neighborhoods. We say that $X \subseteq W$ is **consistent (compatible) with** φ if $X \cap [\![\varphi]\!]_{\mathcal{M}} \neq \emptyset$. Truth of conditional evidence can then be defined as follows:

- $\mathcal{M}, w \models \Box^{\varphi}\psi$ iff there exists an evidence set $X \in E(w)$ which is consistent with φ such that for all worlds $v \in X \cap [\![\varphi]\!]_{\mathcal{M}}$, $\mathcal{M}, v \models \varphi$.

In particular, if there is no evidence consistent with φ, then $\Box^{\varphi}\psi$ is false. This, in turn means that $\Box^{\varphi}\psi$ is not equivalent to $\Box(\varphi \to \psi)$. [11] Indeed, a simple bisimulation argument shows that no definition exists for conditional evidence in the language with absolute evidence and belief. The conditional evidence operators satisfy the monotonicity inference rule (from $\varphi \to \psi$ infer $\Box^{\alpha}\varphi \to \Box^{\alpha}\psi$) and, for example, the axiom scheme $\Box^{\alpha}\varphi \to \Box^{\alpha}\alpha$; however, a complete logic will be left for future work.

Both of these operators are special cases of more general modalities that were discovered in [21] as static counterparts to natural dynamic modalities of adding or removing evidence.

For the conditional evidence operator $\Box^{\varphi}\psi$, we can require the witnessing evidence set to be "compatible" with a sequence of formulas. Let $\overline{\varphi} = (\varphi_1, \ldots, \varphi_n)$ be a finite sequence of formulas. We say that a set of states X is **compatible with** $\overline{\varphi}$ provided that, for each formula φ_i, $X \cap [\![\varphi_i]\!]_{\mathcal{M}} \neq \emptyset$. Then, we define the general conditional evidence operator $\Box^{\alpha}_{\overline{\varphi}}\psi$ as follows:

- $\mathcal{M}, w \models \Box^{\alpha}_{\overline{\varphi}}\psi$ iff there is $X \in E(w)$ compatible with $\overline{\varphi}$, α such that $X \cap [\![\alpha]\!]_{\mathcal{M}} \subseteq [\![\psi]\!]_{\mathcal{M}}$.

---

[11] To see this, consider a model where φ is false at all worlds. Then $\Box^{\varphi}\psi$ is also false at all worlds, but $\Box(\varphi \to \psi)$ will be true at all worlds, since $\varphi \to \psi$ is true everywhere.

The conditional belief operator can be generalized in two ways. The first way is to incorporate the above notion of compatibility with a sequence of formulas. The intended interpretation of $B^{\varphi}_{\overline{\gamma}}\psi$ is "the agent believes χ conditional on φ assuming compatibility with each of the γ_i". The formal definition is a straightforward generalization of the earlier definition of $B^{\varphi}\psi$. A maximal f.i.p. set $\mathcal{X}$ is **compatible with** a sequence of formulas $\overline{\varphi}$ provided for each $X \in \mathcal{X}$, X is compatible with $\overline{\varphi}$. Then,

- $\mathcal{M}, w \models B^{\alpha}_{\overline{\varphi}}\psi$ iff for each maximal α-f.i.p. $\mathcal{X}$ compatible with $\overline{\varphi}$, we have that $\bigcap \mathcal{X}^{\alpha} \subseteq [\![\psi]\!]_{\mathcal{M}}$.

The second generalization focuses on the conditioning operation. Note that $B^{\varphi}\psi$ may be true at a state w without having any w-scenarios that imply φ (i.e., there is no w-scenario $\mathcal{X}$ such that $\bigcap \mathcal{X} \subseteq [\![\varphi]\!]_{\mathcal{M}}$). A more general form of conditioning is $B^{\varphi,\alpha}\psi$ where "the agent believe ψ, after having settled on α and conditional on φ". Formally,

- $\mathcal{M}, w \models B^{\varphi,\psi}\chi$ iff for all maximally φ-compatible sets $\mathcal{X} \subseteq E(w)$, if $\bigcap \mathcal{X} \cap [\![\varphi]\!]_{\mathcal{M}} \subseteq [\![\psi]\!]_{\mathcal{M}}$, then $\bigcap \mathcal{X} \cap [\![\varphi]\!]_{\mathcal{M}} \subseteq [\![\chi]\!]_{\mathcal{M}}$.

Note that $B^{+\varphi}$ can be defined as $B^{\varphi,\top}$.

This splitting of notions shows that neighborhood structures are a good vehicle for exploring finer epistemic and doxastic distinctions than those found in standard relational models for modal logic. Moreover, they support interesting matching forms of reasoning beyond standard axioms. Validities include interesting connections between varieties of conditional belief such as: $B^{\varphi}\psi \to B(\varphi \to \psi)$ and $B(\varphi \to \psi) \to B^{\top,\varphi}\psi$.

At the same time, this richness means that new techniques may be needed in the logical analysis of this richer form of neighborhood semantics. Model-theoretically, we need stronger notions of bisimulation and -morphism matching these more expressive languages, while proof-theoretically, we need to lift our earlier completeness technique to this setting.

6 Conclusion and Future Work

There are two main contributions in this paper. First, our completeness theorems solve an important problem left open in [21] contributing to the general study of basic evidence logic and its dynamics. Second, in doing so, we develop a new perspective on neighborhood models that suggests extensions of the usual systems. For the modal logician, the pleasant surprise of our language extensions is that there is a lot of well-motivated new modal structure to be explored on these simple models.

Clearly, many open problems remain. These start at the base level of the motivations for our framework, touched upon lightly earlier on. For instance, our framework still needs to be related to other modal logics of evidence [18,5], justification [1] and argumentation [4]. Here are a few more specific technical avenues for future research:

- *Further interpretations*: By imposing additional constraints on the evidence relations (i.e., the neighborhood functions), we get evidence models that are topological spaces. Can we give a spatial interpretation to our belief operator and the new modalities discussed in Section 5? This suggests new, richer spatial logics for reasoning about topological spaces (cf. [20]).
- *Computational complexity*: Neighborhood logics are often NP-complete, while basic modal logics on relational models are often Pspace-complete. What about mixtures of the two? By Corollary 4.14 we know that the validity problem for flat and general evidence logic is decidabile. What is the precise complexity?
- *Extended model theory*: Combining existing notions of bisimulation for relational models and neighborhood models takes us only so far. What new notions of bisimulation directly on evidence models will match our extended modal languages? And in this setting, can the key representation method of this paper (cf. Theorem 3.8) be extended to deal with such richer languages, perhaps starting from richer extended evidence models carrying plausibility relations?
- *Extended proof theory*: How can we axiomatize the richer modal logic of neighborhood models that arises in Section 5? Will modifying existing techniques, including our approach in this paper, still work, or do we need a new style of analysis of canonical models?

Appendix

A Proof from Section 3

Lemma A.1 (Lemma 3.7) *Let $\mathfrak{M} = \langle W, R, E, V\rangle$ be a extended evidence model with associated evidence model $\mathfrak{M}^+$. Let $\alpha \in W^+$ and $v \in W$ be arbitrary.*

Then, $\pi(\alpha)\ R\ v$ if and only if there is $\beta \in \pi^{-1}(v)$ such that $\alpha\ R_{E^+}\ \beta$.

Proof. Assume that $\alpha \in W^+$ and $\pi(\alpha) = w$. We first claim that every α-scenario on $\mathfrak{M}^+$ is of the form $\mathcal{B}^Y(\alpha) \cup \{W^+\}$.

For this, suppose that $\alpha\ E^+\ X$ and $\alpha\ E^+\ Y$ with $X \cap Y \neq \varnothing$ and $X, Y \neq W^+$. This means that there is a $\langle v, Z, n\rangle \in X \cap Y$. By inspecting the definition of E^+, this implies $X, Y \in \mathcal{B}^Z(\alpha)$. Hence, any collection $\mathcal{X} \subseteq E^+(\alpha)$ with the finite intersection property must be contained in $\mathcal{B}^Z(\alpha) \cup \{W^+\}$ for some Z.

Meanwhile, it is easy to see that for every $Z \subseteq W$, $\mathcal{B}^Z(\alpha) \cup \{W^+\}$ already has the fip (by case-by-case inspection), hence every scenario is of this form.

Now, if Y is not flat for w, one can check that $\bigcap \mathcal{B}^Y(w) = \varnothing$. If Y is flat for w, then all elements of $\bigcap \mathcal{B}^Y(w)$ are of the form $\langle v, Y, 0\rangle$ with $w\ R\ v$ and $v \in Y$. In either case, $\pi[\bigcap \mathcal{B}^Y(w)] \subseteq R(w)$. It follows that, whenever $\alpha\ R_{E^+}\ \beta$, then $w\ R\ \pi(\beta)$.

It is also straightforward to see that

$$\bigcap \mathcal{B}^W(w) = R(w) \times \{W\} \times \{0\},$$

so that $\pi[\bigcap \mathcal{B}^W(w)] = R(w)$. Thus, if $w \ R \ v$ then there is an α-scenario $\mathcal{U}$ (namely $\mathcal{U} = \mathcal{B}^W(w) \cup \{W^+\}$) with $\langle v, W, 0\rangle \in \bigcap \mathcal{U}$, so that $\alpha \ R_{E^+} \ \beta = \langle v, W, 0\rangle$. □

B Proofs from Section 4

Lemma B.1 (Lemma 4.6) *Let Γ, Δ be sets of formulas.*

(i) *Suppose that $\varphi, A\Gamma, \widehat{A}\Delta \vdash_\lambda \psi$. Then, $\Box\varphi, A\Gamma, \widehat{A}\Delta \vdash_\lambda \Box\psi$.*

(ii) *Suppose that $\Phi, A\Gamma, \widehat{A}\Delta \vdash_\lambda \psi$. Then, $B\Phi, A\Gamma, \widehat{A}\Delta \vdash_\lambda B\psi$.*

Without loss of generality we can assume Γ, Δ, Φ to be finite, since in general we have that $\Theta \vdash_\lambda \alpha$ if and only if for some finite $\Theta' \subseteq \Theta$, $\Theta' \vdash_\lambda \alpha$.

Proof. Note that

$$\bigwedge A\Gamma \wedge \bigwedge \widehat{A}\Delta \leftrightarrow A\left(\bigwedge A\Gamma \wedge \bigwedge \widehat{A}\Delta\right)$$

is derivable in S5, so we can replace $A\Gamma, \widehat{A}\Delta$ by a single formula $A\gamma$.

(i) If $\varphi, A\gamma \vdash_\lambda \psi$, then

$$\vdash_\lambda \varphi \wedge A\gamma \to \psi.$$

By the monotonicity rule

$$\vdash_\lambda \Box(\varphi \wedge A\gamma) \to \Box\psi.$$

Applying the pullout axiom we get

$$\vdash_\lambda \Box\varphi \wedge A\gamma \to \Box\psi,$$

that is,

$$\Box\varphi, A\gamma \vdash_\lambda \Box\psi.$$

(ii) If $\Phi, A\gamma \vdash_\lambda \psi$, since B is a normal operator, we also have

$$B\Phi, BA\gamma \vdash_\lambda B\psi.$$

But since $A\alpha \to BA\alpha$ is an axiom, we get

$$B\Phi, A\gamma \vdash_\lambda B\psi.$$

□

Below, recall that, if $\lambda \in \{\mathsf{Log}, \mathsf{Log}_\flat\}$, we defined $\Gamma^\lambda = A\Gamma^A \cup \widehat{A}\Gamma^{\widehat{A}}$, while if $\lambda \in \{\mathsf{Log}_u, \mathsf{Log}_{\flat u}\}$,

$$\Gamma^\lambda = \bigcup\{\bigcirc\Gamma^\bigcirc : \bigcirc = A, \widehat{A}, B, \widehat{B}, \Box, \Diamond\}.$$

Lemma B.2 (Lemma 4.7) *Let Γ be a set of formulas.*

(i) *If Γ is λ-consistent, $\alpha \in \Gamma^\Box$ and $\delta \in \Gamma^\Diamond$, then $\{\alpha, \delta\} \cup \Gamma^\lambda$ is λ-consistent.*

(ii) *If Γ is λ-consistent and $\psi \in \Gamma^{\widehat{B}}$, then $\{\psi\} \cup \Gamma^B \cup \Gamma^\lambda$ is λ-consistent.*

(iii) *If λ is flat, Γ is λ-consistent and $\psi \in \Gamma^{\Box}$, then $\{\psi\} \cup \Gamma^{B} \cup \Gamma^{\lambda}$ is λ-consistent.*

Proof. Assume Γ is λ-consistent.

Note that we have $\Gamma^{\lambda} \equiv A\Gamma^{\lambda}$ independently of λ; over the non-uniform case, this follows from the axiom $\widehat{A}\varphi \to A\widehat{A}\varphi$, while over the uniform case (i.e., if $\lambda \in \{\mathsf{Log}_u, \mathsf{Log}_{\flat u}\}$), this is because $\bigcirc\Gamma^{\bigcirc}$ is equivalent to $A\bigcirc\Gamma^{\bigcirc}$ over λ for any modality $\bigcirc$.

As before, we may assume Γ is finite. Thus we may replace Γ^{λ} by a single formula $A\gamma$ equivalent to $\bigwedge\Gamma^{\lambda}$ over λ.

(i) Suppose otherwise. Then, we would have

$$\alpha, A\gamma \vdash_{\lambda} \neg\delta.$$

Thus by Lemma 4.6.1,

$$\Box\alpha, A\gamma \vdash_{\lambda} \Box\neg\delta,$$

so that

$$\{\neg\Box\neg\delta, \Box\alpha, A\gamma\}$$

is λ-inconsistent. But this is a subset of Γ, contradicting our assumption that Γ is λ-consistent.

(ii) Suppose $\psi \in \Gamma^{\widehat{B}}$. If the claim was false, then we would have

$$\Gamma^{B}, A\gamma \vdash_{\lambda} \neg\psi,$$

and hence, by Lemma 4.6.2,

$$B\Gamma^{B}, A\gamma \vdash_{\lambda} B\neg\psi.$$

Thus

$$\{\neg B\neg\psi, A\gamma\} \cup B\Gamma^{B}$$

would be λ-inconsistent. If $\psi \in \Gamma^{\widehat{B}}$, this contradicts the consistency of Γ.

(iii) If $\psi \in \Gamma^{\Box}$, the claim follows from item 2 above using the axiom $\Box\psi \to \widehat{B}\psi$.

$\Box$

Lemma B.3 (Lemma 4.8) *Let Γ be a (Φ, λ)-type.*

(i) *If $\Box\alpha, \Diamond\beta \in \Gamma$, there is $\Delta \in \mathcal{N}_{\lambda}^{\alpha}(\Gamma)$ with $\beta \in \Delta$.*

(ii) *If $\lambda \in \{\mathsf{Log}_{\flat}, \mathsf{Log}_{\flat u}\}$ and $\alpha \in \Gamma^{\Box}$, then $\mathcal{N}_{\lambda}^{\alpha}(\Gamma) \cap \mathrm{type}_{\lambda}^{B}(\Gamma)$ is non-empty.*

(iii) *Suppose that $\widehat{B}\alpha \in \Gamma$. Then, there is $\Delta \in \mathrm{type}_{\lambda}^{B}(\Gamma)$ with $\beta \in \Delta$.*

Proof. Let Γ be a (Φ, λ)-type (for some λ-consistent set of formulas Φ).

(i) Since Γ is λ-consistent and $\Box\alpha, \Diamond\beta \in \Gamma$, we have, by Lemma 4.7.1, that

$$\{\alpha, \beta\} \cup \Gamma^{\lambda}$$

is λ-consistent. This can be extended to a Φ-type Δ. Then, by definition we have $\Delta \in \mathcal{N}_{\lambda}^{\alpha}(\Gamma)$ and $\beta \in \Delta$, as desired.

(ii) By Lemma 4.7.3, if Γ is λ-consistent and λ is flat, then $\{\alpha\} \cup \Gamma^B \cup \Gamma^\lambda$ is λ-consistent as well, and hence can be extended to a Φ-type Δ. Obviously,

$$\Delta \in \mathcal{N}_\lambda^\alpha(\Gamma) \cap \mathrm{type}_\lambda^B(\Gamma).$$

(iii) By Lemma 4.7.2, since Γ is λ-consistent, we have that

$$\{\beta\} \cup \Gamma^B \cup \Gamma^\lambda$$

is λ-consistent as well. Thus, it can be extended to a Φ-type Δ. Evidently, $\Delta \in \mathrm{type}_\lambda^B(\Gamma)$ and $\beta \in \Delta$.

□

References

[1] Artemov, S. and E. Nogina, *Introducing justification into epistemic logic*, Journal of Logic and Computation **15** (2005), pp. 1059–1073.

[2] Board, O., *Dynamic interactive epistemology*, Games and Economic Behavior **49** (2004), pp. 49–80.

[3] Gärdenfors, P., "Knowledge in Flux: Modeling the Dynamics of Epistemic States," Bradford Books, MIT Press, 1988.

[4] Grossi, D., *On the logic of argumentation theory*, in: W. van der Hoek, G. Kaminka, Y. Lesperance, M. Luck and S. Sandip, editors, *Proceedings of the 9th International Conference on Autonomous Agents and Multiagent Systems (AAMAS'10)*, 2010, pp. 409–416.

[5] Halpern, J. and R. Pucella, *A logic for reasoning about evidence*, Journal of AI Research **26** (2006), pp. 1–34.

[6] Hansen, H. H., "Monotonic Modal Logic," Master's thesis, Universiteit van Amsterdam (ILLC technical report: PP-2003-24) (2003).

[7] Hansen, H. H., C. Kupke and E. Pacuit, *Neighbourhood structures: Bisimilarity and basic model theory*, Logical Methods in Computer Science **5** (2009), pp. 1–38.

[8] Horty, J. F., *Reasons as defaults*, Philosophers' Imprint **7** (2007), pp. 1–28.

[9] Kratzer, A., *What* must *and* can *must and can mean*, Linguistics and Philosophy **1** (1977), pp. 337–355.

[10] Lismont, L. and P. Mongin, *A non-minimal but very weak axiomatization of common belief*, Artificial Intelligence **70** (1994), pp. 363–374.

[11] List, C. and F. Dietrich, *Reasons for (prior) belief in bayesian epistemology* (2012), unpublished manuscript.

[12] Pacuit, E., *Neighborhood semantics for modal logic: An introduction* (2007), ESSLLI 2007 course notes (`ai.stanford.edu/~epacuit/classes/`).

[13] Parikh, R., *The logic of games and its applications*, in: M. Karpinski and J. v. Leeuwen, editors, *Topics in the Theory of Computation*, Annals of Discrete Mathematics 24, Elsevier, 1985 pp. 111 – 140.

[14] Pauly, M., *A modal logic for coalitional power in games*, Journal of Logic and Computation **12** (2002), pp. 149–166.

[15] Peleg, D., *Concurrent dynamic logic*, Journal of the ACM **34** (1987), pp. 450–479. URL `http://doi.acm.org/10.1145/23005.23008`

[16] Rott, H., "Change, Choice and Inference: A Study in Belief Revision and Nonmonotonic Reasoning," Oxford University Press, 2001.

[17] Segerberg, K., "An Essay in Classical Modal Logic," Filosofisska Stuier, Uppsala Universitet, 1971.

[18] Shafer, G., "A Mathematical Theory of Evidence," Princeton University Press, 1976.

[19] van Benthem, J., "Logical Dynamics of Information and Interaction," Cambridge University Press, 2011.

[20] van Benthem, J. and G. Bezhanishvili, *Modal logics of space*, in: M. Aiello, I. Pratt-Hartmann and J. van Benthem, editors, *Handbook of Spatial Reasoning*, 2007, pp. 217–298.

[21] van Benthem, J. and E. Pacuit, *Dynamic logics of evidence-based beliefs*, Studia Logica **99** (2011), pp. 61–92.

[22] Vardi, M., *On epistemic logic and logical omniscience*, in: J. Halpern, editor, *Theoretical Aspects of Reasoning about Knowledge: Proceedings of the 1986 Conference* (1986), pp. 293–305.

[23] Veltman, F., *Prejudices, presuppositions and the theory of conditionals*, in: J. Groenendijk and M. Stokhof, editors, *Amsterdam Papers in Formal Grammar* (1976), pp. 248–281.

[24] Zvesper, J., "Playing with Information," Ph.D. thesis, ILLC University of Amsterdam Dissertation Series DS-2010-02 (2010).

Distributive Substructural Logics as Coalgebraic Logics over Posets

Marta Bílková, Rostislav Horčík [1]

Institute of Computer Science, Academy of Sciences of the Czech Republic

Jiří Velebil [1]

Faculty of Electrical Engineering, Czech Technical University in Prague

Abstract

We show how to understand frame semantics of distributive substructural logics coalgebraically, thus opening a possibility to study them as coalgebraic logics. As an application of this approach we prove a general version of Goldblatt-Thomason theorem that characterizes definability of classes of frames for logics extending the distributive Full Lambek logic, as e.g. relevance logics, many-valued logics or intuitionistic logic. The paper is rather conceptual and does not claim to contain significant new results. We consider a category of frames as posets equipped with monotone relations, and show that they can be understood as coalgebras for an endofunctor of the category of posets. In fact, we adopt a more general definition of frames that allows to cover a wider class of distributive modal logics. Goldblatt-Thomason theorem for classes of resulting coalgebras for instance shows that frames for axiomatic extensions of distributive Full Lambek logic are modally definable classes of certain coalgebras, the respective modal algebras being precisely the corresponding subvarieties of distributive residuated lattices.

Keywords: Substructural logics, frame semantics, coalgebras, coalgebraic logic, Goldblatt-Thomason theorem.

1 Introduction

Modal logics are coalgebraic, the relational frames of classical modal logics can be seen as Set coalgebras for the powerset functor. Given an endofunctor T on Set, a conceptually clear setting of classical coalgebraic logic of T-coalgebras can be based on an adjunction called *logical connection*, linking categories Set and BA of sets and Boolean algebras [5,6] and capturing syntax and semantics of the propositional part of the language. Such connection can be "lifted" to a connection between categories of T-coalgebras and Boolean algebras with

[1] The authors acknowledge the support of the grant No. P202/11/1632 of the Czech Science Foundation. email: `bilkova@cs.cas.cz`, `horcik@cs.cas.cz`, `velebil@math.feld.cvut.cz`

operators, which is in general "almost" an adjunction, capturing syntax and semantics of the modal part of the language. From certain properties of the lifted connection one automatically obtains soundness, completeness and expressivity of the modal language. One can also explore the connection to obtain the Goldblatt-Thomason definability theorem for classes of T-coalgebras for a reasonable class of Set functors [26].

In this paper, lead by a motivation to approach distributive substructural logics in a coalgebraic way, we use an (enriched) logical connection [27,31] between categories Pos of posets and DL of distributive lattices. We consider a general language of distributive lattices with operators, including the usual language of substructural logics as an instance. We start with requiring no additional axioms the operators should satisfy (not even the residuation laws), obtaining coalgebras for a certain endofunctor T on posets as semantics of this language. As an application of this setting we prove Goldblatt-Thomason definability theorem for classes of T-coalgebras. Classes of T-coalgebras definable by additional axioms of distributive substructural logics then precisely correspond to frames for these logics as surveyed and studied in [32]. Distributive modal logics have been treated coalgebraically before [7,29]. We see the main novelty of this paper in the fact that we use a weaker assumption than a duality of the category of algebras and certain topological spaces, thus resulting in non-topological coalgebras as semantics of distributive modal or substructural logics.

A leading example of a logic, semantics of which we want to cover, is the distributive full Lambek calculus **dFL** [15] in the following language

$$\varphi ::= p \mid \varphi \wedge \varphi \mid \varphi \vee \varphi \mid \varphi \otimes \varphi \mid \varphi \to \varphi \mid \varphi \leftarrow \varphi \mid e \tag{1}$$

where p ranges through a given poset of atomic propositions, $\wedge$ and $\vee$ are tied together by a distributive law, and the remaining four connectives $\otimes$, $\leftarrow$, $\to$, e satisfy additional equational axioms as, for example, the residuation laws. The algebraic semantics of **dFL** are residuated lattices.

We want to take the stance that $\wedge$ and $\vee$ are the only *propositional connectives* of the language, while the remaining four constructions $\otimes$, $\leftarrow$, $\to$, e are *modalities*. To prove that the study of relational models of the above language falls into the realm of *coalgebraic modal logic* it will be essential to start with a weaker setting, with *no* additional requirements on the modalities, apart from being monotone and preserving $\wedge$ or $\vee$, i.e. being operators over distributive lattices.

As it turns out, the natural environment for giving models of the above language is the one of *posets* and *monotone* relations. Namely, a relational model will consist of a poset $\mathscr{W}$ and four monotone relations $P_\otimes$, $P_\leftarrow$, $P_\to$ and P_e on $\mathscr{W}$. For example, $P_\otimes$ will be a monotone relation (i.e., a monotone map $P_\otimes : \mathscr{W}^{op} \times \mathscr{W}^{op} \times \mathscr{W} \longrightarrow \mathbb{2}$, where $\mathbb{2}$ is the two-element chain) that we will denote by $P_\otimes : \mathscr{W} \times \mathscr{W} \longrightarrow\!\!\!\!\!\!\!+\ \ \mathscr{W}$. Hence the "arity" of $P_\otimes$ mirrors the arity of the "modality" $\otimes$. Analogously, P_e will be a monotone relation of the form $P_e : \mathbb{1} \longrightarrow\!\!\!\!\!\!\!+\ \ \mathscr{W}$ where $\mathbb{1}$ denotes the one-element preorder. Hence P_e will appear as a "nullary" monotone relation, mirroring the fact that the "modality" e is nullary. We prove that the above quintuple $\mathbb{W} = (\mathscr{W}, P_\otimes, P_\leftarrow, P_\rightarrow, P_e)$ can be seen as a *coalgebra* for an endofunctor T of the category Pos of posets and monotone maps.

The reasoning does not change much if we incorporate slightly more general languages of the form

$$\varphi ::= p \mid \varphi \wedge \varphi \mid \varphi \vee \varphi \mid \heartsuit(\varphi_0, \dots, \varphi_{n-1}) \mid (\varphi_0, \dots, \varphi_{l-1}) \multimap \psi \mid {\sim}\varphi \qquad (2)$$

where p ranges through a poset At of atomic propositions, the connectives $\wedge$, $\vee$ are tied together by the distributive law, $\heartsuit$ is an n-ary *fusion-like* connective, $\multimap$ is an l-ary *implication-like* connective, and $\sim$ is a *negation-like* connective. These connectives are required to interact with $\wedge$ and $\vee$ in the sense that the following equalities are valid for each $0 \leq i \leq n$:

$$\heartsuit(\dots, \varphi_i \vee \varphi_i' \dots) = \heartsuit(\dots, \varphi_i, \dots) \vee \heartsuit(\dots, \varphi_i', \dots)$$
$$(\dots, \varphi_i \vee \varphi_i', \dots) \multimap \psi = ((\dots, \varphi_i, \dots) \multimap \psi) \wedge ((\dots, \varphi_i', \dots) \multimap \psi)$$
$$(\varphi_0, \dots, \varphi_{l-1}) \multimap (\psi \wedge \psi') = ((\varphi_0, \dots, \varphi_{l-1}) \multimap \psi) \wedge ((\varphi_0, \dots, \varphi_{l-1}) \multimap \psi')$$
$${\sim}(\varphi_1 \vee \varphi_2) = {\sim}\varphi_1 \wedge {\sim}\varphi_2$$

In slogans: $\heartsuit$ should preserve $\vee$ pointwise, $\multimap$ should pointwise transform $\vee$ in its premises to $\wedge$, and it should preserve $\wedge$ in its conclusion, $\sim$ should transform $\vee$ into $\wedge$.[2]

We will prove that:

(i) Relational models of the language (2) are precisely the coalgebras for an endofunctor $T : \mathsf{Pos} \longrightarrow \mathsf{Pos}$. Moreover, the construction of T copies the syntax of the "modalities" $\heartsuit$, $\multimap$, $\sim$ in (2).

(ii) The algebraic semantics of (2) will be given by a variety $\mathsf{DL}_{\heartsuit,\multimap,\sim}$ of distributive lattices with operators $\heartsuit$, $\multimap$ and $\sim$.

(iii) It is essential to start with *no* requirements on the modalities in order to obtain a coalgebraic description. Any additional equational requirements on the modalities $\heartsuit$, $\multimap$ and $\sim$ will result in a *modally definable* class of T-coalgebras. We characterize modally definable classes in the spirit of *Goldblatt-Thomason Theorem* known from the classical modal logic.

As an illustration, we explain how various classes of frames for languages of the type (2) can be perceived as modally definable. In particular, we cover all the frames for the distributive substructural logics as studied in [32], namely:

- The class of frames modelling the *distributive full Lambek calculus* is modally definable by the equations for residuated distributive lattices. The modalities are $\otimes$, $\rightarrow$, $\leftarrow$ and e.

---

[2] The language above, in its greatest generality, allows for finitely many connectives of each kind, all of various arities. In order not to make the notation too heavy, we will assume that there is just one connective of each kind in our signature. The results for the general case are straightforward generalisations of results for our simplification.

- The class of frames modelling the *intuitionistic logic* is modally definable by the equations for Heyting algebras. The modalities are $\otimes$ (coinciding with $\wedge$) and $\rightarrow$.
- The class of frames modelling *relevance logic* is modally definable. The modalities are $\otimes$, $\rightarrow$, $\leftarrow$, e and $\sim$.

Related work: Using relational models on posets for modelling semantics of various nonclassical logics goes back at least to the work of Routley and Meyer [34], and Dunn, see [10,11,12] or [32] for an overview. We see the novelty of our approach in the fact that we can systematically work with such frames as coalgebras, hence one has a canonical notion of a frame morphism as morphism of corresponding coalgebras. [3]

The original Goldblatt-Thomason theorem for modal logics [23] characterizes modally definable classes of Kripke *frames*. For positive modal logic it was proved in [9]. A definability theorem for classes of models is due to Venema [38]. Our version of the theorem is an analogue of coalgebraic Goldblatt-Thomason theorem for Set coalgebras [26, Theorem 3.15(2.)]. Possibilities to generalize the theorem to coalgebras over measurable spaces have been explored in [30]. Coalgebraic Goldblatt-Thomason theorem for classes of *models* can be found in [21] and [26, Theorem 3.15(1.)]. A Goldblatt-Thomason theorem for classes of intuitionistic frames appeared in the thesis [33], a definability theorem for classes of intuitionistic models appeared in [22].

Our approach relates to, but also, due to the coalgebraic formulation, differs from extensive work relating algebraic and frame (or topological) semantics of modal and substructural logics, using dualities and discrete dualities for distributive lattices [17,18,19], distributive lattices with operators [20,35,36,25,29], or posets [13], most of it using canonical extensions: in contrast to this approach we do not use a dual equivalence of distributive lattices and certain topological spaces, a weaker kind of adjunction between DL and posets, called logical connection, is enough. The frames, and thus the coalgebras we consider are not topological as those obtained in [29], [7] or [1], they can however be seen as non-topological analogues of those — the map from algebras to frames can be factored through topological frames.

Organisation of the paper: Section 2 is devoted to fixing the terminology and notation for monotone relations. In Section 3 we briefly recall how the *semantics of the propositional part of coalgebraic logic* is captured by an adjunction of a special kind, called *logical connection*. Relational *frames as coalgebras* are introduced in Section 4. *Complex algebras* and *canonical frames* are studied in Sections 5 and 6. Our main result: the *modal definability theorem* is proved in Section 7. We illustrate this result by examples of distributive *full Lambek calculus*, *relevance logic*, etc. We hint at possible generalizations of our approach in Section 8.

---

[3] The usual notion of morphism for substructural frames is different — it requires equalities $a = f(x), b = f(y)$ in the back condition in 4.6. The same notion of a frame morphism as ours in the special case of frames for fuzzy logics is given in [8].

Remark on the notation we use: We work with posets and monotone relations as with categories enriched over the two-element chain $\mathbb{2}$, see Section 2. Therefore our formulas for manipulation monotone relations use the structure of the complete Boolean algebra $\mathbb{2}$ and are to be computed there. We think that the notation will become convenient in future generalizations to enriched categories, see Section 8.

Due to space limitations we have omitted some of the proofs. §

2 Preliminaries

Recall that a *poset* $\mathscr{W}$ is a set W equipped with a reflexive, transitive and anti-symmetric relation $\leq$. Instead of writing $x \leq x'$ we will often write $\mathscr{W}(x,x') = 1$ (and writing $\mathscr{W}(x,x') = 0$, if $x \leq x'$ does not hold). This is in compliance with the fact that a poset $\mathscr{W}$ can be seen as a small category *enriched* in the two-element chain $\mathbb{2}$. Although we will not use any machinery of enriched category theory explicitly, we find the above notation convenient in the view of further generalizations, see Section 8 below.

An *opposite* $\mathscr{W}^{op}$ of the poset $\mathscr{W}$ has the same set of elements as $\mathscr{W}$, but we put $\mathscr{W}^{op}(x,x') = \mathscr{W}(x',x)$.

Recall further that a *monotone map* $f : \mathscr{W}_1 \longrightarrow \mathscr{W}_2$ consists of an assignment $x \mapsto fx$ such that, for any x and x', the inequality $\mathscr{W}_1(x,x') \leq \mathscr{W}_2(fx,fx')$ holds in $\mathbb{2}$. The poset of all monotone maps from $\mathscr{W}_1$ to $\mathscr{W}_2$, with the order defined pointwise, is denoted by $[\mathscr{W}_1,\mathscr{W}_2]$. A *product* $\mathscr{W}_1 \times \mathscr{W}_2$ of posets $\mathscr{W}_1$, $\mathscr{W}_2$ is an order on the pairs of elements, defined pointwise. We denote by $\mathscr{W}^n$ the product of n-many copies of $\mathscr{W}$ with itself, writing $\mathscr{W}^0 = \mathbb{1}$ — the one-element poset.

Given posets $\mathscr{W}_1$ and $\mathscr{W}_2$, a *monotone relation from* $\mathscr{W}_1$ *to* $\mathscr{W}_2$, denoted by

$$R : \mathscr{W}_1 \longrightarrow\!\!\!\!\!\!\!+\;\; \mathscr{W}_2$$

is a monotone map of the form $R : \mathscr{W}_1^{op} \times \mathscr{W}_2 \longrightarrow \mathbb{2}$. We write $R(x,x') = 1$ to denote that x is related to x'. In what follows we will omit the adjective 'monotone' and speak just of relations. A relation of the form

$$R : \mathscr{W}^n \longrightarrow\!\!\!\!\!\!\!+\;\; \mathscr{W}$$

is called an *n-ary relation* on $\mathscr{W}$, where $n \geq 0$. For $n = 0$ we obtain

$$R : \mathbb{1} \longrightarrow\!\!\!\!\!\!\!+\;\; \mathscr{W}$$

and it is easy to see that such a relation corresponds to an *upperset* of $\mathscr{W}$, i.e., the set $U = \{x \mid Rx = 1\}$ has the property: if $x \in U$ and $x \leq x'$, then $x' \in U$.

Relations compose in the usual way: the composite of the relations $R : \mathscr{W}_1 \longrightarrow\!\!\!\!\!\!\!+\;\; \mathscr{W}_2$ $\quad S : \mathscr{W}_2 \longrightarrow\!\!\!\!\!\!\!+\;\; \mathscr{W}_3$ is a relation $S \cdot R : \mathscr{W}_1 \longrightarrow\!\!\!\!\!\!\!+\;\; \mathscr{W}_3$ given by

§ The full version of the paper including missing proofs can be found at `http://math.feld.cvut.cz/velebil/research/papers.html`.

the formula

$$S \cdot R(x,z) = \bigvee_y S(y,z) \wedge R(x,y)$$

For every poset $\mathscr{W}$, the *identity relation* $id_{\mathscr{W}} : \mathscr{W} \longrightarrow\!\!\!\!\!\!\!\!+\quad \mathscr{W}$ is defined by putting $id_{\mathscr{W}}(x,x') = 1$ iff $x \leq x'$. Hence it is consistent to write $\mathscr{W}$ instead of $id_{\mathscr{W}}$.

It is easy to see that the above composition is associative and that it has identity relations as units, hence we obtain a category (enriched in posets) of posets and relations. The above definitions are specializations of the theory of *profunctors* (also *distributors*, or, *modules*), known from enriched category theory. See, for example, [37] for more details.

3 The logical connection

The semantics of the propositional part of the language, i.e., of the language

$$\varphi ::= p \mid \varphi \wedge \varphi \mid \varphi \vee \varphi \tag{3}$$

where p ranges through a poset At of atomic propositions and $\wedge$ and $\vee$ are tied by the distributive law, will be given by a *logical connection* of the category Pos of *posets* and *monotone maps* and the category DL of *distributive lattices* and *lattice morphisms*. The logical connection

$$\mathit{Stone} \dashv \mathit{Pred} : \mathsf{Pos}^{op} \longrightarrow \mathsf{DL} \tag{4}$$

is given by the two-element chain 2 as a *schizophrenic object*. Recall how the above connection works (we refer to [31] or [27] for more details on logical connections):

(i) *Pred* sends a poset $\mathscr{W}$ to the distributive lattice $([\mathscr{W},2],\cap,\cup)$ of uppersets on $\mathscr{W}$. A monotone map f is sent to $[f,2] : U \mapsto U \cdot f$.

For a poset $\mathscr{W}$, the distributive lattice $\mathit{Pred}(\mathscr{W})$ is to be considered as the "distributive lattice of truth-distributions on $\mathscr{W}$".

(ii) For a distributive lattice $\mathscr{A}$, $\mathit{Stone}(\mathscr{A})$ is the poset $\mathsf{DL}(\mathscr{A},2)$ of *prime filters* on $\mathscr{A}$. The mapping $\mathit{Stone}(h)$ is given by composition: a prime filter F is sent to the prime filter $F \cdot h$.

The poset $\mathit{Stone}(\mathscr{A})$ is the "Stone space" of the distributive lattice $\mathscr{A}$.

(iii) The unit $\eta_{\mathscr{A}} : \mathscr{A} \longrightarrow [\mathsf{DL}(\mathscr{A},2),2]$ is the lattice homomorphism sending x in $\mathscr{A}$ to the upperset of all prime filters on $\mathscr{A}$ that contain x.

(iv) The counit $\varepsilon_{\mathscr{W}} : \mathscr{W} \longrightarrow \mathsf{DL}([\mathscr{W},2],2)$ is the monotone map sending x in $\mathscr{W}$ to the prime filter of those uppersets on $\mathscr{W}$ that contain x.

The semantics of the propositional language (3) is given by the logical connection (4), together with another adjunction

$$F \dashv U : \mathsf{DL} \longrightarrow \mathsf{Pos} \tag{5}$$

where U denotes the obvious forgetful functor and F sends a poset $\mathscr{X}$ to the *free distributive lattice* on $\mathscr{X}$. More in detail, the semantics is given as follows:

(i) Fix a poset At of atomic propositions. The distributive lattice $F(\mathsf{At})$ is then the *Lindenbaum-Tarski* algebra of formulas.

(ii) Observe that $U(\mathit{Pred}(\mathscr{W})) = [\mathscr{W}, 2]$, for every poset $\mathscr{W}$. Hence, due to the adjunction $F \dashv U$, monotone maps of the form $\mathsf{val} : \mathsf{At} \longrightarrow [\mathscr{W}, 2]$ are in bijective correspondence with lattice morphisms $\|{-}\|_{\mathsf{val}} : F(\mathsf{At}) \longrightarrow \mathit{Pred}(\mathscr{W})$.

Of course, as the notation suggests, the monotone map val is the *valuation* of atomic propositions, assigning to every p the upperset $\mathsf{val}(p)$ of those x's in $\mathscr{W}$, where p is valid. The lattice homomorphism $\|{-}\|_{\mathsf{val}}$ is then the free extension of the valuation val. It can be described inductively as follows:

$$\|p\|_{\mathsf{val}} = \mathsf{val}(p),\ \|\varphi_1 \wedge \varphi_2\|_{\mathsf{val}} = \|\varphi_1\|_{\mathsf{val}} \cap \|\varphi_2\|_{\mathsf{val}},\ \|\varphi_1 \vee \varphi_2\|_{\mathsf{val}} = \|\varphi_1\|_{\mathsf{val}} \cup \|\varphi_2\|_{\mathsf{val}}$$

We will later add more connectives (fusion-like, implication-like and negation-like) but we are going to consider them as *modal operators* on distributive lattices. In fact, as we will see, such extension of the language will yield extensions of the above two functors *Pred* and *Stone*.

4 Relational frames as coalgebras

We define structures that we call (relational) frames for the language of the type (2). Frames will consist of a poset of states and various relations reflecting the syntax of "modalities" of the language, compare to frames in [32]. We prove that frames are exactly the coalgebras for a certain endofunctor of the category of posets.

Notation 4.1 We will introduce the following "vector" conventions: for a relation $P : \mathscr{W}^n \longrightarrow\!\!\!\!\!\!\!+\ \mathscr{W}$ we will write $P(\vec{x}; x)$ instead of $P(x_0, \ldots, x_{n-1}; x)$. For $P : \mathscr{W} \longrightarrow\!\!\!\!\!\!\!+\ (\mathscr{W}^{op})^l \times \mathscr{W}$ we will write $P(x; \vec{y}, z)$ instead of $P(x; y_0, \ldots, y_{l-1}, z)$. Analogously we will write $\mathscr{W}_2(\vec{a}, f\vec{x})$ instead of $\mathscr{W}_2(a_0, fx_0) \wedge \cdots \wedge \mathscr{W}_2(a_{n-1}, fx_{n-1})$, etc.

Definition 4.2 *A* relational frame *for the language (2) is a quadruple* $\mathbb{W} = (\mathscr{W}, P_\heartsuit, P_\multimap, P_\sim)$, *consisting of a poset* $\mathscr{W}$, *and relations*

$$P_\heartsuit : \mathscr{W}^n \longrightarrow\!\!\!\!\!\!\!+\ \mathscr{W}, \quad P_\multimap : \mathscr{W} \longrightarrow\!\!\!\!\!\!\!+\ (\mathscr{W}^l)^{op} \times \mathscr{W}, \quad P_\sim : \mathscr{W} \longrightarrow\!\!\!\!\!\!\!+\ \mathscr{W}^{op}$$

A morphism *from* $\mathbb{W}_1 = (\mathscr{W}_1, P^1_\heartsuit, P^1_\multimap, P^1_\sim)$ *to* $\mathbb{W}_2 = (\mathscr{W}_2, P^2_\heartsuit, P^2_\multimap, P^2_\sim)$ *is a monotone map* $f : \mathscr{W}_1 \longrightarrow \mathscr{W}_2$ *such that the following three equations hold:*

$$P^2_\heartsuit(\vec{a}; fy) = \bigvee_{\vec{x}} \mathscr{W}_2(\vec{a}, f\vec{x}) \wedge P^1_\heartsuit(\vec{x}; y) \tag{6}$$

$$P^2_\multimap(fx; \vec{b}, c) = \bigvee_{\vec{y}, z} \mathscr{W}_2(\vec{b}, f\vec{y}) \wedge \mathscr{W}_2(fz, c) \wedge P^1_\multimap(x; \vec{y}, z) \tag{7}$$

$$P^2_\sim(fx; b) = \bigvee_{y} \mathscr{W}_2(b, fy) \wedge P^1_\sim(x; y) \tag{8}$$

We write $f : \mathbb{W}_1 \longrightarrow \mathbb{W}_2$ *to indicate that* f *is a morphism of relational frames.*

Remark 4.3 We have not defined semantics in a relational frame yet, but the following intuitions about the "meaning" of the individual relations $P_{\heartsuit}$, $P_{\multimap}$ and $P_{\sim}$ on $\mathscr{W}$ might be useful (see Notation 4.1).

(i) $P_{\heartsuit}(\vec{x}; y) = 1$ holds, only if $\vec{x} \Vdash \vec{\varphi}$ implies $y \Vdash \heartsuit\vec{\varphi}$.

(ii) $P_{\multimap}(x; \vec{y}, z) = 1$ holds, only if $x \Vdash \vec{\varphi} \multimap \psi$ and $\vec{y} \Vdash \vec{\varphi}$ imply $z \Vdash \psi$.

(iii) $P_{\sim}(x; y) = 1$ holds, only if $y \Vdash \varphi$ implies $x \nVdash {\sim}\varphi$.

See Remark 5.5 below for precising the above intuitions.

Example 4.4 A relational frame $\mathbb{W}$ for the language (1) consists of a poset $\mathscr{W}$, together with fusion-like relations $P_{\otimes} : \mathscr{W} \times \mathscr{W} \longrightarrow\!\!\!\!\!\!\!+\;\; \mathscr{W}$, $P_e : \mathbb{1} \longrightarrow\!\!\!\!\!\!\!+\;\; \mathscr{W}$, and implication-like relations $P_{\to} : \mathscr{W} \longrightarrow\!\!\!\!\!\!\!+\;\; \mathscr{W}^{op} \times \mathscr{W}$ and $P_{\leftarrow} : \mathscr{W} \longrightarrow\!\!\!\!\!\!\!+\;\; \mathscr{W}^{op} \times \mathscr{W}$.

Let us stress that the relations $P_{\otimes}$, P_e, $P_{\to}$ and $P_{\leftarrow}$ are (as of yet) arbitrary. When one needs special properties as, for example, the frame to be the model of a distributive full Lambek calculus (for such frames see [32]), one needs to invoke modal definability theorem. This is shown in Example 7.7 below.

Example 4.5 Relational frames for the language $\wedge$, $\vee$, $\otimes$, $\to$, e and $\sim$ of *relevance logic*, see [12], are posets $\mathscr{W}$ equipped with relations $P_{\otimes} : \mathscr{W} \times \mathscr{W} \longrightarrow\!\!\!\!\!\!\!+\;\; \mathscr{W}$, $P_e : \mathbb{1} \longrightarrow\!\!\!\!\!\!\!+\;\; \mathscr{W}$, $P_{\to} : \mathscr{W} \longrightarrow\!\!\!\!\!\!\!+\;\; \mathscr{W}^{op} \times \mathscr{W}$ and $P_{\sim} : \mathscr{W} \longrightarrow\!\!\!\!\!\!\!+\;\; \mathscr{W}^{op}$. The above relations are as of yet arbitrary. Frames for various classes of relevance logic are modally definable, see Remark 7.8 below.

Remark 4.6 It is very easy to prove that the above equations (6)–(8) can be "split" into six inequalities, giving us the *back & forth* description of morphisms for fusion-like, implication-like and negation-like connectives. More precisely:

(i) The equation (6) is equivalent to the conjunction of the following two inequalities

$$P^1_{\heartsuit}(\vec{x}; y) \leq P^2_{\heartsuit}(f\vec{x}; fy) \tag{9}$$

$$P^2_{\heartsuit}(\vec{a}; fy) \leq \bigvee_{\vec{x}} \mathscr{W}_2(\vec{a}, f\vec{x}) \wedge P^1_{\heartsuit}(\vec{x}; y) \tag{10}$$

(ii) The equation (7) is equivalent to the conjunction of the following two inequalities

$$P^1_{\multimap}(x; \vec{y}, z) \leq P^2_{\multimap}(fx; f\vec{y}; fz) \tag{11}$$

$$P^2_{\multimap}(fx; \vec{b}, c) \leq \bigvee_{\vec{y}, z} \mathscr{W}_2(\vec{b}, f\vec{y}) \wedge \mathscr{W}_2(fz, c) \wedge P^1_{\multimap}(x; \vec{y}, z) \tag{12}$$

(iii) The equation (8) is equivalent to the conjunction of the following two inequalities

$$P^1_{\sim}(x; y) \leq P^2_{\sim}(fx; fy) \tag{13}$$

$$P^2_{\sim}(fx; b) \leq \bigvee_{y} \mathscr{W}_2(b, fy) \wedge P^1_{\sim}(x; y) \tag{14}$$

We define now three functors

$$T_\heartsuit : \mathsf{Pos} \longrightarrow \mathsf{Pos}, \quad T_{\multimap} : \mathsf{Pos} \longrightarrow \mathsf{Pos}, \quad T_\sim : \mathsf{Pos} \longrightarrow \mathsf{Pos}$$

and prove that their product $T = T_\heartsuit \times T_{\multimap} \times T_\sim$ gives rise to relational frames and their morphisms. Namely: frames are T-coalgebras and frame morphisms are T-coalgebra morphisms.

Definition 4.7

(i) *The functor $T_\heartsuit$ sends $\mathscr{W}$ to the poset $[(\mathscr{W}^n)^{op}, 2]$ of lowersets on $\mathscr{W}^n$. For a monotone map $f : \mathscr{W}_1 \longrightarrow \mathscr{W}_2$, the map $T_\heartsuit(f)$ sends $\vec{L} : (\mathscr{W}_1^n)^{op} \longrightarrow 2$ to*

$$\vec{b} \mapsto \bigvee_{\vec{x}} \mathscr{W}_2(\vec{b}, f\vec{x}) \wedge \vec{L}\vec{x}$$

(ii) *The functor $T_{\multimap}$ sends $\mathscr{W}$ to the poset $[(\mathscr{W}^l)^{op} \times \mathscr{W}, 2]^{op}$. For a monotone map $f : \mathscr{W}_1 \longrightarrow \mathscr{W}_2$, the map $T_{\multimap}(f)$ sends $X : (\mathscr{W}_1^l)^{op} \times \mathscr{W}_1 \longrightarrow 2$ to*

$$(\vec{b}, c) \mapsto \bigvee_{\vec{y}, z} \mathscr{W}_2(\vec{b}, f\vec{y}) \wedge \mathscr{W}_2(fz, c) \wedge X(\vec{y}, z)$$

(iii) *The functor $T_\sim$ sends $\mathscr{W}$ to the poset $[\mathscr{W}^{op}, 2]^{op}$. For a monotone map $f : \mathscr{W}_1 \longrightarrow \mathscr{W}_2$, the map $T_\sim(f)$ sends $X : \mathscr{W}_1^{op} \longrightarrow 2$ to*

$$b \mapsto \bigvee_{y} \mathscr{W}_2(b, fy) \wedge Xy$$

Proposition 4.8 *Put $T = T_\heartsuit \times T_{\multimap} \times T_\sim$. The category of relational frames and their morphisms is isomorphic to the category Pos_T of T-coalgebras and their morphisms.*

Proof.

(i) To give a monotone map $\gamma : \mathscr{W} \longrightarrow T(\mathscr{W})$ is to give three monotone maps $\gamma_\heartsuit : \mathscr{W} \longrightarrow T_\heartsuit(\mathscr{W})$, $\gamma_{\multimap} : \mathscr{W} \longrightarrow T_{\multimap}(\mathscr{W})$ and $\gamma_\sim : \mathscr{W} \longrightarrow T_\sim(\mathscr{W})$. Each of the three maps, however, can be uncurried to produce monotone maps $P_\heartsuit : (\mathscr{W}^n)^{op} \times \mathscr{W} \longrightarrow 2$, $P_{\multimap} : \mathscr{W}^{op} \times (\mathscr{W}^l)^{op} \times \mathscr{W} \longrightarrow 2$ and $P_\sim : \mathscr{W}^{op} \times \mathscr{W}^{op} \longrightarrow 2$. To conclude: T-coalgebras are exactly the relational frames.

(ii) To give a monotone map $f : \mathscr{W}_1 \longrightarrow \mathscr{W}_2$ such that the square

$$\begin{array}{ccc} \mathscr{W}_1 & \xrightarrow{\gamma_1} & T(\mathscr{W}_1) \\ {\scriptstyle f}\downarrow & & \downarrow{\scriptstyle T(f)} \\ \mathscr{W}_2 & \xrightarrow[\gamma_2]{} & T(\mathscr{W}_2) \end{array}$$

commutes, is, by Definition 4.7, to give a monotone map f such that equations (6)–(8) hold. To conclude: coalgebra homomorphisms are exactly the morphisms of relational frames.

□

5 Complex algebras

The complex algebra $Pred^\sharp(\mathbb{W})$ of the frame $\mathbb{W}$ will be a distributive lattice $Pred(\mathscr{W})$, equipped with extra operators $\heartsuit$, $\multimap$ and $\sim$.

We prove that taking a complex algebra defines a functor $Pred^\sharp$ from the (opposite of the) category of relational frames and their morphisms to the category $\mathsf{DL}_{\heartsuit,\multimap,\sim}$ of distributive lattices equipped with extra operations. Moreover, this construction extends the predicate functor $Pred : \mathsf{Pos}^{op} \longrightarrow \mathsf{DL}$ in the sense that the square

$$\begin{array}{ccc} (\mathsf{Pos}_T)^{op} & \xrightarrow{Pred^\sharp} & \mathsf{DL}_{\heartsuit,\multimap,\sim} \\ {\scriptstyle (V_T)^{op}}\downarrow & & \downarrow{\scriptstyle U_{\heartsuit,\multimap,\sim}} \\ \mathsf{Pos}^{op} & \xrightarrow[Pred]{} & \mathsf{DL} \end{array} \tag{15}$$

commutes. Above, $V_T\colon \mathsf{Pos}_T \longrightarrow \mathsf{Pos}$ is the forgetful functor sending a coalgebra $(\mathscr{W},\gamma)$ to the poset $\mathscr{W}$.

Definition 5.1 *The category $\mathsf{DL}_{\heartsuit,\multimap,\sim}$ is defined as follows:*

(i) *Objects are distributive lattices $\mathscr{A} = (\mathscr{A}_o, \wedge, \vee)$ (where $\mathscr{A}_o$ denotes the underlying* poset*), together with* monotone *maps*

$$[\![\heartsuit]\!]_{\mathscr{A}} : \mathscr{A}_o^n \longrightarrow \mathscr{A}_o, \quad [\![\multimap]\!]_{\mathscr{A}} : (\mathscr{A}_o^l)^{op} \times \mathscr{A}_o \longrightarrow \mathscr{A}_o, \quad [\![\sim]\!]_{\mathscr{A}} : \mathscr{A}_o^{op} \longrightarrow \mathscr{A}_o$$

called the interpretations *of $\heartsuit$, $\multimap$ and $\sim$. We will usually omit the brackets $[\![-]\!]_{\mathscr{A}}$ and denote $(\mathscr{A},\heartsuit,\multimap,\sim)$ by $\mathbb{A}$.*

The operations are required to satisfy the following axioms, for each $0 \le i \le n$:

$$\heartsuit(\dots,x_i \vee x_i',\dots) = \heartsuit(\dots,x_i,\dots) \vee \heartsuit(\dots,x_i',\dots)$$
$$(\dots,x_i \vee x_i',\dots) \multimap y = ((\dots,x_i,\dots) \multimap y) \wedge ((\dots,x_i',\dots) \multimap y)$$
$$\vec{x} \multimap (y_1 \wedge y_2) = (\vec{x} \multimap y_1) \wedge (\vec{x} \multimap y_2)$$
$$\sim(x_1 \vee x_2) = \sim x_1 \wedge \sim x_2$$

(ii) *A* morphism *from $\mathbb{A}_1$ to $\mathbb{A}_2$ is a lattice morphism $h : \mathscr{A}_1 \longrightarrow \mathscr{A}_2$ preserving the additional operations $\heartsuit$, $\multimap$ and $\sim$ on the nose.*

The obvious underlying functor will be denoted by $U_{\heartsuit,\multimap,\sim} : \mathsf{DL}_{\heartsuit,\multimap,\sim} \longrightarrow \mathsf{DL}$.

Remark 5.2 It is clear that $\mathsf{DL}_{\heartsuit,\multimap,\sim}$ is a finitary variety over Pos in the sense of categorical universal algebra. More precisely: the composite $U \cdot U_{\heartsuit,\multimap,\sim}$: $\mathsf{DL}_{\heartsuit,\multimap,\sim} \longrightarrow \mathsf{Pos}$ of the obvious forgetful functors is a monadic functor. In particular, the forgetful functor $U \cdot U_{\heartsuit,\multimap,\sim} : \mathsf{DL}_{\heartsuit,\multimap,\sim} \longrightarrow \mathsf{Pos}$ has a left adjoint, hence there also exists a left adjoint $F_{\heartsuit,\multimap,\sim} : \mathsf{DL} \longrightarrow \mathsf{DL}_{\heartsuit,\multimap,\sim}$ to $U_{\heartsuit,\multimap,\sim}$. Thus, given a poset At, we can form $F_{\heartsuit,\multimap,\sim}(F(\mathsf{At}))$. This is the *Lindenbaum-Tarski algebra* of formulas for the language (2) and we denote it by $\mathscr{L}(\mathsf{At})$.

Definition 5.3 *The* complex algebra $Pred^\sharp(\mathbb{W}) = (([\mathscr{W},2],\cap,\cup),\heartsuit,\multimap,\sim)$ *is defined as follows:*

(i) *Given a vector $\vec{U}$ of uppersets $U_0, \ldots, U_{n-1}$, the upperset $\heartsuit\vec{U}$ is defined by the formula*

$$y \mapsto \bigvee_{\vec{x}} \vec{U}\vec{x} \wedge P_\heartsuit(\vec{x};y)$$

(ii) *Given a vector $\vec{U}$ of uppersets $U_0, \ldots, U_{l-1}$, and an upperset W, the upperset $\vec{U} \multimap W$ is defined by the formula*

$$x \mapsto \bigwedge_{\vec{y},z} \vec{U}\vec{y} \wedge P_\multimap(x;\vec{y},z) \Rightarrow Wz$$

(iii) *Given an upperset U, the upperset $\sim U$ is defined by the formula*

$$x \mapsto \bigwedge_{y} P_\sim(x;y) \Rightarrow \neg Uy$$

where the $\neg$ sign is negation in 2.

The following result is easy to prove, when one uses the back & forth description of morphism of frames, see Remark 4.6.

Proposition 5.4 *The assignment $\mathbb{W} \mapsto Pred^\sharp(\mathbb{W})$ can be extended to a functor from $(\mathsf{Pos}_T)^{op}$ to $\mathsf{DL}_{\heartsuit,\multimap,\sim}$. Moreover, the square (15) commutes.*

Proof. It is easy to verify that, given a frame $\mathbb{W}$, the algebra $Pred^\sharp(\mathbb{W})$ is an object of $\mathsf{DL}_{\heartsuit,\multimap,\sim}$.

For a frame morphism $f : \mathbb{W}_1 \longrightarrow \mathbb{W}_2$, put $Pred^\sharp(f)$ to be the mapping $[f,2] : [\mathscr{W}_2,2] \longrightarrow [\mathscr{W}_1,2]$. We verify that the three operations are preserved on the nose:

(i) The commutativity of the square

$$\begin{array}{ccc} [\mathscr{W}_2,2]^n & \xrightarrow{[f,2]^n} & [\mathscr{W}_1,2]^n \\ \heartsuit\downarrow & & \downarrow\heartsuit \\ [\mathscr{W}_2,2] & \xrightarrow[{[f,2]}]{} & [\mathscr{W}_1,2] \end{array}$$

is the requirement that the equality

$$\bigvee_{\vec{a}} \vec{U}\vec{a} \wedge P^2_\heartsuit(\vec{a};fy) = \bigvee_{\vec{x}} \vec{U}f\vec{x} \wedge P^1_\heartsuit(\vec{x};y)$$

holds for every y. The inequality $\geq$ is obvious: put $\vec{a} = f\vec{x}$ and use that $P^1_\heartsuit(\vec{x};y) \leq P^2_\heartsuit(f\vec{x};fy)$ holds, see (9). The converse inequality follows from the inequality (10).

(ii) The commutativity of the square

$$\begin{array}{ccc} ([\mathscr{W}_2,2]^l)^{op}\times[\mathscr{W}_2,2] & \xrightarrow{([f,2]^l)^{op}\times[f,2]} & ([\mathscr{W}_1,2]^l)^{op}\times[\mathscr{W}_1,2] \\ {\scriptstyle \multimap}\big\downarrow & & \big\downarrow{\scriptstyle\multimap} \\ [\mathscr{W}_2,2] & \xrightarrow[{[f,2]}]{} & [\mathscr{W}_1,2]\end{array}$$

is the requirement that the equality

$$\bigwedge_{\vec{b},c} \vec{U}\vec{b} \wedge P^2_{\multimap}(fx;\vec{b},c) \Rightarrow Wc = \bigwedge_{\vec{y},z} \vec{U}f\vec{y} \wedge P^1_{\multimap}(x;\vec{y},z) \Rightarrow Wfz$$

holds for every x. The inequality $\leq$ follows from $P^1_{\multimap}(x;\vec{y},z) \leq P^2_{\multimap}(fx;f\vec{y},fz)$, see (11). For the converse inequality, use inequality (12).

(iii) The commutativity of the square

$$\begin{array}{ccc} [\mathscr{W}_2,2]^{op} & \xrightarrow{[f,2]^{op}} & [\mathscr{W}_1,2]^{op} \\ {\scriptstyle \sim}\big\downarrow & & \big\downarrow{\scriptstyle\sim} \\ [\mathscr{W}_2,2] & \xrightarrow[{[f,2]}]{} & [\mathscr{W}_1,2]\end{array}$$

is the requirement that the equality

$$\bigwedge_{b} P^2_{\sim}(fx;b) \Rightarrow \neg Ub = \bigwedge_{y} P^1_{\sim}(x;y) \Rightarrow \neg Ufy$$

holds for every x. The inequality $\leq$ follows from inequality (13). For the converse inequality, use inequality (14).

□

Remark 5.5 The square (15) allows us to give *semantics* of the language. More precisely, we saw in Section 3 that the adjunction $F \dashv U : \mathsf{DL} \longrightarrow \mathsf{Pos}$, together with *Stone* $\dashv$ *Pred* : $\mathsf{Pos}^{op} \longrightarrow \mathsf{DL}$, takes care of the semantics $\|-\|_{\mathsf{val}}$ of the propositional part of the logic.

The adjunction $F_{\heartsuit,\multimap,\sim} \dashv U_{\heartsuit,\multimap,\sim} : \mathsf{DL}_{\heartsuit,\multimap,\sim} \longrightarrow \mathsf{DL}$, together with square (15), allow us to define, for every frame $\mathbb{W}$, a *semantics morphism*

$$\|-\|_{\mathsf{val}} : \mathscr{L}(\mathsf{At}) \longrightarrow \mathit{Pred}^{\sharp}(\mathbb{W})$$

in $\mathsf{DL}_{\heartsuit,\multimap,\sim}$ as the transpose under the composite adjunction

$$\mathsf{DL}_{\heartsuit,\multimap,\sim} \underset{U_{\heartsuit,\multimap,\sim}}{\overset{F_{\heartsuit,\multimap,\sim}}{\overset{\longleftarrow}{\underset{\longrightarrow}{\bot}}}} \mathsf{DL} \underset{U}{\overset{F}{\overset{\longleftarrow}{\underset{\longrightarrow}{\bot}}}} \mathsf{Pos}$$

of a *valuation* $\mathsf{val} : \mathsf{At} \longrightarrow [\mathscr{W},2]$.

It is possible to give an inductive description of $\|-\|_{\mathsf{val}}$. Namely: the equations

$$\begin{array}{ll} \|p\|_{\mathsf{val}} = \mathsf{val}(p), & \|\varphi_1 \wedge \varphi_2\|_{\mathsf{val}} = \|\varphi_1\|_{\mathsf{val}} \cap \|\varphi_2\|_{\mathsf{val}}, \\ \|\varphi_1 \vee \varphi_2\|_{\mathsf{val}} = \|\varphi_1\|_{\mathsf{val}} \cup \|\varphi_2\|_{\mathsf{val}} & \|\heartsuit\vec{\varphi}\|_{\mathsf{val}} = \heartsuit\|\vec{\varphi}\|_{\mathsf{val}}, \\ \|\vec{\varphi} \multimap \psi\|_{\mathsf{val}} = \|\vec{\varphi}\|_{\mathsf{val}} \multimap \|\psi\|_{\mathsf{val}}, & \|{\sim}\varphi\|_{\mathsf{val}} = {\sim}\|\varphi\|_{\mathsf{val}} \end{array}$$

hold. Above, the symbols $\heartsuit$, $\multimap$ and $\sim$ on the right-hand sides are to be interpreted as the operations in the complex algebra $Pred^\sharp(\mathbb{W})$.

Let us call the pair $(\mathbb{W}, \mathsf{val})$, consisting of a frame and a valuation, a *model*. Then the morphism $\|-\|_{\mathsf{val}}$ defines the notion of *local truth* in the model $(\mathbb{W}, \mathsf{val})$ — we write $x \Vdash_{\mathbb{W},\mathsf{val}} \varphi$, if x belongs to the upperset $\|\varphi\|_{\mathsf{val}}$, or, equivalently, if $\|\varphi\|_{\mathsf{val}} x = 1$. If rewritten in terms of $\Vdash$, the above equations give the familiar inductive definition of validity. Namely (omitting the obvious cases of atomic propositions and $\wedge$ and $\vee$):

(i) $x \Vdash_{\mathbb{W},\mathsf{val}} \heartsuit\vec{\varphi}$ holds iff there exists $\vec{y}$ such that both $\vec{y} \Vdash \vec{\varphi}$ and $P_\heartsuit(\vec{y}; x)$ hold.

(ii) $x \Vdash_{\mathbb{W},\mathsf{val}} \vec{\varphi} \multimap \psi$ holds iff for all $\vec{y}$ and z such that $\vec{y} \Vdash \vec{\varphi}$ and $P_\multimap(x; \vec{y}, z)$ hold, $z \Vdash \psi$ holds.

(iii) $x \Vdash_{\mathbb{W},\mathsf{val}} {\sim}\varphi$ iff for all y such that $P_\sim(x; y)$ holds, $y \nVdash \varphi$ holds.

6 Canonical relational frames

The assignment of the *canonical frame* $Stone^\sharp(\mathbb{A})$ to an object $\mathbb{A}$ of $\mathsf{DL}_{\heartsuit,\multimap,\sim}$ is, in a way, dual to the formation of complex algebras. We prove below that $\mathbb{A} \mapsto Stone^\sharp(\mathbb{A})$ is functorial and that the square

$$\begin{array}{ccc} \mathsf{DL}_{\heartsuit,\multimap,\sim} & \xrightarrow{Stone^\sharp} & (\mathsf{Pos}_T)^{op} \\ {\scriptstyle U_{\heartsuit,\multimap,\sim}}\downarrow & & \downarrow{\scriptstyle (V_T)^{op}} \\ \mathsf{DL} & \xrightarrow[Stone]{} & \mathsf{Pos}^{op} \end{array} \qquad (16)$$

commutes.

Definition 6.1 *Suppose $\mathbb{A} = (\mathscr{A}, \heartsuit, \multimap, \sim)$ is in $\mathsf{DL}_{\heartsuit,\multimap,\sim}$. Define $Stone^\sharp(\mathbb{A})$ as follows:*

(i) *The underlying poset of $Stone^\sharp(\mathbb{A})$ is the poset $\mathsf{DL}(\mathscr{A}, 2)$ of prime filters on the distributive lattice $\mathscr{A}$.*

(ii) *The relation $P_\heartsuit$ is defined as follows:*

$$P_\heartsuit(\vec{F}; G) = \bigwedge_{\vec{x}} \vec{F}\vec{x} \Rightarrow G(\heartsuit\vec{x})$$

(iii) *The relation $P_\multimap$ is defined as follows:*

$$P_\multimap(F; \vec{G}, H) = \bigwedge_{\vec{x},y} F(\vec{x} \multimap y) \wedge \vec{G}\vec{x} \Rightarrow Hy$$

(iv) *The relation* $P_\sim$ *is defined as follows:*

$$P_\sim(F;G) = \bigwedge_x Gx \Rightarrow \neg F(\sim x)$$

where the $\neg$ *sign is the negation in* 2.

The above definitions clearly make sense if we work with mere *uppersets* in lieu of prime filters. We will need the following three technical results that slightly generalize the results originating in the work on relevance logic, see Section 6 of [11].

Lemma 6.2 (Squeeze Lemma for $\heartsuit$**)** *Suppose* $P_\heartsuit(\vec{F'};G) = 1$ *holds, where* $\vec{F'}$ *is a vector of filters and* G *a prime filter. Then there is a vector* $\vec{F}$ *of prime filters that extends* $\vec{F'}$ *and* $P_\heartsuit(\vec{F};G) = 1$.

Lemma 6.3 (Squeeze Lemma for $\multimap$**)** *Suppose* $P_\multimap(F;\vec{G'},\overline{I'}) = 1$, *where* F *is a prime filter,* $\vec{G'}$ *is a vector of filters and* $\overline{I'}$ *is a complement of an ideal* I'. *Then there exists a vector* $\vec{G}$ *of prime filters such that* $\vec{G}$ *extends* $\vec{G'}$ *and a prime ideal* I *that extends* I' *and* $P_\multimap(F;\vec{G},\overline{I}) = 1$, *where* $\overline{I}$ *denotes the complement of* I.

Lemma 6.4 (Squeeze Lemma for $\sim$**)** *Suppose* $P_\sim(F;G') = 1$, *where* F *is a prime filter and* G' *is a filter. Then there exists a prime filter* G *extending* G' *such that* $P_\sim(F;G) = 1$.

The above three lemmata allow us to prove that the computation of a canonical frame is a functorial process.

Proposition 6.5 *The assignment* $\mathbb{A} \mapsto \mathit{Stone}^\sharp(\mathbb{A})$ *can be extended to a functor from* $\mathsf{DL}_{\heartsuit,\multimap,\sim}$ *to* $(\mathsf{Pos}_T)^{op}$. *Moreover, the square (16) commutes.*

Proof. Given $h : \mathbb{A}_1 \longrightarrow \mathbb{A}_2$, we define $\mathit{Stone}^\sharp(h)$ as $\mathsf{DL}(h,2) : \mathsf{DL}(\mathscr{A}_2,2) \longrightarrow \mathsf{DL}(\mathscr{A}_1,2)$. We only need to prove that equations (6)–(8) are satisfied. For the purposes of better readability we denote $[h,2]$ by $h^\dagger$ in what follows.

(i) The required equality

$$P^1_\heartsuit(\vec{K};h^\dagger G) = \bigvee_{\vec{F}} \mathsf{DL}(\mathscr{A}_1,2)(\vec{K},h^\dagger\vec{F}) \wedge P^2_\heartsuit(\vec{F};G)$$

can be rewritten, using the definition of $h^\dagger$, to the equation

$$P^1_\heartsuit(\vec{K};Gh) = \bigvee_{\vec{F}} \mathsf{DL}(\mathscr{A}_1,2)(\vec{K},\vec{F}h) \wedge P^2_\heartsuit(\vec{F};G)$$

We prove inequalities (9) and (10):

(a) To prove $P^2_\heartsuit(\vec{F};G) \leq P^1_\heartsuit(\vec{F}h;Gh)$, suppose $\vec{F}hx = 1$. Then $G(\heartsuit(hx)) = Gh(\heartsuit x) = 1$ and we are done.

(b) We prove $P^1_{\heartsuit}(\vec{K};Gh) \leq \bigvee_{\vec{F}} \mathsf{DL}(\mathscr{A}_1,2)(\vec{K},\vec{F}h) \wedge P^2_{\heartsuit}(\vec{F};G)$.

Define a vector $\vec{K}'$ of *filters* on $\mathscr{A}_2$ by putting

$$\vec{K}'\vec{a} = \bigvee_{\vec{x}} \mathscr{A}_2(h\vec{x},\vec{a}) \wedge \vec{K}\vec{x}$$

We prove $P^2_{\heartsuit}(\vec{K}';G) = 1$, supposing $P^1_{\heartsuit}(\vec{K};Gh) = 1$. To that end, suppose $\vec{K}'\vec{a} = 1$ and choose $\vec{x}$ such that $\mathscr{A}_2(h\vec{x},\vec{a}) \wedge \vec{K}\vec{x} = 1$. Then $Gh(\heartsuit\vec{x}) = G(\heartsuit(h\vec{x})) = 1$, hence $G(\heartsuit\vec{a}) = 1$, since $\heartsuit$ is monotone.

Now use Lemma 6.2 to find a vector $\vec{F}$ of *prime filters* such that $\vec{F}$ extends $\vec{K}'$ and $P^2_{\heartsuit}(\vec{F};G) = 1$ holds. It remains to prove the equality $\mathsf{DL}(\mathscr{A}_1,2)(\vec{K},\vec{F}h) = 1$. This follows immediately from the fact that $\vec{F}$ extends $\vec{K}'$: if $\vec{K}\vec{x} = 1$, then $\vec{K}'(h\vec{x}) = 1$, hence $\vec{F}h\vec{x} = 1$.

(ii) The required equality

$$P^1_{\multimap}(h^\dagger F;\vec{L},M) = \bigvee_{\vec{G},H} \mathsf{DL}(\mathscr{A}_1,2)(\vec{L},h^\dagger\vec{G}) \wedge \mathsf{DL}(\mathscr{A}_1,2)(h^\dagger H,M) \wedge P^2_{\multimap}(F;\vec{G},H)$$

can be rewritten to the equality

$$P^1_{\multimap}(Fh;\vec{L},M) = \bigvee_{\vec{G},H} \mathsf{DL}(\mathscr{A}_1,2)(\vec{L},\vec{G}h) \wedge \mathsf{DL}(\mathscr{A}_1,2)(Hh,M) \wedge P^2_{\multimap}(F;\vec{G},H)$$

We prove inequalities (11) and (12):

(a) For proving the inequality $P^2_{\multimap}(F;\vec{G},H) \leq P^1_{\multimap}(Fh;\vec{G}h,Hh)$, assume $P^2_{\multimap}(F;\vec{G},H) = 1$. If $Fh(\vec{x} \multimap y) \wedge \vec{G}h\vec{x} = F(h\vec{x} \multimap hy) \wedge \vec{G}h\vec{x} = 1$, then $Hhy = 1$, which was to be proved.

(b) We prove the inequality

$$P^1_{\multimap}(Fh;\vec{L},M) \leq \bigvee_{\vec{G},H} \mathsf{DL}(\mathscr{A}_1,2)(\vec{L},\vec{G}h) \wedge \mathsf{DL}(\mathscr{A}_1,2)(Hh,M) \wedge P^2_{\multimap}(F;\vec{G},H)$$

Define

$$\vec{G}'\vec{b} = \bigvee_{\vec{y}} \mathscr{A}_2(h\vec{y},\vec{b}) \wedge \vec{L}\vec{y}, \quad I'c = \bigvee_{z} \mathscr{A}_2(c,hz) \wedge \neg Mz$$

and observe that $\vec{G}'$ is a vector of filters and I' is an ideal. Moreover, the complement $\overline{I'}$ of I' is given by the formula

$$\overline{I'}c = \bigwedge_{z} \mathscr{A}_2(c,hz) \Rightarrow Mz$$

We will prove that $P^2_{\multimap}(F;\vec{G}',\overline{I'}) = 1$, if we suppose $P^1_{\multimap}(Fh;\vec{L},M) = 1$.

To that end, suppose $F(\vec{b} \multimap c) \wedge \vec{G'}\vec{b} = 1$ and suppose z is such that $\mathscr{A}_2(c, hz) = 1$ holds. We need to prove $Mz = 1$.

Pick $\vec{y}$ witnessing $\vec{G'}\vec{b} = 1$. Then $F(h\vec{y} \multimap hz) = Fh(\vec{y} \multimap z) = 1$ and $\vec{L}\vec{y} = 1$. Therefore $Mz = 1$, since we assumed $P^1_{\multimap}(Fh; \vec{L}, M) = 1$.

By Lemma 6.3 there exist $\vec{G}$ and I such that $\vec{G}$ is a vector of prime filters extending $\vec{G'}$, I is a prime ideal extending I', and $P^2_{\multimap}(F, \vec{G}, \overline{I}) = 1$ holds. Since a complement of a prime ideal is a prime filter, we can put $H = \overline{I}$.

It remains to show that $\vec{G}h$ extends L and Hh is extended by M. Since $\vec{G'}h$ clearly extends L, so does $\vec{G}h$ (use that $\vec{G}$ extends $\vec{G'}$).

Since $\overline{I'}h$ is extended by M, so is $\overline{I}h$. This follows from the fact that I extends I'.

(iii) The required equality

$$P^1_{\sim}(h^\dagger F; L) = \bigvee_G \mathsf{DL}(\mathscr{A}_1, 2)(L, h^\dagger G) \wedge P^2_{\sim}(F; G)$$

can be rewritten to the equality

$$P^1_{\sim}(Fh; L) = \bigvee_G \mathsf{DL}(\mathscr{A}_1, 2)(L, Gh) \wedge P^2_{\sim}(F; G)$$

We prove inequalities in (13) and (14):

(a) To prove the inequality $P^2_{\sim}(F; G) \leq P^1_{\sim}(Fh; Gh)$, suppose that $P^2_{\sim}(F; G) = 1$ and $Fhx = 1$. Then $\neg G({\sim} hx) = \neg Gh({\sim} x) = 1$, which had to be proved.

(b) We prove the inequality $P^1_{\sim}(Fh; L) \leq \bigvee_G \mathsf{DL}(\mathscr{A}_1, 2)(L, Gh) \wedge P^2_{\sim}(F; G)$.

Define the filter G' by the formula

$$G'b = \bigvee_y \mathscr{A}_2(hy, b) \wedge Ly$$

and observe that $P^2_{\sim}(F; G') = 1$ holds, if we assume $P^1_{\sim}(Fh; L) = 1$.

Indeed: suppose $G'b = 1$ and let y witness this equality. We need to prove $\neg F({\sim} b) = 1$. But we know $\neg Fh({\sim} y) = \neg F({\sim}(hy)) = 1$. Therefore $\neg F({\sim} b) = 1$, since $\mathscr{A}_2(hy, b) = 1$ and F is an upperset.

By Lemma 6.4 we can find a prime filter G extending G' such that $P^2_{\sim}(F; G) = 1$ holds. Moreover, Gh extends L, since $G'h$ does.

□

7 Modal definability

Our modal definability theorem (Theorem 7.6 below) will identify classes **C** of frames such that the image of **C** under $Pred^\sharp$ is an "HSP" class in $\mathsf{DL}_{\heartsuit, \multimap, \sim}$, i.e., it is a variety (compare with the version of Goldblatt-Thomason theorem for modal logics [3, Theorem 5.54] and [26, Theorem 3.15/2.]). Since we work over posets, the notion of HSP-closedness has to take this fact under consideration. Namely, we will use the factorization system $(\mathcal{E}, \mathcal{M})$ on Pos where $\mathcal{E}$ consists of surjective monotone maps and $\mathcal{M}$ of monotone maps reflecting order, i.e., $f : \mathscr{W}_1 \longrightarrow \mathscr{W}_2$ is in $\mathcal{M}$ if $\mathscr{W}_1(x, x') = \mathscr{W}_2(fx, fx')$ holds for every x and x'. That $(\mathcal{E}, \mathcal{M})$ is indeed a factorization system on Pos is proved in [4]. We will use the HSP Theorem w.r.t. a factorization system, see [28]:

A class $\mathbf{A}$ of algebras in a variety $\mathscr{V}$ over Pos is definable by equations in $\mathscr{V}$ iff $\mathbf{A}$ satisfies the following conditions ($U : \mathscr{V} \longrightarrow \mathsf{Pos}$ denotes the underlying functor):

(H) If $e : \mathbb{A}_1 \longrightarrow \mathbb{A}_2$ is such that $U(e)$ is a split epi in Pos and $\mathbb{A}_1$ is in $\mathscr{A}$, then $\mathbb{A}_2$ is in $\mathbf{A}$.

(S) If $m : \mathbb{A}_1 \longrightarrow \mathbb{A}_2$ is such that $U(m)$ is in $\mathcal{M}$ and $\mathbb{A}_2$ is in $\mathscr{A}$, then $\mathbb{A}_1$ is in $\mathbf{A}$.

(P) If $\mathbb{A}_i$, $i \in I$, are in $\mathscr{A}$, then $\prod_{i \in I} \mathbb{A}_i$ is in $\mathbf{A}$.

In fact, since the algebraic semantics of our logic takes place in (distributive) lattices, we may as well replace equationally defined classes by inequationally defined. We prefer to introduce the inequational description, since it is often more useful in applications.

Definition 7.1 *Suppose* $\mathbb{W}$ *is a relational frame. We say that* α entails β, *and denote this fact by* $\alpha \models_{\mathbb{W}} \beta$, *provided that* $\|\alpha\|_{\mathsf{val}} \leq \|\beta\|_{\mathsf{val}}$ *holds, for every valuation* $\mathsf{val} : \mathsf{At} \longrightarrow [\mathscr{W}, 2]$.

Given a class Σ *of pairs of formulas, we denote by* $\mathsf{Mod}(\Sigma)$ *the class of frames* $\mathbb{W}$ *such that* $\alpha \models_{\mathbb{W}} \beta$, *for all* $(\alpha, \beta) \in \Sigma$.

The following result is trivial.

Lemma 7.2 $\alpha \models_{\mathbb{W}} \beta$ *holds iff* $Pred^{\sharp}(\mathbb{W}) \models \alpha \wedge \beta = \alpha$, *where the* $\models$ *sign on the right denotes validity in the sense of universal algebra.*

Although the notation might suggest it, *it is not the case* that the logical connection *Stone* $\dashv$ *Pred* lifts to an adjunction $Stone^{\sharp} \dashv Pred^{\sharp}$. The unit of *Stone* $\dashv$ *Pred* does lift, however, and we will need it in the proof of Theorem 7.6.

This fact, known in modal logic as the Jónsson-Tarski theorem, is itself interesting since it gives us the modal completeness of the resulting logic w.r.t. T-coalgebras: namely, the T-coalgebra (the canonical frame) corresponding to the Lindenbaum-Tarski $\mathsf{DL}_{\heartsuit,\multimap,\sim}$-algebra provides a counterexample to any unprovable formula.

Lemma 7.3 *The unit* η *of Stone* $\dashv$ *Pred is a morphism in* $\mathsf{DL}_{\heartsuit,\multimap,\sim}$, *i.e.,* η *lifts along the functor* $U_{\heartsuit,\multimap,\sim} : \mathsf{DL}_{\heartsuit,\multimap,\sim} \longrightarrow \mathsf{DL}$ *to a natural transformation* $\eta^{\sharp} : Id_{\mathsf{DL}_{\heartsuit,\multimap,\sim}} \longrightarrow Pred^{\sharp} Stone^{\sharp}$.

Another technical result that we need for Theorem 7.6 is the following one.

Lemma 7.4 *The functor Stone sends maps reflecting order to surjective monotone maps.*

Proof. Suppose $m : \mathscr{A} \longrightarrow \mathscr{B}$ is a lattice homomorphism that reflects order. We need to prove that the monotone map $Stone(m) : Stone(\mathscr{B}) \longrightarrow Stone(\mathscr{A})$ is surjective. To that end, fix a prime filter F on $\mathscr{A}$. Define the set

$$\mathbf{E} = \{G \mid G \cdot m = F\}$$

of filters on $\mathscr{B}$, ordered by inclusion. The set $\mathbf{E}$ is nonempty, since m reflects order: put $Gb = \bigvee_a \mathscr{B}(ma, b) \wedge Fa$ and observe that G is in $\mathbf{E}$. Furthermore, the union of a nonempty chain of elements of $\mathbf{E}$ is an element of $\mathbf{E}$. By Zorn's Lemma, $\mathbf{E}$ has a maximal element G_0. It is easy to prove that it is a prime filter. □

Finally, before stating Theorem 7.6, we need to introduce the concept of a *prime extension* of a frame.

Definition 7.5 *The frame* $\mathbb{W}^* = Stone^\sharp Pred^\sharp(\mathbb{W})$ *is called the* prime extension *of* $\mathbb{W}$.

Theorem 7.6 *Suppose* $\mathbf{C}$ *is a class of relational frames that is closed under prime extensions (if* $\mathbb{W}$ *is in* $\mathbf{C}$*, then* $\mathbb{W}^*$ *is in* $\mathbf{C}$*). Then the following are equivalent:*

(i) *There is* Σ *such that* $\mathbf{C} = \mathsf{Mod}(\Sigma)$.

(ii) $\mathbf{C}$ *satisfies the following four conditions:*

  (a) $\mathbf{C}$ *is closed under "surjective coalgebraic quotients", i.e., if* $e : \mathbb{W}_1 \longrightarrow \mathbb{W}_2$ *is surjective and* $\mathbb{W}_1$ *is in* $\mathbf{C}$*, so is* $\mathbb{W}_2$.

  (b) $\mathbf{C}$ *is closed under "subcoalgebras", i.e., if* $m : \mathbb{W}_1 \longrightarrow \mathbb{W}_2$ *reflects order and* $\mathbb{W}_2$ *is in* $\mathbf{C}$*, so is* $\mathbb{W}_1$.

  (c) $\mathbf{C}$ *is closed under coproducts.*

  (d) $\mathbf{C}$ *reflects prime extensions: if* $\mathbb{W}^*$ *is in* $\mathbf{C}$*, so is* $\mathbb{W}$.

Proof. 1 implies 2. Suppose $\mathbf{C} = \mathsf{Mod}(\Sigma)$. We will verify the four conditions for $\mathbf{C}$.

(a) Suppose $e : \mathbb{W}_1 \longrightarrow \mathbb{W}_2$ is a surjective coalgebra morphism. We prove that if $\alpha \models_{\mathbb{W}_1} \beta$, then $\alpha \models_{\mathbb{W}_2} \beta$.

Consider $y \in \mathscr{W}_2$ and a valuation $\mathsf{val} : \mathsf{At} \longrightarrow [\mathscr{W}_2, 2]$. We can define a new valuation $\mathsf{val}' : \mathsf{At} \longrightarrow [\mathscr{W}_1, 2]$ by the composition

$$\mathsf{At} \xrightarrow{\mathsf{val}} [\mathscr{W}_2, 2] \xrightarrow{[e,2]} [\mathscr{W}_1, 2]$$

Then the diagram

$$\mathscr{L}(\mathsf{At}) \xrightarrow{\|-\|_{\mathsf{val}}} Pred^\sharp(\mathbb{W}_2) \xrightarrow{Pred^\sharp(e)} Pred^\sharp(\mathbb{W}_1), \qquad \|-\|^{\mathsf{val}'} : \mathscr{L}(\mathsf{At}) \longrightarrow Pred^\sharp(\mathbb{W}_1)$$

commutes in $\mathsf{DL}_{\heartsuit,\multimap,\sim}$.

Let x be such that $ex = y$. Then, by assumption, $x \Vdash_{\mathsf{val}'} \alpha \leq \beta$, hence

$$\|\alpha \wedge \beta\|_{\mathsf{val}} ex = [e,2](\|\alpha \wedge \beta\|_{\mathsf{val}})x = \|\alpha \wedge \beta\|_{\mathsf{val}'} x = \|\alpha\|_{\mathsf{val}'} x = [e,2](\|\alpha\|_{\mathsf{val}})x$$
$$= \|\alpha\|_{\mathsf{val}} ex$$

Therefore $ex \Vdash_{\mathsf{val}} \alpha \leq \beta$, i.e., $y \Vdash_{\mathsf{val}} \alpha \leq \beta$.

(b) Suppose $m : \mathbb{W}_1 \longrightarrow \mathbb{W}_2$ is a coalgebra morphism with m reflecting order. We prove that if $\alpha \models_{\mathbb{W}_2} \beta$, then $\alpha \models_{\mathbb{W}_1} \beta$.

Observe that $[m,2] : [\mathscr{W}_2, 2] \longrightarrow [\mathscr{W}_1, 2]$ is a split epimorphism in Pos. Indeed: there exists a monotone map $z : [\mathscr{W}_1, 2] \longrightarrow [\mathscr{W}_2, 2]$ such that $[m,2] \cdot z = id$. Given $u : \mathscr{W}_1 \longrightarrow 2$, define $v : \mathscr{W}_2 \longrightarrow 2$ by the formula

$$vy = \bigvee_x \mathscr{W}_2(mx, y) \wedge ux$$

Then $z : u \mapsto v$ is monotone and the equalities

$$vmx' = \bigvee_x \mathscr{W}_2(mx, mx') \wedge ux = \bigvee_x \mathscr{W}_1(x, x') \wedge ux = ux'$$

prove $[m,2] \cdot z = id$ (above, we have used that m reflects order).

Suppose $\mathsf{val} : \mathsf{At} \longrightarrow [\mathscr{W}_1, 2]$ is given. To prove $x \in \|\alpha\|_{\mathsf{val}}$, consider

$$\mathsf{val}' \equiv \mathsf{At} \xrightarrow{\mathsf{val}} [\mathscr{W}_1, 2] \xrightarrow{z} [\mathscr{W}_2, 2]$$

By assumption, $\|\alpha \wedge \beta\|_{\mathsf{val}'} mx = \|\alpha\|_{\mathsf{val}'} mx$. But the diagram

$$\mathscr{L}(\mathsf{At}) \xrightarrow{\|-\|_{\mathsf{val}'}} Pred^{\sharp}(\mathbb{W}_2) \xrightarrow{Pred^{\sharp}(m)} Pred^{\sharp}(\mathbb{W}_1)$$
$$\underbrace{\qquad\qquad\qquad\qquad\qquad}_{\|-\|_{\mathsf{val}}}$$

commutes in $\mathsf{DL}_{\heartsuit, -\circ, \sim}$ due to $[m,2] \cdot z = id$. Hence $\|\alpha \wedge \beta\|_{\mathsf{val}} x = \|\alpha\|_{\mathsf{val}} x$.

(c) Suppose $\alpha \models_{\mathbb{W}_i} \beta$, for all $i \in I$. We prove that $\alpha \models_{\coprod_{i \in I} \mathbb{W}_i} \beta$.

The functor $Pred^{\sharp}$ preserves products (in fact, it preserves all limits). Products in $(\mathsf{Pos}_T)^{op}$ are, of course, coproducts in Pos_T.

Consider x in $\coprod_{i \in I} \mathscr{W}_i$. Since coproducts of frames are formed on the level of posets, there is $i \in I$ such that x is in $\mathscr{W}_i$. Let $\mathsf{val} : \mathsf{At} \longrightarrow \prod_{i \in I} [\mathscr{W}_i, 2]$ be any valuation. Then, by assumption, $x \Vdash_{\mathsf{val}_i} \alpha \wedge \beta = \alpha$, where

$$\mathsf{val}_i \equiv \mathsf{At} \xrightarrow{\mathsf{val}} \prod_{i \in I} [\mathscr{W}_i, 2] \xrightarrow{p_i} [\mathscr{W}_i, 2]$$

and where p_i denotes the i-th projection.

This proves $\|\alpha \wedge \beta\|_{\mathsf{val}} x = \|\alpha\|_{\mathsf{val}} x$.

(d) Suppose $\alpha \models_{\mathscr{W}^*} \beta$. We prove that $\alpha \models_{\mathscr{W}} \beta$.

Take x in $\mathscr{W}$ and $\mathsf{val} : \mathsf{At} \longrightarrow [\mathscr{W}, 2]$. Recall that, by Lemma 7.3, η lifts to $\eta^{\sharp}$, hence we can consider the valuation

$$\mathsf{val}' \equiv \mathsf{At} \xrightarrow{\mathsf{val}} [\mathscr{W}, 2] \xrightarrow{UU_{\heartsuit, -\circ, \sim}(\eta^{\sharp}_{Pred^{\sharp}(\mathbb{W})})} [Stone\, Pred(\mathscr{W}), 2]$$

and therefore the diagram

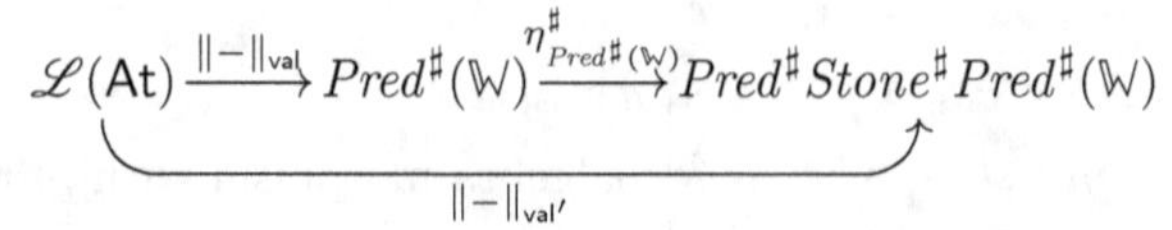

$$\mathscr{L}(\text{At}) \xrightarrow{\|-\|_{\text{val}}} Pred^{\sharp}(\mathbb{W}) \xrightarrow{\eta^{\sharp}_{Pred^{\sharp}(\mathbb{W})}} Pred^{\sharp} Stone^{\sharp} Pred^{\sharp}(\mathbb{W}) \quad (17)$$

$$\underbrace{\qquad\qquad\qquad\qquad\qquad\qquad}_{\|-\|_{\text{val}'}}$$

commutes in $\text{DL}_{\heartsuit,-\circ,\sim}$. Thus, we obtain a commutative diagram

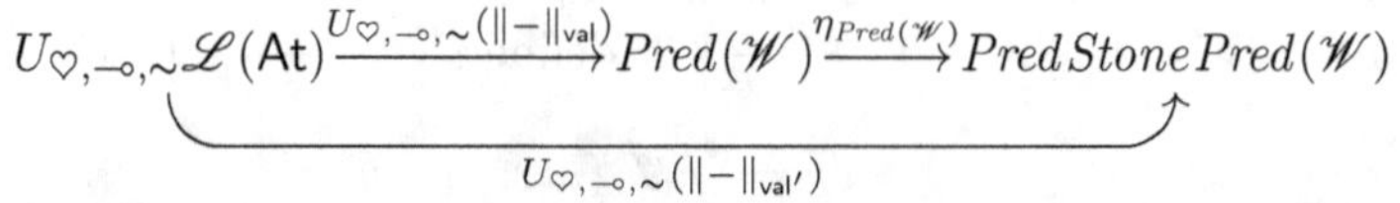

$$U_{\heartsuit,-\circ,\sim}\mathscr{L}(\text{At}) \xrightarrow{U_{\heartsuit,-\circ,\sim}(\|-\|_{\text{val}})} Pred(\mathscr{W}) \xrightarrow{\eta_{Pred(\mathscr{W})}} Pred\, Stone\, Pred(\mathscr{W})$$

$$\underbrace{\qquad\qquad\qquad\qquad}_{U_{\heartsuit,-\circ,\sim}(\|-\|_{\text{val}'})}$$

in DL (apply $U_{\heartsuit,-\circ,\sim}$ to diagram (17) and use that $U_{\heartsuit,-\circ,\sim}(\eta^{\sharp}_{Pred^{\sharp}(\mathbb{W})}) = \eta_{Pred(\mathscr{W})}$).

Hence also the diagram

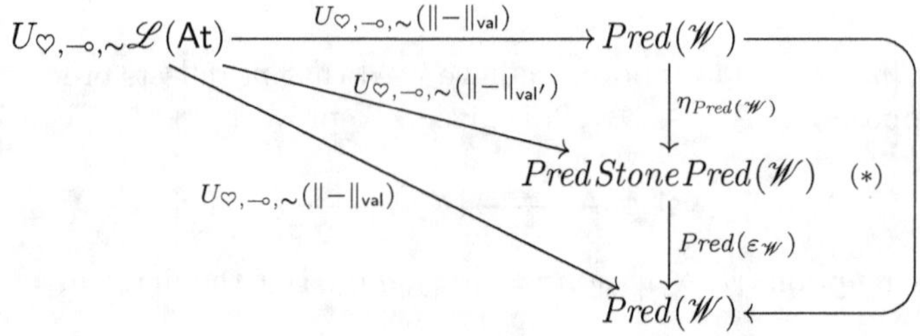

commutes in DL. In fact, the area $(*)$ in the above diagram is just one of the triangle equalities for $Stone \dashv Pred$.

By assumption, $\varepsilon_{\mathscr{W}}(x) \Vdash_{\text{val}'} \alpha \wedge \beta = \alpha$. From the lower triangle in (18) it follows that $x \Vdash_{\text{val}} \alpha \wedge \beta = \alpha$:

$$\|\alpha \wedge \beta\|_{\text{val}} x = [\varepsilon_{\mathscr{W}}, 2](\|\alpha \wedge \beta\|_{\text{val}'} x) = \|\alpha \wedge \beta\|_{\text{val}'} \varepsilon_{\mathscr{W}}(x) = \|\alpha\|_{\text{val}'} \varepsilon_{\mathscr{W}}(x)$$
$$= [\varepsilon_{\mathscr{W}}, 2](\|\alpha\|_{\text{val}'} x) = \|\alpha\|_{\text{val}} x$$

2 implies 1. Denote by Σ the set of pairs (α, β) such that $\alpha \models_{\mathbb{W}} \beta$, for all $\mathbb{W}$ in $\mathbf{C}$. Hence $\mathbf{C} \subseteq \text{Mod}(\Sigma)$ by definition.

Suppose $\mathbb{W}_0$ is in $\text{Mod}(\Sigma)$, we want to prove that $\mathbb{W}_0$ is in $\mathbf{C}$.

Define $\mathbf{A}$ to be the closure of $\{Pred^{\sharp}(\mathbb{W}) \mid \mathbb{W} \in \mathbf{C}\}$ under products, subalgebras along monotone maps reflecting order and images along split epis in Pos. Therefore $Pred^{\sharp}(\mathbb{W}_0)$ is in $\mathbf{A}$ and there is a diagram

$$Pred^{\sharp}(\mathbb{W}_0) \xleftarrow{e} \mathbb{A} \xrightarrow{m} \prod_{i \in I} Pred^{\sharp}(\mathbb{W}_i)$$

in $\text{DL}_{\heartsuit,-\circ,\sim}$, where $\mathbb{A}$ is in $\mathbf{A}$, $\mathbb{W}_i$ are in $\mathbf{C}$, for all $i \in I$, and m reflects orders, and e is split epi in Pos.

Consider the image of the above diagram

$$Stone^{\sharp} Pred^{\sharp}(\mathbb{W}_0) \xleftarrow{Stone^{\sharp}(e)} Stone^{\sharp}(\mathbb{A}) \xrightarrow{Stone^{\sharp}(m)} Stone^{\sharp}(\prod_{i \in I} Pred^{\sharp}(\mathbb{W}_i))$$

under $\mathit{Stone}^\sharp : \mathsf{DL}_{\heartsuit,\multimap,\sim} \longrightarrow (\mathsf{Pos}_T)^{op}$.

When reading the above diagram in Pos_T, i.e., when reversing the arrows, we obtain a diagram

$$\mathit{Stone}^\sharp \mathit{Pred}^\sharp(\mathbb{W}_0) \xrightarrow{\mathit{Stone}^\sharp(e)} \mathit{Stone}^\sharp(\mathbb{A}) \xleftarrow{\mathit{Stone}^\sharp(m)} \mathit{Stone}^\sharp(\mathit{Pred}^\sharp(\textstyle\coprod_{i\in I} \mathbb{W}_i))$$

Then:

(i) $\mathit{Stone}^\sharp(\mathit{Pred}^\sharp(\coprod_{i\in I} \mathbb{W}_i))$ is in **C**, since it is a prime extension of a coproduct of elements of **C**.

(ii) $\mathit{Stone}^\sharp(\mathbb{A})$ is in **C**.
  (a) By Lemma 7.4, $\mathit{Stone}^\sharp(m)$ is a surjective coalgebra homomorphism. Indeed, the underlying map of $\mathit{Stone}^\sharp(m)$ is $\mathit{Stone}(m)$ by (16).
  (b) Since $\mathit{Stone}^\sharp(\mathit{Pred}^\sharp(\coprod_{i\in I} \mathbb{W}_i))$ is in **C**, so is $\mathit{Stone}^\sharp(\mathbb{A})$. Use properties of **C**.

(iii) $\mathit{Stone}^\sharp \mathit{Pred}^\sharp(\mathbb{W}_0)$ is in **C**.

This will follow after we prove that $\mathit{Stone}^\sharp(e)$ reflects orders. Its underlying map is restriction along e from the poset of prime filters on $\mathbb{A}$ to the poset of prime filters on $\mathit{Pred}(\mathscr{W}_0)$. Recall that e is a split epimorphism, denote by z the monotone map satisfying $e \cdot z = id$. Consider two prime filters u, u' on $\mathbb{A}$ such that $u \cdot e \leq u' \cdot e$ holds. Then $u = u \cdot e \cdot z \leq u' \cdot e \cdot z = u'$ holds.

Since we proved that $\mathit{Stone}^\sharp \mathit{Pred}^\sharp(\mathbb{W}_0)$ is in **C**, we know that $\mathbb{W}_0$ is in **C**, since **C** reflects ultrafilter extensions. □

Example 7.7 The *distributive and associative full Lambek calculus* (denoted by **dFL**) is given by the grammar (1), where $\otimes$ is required to be associative, to have e as a unit and to satisfy the residuation laws $\varphi \otimes \psi \leq \chi$ iff $\psi \leq \varphi \to \chi$ iff $\varphi \leq \chi \leftarrow \psi$. Thus, the subvariety of $\mathsf{DL}_{\otimes,e,\to,\leftarrow}$ that we want to deal with is exactly that of distributive residuated lattices.

The frames that are definable by the above (in)equations are precisely the quintuples $(\mathscr{W}, P_\otimes, P_\to, P_\leftarrow, P_e)$ that satisfy the following conditions (for details see [32, Chapter 11]):

(i) $P_\otimes$ is associative: $\bigvee_z (P_\otimes(x, y; z) \wedge P_\otimes(z, u; v)) = \bigvee_w (P_\otimes(y, u; w) \wedge P_\otimes(x, w; v))$

(ii) and has P_e as a (left and right) unit:
$W(x, y) = \bigvee_z (P_e(z) \to P_\otimes(z, x; y)) = \bigvee_z (P_e(z) \to P_\otimes(x, z; y))$

(iii) The equalities $P_\otimes(x_0, x_1; y) = P_\to(x_1; x_0, y) = P_\leftarrow(x_0; x_1, y)$ hold.

Class **C** of frames satisfying the above conditions is easily seen to verify the conditions in Theorem 7.6.

Example 7.8 Many interesting examples can be found among the extensions of (associative) **dFL** with, e.g., the structural rules, or when expanding the language by negation. Instances of the first possibility are: **dFL** extended with any combination of: exchange, weakening, contraction. See [32] for details on what follows.

(i) The exchange rule corresponds to the commutativity of $P_\otimes$, i.e. to the equality $P_\otimes(x, y; z) = P_\otimes(y, x; z)$.

(ii) Weakening corresponds to: $P_\otimes(x_0, x_1; y)$ implies $x_0 \le y$ and $x_1 \le y$.

(iii) Contraction corresponds to the equality $P_\otimes(x, x; x) = 1$.

This includes, for example, intuitionistic logic, obtained as an extension of **dFL** with all the three structural rules.[4] Instances of the second possibility include, e.g., the relevance logic **R**, see [12] or [32]. Here the language $\otimes$, $\to$, $\leftarrow$, e is extended by a negation connective $\sim$.

The frames $(\mathscr{W}, P_\otimes, P_\to, P_\leftarrow, P_e, P_\sim)$ for the relevance logic **R** are the frames for **dFL** satisfying, in addition, the contraction equality together with the following three axioms ([32]):

(a) $P_\sim(x; y) = P_\sim(y; x)$,

(b) $\bigvee_y P_\otimes(x_0, x_1; y) \wedge P_\sim(y; u) \le \bigvee_s P_\otimes(u, x_0; s) \wedge P_\sim(x_1; s)$,

(c) $\bigvee_y (P_\sim(x; y) \wedge \bigwedge_z (P_\sim(y; z) \Rightarrow \mathscr{W}(z, x))) = 1$.

The class **C** of frames satisfying these axioms is easily seen to verify the conditions of Theorem 7.6. It is modally definable by corresponding axioms of **R**.

8 Conclusions and further work

We have shown that frames for various kinds of distributive substructural logic can be perceived naturally as modally definable classes of poset coalgebras. It seems natural to construct first frames for logics that have minimal necessary restrictions on the modalities — these frames are exactly the coalgebras for a certain endofunctor of the category of posets. Such an approach yields the notion of frame morphisms for free: the morphisms of frames are exactly the coalgebra morphisms. Any (in)equational requirement on the modalities results in singling out a subclass of frames that is modally definable in the sense of Goldblatt-Thomason Theorem. Hence any subvariety of modal algebras (= distributive lattices with operators) defines a Goldblatt-Thomason subclass of frames, and vice versa, which has been illustrated by well-known examples of frames for distributive full Lambek calculus, relevance logic, etc.

The limitation of our result lies certainly in the presence of the distributive law for the propositional part of the logic since it leaves out nondistributive substructural logics. We believe that this can be easily overcome by passing to general lattices and using a two-sorted representation of lattices in the sense of [24]. The underlying logical connection will be two-sorted, hence the "state space" will consist of two posets connected with a monotone relation. This

[4] A usual frame $(X, \le)$ for intuitionistic logic can be perceived as a relational frame defining $P(x, y; z) = x \le z \wedge y \le z$. Then coalgebraic morphisms correspond precisely to bounded morphisms.

is in compliance with various notions of generalized frames, as studied, e.g., in [16] and [14]. Furthermore, this approach will also allow to pass naturally from posets to categories enriched in a general commutative quantale. In the latter framework, we believe to be able to study, e.g., many-valued modal and substructural logics in a rather conceptual way.

A natural further direction would be to prove a more general Goldblatt-Thomason theorem for coalgebras over posets or categories enriched in a general commutative quantale, obtaining an analogue of [26, Theorem 3.15]. Another line of research explores the fact that the coalgebraic functor we obtained is easily seen to satisfy the Beck-Chevalley Condition in the sense of [2]. Hence it will be possible to develop the theory of cover modalities over coalgebras for distributive substructural logics.

References

[1] Bezhanishvili, N. and M. Gehrke, *Finitely generated free Heyting algebras via Birkhoff duality and coalgebra*, Logical Methods in Computer Science **7** (2011), pp. 1–24.

[2] Bílková, M., A. Kurz, D. Petrişan and J. Velebil, *Relation liftings on preorders*, in: A. Corradini, B. Klin and C. Cirstea, editors, *Proceedings of CALCO 2011*, LNCS **6859** (2011), pp. 115–129.

[3] Blackburn, P., M. de Rijke and Y. Venema, "Modal Logic," Cambridge University Press, 2001.

[4] Bloom, S. and J. Wright, *P-varieties: a signature free characterization of ordered algebras*, Jour. Pure Appl. Alg. **29** (1983), pp. 13–58.

[5] Bonsangue, M. M. and A. Kurz, *Duality for logics of transition systems*, in: *Proceedings of FoSaCS'05*, LNCS **3441** (2005), pp. 455–469.

[6] Bonsangue, M. M. and A. Kurz, *Presenting functors by operations and equations*, in: *Proceedings of FoSaCS'06*, LNCS **3921** (2006), pp. 455–469.

[7] Bonsangue, M. M., A. Kurz and I. M. Rewitzky, *Coalgebraic representation of distributive lattices with operators*, Topology and its Applications **154** (2007), pp. 778–791.

[8] Cabrer, L. M. and S. A. Celani, *Priestley dualities for some lattice ordered algebraic structures, including MTL, IMTL and MV-algebras*, Cent. Eur. J. Math. **4** (2006), pp. 600–623.

[9] Celani, S. and R. Jansana, *Priestley duality, a Sahlqvist theorem and a Goldblatt-Thomason theorem for positive modal logic*, Logic Jnl IGPL **7** (1999), pp. 683–715.

[10] Dunn, J. M., *Gaggle theory: An abstraction of Galois connections and residuation, with applications to negation, implication and various logical operators*, in: J. van Eijck, editor, *Proceedings of European Workshop JELIA '90*, LNCS **478** (1991), pp. 31–51.

[11] Dunn, J. M., *A representation of relation algebras using Routley-Meyer frames*, in: *Logic, Meaning and Computation. Essays in memory of Alonzo Church*, Kluwer, 2001 pp. 77–108.

[12] Dunn, J. M. and G. Restall, *Relevance logic*, in: D. Gabbay and F. Guenther, editors, *Handbook of Philosophical Logic vol. 5*, Kluwer, 2002 pp. 1–128.

[13] Dunn, M., M. Gehrke and A. Palmigiano, *Canonical extensions and relational completeness of some substructural logics*, Journal of Symbolic Logic **70** (2005), pp. 713–740.

[14] Galatos, N. and P. Jipsen, *Residuated frames with applications to decidability* (2011), to appear in *Trans. Amer. Math. Soc.*.

[15] Galatos, N., P. Jipsen, T. Kowalski and H. Ono, "Residuated Lattices: An Algebraic Glimpse at Substructural Logics," Elsevier, 2007.

[16] Gehrke, M., *Generalized Kripke frames*, Studia Logica **84** (2006), pp. 241–275.

[17] Gehrke, M. and B. Jónsson, *Bounded distributive lattices with operators*, Mathematica Japonica **40** (1994), pp. 207–215.
[18] Gehrke, M. and B. Jónsson, *Monotone bounded distributive lattice expansions*, Mathematica Japonica **52** (2000), pp. 197–213.
[19] Gehrke, M. and B. Jónsson, *Canonical extensions of bounded distributive lattice expansions*, Mathematica Scandinavica **94** (2004), pp. 13–45.
[20] Goldblatt, R., *Varieties of complex algebras*, Annals of Pure and Applied Logic **44** (1989), pp. 173–242.
[21] Goldblatt, R., *What is the coalgebraic analogue of Birkhoff's variety theorem?*, Theoretical Computer Science **266** (2001), pp. 853–886.
[22] Goldblatt, R., *Axiomatic classes of intuitionistic models*, Journal of Universal Computer Science **11** (2005), pp. 1945–1962.
[23] Goldblatt, R. I. and S. K. Thomason, *Axiomatic classes in propositional modal logic*, in: *Algebra and Logic*, Lecture Notes in Mathematics **450** (1975), pp. 163–173.
[24] Hartonas, C., *Duality for lattice-ordered algebras and for normal algebraizable logics*, Studia Logica **58** (1997), pp. 403–450.
[25] Kupke, C., A. Kurz and Y. Venema, *Stone coalgebras*, Theoretical Computer Science **327** (2004), pp. 109–134.
[26] Kurz, A. and J. Rosický, *The Goldblatt-Thomason theorem for coalgebras*, in: *Proceedings of CALCO 2007*, LNCS **4624** (2007), pp. 342–355.
[27] Kurz, A. and J. Velebil, *Enriched logical connections* (September 2011), *Appl. Categ. Structures* online first.
[28] Manes, E. G., "Algebraic theories," Graduate Texts in Mathematics, Springer, 1976.
[29] Palmigiano, A., *Coalgebraic semantics for positive modal logic*, Theoretical Computer Science **327** (2004), pp. 175–197.
[30] Petrişan, D., L. Nentvich and J. Velebil, *The Goldblatt-Thomason theorem beyond sets*, in: *Workshop CTU in Prague*, 2011.
[31] Porst, H.-E. and W. Tholen, *Concrete dualities*, in: H. Herrlich and H.-E. Porst, editors, *Category theory at work*, Heldermann Verlag, Berlin, 1991 pp. 111–136.
[32] Restall, G., "An introduction to substructural logics," Routledge, 2000.
[33] Rodenburg, P., "Intuitionistic Correspondence Theory," Ph.D. thesis, University of Amsterdam (1986).
[34] Routley, R. and R. K. Meyer, *The semantics of entailment I*, in: H. Leblanc, editor, *Truth, syntax and modality*, North Holland, Amsterdam, 1972 pp. 199–243.
[35] Sofronie-Stokkermans, V., *Duality and canonical extensions of bounded distributive lattices with operators, and applications to the semantics of non-classical logics I*, Studia Logica **64** (2000), pp. 93–132.
[36] Sofronie-Stokkermans, V., *Duality and canonical extensions of bounded distributive lattices with operators, and applications to the semantics of non-classical logics II*, Studia Logica **64** (2000), pp. 151–172.
[37] Stubbe, I., *Categorical structures enriched in a quantaloid: categories, distributors and functors*, Theory Appl. Categ. **14** (2005), pp. 1–45.
[38] Venema, Y., *Model definability, purely modal*, in: *JFAK. Essays Dedicated to Johan van Benthem on the Occasion of his 50th Birthday* (1999).

First Steps in Synthetic Guarded Domain Theory

Lars Birkedal [1]

IT University of Copenhagen
Denmark

We present the topos $\mathcal{S}$ of trees as a model of guarded recursion. We study the internal dependently-typed higher-order logic of $\mathcal{S}$ and show that $\mathcal{S}$ models two modal operators, on predicates and types, which serve as guards in recursive definitions of terms, predicates, and types. In particular, we show how to solve recursive type equations involving *dependent* types. We propose that the internal logic of $\mathcal{S}$ provides the right setting for the synthetic construction of abstract versions of step-indexed models of programming languages and program logics. As an example, we show how to construct a model of a programming language with higher-order store and recursive types entirely inside the internal logic of $\mathcal{S}$. Moreover, we give an axiomatic categorical treatment of models of synthetic guarded domain theory and prove that, for any complete Heyting algebra A with a well-founded basis, the topos of sheaves over A forms a model of synthetic guarded domain theory, generalizing the results for $\mathcal{S}$.

This talk is based on joint work with Rasmus Møgelberg, Jan Schwinghammer, and Kristian Støvring.

[1] `birkedal@itu.dk`

Indexical Hybrid Tense Logic

Patrick Blackburn

Section for Philosophy and Science Studies
Roskilde University

Klaus Frovin Jørgensen

Section for Philosophy and Science Studies
Roskilde University

Abstract

In this paper we explore the logic of *now*, *yesterday*, *today* and *tomorrow* by combining the semantic approach to indexicality pioneered by Hans Kamp [9] and refined by David Kaplan [10] with hybrid tense logic. We first introduce a special *now* nominal (our $@_{now}$ corresponds to Kamp's original now operator N) and prove completeness results for both logical and contextual validity. We then add propositional constants to handle *yesterday*, *today* and *tomorrow*; our system correctly treats sentences like "Niels will die yesterday" as contextually unsatisfiable. Building on our completeness results for *now*, we prove completeness for the richer language, again for both logical and contextual validity.

Keywords: Hybrid logic, two-dimensional logic, nominals, indexicals, now

Human languages are rife with indexicals. Words like *now*, *here*, and *I* are context sensitive: when uttered at different times and places by different speakers they denote different times, places and people, and they do so in a constrained way. The indexical *now* picks out the time of utterance, *here* picks out the place, and *I* the speaker. These semantical constraints mean that they possess an interesting logic. The expression

I am here now,

for example, cannot be uttered falsely. We might not want to call it a logical validity, but it clearly is some kind of validity. We shall use the term *contextual validity* to distinguish such sentences from ordinary logical validities such as *Either it is raining or it is not.*

In this paper, which builds on Blackburn [2], we are going to examine the logic of four temporal indexicals, *now*, *yesterday*, *today* and *tomorrow* within the setting of hybrid tense logic. The distinguishing feature of hybrid tense logic is that it uses special propositional symbols called nominals to refer to times. A nominal i is true at a unique time; it names the time it is true at. We will treat *now* as a special nominal that is true at a contextually determined utterance time, and view *yesterday*, *today* and *tomorrow* as propositional constants true at unbroken stretches of time correctly aligned around this special point. The contextual semantics of these indexicals will be handled by the method pioneered by Hans Kamp in *Formal Properties of 'Now'* [9], and refined and extended by David Kaplan in *Demonstratives* [10]. Indeed, the present paper could be described as "Hybrid Logic meets Kamp-Kaplan semantics".

This meeting has two main advantages. The first is semantic. Hybridization enables us to perspicuously capture a number of important facts about indexicals and their interaction with the tenses. Consider, for example

Niels will die yesterday.

This is contextually incoherent. It cannot be truthfully uttered in any context because of the clash between the indexical *yesterday* (which places the dying in the past, namely sometime during yesterday) and the tensed verb *will die* which places the dying in the future. We shall represent this sentence as

$$F(\textit{yesterday} \wedge \textit{Niels-die})$$

and our semantics will guarantee that this formula cannot be satisfied in any context of utterance in any model, which is just as it should be.

The second advantage is logical. The literature on hybrid logic contains many general results on completeness and other topics; these results (which were proved for ordinary nominals) can be straightforwardly adapted to deal with the logic of *now*, and doing so sheds interesting light on results such as Kamp's eliminability result for his N operator. Moreover, once the logics of *now* have been captured, it is straightforward to build on them to capture the logics of *yesterday*, *today* and *tomorrow*. Indeed, by the end of the paper it should be clear that what we are presenting is not so much a particular logic of indexicality as a framework for modeling indexicality in a wide range of modal logics and languages.

1 Basic hybrid tense logic

The basic idea behind hybrid logic is to introduce a second sort of atomic symbol, nominals. Thus the point of departure for our work is a two-sorted language $\mathcal{L}$ which contains a countable set $\Phi = \{p, q, r, \ldots\}$ of propositional symbols and another (disjoint) countable set of nominals $\Omega = \{i, j, k, \ldots\}$. These are our atomic symbols. Our tense logic will be diamond-based: there is a diamond P for looking backwards in time, a diamond F for looking forwards, and for each nominal i an $@_i$-operator. Formulas of $\mathcal{L}$ are built as follows:

$$\phi ::= i \mid p \mid \bot \mid \neg\phi \mid \phi \wedge \psi \mid P\phi \mid F\phi \mid @_i\phi.$$

We define $H\phi$ to be $\neg P \neg\phi$ and $G\phi$ to be $\neg F \neg \phi$. A formula is said to be *pure* if its only atomic subformulas are nominals. Note that nominals can occur either as subscripts to @ ("in operator position") or as formulas in their own right ("in formula position").

The semantics for the basic hybrid tense logic is given by interpreting formulas of $\mathcal{L}$ in models based on a frame (T, R) together with a valuation V. Here T is a non-empty set and R is a binary relation. The elements of T are thought of as points of time and the relation R as the temporal flow. P searches backwards along this relation (into the past) whereas F searches forward (into the future). The valuation V distributes information over the frame, thus V takes atomic formulas to subsets of T and it satisfies the following two conditions:

(i) $V(p)$ is a subset of T, when $p \in \Phi$,

(ii) $V(i)$ is a singleton subset of T, when $i \in \Omega$.

We say that t is the denotation of i under V iff $t \in V(i)$.

Satisfiability in a model is defined in the usual way as a relation which obtains between a model $\mathfrak{M}$, a point t in the model, and a formula ϕ:

$$\begin{array}{ll}
\mathfrak{M}, t \models \bot & \text{never} \\
\mathfrak{M}, t \models a & \text{iff } a \text{ is atomic and } t \in V(a) \\
\mathfrak{M}, t \models \neg\phi & \text{iff } \mathfrak{M}, t \not\models \phi \\
\mathfrak{M}, t \models \phi \wedge \psi & \text{iff } \mathfrak{M}, t \models \phi \text{ and } \mathfrak{M}, t \models \psi \\
\mathfrak{M}, t \models P\phi & \text{iff for some } t',\ t'Rt \text{ and } \mathfrak{M}, t' \models \phi \\
\mathfrak{M}, t \models F\phi & \text{iff for some } t',\ tRt' \text{ and } \mathfrak{M}, t' \models \phi \\
\mathfrak{M}, t \models @_i\phi & \text{iff } \mathfrak{M}, t' \models \phi \text{ and } t' \in V(i).
\end{array}$$

A formula ϕ is *true in* $\mathfrak{M} = (T, R, V)$ when for all points $t \in T$ we have that $\mathfrak{M}, t \models \phi$. A formula is *logically valid* if it is true in all models.

The set of logically valid formulas (the basic hybrid tense logic) is called K_h^t. This logic can be proof theoretically characterized in a number of ways, for example via Hilbert systems [4], via natural deduction systems [6], or via tableau systems [3,5]. In this paper we shall use a tableau system; see the Appendix for details. For this system we have:

Theorem 1.1 (Basic Completeness) *Any set of formulas in $\mathcal{L}$ that is K_h^t-consistent is satisfiable in a model. Moreover, if Π is some set of pure $\mathcal{L}$-axioms, then any set of formulas which is $K_h^t + \Pi$-consistent is satisfiable in a model based on a frame satisfying the frame properties defined by Π.*

Note that this is not one, but many, completeness theorems. In our definition of the semantics, we imposed no restrictions on the relation R. However, given that we think of R as the flow of time, it would be natural to impose additional demands such as transitivity and irreflexivity. Theorem 1.1 tells us that if the required constraints can be expressed by a pure formula, then adding that formula as an axiom to K_h^t automatically yields completeness with respect to these properties. For example, $@_i\neg Fi$ and $FFi \to Fi$ express, respectively, the properties of irreflexivity and transitivity. Hence (as both formulas are pure) adding them as axioms to K_h^t yields a tableau system that is complete with respect to the class of frames in which R is a strict partial order.

2 Adding *now*

Our first step is to extend $\mathcal{L}$ with the atomic symbol *now* to obtain $\mathcal{L}(now)$. The new symbol is in essence a nominal, but it is a very special one with a very special (indexical) meaning. So with the introduction of *now* we have three sorts of atomic formulas: the propositional symbols Φ, the ordinary nominals Ω, and the indexical *now*. The formulas of $\mathcal{L}(now)$ are built as follows:

$$\phi ::= now \mid i \mid p \mid \bot \mid \neg\phi \mid \phi \wedge \psi \mid P\phi \mid F\phi \mid @_{now}\phi \mid @_i\phi.$$

Models for $\mathcal{L}(now)$ will be contextualized versions of the ordinary models presented in the previous section. A contextual model $\mathfrak{M}$ is a 5-tuple

$$\mathfrak{M} = (T, R, V, C, \eta)$$

where (T, R) is an ordinary frame, V is a valuation function (to be defined below), C is a non-empty set of contexts, and η is a mapping from contexts to points in T. The function η is crucial: it specifies, for any context $c \in C$, what the time (or temporal location) of any utterance in that context is. That is, it tells you, for any context, what your "now" moment is. Thus η is exactly what Kaplan [10] calls the *character* of "now". As before, the valuation V interprets atomic formulas, but now it does so relative to contexts. Therefore a contextual valuation V takes a pair (c, a) consisting of a context c and an atom a and assigns a subset of T to the pair subject to the following restrictions:

(i) $V(c, p)$ is a subset of T, when $p \in \Phi$,

(ii) $V(c, i)$ is a singleton subset of T, when $i \in \Omega$,

(iii) $V(c, now) = \{\eta(c)\}$.

Note the restriction hard-wired into the semantics of *now*. This special nominal denotes a singleton, but not just any singleton: in any context c it denotes the utterance time $\eta(c)$ in that context. Satisfiability in contextual models is construed as a relation obtaining between four elements: a model $\mathfrak{M} = (T, R, V, C, \eta)$, a context c, a point t, and a formula ϕ:

$\mathfrak{M}, c, t \models \bot$	never
$\mathfrak{M}, c, t \models a$	iff a is atomic and $t \in V(c, a)$
$\mathfrak{M}, c, t \models \neg\phi$	iff $\mathfrak{M}, c, t \not\models \phi$
$\mathfrak{M}, c, t \models \phi \wedge \psi$	iff $\mathfrak{M}, c, t \models \phi$ and $\mathfrak{M}, c, t \models \psi$
$\mathfrak{M}, c, t \models P\phi$	iff for some t', $t'Rt$ and $\mathfrak{M}, c, t' \models \phi$
$\mathfrak{M}, c, t \models F\phi$	iff for some t', tRt' and $\mathfrak{M}, c, t' \models \phi$
$\mathfrak{M}, c, t \models @_a\phi$	iff $\mathfrak{M}, c, t' \models \phi$ and $t' \in V(c, a)$, where a is *now* or $a \in \Omega$.

Let's think carefully about what the last clause means when working with the $@_{now}$-operator. When we evaluate $@_{now}\phi$ at (c, t) (that is, when we ask whether $\mathfrak{M}, c, t \models @_{now}\phi$ holds) we jump to the point $\eta(c) \in T$ associated with c and ask whether ϕ is satisfied there relative to context c (recall the intuition that $\eta(c)$ is the temporal location of any utterance in c). That is:

$$\mathfrak{M}, c, t \models @_{now}\phi \quad \text{iff} \quad \mathfrak{M}, c, \eta(c) \models \phi.$$

What about validity? Well, clearly we can generalize our previous definition. We say that ϕ is *true in* $\mathfrak{M}$ if for all pairs c, t from $\mathfrak{M}$ have $\mathfrak{M}, c, t \models \phi$. And then we say that ϕ is *logically valid* if ϕ is true in all models. This is essentially the regular notion of validity familiar from modal logic.

But the whole point of working with contextualized models is that they support a second notion of validity, *contextual validity*. We are interested in characterising not only the set of logically valid formulas, but also *the set of formulas that are always true whenever they are uttered.* To put this more precisely: we say, with respect to $\mathfrak{M}$, that ϕ *is satisfied in the context* c precisely when ϕ in $\mathfrak{M}$ is satisfied in c at the utterance time of c, that is, whenever

$$\mathfrak{M}, c, \eta(c) \models \phi.$$

We say that ϕ is *contextually true in* $\mathfrak{M}$ when ϕ is satisfied in every context c in $\mathfrak{M}$. And ϕ is *contextually valid* if it is contextually true in every model. [1]

Let us consider some examples of contextual validities. Clearly ordinary tautologies from propositional logic are logically valid. And trivially, any formula which is logically valid is also contextually valid. On the other hand, *now* is not logically valid, but it is contextually valid: given any $\mathfrak{M}$, and any c, $\eta(c)$ is the denotation of *now* under V in c, That is, for any context c, $\eta(c) \in V(c, now)$ and so we always have that:

$$\mathfrak{M}, c, \eta(c) \models now$$

In Kaplan's words: *now* "cannot be uttered falsely" (Kaplan, 1978, p. 402).

Now for two simple but important model-theoretic properties of contextual semantics with respect to $\mathcal{L}$ (recall that $\mathcal{L}$ is $\mathcal{L}(now)$ without *now*):

Lemma 2.1 (Playing with Contexts) *All $\mathcal{L}$-formulas are semantically insensitive to contexts:*

(i) *Any ordinary model can be extended to a contextual model which, with respect to $\mathcal{L}$-formulas, is semantically equivalent to the original model.*

(ii) *Moreover, $\mathcal{L}$-formulas are insensitive to how contexts are mapped to moments of time.*

Proof. Part (i). Let $\mathfrak{M}$ be an ordinary model (T, R, V) and let $\mathfrak{M}'$ be $\mathfrak{M}$ contextualised by $C = \{c\}$ and η defined by $\eta(c) = t_1$, where t_1 is the denotation of i_1 (the first nominal in some enumeration of the nominals) under V. Moreover, let V' be defined by $V'(c, a) = V(a)$ for any atomic a. With $\mathfrak{M}' = (T, R, V', C, \eta)$ we have for any formula ϕ of $\mathcal{L}$ that:

$$\mathfrak{M}, t \models \phi \ \text{ iff } \ \mathfrak{M}', c, t \models \phi.$$

[1] Contextually validity is what Kaplan calls *validity* [10, p. 547]. This strikes us as confusing. We prefer to use the more explicit *contextual validity* to clearly signal when we are talking about the Kamp-Kaplan notion of validity, and *logical validity* for regular modal validity.

The proof is a trivial induction on the complexity of ϕ.

Part (ii). Given any model $\mathfrak{M} = (T, R, V, C, \eta)$ we can replace η with an arbitrary η' having C as domain, resulting in a model $\mathfrak{M}' = (T, R, V', C, \eta')$, where V' is defined by the following:

$$V'(c,a) = \begin{cases} \{\eta'(c)\}, & \text{if } a \text{ is } now, \\ V(c,a), & \text{otherwise.} \end{cases}$$

Then $\mathfrak{M}$ and $\mathfrak{M}'$ have the same semantical behaviour with respect to $\mathcal{L}$-formulas:

$$\mathfrak{M}, c, t \models \phi \quad \text{iff} \quad \mathfrak{M}', c, t \models \phi.$$

Once again, the proof is a trivial induction on the structure of $\mathcal{L}$-formulas. □

To conclude this section we briefly discuss Hans Kamp's [9] eliminability result for his now operator N (for a succinct overview, see Burgess [7]). Expressed in our notation, Kamp's result is for the language of tense logic together with $@_{now}$ (that is, his language contains no ordinary nominals, no ordinary @-operators, and no *now* nominal). He proved that if a formula containing occurrences of $@_{now}$ is satisfiable at the designated now point (that is, $\eta(c)$), then there is an equivalent $@_{now}$ free formula satisfiable at that point.

Kamp's result can be viewed as a special case of a more general observation, due to Balder ten Cate [8], concerning arbitrary @-operators. Let's first consider the special case of Kamp's language. Suppose that ϕ is a formula in the restricted language just described, and suppose that $@_{now}\psi$ is some subformula occurrence in ϕ. Then (following ten Cate) we observe that:

$$@_{now}\psi \wedge \big(\phi[@_{now}\psi \leftarrow \top] \vee \phi[@_{now}\psi \leftarrow \bot]\big)$$

is equivalent to ϕ. By iteratively continuing this process we eventually arrive at a formula where no occurrence of $@_{now}$ lies in the scope of any other $@_{now}$ or any tense operator. This longer formula (call it ϕ^+) is equivalent to ϕ, and hence satisfiable at the utterance time iff ϕ is. And now a simple observation yields Kamp's result: at the utterance time, for any formula θ we have $@_{now}\theta \leftrightarrow \theta$. So replacing each subformula of the form $@_{now}\theta$ in ϕ^+ by θ, yields an equivalent $@_{now}$-free formula ϕ^k which is satisfiable at the utterance time iff ϕ^+ is.

That gives us the classic Kamp result, but ten Cate's simple argument works for *arbitrary* $@_i$-operators, not just for the special case of $@_{now}$. So it is easy to generalize Kamp's result to the full language of this paper. Suppose we have a formula ϕ of this language. Following ten Cate, we expand our original ϕ to an equivalent ϕ^+ with the property that no $@_i$-operator or $@_{now}$ lies under the scope of any other @-operator, or $@_{now}$, or any tense operator. And then we argue as above: any $@_i$-operator can be eliminated at the denotation of i using the equivalence $@_i\phi \leftrightarrow \phi$. Note that this argument does *not* eliminate nominals (including *now*) that occur in formula position; it simply shows that we can select any operator $@_i$ and eliminate it at the denotation of i.

3 Completeness for *now*

In this section we are going to look at the two languages $\mathcal{L}$ and $\mathcal{L}(now)$ with respect to the two different notions of validity given by contextual semantics. That is, we are going to examine four logics and establish the results given in the following matrix:

	$\mathcal{L}$	$\mathcal{L}(now)$
Logical validity	**Logic 1:** K_h^t	**Logic 3:** $K_h^t(now)$
Contextual validity	**Logic 2:** K_h^t	**Logic 4:** $K_h^t(now) + now$

Recall that K_h^t is the basic hybrid tense logic in language $\mathcal{L}$. The logic $K_h^t(now)$ is obtained by applying the tableau rules of K_h^t to the formulas of $\mathcal{L}(now)$ treating *now* as if it were an ordinary nominal. We'll explain what $K_h^t(now) + now$ is later in this section.

Our arguments will be semantic. In essence, in the work that follows we are trying to pin down when *now* behaves like a regular nominal, and when it does not (that is, when its special contextual properties kick in). This information will be valuable when we later add *yesterday*, *today* and *tomorrow*. As a preliminary step, we state the soundness of our tableau system.

A formula is called a *satisfaction statement* if it is either of the form $@_a\phi$ or $\neg @_a\phi$, where a is *now* or an ordinary nominal. If Σ is a set of satisfaction statements and $\mathfrak{M}$ is a contextual model, then we say that Σ is *satisfied by label* in $\mathfrak{M}$ if there is a context c such that for every formula in Σ:

(i) If $@_a\phi \in \Sigma$ then $\mathfrak{M}, c, \overline{a} \models \phi$,

(ii) If $\neg @_a\phi \in \Sigma$ then $\mathfrak{M}, c, \overline{a} \not\models \phi$,

where $\overline{a}$ is the denotation of a under V in c.

Lemma 3.1 (Soundness) *For any set of satisfaction statements Σ in $\mathcal{L}(now)$, if Σ is satisfiable by label, then so is at least one of the sets obtained by applying any rule of $K_h^t(now)$ to Σ. Hence both $K_h^t(now)$ and its subsystem K_h^t are sound.*

Proof. Proving this requires essentially nothing beyond the soundness proof for ordinary semantics given in Blackburn [3]. The basic point is that as far as tableaux rules are concerned, *now* behaves much like an ordinary nominal. □

Logically and contextually valid formulas in $\mathcal{L}$

Let's turn to completeness. Our first result should not be a surprise. It says that K_h^t captures logical validity in $\mathcal{L}$ with respect to contextual semantics.

Theorem 3.2 K_h^t *is complete with respect to logically valid $\mathcal{L}$-formulas.*

Proof. Let ϕ be some K_h^t-consistent $\mathcal{L}$-formula. By Theorem 1.1, ϕ is satisfiable in some ordinary model. But then, by Lemma 2.1, ϕ is satisfiable in a contextual model too. □

But we can also show that, for $\mathcal{L}$, K_h^t is still the complete tableau system when we turn from logical validity to contextual validity (recall that a formula ϕ is contextually valid iff for all $\mathfrak{M}$ and all c, $\mathfrak{M}, c, \eta(c) \models \phi$). And this should not be surprising either—after all, without *now* in the language, we have nothing that gets to grips with contexts. And that's indeed how things turn out:

Theorem 3.3 *K_h^t is complete with respect to contextually valid $\mathcal{L}$-formulas.*

Proof. We show that the set of contextually valid $\mathcal{L}$-formulas equals the set of logically valid $\mathcal{L}$-formulas. Given that, the result follows from Theorem 3.2.

Suppose for the sake of contradiction that ϕ in $\mathcal{L}$ is contextually valid but not logically valid. Then there is a model $\mathfrak{M} = (T, R, V, C, \eta)$, a context c, and a point t such that $\mathfrak{M}, c, t \not\models \phi$. Define $\eta' : C \to T$ to be constantly t and replace η in the definition of V by η' to obtain $\mathfrak{M}' = (T, R, V', C, \eta')$. As formulas in $\mathcal{L}$ are insensitive to how contexts are mapped to moments in time (Lemma 2.1.ii) we have $\mathfrak{M}', c, \eta'(c) \not\models \phi$ which contradicts the contextual validity of ϕ. □

Summing up, we have established the results in the left-hand column of our matrix: K_h^t is both **Logic 1** and **Logic 2**. Time to turn to $\mathcal{L}(now)$.

Logically and contextually valid formulas in $\mathcal{L}(now)$

The tableau system $K_h^t(now)$ in $\mathcal{L}(now)$—where *now* is treated as an ordinary nominal—will characterise the logically valid $\mathcal{L}(now)$-formulas. The contextually valid formulas, however, will be characterised by $K_h^t(now)$ together with the formula *now* regarded as a single axiom; we call the tableau system embodying this idea $K_h^t(now) + now$. More precisely, an $\mathcal{L}(now)$-formula ϕ is provable in $K_h^t(now) + now$ if and only if there is a finite closed tableau whose root node is $@_i(now \wedge \neg\phi)$ where i does not occur in ϕ. The idea is to enforce i to denote both the utterance time and the point where ϕ must be falsified; if we can't falsify ϕ at the utterance time then it must be contextually valid.[2] In order to prove that these two systems characterise the two different notions of validities we need some technicalities regarding substitutions. Moreover, until now, we have worked with fixed sets of propositional symbols Φ and nominals Ω, but in the proofs that follow we shall need a nominal not in Ω. Thus, for the rest of this section, let j be a fresh nominal not in this set, and let our tableau system in $\mathcal{L}$ extended with nominal j be called $K_h^t(j)$. Finally, we'll use the following notation: if in ϕ we uniformly substitute ρ for ψ we obtain $\phi[\psi \leftarrow \rho]$.

Lemma 3.4 (Playing with Substitutions) *With regard to satisfiability and consistency, now is just a nominal:*

(i) *If ϕ in $\mathcal{L}$ is satisfiable at a pair (c, t) from a model $\mathfrak{M}$, then now can be uniformly substituted for any nominal i in ϕ and the resulting formula $\phi[i \leftarrow now]$ is also satisfiable at (c, t) at a model $\mathfrak{M}'$ which differs from $\mathfrak{M}$ only in the value it assigns to now.*

(ii) *Given any formula ϕ of $\mathcal{L}(now)$, ϕ is $K_h^t(now) + now$-consistent if and only if $j \wedge \phi[now \leftarrow j]$ is $K_h^t(j)$-consistent.*

[2] In practice, when proving formulas in $K_h^t(now) + now$ we build a tableau whose root node is $@_i now$ and whose second node is $\neg @_i \phi$, thereby saving two steps.

(iii) *Given any formula* ϕ *of* $\mathcal{L}(now)$, *if* $\mathfrak{M}, c, \eta(c) \models j \wedge \phi[now \leftarrow j]$ *then* $\mathfrak{M}, c, \eta(c) \models now \wedge \phi$.

Proof. Part (i). Suppose ϕ in $\mathcal{L}$ is satisfiable in $\mathfrak{M} = (T, R, V, C, \eta)$ and suppose i is a nominal in ϕ. As ϕ is insensitive to contexts we simply take some context c and ask what the denotation of i under V in context c is. Suppose it is t'. We let η' be constantly t' and let V' just be V with η replaced by η'. For $\mathfrak{M}' = (T, R, V', C, \eta')$ it can be proved by an easy induction that for any $t \in T$ and any subformula ψ of ϕ:

$$\mathfrak{M}, c, t \models \psi \ \text{ iff } \ \mathfrak{M}', c, t \models \psi[i \leftarrow now].$$

Part (ii). We first note two simple but useful facts:

- For any ψ in $\mathcal{L}(now)$ the formula $\psi[now \leftarrow j]$ is in $\mathcal{L}(j)$ and j only occurs where it has replaced now.
- If ψ is a formula in $\mathcal{L}(j)$ then $\psi[j \leftarrow now]$ is in $\mathcal{L}(now)$, and now occurs only in $\psi[j \leftarrow now]$ where it has replaced j.

Hence part (ii) follows easily. For suppose $j \wedge \phi[now \leftarrow j]$ is $K_h^t(j)$-consistent. Then any tableau having $@_i(j \wedge \phi[now \leftarrow j])$ as root node will contain an open branch. Therefore there cannot be a closed tableau in $K_h^t(now) + now$ with $@_i(now \wedge \phi)$ at the root, as this could be turned into a closed tableau for $@_i(j \wedge \phi[now \leftarrow j])$ by uniformly substituting j for now. And conversely, if ϕ in $\mathcal{L}(now)$ is $K_h^t(now) + now$-consistent, then any tableau for $@_i(now \wedge \phi)$ will contain an open branch. So there cannot be closed tableau in $K_h^t(j)$ with $@_i(j \wedge \phi[now \leftarrow j])$ at the root either.

Part (iii). Given the assumption that $\mathfrak{M}, c, \eta(c) \models j \wedge \phi[now \leftarrow j]$ it is easy to prove, that for any subformula ψ of ϕ

$$\mathfrak{M}, c, \eta(c) \models \psi[now \leftarrow j] \ \text{ iff } \ \mathfrak{M}, c, \eta(c) \models \psi.$$

For the atomic case the claim is completely trivial for all atomic formulas except when ψ is now. But then $\mathfrak{M}, c, \eta(c) \models j$ iff $\mathfrak{M}, c, \eta(c) \models now$ as we are working under the assumption that j denotes $\eta(c)$ under V in context c. The induction step then follows trivially except for when ψ is $@_{now}\theta$. But again, this follows by assumption as j and now both denote $\eta(c)$ under V in context c. □

Theorem 3.5 $K_h^t(now)$ *is complete with respect to the logically valid formulas in contextual semantics.*

Proof. Let ϕ be a formula of $\mathcal{L}(now)$ and suppose ϕ is $K_h^t(now)$-consistent. Let i be the first nominal not occurring in ϕ and substitute i for now in ϕ. Clearly $\phi[now \leftarrow i]$ is $K_h^t(now)$-consistent, thus by Theorem 3.2 $\phi[now \leftarrow i]$ has a model. We then apply Lemma 3.4.i to get the required model for ϕ. □

Theorem 3.6 $K_h^t(now)$+now *is complete with respect to the contextually valid formulas in contextual semantics.*

Proof. Suppose ϕ in $\mathcal{L}(now)$ is $K_h^t(now) + now$-consistent. By Lemma 3.4.ii, $j \wedge \phi[now \leftarrow j]$ is $K_h^t(j)$-consistent, hence by Theorem 3.2 there is a model $\mathfrak{M} = (T, R, V, C, \eta)$ such that:

$$\mathfrak{M}, c, t \models j \wedge \phi[now \leftarrow j].$$

As $\phi[now \leftarrow j]$ does not contain *now*, it is insensitive to the way C is mapped on T (Lemma 2.1.ii). We then do the usual trick of defining η' to be constantly t, and this induces the usual V' obtained from V by replacing η by η'. We let $\mathfrak{M}' = (T, R, V', C, \eta')$ and obtain:

$$\mathfrak{M}', c, \eta'(c) \models j \wedge \phi[now \leftarrow j].$$

By Lemma 3.4.iii, $now \wedge \phi$, and thus ϕ, is satisfiable at $c, \eta'(c)$ in $\mathfrak{M}'$ too. □

We end our discussion of *now* with a general theorem.

Theorem 3.7 *Let* Λ *be* K_h^t *extended with pure axioms, and* $\Lambda + now$ *be its contextualized counterpart. Then:*

(i) $\Lambda + now$ *is contextually complete with respect to the same classes of models as that* Λ *is logically complete for.*

(ii) Λ*-satisfiability has the same complexity as* $\Lambda + now$ *satisfiability.*

(iii) *There is a terminating tableau system for* $\Lambda + now$ *iff there is a terminating tableau system for* Λ.

Proof. Part (i) is essentially the standard result, Theorem 1.1, from hybrid logic: pure axioms restrict us to the appropriate model classes, and adding *now* does not effect this, as satisfiability in a contextual model for any ϕ in $\mathcal{L}(now)$ can be reduced to satisfiability of $\phi[now \leftarrow j]$ in an *ordinary* model (where j is fresh). As for Parts (ii) and (iii), it follows by our results above that finding a satisfying model (or a terminating tableau) for $@_i(now \wedge \phi)$ amounts to finding a satisfying model (or a terminating tableau) for $@_j(j \wedge \phi[now \leftarrow j])$, where j is a fresh nominal. □

4 Adding *yesterday*, *today* and *tomorrow*

Let's extend $\mathcal{L}(now)$ with three symbols: *yesterday*, *today* and *tomorrow*. Like *now* these are indexicals, but they are not nominals. Rather, they are special propositional symbols: each denotes a "daylike" set of points correctly positioned in the model with respect to the utterance time. So the formulas of $\mathcal{L}(now, yesterday, today, tomorrow)$ are formed like those of $\mathcal{L}(now)$, except that we can use *yesterday*, *today* and *tomorrow* as atomic symbols. It's perhaps worth emphasizing that these three new symbols only occur in formula position, never in operator position; they are not nominals, and @ requires nominals as subscripts.

What about models? The key point is to add further structure to contexts. Until now, the only structure on contexts has been the function η which returns the utterance time. Here we will add three further functions: YESTERDAY,

TODAY, and TOMORROW. These map contexts to sets of times, and we impose a number of constraints. These constraints ensure that the three sets are correctly "contextually placed", with respect to the utterance time $\eta(c)$ and each other, and that they are sufficiently "daylike". In essence we are specifying what Kaplan calls the character of *yesterday*, *today* and *tomorrow*, and the following diagram shows what we require:

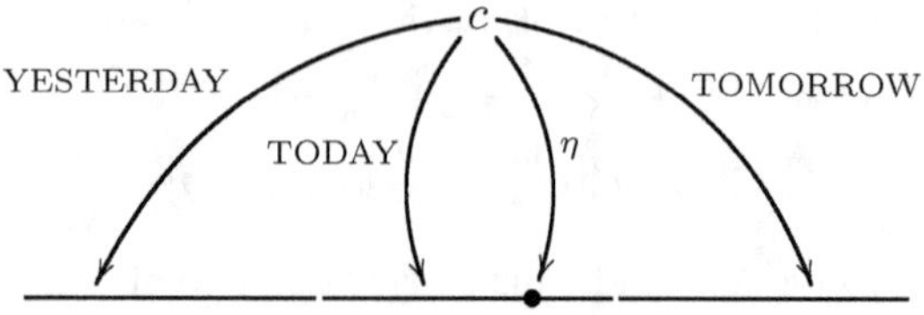

An important remark: we are *not* going to impose any global requirements on R. It would be easy to insist that R be irreflexive, transitive, or linear, but we won't do this. Instead, we impose structure *locally*, that is, on only these three sets of times. Why work this way? For a number of reasons. For a start, working locally means that our approach can be used with very weak tense logics. Moreover, there is no one class of temporal models suitable for every application: philosophers may be torn between Ockhamist and Peircean branching time, semanticists may demand linear time, while computer scientists may want discrete time for some applications and dense or even continuous time for others. We want our analysis to adapt easily to all such demands.

Let's make the pictorial constraints on character explicit. For all $c \in C$, YESTERDAY(c), TODAY(c) and TOMORROW(c) are subsets of T such that:

(i) $\eta(c) \in$ TODAY(c).

(ii) $\eta(c)$ is an R-successor of every point in YESTERDAY(c).

(iii) $\eta(c)$ is an R-predecessor of every point in TOMORROW(c).

(iv) YESTERDAY(c), TODAY(c), and TOMORROW(c) are pairwise disjoint.

(v) Every point in YESTERDAY(c) R-precedes every point in TODAY(c).

(vi) Every point in TODAY(c) R-precedes every point in TOMORROW(c).

(vii) Every point in YESTERDAY(c) R-precedes every point in TOMORROW(c).

(viii) YESTERDAY(c), TODAY(c), and TOMORROW(c) are all R-convex.

(ix) If $t \in$ YESTERDAY(c) and $t' \in$ TODAY(c) and tRs and sRt', then either $s \in$ YESTERDAY(c) or $s \in$ TODAY(c).

(x) If $t \in$ TODAY(c) and $t' \in$ TOMORROW(c) and tRs and sRt', then either $s \in$ TODAY(c) or $s \in$ TOMORROW(c).

In the presence of global assumptions about R (such as irreflexivity and transitivity) this list contains redundancies; for example, we don't need item vii if R is transitive. Its virtue (as we shall see below) is that even in the absence of such assumptions it imposes enough constraints to support interesting local inferences about contextuality.

Models for our expanded language simply build in this extra structure. That is, a contextual model is now an 8-tuple

$$\mathfrak{M} = (T, R, V, C, \eta, \text{YESTERDAY}, \text{TODAY}, \text{TOMORROW}),$$

and it only remains to specify the valuation for our three new atomic symbols:

(i) $V(c, \textit{yesterday}) = \text{YESTERDAY}(c)$,

(ii) $V(c, \textit{today}) = \text{TODAY}(c)$,

(iii) $V(c, \textit{tomorrow}) = \text{TOMORROW}(c)$.

So what about completeness? Let us us first deal with logical validity. We define $K_h^t(now, yesterday, today, tomorrow)$ to be $K_h^t(now)$ augmented with all instances of the following axioms. That is, we extend our tableau system by allowing any of the following formulas to be freely introduced in the course of tableau construction:

Now Placement	**Disjointness**
$now \to today$	$today \to \neg tomorrow$
$yesterday \to F now$	$today \to \neg yesterday$
$tomorrow \to P now$	$yesterday \to \neg tomorrow$
One Step Alignment	**Two Step Alignment**
$today \to G\neg yesterday$	
$tomorrow \to G\neg today$	$tomorrow \to G\neg yesterday$
Convexity	**No Gaps**
$P yesterday \wedge F yesterday \to yesterday$	$P yesterday \wedge F today \to yesterday \vee today$
$P today \wedge F today \to today$	$P today \wedge F tomorrow \to today \vee tomorrow$
$P tomorrow \wedge F tomorrow \to tomorrow$	

These axioms correspond in an obvious way to the ten requirements we demand of our character functions. Indeed—because of the clarity of the correspondences involved—it would be straightforward to impose even *less* structure simply by dropping suitable axioms. For example, for some applications we might want to think of all three days as points, in which case we would simply drop the Convexity and No Gaps axioms. But in any case, adding all the above axioms results in *logical* completeness with respect to context structures as defined by the ten constraints on character.

But now for the key issue: how do we get *contextual* completeness? Exactly as we did before. Let

$$K_h^t(now, yesterday, today, tomorrow) + now$$

be $K_h^t(now, yesterday, today, tomorrow)$ extended by insisting that the tableau construction for a formula ϕ starts with $@_k(now \wedge \phi)$ at the root node, or equivalently, with $@_k now$ at the root node and $@_k\phi$ immediately afterwards (as before, k is a nominal not occuring in ϕ). Informally, just as with $K_h^t(now)+$ *now*, contextual validity is captured by asserting *Now!* at the start of tableaux construction. And a little reflection shows why this must be so. If we utter *Now!* in a context, then modus ponens fires the Now Placement axioms and we immediately infer that *It's Today!*, that *It's not Yesterday!* and that *It's not Tomorrow!* And (just glance through the axiom list) modus ponens and modus tollens keep firing until we have full information about the relative location and structure of the three days. To put it another way: *now* is the only bridge we require between realms of logical and contextual validity. Uttering *now* nails $\eta(c)$ firmly into the temporal flow, and *yesterday*, *today*, and *tomorrow* line up obediently around it.

Here is an example of the system in action. We shall show that in any context of utterance, it is impossible to satisfy

$$F(yesterday \wedge Niels\text{-}die).$$

Here is the required tableau:

1 $@_k now$	
2 $@_k F(yesterday \wedge Niels\text{-}die)$	
3 $@_k Fi$	2, *F Elimination*
4 $@_i(yesterday \wedge Niels\text{-}die)$	2, *F Elimination*
5 $@_i yesterday$	4, $\wedge$ *Elimination*
6 $@_i Niels\text{-}die$	4, $\wedge$ *Elimination*
7 $@_k(now \rightarrow today)$	*Now Placement axiom*
8 $@_k today$	1, 7, *Modus Ponens*
9 $@_k(today \rightarrow G\neg yesterday)$	*One Step Alignment axiom*
10 $@_k G\neg yesterday$	8, 9, *Modus Ponens*
11 $@_i \neg yesterday$	3, 10, *G Propogation*
12 $\perp$	5, 11

Theorem 4.1 *Let* Λ *be* $K_h^t(now, yesterday, today, tomorrow)$ *extended with pure axioms, and let* Λ+*now be its contextualized counterpart. Then* Λ+*now is contextually complete with respect to the same class of models that* Λ *is logically complete for.*

Proof. As we have already informally discussed, beginning the tableau construction with $@_k now$ and $\neg @_k \phi$ is all that needs to be done to test ϕ for contextual rather than logical validity. The claim about completeness for pure logics Λ is just the familiar point that properly designed hybrid proof systems are complete for any pure extension of the minimal logic. □

But the previous theorem is only a partial analog of Theorem 3.7. In our earlier work on $K_h^t(now) + now$ we obtained not merely general completeness results but also general results concerning complexity and tableau termination. With the richer language, however, matters are more complicated: the computational consequences of our use of axioms for character is unclear. But for some classes of commonly used models (such as those based on linear frames) more can be said. Here's an example.

Nowadays modal logicians often add A, the universal modality, as a new primitive, to logics they find interesting: $A\phi$ asserts that ϕ holds at *all* points in a model. But on some classes of model A may be definable using the tense operators. For example, when working with transitive linear models, $A\phi$ is definable as $H\phi \vee \phi \vee G\phi$. This leads to the following:

Theorem 4.2 *Let* M *be a class of models on which the universal modality is definable in terms of the tense operators. Then the complexity of the satisfiability problem for* $K_h^t(now, yesterday, today, tomorrow)$ *is the same as the complexity satisfiability problem for* K_h^t *over* M*, and so is the satisfiability problem for* $K_h^t(now, yesterday, today, tomorrow) + now$.

Proof. Let's first treat $K_h^t(now, yesterday, today, tomorrow)$ satisfiability. Let **Character** be the conjunction of the character axioms listed above. Then observe that $K_h^t(now, yesterday, today, tomorrow)$ satisfying a formula ϕ is the same as satisfying the conjunction

$$\phi \wedge A(\textbf{Character}).$$

That is, we are relying on the universal modality to appropriately constrain the denotations of the day-indexicals, and clearly it is strong enough do this.

Now, by assumption, A is definable using the tense operators, but we said nothing about how complex this definition was (some definitions might lead to blowups in formula size). But this is irrelevant: **Character** is a fixed formula, thus $A(\textbf{Character})$ has constant size, and so we have that the length of $\phi \wedge A(\textbf{Character})$ depends only on the length of ϕ. So we have polynomial time reduced satisfiability for $K_h^t(now, yesterday, today, tomorrow)$ to satisfiability for $K_h^t(now)$, and (by replacing all occurrences of now and $@_{now}$ using some fresh nominal k, as in the proof of Theorem 3.7) this in turn reduces essentially to K_h^t satisfiability over M.

As for $K_h^t(now, yesterday, today, tomorrow) + now$ satisfiability, here the task is to satisfy ϕ in some model at the utterance time. But this simply means: find a model for $now \wedge \phi$. But this is a $K_h^t(now, yesterday, today, tomorrow)$ satisfiability task. □

5 Concluding Remarks

In this paper we have shown how to incorporate the temporal indexicals *now*, *yesterday*, *today* and *tomorrow* into hybrid tense logic and have provided completeness theorems for both ordinary logical validity and the Kamp-Kaplan notion of contextual validity. Our analyses have been modular. Perhaps most importantly, the bridge between logical and contextual validity has been provided in a uniform way for both the *now* language, and the stronger language containing *yesterday*, *today* and *tomorrow*; in both cases, simply adding the utterance *now* as an additional contextual axiom is enough to lift us from logical to contextual validities. Furthermore, the local approach we adopted when introducing the character constraints on *yesterday*, *today* and *tomorrow* means that our results are compatible with a wide range of global assumptions about the structure of time. Finally, the fact we are working in hybrid logic means that our tableau systems can easily be strengthened (with the help of pure axioms) from the minimal tense logic (no conditions on R) to more temporally interesting classes of models, such as strict partial orders or strict total orders—or, indeed, even to the richer setting of interval-based logic.

But in our view hybrid logic has helped in a less obvious but no less important way. We believe that some work in the Kamp-Kaplan tradition has gone astray by using contextual semantics to simulate temporal reference in a rather artificial way; see van Benthem [1] for a detailed critique of this tendency. Using hybrid logic avoids this pitfall: it provides the tools required to tackle temporal reference head on. But this means that we can use Kamp-Kaplan semantics for the purpose for which it was originally intended, and for which it is best suited: *handling indexicality.* In our view, hybrid logic is capable of bringing clarity and simplicity to this interesting but complex area, and we hope to substantiate this claim in future work.

Acknowledgments

We would like thank the three anonymous referees for their comments on an earlier version of this paper.

References

[1] Benthem, J. V., *Tense logic and standard logic*, Logique et Analyse **20** (1977), pp. 395–437.

[2] Blackburn, P., *Tense, Temporal Reference and Tense Logic*, Journal of Semantics **11** (1994), pp. 83–101.

[3] Blackburn, P., *Internalizing labelled deduction*, Journal of Logic and Computation **10** (2000), pp. 137–168.

[4] Blackburn, P. and B. T. Cate, *Pure Extensions, Proof Rules, and Hybrid Axiomatics*, Studia Logica **84** (2006), pp. 277–322.

[5] Bolander, T. and P. Blackburn, *Termination for hybrid tableaus*, Journal of Logic and Computation **17** (2007), pp. 517–554.

[6] Bräuner, T., "Hybrid Logic and Its Proof-Theory," Springer, Heidelberg, 2011.

[7] Burgess, J., *Basic tense logic*, in: Gabbay and Guenthner, editors, *Handbook of Philosophical Logic, Volume 2*, Reidel, 1984 pp. 89–133.

[8] Cate, B. T., "Model Theory for Extended Modal Languages," Ph.D. thesis, ILLC, University of Amsterdam (2004).

[9] Kamp, H., *Formal Properties of 'Now'*, Theoria **37** (1971), pp. 237–273.

[10] Kaplan, D., *Demonstratives: An Essay on the Semantics, Logic, Metaphysics, and Epistemology of Demonstratives and Other Indexicals*, in: J. Almog, J. Perry and H. Wettstein, editors, *Themes from Kaplan*, Oxford University Press, Oxford, New York, 1989 pp. 481–564.

Appendix

Below are the hybrid tableau rules assumed in this paper; they are tense-logical versions of those introduced in Blackburn [3]. Here s, t, and u range over ordinary nominals and *now*, while a is used to indicate that a new (ordinary) nominal is being introduced. A formula ϕ is (logically) provable using this tableau system iff a closed tableau for $\neg @_i\phi$ can be constructed, where i is a nominal not occurring in ϕ. When axioms are introduced onto a tableau branch they are first prefixed by $@_i$, where i can be any nominal already occurring on that branch. While we hope this appendix is reasonably self-contained, some readers may find it useful to consult Blackburn [3], which contains several examples of tableau proofs, and a completeness proof for the minimal (modal) logic and all pure axiomatic extensions.

$$\frac{@_s\neg\phi}{\neg @_s\phi}\ [\neg] \qquad\qquad \frac{\neg @_s\neg\phi}{@_s\phi}\ [\neg\neg]$$

$$\frac{@_s(\phi\wedge\psi)}{\begin{array}{c}@_s\phi\\ @_s\psi\end{array}}\ [\wedge] \qquad\qquad \frac{\neg @_s(\phi\wedge\psi)}{\neg @_s\phi \mid \neg @_s\psi}\ [\neg\wedge]$$

$$\frac{@_s @_t\phi}{@_t\phi}\ [@] \qquad\qquad \frac{\neg @_s @_t\phi}{\neg @_t\phi}\ [\neg @]$$

$$\frac{@_s P\phi}{\begin{array}{c}@_s Pa\\ @_a\phi\end{array}}\ [P] \qquad\qquad \frac{\neg @_s P\phi \quad @_s Pt}{\neg @_t\phi}\ [\neg P]$$

$$\frac{@_s H\phi \quad @_s Pt}{@_t\phi}\ [H] \qquad\qquad \frac{\neg @_s H\phi}{\begin{array}{c}@_s Pa\\ \neg @_a\phi\end{array}}\ [\neg H]$$

$$\frac{@_s F\phi}{\begin{array}{c}@_s Fa\\ @_a\phi\end{array}}\ [F] \qquad\qquad \frac{\neg @_s F\phi \quad @_s Ft}{\neg @_t\phi}\ [\neg F]$$

$$\frac{@_s G\phi \quad @_s Ft}{@_t\phi}\ [G] \qquad\qquad \frac{\neg @_s G\phi}{\begin{array}{c}@_s Fa\\ \neg @_a\phi\end{array}}\ [\neg G]$$

$$\frac{@_s Pt \quad @_t u}{@_s Pu}\ [P\text{-Bridge}] \qquad \frac{@_s Ft \quad @_t u}{@_s Fu}\ [F\text{-Bridge}] \qquad \frac{@_s Pt}{@_t Fs}\ [P\text{-Trans}] \qquad \frac{@_s Ft}{@_t Ps}\ [F\text{-Trans}]$$

$$\frac{[s \text{ on branch}]}{@_s s}\ [\text{Ref}] \qquad\qquad \frac{@_t s}{@_s t}\ [\text{Sym}] \qquad\qquad \frac{@_s t \quad @_t\phi}{@_s\phi}\ [\text{Nom}]$$

Here's a first example of the system in action. In this paper, when discussing *yesterday*, *today*, and *tomorrow* we introduced two axioms governing what we call One Step Alignment: *today* $\rightarrow G\neg$*yesterday* and *tomorrow* $\rightarrow G\neg$*today*. Conspicuous by their absence are their backward-looking counterparts, namely *today* $\rightarrow H\neg$*tomorrow* and *yesterday* $\rightarrow H\neg$*today*. But both are provable (by essentially identical proofs). Here's an example:

1	$\neg @_i(today \to H\neg tomorrow)$	
2	$@_i today$	$1, \neg\!\to$ *Elimination*
3	$\neg @_i H\neg tomorrow$	$1, \neg\!\to$ *Elimination*
4	$@_i Pj$	$3, \neg H$ *Elimination*
5	$\neg @_j \neg tomorrow$	$3, \neg H$ *Elimination*
6	$@_j tomorrow$	$5, \neg\neg$ *Elimination*
7	$@_j(tomorrow \to G\neg today)$	*One Step Alignment axiom*
8	$@_j G\neg today$	6, 7, *Modus Ponens*
9	$@_j Fi$	4, *P Transpose*
10	$@_i \neg today$	8, 9, *G Propogation*
11	$\bot$	2, 10

Note that when we introduced the needed axiom at line 7, we prefixed it with an @-operator that already occurred on the branch. Also note that we applied Modus Ponens at line 8 rather than following strict tableau procedure and disjunctively splitting the branch; clearly this shortcut is harmless.

Let's look at a more interesting example. In this paper we introduced the Two Step Alignment axiom, $tomorrow \to G\neg yesterday$. We remarked that this axiom is superfluous when working with transitive models. The following tableau proof shows that, assuming transitivity, we can derive it using One Step Alignment and Now Placement:

1	$\neg @_i(tomorrow \to G\neg yesterday)$	
2	$@_i tomorrow$	$1, \neg\!\to$ *Elimination*
3	$\neg @_i G\neg yesterday$	$1, \neg\!\to$ *Elimination*
4	$@_i Fj$	$3, \neg G$ *Elimination*
5	$\neg @_j \neg yesterday$	$3, \neg G$ *Elimination*
6	$@_j yesterday$	$5, \neg\neg$ *Elimination*
7	$@_j(yesterday \to Fnow)$	*Now Placement axiom*
8	$@_j Fnow$	6, 7, *Modus Ponens*
9	$@_j Fk$	8, *F Elimination*
10	$@_k now$	8, *F Elimination*
11	$@_k(now \to today)$	*Now Placement axiom*
12	$@_k today$	10, 11, *Modus Ponens*
13	$@_i(tomorrow \to G\neg today)$	*One Step Alignment axiom*
14	$@_i G\neg today$	12, 13, *Modus Ponens*
15	$@_i Fk$	4, 9, *Transitivity Rule*
16	$@_k \neg today$	14, 15, *G Propogation*
17	$\bot$	12, 16

One of the key points made in this paper is that we obtain complete proof systems for contextual validity simply by assuming *now* as an extra axiom at the start of tableau construction. More precisely, whereas to prove that ϕ is *logically* valid we attempt to construct a closed tableau starting with $\neg @_i \phi$; to prove that ϕ is *contextually* valid we attempt to construct a closed tableau starting with $@_i now$ and $\neg @_i \phi$ (in both cases, i is a nominal not occurring in ϕ). Here's a simple example. We shall show that *today* is contextually valid:

1	$@_i now$	
2	$\neg @_i today$	
3	$@_i(now \to today)$	*Now Placement axiom*
4	$@_i today$	1, 3, *Modus Ponens*
5	$\bot$	2, 4

Beyond Regularity for Presburger Modal Logics

Facundo Carreiro [1]

ILLC, University of Amsterdam, the Netherlands

Stéphane Demri

LSV, ENS Cachan, CNRS, INRIA, France

Abstract

Satisfiability problem for modal logic K with quantifier-free Presburger and regularity constraints (EML) is known to be PSPACE-complete. In this paper, we consider its extension with nonregular constraints, and more specifically those expressed by visibly pushdown languages (VPL). This class of languages behaves nicely, in particular when combined with Propositional Dynamic Logic (PDL). By extending EML, we show that decidability is preserved if we allow at most one positive VPL-constraint at each modal depth. However, the presence of two VPL-contraints or the presence of a negative occurrence of a single VPL-constraint leads to undecidability. These results contrast with the decidability of PDL augmented with VPL-constraints.

Keywords: Presburger constraint, context-free constraint, decidability

1 Introduction

Presburger modal logics. Graded modal logics are extensions of modal logic K in which the modality $\Diamond$ (see e.g. [5]) is replaced by $\Diamond_{\geqslant n}$; the formula $\Diamond_{\geqslant n}\, p$ states that there are at least n successor worlds satisfying the proposition p, early works about such counting operators can be found in [15,4,10]. In [30], the minimal graded modal logic, counterpart of the modal logic K, is shown decidable in PSPACE. Recent complexity results can be found in [20], see also enriched logics with graded modalities in [21,6]. Nevertheless, richer arithmetical constraints about successor worlds are conceivable. For instance, in the majority logic [25], one can express that more than half of the successors satisfy a given formula. All of the above constraints, and many more, are expressible in Presburger arithmetic (PrA), the decidable first-order theory of natural numbers with addition. In the sequel, Presburger modal logics refer to logics admitting arithmetical constraints from PrA. Not only arithmetical constraints from graded modal logics have been considered in several classes of

[1] F. Carreiro has been supported by INRIA at LSV, ENS Cachan and by CONICET Argentina at the Department of Computer Science of the University of Buenos Aires.

non-classical logics, including epistemic logics [31] and description logics [18,9,3] but also PSPACE-complete logics with richer arithmetical constraints have been designed [27,13,11]. If the PSPACE upper bound is given up, modal-like logics with more expressive Presburger constraints on the number of children can be found in [21,34,28].

Assuming that the children of a node are ordered, it is possible to enrich Presburger modal logics with regularity constraints related to the ordering of siblings. This is particularly meaningful to design logical formalisms to query XML documents viewed as finite ordered unranked labeled trees; such logical and automata-based formalisms can be found, e.g., in [34,28]. For instance, a logic with fixpoint operators, arithmetical and regularity constraints is introduced in [28] and shown decidable with an exponential-time complexity, which improves results for description logics with qualified number restrictions [9]. Similarly, the main goal of [13] was to introduce a Presburger modal logic EML with regularity constraints as in the logical formalisms from [33,34,28] with a satisfiability problem that could still be solved in polynomial space.

Our motivations: beyond regularity. The PSPACE upper bound established in [13] for the satisfiability problem for EML is based on a Ladner-like algorithm [22]. Moreover it takes advantage of the properties that the commutative images of context-free languages (CFLs) are effectively semilinear [26] and satisfiability for existential Presburger formulae can be solved in NP thanks to the existence of small solutions [7]. In this paper, our goal is to revisit decidability and complexity results for Presburger modal logics when context-free constraints are considered instead of regular ones. Usefulness of such contraints is best witnessed by their use to define Document Type Definitions (DTDs) for XML documents or to express nonregular programs as done in [23], but the paper focuses on decidability issues. However, it is not reasonable to expect that adding all context-free languages (CFLs) would preserve decidability. We are aware of several situations in which such an extension does not preserve decidability. For instance, PDL with regular programs augmented with the context-free program $\{\mathtt{a}^i\mathtt{b}\mathtt{a}^i : i \geqslant 0\}$ is known to be undecidable [17]. On the other hand, PDL augmented with context-free programs definable by *semi-simple minded pushdown automata* is decidable [16]. In such pushdown automata, the input symbol determines the stack operation to be performed, the next control state and the symbol to be pushed in case of a push operation. The ultimate decidability result has been obtained for PDL augmented by programs definable by visibly pushdown automata (VPA) [23]. These automata are introduced in [1] (see also the equivalent class of *input-driven pushdown automata* in [24]) and are motivated by verification problems for recursive state machines. They generalize the semi-simple-minded pushdown automata since we only require that the input symbol determines the stack operation.

The initial motivation for this work is to understand the decidability status of EML augmented with context-free constraints from VPAs (*VPL-constraints*). Also, the decidability proof for EML takes advantage of the Boolean closure of regular languages, which is also the case for visibly pushdown languages [1] (complementation can be performed in exponential time and causes -only- an exponential blow-up [1]). Moreover, by [32], from a VPA, one can compute in linear time a simple Presburger formula whose solutions are the Parikh image of its language. This may allow us to extend [13]. Hence, boosted by the decidability and complexity results from [23] for PDL with VPA and by the nice properties of VPLs, we aim at understanding how much of the power of context-free constraints can be added to EML while still preserving decidability.

Our contributions. We can distinguish three types of contributions: showing that different extensions of EML have identical expressive power, proving undecidability results when context-free constraints are present in formulae and establishing decidability for the satisfiability problem of EML extended with restricted context-free constraints.

- EML extended with VPL-constraints such that at each depth, there are at most i such constraints (written $\mathsf{EML}_i(\mathrm{VPL})$) is expressively equivalent to EML extended with CFL-constraints such that at each depth, there are at most i such constraints (written $\mathsf{EML}_i(\mathrm{CFL})$).
- We show that a strict fragment of $\mathsf{EML}_2(\mathrm{CFL})$, with only positive CFL-constraints, and a strict fragment of $\mathsf{EML}_1(\mathrm{CFL})$, with a single CFL-constraint occurring negatively, have undecidable satisfiability problems.
- Last, but not least, we show that $\mathsf{EML}_1^+(\mathrm{CFL})$, a fragment of $\mathsf{EML}_1(\mathrm{CFL})$ but a substantial extension of EML, has a decidable satisfiability problem by extending the proof technique from [13]. Herein, $\mathsf{EML}_1^+(\mathrm{CFL})$ denotes the fragment of $\mathsf{EML}_1(\mathrm{CFL})$ in which at each depth, there is at most one context-free constraint and it occurs positively plus additional more technical conditions. Positive occurrences of constraints shall be defined in Section 2 and this notion slightly differs from the one that only counts the number of negations. This is the best we can hope for in view of the previous results.

It was unexpected that $\mathsf{EML}_2(\mathrm{VPL})$ is already undecidable. Decidability can be regained if only one VPL-constraint occurs positively at each depth. If only one VPL-constraint may occur negatively at each depth, then undecidability is back. This contrasts with the decidability of PDL extended with VPAs [23] or with the undecidability of LTL extended with a fixed VPA [12]. Since modal logic K is a fragment of PDL, this seems to contradict the undecidability of $\mathsf{EML}_2(\mathrm{CFL})$. However, language-based constraints on PDL are related to paths whereas in EML it is related to the ordering of successor worlds. Moreover, a CFL-constraint $\mathrm{L}(p_1, \ldots, p_n)$ on the children can be turned into an arithmetical constraint by using Parikh's Theorem [26]. In a sense, CFL-constraints on the successor worlds can be handled by arithmetical constraints. Again, this seems to contradict the undecidability of EML(CFL) fragments; this contradiction vanishes if we recall that the Parikh image of $\overline{\mathrm{L}}$ is not equal to the complement of the Parikh image of L and the Parikh image of the intersection of two CFLs is not equal to the intersection of the Parikh images of the languages. Hence, only isolated CFL-constraints (without other language-based constraints) can be safely replaced by arithmetical constraints.

2 Preliminaries

In this section, we recall the extended modal logic EML and we present extensions based on CFLs. In particular, we shall recall what VPLs are.

2.1 Extended Modal Logic EML

The logics studied in this paper extend the logic EML that has Presburger and regularity constraints. Given a countably infinite set $\mathrm{AT} = \{p_1, p_2, \dots\}$ of propositional variables, the set of terms and EML formulae is inductively defined as follows

$$t ::= a \times \sharp\varphi \mid t + a \times \sharp\varphi$$

$$\varphi ::= p \mid \neg\varphi \mid \varphi \wedge \varphi \mid \varphi \vee \varphi \mid t > b \mid t \equiv_k c \mid \mathrm{L}(\varphi_1, \dots, \varphi_n)$$

where

- $b, c \in \mathbb{N}$, $k \in \mathbb{N} \setminus \{0, 1\}$, $a \in \mathbb{Z} \setminus \{0\}$.
- L is any regular language specified by a finite-state automaton.

A model for EML is a structure $\langle T, R, (<_s)_{s \in T}, l \rangle$ where T is the (possibly infinite) set of nodes and $R \subseteq T^2$ such that the set $\{t \in T : \langle s, t \rangle \in R\}$ is finite for each $s \in T$. Each relation $<_s$ is a total ordering on the R-succesors of s and $l : T \to 2^{\mathrm{AT}}$ is the valuation function. We write $R(s) = s_1 < \cdots < s_m$ if the children of s are $s_1, \dots, s_m$ and they are ordered this way. The satisfaction relation is inductively defined as follows (we omit Boolean clauses):

- $\mathcal{M}, s \models p$ iff $p \in l(s)$,
- $\mathcal{M}, s \models \sum_i a_i \sharp\varphi_i \equiv_k c$ iff there is $n \in \mathbb{N}$ such that $\sum_i a_i R^{\sharp}_{\varphi_i}(s) = nk + c$ with $R_{\varphi_i} = \{\langle s', s'' \rangle \in T^2 : \langle s', s'' \rangle \in R$, and $\mathcal{M}, s'' \models \varphi_i\}$ and $R^{\sharp}_{\varphi_i}(s) = \mathrm{card}(R_{\varphi_i}(s))$,
- $\mathcal{M}, s \models \sum_i a_i \sharp\varphi_i > b$ iff $\sum_i a_i R^{\sharp}_{\varphi_i}(s) > b$,
- $\mathcal{M}, s \models \mathrm{L}(\varphi_1, \dots, \varphi_n)$ iff the finite sequence of children of the node s induces a finite pattern from L. More precisely, iff there is $\mathsf{a}_{i_1} \cdots \mathsf{a}_{i_m} \in \mathrm{L}$ such that given the children of s, $R(s) = s_1 < \cdots < s_m$, for every $j \in [1, m]$, we have $\mathcal{M}, s_j \models \varphi_{i_j}$ (L is any regular language).

A formula φ is *satisfiable* iff there is a model $\mathcal{M}$ and a node s such that $\mathcal{M}, s \models \varphi$. The satisfiability problem is defined accordingly. Observe that although arithmetical constraints are only allowed to use $>$, the operators $\{=, <, \leqslant\}$ can be easily defined using Boolean combinations and constants (no exponential blow-up occurs if renaming is used). Hence, we will use them as abbreviations.

Modalities from K and graded modal logic can also be defined in EML. We use the abbreviation $\Box\varphi \stackrel{\text{def}}{=} \neg(\sharp\neg\varphi > 0)$. Despite the fact that EML extends graded modal logics with richer arithmetical constraints but also with regularity constraints, the satisfiability problem for EML is PSPACE-complete [13].

Now, let us recall what formula trees are. They will be used to define the different fragments that we will be working with. A *finite tree* $\mathcal{T}$ is a finite subset of $(\mathbb{N} \setminus \{0\})^*$ such that $\varepsilon \in \mathcal{T}$ and if $s{\cdot}i \in \mathcal{T}$, then $s \in \mathcal{T}$ and $s{\cdot}j \in \mathcal{T}$ for $j \in [1, i-1]$. The nodes of $\mathcal{T}$ are its elements. The root of $\mathcal{T}$ is the empty word ε. All notions such as parent, child, subtree and leaf, have their standard meanings. A *formula tree* is a labelled tree $\langle \mathcal{T}, \ell \rangle$ representing a formula φ (the label set being a set of subformulae), i.e.

- $\ell(\varepsilon) = \varphi$. If $\ell(s) = \psi_1 \oplus \psi_2$ with $\oplus \in \{\vee, \wedge\}$, then s has two children, $\ell(s{\cdot}1) = \psi_1$ and $\ell(s{\cdot}2) = \psi_2$ ($\neg$ has a similar clause).
- If $\ell(s) = \mathrm{L}(\varphi_1, \ldots, \varphi_n)$, then s has n children and for all $j \in [1, n]$, we have $\ell(s{\cdot}j) = \varphi_j$ (formulae $\sum_i a_i \sharp \varphi_i > b$ and $\sum_i a_i \sharp \varphi_i \equiv_k c$ have similar clauses).

EML formulae are encoded as formula trees and occurrences of ψ in φ correspond to nodes of the formula tree for φ that are labelled by ψ. Given a formula tree $\langle \mathcal{T}, \ell \rangle$, we define its *polarity tree* $\langle \mathcal{T}, \ell_{\mathrm{POL}} \rangle$ as follows:

- $\ell_{\mathrm{POL}}(\varepsilon) = 1$. If $\ell(s) = \neg\psi$, then $\ell_{\mathrm{POL}}(s{\cdot}1) = (1 - \ell_{\mathrm{POL}}(s))$.
- If $\ell(s) = \psi_1 \oplus \psi_2$ with $\oplus \in \{\vee, \wedge\}$, then $\ell_{\mathrm{POL}}(s{\cdot}1) = \ell_{\mathrm{POL}}(s{\cdot}2) = \ell_{\mathrm{POL}}(s)$.
- If $\ell(s) = \sum_i a_i \sharp \varphi_i > b$ or $\ell(s) = \sum_i a_i \sharp \varphi_i \equiv_k c$ or $\ell(s) = \mathrm{L}(\varphi_1, \ldots, \varphi_n)$, then for all j, we have $\ell_{\mathrm{POL}}(s{\cdot}j) = \ell_{\mathrm{POL}}(s)$.

Given a formula tree $\langle \mathcal{T}, \ell \rangle$, we define its *depth tree* $\langle \mathcal{T}, \ell_{\mathrm{DEP}} \rangle$ as follows:

- $\ell_{\mathrm{DEP}}(\varepsilon) = 0$. If $\ell(s) = \neg\psi$, then $\ell_{\mathrm{DEP}}(s{\cdot}1) = \ell_{\mathrm{DEP}}(s)$.
- If $\ell(s) = \psi_1 \oplus \psi_2$ with $\oplus \in \{\vee, \wedge\}$, then $\ell_{\mathrm{DEP}}(s{\cdot}1) = \ell_{\mathrm{DEP}}(s{\cdot}2) = \ell_{\mathrm{DEP}}(s)$.
- If $\ell(s) = \sum_i a_i \sharp \varphi_i \equiv_k c$ or $\ell(s) = \mathrm{L}(\varphi_1, \ldots, \varphi_n)$ or $\ell(s) = \sum_i a_i \sharp \varphi_i > b$ then for all j, we have $\ell_{\mathrm{DEP}}(s{\cdot}j) = 1 + \ell_{\mathrm{DEP}}(s)$.

Intuitively, this last definition assigns the modal-arithmetical depth to the subformula in each node. By way of example, we present the formula tree for $\neg(\sharp p > 3) \vee \neg\mathrm{L}(p, q)$ together with its polarity tree and its depth tree.

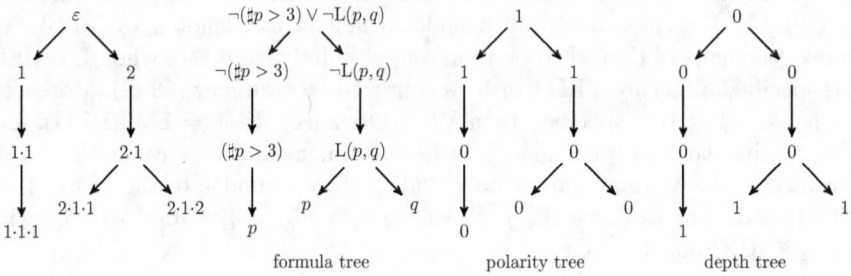

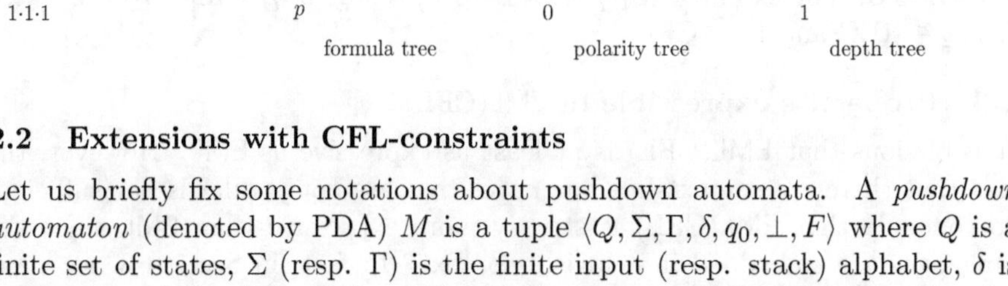

2.2 Extensions with CFL-constraints

Let us briefly fix some notations about pushdown automata. A *pushdown automaton* (denoted by PDA) M is a tuple $\langle Q, \Sigma, \Gamma, \delta, q_0, \perp, F \rangle$ where Q is a finite set of states, Σ (resp. Γ) is the finite input (resp. stack) alphabet, δ is a mapping of $Q \times (\Sigma \cup \{\varepsilon\}) \times \Gamma$ into finite subsets of $Q \times \Gamma^*$, $q_0 \in Q$ is the initial state, $\perp$ is the initial stack symbol, $F \subseteq Q$ is the set of accepting states. A configuration $\langle q, \alpha, \gamma \rangle$ is an element of $Q \times \Sigma^* \times \Gamma^*$. Initial configurations are of the form $\langle q_0, \varepsilon, \perp \rangle$. The one-step relation $\vdash$ (with respect to M) between configurations is defined as follows: $\langle q, \alpha, \mathtt{b}\gamma \rangle \vdash \langle q', \alpha\mathtt{a}, \gamma'\gamma \rangle$ whenever there is $\langle q', \gamma' \rangle \in \delta(q, \mathtt{a}, \mathtt{b})$. As usual, $\vdash^*$ is the reflexive and transitive closure of $\vdash$. We write $\mathcal{L}(M)$ to denote the language accepted by the pushdown automaton M when acceptance is by final states, i.e. $\mathcal{L}(M) = \{\alpha \in \Sigma^* : \langle q_0, \varepsilon, \perp \rangle \vdash^* \langle q, \alpha, \gamma \rangle \text{ with } q \in F\}$.

A *pushdown alphabet* Σ is an alphabet equipped with a partition $\langle \Sigma_c, \Sigma_i, \Sigma_r \rangle$ such that the letters in Σ_c are *calls* (corresponding to push actions on the stack), the letters in Σ_i are *internal actions* (no action on the stack) and the letters in Σ_r are *returns* (corresponding to pop actions on the stack). A *visibly pushdown automaton* (VPA) over Σ is defined as a PDA except that the input letters in Σ determine which actions are performing on the stack according to the partition $\langle \Sigma_c, \Sigma_i, \Sigma_r \rangle$. More precisely, a VPA M over a pushdown alphabet $\langle \Sigma_c, \Sigma_i, \Sigma_r \rangle$ is a PDA without ε-transitions and such that for all states q, input symbols $\mathtt{a} \in \Sigma$ and stack symbols $\mathtt{b} \in \Gamma$,

- $\mathtt{a} \in \Sigma_c$ and $\langle q', \gamma \rangle \in \delta(q, \mathtt{a}, \mathtt{b})$ imply $\gamma = \mathtt{b}'\mathtt{b}$ for some $\mathtt{b}' \in \Gamma$ (call),
- $\mathtt{a} \in \Sigma_i$ and $\langle q', \gamma \rangle \in \delta(q, \mathtt{a}, \mathtt{b})$ imply $\gamma = \mathtt{b}'$ (internal action),
- $\mathtt{a} \in \Sigma_r$ and $\langle q', \gamma \rangle \in \delta(q, \mathtt{a}, \mathtt{b})$ imply $\gamma = \varepsilon$ (return).

A *visibly pushdown language* (VPL) is a language accepted by a VPA. By [1], given two VPLs L_1 and L_2 over the same pushdown alphabet Σ, $\mathrm{L}_1 \cup \mathrm{L}_2$, $\mathrm{L}_1 \cap \mathrm{L}_2$ and $\Sigma^* \setminus \mathrm{L}_1$ are also VPLs and one can effectively compute the corresponding VPA.

Given a class $\mathcal{C}$ of languages (made of finite words from finite alphabets), we define the modal logic $\mathsf{EML}(\mathcal{C})$ interpreted over finitely-branching Kripke structures such that $\mathsf{EML}(\mathcal{C})$ is an extension of EML by allowing formula $\mathrm{L}(\varphi_1, \ldots, \varphi_n)$ with $\mathrm{L} \in (\mathcal{C} \cup \mathrm{REG})$ where REG is the class of regular languages. In this paper, $\mathcal{C}$ is either the class of context-free languages (denoted by CFL) or the class of visibly pushdown languages (denoted by VPL). Of course, elements of $\mathcal{C}$ need to be represented finitely; to do so, when $\mathrm{L} \in \mathrm{CFL}$, L is specified either by a PDA or by a context-free grammar (CFG). Moreover, when $\mathrm{L} \in \mathrm{VPL}$, L is specified by a VPA. Obviously, $\mathsf{EML} = \mathsf{EML}(\mathrm{REG})$. Let $\mathsf{EML}_i(\mathcal{C})$ be the set of formulae φ in $\mathsf{EML}(\mathcal{C})$ such that for any depth j in its depth tree, there are at most i nodes labelled by formulae of the form $\mathrm{L}(\cdots)$ with $\mathrm{L} \in \mathcal{C}$. For example, $\mathrm{L}_1(p,q) \wedge \mathrm{L}_2(p,q) \wedge \neg\mathrm{L}_3(p,q) \in \mathsf{EML}_1(\mathrm{CFL})$ with $\mathrm{L}_1, \mathrm{L}_2 \in \mathrm{REG}$ and $\mathrm{L}_3 \in \mathrm{CFL}$.

2.3 Properties expressible in EML(CFL)

It is obvious that EML(CFL) is at least as expressive as EML. However, the fact that there are context-free languages that are not regular is not sufficient to conclude that EML(CFL) is strictly more expressive than EML since the languages are embedded in a logical context. In fact, in Section 3, we show that EML(VPL) is as expressive as EML(CFL) even though VPL is strictly included in CFL. In the paragraphs below, we explore the relationship between properties expressible in EML and those expressible in EML(CFL). Towards the end of the section, we finally show that EML(CFL) is strictly more expressive.

Let us express in EML(CFL) that there are no children of distance exactly 2^{N+1} such that the first child satisfies p and the second one satisfies q. This property can be instrumental for reducing tiling problems and, with it, establish complexity lower bounds. Let L be the language defined by the context-free grammar below (S is the axiom):

$$S \to T\mathtt{a}S_N S_N^\star \mathtt{b}T,\ T \to \mathtt{c},\ T \to TT,\ T \to \varepsilon,$$

$$S_N \to S_{N-1}S_{N-1},\ S_N^\star \to S_{N-1}S_{N-1}^\star, \cdots,\ S_2 \to S_1 S_1,\ S_2^\star \to S_1 S_1^\star,$$
$$S_1 \to \mathtt{cc},\ S_1^\star \to \mathtt{c}$$

One can check that $\neg\mathrm{L}(p, q, \top)$ whenever there are no children of distance exactly 2^{N+1} such that the first child satisfies p and the second one satisfies q. Note that L is a regular language (even though it is defined with a CFG) and therefore such a property can be also expressed in EML. However, this can be done much more succinctly in EML(CFL) since the context-free grammar is of linear size in N. So, we can conclude that EML(CFL) is at least a succinct version of EML.

Now, let us turn to a genuine context-free property: the sequence of children can be divided into two sequences of equal cardinality such that every child of the first part satisfies p and every child of the second part satisfies q which is a quite natural property to express. In EML(CFL), this statement can be expressed by $\mathrm{L}(p \wedge \neg q, q \wedge \neg p)$ with the context-free language

$$\mathrm{L} = \{\mathtt{a}^i\mathtt{b}^i : i \in \mathbb{N}\}$$

However, it is not difficult to show that the formula $(\sharp p = \sharp q) \wedge \mathrm{L}'(p \wedge \neg q, q \wedge \neg p)$ in EML can also express exactly the same property where $\mathrm{L}' = \mathtt{a}^*\mathtt{b}^*$. In a sense, formulae in EML(CFL) can be viewed as macros for formulae from EML. This is analogous to the situation in modal logic K for which we may consider only one of the operators $\Box$ and $\Diamond$ or both in order to make the formulae clearer. So, EML(CFL) is at least a more friendly version of EML.

The key question remains whether EML(CFL) is strictly more expressive than EML. By generalizing the above reasoning, one can show that context-free constraints of the form $\mathrm{L}(\cdots)$ can be expressed in EML if there are a formula ψ from PrA and a regular language L' such that for every $\alpha \in \Sigma^*$, we have $\alpha \in \mathrm{L}$ iff the Parikh image of α satisfies ψ and $\alpha \in \mathrm{L}'$. Note that $\{\mathtt{a}^i\mathtt{b}^i : i \in \mathbb{N}\}$ precisely satisfies this condition. Such context-free languages L definable by a Presburger formula and by a regular language are exactly the context-free languages definable by so-called *Parikh automata*, see e.g. [8].

The constraints on children that are expressible in EML are Boolean combinations of arithmetical constraints and regular constraints. Such formulae can be reduced equivalently to a disjunction such that each disjunct is a conjunction made of a Boolean combination of arithmetical constraints and made of a regular constraint of the form $\mathcal{A}(\cdots)$. Indeed, a conjunction of positive or negative regular constraints can be equivalently turned into a single regular constraint. Hence, EML(CFL) is strictly more expressive than EML if there is a context-free language that cannot be recognized by a Parikh automaton. Let us consider the language

$$\mathrm{L}_{REV} = \{\alpha \mathsf{c} \alpha^{\mathrm{rev}} : \alpha \in \{\mathsf{a}, \mathsf{b}\}\}$$

where α^{rev} is the reverse of α. For instance

$$\mathrm{L}_{REV}(p \wedge \neg q \wedge \neg r, q \wedge \neg p \wedge \neg r, r \wedge \neg p \wedge \neg q)$$

roughly states that the sequence of children can be divided into three parts such that the sequence of the first part is the mirror image of the sequence of the third part and the second part is reduce to a single child satisfying $r \wedge \neg p \wedge \neg q$. By adapting the proof of [8, Proposition 3], one can show that L_{REV} is a context-free language that is not definable by a Parikh automaton and therefore EML(CFL) is stricly more expressive that EML, which is after all not a big surprise.

3 On the Expressive Power of EML(VPL)

We show that the logic EML(VPL) has the same expressive power as EML(CFL). It is clear that every VPL is a CFL and therefore EML(CFL) is at least as expressive as EML(VPL). For the other direction, we state that every CFL L over an alphabet Σ can be transformed into a VPL L$'$ over the pushdown alphabet $\langle \Sigma \times \{c\}, \Sigma \times \{i\}, \Sigma \times \{r\} \rangle$ such that $\pi_1(\mathrm{L}') = \mathrm{L}$ where π_1 is the projection over the first component, i.e. π_1 erases the second component whose value is among $\{c, i, r\}$. With this 'embedding' we will be able to show that EML(VPL) is as expressive as EML(CFL) using a truth-preserving reduction.

Theorem 3.1 ([2, Theorem 5.2]) *For any CFL* L *over the alphabet* Σ*, we can effectively build a VPA* M *over the alphabet* $\Sigma_M = \langle \Sigma \times \{c\}, \Sigma \times \{i\}, \Sigma \times \{r\} \rangle$ *such that for all* $\alpha \in \Sigma^*$*, we have* $\alpha \in \mathrm{L}$ *iff there is* $\alpha' \in \mathcal{L}(M)$ *s.t.* $\pi_1(\alpha') = \alpha$.

Proof. Let $M' = \langle Q, \Sigma, \Gamma, q_0, \delta, \bot, F \rangle$ be a PDA such that $\mathcal{L}(M') = \mathrm{L}$. Without loss of generality we can also assume that (i) no ε-transitions are used and (ii) the net effect (on the stack) of any transition is null or it pushes or pops one symbol. Define the VPA $M = \langle Q, \Sigma_M, \Gamma, q_0, \delta', \bot, F \rangle$ with $\Sigma_M = \langle \Sigma \times \{c\}, \Sigma \times \{i\}, \Sigma \times \{r\} \rangle$ such that δ' is the minimum relation satisfying

	for every transition $q \to q'$ in M'			there is a transition $q \to q'$ in M		
	reading	*w/stack top*	*action*	*reading*	*w/stack top*	*pushing*
Call	a	b$'$	push b	$\langle \mathsf{a}, c \rangle$	b$'$	b$'$b
Return	a	b$'$	pop b$'$	$\langle \mathsf{a}, r \rangle$	b$'$	ε
Internal	a	b$'$	−	$\langle \mathsf{a}, i \rangle$	b$'$	b$'$

Intuitively, we simulate the associated PDA by annotating the input word with information on how the stack is managed. The proof follows from the fact that there is a 1-to-1 correspondence between the transitions of M and M'. $\square$

Now we can show that the fragment $\mathsf{EML}(\text{VPL})$ is at least as expressive as $\mathsf{EML}(\text{CFL})$. This result is surprising because, as we said before, the class of VPLs is strictly included in the class of CFLs. To prove this claim, we give a truth-preserving translation $\mathsf{T} : \mathsf{EML}(\text{CFL}) \to \mathsf{EML}(\text{VPL})$ defined inductively over the terms and formulae. T is defined homomorphically for all Boolean connectives; arithmetical and periodicity constraints are translated as $\mathsf{T}(t > b) \stackrel{\text{def}}{=} \mathsf{T}(t) > b$, $\mathsf{T}(t \equiv_k c) \stackrel{\text{def}}{=} \mathsf{T}(t) \equiv_k c$ and the terms are defined as follows $\mathsf{T}(a \times \sharp\varphi) \stackrel{\text{def}}{=} a \times \sharp\mathsf{T}(\varphi)$, $\mathsf{T}(t + a \times \sharp\varphi) \stackrel{\text{def}}{=} \mathsf{T}(t) + a \times \sharp\mathsf{T}(\varphi)$. The translation is handled differently for regular and context-free languages.

$$\mathsf{T}(\mathrm{L}(\varphi_1, \ldots, \varphi_n)) \stackrel{\text{def}}{=} \mathrm{L}(\mathsf{T}(\varphi_1), \ldots, \mathsf{T}(\varphi_n)) \text{ if } \mathrm{L} \in \mathrm{REG}$$
$$\mathsf{T}(\mathrm{L}(\varphi_1, \ldots, \varphi_n)) \stackrel{\text{def}}{=} \mathrm{L}'(\mathsf{T}(\varphi_1), \mathsf{T}(\varphi_1), \mathsf{T}(\varphi_1), \ldots, \mathsf{T}(\varphi_n), \mathsf{T}(\varphi_n), \mathsf{T}(\varphi_n)) \text{ if } \mathrm{L} \in \mathrm{CFL}.$$

where L' is a VPL built thanks to Theorem 3.1. Note that this is only needed if L is a CFL (hence represented by a PDA) whereas when L is in REG, the translation is homomorphic. The different numbers of arguments are due to the fact that the cardinal of the alphabet of M' is equal to three times the cardinal of Σ. Assuming that the alphabet of M' is grouped by their first component, the above translation assigns the same formula to symbols sharing the same first component. So, $\mathsf{T}(\varphi)$ can be effectively computed from φ. Moreover, an exponential blow-up is observed if formulae are encoded as formula trees, unlike the case when formulae are encoded as directed acyclic graphs.

Lemma 3.2 *Let $\varphi \in \mathsf{EML}(\text{CFL})$. For all models $\mathcal{M} = \langle T, R, (<_s)_{s \in T}, l\rangle$ and nodes $s \in T$, we have $\mathcal{M}, s \models \varphi$ iff $\mathcal{M}, s \models \mathsf{T}(\varphi)$.*

Proof. We prove the result by structural induction. The base case as well as the cases in induction step for Boolean connectives, regularity constraints and arithmetical constraints are by an easy verification. We focus on the remaining case with context-free constraints.

Left to right: Suppose that $\mathcal{M}, s \models \mathrm{L}(\varphi_1, \ldots, \varphi_n)$ and the alphabet of L is $\Sigma = \{\mathtt{a}_1, \ldots, \mathtt{a}_n\}$. There is a word $\alpha = \mathtt{a}_{j_1} \ldots \mathtt{a}_{j_m} \in \mathrm{L}$ and children $s_1, \ldots, s_m$ such that $s_i \models \varphi_{j_i}$ for $i \in [1, m]$. Using the properties satisfied by $\mathrm{L}' \in \mathrm{VPL}$, we know that there is $\alpha' \in \mathrm{L}'$ such that $\pi_1(\alpha') = \alpha$. It is enough to show that for all i, if $\pi_1(\alpha'(i)) = \mathtt{a}_j$ then $s_i \models \mathsf{T}(\varphi_j)$ and this is clearly satisfied by $\mathcal{M}$ by induction hypothesis.

For the other direction, suppose that $\mathcal{M}, s \models \mathsf{T}(\mathrm{L}(\varphi_1, \ldots, \varphi_n))$. Hence, there exist $\alpha' \in \mathrm{L}'$ and children $s_1, \ldots, s_m$. Using again the properties satisfied by $\mathrm{L}' \in \mathrm{VPL}$, it is clear that $\pi_1(\alpha') \in \mathrm{L}$. Observe also that any child s_{j_k} corresponding to a symbol $(\mathtt{a}_{j_k}, \mathtt{action})$ satisfies $s_{j_k} \models \mathsf{T}(\varphi_{j_k})$. With this observation we conclude, by induction hypothesis, that $s_1, \ldots, s_m$ satisfy the needed formulae and thus $\mathcal{M}, s \models \mathrm{L}(\varphi_1, \ldots, \varphi_n)$. □

Observe that the translation T preserves the number of context-free constraints at each depth. Furthermore, polarity of the subformulae is preserved. Consequently, we can state the following result for several fragments.

Theorem 3.3 *There is a reduction from $\mathsf{EML}(\text{CFL})$ [resp. $\mathsf{EML}_i(\text{CFL})$] satisfiability to $\mathsf{EML}(\text{VPL})$ [resp. $\mathsf{EML}_i(\text{VPL})$] satisfiability.*

The exponential blow-up in T for EML(CFL) formulae can be avoided using the renaming technique, whence by introducing new propositional variables. The formula and its translation would not be logically equivalent but they would be equi-satisfiable. Theorem 3.3 states relationships between fragments that will have an important impact on the expressive power and decidability of the logic EML(VPL). Recall that the class of VPLs forms a Boolean algebra and they behave nicely with other logics. Hence, we expected to be able to get numerous positive decidability results as well. From Theorem 3.3, we can conclude that adding VPLs to EML is as powerful as adding the full class of CFLs. In the forthcoming section, we investigate the decidability status of several extensions of EML with CFL-constraints or VPL-constraints.

4 Undecidable Extensions of Presburger Modal Logics

We start by extending EML with CFLs and VPLs and show that they are already undecidable when we allow two positive language constraints to appear simultaneously. Moreover, we show that if we allow negative occurrences of language constraints, we already get undecidability with just one language constraint.

Undecidability of WEML_2^+(CFL). Let $\mathsf{WEML}_2^+(\mathcal{C})$ be the fragment of $\mathsf{EML}_2(\mathcal{C})$ defined as the restriction of the set of formulae specified by

$$\varphi ::= p \mid \neg p \mid \varphi \wedge \varphi \mid \mathrm{L}(\psi_1, \ldots, \psi_n)$$

where $\mathrm{L} \in \mathcal{C}$, ψ_i are propositional formulae *and* φ has at most two language constraints from $\mathcal{C}$.Observe that this fragment does not include arithmetical constraints.

Theorem 4.1 *The satisfiability problem for* WEML_2^+(CFL) *is undecidable.*

Proof. The proof is by reduction from the nonemptiness problem for the intersection of CFLs. Let $\mathrm{L}_1, \mathrm{L}_2$ be two CFLs. We assume that both languages share the same alphabet $\Sigma = \{\mathsf{a}_1, \ldots, \mathsf{a}_n\}$ (if it is not the case, we take Σ as the union of their alphabets). Let φ_i be a formula stating that p_i is the only propositional variable holding among $\{p_1, \ldots, p_n\}$. Let $\psi = \mathrm{L}_1(\varphi_1, \ldots, \varphi_n) \wedge \mathrm{L}_2(\varphi_1, \ldots, \varphi_n)$. We prove that ψ is satisfiable iff $\mathrm{L}_1 \cap \mathrm{L}_2 \neq \emptyset$.

For the 'if' direction suppose that $\mathcal{M}, s \models \psi$ and there exist a sequence of children $s_1, \ldots, s_k$ and words $\alpha_1 \in \mathrm{L}_1$, $\alpha_2 \in \mathrm{L}_2$ such that $|\alpha_1| = |\alpha_2| = k$. Let $w_1 = \mathsf{a}_{i_1} \ldots \mathsf{a}_{i_k}$ and $w_2 = \mathsf{a}_{j_1} \ldots \mathsf{a}_{j_k}$. Note that for $\ell \in \{1, 2\}$, $\alpha_\ell(i) = \mathsf{a}_j$ iff $s_i \models p_j$. As only one propositional variable can be true of a given child then we can conclude that $\alpha_1 = \alpha_2$ and therefore $\mathrm{L}_1 \cap \mathrm{L}_2 \neq \emptyset$. For the 'only if' direction: Let $\alpha \in \mathrm{L}_1 \cap \mathrm{L}_2$ and $|\alpha| = k$. Take the model $\mathcal{M}, s$ with children $s_1, \ldots, s_k$ such that $s_i \models p_j$ iff $\alpha(i) = \mathsf{a}_j$. It is easy to check that this model satisfies ψ. □

Undecidability of EML_2(VPL). We show that the satisfiability problem for EML(VPL) is undecidable. Moreover, the EML_2(VPL) fragment is already undecidable. In contrast with CFLs, there is one more parameter to be taken into account in our analysis: every VPL is defined over a pushdown alphabet. Therefore, given many VPLs occurring in a formula, we analyze how the relationship among their partitions relates to the undecidability result.

Corollary 4.2 *The satisfiability problem for* WEML_2^+(VPL) *is undecidable, even restricted to formulae such that every two distinct VPLs occurring at the same depth share the same pushdown alphabet.*

Proof. Undecidability is by reduction from the satisfiability problem for the fragment WEML_2^+(CFL) (see Theorem 4.1). Given a formula $\varphi \in \mathsf{WEML}_2^+$(CFL), we can use the translation T to obtain a formula $\mathsf{T}(\varphi)$ in WEML_2^+(VPL). By truth-preservation of T, we obtain that the satisfiability problem for WEML_2^+(VPL) is undecidable. Observe that we can always assume that for $\varphi \in \mathsf{WEML}_2^+$(CFL), if two distinct context-free languages occur at the same modal depth, then they share the same pushdown alphabet. In this way, we get a translation $\mathsf{T}(\varphi)$ in the fragment mentioned in the statement of the theorem. □

Considering T, the translation of a CFL provides the same pushdown alphabet and formula arguments. One could argue that it is exactly this connection through the pushdown alphabet that allows us to encode the nonemptiness problem for the intersection of CFLs. In the following theorem we show that, even if the VPLs do not share the pushdown alphabet, we can still encode the same problem using propositional variables that play the role of 'binders'.

Theorem 4.3 *The satisfiability problem for* WEML_2^+(VPL) *is undecidable even restricted to formulae such that every two distinct VPLs occuring at the same modal depth do not share the pushdown alphabet.*

Proof. Let $\mathrm{L}_1, \mathrm{L}_2$ be CFLs over the alphabet Σ. By Theorem 3.1, one can build two VPA M_1 and M_2 over the pushdown alphabet $\langle \Sigma \times \{c\}, \Sigma \times \{i\}, \Sigma \times \{r\} \rangle$ such that for all words $\alpha \in \Sigma^*$, we have (1) $\alpha \in \mathrm{L}_1$ iff there is $\alpha' \in \mathcal{L}(M_1)$ such that $\pi_1(\alpha') = \alpha$ and (2) $\alpha \in \mathrm{L}_2$ iff there is $\alpha' \in \mathcal{L}(M_2)$ such that $\pi_1(\alpha') = \alpha$.

Let Σ_1 and Σ_2 be two distinct alphabets of cardinality $3 \times \mathrm{card}(\Sigma)$ with $\Sigma_1 \cap \Sigma_2 = \emptyset$ and, $\sigma_1 : \Sigma \times \{c, r, i\} \to \Sigma_1$ and $\sigma_2 : \Sigma \times \{c, r, i\} \to \Sigma_2$ be two bijective renamings. Recall that VPLs are closed under renamings and one can easily compute VPA $M_1^\star$ and $M_2^\star$ such that $\mathcal{L}(M_1^\star) = \sigma_1(\mathcal{L}(M_1))$ and $\mathcal{L}(M_2^\star) = \sigma_2(\mathcal{L}(M_2))$. Note that $M_1^\star$ and $M_2^\star$ are defined over distinct pushdown alphabets. Again, for all words $\alpha \in \Sigma^*$, we have (1′) $\alpha \in \mathrm{L}_1$ iff there is $\alpha' \in \mathcal{L}(M_1^\star)$ such that $\pi_1(\sigma_1^{-1}(\alpha')) = \alpha$ and (2′) $\alpha \in \mathrm{L}_2$ iff there is $\alpha' \in \mathcal{L}(M_2^\star)$ such that $\pi_1(\sigma_2^{-1}(\alpha')) = \alpha$.

Let $\Sigma = \{\mathtt{a}_1, \ldots, \mathtt{a}_n\}$, $\Sigma_1 = \{\mathtt{s}_1, \ldots, \mathtt{s}_{3n}\}$ and $\Sigma_2 = \{\mathtt{s}'_1, \ldots, \mathtt{s}'_{3n}\}$. Let φ_i be a formula that states that the only propositional variable true among $\{p_1, \ldots, p_{3n}\}$ at a given node is p_i. We define a function which identifies each symbol from the original alphabet Σ with a propositional variable. Let $g : \Sigma \times \{c, r, i\} \to \{\varphi_1, \ldots, \varphi_{3n}\}$ be defined as $g((\mathtt{a}_i, \mathtt{action})) = \varphi_i$ and let $\theta_j^i = g(\sigma_i^{-1}(\mathtt{s}_j))$ then we define $\psi = \mathcal{L}(M_1^\star)(\theta_1^1, \ldots, \theta_{3n}^1) \wedge \mathcal{L}(M_2^\star)(\theta_1^2, \ldots, \theta_{3n}^2)$. It is easy to prove that ψ is satisfiable iff $\mathcal{L}(M_1) \cap \mathcal{L}(M_2) \neq \emptyset$ using an argument similar to that of Theorem 4.1. We conclude that $\mathrm{L}_1 \cap \mathrm{L}_2 \neq \emptyset$ iff ψ is satisfiable. □

Undecidability of WEML_1^-(CFL). Let WEML_1^-(CFL) be the restriction of EML_1(CFL) defined as a fragment specified by

$$\varphi ::= p \mid \neg p \mid \varphi \wedge \varphi \mid \varphi \vee \varphi \mid \neg \mathrm{L}(p_1, \ldots, p_n) \mid \Box\psi$$

where $\mathrm{L} \in$ CFL, ψ is a propositional formula *and* φ has at most one context-free constraint. Given a CFL L over the alphabet $\Sigma = \{\mathsf{a}_1, \ldots, \mathsf{a}_k\}$, we build a formula φ_{L} in WEML_1^-(CFL) such that $\mathrm{L} \neq \Sigma^*$ iff φ_{L} is satisfiable. Since the universality problem for CFLs is undecidable (see e.g. [19]) we conclude that the satisfiability problem for WEML_1^-(CFL) is undecidable too. Let $\varphi_{\mathrm{uni}}(p_1, \ldots, p_k)$ be a formula in WEML_1^-(CFL) enforcing that any child satisfies exactly one propositional variable among $\{p_1, \ldots, p_k\}$, we define

$$\varphi_{\mathrm{L}} = \varphi_{\mathrm{uni}}(p_1, \ldots, p_k) \wedge \neg \mathrm{L}(p_1, \ldots, p_k)$$

Theorem 4.4 *The satisfiability problem for* WEML_1^-(CFL) *is undecidable.*

Proof. First, suppose that $\mathrm{L} \neq \Sigma^*$; hence there is a word $w = \mathsf{a}_{i_1} \cdots \mathsf{a}_{i_n} \notin \mathrm{L}$. Let us consider the model $\mathcal{M}$ with a root node s and children $s_1 < \cdots < s_n$ such that for every $j \in [1, n]$, $s_j \models \varphi_{i_j}$. It is clear that $\mathcal{M}, s \models \varphi_{\mathrm{uni}}(p_1, \ldots, p_k)$. The crucial observation is that, given a model where exactly one propositional variable holds at each child, forming the sequence $p_{i_1}, \ldots, p_{i_n}$ then $\mathcal{M}, s \models \mathrm{L}(p_1, \ldots, p_k)$ iff $\mathsf{a}_{i_1} \cdots \mathsf{a}_{i_n} \in \mathrm{L}$. Hence, $\mathcal{M}, s \models \neg\mathrm{L}(p_1, \ldots, p_k)$.

Now suppose that there is a model $\mathcal{M}$ and a node s such that $\mathcal{M}, s \models \varphi_{\mathrm{L}}$. If s has no successor, then $\epsilon \notin \mathrm{L}$, otherwise $\mathcal{M}, s \models \mathrm{L}(p_1, \ldots, p_k)$. In that case $\mathrm{L} \neq \Sigma^*$. Now suppose that s has children $s_1 < \ldots < s_n$ with $n \geqslant 1$. As before, the propositional variables true at the children form a sequence $p_{i_1}, \ldots, p_{i_n}$ and $\mathcal{M}, s \models \mathrm{L}(p_1, \ldots, p_k)$ iff $\mathsf{a}_{i_1} \cdots \mathsf{a}_{i_n} \in \mathrm{L}$. Therefore there exists a word $\mathsf{a}_{i_1} \cdots \mathsf{a}_{i_n} \notin \mathrm{L}$ which implies $\mathrm{L} \neq \Sigma^*$. □

As a corollary, the satisfiability problem for EML_1^-(VPL) is undecidable too.

5 Decidability of EML_1^+(CFL)

We saw that it is enough to have two positive occurrences (i.e., with nonzero polarity) of CFL-constraints or one negative occurrence (i.e., with zero polarity) of a CFL-constraint to have an undecidable satisfiability problem. It remains one interesting extension of EML to be studied: let EML_1^+(CFL) be the fragment in which there is at most one CFL-constraint at each depth, occurring positively. $\mathsf{EML}_i^+(\mathcal{C})$ is defined as the subset of $\mathsf{EML}_i(\mathcal{C})$ such that every formula's tree node labelled by a formula of the form

1. $\mathrm{L}(\cdots)$ with $\mathrm{L} \in \mathcal{C}$ has a positive polarity,
2. $\sum_i a_i \sharp\varphi_i > b$ with a positive polarity has all $a_i, b > 0$ or all $\varphi_i \in$ EML(REG),
3. $\sum_i a_i \sharp\varphi_i > b$ with a negative polarity has each $\varphi_i \in$ EML(REG),

4. $\sum_i a_i \sharp \varphi_i \equiv_k c$ has $\varphi_i \in \mathsf{EML}(\mathrm{REG})$.

For example, $\mathrm{L}_3(p,q) \wedge \mathrm{L}_1(\mathrm{L}_3(p,q),q) \in \mathsf{EML}_1^+(\mathrm{CFL})$ with $\mathrm{L}_1 \in \mathrm{REG}$, $\mathrm{L}_3 \in \mathrm{CFL}$. We show that the satisfiability problem for $\mathsf{EML}_1^+(\mathrm{CFL})$ is decidable, which is the main decidability result of the paper. Even though the above definition has been finely tailored for our proof to work, we think that it is probably the only potentially interesting fragment left to be analyzed. Nonetheless, this fragment adds extra power over EML: suppose you have an XML file containing 'buy' and 'sell' records. With $\mathsf{EML}_1^+(\mathrm{CFL})$ you can require that for every 'sell' there is a matching 'buy' record before it whereas that cannot be done in EML.

We provide a decision procedure for $\mathsf{EML}_1^+(\mathrm{CFL})$ generalizing the one for EML [13]. A major difference is that we have to deal with CFL-constraints with a fine-tuned procedure. Whereas EML satisfiability is in PSPACE, our algorithm for $\mathsf{EML}_1^+(\mathrm{CFL})$ requires strictly more space since we need to store the stack content which can be of double exponential length in the worst-case, preventing us from using an on-the-fly algorithm as done for EML. Our decidability result is established thanks to a Ladner-like algorithm [22] (see also [29,13]). The original algorithm from [22] allows to show that the satisfiability problem for modal logic K is in PSPACE and it can be viewed as a tableaux-like procedure with non-determinism but without tableaux rules. The beauty of Ladner-like algorithms rests on their simple structure with fine-tuned recursivity; moreover it does not use any specific formalism such as tableaux, automata, sequents etc. Our algorithm below extends the one presented in [13] for the satisfiability of the basic EML. We start by defining the closure for finite sets of formulae. Intuitively, the closure $\mathrm{cl}(X)$ of X contains all formulae useful to evaluate the truth of formulae in X.

Definition 5.1 *Let X be a finite set of formulae, $\mathrm{cl}(X)$ is the smallest set satisfying*

- $X \subseteq \mathrm{cl}(X)$, $\mathrm{cl}(X)$ *is closed under subformulae,*
- *if* $\psi \in \mathrm{cl}(X)$, *then* $\neg\psi \in \mathrm{cl}(X)$ *(we identify* $\neg\neg\psi$ *with* ψ*),*
- *let* $\mathtt{K}$ *be the least common multiple (lcm) of all the constants* k *occurring in subformulae of the form* $t \equiv_k c$. *If* $t \equiv_k c \in \mathrm{cl}(X)$, *then* $t \equiv_{\mathtt{K}} c' \in \mathrm{cl}(X)$ *for every* $c' \in [0, \mathtt{K}-1]$.

Observe that, since $\mathsf{EML}_1^+(\mathrm{CFL})$ is not closed under negation, formulae in $\mathrm{cl}(X)$ may not belong to $\mathsf{EML}_1^+(\mathrm{CFL})$ even if X is a set of $\mathsf{EML}_1^+(\mathrm{CFL})$ formulae. A set X of formulae is said to be closed iff $\mathrm{cl}(X) = X$. We refine the notion of closure by introducing a new parameter n: each set $\mathrm{cl}(n,\varphi)$ is a subset of $\mathrm{cl}(\{\varphi\})$ that corresponds to a closed set obtained from subformulae of depth n. We have $\mathrm{cl}(0,\varphi) \supseteq \mathrm{cl}(1,\varphi) \supseteq \mathrm{cl}(2,\varphi) \cdots$ and the sets are defined by peeling of modalities layer by layer.

Definition 5.2 *Let* $\varphi \in \mathsf{EML}_1^+(\mathrm{CFL})$; *for* $n \in \mathbb{N}$, $\mathrm{cl}(n,\varphi)$ *is the smallest set such that*

- $\mathrm{cl}(0,\varphi) = \mathrm{cl}(\{\varphi\})$, *for every* $n \in \mathbb{N}$, $\mathrm{cl}(n,\varphi)$ *is closed,*

- *for all* $n \in \mathbb{N}$ *and* $\sharp\psi$ *occurring in some formula of* $\mathrm{cl}(n, \varphi)$, $\psi \in \mathrm{cl}(n+1, \varphi)$,
- *for all* $n \in \mathbb{N}$ *and* $\mathrm{L}(\varphi_1, \ldots, \varphi_m) \in \mathrm{cl}(n, \varphi)$, *then* $\{\varphi_1, \ldots, \varphi_m\} \subseteq \mathrm{cl}(n+1, \varphi)$.

We will concentrate on subsets of $\mathrm{cl}(n, \varphi)$ whose conjunction of elements is $\mathsf{EML}_1^+(\mathrm{CFL})$ satisfiable. A necessary condition to be satisfiable is to be consistent locally, i.e. at the propositional level and at the level of arithmetical constraints. Let $\mathtt{K}$ be the lcm of all the constants k occurring in subformulae of φ of the form $t \equiv_k c$.

Definition 5.3 *A set* $X \subseteq \mathrm{cl}(n, \varphi)$ *is said to be* n*-locally consistent iff:*

- *if* $\neg\psi \in \mathrm{cl}(n, \varphi)$, *then* $\neg\psi \in X$ *iff* $\psi \notin X$,
- *if* $\psi_1 \wedge \psi_2 \in \mathrm{cl}(n, \varphi)$, *then* $\psi_1 \wedge \psi_2 \in X$ *iff* $\psi_1, \psi_2 \in X$,
- *if* $\psi_1 \vee \psi_2 \in \mathrm{cl}(n, \varphi)$, *then* $\psi_1 \vee \psi_2 \in X$ *iff* $\psi_1 \in X$ *or* $\psi_2 \in X$,
- *if* $t \equiv_k c \in \mathrm{cl}(n, X)$, *then there is a unique* $c' \in [0, \mathtt{K}-1]$ *such that* $t \equiv_{\mathtt{K}} c' \in X$,
- *if* $t \equiv_k c \in \mathrm{cl}(n, X)$, *then* $\neg t \equiv_k c \in X$ *iff there is* $c' \in [0, \mathtt{K}-1]$ *such that* $t \equiv_{\mathtt{K}} c' \in X$ *and not* $c' \equiv_k c$,

The *kernel* of an n-locally consistent set X is $\ker(X) = X \cap \mathsf{EML}_1^+(\mathrm{CFL})$. As we have observed, the closure operation may introduce formulae outside $\mathsf{EML}_1^+(\mathrm{CFL})$. Negation of EML formulae is not a problem (since they are also in EML) whereas negation of $\mathsf{EML}_1^+(\mathrm{CFL})$ formulae may lead to undecidability (e.g., because of a negative polarity). That is why, in our algorithm, we shall only try to satisfy formulae from $\ker(X)$ and this shall be sufficient from the way the fragment $\mathsf{EML}_1^+(\mathrm{CFL})$ is designed. A slight variation in the definition of $\mathsf{EML}_1^+(\mathrm{CFL})$ may lead to undecidability of the satisfiability problem. Regarding size, observe that $\mathrm{card}(\mathrm{cl}(X))$ is exponential in $\mathrm{card}(X)$. Nevertheless, consistent sets of formulae that are satisfiable contain exactly one formula from the set $\{t \equiv_{\mathtt{K}} c : c \in [0, \mathtt{K}-1]\}$ for each constraint $t \equiv_k c'$ in X. Hence, as explained in [13], encoding consistent sets will only require polynomial space.

Definition 5.4 *Let* φ *be an* $\mathsf{EML}_1^+(\mathrm{CFL})$ *formula,* N *be a natural number and* $\mathcal{M}$ *be a finite tree model such that* $\mathcal{M}, s \models \varphi$ *for some node* s. *We say that* $\langle \mathcal{M}, s\rangle$ *is* N-bounded *for* φ *iff for every node* s' *of distance* d *from* s, *the number of successors of* s' *is bounded by* $nb(d+1) \times N$ *where* $nb(d+1)$ *is the number of distinct* $(d+1)$*-locally consistent sets (with respect to* φ*).*

We define the function SAT (see below) such that φ is $\mathsf{EML}_1^+(\mathrm{CFL})$ satisfiable in some N-bounded model iff there is $X \subseteq \mathrm{cl}(0, \varphi)$ such that $\varphi \in \ker(X)$ and $\mathrm{SAT}(X, 0)$ has a computation returning `true`. Indeed, the function $\mathrm{SAT}(X, d)$ is globally parameterized by some natural number N and by the formula φ. These two parameters should be understood as global variables. The main difference with the algorithm in [13] is related to the update of the stack and we restrict ourselves to formulae in $\ker(X)$.

`function` SAT(X, d)

(consistency) if X is not d-locally consistent then `abort`;

(base case) if X contains only propositional formulae then return `true`;

(initialization-counters) for every $\psi \in \ker(\mathrm{cl}(d+1, \varphi))$ that is not a periodicity constraint of the form $t \equiv_K c$, $C_\psi := 0$;

(initialization-fsa) for every FSA $\mathcal{A}(\psi_1, \ldots, \psi_\alpha) \in \ker(X)$, the state variable $q_{\mathcal{A}(\psi_1,\ldots,\psi_\alpha)} := q_0$ for some initial state q_0 of $\mathcal{A}$;

(initialization-fsa-complement) for every FSA $\neg\mathcal{A}(\psi_1, \ldots, \psi_\alpha) \in \ker(X)$, $Z_{\neg\mathcal{A}(\psi_1,\ldots,\psi_\alpha)} := I$ where I is the set of initial states of $\mathcal{A}$;

(initialization-pda) for every CFL-constraint $\mathcal{C}(\psi_1, \ldots, \psi_\alpha) \in \ker(X)$ where $\mathcal{C}$ is a PDA, the state variable $q_{\mathcal{C}(\psi_1,\ldots,\psi_\alpha)} := q_0$ for some initial state q_0 of $\mathcal{C}$ and the stack variable $S_{\mathcal{C}(\psi_1,\ldots,\psi_\alpha)} := \perp$ for the initial stack symbol $\perp$ of $\mathcal{C}$;

(guess-number-children) guess NB in $\{0, \ldots, nb(d+1) \times N\}$;

(guess-children-from-left-to-right) for $i = 1$ to NB do

- (i) guess $x \in \{1, \ldots, nb(d+1)\}$;
- (ii) if not SAT$(Y_x, d+1)$ then `abort`;
- (iii) for every $\psi \in \ker(\mathrm{cl}(d+1, \varphi))$ different from some $t \equiv_K c$ such that $\psi \in Y_x$ do $C_\psi := C_\psi + 1$;
- (iv) for every finite state automaton $\mathcal{A}(\psi_1, \ldots, \psi_\alpha) \in \ker(X)$,
  - *a.* guess a transition $q_{\mathcal{A}(\psi_1,\ldots,\psi_\alpha)} \xrightarrow{a_i} q'$ in $\mathcal{A}$ with $\Sigma_{\mathcal{A}} = \mathtt{a}_1, \ldots, \mathtt{a}_\alpha$;
  - *b.* if $\psi_i \in Y_x$, then $q_{\mathcal{A}(\psi_1,\ldots,\psi_\alpha)} := q'$, otherwise `abort`;
- (v) for every finite state automaton $\neg\mathcal{A}(\psi_1, \ldots, \psi_\alpha) \in \ker(X)$, $Z_{\neg\mathcal{A}(\psi_1,\ldots,\psi_\alpha)} := \{q : \exists\, q' \in Z_{\neg\mathcal{A}(\psi_1,\ldots,\psi_\alpha)}, q' \xrightarrow{\mathtt{a}_i} q, \psi_i \in Y_x\}$;
- (vi) for every CFL-constraint $\mathcal{C}(\psi_1, \ldots, \psi_\alpha) \in \ker(X)$,
  - *a.* let $S_{\mathcal{C}(\psi_1,\ldots,\psi_\alpha)} = \mathtt{b}_1 \ldots \mathtt{b}_k$ be the stack content of $\mathcal{C}$, guess a transition $q_{\mathcal{C}(\psi_1,\ldots,\psi_\alpha)} \xrightarrow{\mathtt{a}_i} q'$ poping $\mathtt{b}_1$ and pushing γ in $\mathcal{C}$ with $\Sigma_{\mathcal{C}} = \mathtt{a}_1, \ldots, \mathtt{a}_\alpha$;
  - *b.* if $\psi_i \in Y_x$, then $q_{\mathcal{A}(\psi_1,\ldots,\psi_\alpha)} := q'$ and $S_{\mathcal{C}(\psi_1,\ldots,\psi_\alpha)} := \gamma\mathtt{b}_2 \ldots \mathtt{b}_k$, otherwise `abort`;

(final-checking)

- (i) for every $\sum_i a_i \sharp\psi_i \sim b \in X$, if $\sum_i a_i \times C_{\psi_i} \sim b$ does not hold, then `abort`,
- (ii) for every $\sum_i a_i \sharp\psi_i \equiv_k c \in X$, if $\sum_i a_i \times C_{\psi_i} \equiv_k c$ does not hold, `abort`,
- (iii) for every (either finite-state or pushdown) automaton $\mathcal{A}(\psi_1, \ldots, \psi_\alpha) \in X$, if $q_{\mathcal{A}(\psi_1,\ldots,\psi_\alpha)}$ is not a final state of $\mathcal{A}$, then `abort`;
- (iv) for every FSA $\neg\mathcal{A}(\psi_1, \ldots, \psi_\alpha) \in X$, if $Z_{\neg\mathcal{A}(\psi_1,\ldots,\psi_\alpha)}$ contains a final state of $\mathcal{A}$, then `abort`;

(return-true) return `true`.

We shall later fix N which will be doubly exponential in $|\varphi|$ (see Lemma 5.9). The first argument X is intended to be a subset of $\mathrm{cl}(d, \varphi)$ and the $(d+1)$-locally consistent sets are denoted by Y_i for some $1 \leqslant i \leqslant nb(d+1)$. A call to SAT$(X, d)$ performs the following actions. First it checks whether X is d-locally consistent and if the modal degree is zero, then it returns `true` in case of d-locally consistency. In order to check that $\ker(X)$ is satisfiable, children of the node are guessed from left to right (providing an ordering of the successors). For each $\psi \in \ker(X)$, there is a counter C_ψ which contains the current number of children that should satisfy ψ. Similarly, we keep track of periodicity constraints. For each subformula in $\ker(X)$ whose outermost connective is automata-based, we introduce a variable that encodes the current state and stack content. At the end of the guess of the children, this variable should be equal to a final state of the automaton. By contrast, for each formula in $\ker(X)$ whose outermost connective is the negation of some automata-based formula, we introduce a variable that encodes the set of states that could be reached so far in the automaton (simulating a subset construction of the underlying automaton). We never have a negation of a CFL-constraint. At the end of the guess of the children, this variable should not contain any final state of the automaton. After guessing at most $N \times nb(d+1)$ children, there is a final checking which verifies that periodicity and arithmetical constraints are satisfied. Keeping track of the stack may hurt a *lot*, i.e., the resulting size could be linear on the number of children. Therefore, of exponential space.

Lemma 5.5 *For all 0-locally consistent sets X and computations of* $\mathrm{SAT}(X,0)$

(i) *the recursive depth is linear in $|\varphi|$,*

(ii) *each call requires space polynomial in the sum of the space for encoding 0-locally consistent sets and $N + 2^{|\varphi|}$,*

Proof. Since $\mathrm{cl}(|\varphi|,\varphi) = \emptyset$, the size of the stack of recursive calls to SAT is at most $|\varphi|$. In the function SAT, the steps (consistency), (base case), (initialization-counters), (initialization-fsa), (initialization-fsa-complement) and (initialization-pda) can be obviously checked in polynomial time in φ (and therefore in polynomial space). Indeed, every n-locally consistent set has cardinal at most $2 \times |\varphi|$ and can be encoded with a polynomial amount of bits with respect of $|\varphi|$; moreover given $X \subseteq \mathrm{cl}(0,\varphi)$ of cardinal at most $2 \times |\varphi|$ and $n \in \mathbb{N}$, one can decide in polynomial-time in $|\varphi|$ whether X is n-locally consistent. In the step (guess-children-from-left-to-right), one needs a counter to count at most until $nb(d+1) \times N$. A polynomial amount of bits in $|\varphi| + log(N)$ suffices. All the non-recursive instructions in (guess-children-from-left-right) can be done in time polynomial in $|\varphi| + log(N)$. Since at the end of the step (guess-children-from-left-right), the values of the counters are less than or equal to $nb(d+1) \times N$, checking the points (i) and (ii) in (final-checking) can be done in polynomial space in $|\varphi| + log(N)$ (remember that the encoding of constants a_i, b and c and k are already in linear space in $|\varphi|$). The only variables that requires polynomial space in $2^{|\varphi|} + N$ are those containing stack contents since the number of children NB is bounded by $nb(d+1) \times N$.□

An analogous of [13, Lemma 5] is shown below.

Lemma 5.6 *Let $X \subseteq \mathrm{cl}(0,\varphi)$, if* $\mathrm{SAT}(X,0)$ *has successful computation and $\varphi \in \ker(X)$, then φ is* $\mathsf{EML}_1^+(\mathrm{CFL})$ *satisfiable in an N-bounded model.*

Proof. Assume that $\mathrm{SAT}(X,0)$ has an accepting computation with $\varphi \in \ker(X)$. Let us build a model $\mathcal{M} = \langle T, R, (<_s)_{s \in T}, l\rangle$ for which there is $s \in T$ such that for every $\psi \in \ker(X)$ we have $\mathcal{M}, s \models \psi$.

From an accepting computation of $\mathrm{SAT}(X,0)$, we consider the following finite ordered tree $\langle T, R, (<_s)_{s\in T}, \ell\rangle$ that corresponds to the calls tree of $\mathrm{SAT}(X,0)$.

- $\langle T, R, (<_s)_{s\in T}\rangle$ is a finite ordered tree,
- for each $s \in T$, $\ell(s) = \langle Y, d\rangle$ for some d-consistent set Y,
- the root node s_0 is labelled by $\langle X, 0\rangle$,
- for each node s with $s_1 <_s \cdots <_s s_n$, the call related to $\ell(s)$ recursively calls SAT with the respective arguments $\ell(s_1)$, ..., $\ell(s_n)$ and in this very ordering.

The model $\mathcal{M}$ we are looking for, is precisely $\mathcal{M} = \langle T, R, (<_s)_{s\in T}, l\rangle$ for which $l(s) = Y \cap \mathrm{AT}$ where $\ell(s) = \langle Y, d\rangle$ for each s.

By structural induction on ψ, we shall show that for all $s \in T$ with labeling $\ell(s) = \langle Y, d\rangle$ we have, $\psi \in \ker(Y)$ implies $\mathcal{M}, s \models \psi$. Consequently, we then get $\mathcal{M}, s_0 \models \varphi$. The case when ψ is a propositional variable is by definition of l.

Induction hypothesis: for all ψ such that $|\psi| \leqslant n$, and for all $s \in T$ with $\ell(s) = \langle Y, d\rangle$, if $\psi \in \ker(Y)$ then $\mathcal{M}, s \models \psi$.

Let ψ be a formula such that $|\psi| = n + 1$.

Basic boolean cases: $\psi = \psi_1 \wedge \psi_2$ and $\psi = \psi_1 \vee \psi_2$.
Observe that if $\psi \in \ker(Y)$ then $\psi_1, \psi_2 \in \ker(Y)$. These cases follow easily with this observation and the definition of n-locally consistent set.

Negation of boolean operators: $\psi = \neg(\psi_1 \wedge \psi_2)$ and $\psi = \neg(\psi_1 \vee \psi_2)$
For the first case, by definition of n-locally consistent set, at least one of $\neg\psi_1$ or $\neg\psi_2$ belongs to Y. Suppose that $\neg\psi_1 \in Y$. It is easy to check that, as $\psi \in \ker(Y)$ we also have $\neg\psi_1 \in \ker(Y)$. By the induction hypothesis, we get that $\mathcal{M}, s \models \neg\psi_1$ and therefore $\mathcal{M}, s \models \psi$. The case for $\psi = \neg(\psi_1 \vee \psi_2)$ is similar.

Positive language: $\psi = \mathrm{L}(\psi_1, \ldots, \psi_k)$ with $\mathrm{L} \in \mathrm{REG} \cup \mathrm{CFL}$.
Let $s \in T$ with $\ell(s) = \langle Y, d\rangle$ such that $\psi \in \mathrm{cl}(d, \varphi)$. By definition of T, $\mathrm{SAT}(Y, d)$ has an accepting computation. If $\psi \in \ker(Y)$, then each call in the sequence $\mathrm{SAT}(Y_{x_1}, d+1), \ldots, \mathrm{SAT}(Y_{x_{\mathrm{NB}}}, d+1)$ has an accepting computation. Hence the children of s are, from left to right: $s_1, \ldots, s_{\mathrm{NB}}$ such that $\ell(s_i) = \langle Y_{x_i}, d+1\rangle$. Then, it is not difficult to show that the steps (initialization-fsa), (initialization-pda), (guess-children-from-left-to-right) and (final-checking)(iii) guarantee that $\mathcal{M}, s \models \psi$ by using the induction hypothesis and the fact that each ψ_i belongs to $\ker(Y)$ too.

Negation of regular language: $\psi = \neg\mathrm{L}(\psi_1, \ldots, \psi_k)$ with $\mathrm{L} \in \mathrm{REG}$.
Let $s \in T$ with $\ell(s) = \langle Y, d\rangle$ such that $\psi \in \mathrm{cl}(d, \varphi)$. By definition of T, $\mathrm{SAT}(Y, d)$ has an accepting computation. Suppose s has no children, then $\varepsilon \notin \mathrm{L}$, otherwise (final-checking) would fail. This implies $\mathcal{M}, s \not\models \mathrm{L}(\psi_1, \ldots, \psi_k)$.

Suppose s has children $s_1, \ldots, s_{\mathrm{NB}}$ such that $\ell(s_i) = \langle Y_{x_i}, d+1\rangle$. Given an arbitrary word $\gamma = \psi_{j_1} \ldots \psi_{j_{\mathrm{NB}}} \in \mathrm{L}$ suppose that $\psi_{j_i} \in Y_{j_i}$ for all i. In this case the step (guess-children-from-left-to-right) would finish with a set of states including a final state and therefore (final-checking) would abort. Hence, there exists $\psi_{j_k} \notin Y_{j_k}$. By the definition of n-locally consistent set, this implies that $\neg\psi_{j_k} \in Y_{j_k}$. As $\psi \in \mathsf{EML}_1^+(\mathrm{CFL})$ and going through language constraints preserve polarity we know that $\neg\psi_{j_k} \in \mathsf{EML}_1^+(\mathrm{CFL})$ too. We conclude that $\neg\psi_{j_k} \in \ker(Y_{j_k})$ and using the induction hypothesis it is easy to see that $\mathcal{M}, s_k \not\models \psi_{j_k}$ and therefore $\mathcal{M}, s \not\models \mathrm{L}(\psi_1, \ldots, \psi_k)$.

Arithmetical constraint: $\psi = \sum_{i=1}^{i=\alpha} a_i \sharp\psi_i > b$.
Let $s \in T$ such that $\ell(s) = \langle Y, d\rangle$ and $\psi \in \mathrm{cl}(d, \varphi)$. By definition of T, $\mathrm{SAT}(Y, d)$ has an accepting computation. If $\psi \in \ker(Y)$, then each call in the sequence $\mathrm{SAT}(Y_{x_1}, d+1)$, …, $\mathrm{SAT}(Y_{x_{NB}}, d+1)$ has an accepting computation. For every $i \in [1, \alpha]$, there are exactly C_{ψ_i} elements in $Y_{x_1}, \ldots, Y_{x_{NB}}$ that contain ψ_i where C_{ψ_i} is the value of the counter after the step (guess-children-from-left-to-right) in the above-mentioned successful computation for $\mathrm{SAT}(Y, d)$. Hence the children of s in $\mathcal{M}$ are the following (from left to right): $s_1, \ldots, s_{NB}$ with $\ell(s_i) = \langle Y_{x_i}, d+1\rangle$.

Case 1: All the subformulae ψ_i belong to EML.
There are exactly C_{ψ_i} children satisfying ψ_i by induction hypothesis and by the fact that either ψ_i or $\neg\psi_i$ belongs to Y_{x_i}. The sum of the terms will be at least $b+1$ and together with the steps (initialization-counters), (guess-children-from-left-to-right) and (final-checking) guarantee that $\mathcal{M}, s \models \psi$.

Case 2: Some formula ψ_i does not belong to EML.
There are at least C_{ψ_i} children satisfying ψ_i by induction hypothesis. Observe that, given a formula ψ_i belonging to some set $\ker(Y_{x_j})$ we know that $\mathcal{M}, s_j \models \psi_i$. On the contrary, if $\psi_i \notin \ker(Y_{x_j})$ we cannot say that $\mathcal{M}, s_j \not\models \psi_i$. Therefore, the values C_{ψ_i} are *lower bounds* for the number of children satisfying ψ_i. Since each a_i is strictly greater than zero, the sum of the terms will be at least $b+1$ and together with the steps (initialization-counters), (guess-children-from-left-to-right) and (final-checking) guarantee that $\mathcal{M}, s \models \psi$.

Negation of arithmetical constraint: $\psi = \neg\sum_{i=1}^{i=\alpha} a_i \sharp\psi_i > b$.
First, observe that if $\psi \in \mathsf{EML}_1^+(\mathrm{CFL})$, then all $\psi_i \in \mathsf{EML}(\mathrm{REG})$. In this case we need to show that $\mathcal{M}, s \models \sum_{i=1}^{i=\alpha} a_i \sharp\psi_i \leqslant b$, which can be done with an argument similar to the case of positive constraints.

Periodicity constraints: $\psi = \sum_{i=1}^{i=\alpha} a_i \sharp\psi_i \equiv_k c$ and $\psi = \neg\sum_{i=1}^{i=\alpha} a_i \sharp\psi_i \equiv_k c$.
In this case, all $\psi_i \in \mathsf{EML}(\mathrm{REG})$, therefore, $\psi_i \in \mathsf{EML}_1^+(\mathrm{CFL})$ which lets us use the induction hypothesis on them. With this observation, these cases go through exactly as in the original proof.

The current model $\mathcal{M}$ is of double exponential size in $|\varphi|$ and it is easy to show that it is N-bounded. □

Lemma 5.7 *If φ is $\mathsf{EML}_1^+(\mathrm{CFL})$ satisfiable in some N-bounded model then for some $X \subseteq \mathrm{cl}(0, \varphi)$, $\mathrm{SAT}(X, 0)$ has an accepting computation with $\varphi \in \ker(X)$.*

Proof. Assume that φ is EML_1^+(CFL) satisfiable in some N-bounded model $\mathcal{M} = \langle T, R, (<_s)_{s \in T}, l \rangle$. So there is $s \in T$ such that $\mathcal{M}, s \models \varphi$ and $\langle \mathcal{M}, s \rangle$ is N-bounded.

Given a formula φ and an EML_1^+(CFL) model $\mathcal{M}'$ we use $X[s', d]$ with $s' \in T'$ and $d \in [0, |\varphi|]$ to denote the set $\{\psi \in \mathrm{cl}(d, \varphi) : \mathcal{M}', s' \models \psi\}$. We shall show that whenever $\langle \mathcal{M}', s' \rangle$ is N-bounded then $\mathrm{SAT}(X[s', d], d)$ has an accepting computation. We recall that $X[s', d]$ is d-locally consistent. Consequently, we get that $\mathrm{SAT}(X[s, 0], 0)$ has an accepting computation and, by definition, $\varphi \in \ker(X[s, 0])$. The proof is by induction on $d_{max} - d$ where d_{max} is the maximum value such that $\mathrm{cl}(d_{max}, \varphi) \neq \emptyset$.

Base case: $d = d_{max}$.
Any satisfiable set of literals included in $\mathrm{cl}(d_{max}, \varphi)$ is consistent and leads to an accepting computation.

Induction hypothesis: for all $1 \leqslant n \leqslant d' \leqslant |\varphi|$ and $X \subseteq \mathrm{cl}(d', \varphi)$ such that there exist an model $\mathcal{M}' = \langle T', R', (<'_s)_{s \in T'}, l' \rangle$ and $s' \in T'$ with $X_{d'} = \{\psi \in \mathrm{cl}(d', \varphi) : \mathcal{M}', s' \models \psi\}$ and $\langle \mathcal{M}', s' \rangle$ is N-bounded, $\mathrm{SAT}(X[s', d'], d')$ has an accepting computation.

Let $d = n - 1$ and $\mathcal{M}' = \langle T', R', (<'_s)_{s \in T'}, l' \rangle$, with $s' \in T'$ such that $\langle \mathcal{M}', s' \rangle$ is N-bounded. The set $X[s', d]$ is, by definition, d-locally consistent and EML_1^+(CFL) satisfiable.

For $i \in \{1, \ldots, nb(d+1)\}$, let Y_i be the ith (d+1)-locally consistent set. We write n_i to denote the number of times that the set Y_i holds in the children of s', i.e., $n_i = \mathrm{card}(\{Y_i = X[s'', d+1] : s'' \in T' \text{ and } R'(s', s'')\})$. Observe that, since $\mathcal{M}'$ is N-bounded, $\sum_i n_i \leqslant nb(d+1) \times N$.

This is sufficient to establish that $\mathrm{SAT}(X[s', d], d)$ has an accepting computation. Indeed, the step (consistency) is successful because $X[s', d]$ is d-locally consistent. The guessed number NB is obviously $\sum_i n_i$ and each set Y_i is guessed n_i times in the step (guess-children-from-left-to-right). Additionally, the order in which the sets Y_i are guessed is precisely given by the ordering of the children of the root of $\mathcal{M}'$. Since $\mathcal{M}'$ is a model for $X[s', d]$, for every $i \in [1, nb(d'+1)]$, if $n_i \neq 0$, then the set Y_i is satisfiable in some N-bounded model (namely $\langle \mathcal{M}, s'' \rangle$). By the induction hypothesis, $\mathrm{SAT}(Y_i, d+1)$ returns `true`. Each passage to (guess-children-from-left-to-right) as well as the passage to (final-checking) are successful steps because the numbers of children is computed from $\mathcal{M}'$. Hence, $\mathrm{SAT}(X[s', d], d)$ has an accepting computation. □

Theorem 5.8 (Correctness) *A formula φ is EML_1^+(CFL) satisfiable in some N-bounded model iff for some $X \subseteq \mathrm{cl}(0, \varphi)$, $\mathrm{SAT}(X, 0)$ has an accepting computation and $\varphi \in \ker(X)$.*

We are in position to establish that N is at most doubly exponential in $|\varphi|$.

Lemma 5.9 *There is a polynomial $q(\cdot)$ such that for every formula φ, φ is EML_1^+(CFL) satisfiable iff φ is satisfiable in some $2^{2^{q(|\varphi|)}}$-bounded model.*

Proof. The proof extends that in [13, Lemma 7]. Only the main change is explained below. Suppose that $\mathcal{C}, \mathcal{B}_1, \ldots, \mathcal{B}_\ell, \neg\mathcal{B}'_1, \ldots, \neg\mathcal{B}'_m$ are the automata-based formulae (or their negation) occurring in some set $\ker(X) \subseteq X \subseteq cl(d, \varphi)$ ($\mathcal{C}$ is the unique CFL-constraint, if any). First, we build a new PDA $\mathcal{C}^+$, over the alphabet $\Sigma = \{Y_1, \ldots, Y_{nb(n+1)}\}$ where $Y_1, \ldots, Y_{nb(n+1)}$ are the only $(n+1)$-locally consistent sets. The automata $\mathcal{C}^+$ and $\mathcal{C}$ have the same sets of states, initial states, stack symbols and final states and $q \stackrel{Y}{\to} q'$ in $\mathcal{C}^+$ iff $q \stackrel{\psi}{\to} q'$ in $\mathcal{C}$ for some $\psi \in Y$. For each transition, the action on the stack in the new PDA should be the same as in the original one. The FSAs are also converted into new FSA $\mathcal{B}_1^+, \ldots, \mathcal{B}_\ell^+, \mathcal{B}_1^-, \ldots, \mathcal{B}_m^-$ in the same way as before (obviously, without handling the stack). Next we synchronize all the automata $\mathcal{C}^+, \mathcal{B}_1^+, \ldots, \mathcal{B}_\ell^+$ and the *complement* of $\mathcal{B}_1^-, \ldots, \mathcal{B}_m^-$ into a product PDA $\mathcal{C}^*$ over the alphabet Σ. Observe that this can be done because the intersection of all the FSA yields a FSA and the intersection of a PDA and a FSA provides a PDA. The automaton $\mathcal{C}^*$ is such that for every word $\alpha = Y_{j_1} \cdots Y_{j_\alpha} \in \Sigma^*$, $\alpha \in \mathcal{L}(\mathcal{C}^*)$ iff (i) there are $\psi_1 \in Y_{j_1}, \ldots, \psi_\alpha \in Y_{j_\alpha}$ such that $\psi_1 \cdots \psi_\alpha \in \mathcal{L}(\mathcal{C})$; (ii) for all $i \in [1, \ell]$, there are $\psi_1 \in Y_{j_1}, \ldots, \psi_\alpha \in Y_{j_\alpha}$ s.t. $\psi_1 \cdots \psi_\alpha \in \mathcal{L}(\mathcal{B}_i)$; (iii) for all $i \in [1, m]$, there are no $\psi_1 \in Y_{j_1}, \ldots, \psi_\alpha \in Y_{j_\alpha}$ s.t. $\psi_1 \cdots \psi_\alpha \in \mathcal{L}(\mathcal{B}'_i)$.

Observe first that there is a CFG G such that $\mathcal{L}(G) = \mathcal{L}(\mathcal{A}^*)$ and the size of G is polynomial in $|\mathcal{A}^*|$. Moreover, there exists a FSA $\mathcal{A}^*$ such that $\mathcal{L}(\mathcal{A}^*)$ and $\mathcal{L}(G)$ have the same Parikh image (see e.g. [26]) and $|\mathcal{A}^*|$ is exponential in $|G|$, see e.g. [14]. Consequently, there is a FSA $\mathcal{A}^*$ whose size is at most of double exponential in $|\varphi|$, which is one exponential higher than the bound obtained in [13, Lemma 7] for that part of the argument. The rest of the proof follows that of [13, Lemma 7] *except that we carry values with one exponential higher.* □

Note that for CFGs in Chomsky normal form (every CFG can be converted to such form in polynomial time) a lower bound for the size of an equivalent FSA with identical Parikh image is $\Omega(2^n)$, see e.g. [14]. Hence, the reasoning performed in the proof sketch of Lemma 5.9 can be hardly improved. Another technique would be needed for substantial improvement.

Theorem 5.10 *The satisfiability problem for* EML_1^+(CFL) *is in* 2EXPSPACE.

Proof. By Theorem 5.8 and Lemma 5.9, φ is EML_1^+(CFL) satisfiable iff SAT$(X,0)$ has an accepting computation and $\varphi \in \ker(X)$ with $N = 2^{2^{q(|\varphi|)}}$. By Lemma 5.5, in that case SAT runs in space polynomial in $2^{|\varphi|} + N$, whence the complexity upper bound 2EXPSPACE by Savitch's Theorem. □

Remark 5.11 The 2EXPSPACE bound can be shown alternatively. Indeed, by Lemma 5.9, φ is satisfiable iff φ is satisfiable by a tree-like model whose depth is bounded by $|\varphi|$ and branching factor is at most $2^{2^{q(|\varphi|)}}$. A nondeterministic 2EXPSPACE algorithm consists in guessing a model of double exponential size and then in performing a model-checking call on it, which can be done in 2EXPSPACE. Savitch's Theorem allows to regain the 2EXPSPACE bound. Note that running SAT consists in guessing the very large model by pieces while checking satisfaction of subformulae on-the-fly. The optimal complexity lower bound remains unknown.

A natural question about the design of $\mathsf{EML}_1^+(\text{CFL})$ is if we need to simultaneously restrict the positive occurrences of CFL-constraints and the positive occurrences of $\sum_i a_i \sharp\varphi_i > b$ by imposing that either each $a_i > 0$ and $b > 0$ or each $\varphi_i \in \mathsf{EML}(\text{REG})$. Indeed, if we give up this last requirement decidability is lost again. As done in Section 4, we can show that a CFL L over Σ is different from Σ^* iff $\neg\Box\perp \wedge \Box\varphi_{\text{uni}}(p_1,\ldots,p_k) \wedge \sharp\text{L}(p_1,\ldots,p_k) \leqslant 0$ is satisfiable where Σ contains k letters (this formula is not in $\mathsf{EML}_1^+(\text{CFL})$).

6 Conclusion

In this paper, we have investigated the decidability status of EML extensions by adding CFL-constraints, especially those definable from VPLs. For instance, VPLs have been used to express nonregular programs [23] for PDL. Whereas PDL augmented with VPA is decidable, we have shown that the satisfiability problem for $\mathsf{EML}_2(\text{VPL})$ is already undecidable. Also, almost every interesting restriction of this logic still leads to an undecidable fragment. By contrast, in Section 5, we establish that the satisfiability problem for $\mathsf{EML}_1^+(\text{CFL})$ can be solved in 2EXPSPACE by extending proof techniques from [13] that essentially use a Ladner-like algorithm, see e.g. [22], combined with the existence of small solutions for constraint systems. In this fragment, only one CFL-constraint may occur at each modal depth and it should have a positive polarity. Surprisingly, if only one CFL-constraint may occur negatively at each modal depth, then undecidability is back. Schema below contains the decidability status of fragments for $\mathcal{C} \in \{\text{CFL}, \text{VPL}\}$; decidable fragments appear in frames. Our work can be pursued by considering subclasses of VPA such as those definable by semi-simple minded pushdown automata [16]. Such extension are still of interest and our techniques to prove undecidability cannot be applied in that case. Also, the exact characterization of the complexity for the satisfiability problem for $\mathsf{EML}_1^+(\text{CFL})$ is still open.

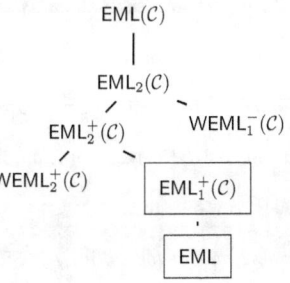

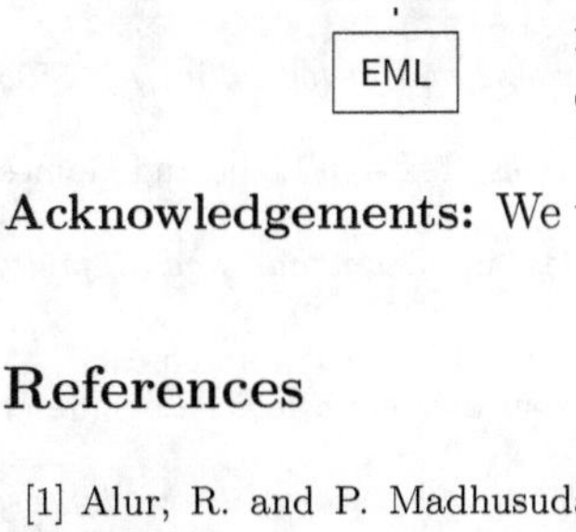

Acknowledgements: We thank the anonymous referees for their suggestions.

References

[1] Alur, R. and P. Madhusudan, *Visibly pushdown languages*, in: *STOC'04* (2004), pp. 202–211.

[2] Alur, R. and P. Madhusudan, *Adding nesting structure to words*, JACM **56** (2009), pp. 1–43.

[3] Areces, C., G. Hoffmann and A. Denis, *Modal Logics with Counting*, in: *WOLLIC'10*, LNCS **6188** (2010), pp. 98–109.

[4] Barnaba, M. F. and F. D. Caro, *Graded modalities*, Studia Logica **44** (1985), pp. 197–221.
[5] Blackburn, P., M. de Rijke and Y. Venema, "Modal Logic," CUP, 2001.
[6] Bonatti, P., C. Lutz, A. Murano and M. Vardi, *The complexity of enriched mu-calculi*, LMCS **4** (2008).
[7] Borosh, I. and L. Treybig, *Bounds on positive integral solutions of linear diophantine equations*, American Mathematical Society **55** (1976), pp. 299–304.
[8] Cadilhac, M., A. Finkel and P. McKenzie, *On the expressiveness of Parikh automata and related models*, in: *NCMA'11*, books@ocg.at **282** (2011), pp. 103–119.
[9] Calvanese, D. and G. D. Giacomo, *Expressive description logics*, in: *Description Logics Handbook* (2005), pp. 178–218.
[10] Cerrato, C., *General canonical models for graded normal logics*, Studia Logica **49** (1990), pp. 242–252.
[11] Cîrstea, C., A. Kurz, D. Pattinson, L. Schröder and Y. Venema, *Modal logics are coalgebraic*, The Computer Journal **54** (2011), pp. 31–41.
[12] Demri, S. and P. Gastin, "Chapter 'Specification and Verification using Temporal Logics' in Modern Applications of Automata Theory," IIsc Research Monographs, World Scientific, 2011 To appear.
[13] Demri, S. and D. Lugiez, *Complexity of modal logics with Presburger constraints*, JAL **8** (2010), pp. 233–252.
[14] Esparza, J., P. Ganty, S. Kiefer and M. Luttenberger, *Parikh's Theorem: A simple and direct automaton construction*, IPL **111** (2011), pp. 614–619.
[15] Fine, K., *In so many possible worlds*, NDJFL **13** (1972), pp. 516–520.
[16] Harel, D. and M. Kaminsky, *Strengthened Results on Nonregular PDL*, Technical Report MCS99-13, The Weizmann Institute of Science (1999).
[17] Harel, D., A. Pnueli and J. Stavi, *Propositional dynamic logic of nonregular programs*, JCSS **26** (1983), pp. 222–243.
[18] Hollunder, B. and F. Baader, *Qualifying number restrictions in concept languages*, in: *KR'91*, 1991, pp. 335–346.
[19] Ibarra, O., *Restricted one-counter machines with undecidable universe problems*, Mathematical Systems Theory **13** (1979), pp. 181–186.
[20] Kazakov, Y. and I. Pratt-Hartmann, *A note on the complexity of the satisfiability problem for graded modal logic*, in: *LICS'09* (2009), pp. 407–416.
[21] Kupferman, O., U. Sattler and M. Vardi, *The complexity of the graded μ-calculus*, in: *CADE'02*, LNCS **2392** (2002), pp. 423–437.
[22] Ladner, R., *The computational complexity of provability in systems of modal propositional logic*, SIAM Journal of Computing **6** (1977), pp. 467–480.
[23] Löding, C., C. Lutz and O. Serre, *Propositional dynamic logic with recursive programs*, Journal of Logic and Algebric Programming **73** (2007), pp. 51–69.
[24] Mehlhorn, K., *Pebbling mountain ranges and its application to DCFL-recognization*, in: *ICALP'80*, LNCS **85** (1980), pp. 422–435.
[25] Pacuit, E. and S. Salame, *Majority logic*, in: *KR'04* (2004), pp. 598–605.
[26] Parikh, R., *On context-free languages*, JACM **13** (1966), pp. 570–581.
[27] Schröder, L. and D. Pattinson, *PSPACE bounds for rank-1 modal logics*, in: *LICS'06* (2006), pp. 231–240.
[28] Seidl, H., T. Schwentick and A. Muscholl, *Counting in trees*, Texts in Logic and Games **2** (2007), pp. 575–612.
[29] Spaan, E., *The complexity of propositional tense logics*, in: *Diamonds and Defaults* (1993), pp. 287–309.
[30] Tobies, S., *PSPACE reasoning for graded modal logics*, JLC **11** (2001), pp. 85–106.
[31] van der Hoek, W. and J.-J. Meyer, *Graded modalities in epistemic logic*, Logique et Analyse **133–134** (1991), pp. 251–270.
[32] Verma, K. N., H. Seidl and T. Schwentick, *On the complexity of equational Horn clauses*, in: *CADE'05*, LNCS **3632**, 2005, pp. 337–352.
[33] Wolper, P., *Temporal logic can be more expressive*, I & C **56** (1983), pp. 72–99.
[34] Zilio, S. D. and D. Lugiez, *XML schema, tree logic and sheaves automata*, Applicable Algebra in Engineering, Communication and Computing **17** (2006), pp. 337–377.

Guarded Negation

Balder ten Cate

UC Santa Cruz

In recent work with Vince Barany and Luc Segoufin [4], we introduce fragments of first-order logic and of fixed-point logic in which all occurrences of negation are required to be guarded by an atomic predicate. In terms of expressive power, the logics in question, called GNFO and GNFP, properly extend the guarded fragment of first-order logic [1] and guarded fixed-point logic [6], respectively. GNFO can be viewed as a natural common generalization of the guarded fragment and the language of unions of conjunctive queries, while GNFP can be viewed as a natural common generalization of guarded fixpoint logic and monadic Datalog. In this sense, the concept of guarded negation provides a unifying perspective on decidable formalisms arising in the areas of modal logic and database theory. The results in [4] (which build on recent results for the guarded fragment and guarded fixed point logic [2,3]) show that GNFO and GNFP are well behaved logics, both computationally and model theoretically.

The aim of the talk is to present the guarded negation logics GNFO and GNFP, and also to describe some of the technical constructions used in the proofs of our main results. In particular, I will discuss two important (known) concepts that are frequently used in this context, and that deserve to belong to the toolkit of any modern modal logician, namely the concepts of *bounded treewidth*, and of *locally acyclic covers*.

Bounded treewidth provides, in some sense, a precise characterization of the dividing line between the decidable world of trees and the undecidable world of grids and tilings. If a class of structures has bounded treewidth, then the first-order theory, and in fact even the guarded second order theory, of that class is decidable, as can be shown by a reduction to the monadic second order theory of trees. If, on the other hand, a class of structures has unbounded treewidth, then it follows from results in graph theory due to Robinson and Seymour, that we can find, inside this class, arbitrarily large grids as graph minors, from which it then follows that the guarded second order theory of the class is undecidable. The decidability of the guarded fragment of first-order logic, as well the decidability of guarded fixed point logic, stems from the fact that these logics are invariant under guarded bisimulations, and every structure is guarded bisimilar to a structure of bounded treewidth namely its "guarded unraveling" [1,5]. A similar analysis applies more generally to the guarded-negation logics GNFO and GNFP.

Locally acyclic structures form finite "approximations" of infinite unravelings. In general, unraveling a structure into a tree-like structure is only possible if one allows for infinite structures (as can be illustrated by the fact that, if a finite Kripke structure contains a cycle, then every finite Kripke structure bisimilar to it must contain a cycle as well, and therefore cannot be a tree). Nevertheless, it is often possible to construct a finite structure that behaves "similarly enough" to an infinite unraveling. In particular, in the case of the basic modal language, every finite Kripke structure is bimilar to a Kripke structure that contains only cycles of length at least k, where k is a natural number that may be chosen arbitrarily large. It follows from locality theorems for first-order logic that such structures cannot be distinguished by means of a first-order formula from structures that are fully acyclic (picking k large enough, depending on the first-order formula in question). Such locally acyclic structures can then be used as a substitute for infinite unravelings in proving finite model-theoretic results [7]. Similar, but more complicated, constructions exist for the guarded fragment [3,8,2], and they are used also to the study of the guarded negation logics GNFO and GNFP.

References

[1] Andréka, H., J. Benthem and I. Németi, *Modal languages and bounded fragments of predicate logic*, Journal of Philosophical Logic **27** (1998), pp. 217–274.

[2] Bárány, V. and M. Bojanczyk, *Finite satisfiability for guarded fixpoint logic*, IPL **112** (2012), pp. 371–375.

[3] Bárány, V., G. Gottlob and M. Otto, *Querying the guarded fragment*, in: *Proc. LICS*, 2010.
URL `http://www.mathematik.tu-darmstadt.de/~otto/papers/BGOlics2010.pdf`

[4] Bárány, V., B. ten Cate and L. Segoufin, *Guarded negation*, in: *Proc. ICALP*, 2011.

[5] Grädel, E., *On the restraining power of guards*, Journal of Symbolic Logic **64** (1999), pp. 1719–1742.

[6] Grädel, E. and I. Walukiewicz, *Guarded fixed point logic*, in: *LICS*, 1999, pp. 45–54.

[7] Otto, M., *Modal and guarded characterisation theorems over finite transition systems*, in: *LICS* (2002), pp. 371–.

[8] Otto, M., *Highly acyclic groups, hypergraph covers and the guarded fragment*, Logic in Computer Science, Symposium on **0** (2010), pp. 11–20.

Kripke Models of Transfinite Provability Logic

David Fernández-Duque [1]

Universidad de Sevilla

Joost J. Joosten [2]

Universitat de Barcelona

Abstract

For any ordinal Λ, we can define a polymodal logic GLP_Λ, with a modality $[\xi]$ for each $\xi < \Lambda$. These represent provability predicates of increasing strength. Although GLP_Λ has no non-trivial Kripke frames, Ignatiev showed that indeed one can construct a universal Kripke frame for the variable-free fragment with natural number modalities, denoted GLP^0_ω.

In this paper we show how to extend these constructions for arbitrary Λ. More generally, for each ordinals Θ, Λ we build a Kripke model $\mathfrak{I}^\Theta_\Lambda$ and show that GLP^0_Λ is sound for this structure. In our notation, Ignatiev's original model becomes $\mathfrak{I}^{\varepsilon_0}_\omega$.

Keywords: proof theory, modal logic, provability logic

1 Introduction

It was Gödel who first suggested interpreting the modal $\Box$ as a provability predicate, which as he observed should satisfy $\Box(\phi \to \psi) \to (\Box\phi \to \Box\psi)$ and $\Box\phi \to \Box\Box\phi$. In this way, the Second Incompleteness Theorem could be expressed succinctly as $\Diamond\top \to \Diamond\Box\bot$.

More generally, Löb's axiom $\Box(\Box\phi \to \phi) \to \Box\phi$ is valid for this interpretation, and with this we obtain a complete characterization of the propositional behavior of provability in Peano Arithmetic [11]. The modal logic obtained from Löb's axiom is called GL (for Gödel-Löb) and is rather well-behaved; it is decidable and has finite Kripke models, based on transitive, well-founded frames [10].

Japaridze [5] then suggested extending GL by a sequence of provability modalities $[n]$, for $n < \omega$, where $[n]\phi$ could be interpreted (for example) as ϕ *is derivable using* n *instances of the* ω*-rule.* We shall refer to this extension as GLP_ω. GLP_ω turns out to be much more powerful than GL, and indeed Beklemishev has shown how it can be used to perform ordinal analysis of Peano Arithmetic and its natural subtheories [1].

[1] dfduque@us.es

[2] jjosten@ub.edu

However, as a modal logic, it is much more ill-behaved than GL. Most notably, over the class of GLP Kripke frames, the formula $[1]\bot$ is valid! This is clearly undesirable. There are ways to get around this, for example using topological semantics. However, Ignatiev in [7] showed how one can still get Kripke frames for the *closed* fragment of GLP_ω, which contains no propositional variables (only $\bot$). This fragment, which we denote GLP^0_ω, is still expressive enough to be used in Beklemishev's ordinal analysis.

Our goal is to extend Ignatiev's construction for GLP^0_ω to GLP^0_Λ, where Λ is an arbitrary ordinal (or, if one wishes, the class of all ordinals). To do this we build upon known techniques, but dealing with transfinite modalities poses many new challenges. In particular, frames will now have to be much 'deeper' if we wish to obtain non-empty accessibility relations.

Our structures naturally extend the model which was first defined and studied by Ignatiev for GLP^0_ω in [7], and in our notation becomes $\mathfrak{I}^{\varepsilon_0}_\omega$. Originally, Ignatiev's study was an amalgamate of modal, arithmetical and syntactical methods. In [8] the model was first submitted to a purely modal analysis and [3] built forth on this work. In this paper, we prove soundness and non-triviality of the accessibility relations using purely semantic techniques.

The layout of the paper is as follows. In Section 2 we give a quick overview of the logics GLP_Λ. Section 3 then gives some motivation for the constructions we shall present, and Section 4 reviews some operations on ordinals and the notation we will use.

In Section 5, we introduce our generalized Ignatiev models, denoted $\mathfrak{I}^\Theta_\Lambda$, where Θ, Λ are ordinal parameters. Section 6 then defines some operations on the points of our model, which are called *ℓ-sequences*. With these operations we prove soundness in Section 7.

Finally, Section 8 shows that, for an arbitrarily large ordinal ξ with $\xi < \Lambda$, if Θ is large enough, then $<_\xi$ is non-empty on $\mathfrak{I}^\Theta_\Lambda$. This result is not a full completeness proof, however it is a crucial step; in [4], we shall show how one deduces completeness of GLP^0_Λ for a Kripke frame $\mathfrak{F}$ from non-triviality of the accessibility relations. The latter result is syntactical and does not depend much on the structure $\mathfrak{F}$.

2 The logic GLP^0_Λ

Let Λ be either an ordinal or the class of all ordinals. Formulas of GLP^0_Λ are built from $\bot$ using Boolean connectives $\neg, \wedge$ and a modality $[\xi]$ for each $\xi < \Lambda$. As is customary, we use $\langle\xi\rangle$ as a shorthand for $\neg[\xi]\neg$.

Note that there are no propositional variables, as we are concerned here with the *closed fragment* of GLP_Λ.

The logic GLP^0_Λ (see [2]) is given by the following axioms:

(i) all propositional tautologies,

(ii) $[\xi](\phi \to \psi) \to ([\xi]\phi \to [\xi]\psi)$ for all $\xi < \Lambda$,

(iii) $[\xi]([\xi]\phi \to \phi) \to [\xi]\phi$ for all $\xi < \Lambda$,

(iv) $\langle\xi\rangle\,\phi \to \langle\zeta\rangle\,\phi$ for $\zeta < \xi < \Lambda$,

(v) $\langle\zeta\rangle\,\phi \to [\xi]\,\langle\zeta\rangle\,\phi$ for $\zeta < \xi < \Lambda$.

The rules of our logic are Modus Ponens and Necessitation for each modality. Although no full completeness result has yet been published for (hyper)arithmetical interpretations of GLP_Λ with $\Lambda > \omega$ the community is confident that such interpretations exist. One such example would be to interpret $[\alpha]$ as "provable in some base theory using α many nested iterations of the ω rule" in some infinitary calculus. In this paper our focus is on the modal aspects of the logics GLP_Λ only.

A *Kripke frame*[3] is a structure $\mathfrak{F} = \langle W, \langle R_i\rangle_{i<I}\rangle$, where W is a set and $\langle R_i\rangle_{i<I}$ a family of binary relations on W. To each formula ψ in the closed modal language with modalities $\langle i\rangle$ for $i < I$ we assign a set $[\![\psi]\!]_{\mathfrak{F}} \subseteq W$ inductively as follows:

$$[\![\bot]\!]_{\mathfrak{F}} = \varnothing$$

$$[\![\neg\phi]\!]_{\mathfrak{F}} = W \setminus [\![\phi]\!]_{\mathfrak{F}}$$

$$[\![\phi \wedge \psi]\!]_{\mathfrak{F}} = [\![\phi]\!]_{\mathfrak{F}} \cap [\![\psi]\!]_{\mathfrak{F}}$$

$$[\![\langle i\rangle\,\phi]\!]_{\mathfrak{F}} = R_i^{-1}\,[\![\phi]\!]_{\mathfrak{F}}\,.$$

As always, for a binary relation S on W, if $X \subseteq W$ we denote by $S^{-1}X$ the set $\{y \in W \mid \exists x{\in}X\ ySx\}$. Often we will write $\langle\mathfrak{F}, x\rangle \Vdash \psi$ instead of $x \in [\![\psi]\!]_{\mathfrak{F}}$.

It is well-known that GL is sound for $\mathfrak{F}$ whenever R_i^{-1} is well-founded and transitive, in which case we write it $<_i$. However, constructing models of GLP_Λ is substantially more difficult than constructing models of GL, as we shall see.

3 Motivation for our models

The full logic GLP_Λ cannot be sound and complete with respect to any class of Kripke frames. Indeed, let $\mathfrak{F} = \langle W, \langle <_\xi\rangle_{\xi<\lambda}\rangle$ be a polymodal frame.

Then, it is not too hard to check that in $\mathfrak{F}$ we have the following correspondences

(i) Löb's axiom $[\xi]([\xi]\phi \to \phi) \to [\xi]\phi$ is valid if and only if $<_\xi$ is well-founded and transitive,

(ii) the axiom $\langle\zeta\rangle\phi \to \langle\xi\rangle\phi$ for $\xi \le \zeta$ is valid if and only if, whenever $w <_\zeta v$, then $w <_\xi v$, and

(iii) $\langle\xi\rangle\phi \to [\zeta]\langle\xi\rangle\phi$ for $\xi < \zeta$ is valid if, whenever $v <_\zeta w$, $u <_\xi w$ and $\xi < \zeta$, then $u <_\xi v$.

Suppose that for $\xi < \zeta$, there are two worlds such that $w <_\zeta v$. Then from Correspondence (ii) above we see that $w <_\xi v$, while from (iii) this implies that $w <_\xi w$. But this clearly violates (i). Hence if $\mathfrak{F} \models \mathsf{GLP}$, it follows that all accessibility relations (except possibly $<_0$) are empty.

[3] Since we are restricting to the closed fragment we make no distinction between Kripke frames and Kripke *models*.

However, this does not rule out the possibility that the closed fragments GLP^0_Λ have Kripke frames for which they are sound and complete. This turned out to be the case for GLP^0_ω and in the current paper we shall extend this result to GLP^0_Λ, with Λ arbitrary.

More precisely, given ordinals Λ, Θ, we will construct a Kripke frame $\mathfrak{J}^\Theta_\Lambda$ with 'depth' Θ (i.e., the order-type of $<_0$) and 'length' Λ (the set of modalities it interprets). $\mathfrak{J}^\Theta_\Lambda$ validates all frame conditions except for condition (ii). We shall only approximate it in that we require, for $\zeta < \xi$,

$$v <_\xi w \Rightarrow \exists v' <_\zeta w \text{ such that } v' \leftrightarrows_{\mathsf{p}} v.$$

Here p will be a set of parameters and $u' \leftrightarrows_{\mathsf{p}} u$ denotes that u' is p-*bisimilar* to u. The parameters p can be adjusted depending on ϕ in order to validate each instance of the axiom.

One convenient property of the closed fragment is that it is not sensitive to 'branching'. Indeed, consider any Kripke frame $\langle W, < \rangle$ for GL^0. To each $w \in W$ assign an ordinal $o(w)$ as follows: if w is minimal, $o(w) = 0$. Otherwise, $o(w)$ is the supremum of $o(v) + 1$ over all $v < w$.

The map o is well-defined because models of GL are well-founded. Further, because there are no variables, it is easy to check that $o : W \to \Lambda$ (where Λ is a sufficiently large ordinal) is a bisimulation.

Thus to describe the modal logic of W it is enough to describe $o(W)$. We can extend this idea to GLP_Λ; if we have a well-founded frame $\mathfrak{F} = \langle W, \langle <_\xi \rangle_{\xi<\Lambda} \rangle$, we can represent a world w by the sequence $\boldsymbol{o}(w) = \langle o_\xi(w) \rangle_{\xi<\Lambda}$, where o_ξ is defined analogously to o. Thus we can identify elements of our model with sequences of ordinals. It is a priori not clear that this representation suffices also for the polymodal case, and one of the main purposes of this paper is to see that it actually does.

Meanwhile, there are certain conditions these sequences must satisfy. They arise from considering *worms*, which are formulas of the form $\langle \xi_0 \rangle \ldots \langle \xi_n \rangle \top$. In various ways we can see worms as the backbone of the closed fragment of GLP. It is known that each formula of GLP^0_Λ is equivalent to a Boolean combination of worms.

Given worms A, B and an ordinal ξ, we define $A \prec_\xi B$ if $\vdash B \to \langle \xi \rangle A$. This gives us a well-founded partial order.

In [6], we study $\boldsymbol{\Omega}(A)$, where

$$\Omega_\xi(A) = \sup_{B \prec_\xi A} \Omega_\xi(B);$$

this gives us a good idea of what sequences may be included in the model. As it turns out, $\boldsymbol{\Omega}(A)$ is a 'local bound' for $\boldsymbol{o}(w)$ (see Definition 5.1).

4 Some ordinal arithmetic

As mentioned in the previous section, a world f in our model will be coded by a sequence that for each ξ tells us the order-type of f with respect $<_\xi$.

These order-types are ordinals and it will be convenient to review some basic properties of ordinals that shall be used throughout this paper. We dedicate this section to this purpose.

We shall simply state the main properties without proof. For further details, we refer the reader to [9]. Ordinals are canonical representatives for well-orders. The first infinite ordinal is as always denoted by ω.

Most operations on natural numbers can be extended to ordinal numbers, like addition, multiplication and exponentiation (see [9]). However, in the realm of ordinal arithmetic things become often more subtle; for example, $1+\omega = \omega \neq \omega + 1$. Other operations differ considerably from ordinary arithmetic as well.

However, there are also various similarities. In particular we have a form of subtraction available in ordinal arithmetic.

Lemma 4.1

(i) *Given ordinals $\zeta<\xi$, there exists a unique ordinal $\eta = -\zeta + \xi$ such that $\zeta + \eta = \xi$.*

(ii) *Given $\eta > 0$, there exist α, β such that $\eta = \alpha + \omega^\beta$. The value of β is uniquely defined and we denote it $\ell\eta$, the 'last exponent' of η.*

(iii) *Given $\eta > 0$, there exist unique values of α, β such that $\eta = \omega^\alpha + \beta$ and $\beta < \omega^\alpha + \beta$.*

It is convenient to have representation systems for ordinals. One of the most convenient is given by Cantor Normal Forms (CNFs).

Theorem 4.2 (Cantor Normal Form Theorem)
For each ordinal α there are unique ordinals $\alpha_1 \geq \ldots \geq \alpha_n$ such that

$$\alpha = \omega^{\alpha_1} + \ldots + \omega^{\alpha_n}.$$

Another difference between ordinal and ordinary arithmetic is that various increasing functions in ordinal arithmetic have fixpoints where the ordinary counterparts do not. Let us make this precise. We call a function f *increasing* if $\alpha < \beta$ implies $f(\alpha) < f(\beta)$. An ordinal function is called *continuous* if $\bigcup_{\zeta<\xi} f(\zeta) = f(\xi)$ for all limit ordinals[4] ξ. Functions which are both increasing and continuous are called *normal.*

It is not hard to see that each normal function has an unbounded set of fixpoints. For example the first fixpoint of the function $x \mapsto \omega^x$ is

$$\sup\{\omega, \omega^\omega, \omega^{\omega^\omega}, \ldots\}$$

and is denoted ε_0. Clearly for these fixpoints, CNFs are not too informative as, for example, $\varepsilon_0 = \omega^{\varepsilon_0}$. Here it is convenient to pass to normal forms capable of representing fixed points of the ω-exponential: Veblen Normal Forms (VNF).

[4] Henceforth we shall write $\lim_{\zeta\to\xi} f(\zeta)$ instead of $\bigcup_{\zeta<\xi} f(\zeta)$.

In his seminal paper [12], Veblen considered for each normal function f its derivative f' that enumerates the fixpoints of f. Taking derivatives can be transfinitely iterated for unbounded f.

Each closed (under taking suprema) unbounded set X is enumerated by a normal function. The derivative X' of a closed unbounded set X is defined to be the set of fixpoints of the function that enumerates X and likewise for transfinite progressions:

$$\begin{array}{lll} X_{\alpha+1} & := & (X_\alpha)'; \\ X_\lambda & := & \bigcap_{\alpha<\lambda} X_\alpha \ \text{ for limit } \lambda. \end{array}$$

By taking $F_0 := \{\omega^\alpha \mid \alpha \in \mathsf{On}\}$ one obtains Veblen's original hierarchy and the φ_α denote the corresponding enumeration functions of the thus obtained F_α.

Beklemishev noted in [2] that in the setting of GLP it is desirable to have $1 \notin F_0$. Thus he considered the progression that started with $F_0^B := \{\omega^{1+\alpha} \mid \alpha \in \mathsf{On}\}$. We denote the corresponding enumeration functions by $\hat{\varphi}_\alpha$.

In [6] the authors realized that, moreover it is desirable to have 0 in the initial set, whence they departed from $E_0 = \{0\} \cup \{\omega^{1+\alpha} \mid \alpha \in \mathsf{On}\}$. We shall denote the corresponding enumeration functions by e_α.

One readily observes that

$$\begin{array}{rclcll} e_\alpha(0) & = & 0 & & & \text{for all } \alpha; \\ e_0(1+\beta) & = & \varphi_0(1+\beta) & = & \hat{\varphi}_0(\beta) & \text{for all } \beta; \\ e_{1+\alpha}(1+\beta) & = & \varphi_{1+\alpha}(\beta) & = & \hat{\varphi}_{1+\alpha}(\beta) & \text{for all } \alpha, \beta. \end{array}$$

Often, we can write an ordinal ω^α in many ways as $\varphi_\xi(\eta)$. However, if we require that $\eta < \varphi_\xi(\eta)$, then both ξ and η are uniquely determined. In other words, for every ordinal α, there exist unique η, ξ such that $\omega^\alpha = \varphi_\xi(\eta)$ and $\eta < \varphi_\xi(\eta)$.

Combining this fact with the CNF Theorem one obtains so-called Veblen Normal Forms for ordinals.

Theorem 4.3 (Veblen Normal Form Theorem) *For all α there exist unique $\alpha_1, \beta_1, \ldots \alpha_n, \beta_n$ $(n \geq 0)$ such that*

(i) $\alpha = \varphi_{\alpha_1}(\beta_1) + \ldots + \varphi_{\alpha_n}(\beta_n)$,

(ii) $\varphi_{\alpha_i}(\beta_i) \geq \varphi_{\alpha_{i+1}}(\beta_{i+1})$ *for* $i < n$,

(iii) $\beta_i < \varphi_{\alpha_i}(\beta_i)$ *for* $i \leq n$.

Note that $\alpha_i \geq \alpha_{i+1}$ does not in general hold in the VNF of α. For example, $\omega^{\varepsilon_0+1} + \varepsilon_0 = \varphi_0(\varphi_{\varphi_0(0)}(0) + \varphi_0(0)) + \varphi_{\varphi_0(0)}(0)$.

5 Generalized Ignatiev models

In this section we will generalize Ignatiev's universal model for GLP^0_ω to obtain models for GLP^0_Λ, for arbitrary Λ. Our model combines ideas from Ignatiev's construction with some new methods for dealing with limit modalities. The model is universal in that it validates all theorems and refutes all non-theorems of GLP^0_Λ. It also has some minimality properties that we shall not discuss in this paper.

We use the 'last exponent' operation ℓ described above to define the 'worlds' of our model. They will be (typically infinite) sequences of ordinals which we call *ℓ-sequences.*

Definition 5.1 *Let Θ, Λ be ordinals. We define an* ℓ-sequence *(of depth Θ and length Λ) to be a function*

$$f : \Lambda \to \Theta$$

such that, for every $\zeta < \xi < \Lambda$,

$$\ell f(\zeta) \geq \ell e_{\ell\xi} f(\xi). \tag{1}$$

Note that in ℓ-sequences, for $\xi = \zeta + 1$ we have

$$\ell f(\zeta) \geq \ell e_0 f(\xi) = f(\xi)$$

which is as in the original Ignatiev model for GLP^0_ω. We shall now see that ℓ-sequences can be described either globally, as above, or locally.

Definition 5.2 *Let Θ, Λ be ordinals. A function $f : \Lambda \to \Theta$ is a* local ℓ-sequence *if and only if, given $\xi \in (0, \Lambda)$ there is $\vartheta < \xi$ such that*

$$\ell f(\zeta) \geq \ell e_{\ell\xi} f(\xi)$$

for all $\zeta \in [\vartheta, \xi)$.

If one requires equality in the above definition, i.e., $\ell f(\zeta) = \ell e_{\ell\xi} f(\xi)$ then one exactly gets the sequences $\mathbf{\Omega}(A)$ for worms A.

Lemma 5.3 *A function $f : \Lambda \to \Theta$ is an ℓ-sequence if and only if it is a local ℓ-sequence.*

Proof. Clearly every ℓ-sequence is a local ℓ-sequence.

Now, if f is a local ℓ-sequence, towards a contradiction suppose that it is not an ℓ-sequence, and let $\xi \in (0, \Lambda)$ be least with the property that, for some $\zeta < \xi$,

$$\ell f(\zeta) < \ell e_{\ell\xi} f(\xi). \tag{2}$$

Now pick $\vartheta < \xi$ such that, for all $\zeta' \in [\vartheta, \xi)$,

$$\ell f(\zeta') \geq \ell e_{\ell\xi} f(\xi).$$

Such a ϑ exists, since f is a local ℓ-sequence.

Evidently $\zeta < \vartheta < \xi$, whence by minimality of ξ, $\ell f(\zeta) \geq \ell e_{\ell\vartheta} f(\vartheta)$ and

$$\ell f(\zeta) \geq \ell e_{\ell\vartheta} f(\vartheta) \geq \ell f(\vartheta) \geq \ell e_{\ell\xi} f(\xi).$$

This contradicts (2). □

Now rather than considering an ℓ-sequence in isolation, we will be interested in the structure of all ℓ-sequences:

Definition 5.4 *Given ordinals Θ, Λ, define a structure*

$$\mathfrak{I}^{\Theta}_{\Lambda} = \left\langle D^{\Theta}_{\Lambda}, \langle <_{\xi} \rangle_{\xi<\Lambda} \right\rangle$$

by setting D^{Θ}_{Λ} to be the set of all ℓ-sequences of depth Θ and length Λ. Define $f <_{\xi} g$ if and only if $f(\zeta) = g(\zeta)$ for all $\zeta < \xi$ and $f(\xi) < g(\xi)$.

One can check that Ignatiev's original model is precisely $\mathfrak{I}^{\varepsilon_0}_{\omega}$ in our notation. The novelty is that now Λ could be much, much bigger than ω.

6 Operations on ℓ-sequences

$\mathfrak{I}^{\Theta}_{\Lambda}$ is not a genuine GLP_{Λ} frame. However, we shall show that indeed it is a model of $\mathsf{GLP}^{0}_{\Lambda}$. In this section we shall develop some tools which will be useful for proving this fact.

6.1 Simple sequences

A useful elementary notion will be that of *simple sequences.* These are finite increasing sequences of ordinals that only make 'jumps' of the form ω^{β}. When analizing a formula ψ, it will be easier to extend the modalities appearing in ψ to a simple sequence and treat them, to some extent, as if they appeared in ψ.

Definition 6.1 *A finite sequence of ordinals $\langle \sigma_i \rangle_{i \leq I}$ is* simple *if $\sigma_0 = 0$ and for every $i < I$ there exists β_i such that $\sigma_{i+1} = \sigma_i + \omega^{\beta_i}$.*

Lemma 6.2 *Every finite increasing sequence of ordinals can be extended to a simple sequence.*

Proof. Induction on I. Suppose $\langle \sigma_i \rangle_{i \leq I+1}$ is a finite increasing sequence of ordinals. We assume that there is a simple sequence $\langle \alpha_i \rangle_{i \leq J}$ extending $\langle \sigma_0, ..., \sigma_I \rangle$ with $\alpha_J = \sigma_I$.

Since $\sigma_I < \sigma_{I+1}$, there exists a unique ordinal η such that $\sigma_{I+1} = \sigma_I + \eta$. Write

$$\eta = \sum_{k<K} \omega^{\gamma_k}$$

in Cantor Normal Form.

Then define, for each $k \leq K$,

$$\beta_k = \sigma_I + \sum_{i<k} \omega^{\gamma_i}.$$

Finally, setting

$$\boldsymbol{\delta} = \langle \alpha_0, \alpha_1, ..., \alpha_J, \beta_1, \beta_2, ..., \beta_K \rangle$$

gives us the desired simple extension of $\boldsymbol{\sigma}$. □

6.2 Approximations of ℓ-sequences

Given a formula ϕ and an ℓ-sequence $f \in D^{\Theta}_{\Lambda}$ with $\langle \mathfrak{I}^{\Theta}_{\Lambda}, f\rangle \Vdash \phi$, there is a sense in which every ℓ-sequence g that is 'close enough' to f also satisfies ϕ. To make this precise, we will define *$\langle p, \boldsymbol{\sigma}\rangle$-approximations* of f.

Below, given an ordinal ξ in Veblen Normal Form $\sum_{i<I} \varphi_{\alpha_i}(\beta_i)$, we define the *width* of ξ recursively as the maximal sum-size in the VNF of ξ:

$$\mathrm{wdt}(\xi) := \max(\{I\} \cup \{\mathrm{wdt}(\beta_i) : i < I\}).$$

Similarly, the *height* of ξ is defined as the maximal number of nested φ's in the VNF of ξ:

$$\mathrm{hgt}(\xi) := 1 + \max_{i<I} \mathrm{hgt}(\beta_i).$$

Both the width and height of 0 are stipulated to be zero.

Note that α_i will not be used in computing the height or width of ξ; these are seen as atomic symbols and will take on only finitely many possible values.

More specifically, we say α is a *subindex* of ξ if, when writing

$$\xi = \sum_{j<J} \varphi_{\alpha_j}(\beta_j)$$

in Veblen Normal Form, we have that either $\alpha = \alpha_j$ for some $j < J$ or α is, inductively, a subindex of some β_j.

Definition 6.3 *Given a natural number p and a finite sequence of ordinals $\boldsymbol{\sigma} = \langle \sigma_0, ..., \sigma_I\rangle$, we say β is a $\langle p, \boldsymbol{\sigma}\rangle$-*approximation *of α if*

(i) $\beta < \alpha$,

(ii) $\mathrm{wdt}(\beta)$ *and* $\mathrm{hgt}(\beta)$ *are both at most* p,

(iii) *every subindex of β is of the form $\ell\sigma_i$.*

Clearly, for fixed α, p and $\boldsymbol{\sigma}$ one can only make finitely many syntactical expressions of nested width and height with subindices in $\boldsymbol{\sigma}$. Thus, there are only finitely many $\langle p, \boldsymbol{\sigma}\rangle$-approximations of a given α, and hence there is a maximum one: we denote it by $\lfloor \alpha \rfloor^p_{\boldsymbol{\sigma}}$. It will be convenient to stipulate $\lfloor 0 \rfloor^p_{\boldsymbol{\sigma}} = -1$. Clearly $\lfloor \alpha \rfloor^p_{\boldsymbol{\sigma}}$ is weakly monotone in all of its arguments.

We are not interested in approximating only ordinals, but rather entire ℓ-sequences:

Definition 6.4 *Let $\boldsymbol{\sigma}$ be a simple sequence $\langle 0 = \sigma_0, \ldots, \sigma_I\rangle$. We extend the use of $\lfloor \cdot \rfloor^p_{\boldsymbol{\sigma}}$ to sequences $f : \Lambda \to \Theta$ as follows:*

$$\lfloor f \rfloor^p_{\boldsymbol{\sigma}}(\xi) = \begin{cases} 0 & \text{for } \xi > \sigma_I \\ \lfloor f(\sigma_I)\rfloor^p_{\boldsymbol{\sigma}} + 1 & \text{for } \xi = I \\ \lfloor f(\sigma_i)\rfloor^p_{\boldsymbol{\sigma}} + 1 + e_{\ell\sigma_{i+1}} \lfloor f \rfloor^p_{\boldsymbol{\sigma}}(\sigma_{i+1}) & \text{for } \xi = \sigma_i \text{ with } i < I \\ e_{\ell\sigma_{i+1}} \lfloor f \rfloor^p_{\boldsymbol{\sigma}}(\sigma_{i+1}) & \text{for } \sigma_i < \xi < \sigma_{i+1} \end{cases}$$

Note that it is nearly never the case that $\lfloor f \rfloor^p_{\boldsymbol{\sigma}}(\xi) = \lfloor f(\xi) \rfloor^p_{\boldsymbol{\sigma}}$; however as we will see later in Lemma 6.8, they cannot be too different. Before this, we observe that this operation always produces ℓ-sequences.

Lemma 6.5 *Given any $f : \Lambda \to \Theta$ and parameters $p, \boldsymbol{\sigma}$, $g = \lfloor f \rfloor^p_{\boldsymbol{\sigma}}$ is an ℓ-sequence.*

Further, it has the property that for all $i \leq I$

$$\ell g(\sigma_i) = \ell e_{\ell \sigma_{i+1}} g(\sigma_{i+1}). \tag{3}$$

Proof. Using Lemma 5.3 it suffices to see that g is a local ℓ -sequence. To see this, we make a few case distinctions.

$g(\xi) = 0$. Note that this covers the case where $\xi > \sigma_I$. In this case, the inequality $\ell g(\zeta) \geq e_{\ell\xi} g(\xi)$ holds trivially for all $\zeta < \xi$ since the right-hand side is zero.

$g(\xi) > 0$ **and** $\xi = \sigma_i$ **for some** i.
Then, $\ell g(\zeta) = \ell e_{\ell\xi} g(\xi)$, for all $\zeta \in [\sigma_i, \xi)$.

$g(\xi) > 0$ **and** $\sigma_i < \xi < \sigma_{i+1}$ **for some** i. We first claim that $\ell\xi < \ell\sigma_{i+1}$. Indeed, since $\boldsymbol{\sigma}$ is simple, we have that

$$\sigma_{i+1} = \sigma_i + \omega^{\ell \sigma_{i+1}}.$$

Meanwhile, since $\xi > \sigma_i$ we can write

$$\xi = \sigma_i + \sum_{j \leq J} \omega^{\beta_j}.$$

Now, clearly if $\beta_J = \ell\xi$ were greater or equal to $\ell\sigma_{i+1}$, we would have $\xi \geq \sigma_{i+1}$, contrary to our assumption. In particular, note that this implies $\ell\sigma_{i+1} > 0$.

But then we know that $e_{\ell\sigma_{i+1}} g(\sigma_{i+1})$ is a fixpoint of $e_{\ell\xi}$, and thus for all $\zeta \in [\sigma_i, \xi)$,

$$\begin{aligned} \ell g(\zeta) &= \ell e_{\ell\sigma_{i+1}} g(\sigma_{i+1}) \\ &= \ell e_{\ell\xi} e_{\ell\sigma_{i+1}} g(\sigma_{i+1}) \\ &= \ell e_{\ell\xi} g(\xi). \end{aligned}$$

This covers all cases and shows that g is an ℓ-sequence satisfying (3), as desired. □

Lemma 6.6 *If $\boldsymbol{\sigma} = \langle \sigma_i \rangle_{i \leq I}$ is a simple sequence, $f \in D^{\Theta}_{\Lambda}$ and $\xi < \Lambda$, then $\mathrm{wdt}(\lfloor f \rfloor^p_{\boldsymbol{\sigma}}(\xi))$ and $\mathrm{hgt}(\lfloor f \rfloor^p_{\boldsymbol{\sigma}}(\xi))$ are both at most[5] $p + I$.*

[5] Actually, $\mathrm{wdt}(\lfloor f \rfloor^p_{\boldsymbol{\sigma}}(\xi)) \leq p + 1$, but it seems more convenient to bound the height and width uniformly.

Proof. One can check easily that $\mathrm{wdt}(\lfloor f \rfloor^p_{\boldsymbol{\sigma}}(\sigma_i)) \leq \mathrm{wdt}(\lfloor f \rfloor^p_{\boldsymbol{\sigma}}(\sigma_{i+1})) + 1$ and

$$\mathrm{hgt}(\lfloor f \rfloor^p_{\boldsymbol{\sigma}}(\sigma_i)) \leq \mathrm{hgt}(\lfloor f \rfloor^p_{\boldsymbol{\sigma}}(\sigma_{i+1})) + 1.$$

Thus the width and height of all terms is bounded by $I+p$; intermediate terms (i.e., $\lfloor f \rfloor^p_{\boldsymbol{\sigma}}(\xi)$ for $\sigma_i < \xi < \sigma_{i+1}$) obviously have width and height bounded by that of $\lfloor f \rfloor^p_{\boldsymbol{\sigma}}(\sigma_i)$. □

The following simple observation will be quite useful later:

Lemma 6.7 *If $\alpha < \xi$, then*

(i) *If $\beta \leq \ell\xi$, then $\alpha + \omega^\beta \leq \xi$;*

(ii) *If $\beta < \ell\xi$, then $\alpha + \omega^\beta < \xi$.*

Proof. By observations on the Cantor normal form of ξ. □

Lemma 6.8 *For every $f \in D^\Theta_\Lambda$ and $i \leq I$,*

$$\lfloor f(\sigma_i) \rfloor^p_{\boldsymbol{\sigma}} < \lfloor f \rfloor^p_{\boldsymbol{\sigma}}(\sigma_i) \leq f(\sigma_i).$$

Moreover, if $\lfloor f \rfloor^p_{\boldsymbol{\sigma}}(\sigma_i) < f(\sigma_i)$ then $\lfloor f \rfloor^p_{\boldsymbol{\sigma}}(\sigma_j) < f(\sigma_j)$ for all $j < i$.

Proof. That $\lfloor f(\sigma_i) \rfloor^p_{\boldsymbol{\sigma}} < \lfloor f \rfloor^p_{\boldsymbol{\sigma}}(\sigma_i)$ is obvious from the definition of $\lfloor f \rfloor^p_{\boldsymbol{\sigma}}(\sigma_i)$, since it is always of the form

$$\lfloor f(\sigma_i) \rfloor^p_{\boldsymbol{\sigma}} + \omega^\rho \tag{4}$$

for some ordinal ρ. In particular we see $\lfloor f \rfloor^p_{\boldsymbol{\sigma}}(\sigma_i) > 0$.

To see the other inequality, we use backwards induction on i; clearly

$$\lfloor f \rfloor^p_{\boldsymbol{\sigma}}(\sigma_I) \leq f(\sigma_I),$$

since $\lfloor f \rfloor^p_{\boldsymbol{\sigma}}(\sigma_I) = \lfloor f(\sigma_I) \rfloor^p_{\boldsymbol{\sigma}} + 1$ and $\lfloor f(\sigma_I) \rfloor^p_{\boldsymbol{\sigma}} < f(\sigma_I)$.

Now, assume inductively that

$$\lfloor f \rfloor^p_{\boldsymbol{\sigma}}(\sigma_{i+1}) \leq f(\sigma_{i+1}),$$

and once again write $\lfloor f \rfloor^p_{\boldsymbol{\sigma}}(\sigma_i)$ in the form (4).

First we note that the function ℓe_α is increasing independently of α: if $\alpha = 0$ it is the identity; otherwise, $\ell e_\alpha = e_\alpha$, which is a normal function.

Thus we have that, if $\sigma_{i+1} = \sigma_i + \omega^\alpha$,

$$\rho = \ell e_\alpha \lfloor f \rfloor^p_{\boldsymbol{\sigma}}(\sigma_{i+1}) \overset{\mathrm{IH}}{\leq} \ell e_\alpha f(\sigma_{i+1}) = \ell f(\sigma_i),$$

where the last equality is by Lemma 6.5.

In either case we get $\rho \leq \ell f(\sigma_i)$ so by Lemma 6.7.1,

$$\lfloor f \rfloor^p_{\boldsymbol{\sigma}}(\sigma_i) = \lfloor f(\sigma_i) \rfloor^p_{\boldsymbol{\sigma}} + \omega^\rho \leq f(\sigma_i).$$

Moreover, if $\lfloor f \rfloor^p_{\boldsymbol{\sigma}}(\sigma_i) < f(\sigma_i)$ then we use Lemma 6.7.2 to conclude $\lfloor f \rfloor^p_{\boldsymbol{\sigma}}(\sigma_{i+1}) < f(\sigma_{i+1})$. □

6.3 Concatenation

Definition 6.9 *Given sequences* $f, g : \Lambda \to \Theta$, *we define their* λ-concatenation $f \overset{\lambda}{*} g : \Lambda \to \Theta$ *by*

$$f \overset{\lambda}{*} g(\xi) = \begin{cases} f(\xi) & \text{if } \xi < \lambda \\ g(\xi) & \text{otherwise.} \end{cases}$$

Lemma 6.10 *If* $f, g \in D^{\Theta}_{\Lambda}$ *and* $g(\lambda) \leq f(\lambda)$, *then* $f \overset{\lambda}{*} g$ *is an* ℓ*-sequence.*

If, further, $g(\lambda) < f(\lambda)$, *then* $f \overset{\lambda}{*} g <_{\lambda} f$.

Proof. Immediate from the definition of $f \overset{\lambda}{*} g$ and Lemma 5.3. □

7 Soundness

The sequence $\lfloor f \rfloor^{p}_{\boldsymbol{\sigma}}$ does not satisfy the same formulas of the modal language as f, but it does satisfy the same formulas that are 'simple enough'. To see this we extend the notion of *n-bisimulation* to the slightly more general notion of $\langle p, \boldsymbol{\sigma} \rangle$*-bisimulation*:

Definition 7.1 *Given* $f, g \in D^{\Theta}_{\Lambda}$, *a sequence of ordinals* $\boldsymbol{\sigma}$ *and* $p < \omega$, *we say* f *is* $\langle p, \boldsymbol{\sigma} \rangle$-bisimilar *to* g *(in symbols,* $f \leftrightarrows^{p}_{\boldsymbol{\sigma}} g$*) by induction on* p *as follows:*

For $p = 0$, *any two* ℓ*-sequences are* $\langle p, \boldsymbol{\sigma} \rangle$*-bisimilar.*

For $p = q + 1$, $f \leftrightarrows^{p}_{\boldsymbol{\sigma}} g$ *if and only if, for every* ξ *of the form* σ_i*:*

Forth. Whenever $f' <_{\xi} f$, *there is* $g' <_{\xi} g$ *with* $f' \leftrightarrows^{q}_{\boldsymbol{\sigma}} g'$.

Back. Whenever $g' <_{\xi} g$, *there is* $f' <_{\xi} f$ *with* $f' \leftrightarrows^{q}_{\boldsymbol{\sigma}} g'$.

The following lemma is standard in modal logic.

Lemma 7.2 *If* $f \leftrightarrows^{p}_{\boldsymbol{\sigma}} g$, *then* f *and* g *validate the same formulas* ψ *of modal depth* p *where all the modalities in* ψ *are among* $\boldsymbol{\sigma}$.

There is a close relation between $\langle p, \boldsymbol{\sigma} \rangle$-approximation and $\langle p, \boldsymbol{\sigma} \rangle$-bisimulation.

The following lemma will be quite useful in making this precise. Say that two ℓ-sequences are $\langle p, \langle \sigma_k \rangle_{k \leq I} \rangle$*-close* if, for all $k \leq I$,

(i) $\lfloor f(\sigma_k) \rfloor^{p}_{\boldsymbol{\sigma}} < g(\sigma_k)$ and

(ii) $\lfloor g(\sigma_k) \rfloor^{p}_{\boldsymbol{\sigma}} < f(\sigma_k)$.

We will write $f \sim^{p}_{\boldsymbol{\sigma}} g$.

Lemma 7.3 *Let* g, f, f' *be* ℓ*-sequences and* $\langle p, \langle \sigma_k \rangle_{k \leq I} \rangle$ *be parameters.*

Suppose that for some $i \leq I$, $f' <_{\sigma_i} f$, *and* $f \sim^{p+I}_{\boldsymbol{\sigma}} g$.

For $j \leq i$, *let*

$$g_j = g \overset{\sigma_j}{*} \lfloor f' \rfloor^{p}_{\boldsymbol{\sigma}}.$$

Then, we have that g_j *is an* ℓ*-sequence,* $g_j <_{\sigma_j} g$ *and* $f' \sim^{p}_{\boldsymbol{\sigma}} g_j$.

Proof. To see that g_j is an ℓ-sequence, by Lemmas 6.10 and 6.5, it suffices to show that $g_j(\sigma_j) = \lfloor f' \rfloor^{p}_{\boldsymbol{\sigma}}(\sigma_j) < g(\sigma_j)$. By Lemma 6.8 we see $\lfloor f' \rfloor^{p}_{\boldsymbol{\sigma}}(\sigma_i) \leq f'(\sigma_i)$, and since $f' <_{\sigma_i} f$ we have that $g_j(\sigma_i) < f(\sigma_i)$. Thus, by Lemma

6.8 again, we see that also $g_j(\sigma_j) < f(\sigma_j)$. Meanwhile, by Lemma 6.6, the height and width of $g_j(\sigma_j)$ are bounded by $p+I$, so that $g_j(\sigma_j)$ is a $\langle p+I, \boldsymbol{\sigma}\rangle$-approximation of $f(\sigma_j)$ and thus $g_j(\sigma_j) \leq \lfloor f(\sigma_j) \rfloor_{\boldsymbol{\sigma}}^{p+I}$. Now, by assumption $\lfloor f(\sigma_j) \rfloor_{\boldsymbol{\sigma}}^{p+I} < g(\sigma_j)$, so $g_j(\sigma_j) < g(\sigma_j)$ as required and g' is indeed an ℓ-sequence.

As $g_j(\xi) = g(\xi)$ for $\xi < \sigma_j$ we also conclude that $g_j <_{\sigma_j} g$. Thus, it remains to see that 1 and 2 hold for g_j and f'.

For $k < j$ we see that

$$\lfloor g_j(\sigma_k) \rfloor_{\boldsymbol{\sigma}}^{p} \leq \lfloor g_j(\sigma_k) \rfloor_{\boldsymbol{\sigma}}^{p+I} < f(\sigma_k) = f'(\sigma_k),$$

and by a symmetric argument, $\lfloor f'(\sigma_k) \rfloor_{\boldsymbol{\sigma}}^{p} < g_j(\sigma_k)$.

For $k \geq j$ we use Lemma 6.8 to obtain

$$\lfloor f'(\sigma_k) \rfloor_{\boldsymbol{\sigma}}^{p} < \lfloor f' \rfloor_{\boldsymbol{\sigma}}^{p}(\sigma_k) = g_j(\sigma_k)$$

and

$$\lfloor g_j(\sigma_k) \rfloor_{\boldsymbol{\sigma}}^{p} = \lfloor \lfloor f' \rfloor_{\boldsymbol{\sigma}}^{p}(\sigma_k) \rfloor_{\boldsymbol{\sigma}}^{p} \leq \lfloor f'(\sigma_k) \rfloor_{\boldsymbol{\sigma}}^{p} < f'(\sigma_k).$$

□

Lemma 7.4 *Let $\boldsymbol{\sigma} = \langle 0 = \sigma_0, \ldots, \sigma_I \rangle$ be a simple sequence. If $f, g \in D_\Lambda^\Theta$ are such that $f \sim_{\boldsymbol{\sigma}}^{Ip} g$, then $g \leftrightarrows_{\boldsymbol{\sigma}}^{p} f$.*

Proof. We prove the claim by induction on p. By symmetry it is enough to consider the 'forth' condition.

Thus, we suppose that $f \sim_{\boldsymbol{\sigma}}^{I(p+1)} g$, and $f' <_{\sigma_i} f$. We must find $g' <_{\sigma_i} g$ such that $g' \leftrightarrows_{\boldsymbol{\sigma}}^{p} f'$.

But by Lemma 7.3, $g' = g \overset{\sigma_i}{*} \lfloor f' \rfloor_{\boldsymbol{\sigma}}^{Ip}$ satisfies $g' <_{\sigma_i} g$ and $g' \sim_{\boldsymbol{\sigma}}^{p} f'$. By induction hypothesis we can conclude that also $g' \leftrightarrows_{\boldsymbol{\sigma}}^{p} f'$, as required. □

Theorem 7.5 (Soundness) GLP_Λ^0 *is sound for $\mathfrak{I}_\Lambda^\Theta$.*

Proof. That each of the modalities satisfy the GL axioms is a consequence of the well-foundedness and transitivity of $<_\xi$.

Let us see that the axiom $\langle \zeta \rangle \phi \to [\xi] \langle \zeta \rangle \phi$, for $\xi > \zeta$, is valid. Thus, suppose f satisfies $\langle \zeta \rangle \phi$, so that there is $g <_\zeta f$ which satisfies ϕ. Then, if $h <_\xi f$ with $\zeta < \xi$, we have that $h(\eta) = f(\eta)$ for all $\eta \leq \zeta$, so it is also the case that $g(\zeta) < h(\zeta)$ and hence h satisfies $\langle \zeta \rangle \phi$. Since h was arbitrary we conclude that f satisfies $[\xi] \langle \zeta \rangle \phi$.

The validity of any instance of $\psi = \langle \xi \rangle \phi \to \langle \zeta \rangle \phi$ follows from Lemma 7.2. Let $\boldsymbol{\sigma}$ be a simple saturation of all the ordinals appearing in ψ so that $\xi = \sigma_i$ and $\zeta = \sigma_j$ for some $j \leq i$. Let p be the modal depth of ψ. If for some ℓ-sequence f we have that $\langle \mathfrak{I}_\Lambda^\Theta, f \rangle \Vdash \langle \sigma_i \rangle \phi$, then there is some $f' <_{\sigma_i} f$ such that $\langle \mathfrak{I}_\Lambda^\Theta, f' \rangle \Vdash \phi$. Now we note that $f \sim_{\boldsymbol{\sigma}}^{Ip+I} f$ and apply Lemma 7.3 to see that for $g = f \overset{\sigma_j}{*} \lfloor f' \rfloor_{\boldsymbol{\sigma}}^{Ip}$, we have that $g \sim_{\boldsymbol{\sigma}}^{Ip} f'$ and $g <_{\sigma_j} f$. By Lemmata 7.4 and 7.2 we see that $\langle \mathfrak{I}_\Lambda^\Theta, g \rangle \Vdash \phi$ and thus $\langle \mathfrak{I}_\Lambda^\Theta, f \rangle \Vdash \langle \sigma_j \rangle \phi$, as required. □

8 Non-triviality: the first step towards completeness

In this section we will show that, for arbitrary λ, if $\lambda < \Lambda$, then the relation $<_\lambda$ is non-empty on $\mathfrak{I}^\Theta_\Lambda$, provided Θ is large enough. For this, it suffices to find an ℓ-sequence f with $f(\lambda) > 0$.

This, of course, shows that our structures $\mathfrak{I}^\Theta_\Lambda$ indeed give interesting models of GLP^0_Λ. However, as we shall prove in the upcoming [6], it implies much more; for indeed, if Λ is a limit ordinal and $\mathfrak{F}$ is a Kripke frame such that, for all $\lambda < \Lambda$, $<_\lambda$ is non-empty, it immediately follows that GLP^0_Λ is complete for $\mathfrak{F}$. Moreover, this can be shown by purely syntactical methods and does not depend substantially on the structure of $\mathfrak{F}$.

However, said syntactical considerations require some care and lie beyond the scope of the current paper. For now, we shall limit ourselves to establishing non-triviality of the accessibility relations.

Lemma 8.1 *Let α, ϑ be ordinals and let the CNF of α be given by*

$$\alpha = \omega^{\alpha_0} + ... + \omega^{\alpha_N}.$$

Then, there exists an ℓ-sequence f with $f(0) = e_{\alpha_0} \cdots e_{\alpha_N}(\vartheta)$ and $f(\alpha) = \vartheta$.

Proof. Let $\alpha' = \omega^{\alpha_0} + ... + \omega^{\alpha_{N-1}}$. By induction on α, there is an ℓ-sequence f' with $f'(\alpha') = e_{\alpha_N}(\vartheta)$ and $f'(0) = e_{\alpha_0} \cdots e_{\alpha_N}(\vartheta)$. Consider f given by

$$f(\gamma) = \begin{cases} f'(\gamma) & \text{if } \gamma \leq \alpha', \\ e_{\alpha_N}(\vartheta) & \text{if } \gamma \in (\alpha', \alpha), \\ \vartheta & \text{if } \gamma = \alpha, \\ 0 & \text{otherwise.} \end{cases}$$

It is very easy to check that $f : \Lambda \to e_{\alpha_0} \cdots e_{\alpha_N}(\vartheta)$ is an ℓ-sequence with all the desired properties for any $\Lambda \geq \alpha$. □

Corollary 8.2 *Let Λ, Θ be ordinals, and write*

$$\Lambda = \omega^{\alpha_0} + \ldots + \omega^{\alpha_N}.$$

GLP^0_Λ *is sound for $\mathfrak{I}^\Theta_\Lambda$ independently of Θ. If, further, we have that*

$$\Theta > e_{\alpha_0} \cdots e_{\alpha_N}(1),$$

then $<_\lambda$ is non-empty for all $\lambda < \Lambda$.

Proof. Immediate from Theorem 7.5 and Lemma 8.1. □

References

[1] Beklemishev, L., *Provability algebras and proof-theoretic ordinals, I*, Annals of Pure and Applied Logic **128** (2004), pp. 103–124.

[2] Beklemishev, L., *Veblen hierarchy in the context of provability algebras*, in: *Logic, Methodology and Philosophy of Science, Proceedings of the Twelfth International Congress*, Kings College Publications, 2005 .

[3] Beklemishev, L., J. Joosten and M. Vervoort, *A finitary treatment of the closed fragment of japaridze's provability logic*, Journal of Logic and Computation **15** (2005), pp. 447–463.

[4] Beklemishev, L. D., D. Fernández-Duque and J. J. Joosten, *On transfinite provability logics* (2012, in preparation).

[5] Dzhaparidze, G., *The polymodal provability logic (in Russian)*, in: *Intensional logics and the logical structure of theories: material from the Fourth Soviet-Finnish Symposium on Logic*, Telavi, 1988 pp. 16–48.

[6] Fernández-Duque, D. and J. J. Joosten, *Well-orders in the transfinite Japaridze algebra* (2011, in preparation).

[7] Ignatiev, K., *On strong provability predicates and the associated modal logics*, The Journal of Symbolic Logic **58** (1993), pp. 249–290.

[8] Joosten, J., "Intepretability Formalized," Department of Philosophy, University of Utrecht, 2004, ph.D. thesis.

[9] Pohlers, W., "Proof Theory, The First Step into Impredicativity," Springer-Verlag, Berlin Heidelberg, 2009.

[10] Segerberg, K., *An essay in classical modal logic*, Filosofiska Föreningen och Filosofiska Institutionen vid Uppsala Universitet (1971).

[11] Solovay, R. M., *Provability interpretations of modal logic*, Israel Journal of Mathematics **28** (1976), pp. 33–71.

[12] Veblen, O., *Continuous increasing functions of finite and transfinite ordinals*, Transactions of the American Mathematical Society **9** (1908), pp. 280–292.

Non-finite Axiomatizability of Dynamic Topological Logic

David Fernández-Duque [1]

Department of Computer Science and Artifical Intelligence
Universidad de Sevilla

Abstract

Dynamic topological logic ($\mathcal{DTL}$) is a polymodal logic designed for reasoning about *dynamic topological systems.* These are pairs $\langle X, f\rangle$, where X is a topological space and $f : X \to X$ is continuous. $\mathcal{DTL}$ uses a language L which combines the topological $\mathsf{S4}$ modality $\Box$ with temporal modalities from linear temporal logic.
Recently, I gave a sound and complete axiomatization DTL^* for an extension of the logic to the language L^*, where $\Diamond$ is allowed to act on finite sets of formulas and is interpreted as a tangled closure operator. No complete axiomatization is known over L, although one proof system, which we shall call KM, was conjectured to be complete by Kremer and Mints.
In this paper we show that, given any language L' such that $\mathsf{L} \subseteq \mathsf{L}' \subseteq \mathsf{L}^*$, the set of valid formulas of L' is not finitely axiomatizable. It follows, in particular, that KM is incomplete.

Keywords: dynamic topological logic, topological semantics, temporal logic, spatial logic

1 Introduction

Finding a transparent axiomatization for Dynamic Topological Logic ($\mathcal{DTL}$) has been an elusive open problem since 2005, when one (which we shall call KM) was proposed by Kremer and Mints in [12] without establishing its completeness. In [7] I offered a complete axiomatization, not over the language L used in [12], but rather in an extended language L^* which allowed the modal $\Diamond$ to be applied to finite sets of formulas. It was then interpreted as a 'tangled closure' operator (see Section 3). The resulting logic is called DTL^*.

However, the fact that DTL^* used the unfamiliar 'tangled closure' operation and was substantially less intuitive than KM left the completeness of the latter as a relevant open problem. Actually, the only motivation given in [7] for passing to an extended language was, *There is a completeness proof which*

[1] E-mail: `dfduque@us.es`. This research was supported by grants from the projects MTM2011-26840 and FFI2011-29609-C02-01 of the Spanish Ministry of Science and Innovation and HUM-5844 of the Junta de Andalucía.

works in the extended language but not in the original one; a valid, but not terribly compelling, reason.

The results in this paper will show that indeed the use of the tangled closure is an essential part of this axiomatization, and cannot be removed without extending KM (although it is not clear what such an extension should look like). In fact, we prove more. We show that, given $k < \omega$, there is a formula $\mathtt{Trouble}^k \in \mathsf{L}$ such that $\mathtt{Trouble}^k$ is derivable in DTL^* only by using formulas of the form $\Diamond\Gamma$ where Γ has at least k elements. This shows that $\mathcal{DTL}$ can be written as a strictly increasing sequence of theories and hence is not finitely axiomatizable; it follows, in particular, that KM is incomplete.

1.1 Previous work on $\mathcal{DTL}$

Dynamic topological logic ($\mathcal{DTL}$) combines the topological S4 with Linear Temporal Logic. The 'topological interior' interpretation of modal logic was already studied by Tarski, McKinsey and others around the 1940s [16] and more recently in works like [2,11,14]. Temporal logic also has a long history, having been studied by Prior before 1960 [15] and received substantial attention since; see [13] for a nice overview.

The purpose of $\mathcal{DTL}$ is to reason about *dynamic topological systems* (dts's); these are pairs $\langle X, f\rangle$, where X is a topological space and $f : X \to X$ is a continuous function, and shall be discussed in greater detail in Section 4. Dynamic Topological Logic was originally introduced as a bimodal logic in [1], where it was called S4C. In our notation, it uses the 'interior' modality $\Box$, interpreted topologically, and 'next-time' modality f, interpreted as a preimage operator; see Section 4 for details. The logic S4C is a rather well-behaved modal logic; it is decidable, axiomatizable and has the finite model property, all of which was established in [1]. Later, [12] showed that a variant, called S4H, was complete for the class of dynamical systems where f is a homeomorphism.

Also in [12], it was noted that by adding the infinitary temporal modality 'henceforth' (here denoted $[f]$), one could reason about long-term behavior of dts's, capturing phenomena such as topological recurrence. Thus they introduced an extension of S4C, which we denote $\mathcal{DTL}$. $\mathcal{DTL}$ turned out to behave much worse than S4C; it was proven to be undecidable in [9], and in [10] it was also shown that, if we restrict to the case where f is a homeomorphism, then the logic becomes non-axiomatizable. Fortunately, with arbitrary continuous functions the logic turned out to be recursively enumerable [4], but the only currently known axiomatization is over L^* [7].

This axiomatization uses the fact, first observed in [3], that L^* is more expressive than L. There, it is shown that, over the class of finite S4 models, $\mathsf{L}^*_\Diamond$ (i.e., the fragment of L^* without temporal modalities) is equally expressive to the bisimulation-invariant fragment of both first-order logic and monadic second-order logic, while $\mathsf{L}_\Diamond$ is weaker. This added expressive power is used in an important way in [7], although it is not proven that such an extension is *necessary*.

1.2 Layout of the paper

Sections 2-5 give a general review of dynamic topological logic and the known results relevant for this paper. In Section 2, we introduce topological spaces and show how one can see preorders as a special case. Section 3 gives the main properties of the tangled closure operator, an important part of $\mathcal{DTL}^*$, introduced in Section 4. Section 5 then reviews the axiomatization from [7] and defines some important sublogics.

Section 6 introduces tangled bisimulations, which are based on those presented in [3]. These are used to show that $\Diamond_{i=1}^{k+1} \gamma_i$ cannot in general be defined using exclusively k-adic occurrences of $\diamond$.

Section 7 defines the formulas $\texttt{Trouble}^k$ which are derivable in DTL^k (see also Appendix A), as well as other formulas which are useful for our purposes. Finally, Section 8 shows that, indeed, DTL^k is consistent with $\neg\texttt{Trouble}^{k+1}$, thus establishing our main result.

2 Topologies and preorders

In this section we shall very briefly review some basic notions from topology. As is well-known, topological spaces provide an interpretation of the modal logic S4, generalizing its well-known Kripke semantics.

Let us recall the definition of a *topological space*:

Definition 2.1 *A* topological space *is a pair* $\mathfrak{X} = \langle |\mathfrak{X}|, \mathcal{T}_{\mathfrak{X}} \rangle$, *where* $|\mathfrak{X}|$ *is a set and* $\mathcal{T}_{\mathfrak{X}}$ *a family of subsets of* $|\mathfrak{X}|$ *satisfying*

(i) $\varnothing, |\mathfrak{X}| \in \mathcal{T}_{\mathfrak{X}}$;

(ii) *if* $U, V \in \mathcal{T}_{\mathfrak{X}}$ *then* $U \cap V \in \mathcal{T}_{\mathfrak{X}}$ *and*

(iii) *if* $\mathcal{O} \subseteq \mathcal{T}_{\mathfrak{X}}$ *then* $\bigcup \mathcal{O} \in \mathcal{T}_{\mathfrak{X}}$.

The elements of $\mathcal{T}_{\mathfrak{X}}$ *are called* open sets. *Complements of open sets are* closed sets.

Given a set $A \subseteq |\mathfrak{X}|$, its *interior*, denoted A°, is defined by

$$A^\circ = \bigcup \{U \in \mathcal{T}_{\mathfrak{X}} : U \subseteq A\}.$$

Dually, we define the closure $\overline{A}$ as $|\mathfrak{X}| \setminus (|\mathfrak{X}| \setminus A)^\circ$; this is the smallest closed set containing A.

Topological spaces generalize transitive, reflexive Kripke frames. Recall that these are pairs $\mathfrak{W} = \langle |\mathfrak{W}|, \preccurlyeq_{\mathfrak{W}} \rangle$, where $\preccurlyeq_{\mathfrak{W}}$ is a preorder on the set $|\mathfrak{W}|$. We will write $\preccurlyeq$ instead of $\preccurlyeq_{\mathfrak{W}}$ whenever this does not lead to confusion.

To see a preorder as a special case of a topological space, define

$$\downarrow w = \{v : v \preccurlyeq w\}.$$

Then consider the topology $\mathcal{T}_{\preccurlyeq}$ on $|\mathfrak{W}|$ given by setting $U \subseteq |\mathfrak{W}|$ to be open if and only if, whenever $w \in U$, we have $\downarrow w \subseteq U$ (so that all sets of the form $\downarrow w$

provide a basis for $\mathcal{T}_{\preccurlyeq}$). A topology of this form is a *preorder topology*[2].

Throughout this text we will often identify preorders with their corresponding topologies, and many times do so tacitly.

We will also use the notation

- $w \prec v$ for $w \preccurlyeq v$ but $v \not\preccurlyeq w$ and
- $w \approx v$ for $w \preccurlyeq v$ and $v \preccurlyeq w$.

The relation $\approx$ is an equivalence relation; the equivalence class of a point $x \in |\mathfrak{W}|$ is usually called a *cluster*, and we will denote it by $[x]$.

3 The tangled closure

The *tangled closure* is an important component of DTL^*. It was introduced in [3] for Kripke frames and has also appeared in [5,6,7,8].

Definition 3.1 *Let $\mathfrak{X}$ be a topological space and $\mathcal{S} \subseteq 2^{|\mathfrak{X}|}$.*

Given $E \subseteq |\mathfrak{X}|$, we say $\mathcal{S}$ is tangled *in E if, for all $A \in \mathcal{S}$, $A \cap E$ is dense in E.*

We define the tangled closure *of $\mathcal{S}$, denoted $\mathcal{S}^*$, to be the union of all sets E such that $\mathcal{S}$ is tangled in E.*

It is important for us to note that the tangled closure is defined over any topological space; however, we will often be concerned with locally finite preorders in this paper. Here, the tangled closure is relatively simple.

Lemma 3.1 *Let $\langle S, \preccurlyeq \rangle$ be a finite preorder, $x \in S$ and $\mathcal{O} \subseteq 2^S$. Then, $x \in \mathcal{O}^*$ if and only if there exist $\langle y_A \rangle_{A \in \mathcal{O}}$ such that $y_A \in A$, $y_A \preccurlyeq x$ for all $A \in \mathcal{O}$ and $y_A \approx y_B$ for all $A, B \in \mathcal{O}$.*

Proof. A proof can be found in any of [5,6,7,8]. □

4 Dynamic Topological Logic

The language L^* is built from propositional variables in a countably infinite set PV using the Boolean connectives $\wedge$ and $\neg$ (all other connectives are to be defined in terms of these), the unary modal operators f ('next') and $[f]$ ('henceforth'), along with a polyadic modality $\diamond$ which acts on finite sets, so that if Γ is a finite set of formulas then $\diamond\Gamma$ is also a formula. Note that this is a modification of the usual language of $\mathcal{DTL}$, where $\diamond$ acts on single formulas only. We write $\Box$ as a shorthand for $\neg\diamond\neg$; similarly, $\langle f \rangle$ denotes the dual of $[f]$. We also write $\diamond\gamma$ instead of $\diamond\{\gamma\}$; its meaning is identical to that of the usual $\mathsf{S4}$ modality [8]. We will often write $\bigdiamond_{i \leq I} \gamma_i$ instead of $\diamond\{\gamma_i\}_{i \leq I}$.

Given a formula ϕ, the *depth* of ϕ, denoted $\mathrm{dpt}(\phi)$, is the modal nesting depth of ϕ, while its *width*, $\mathrm{wdt}(\phi)$, denotes the maximal k such that ϕ has a subformula of the form $\bigdiamond_{i=1}^{k} \gamma_i$. For $k < \omega$, L^k denotes the sublanguage of L^* where all formulas have width at most k. Thus, in particular, $\mathsf{L} = \mathsf{L}^1$.

---

[2] Or, more specifically, a *downset topology*. Note that I stray from convention, since most authors use the upset topology here, but I personally find this presentation more natural. This will later be reflected in the semantics for □.

Formulas of L^* are interpreted on dynamical systems over topological spaces, or *dynamic topological systems.*

Definition 4.1 *A* weak dynamic topological system (dts) *is a triple*

$$\mathfrak{X} = \langle |\mathfrak{X}|, \mathcal{T}_{\mathfrak{X}}, f_{\mathfrak{X}} \rangle ,$$

where $\langle |\mathfrak{X}|, \mathcal{T}_{\mathfrak{X}} \rangle$ *is a topological space and*

$$f_{\mathfrak{X}} : |\mathfrak{X}| \to |\mathfrak{X}|.$$

If further $f_{\mathfrak{X}}$ *is continuous*[3], *we say* $\mathfrak{X}$ *is a* dynamical system.

Definition 4.2 *Given a (weak) dynamic topological system* $\mathfrak{X}$, *a* valuation *on* $\mathfrak{X}$ *is a function*

$$[\![\cdot]\!] : \mathsf{L}^* \to 2^{|\mathfrak{X}|}$$

satisfying

$$\begin{aligned}
[\![\alpha \wedge \beta]\!]_{\mathfrak{X}} &= [\![\alpha]\!]_{\mathfrak{X}} \cap [\![\beta]\!]_{\mathfrak{X}} \\
[\![\neg\alpha]\!]_{\mathfrak{X}} &= |\mathfrak{X}| \setminus [\![\alpha]\!]_{\mathfrak{X}} \\
[\![f\alpha]\!]_{\mathfrak{X}} &= f^{-1} [\![\alpha]\!]_{\mathfrak{X}} \\
[\![[f]\alpha]\!]_{\mathfrak{X}} &= \bigcap_{n \geq 0} f^{-n} [\![\alpha]\!]_{\mathfrak{X}} \\
[\![\diamond \{\alpha_1, ..., \alpha_n\}]\!]_{\mathfrak{X}} &= \{[\![\alpha_1]\!]_{\mathfrak{X}}, ..., [\![\alpha_n]\!]_{\mathfrak{X}}\}^* .
\end{aligned}$$

A *(weak) dynamic topological model* (wdtm/dtm) is a (weak) dynamic topological system $\mathfrak{X}$ equipped with a valuation $[\![\cdot]\!]_{\mathfrak{X}}$. We say a formula ϕ is *valid* on $\mathfrak{X}$ if $[\![\phi]\!]_{\mathfrak{X}} = |\mathfrak{X}|$, and write $\mathfrak{X} \models \phi$. If a formula ϕ is valid on every dynamic topological model, then we write $\models \phi$. $\mathcal{DTL}$ is the set of valid formulas of L under this interpretation, while $\mathcal{DTL}^*$ denotes the set of valid formulas of L^*.

We will often write $\langle \mathfrak{X}, x \rangle \models \phi$ instead of $x \in [\![\phi]\!]_{\mathfrak{X}}$.

5 The axiomatization

We shall distinguish $\mathcal{DTL}^*$, which is defined semantically, from DTL^*, which is a proof system. The two have the same set of theorems, but we will be interested in natural subsystems of DTL^* which are defined syntactically.

Below, note that the modality f is unary, and $f\Gamma$ is merely a shorthand for $\{f\gamma : \gamma \in \Gamma\}$; p denotes a propositional variable and P a finite set of propositional variables. Then, the axiomatization DTL^* consists of the following:

Taut All propositional tautologies.

Topological axioms

K $\Box(p \to q) \to (\Box p \to \Box q)$

T $\Box p \to p$

---

[3] That is, whenever $U \subseteq |\mathfrak{X}|$ is open, then so is $f^{-1}(U)$

4 $\Box p \to \Box\Box p$

Fix$_\Diamond$ $\Diamond P \to \bigwedge_{q\in P} \Diamond(q \wedge \Diamond P)$

Ind$_\Diamond$ $\bigwedge_{i\in I}(p \to \Diamond(q_i \wedge p)) \to (p \to \bigDiamond_{i\in I} q_i)$

Temporal axioms

Neg$_f$ $\neg fp \leftrightarrow f\neg p$

And$_f$ $f(p \wedge q) \leftrightarrow fp \wedge fq$

Fix$_{[f]}$ $[f]p \to p \wedge f[f]p$

Ind$_{[f]}$ $[f](p \to fp) \to (p \to [f]p)$

Cont* $\Diamond fP \to f\Diamond P$.

Rules

MP Modus ponens

Subs $\dfrac{\phi}{\phi[\boldsymbol{p}/\boldsymbol{\psi}]}$

N$_\Box$ $\dfrac{\phi}{\Box\phi}$ N$_f$ $\dfrac{\phi}{f\phi}$ N$_{[f]}$ $\dfrac{\phi}{[f]\phi}$

This axiomatization is sound and complete, as proven in [7]:

Theorem 5.1 DTL* *is sound and complete for the class of dynamic topological models.*

There are many subtleties in our proof system, so before continuing we should make a few remarks.

First, let us say a few words about the substitution rule. It is to be understood as 'simulataneous substitution', where $\boldsymbol{p}$ represents a finite sequence of variables, $\boldsymbol{\psi}$ a finite sequence of formulas and each variable is replaced by the respective formula. By standard arguments, this rule preserves validity, as there is nothing in our semantics distinguishing atomic facts from complex propositions; this much is not problematic.

Further, since we are concerned with finite axiomatizability of a logic it is important to include it; otherwise, each substitution instance of any of the axioms would have to be regarded as a new axiom and the finite axiomatizability would fail for obvious reasons. Of course this is not the only possible presentation, as one can also consider axiomatizations by finitely many *schemas*, but here we shall consider different formulas to be different also as axioms.

With this in mind, we should note that the above axiomatization is not finite, nor can it be modified into a finite version in an obvious way. Evidently the set of all propositional tautologies can be replaced by finitely many axioms, but this is not what concerns us. Much more importantly, we need infinitely many axioms for $\Diamond$, and it is only in the metalanguage that we can give them a uniform presentation. In fact, the symbol P representing a finite set of propositional variables is not a symbol of L*, where we would have to write out explicitly $\{p_1, \ldots, p_k\}$ for each given value of k.

Of particular interest is the schema Cont*. This was originally named TCont; we adopt the new notation to stress that the standard 'continuity' axiom,

$$\mathsf{Cont}^1 : \Diamond fp \to f\Diamond p,$$

is indeed a special case.

Cont^* is really an infinite collection of axioms. To be precise, for $k < \omega$ let

$$\mathsf{Cont}^k = \bigDiamond_{i\in[1,k]} fp_i \to f \bigDiamond_{i\in[1,k]} p_i.$$

note that Cont^{k+1} extends Cont^k since we can always substitute p_{k+1} by p_k.

We then let DTL^k be the variant of DTL^* where Cont^* is replaced by Cont^k. We denote derivability in DTL^k by $\vdash^k$. DTL^0 denotes the system with no continuity axiom.

Our goal will be to show that $\langle \mathsf{DTL}^k \rangle_{k<\omega}$ gives a sequence of theories of strictly increasing strength. Since DTL^* is the union of these theories, it will follow as a straightforward consequence that DTL^* is not finitely axiomatizable. However, to do this we will need a second refinement, this time of each DTL^k.

For $n, k < \omega$, we let DTL^k_n be the subtheory of DTL^k which restricts the substitution rule in the following ways:

(i) Subs may only be applied immediately to axioms and

(ii) if Subs is applied to Cont^k, then each p_i must be replaced by a formula with modal depth at most n.

A very easy induction on derivations shows that any proof in DTL^k may be transformed into one satisfying the above two conditions for some value of n and hence $\mathsf{DTL}^k = \bigcup_{n<\omega} \mathsf{DTL}^k_n$. We denote derivability in DTL^k_n by $\vdash^k_n$.

The reason for passing to DTL^k_n is that the substitution rule, while preserving validity, does not preserve *model* validity; if $\mathfrak{M} \models \phi$, it does not always follow that $\mathfrak{M} \models \phi[p/\psi]$. Later we wish to build specific models of fragments of DTL^*, and to check soundness for these models, DTL^k_n has the advantage that we only need to focus on substitution instances of axioms. This will become relevant in Section 8.

DTL^* is an extension of KM, which can be defined as follows:

Definition 5.2 *The calculus* KM *is the restriction of* DTL^* *to* L^1.

In KM, all appearances of $\Diamond$ must be applied to a single formula; in particular, the axioms $\mathsf{Fix}_\Diamond$ and $\mathsf{Ind}_\Diamond$ are not present, and Cont^* becomes Cont^1. We should note that DTL^1 is very similar, but not identical, to KM. DTL^1 allows formulas of the form $\Diamond\Gamma$ within derivations for Γ arbitrarily large, but Cont^* is also replaced by Cont^1. We do have, however, that $\mathsf{KM} \subseteq \mathsf{DTL}^1$.

Later we shall show that the sequence $\langle \mathsf{DTL}^k \rangle_{k<\omega}$ is strictly increasing in strength, even over L; i.e., there are formulas $\mathtt{Trouble}^k \in \mathsf{L}$ such that $\vdash^{k+1} \mathtt{Trouble}^{k+1}$ but $\nvdash^k \mathtt{Trouble}^{k+1}$. These are defined in Section 7; but first, we need to define *partial tangled bisimulations*, the fundamental tool we shall use to prove our main results.

6 Tangled bisimulations

Many of our proofs are based on partial bisimulation techniques. As we will be working in a polyadic system, we shall need a notion of partial bisimulation which preserves the polyadic $\Diamond$, originally introduced in [3].

We shall only define these partial bisimulations for finite Kripke models, for two reasons. First, this will be sufficient for our purposes, as all the constructions we shall use involve only finite models. Second, our definition exploits the fact that the tangled closure operation is much, much simpler in the finite setting, and would not readily generalize to infinite frames, much less arbitrary topological spaces.

Definition 6.1 *Given models $\mathfrak{X}, \mathfrak{Y}$, $n < \omega$ and $k \leq \omega$, we define a binary relation $\leftrightarroweq^n_* \subseteq |\mathfrak{X}| \times |\mathfrak{Y}|$ by inducion on n as follows.*

If $n = 0$, then $x \leftrightarroweq^n_* y$ if and only if x and y satisfy the same set of atoms.

Otherwise, $x \leftrightarroweq^{n+1}_* y$ if they satisfy the same atoms and

$\mathsf{Forth}_{\preccurlyeq}$ *whenever $m < \omega$ and $x_1 \approx x_2 \approx \ldots \approx x_m \preccurlyeq x$, there are $y_1 \approx y_2 \approx \ldots y_m \preccurlyeq y$ such that $x_i \leftrightarroweq^n_* y_i$ for all $i \leq m$,*

$\mathsf{Back}_{\preccurlyeq}$ *whenever $m < \omega$ and $y_1 \approx y_2 \approx \ldots \approx y_m \preccurlyeq y$, there are $x_1 \approx x_2 \approx \ldots \approx x_m \preccurlyeq x$ such that $x_i \leftrightarroweq^n_* y_i$ for all $i \leq m$,*

Forth_f $f_{\mathfrak{X}}(x) \leftrightarroweq^n_* f_{\mathfrak{Y}}(y)$

$\mathsf{Forth}_{[f]}$ *for every $m < \omega$ there is $m' < \omega$ such that $f^m_{\mathfrak{X}}(x) \leftrightarroweq^n_* f^{m'}_{\mathfrak{Y}}(y)$ and*

$\mathsf{Back}_{[f]}$ *for every $m < \omega$ there is $m' < \omega$ such that $f^m_{\mathfrak{Y}}(y) \leftrightarroweq^n_* f^{m'}_{\mathfrak{X}}(x)$*

Note there is no 'back' clause for f as it would be identical to Forth_f. When the respective structures are clear from context, we may write $x \leftrightarroweq^n_* y$ instead of $\langle \mathfrak{X}, x\rangle \leftrightarroweq^n_* \langle \mathfrak{Y}, y\rangle$.

Lemma 6.1 *If ϕ is a formula such that* $\mathrm{dpt}(\phi) \leq n$, *then given $x \leftrightarroweq^n_* y$, we have that $x \in [\![\phi]\!]_{\mathfrak{X}}$ if and only if $y \in [\![\phi]\!]_{\mathfrak{Y}}$.*

Proof. The proof proceeds by a standard induction on $\mathrm{dpt}(\phi)$ and we omit it. □

7 Trouble formulas

In this section we shall introduce a sequence of formulas $\langle \mathtt{Trouble}^k \rangle_{k<\omega}$ with the property that $\vdash^k \mathtt{Trouble}^k$. As we shall see later, $\nvdash^k \mathtt{Trouble}^{k+1}$, thus establishing that DTL^{k+1} is stronger than DTL^k. The formulas $\mathtt{Trouble}^k$ will all be in L^1.

Before continuing, let us establish some notational conventions. Given natural numbers n, k, we will denote by $|n|_k$ the unique element m of $\{1, \ldots, k\}$ such that $n \equiv m \pmod{k}$. Note that this strays from the standard remainder in that $|k|_k = k$, but it shall simplify several expressions later on. Intervals shall be assumed to be intervals of natural numbers, i.e.

$$[a, b] = \{n \in \mathbb{N} : a \leq n \leq b\}.$$

Further, we shall assume that the set of propositional variables is enumerated by $\{p_k\}_{k<\omega}$.

Definition 7.1 *The following abbreviations shall be used throughout the text.*

$$\begin{aligned}
\mathtt{Cycle}^k &= \Diamond p_k \to \bigwedge_{i=1}^{k} (p_i \to f p_{|i+1|_k}) \\
\mathtt{Start}_i^k &= p_i \wedge [f]\mathtt{Cycle}^k \\
\mathtt{Bundle}^k &= \Box \bigwedge_{i=1}^{k} \Diamond \mathtt{Start}_i^k \\
\mathtt{Tangle}^k &= \bigotimes_{i\in[1,k]} \mathtt{Start}_i^k \\
\mathtt{Trouble}^k &= \mathtt{Bundle}^k \to [f]\Diamond p_k
\end{aligned}$$

Before continuing, let us give some intuition for these formulas.

The formula $\mathtt{Cycle}^k$ states that f 'cycles' the values of $p(x)$; if $p(x) = p_i$, $p(f(x)) = p_{i+1}$, unless $i = k$ in which case $p(f(x)) = p_1$. The formula $\Diamond p_k$ is used as a sort of trigger for this cycling behavior; when $\Diamond p_k$ fails, $p(f(x))$ is unspecified.

$\mathtt{Start}_i^k$ is used to begin the cycling behavior described by $\mathtt{Cycle}^k$ at p_i; it says that, initially, p_i holds, and from then on, f cycles the values of $p(f^n(x))$, provided that $\Diamond p_k$ holds at each step.

$\mathtt{Bundle}^k$ and $\mathtt{Tangle}^k$ are similar, but $\mathtt{Bundle}^k$ is stronger. As we will mainly be interpreting these formulas over finite Kripke models, let us restrict the discussion to this setting. Here, the meaning of $\mathtt{Tangle}^k$ should be familiar; it says there is a cluster where there is a point x_i satisfying each $\mathtt{Start}_i^k$.

Proposition 7.2 *Given $k < \omega$, $\vdash^k \mathtt{Trouble}^k$.*

Proof. See Appendix A. □

8 Incompleteness of finite fragments

The formula $\mathtt{Trouble}^k$ is derivable in DTL^k; let us now see that $\mathtt{Trouble}^{k+1}$ is not. To prove this, we shall introduce models $\mathfrak{D}(N, K)$. They will be composed of two submodels; $\mathfrak{C}(K)$, defined later, and $\mathfrak{B}(N, K)$, defined below.

The general idea is that the models $\mathfrak{D}(n+1, k+1)$ will satisfy

$$\mathsf{DTL}_n^k \cup \{\neg\mathtt{Trouble}^{k+1}\},$$

thus showing that $\nvdash_n^k \mathtt{Trouble}^{k+1}$ for all n. From this we may conclude that $\nvdash^k \mathtt{Trouble}^{k+1}$.

Below, $\coprod$ denotes a disjoint union.

Definition 8.1 *Let $k < \omega$.*

*A preordered model $\mathfrak{S}$ is k-*simple *if*

$$|\mathfrak{S}| = \coprod_{i=1}^{k} [\![p_i]\!]_{\mathfrak{S}} \,.$$

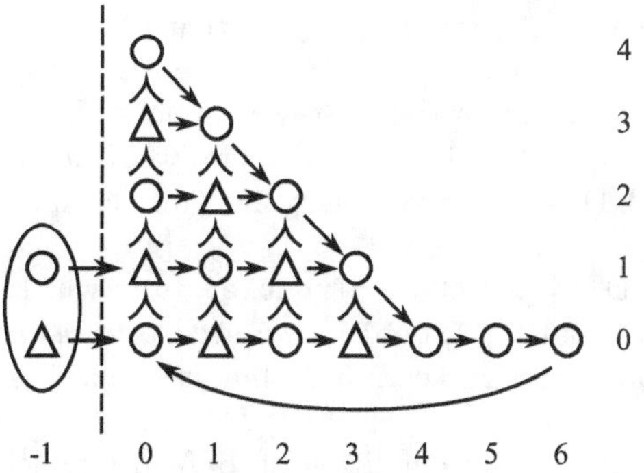

Fig. 1. The model $\mathfrak{D}(2,2)$, described in Definition 8.5. $\mathfrak{B}(2,2)$, as in Definition 8.2, is the submodel on the right-hand side of the dotted line and $\mathfrak{C}(2)$ is the submodel on its left. Arrows indicate $f_{\mathfrak{D}}$, while $\preccurlyeq$ is the transitive, reflexive closure of the relation represented by $\prec$ together with the ellipse on the left, which represents $\approx$. Points represented by a circle satisfy p_1, by a triangle, p_2.

If $\mathfrak{S}$ is simple and $x \in |\mathfrak{S}|$, we write $p_{\mathfrak{S}}(x)$ for the unique $p \in \{p_1, \ldots, p_k\}$ such that $x \in [\![p]\!]_{\mathfrak{S}}$.

As always, we will drop subindices when it does not lead to confusion, writing $p(x)$ instead of $p_{\mathfrak{S}}(x)$. All of the models we shall work with in the rest of the text shall be k-simple for some k.

Before defining our structures formally, let us give a general idea. Consider the model $\mathfrak{D} = \mathfrak{D}(3,2)$ depicted in Figure 1. We will name a point x using triples $(h(x), t(x), k(x))$, where $h(x)$ is the 'spatial' (vertical) coordinate, $t(x)$ the 'temporal' (horizontal) coordinate and $k(x)$ is the index of $p(x)$, which in this case is 1 for points represented by a circle and 2 for triangles. The points on the left of the dotted line will be written $(0, -1, k(x))$.

First, let us observe that $\mathfrak{D} \models \texttt{Cycle}^2$, since $f_{\mathfrak{D}}$ alternates between circles (which satisfy p_1) and triangles (which satisfy p_2). The exception for this are the points on the main diagonal $h + t = 6$ and on the 'tail' $t \geq 4$, but these points do not satisfy $\Diamond p_2$ and thus they also satisfy $\texttt{Cycle}^2$. From this, one can easily check that $(0, -1, 1)$ satisfies $\neg\texttt{Trouble}^2$.

Meanwhile, the key aspect of the model is that $f_{\mathfrak{D}}$ is discontinuous, since $(0, -1, 2) \preccurlyeq (0, -1, 1)$ yet

$$f_{\mathfrak{D}}(0, -1, 2) = (0, 0, 1) \not\preccurlyeq (1, 0, 2) = f_{\mathfrak{D}}(0, -1, 1).$$

This discontinuity is easily seen to make the following instance of Cont^2 fail on $(0, -1, 1)$:

$$\Diamond f\{p_1, p_2\} \to f\Diamond\{p_1, p_2\}.$$

However, instances of Cont^1 of small modal depth do hold.

Consider, for example,

$$\Diamond f p_1 \to f \Diamond p_1.$$

Here we see that $f_{\mathfrak{D}}(0,-1,2)$ satisfies fp_1, so that $(0,-1,1)$ satisfies $\Diamond fp_1$. If $f_{\mathfrak{D}}$ were continuous, we would be able to use $f_{\mathfrak{D}}(0,-1,2)$ as a witness that $(0,-1,1)$ satisfies $f\Diamond p_1$, but in this case we cannot. However, we do have a different witness, namely $(2,0,1)$. More generally, as we shall see in Lemma 8.4, $(2,0,1) \leftrightarrows^1_* (0,0,1)$ so the two satisfy the same formulas of modal depth one.

Thus $\mathfrak{D}$ satisfies DTL^1_1 as well as $\neg\texttt{Trouble}^2$, from which we conclude that $\nvdash^1_1 \texttt{Trouble}^2$. To see that $\nvdash^1_n \texttt{Trouble}^2$, we need to consider a larger model, $\mathfrak{D}(n,2)$, which is built much like $\mathfrak{D}(3,2)$ but is deeper. By varying n, we conclude that $\nvdash^1 \texttt{Trouble}^2$.

Now, let us give the formal definition of $\mathfrak{B}(N,K)$, which is the submodel of $\mathfrak{D}(N,K)$ on the right of the dotted lines in Figure 1.

Definition 8.2 *Given $N,K<\omega$, we define a K-simple dynamic model $\mathfrak{B} = \mathfrak{B}(N,K)$ by letting*

(i) *$|\mathfrak{B}|$ be the set of all triples of natural numbers (h,t,k) such that either*
  (a) *$h+t \le NK$, $k \in [1,K]$ and $k \not\equiv h+t \pmod K$ or*
  (b) *$h=0$, $t \in [KN+1,(K+1)N]$ and $k \neq K$.*

(ii) *$(h_1,t_1,k_1) \preccurlyeq_{\mathfrak{B}} (h_2,t_2,k_2)$ if and only if $t_1=t_2$ and $h_1 \ge h_2$;*

(iii)
$$f_{\mathfrak{B}}(h,t,k) = \begin{cases} (h,t+1,|k+1|_K) & \text{if } h+t<NK \\ (h-1,t+1,k) & \text{if } h+t=NK \text{ and } h>0 \\ (h,t+1,k) & \text{if } h=0 \text{ and } t<N(K+1) \\ (0,0,|k+1|_K) & \text{if } h=0 \text{ and } t=N(K+1) \end{cases}$$

(iv) *$p(h,t,k)=p_k$.*

We will write $x=(h(x),k(x),t(x))$. We will also write $s(x)=h(x)+t(x)$.

It will be convenient to describe the $\leftrightarrows^m_*$-equivalence classes over $\mathfrak{B}(N,K)$. We shall do this using the relations $\sim^m$, defined below.

Definition 8.3 *For $m<N$, say $x \sim^m y$ if $p(x)=p(y)$ and one (or more) of the following occurs:*

(i) *$s(x)=s(y)$,*

(ii) *$s(x),s(y) \le K(N-m)$ or*

(iii) *$s(x),s(y) \in [NK, N(K+1)-m]$.*

The models $\mathfrak{B}(k,n)$ are designed to be very homogeneous, so that different points are hard to distinguished using L^*. The relations $\sim^m$ are representative of this.

Lemma 8.1 *For every $x \in |\mathfrak{B}(N,K)|$ and $m<N$,*

(i) *there is $y \sim^m x$ with $h(y)=0$ and*

(ii) *if* $h(y) = 0$ *there is* $n < \omega$ *such that* $f^n_{\mathfrak{B}}(x) = y$.

Proof. The first claim is obvious if we notice that

$$(h, t, k) \sim^m (0, h + t, k).$$

For the second, first we note that $h(f^{N(K+1)+1}(x)) = 0$ independently of x; then note that $f_{\mathfrak{B}}$ is clearly transitive on those elements z with $h(z) = 0$, given that

$$f^{N(K+1)+1}(0, 0, k) = (0, 0, |k + 1|_K),$$

thus 'rotating' $k(z)$. □

Now, let us see that $\sim^m$ does, indeed, guarantee partial bisimulation.

Proposition 8.4 *If* $x \sim^m y$ *then* $x \leftrightarroweq^m_* y$.

Proof. We work by induction on m, considering each clause of a tangled bisimulation.

Note that $\sim^m$ preserves atoms, in particular covering the case $m = 0$.

Otherwise, suppose $x \sim^{m+1} y$. Clearly we only need to prove the 'forth' clauses, since the 'back' clauses are symmetric.

$\mathsf{Forth}_{\preccurlyeq}$ We shall only consider the case where $s(x), s(y) < NK$; the other case is similar and easier.

Suppose $x_0 \approx x_1 \approx \ldots \approx x_{I-1} \preccurlyeq x$; note that we can assume $I \leq K$, since $\mathfrak{B}$ has cluster width K. Note also that each x_i has $h(x_i) \geq h(x)$ and $t(x_i) = t(x)$.

Consider $h' = h(x_i) + t(y) - t(x)$. If $h' \geq h(y)$, set $h = h'$; otherwise, let h be the least value such that $h \geq h(y)$ and $h + t(y) \equiv h(x_i) + t(x) \pmod{K}$. Then, set $y_i = (h, t(y), k(x_i))$.

First, note that $s(y_i) \equiv s(x_i) \pmod{K}$, so that all y_i are elements of $|\mathfrak{B}|$. Now, we further have that $s(y_i) = s(x_i)$ except in the case that $h' < h(y)$, in which it easily follows that $s(x) \neq s(y)$, so $s(x), s(y) < K(N - (m+1))$ and thus $s(x_i), s(y_i) < K(N - m)$.

In either case we use our induction hypothesis to see that $y_i \leftrightarroweq^m x_i$, as claimed.

Forth_f This follows from observing that the required (in)equalities are preserved by $f_{\mathfrak{B}}$ and we skip it.

$\mathsf{Forth}_{[f]}$ Let $n < \omega$ and consider $z = f^n_{\mathfrak{B}}(x)$. Then, by Lemma 8.1(i), there is $z' \sim^m z$ with $h(z') = 0$, while by Lemma 8.1(ii), there is n' such that $f^{n'}_{\mathfrak{B}}(y) = z'$, as required.

□

Now that we have studied the models $\mathfrak{B}(N, K)$, let us add the 'head' $\mathfrak{C}(K)$, which is where the trouble really lurks. The resulting model will be called $\mathfrak{D}(N, K)$, where points in $\mathfrak{C}(K)$ will map discontinuously onto $\mathfrak{B}(N, K)$. However, these discontinuities will require large formulas to capture in L^{K-1}, given that $\mathfrak{C}(K)$ will consist of a cluster with K points.

Definition 8.5 *We define a model $\mathfrak{C} = \mathfrak{C}(K)$ where*

- $|\mathfrak{C}| = \{0\} \times \{-1\} \times [1, K]$
- $\preccurlyeq_{\mathfrak{C}}$ *is total (i.e.,* $\mathfrak{C}$ *consists of a single cluster)*
- $p(0, -1, k) = p_k$.

We define a model $\mathfrak{D} = \mathfrak{D}(N, K)$ *based on* $|\mathfrak{C}(K)| \cup |\mathfrak{B}(N, K)|$ *with*

$$\preccurlyeq_{\mathfrak{D}} = \preccurlyeq_{\mathfrak{C}(K)} \cup \preccurlyeq_{\mathfrak{B}(N,K)}$$

and

$$f_{\mathfrak{D}}(0, -1, k) = \begin{cases} (0, 0, |k+1|_K) & \text{if } k \neq K-1 \\ \\ (1, 0, K) & \text{if } k = K-1. \end{cases}$$

Our strategy now is to show that $\mathfrak{D}(N+1, K+1)$ is a model of $\mathsf{DTL}^K_N \cup \{\neg \mathtt{Trouble}^{K+1}\}$; from this we may conclude that $\not\vdash^K \mathtt{Trouble}^{K+1}$, given that $\mathsf{DTL}^K = \bigcup_{n<\omega} \mathsf{DTL}^K_n$.

Lemma 8.2 $\mathfrak{D}(N+1, K+1) \models \mathsf{DTL}^K_N$.

Proof. All the rules of DTL^K_N preserve model validity, so it suffices to check that $\mathfrak{D}(N+1, K+1)$ satisfies all axioms of DTL^K_N; that is, all permitted substitution instances of axioms of DTL^K.

Since $\mathfrak{D}(N+1, K+1)$ is a weak dynamical system, it satisfies every axiom of DTL^K_N except possibly for instances of Cont^K.

So, let

$$\sigma = \bigdiamond_{i \leq K} f\delta_i \to f \bigdiamond_{i \leq K} \delta_i$$

be a substitution instance of Cont^K where each δ_i has modal depth at most N.

Let $x \in |\mathfrak{D}|$ and assume that

$$x \in \left[\!\left[\bigdiamond_{k \leq K} f\delta_i \right]\!\right]_{\mathfrak{D}};$$

since $f_{\mathfrak{D}} \upharpoonright |\mathfrak{B}|$ is continuous, we can suppose that $x \in |\mathfrak{C}|$, for otherwise $x \in [\![\sigma]\!]_{\mathfrak{D}}$.

Then, given $i < K$ there is $x_i = (0, -1, k_i) \approx x$ such that $x_i \in [\![f\delta_i]\!]_{\mathfrak{D}}$. For at least one value of $k_* \in [1, K+1]$ we have that $k_* \neq |k_i + 1|_K$ for all i; we then have that $y_i = (k_*, 0, |k_i + 1|_K)$ is an element of $|\mathfrak{B}|$ and by Lemma 8.4

$$y_i \leftrightarrows^N_* f_{\mathfrak{D}}(x_i).$$

Meanwhile, $y_i \approx y_j \preccurlyeq f_{\mathfrak{D}}(x)$ for all i, j, so that

$$x \in \left[\!\left[f \bigdiamond_{i \leq K} \delta_i \right]\!\right]_{\mathfrak{D}},$$

as required. □

Lemma 8.3 *Given $K, N < \omega$ and $k \in [1, k]$,*

$$\langle \mathfrak{D}(N,K), (0,-1,k) \rangle \models \neg \mathtt{Trouble}^K.$$

Proof. Let $\mathfrak{D} = \mathfrak{D}(N,K)$.

First, let us show that every $x \in |\mathfrak{D}|$ satisfies

$$\mathtt{Cycle}^K = \Diamond p_K \to \bigwedge_{k \leq K} (p_k \to f p_{|k+1|_K}).$$

If $s(x) \geq NK$, then $x \notin [\![\Diamond p_K]\!]_{\mathfrak{D}}$ and thus $x \notin [\![\Diamond p_K]\!]_{\mathfrak{D}}$. This shows that $x \in [\![\mathtt{Cycle}^K]\!]_{\mathfrak{D}}$, as required.

Otherwise, letting $y = f_{\mathfrak{D}}(x)$, we note by case-by-case inspection that $k(y) = |k(x)+1|_K$, so that x satisfies $p_{k(x)} \to f p_{|k(x)+1|_K}$, whereas for $k \neq k(x)$, x satisfies $p_k \to f p_{|k+1|_K}$ trivially. Thus $\mathtt{Cycle}^K$ holds everywhere, as claimed.

It follows from this, in particular, that $(0,-1,k)$ satisfies $p_k \wedge [f]\mathtt{Cycle}^K$, i.e. $\mathtt{Start}^K_k$; this shows that $(0,-1,k)$ satisfies $\bigwedge_{1 \leq i \leq K} \Diamond \mathtt{Start}^K_i$ and, given that $k \in [1, K]$ was arbitrary,

$$\langle \mathfrak{D}, (0,-1,k) \rangle \models \Box \bigwedge_{i=1}^{k} \Diamond \mathtt{Start}^K_i = \mathtt{Bundle}^K.$$

It remains to show that $(0,-1,k)$ satisfies $\langle f \rangle \Box \neg p_K$; but this follows from the observation that

$$f^{NK+1}_{\mathfrak{D}}(0,-1,k) = (0, NK, k') \notin [\![\Diamond p_K]\!]_{\mathfrak{D}}.$$

We conclude that

$$\langle \mathfrak{D}, (0,-1,k) \rangle \models \mathtt{Bundle}^K \wedge \neg [f] \Diamond p_K \equiv \neg \mathtt{Trouble}^K,$$

as claimed. □

The following lemma summarizes our results so far:

Lemma 8.6 *For all $k < \omega$, the formula $\mathtt{Trouble}^{k+1} \in \mathsf{L}^1$ is derivable in DTL^{k+1}, but not in DTL^k.*

Proof. By Proposition 7.2, $\vdash^{k+1} \mathtt{Trouble}^{k+1}$; meanwhile, if $\vdash^k \mathtt{Trouble}^{k+1}$, we would have that $\vdash^k_n \mathtt{Trouble}^{k+1}$ for some n.

But this cannot be, since we have seen that

$$\mathfrak{D}(n+1, k+1) \models \mathsf{DTL}^k_n \cup \{\neg \mathtt{Trouble}^{k+1}\},$$

and thus $\not\vdash^k_n \mathtt{Trouble}^{k+1}$. □

With this, we may easily prove our main result.

Theorem 8.7 *Let λ be any language such that $\mathsf{L} \subseteq \lambda \subseteq \mathsf{L}^*$, and let $\mathcal{DTL}[\lambda] = \mathcal{DTL}^* \cap \lambda$.*

Similarly, for $k < \omega$, define $\mathsf{DTL}^k[\lambda] = \mathsf{DTL}^k \cap \lambda$.

Then, given any natural number k, $\mathcal{DTL}[\lambda]$ is not finitely axiomatizable[4] *over $\mathsf{DTL}^k[\lambda]$.*

Proof. Let T be any sound, finite extension of $\mathsf{DTL}^k[\lambda]$, so that without loss of generality we may assume $\mathsf{T} = \mathsf{DTL}^k[\lambda] + \phi$ for some valid formula ϕ.

Since DTL^* is complete, we would have that that $\mathsf{DTL}^* \vdash \phi$, and hence, for some value of K, $\mathsf{DTL}^K \vdash \phi$; obviously, we may take $K \geq k$.

But then, we have by Lemma 8.6 that $\mathsf{DTL}^K \nvdash \mathtt{Trouble}^{K+1}$, and hence

$$\mathsf{DTL}^k[\lambda] + \phi \nvdash \mathtt{Trouble}^{K+1} \in \mathcal{DTL}[\lambda].$$

Meanwhile, T was arbitrary, so we conclude that $\mathcal{DTL}[\lambda]$ is not finitely axiomatizable over $\mathsf{DTL}^k[\lambda]$. □

This result is quite general, so it may be convenient to explicitly mention some special cases. The following corollary states some immediate consequences of Theorem 8.7; below, recall that $\mathsf{KM} \subseteq \mathsf{DTL}^1[\mathsf{L}]$.

Corollary 8.8 *$\mathcal{DTL}$ and $\mathcal{DTL}^*$ are not finitely axiomatizable. In particular, KM is incomplete for the class of dynamic topological models.*

9 Concluding remarks

The axiomatization DTL^* introduced the tangled modality to Dynamic Topological Logic as a sort of scaffolding, to be removed once the appropriate techniques were available. However, I believe the present work to be a convincing argument that indeed it is a central element of the logic; tangled sets affect the behavior of dynamical systems and to be unable to reason about them directly gives a logical formalism an unnecessary handicap.

Of course, none of the results presented here show that a reasonable axiomatization within L^1 is impossible to find. I am not sure how relevant such an axiomatization would be at this point, but it remains an interesting problem.

Meanwhile, I believe a more fruitful direction is to analyze other logics which are hard to axiomatize because of a similar lack in expressive power. In particular, there are many products of modal logics which have very similar models to those of Dynamic Topological Logic; perhaps they too would benefit from a polyadic variant?

Acknowledgements

I would like to thank the anonymous referees for their useful comments; in particular, the current statement of Theorem 8.7 and even the final title of the paper are due to a referee who observed that non-finite axiomatizability

[4] Observe that DTL^{k+1} is finitely axiomatizable over DTL^k, but it does not necessarily follow from this that $\mathsf{DTL}^{k+1}[\lambda]$ is finitely axiomatizable over $\mathsf{DTL}^k[\lambda]$ for all λ.

followed from the results presented in the submitted version. The original version only observed that DTL^k was incomplete for all k.

Appendix

A Derivability of Trouble formulas

In this Appendix, we shall give a syntactic proof of Proposition 7.2. Recall that this proposition states that, for all $k < \omega$, $\vdash^k \texttt{Trouble}^k$.

Proof. Reasoning within S4 one readily sees that, for any $i < k$, $\vdash^0 \texttt{Bundle}^k \to \Diamond(\texttt{Start}_i^k \wedge \texttt{Bundle}^k)$; thus we may apply the rule $\mathsf{Ind}_\Diamond$ to derive $\texttt{Bundle}^k \to \bigDiamond_{i=1}^k \texttt{Start}_i^k$ and obtain

$$\vdash^0 \texttt{Bundle}^k \to \texttt{Tangle}^k. \tag{A.1}$$

Further, we note that

$$\vdash^0 \texttt{Tangle}^k \to \Diamond p_k, \tag{A.2}$$

since this is a consequence of the axiom $\mathsf{Fix}_\Diamond$.

For any $i \in [1, k]$ we may use $\mathsf{Fix}_\Diamond$ to see that

$$\vdash^0 \texttt{Tangle}^k \to \Diamond(\texttt{Start}_i^k \wedge \texttt{Tangle}^k).$$

Using (A.2), this imples

$$\vdash^0 \texttt{Tangle}^k \to \Diamond(\texttt{Start}_i^k \wedge \Diamond p_k \wedge \texttt{Tangle}^k),$$

i.e.

$$\vdash^0 \texttt{Tangle}^k \to \Diamond(p_i \wedge [f]\texttt{Cycle}^k \wedge \Diamond p_k \wedge \texttt{Tangle}^k).$$

Now, by $\mathsf{Fix}_{[f]}$, $\vdash^0 [f]\texttt{Cycle}^k \to (\texttt{Cycle}_i^k \wedge f[f]\texttt{Cycle}^k)$, whereas

$$\vdash^0 \Diamond p_k \wedge \texttt{Cycle}^k \to (p_i \to f p_{|i+1|_k}),$$

i.e. $\vdash^0 p_i \wedge \Diamond p_k \wedge \texttt{Cycle}^k \to f p_{|i+1|_k}$. From this we conclude that

$$\vdash^0 \texttt{Tangle}^k \to \Diamond(f p_{|i+1|_k} \wedge f[f]\texttt{Cycle}^k \wedge \texttt{Tangle}^k),$$

and since it holds for all $i \in [1, k]$ we can use $\mathsf{Ind}_\Diamond$ to obtain

$$\vdash^0 \texttt{Tangle}^k \to \bigDiamond_{i\in[1,k]} (f p_{|i+1|_k} \wedge f[f]\texttt{Cycle}^k),$$

which, rearranging indices and pulling out f, shows that

$$\vdash^0 \texttt{Tangle}^k \to \bigDiamond_{i\in[1,k]} f(p_i \wedge [f]\texttt{Cycle}^k).$$

Now, we may use Cont^k to obtain

$$\vdash^k \texttt{Tangle}^k \to f \bigDiamond_{i\in[1,k]} (p_i \wedge [f]\texttt{Cycle}^k);$$

by $\mathsf{Ind}_{[f]}$ this yields

$$\vdash^k \mathtt{Tangle}^k \to [f] \bigdiamond_{i\in[1,k]} (p_i \wedge [f]\mathtt{Cycle}^k),$$

i.e. $\vdash^k \mathtt{Tangle}^k \to [f]\mathtt{Tangle}^k$.

Putting this together with A.2 we see that

$$\vdash^k \mathtt{Tangle}^k \to [f]\Diamond p_k, \tag{A.3}$$

which, together with (A.1) gives us

$$\vdash^k \mathtt{Bundle}^k \to [f]\Diamond p_k,$$

i.e. $\vdash^k \mathtt{Trouble}^k$, as claimed. □

References

[1] Artemov, S., J. Davoren and A. Nerode, *Modal logics and topological semantics for hybrid systems*, Technical Report MSI 97-05 (1997).

[2] Bezhanishvili, G. and M. Gehrke, *Completeness of s4 with respect to the real line: revisited*, Annals of Pure and Applied Logic **131** (2005), pp. 287 – 301.
URL `http://www.sciencedirect.com/science/article/pii/S0168007204000843`

[3] Dawar, A. and M. Otto, *Modal characterisation theorems over special classes of frames*, Annals of Pure and Applied Logic **161** (2009), pp. 1–42, extended journal version LICS 2005 paper.

[4] Fernández-Duque, D., *Non-deterministic semantics for dynamic topological logic*, Annals of Pure and Applied Logic **157** (2009), pp. 110–121, kurt Gödel Centenary Research Prize Fellowships.

[5] Fernández-Duque, D., *On the modal definability of simulability by finite transitive models*, Studia Logica **98** (2011), pp. 347–373.
URL `http://dx.doi.org/10.1007/s11225-011-9339-x`

[6] Fernández-Duque, D., *Tangled modal logic for spatial reasoning*, in: T. Walsh, editor, *Proceedings of IJCAI*, 2011, pp. 857–862.

[7] Fernández-Duque, D., Journal of Symbolic Logic (2012), forthcoming.

[8] Fernández-Duque, D., *Tangled modal logic for topological dynamics*, Annals of Pure and Applied Logic (2012).

[9] Konev, B., R. Kontchakov, F. Wolter and M. Zakharyaschev, *Dynamic topological logics over spaces with continuous functions*, , **6** (2006), pp. 299–318.

[10] Konev, B., R. Kontchakov, F. Wolter and M. Zakharyaschev, *On dynamic topological and metric logics*, Studia Logica **84** (2006), pp. 129–160.

[11] Kremer, P., *Strong completeness of* S4 *wrt the real line* (2012).

[12] Kremer, P. and G. Mints, *Dynamic topological logic*, Annals of Pure and Applied Logic **131** (2005), pp. 133–158.

[13] Lichtenstein, O. and A. Pnueli, *Propositional temporal logics: Decidability and completeness*, Logic Jounal of the IGPL **8** (2000), pp. 55–85.

[14] Mints, G. and T. Zhang, *Propositional logic of continuous transformations in Cantor space*, Archive for Mathematical Logic **44** (2005), pp. 783–799.

[15] Prior, A., *Time and modality* (1957).

[16] Tarski, A., *Der Aussagenkalkül und die Topologie*, Fundamenta Mathematica **31** (1938), pp. 103–134.

Synthesis for Temporal Logic over the Reals

Tim French John McCabe-Dansted Mark Reynolds [1]

School of Computer Science and Software Engineering, The University of Western Australia
35 Stirling Highway, Crawley WA 6009
Perth, Australia.

Abstract

We develop the notion of synthesizing, or constructing, a temporal structure over the real numbers flow of time, from a given temporal or first-order specification. We present a new notation for giving a manageable description of the compositional construction of such a model and an efficient procedure for finding it from the specification.

Keywords: Compositional Models, RTL, Continuous Time.

1 Introduction

Standard temporal logics are based on a discrete, natural numbers model of time [8]. However, a dense, continuous or specifically real-numbers model of time may be better for many applications, ranging from philosophical, natural language and AI modelling of human reasoning to computing and engineering applications of concurrency, refinement, open systems, analogue devices and metric information.

The most natural and well-established such temporal logic is RTL, propositional temporal logic over real-numbers time using the Until and Since connectives introduced in [5]. We know from [5] that, as far as defining properties is concerned, this logic is as expressive as the first-order monadic logic of the real numbers order, and so RTL is at least as expressive as any other standard temporal logic which could be defined over real-numbers time.

Reasoning in RTL is fairly well understood: complete, Hilbert-style axioms systems for RTL are given in [4] and [11]. Satisfiability and validity in RTL is decidable [1]. However, the decision procedure in [1] uses Rabin's non-elementarily complex decision procedure for the second-order monadic logic of two successors, and so is far from practical. Furthermore, deciding validity in the equally expressive first-order monadic logic of the real order is a non-elementary problem [17]. More recently, there has been some more pos-

[1] The work was partially supported by the Australian Research Council.

itive news as [15] showed that deciding (validity or satisfiability in) RTL is PSPACE-complete.

Our task here, synthesis, is a harder problem than satisfiability checking. It requires an algorithm which can output a complete description of a specific model of the input formula, whenever the input formula is satisfiable. In this paper we give the first synthesis result for RTL.

Towards this end, we first present a new suitable notation for describing models in a concrete way. The compositional approach presented here was hinted at in [12], and traces back to pioneering work in [6] and [1]. It uses a small number of distinct operations for putting together a larger model from one or more smaller ones, or copies thereof. For example, the *shuffle* construct makes a new linear structure from a dense mixture of copies of a finite number of simpler ones. A good overview of the mathematics of linear orders may be found in [16].

We introduce a formal model expression language for defining a model via these inductive operations. In fact, we first give a language for making general linear structures in this way and then define a restricted sub-language (the real model expression language) capable of specifying structures with the real number flow of time. Having a formal model building language opens up the possibility for workable definitions of such tasks as synthesis and model checking for real-flowed structures. It also allows us to formalise questions of expressibility and to assess the computational complexity of these reasoning tasks.

One of our results here, echoing the earlier work of [6], [1] and others is that the real model expression language is able to describe some real-flowed model of every satisfiable RTL formula.

The major advance of this paper on that previously-known expressiveness result, is that we also present an EXPTIME procedure for finding the real model expression of a model from any given satisfiable RTL formula. EXPTIME is best possible. The real model expression tells us exactly how to make a specific real-flowed model of the formula[2]. This is our synthesis result.

Some of the proofs here use the mosaic techniques for temporal logic developed in [15]. These mosaics are small pieces of a real-flowed structure. In that paper we try to find a finite set of small pieces which is sufficient to be used to build a real-numbers model of a given formula. We showed that if a formula was satisfiable then we could find a sufficient set of mosaics. In this paper, we go further and show how to build a compositional model (i.e. one corresponding to an expression in our model language) when there is such a set of mosaics.

The extension here is built on a series of lemmas mirroring some of the earlier results but keeping a much closer track on a relationship between mosaics and some compositionally built structures which witness them. Thus there is a series of quite long and detailed lemmas needed for this result. In the short

[2] Of course, any isomorphic real-flowed structure will also be a model

version of this paper we only sketch some of the new proofs but the full proofs can be found in [2].

In section 2 we present RTL and monadic logic. In section 3 we introduce the compositional approach to building linear models. In section 4 we remind ourselves of useful properties of mosaics from [15]. In section 5 we use mosaics to give a proof of the previously-known expressiveness result for RTL: satisfiability implies satisfiability in a compositional model. We apply this technique to present the new synthesis result and we conclude in section 6.

2 The logic

Fix a countable set $\mathbf{L}$ of atoms. Here, frames $(T, <)$, or flows of time, will be irreflexive linear orders. Structures $\mathbf{T} = (T, <, h)$ will have a frame $(T, <)$ and a valuation h for the atoms i.e. for each atom $p \in \mathbf{L}$, $h(p) \subseteq T$. Of particular importance will be *real* structures $\mathbf{T} = (\mathbb{R}, <, h)$ which have the real numbers flow (with their usual irreflexive linear ordering). We will also introduce structures over sub-orders of this standard model using $<$ to mean the usual ordering. For example, $(]0, 1[, <, h)$ is some structure over the open unit interval of the reals.

The language $L(U, S)$ is generated by the 2-place connectives U and S along with classical $\neg$ and $\wedge$. That is, we define the set of formulas recursively to contain the atoms and for formulas α and β we include $\neg\alpha$, $\alpha \wedge \beta$, $U(\alpha, \beta)$ and $S(\alpha, \beta)$.

Formulas are evaluated at points in structures $\mathbf{T} = (T, <, h)$. We write $\mathbf{T}, x \models \alpha$ when α is true at the point $x \in T$. This is defined recursively as follows. Suppose that we have defined the truth of formulas α and β at all points of $\mathbf{T}$. Then for all points x:

$\mathbf{T}, x \models p$	iff	$x \in h(p)$, for p atomic;
$\mathbf{T}, x \models \neg\alpha$	iff	$\mathbf{T}, x \not\models \alpha$;
$\mathbf{T}, x \models \alpha \wedge \beta$	iff	both $\mathbf{T}, x \models \alpha$ and $\mathbf{T}, x \models \beta$;
$\mathbf{T}, x \models U(\alpha, \beta)$	iff	there is $y > x$ in T such that $\mathbf{T}, y \models \alpha$ and for all $z \in T$ such that $x < z < y$ we have $\mathbf{T}, z \models \beta$; and
$\mathbf{T}, x \models S(\alpha, \beta)$	iff	there is $y < x$ in T such that $\mathbf{T}, y \models \alpha$ and for all $z \in T$ such that $y < z < x$ we have $\mathbf{T}, z \models \beta$.

The logic is discussed more fully in [13], [15] and [14], for example. See those references for investigations of the "strict" versus "non-strict" connectives, infix versus postfix operators, etc. We use the following abbreviations in illustrating the logic: $F\alpha = U(\alpha, \top)$, "alpha will be true (sometime in the future)"; $G\alpha = \neg F(\neg\alpha)$, "alpha will always hold (in the future)"; and their mirror images P and H. Particularly for dense time applications we also have: $\Gamma^+\alpha = U(\top, \alpha)$, "alpha will be constantly true for a while after now"; and $K^+\alpha = \neg\Gamma^+\neg\alpha$,

"alpha will be true arbitrarily soon". They have mirror images Γ^- and K^-.

2.1 Reasoning with RTL

A formula ϕ is $\mathbb{R}$-*satisfiable* if it has a real model: i.e. there is a real structure $\mathbf{S} = (\mathbb{R}, <, h)$ and $x \in \mathbb{R}$ such that $\mathbf{S}, x \models \phi$. A formula is $\mathbb{R}$-*valid* iff it is true at all points of all real structures. Of course, a formula is $\mathbb{R}$-valid iff its negation is not $\mathbb{R}$-satisfiable. We will refer to the logic of L(U,S) over real structures as RTL.

Let RTL-SAT be the problem of deciding whether a given formula of $L(U, S)$ is $\mathbb{R}$-satisfiable or not. The main result of [15] is:

Theorem 2.1 *RTL-SAT is PSPACE-complete.*

In order to help get a feel for the sorts of formulas which are valid in RTL it is worth considering a few formulas in the language. $U(\top, \bot)$ is a formula which only holds at a point with a discrete successor point so $G\neg U(\top, \bot)$ is valid in RTL. $Fp \to FFp$ is a formula which can be used as an axiom for density and is also a valid in RTL.

$(\Gamma^+ p \wedge F\neg p) \to U(\neg p \vee K^+(\neg p), p)$ was used as an axiom for Dedekind completeness (in [11]) and is valid. Recall that a linear order is Dedekind complete if and only if each non-empty subset which has an upper bound has a least upper bound. The formula says that if p is true constantly for a while but not forever then there is an upper bound on the interval in which it remains true. This formula is not valid in the temporal logic with until and since over the rational numbers flow of time.

One of the most interesting valid formulas of RTL is Hodkinson's axiom "Sep" (see [11]). It is

$$K^+ p \wedge \neg K^+(p \wedge U(p, \neg p)) \to K^+(K^+ p \wedge K^- p).$$

This can be used in an axiomatic completeness proof to enforce the *separability* of the linear order:

Definition 2.2 *A linear order is separable iff it has a countable suborder which is spread densely throughout the order: i.e. between every two elements of the order lies an element of the suborder.*

The fact that the rationals are dense in them shows that the reals are separable. There are dense, Dedekind complete linear orders with end points which are not separable (e.g. , see [11]). The negation of Sep will be satisfiable over them but not over the reals.

As we have noted in the introduction, there are complete axiom systems for RTL in [4] and in [11]: the former using a special rule of inference and the latter just using orthodox rules.

Rabin's decision procedure for the second-order monadic logic of two successors [10] is used in [1] to show that RTL is decidable. One of the two decision procedures in that paper just gives us a non-elementary upper bound on the complexity of RTL-SAT.

2.2 Monadic Logic

The first-order monadic language of order, FOMLO, is a first order language which can describe the structures we are dealing with and it is useful to translate between it and the temporal language.

The relation symbols of FOMLO are 2-ary $<$ and 1-ary, or monadic, P_0, P_1, P_2, ... each corresponding respectively to the atoms $p_0, p_1, p_2, \ldots$ of L. So atomic propositions are $x_i < x_j$ and $P_k(x_j)$ for each variable symbol x_i and each 1-ary relation symbol P_k. Formulas of the language are built up from the atoms as follows: $\neg\alpha$, $\alpha \wedge \beta$, and $\forall x_i \alpha$.

The notions of free and bound variables and sentences are as usual.

Given a temporal structure $(T, <, g)$ we can evaluate monadic formulas in it by interpreting the 1-ary predicates P_i as 1-ary relations on (i.e. subsets of) T using the valuation $g(p_i)$ to tell us where the interpretation of P_i holds as follows:

$(T, <, g), \mu \models P_i(x_j)$ iff $t_j \in g(p_i)$
$(T, <, g), \mu \models x_i < x_j$ iff $t_i < t_j$
$(T, <, g), \mu \models \neg\alpha$ iff it is not the case that $(T, <, g), \mu \models \alpha$
$(T, <, g), \mu \models \alpha_1 \wedge \alpha_2$ iff $(T, <, g), \mu \models \alpha_1$ and $(T, <, g), \mu \models \alpha_2$
$(T, <, g), \mu \models \forall x_i \alpha$ iff for every $d \in T$, $(T, <, g), \mu[x_i \mapsto d] \models \alpha$

Here μ is a (possibly partial) map from $\{x_1, x_2, \ldots\}$ to T and $\mu[x_i \mapsto d]$ is the map which is the same as μ except that x_i is mapped to d. We require that μ is defined on all the free variables of α. The truth of $(T, <, g), \mu \models \alpha$ does not depend on the value of $\mu(x_i)$ if x_i is not free in α.

Definition 2.3 *We say that the temporal language $L(B)$ is* expressively complete *over class K of linear orders iff for every FOMLO formula $\alpha(t)$, there is some ϕ of the temporal language such that ϕ is equivalent to α over K.*

Kamp showed in [5] that $L(U, S)$ is expressively complete over $\mathbb{R}$ and over $\mathbb{N}$.

2.3 Isomorphisms

An isomorphism is a bijective mapping from one structure to another that preserves the temporal relation $<$ and the valuation h. This is an important notion of equivalence for us, as we will show that equivalent structures satisfy the same set of formulas in $L(U, S)$.

Definition 2.4 *We say two structures $\mathbf{T} = (T, <, h)$ and $\mathbf{T}' = (T', <', h')$ are isomorphic (written $\mathbf{T} \cong \mathbf{T}'$) if and only if there is a bijection $f : T \longrightarrow T'$ where for all $x, y \in T$ $x < y$ if and only if $f(x) <' f(y)$, and for all $p \in P$ $x \in h(p)$ if and only if $f(x) \in h'(p)$.*

It is well known that isomorphisms between structures preserve the truth of formulas of temporal logic.

Lemma 2.5 *Suppose that $\mathbf{T} = (T, <, h)$, $\mathbf{T}' = (T', <', h')$ and $\mathbf{T} \cong \mathbf{T}'$, via the bijection f. Then for any $\alpha \in L(U, S)$, for any $t \in T$, $\mathbf{T}, t \models \alpha$ if and only if $\mathbf{T}', f(t) \models \alpha$.*

3 Building Structures

We introduce a notation which allows the description of a temporal structure in terms of simple basic structures via a small number of ways of putting structures together to form larger ones.

The general idea is simple: using singleton structures (the flow of time is one point), we build up to more complex structures by the recursive application of four operations. They are:

- concatenation or sum of two structures, consisting of one followed by the other;
- ω repeats of some structure laid end to end towards the future;
- ω repeats laid end to end towards the past;
- and making a densely thorough *shuffle* of copies from a finite set of structures.

These operations are well-known from the study of linear orders (see, for example, [6,16,1]).

Model Expressions are an abstract syntax for defining models that are constructed using the follow set of primitive operators:

$$\mathcal{I} ::= \; a \mid \mathcal{I} + \mathcal{J} \mid \overleftarrow{\mathcal{I}} \mid \overrightarrow{\mathcal{I}} \mid \langle \mathcal{I}_0, \ldots, \mathcal{I}_n \rangle$$

where $a \in \Sigma$, where Σ is some alphabet [3]. We refer to these operators, respectively, as *a letter*, *concatenation*, *lead*, *trail*, and *shuffle*.

Definition 3.1 (Correspondence) *Given* $\Sigma = 2^P$, *a model expression* $\mathcal{I}$ corresponds *to a structure as follows. A letter* a *corresponds to any single point model* $(\{x\}, <, h)$ *where* $<$ *is the empty relation and* $h(p) = \{x\}$ *if and only if* $p \in a$. *For the inductive cases we require the notion of an isomorphism (Definition 2.4). Then:*

- $\mathcal{I} + \mathcal{J}$ *corresponds to a structure* $(T, <, h)$ *if and only if* T *is the disjoint union of two sets* U *and* V *where* $\forall u \in U$, $\forall v \in V$, $v < V$ *and* $\mathcal{I}$ *corresponds to* $(U, <^U, h^U)$ *and* $\mathcal{J}$ *corresponds to* $(V, <^V, h^V)$. *(*$<^U, h^U$ *refers to the restriction of the relations* $<$ *and* h *to apply only to elements of* U*).*
- $\overleftarrow{\mathcal{I}}$ *corresponds to the structure* $(T, <, h)$ *if and only if* T *is the disjoint union of sets* $\{U_i | i \in \omega\}$ *where for all* i, *for all* $u \in U_i$, *for all* $v \in U_{i+1}$, $v < u$, *and* $\mathcal{I}$ *corresponds to* $(U_i, <^{U_i}, h^{U_i})$.
- $\overrightarrow{\mathcal{I}}$ *corresponds to the structure* $(T, <, h)$ *if and only if* T *is the disjoint union of sets* $\{U_i | i \in \omega\}$ *where for all* i, *for all* $u \in U_i$, *for all* $v \in U_{i+1}$, $u < v$, *and* $\mathcal{I}$ *corresponds to* $(U_i, <^{U_i}, h^{U_i})$.
- $\langle \mathcal{I}_0, \ldots, \mathcal{I}_n \rangle$ *corresponds to the structure* $(T, <, h)$ *if and only if* T *is the disjoint union of sets* $\{U_i | i \in \mathbb{Q}\}$ *where*

(i) *for all* $i \in \mathbb{Q}$ $(U_i, <^{U_i}, h^{U_i})$ *corresponds to some* $\mathcal{I}_j$ *for* $j \leq n$,

[3] Typically, we will let $\Sigma = \wp(\mathbf{L})$ so the letter indicates the atoms true at a point

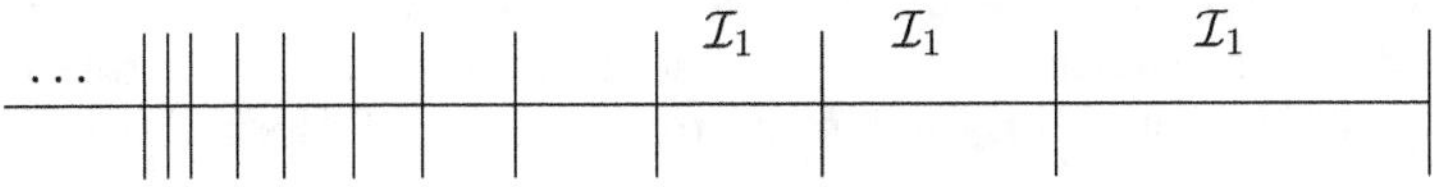

Fig. 1. The lead operation, where $\mathcal{I} = \overleftarrow{\mathcal{I}_1}$

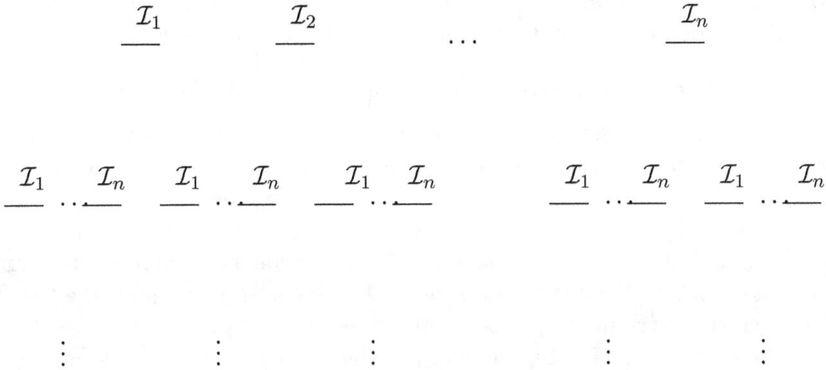

Fig. 2. The shuffle operation, where $\mathcal{I} = \langle \mathcal{I}_1, \ldots, \mathcal{I}_n \rangle$

(ii) *for every* $j \leq n$, *for every* $a \neq b \in \mathbb{Q}$, *there is some* $k \in (a, b)$ *where* $\mathcal{I}_j$ *corresponds to* $(U_k, <^{U_k}, h^{U_k})$,

(iii) *for every* $a < b \in \mathbb{Q}$ *for all* $u \in U_a$, *for all* $v \in U_b$, $u < v$.

We will give an illustration of the non-trivial operations below. The *lead* operation, $\mathcal{I} = \overleftarrow{\mathcal{I}_1}$ corresponds to ω submodels, each corresponding to $\mathcal{I}$, and each preceding the last, as illustrated in Figure 1.

The *trail* operator is the mirror image of the *lead* operation, whereby $\mathcal{I} = \overrightarrow{\mathcal{I}_1}$ corresponds to ω structures, each corresponding to $\mathcal{I}_1$ and each proceeding the earlier structures.

The *shuffle* operator is harder to represent with a diagram. The model expression $\mathcal{I} = \langle \mathcal{I}_1, \ldots \mathcal{I}_n \rangle$ corresponds to a dense, thorough mixture of intervals corresponding to $\mathcal{I}_1, \ldots, \mathcal{I}_n$, without endpoints.

Model expressions give us a grammar that corresponds to structures over general linear frames in a similar manner to the way regular expressions corresponds to words over a given alphabet. Our particular interest in this paper is for frames that are isomorphic to the real numbers so we are required to identify a sublanguage of model expressions. However, the recursive definition of correspondence given in Definition 3.1 is restricted to countable frames. To address this we:

(i) define a Dedekind closure of a structure.

(ii) show that there is a sublanguage of model expressions that correspond to dense, separable structures without endpoints, which agree with their Dedekind closures on the interpretation of $L(U, S)$ formulas.

As every real valued structure is isomorphic to a dense, separable, Dedekind complete structure without end-points, and vice-versa (see [11]) this is sufficient to justify the use of (some) model expressions as the base artefact for synthesis and model checking results.

To define a sublanguage of separable, dense structures without endpoints, we must address the fact that some of the operators of model expressions, such as concatenation, naturally imply a discrete gap in the linear order. We build *real model expressions* via induction using the definitions above:

$$\mathcal{R} ::= \langle a_0, \ldots, a_m, \mathcal{R}_1, \ldots, \mathcal{R}_n \rangle \mid \mathcal{R}_0 + a + \mathcal{R}_1 \mid \overleftarrow{a + \mathcal{R}} \mid \overrightarrow{\mathcal{R} + a}$$

where $a, a_i \in \Sigma$, and $m, n \geq 0$. The letter a_0 is used as a sort of background filler to ensure that the shuffle is Dedekind complete. The abstract syntax for real model expressions is a direct sub-language for the abstract syntax for general model expressions. We note that their syntax will always define open intervals and that the base element of this recursion is a shuffle containing only points. This will define a dense, separable linear order with all the letters homogeneously distributed across the linear order.

Such a sublanguage is suggested in [1] where similar refinements of the [6] operations were applied to provide a decidability result for the monadic theory of the reals. The following lemmas are implicit in that work, but we include them here for completeness.

Lemma 3.2 *Every real model expression corresponds to some structure whose frame is dense, separable and without end-points.*

It is important to note that the corresponding structures are *not* based on a real frame. In fact, any structure corresponding to a model expression $\mathcal{I}$ must be countable and therefore cannot be isomorphic to the reals. However, real model expressions are sufficient for our purposes as the set of formulas satisfiable over the reals is exactly the set of formulas satisfiable over dense, separable, Dedekind complete linear orders without endpoints [3]. As real model expressions correspond to dense, separable linear orders without endpoints, we must show that we can take a further step to a related Dedekind complete order without affecting the interpretation of $L(U, S)$ formulas.

To address this we define a *Dedekind* closure of a structure, and show that any model corresponding to a real model expression agrees with its Dedekind closures on the interpretation of $L(U, S)$ formulae.

Definition 3.3 *Given a structure* $\mathbf{T} = (T, <, h)$*, we say a* Dedekind gap *is pair of sequences in* T*,* $\ell_0, \ell_1, \ldots$ *and* $u_0, u_1, \ldots$*, where for all* i*,* $\ell_i < \ell_{i+1}$*,* $u_i > u_{i+1}$ *and for all* j*,* $\ell_i < u_j$*, and there is no point* $x \in T$ *where for all* i*,* $\ell_i < x < u_i$*.*
Given a point $x \in T$*, we say a* context of x *is the triple* (a, A, B) *where* $a \in \Sigma$ *and* $A, B \subseteq \Sigma$ *where:*

(i) $x \in h(a)$,

(ii) $A = \{b \mid \forall t < x \exists u, t < u < x, \text{ and } u = h(b)\}$

(iii) $B = \{b \mid \forall t > x \exists u, x < u < t, \text{ and } u = h(b)\}$.

A Dedekind gap is curable *if there is a context (a, A, A) such that for all i there are points x_i and y_i where for all j, $\ell_i < x_i < u_j$, $\ell_j < y_i < u_i$, and the context of x_i and y_i is (a, A, A).*

If every Dedekind gap in $\mathbf{T}$ is curable, the Dedekind closure *of $\mathbf{T}$ is the structure $\mathbf{T}^* = (T \cup X, <^*, h^*)$ where:*

(i) *X is a set of new points, one for each Dedekind gap of $\mathbf{T}$,*

(ii) *$<^*$ is the extension of $<$ such that if x is a new point corresponding to a gap defined by $\ell_0, \ell_1, \ldots$ and $u_0, u_1, \ldots$ then for all i, $\ell_i <^* x$ and $x <^* u_i$.*

(iii) *h^* is the extension of h such that for each new point x, $x \in h^*(a)$ where (a, A, A) is some context for x.*

Note that not every structure has a Dedekind closure. However, we have defined real model expressions in such a way that they guarantee that every Dedekind gap will be curable, and furthermore, the cure will not affect the interpretation of any formula.

Lemma 3.4 *Every structure corresponding to a real model expression agrees with its Dedekind closures on the interpretation of $L(U, S)$ formulae.*

Proof. (Sketch)
We show a structure corresponding to a real model expression agrees with its Dedekind closure on the interpretation of $L(U, S)$ formulae by induction over the complexity of formulae. We must show that the addition of the new points in a Dedekind closure does not affect to interpretation of any $L(U, S)$ formula at any of the original points of the structure. This is clearly only relevant in the case of the U and S operators.

Suppose $\mathbf{T} = (T, <, h)$ is a structure and $\mathbf{T}^* = (T \dot\cup X, <^*, h^*)$ is its Dedekind closure. Let $U(\alpha, \beta)$ be given, and suppose that for all $x \in T$ $\mathbf{T}^*, x \models \alpha$ if and only if $\mathbf{T}, x \models \alpha$, and for all $x \in X$, $\mathbf{T}^*, x \models \alpha$ if and only if both for all $t < x$ there is some $\ell \in T$ where $t < l < x$ and $\mathbf{T}, \ell \models \alpha$ and for all $t > x$ there is some $u \in T$ where $x < u < t$ and $\mathbf{T}, u \models \alpha$, (and the same conditions hold for β). Where α and β are propositional, these conditions follow from the definition of Dedekind closure.

Suppose that $x \in T$. If $\mathbf{T}^*, x \models U(\alpha, \beta)$, then there is some point $y > x$ and $\mathbf{T}^*, y \models \alpha$, and for all z where $x < z < y$, $\mathbf{T}^*, z \models \beta$. Therefore, there must be some point $y' \in T$ where $x < y' < z$ where $\mathbf{T}, y' \models \alpha$, and for every point, $t \in T$, where $x < t < y'$ where must have $\mathbf{T}, t \models \beta$ by the induction hypothesis. Conversely, suppose that $\mathbf{T}, x \models U(\alpha, \beta)$. Therefore there is some $y \in T$ where $y > x$ and $\mathbf{T}, y \models \alpha$, and for all $z \in T$ where $x < z < y$, $\mathbf{T}, z \models \beta$. By the induction hypothesis, we have $\mathbf{T}^*, y \models \alpha$ and for all $z \in T \cup X$ where $x <^* z <^* y$, we have $\mathbf{T}^*, z \models \beta$, so $\mathbf{T}^*, x \models U(\alpha, \beta)$.

Suppose now that $x \notin T$. If $\mathbf{T}^*, x \models U(\alpha, \beta)$, then there is some point

$y > x$ and $\mathbf{T}^*, y \models \alpha$, and for all z where $x < z < y$, $\mathbf{T}^*, z \models \beta$. By the induction hypothesis there must be some $x' \in T$ where $x' <^* x$ and for all z where $x' <^* z < y$, $\mathbf{T}^*, z \models \beta$. Hence there must be some $y' \in T$, where $y' <^* y$, $\mathbf{T}, y \models \alpha$ and for all $z \in T$ where $x' \leq z' < y'$, $\mathbf{T}, z \models U(\alpha, \beta)$. Conversely, if for all $t < x$ there is some $\ell \in T$ where $\mathbf{T}, \ell \models U(\alpha, \beta)$ then there must be some $y > x$ where $\mathbf{T}, y \models \alpha$ and for all $z \in T$, $\ell < z < y$, $\mathbf{T}, z \models \beta$. By the induction hypothesis, $\mathbf{T}^*, y \models \alpha$, and for all $z \in T \cup X$, $\ell < z < y$, $\mathbf{T}^*, z \models \beta$ so we must have $\mathbf{T}^*, x \models U(\alpha, \beta)$.

As the case for S is symmetric, we can see that all points $x \in T$ maintain their interpretation of $L(U, S)$ formulas in the Dedekind closure of $\mathbf{T}$, as required. □

Finally we must show that every real model expression corresponds to some real valued structure.

Lemma 3.5 *Every structure corresponding to a real model expression is dense, separable, without endpoints and agrees with its Dedekind closure on the interpretation of $L(U, S)$ formulae.*

Proof. We can see every structure, $\mathbf{T}$, corresponding to a real model expression has a Dedekind closure by construction. Every concatenation, lead and trail operation in a real model expression explicitly includes a single point between the two sub-expressions, so the only place a Dedekind gap may occur is in the shuffle operation. As every shuffle must include at least one single point structure, and the shuffle is dense, then there is a dense set of points in a structure corresponding to the shuffle, where each point has a consistent context. These points can be used to cure all Dedekind defects in $\mathbf{T}$ without affecting the interpretation of any $L(U, S)$ formulae. From Lemma 3.2 we have that $\mathbf{T}$ is dense separable and without end-points so the result follows. □

It is straightforward to make the notation completely formal in the case of a finite set of atoms, and this is the case when we are considering a particular temporal formula. For example, let $[p, \neg q]$ represent a singleton structure with the obvious valuation. We might then suggest $\langle(([p, q]))\rangle + [p, q] + \langle(([p, q], [p, \neg q]))\rangle + [p, q] + \langle(([p, q]))\rangle$, as a model expression for $Gp \wedge U(q, \neg U(q, \neg q) \wedge \neg U(q, q))$.

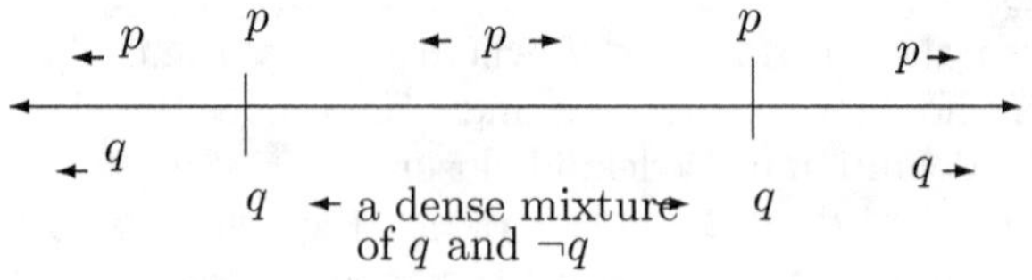

Fig. 3. Diagram representing: $\langle(([p, q]))\rangle + [p, q] + \langle(([p, q], [p, \neg q]))\rangle + [p, q] + \langle(([p, q]))\rangle$

Definition 3.6 *We say that a real-flowed structure* $(\mathbb{R}, <, h)$ *is a* compositional real structure *(or* model*) iff it is isomorphic to the Dedekind closure of a structure which corresponds to some real model expression. In that case, we say that it* realizes *the expression.*

Thus, a compositional real structure is real-flowed by definition. Note also that the real model expression tells us exactly what the model looks like (up to isomorphism).

One of our main results in this paper is that an RTL formula has a real-flowed model iff it has a compositional real model. In the next two sections we briefly describe the proof which uses the mosaic technique for temporal logics.

Before we do so it may be worth noting that there is a similar sort of result in [1] where it is shown that a RTL formula has a real-flowed model iff it has a model with the valuation of each atom being a Borel set, i.e. one obtained from open sets by iterated application of complementation and countable union.

The second half of that paper [1] presents a series of operations corresponding to those of the real model expressions, to show the decidability of the monadic theory of the reals.

The important advantages of our new result are that we provide an explicit notation that is adequate for representing real structures, we are able to give a finite representation in this notation for a model that supports a given satisfiable $L(U, S)$ formula, and we are able to give an efficient means for finding it.

4 Mosaics for U and S

Much of the hard work for us is done by a theorem (4.15 below) from [15] which, unfortunately, must come after a large host of definitions. Our new work also needs these definitions. Due to space considerations the definitions (from that 40 page paper) are selectively presented and only sketched.

In [15], we decided the satisfiability of formulas by considering sets of small pieces of real structures. The idea is based on the mosaics seen in [7] and applied to modal logics. Satisfiability can be decided by checking to see if there exists a finite set of mosaics sufficient to build a model of the formula.

For us, a mosaic is a small piece of a model consisting of three sets of formulas representing those true at each of two points (called the start and end of the mosaic) and those true at all points in-between (called the cover of the mosaic). There are *coherence* conditions on the mosaic which are necessary for it to be part of a model. Note that in the context of a particular formula, ϕ say, (whose satisfiability we might be investigating) we can limit our attention to a finite closure set of formulas and so make these mosaics finite in size. The set of subformulas of ϕ and their negations are a sufficient closure set and will be denoted $\mathrm{Cl}\phi$.

Definition 4.1 *Suppose ϕ is from $L(U, S)$. A ϕ-*mosaic *is a triple (A, B, C) of subsets of the closure set of ϕ such that A and C are maximally propositionally consistent, and B is closed under adding or removing double negations (within the closure set) and the following four* coherency *conditions hold:*

C1. *if* $\neg U(\alpha, \beta) \in A$ *and* $\beta \in B$ *then we have:*
C1.1. $\neg\alpha \in C$;
C1.2. either $\neg\beta \in C$ *or* $\neg U(\alpha, \beta) \in C$ *(or both);*
C1.3. $\neg\alpha \in B$; *and*
C1.4. $\neg U(\alpha, \beta) \in B$.
C2. *if* $U(\alpha, \beta) \in A$ *and* $\neg\alpha \in B$ *then we have:*
C2.1. $\alpha \notin C$ *implies both* $\beta \in C$ *and* $U(\alpha, \beta) \in C$;
C2.2. $\beta \in B$; *and*
C2.3. $U(\alpha, \beta) \in B$.
C3-4 *mirror images of C1-C2.*

Conceptually, A represents the start of the mosaic, C represents the end, and B represents all the points between. We want to show the equivalence of the existence of a model to the existence of a certain set of mosaics: enough mosaics to build a whole model. So the whole set of mosaics also has to obey some conditions. Such conditions are often called *saturation* conditions.

We say that two mosaics compose if the end of the first is the same set as the start of the second. We can then define their composition as the mosaic corresponding to the first point of the first mosaic through to the second point of the second mosaic. Composition is associative.

Definition 4.2 *We say that* ϕ*-mosaics* (A', B', C') *and* (A'', B'', C'') compose *iff* $C' = A''$. *In that case, their* composition *is* $(A', B' \cap C' \cap B'', C'')$.

4.1 Defects

Most of the hard work of finding a saturated set of mosaics is done by breaking up, or decomposing, mosaics into composing sequences of other mosaics. We use a notion of a full decomposition which means that the sequence includes witnesses to all the appropriate formulas in the starts and end of the decomposed mosaic. Eg, if $U(\alpha, \beta)$ is in the start of a mosaic but β is not in its cover then in any full decomposition of the mosaic we should find an initial sequence of mosaics with β in their covers and ends followed by a mosaic with β in its cover and α in its end.

Definition 4.3 *A* defect *in a mosaic* (A, B, C) *is either*

1. *a formula* $U(\alpha, \beta) \in A$ *with either*
   1.1 $\beta \notin B$,
   1.2 ($\alpha \notin C$ *and* $\beta \notin C$), *or*
   1.3 ($\alpha \notin C$ *and* $U(\alpha, \beta) \notin C$);
2. *mirror for* $S(\alpha, \beta) \in C$; *or*
3. *a formula* $\beta \in \mathrm{Cl}\phi$ *with* $\neg\beta \notin B$.

We refer to defects of type 1 to 3 (as listed here). Note that the same formula may be both a type 1 or 2 defect and a type 3 defect in the same mosaic. In that case we count it as two separate defects.

We can talk of sequences of mosaics composing and then find their composition. We define the composition of a sequence of length one to be just the mosaic itself. We leave the composition of an empty sequence undefined.

Definition 4.4 *A* decomposition *for a mosaic (A, B, C) is any finite sequence of mosaics $(A_1, B_1, C_1), (A_2, B_2, C_2), \ldots, (A_n, B_n, C_n)$ which composes to (A, B, C).*

It will be useful to introduce an idea of fullness of decompositions. This is intended to be a decomposition which provides witnesses to the cure of every defect in the decomposed mosaic.

Definition 4.5 *The decomposition above is* full *iff the following three conditions all hold:*

1. *for all $U(\alpha, \beta) \in A$ we have*
   1.1. $\beta \in B$ and either
   $(\beta \in C$ and $U(\alpha, \beta) \in C)$ or $\alpha \in C$,
   1.2. or there is some i with $1 \leq i < n$ such that
   $\alpha \in C_i$, $\beta \in B_j$ (all $j \leq i$)
   and $\beta \in C_j$ (all $j < i$);
2. *the mirror image of 1.; and*
3. *for each $\beta \in \mathrm{Cl}\phi$ such that $\neg\beta \notin B$ there is some i*
   such that $1 \leq i < n$
   and $\beta \in C_i$.

If 1.2 above holds in the case that $U(\alpha, \beta) \in A$ is a type 1 defect in (A, B, C) then we say that *a cure for the defect is witnessed* (in the decomposition) by the end of (A_i, B_i, C_i) (or equivalently by the start of $(A_{i+1}, B_{i+1}, C_{i+1})$). Similarly for the mirror image for $S(\alpha, \beta) \in C$. If $\beta \in C_i$ is a type 3 defect in (A, B, C) then we also say that *a cure for this defect is witnessed* (in the decomposition) by the end of (A_i, B_i, C_i). If a cure for any defect is witnessed then we say that the defect is cured.

Lemma 4.6 *If $m_1, \ldots, m_n$ is a full decomposition of m then every defect in m is cured in the decomposition.*

4.2 Tactics

We would like to introduce an idea of mosaics being fully decomposed in terms of simpler ones. However, sometimes it is allowed that mosaics are decomposed in terms of themselves or equally complicated mosaics. Some recursion is allowed. In order to specify which types of recursion are allowed, we introduce several "tactics" as a sort of meta-level description of how mosaics can be decomposed. There are three tactics with the familiar names of lead, trail and shuffle.

We will see in the next section that by repeated use of a particular tactic to decompose a mosaic we end up showing more or less that it has a model built compositionally via the construction technique with the same name. For example, a mosaic which can be decomposed via a lead tactic into simpler mosaics has a model which is built from simpler structures via a lead construction.

We shall write $\langle p_1, \ldots, p_n \rangle$ for the sequence of mosaics containing $p_1, \ldots, p_n$ in that order. We shall write $\pi^\wedge \rho$ for the sequence resulting from the concatenation of sequences π and ρ in that order. Sequences will always be finite.

Definition 4.7 *We say that m is fully decomposed by the* tactic lead σ, *for some sequence σ of mosaics iff $\langle m\rangle^\wedge\sigma$ is a full decomposition of m.*

The trail σ tactic is mirror.

4.3 Shuffles

The term shuffle has been used in the literature (see, for example, [6] or [1]) to refer to a certain method of constructing a linear structure (often a monadic one) from a thorough mixture of smaller linear structures. A formal definition was given in Section 3.

The intention here is similar except now we need to deal with mosaics corresponding to linear structures. We consider a shuffle S of linear structures $U_0, U_1, \ldots, U_s, V_1, V_2, \ldots, V_r$ where each U_i is a singleton structure and each V_i is a non-singleton structure consisting of a finite sequence of other structures. Thus, we actually only consider an MPC set P_i instead of U_i and a non-empty composing sequence λ_i of mosaics instead of V_i. In this case it is possible to construct a certain set $\{o, m', m'', x_0, \ldots, x_r, y_0, \ldots, y_s\}$ of mosaics such that one, o, corresponds to S and each one in the set has a full decomposition in terms of others in the set and/or the mosaics which decompose each λ_i.

The mosaic m' corresponds to any proper initial interval of S ending at a copy of U_0 while m'' corresponds to any proper final interval of S beginning at a copy of U_0. Each x_i is satisfied by any interior interval of S starting at the end of a copy of V_i (or a copy of U_0 if $i = 0$) and ending at the start of a copy of V_{i+1} (or a copy of U_0 if $i = r$). There are a lot of intervals of this form, for each i, but each one satisfies x_i. Each y_i is satisfied by any interior interval of S starting at a copy of U_i and ending at a copy of U_{i+1} (or a copy of U_0 if $i = s$).

It can be shown that these mosaics can be used to mutually decompose each other according to the patterns described in F1 to F6 in the following definition. We do not need to prove this so we will not. However, this observation does provide the intuition behind this rather involved construct. The fact that we have these mutual full decompositions, means that the mosaic o can be fully decomposed by the others mentioned and they in turn can be fully decomposed and so on: i.e. this provides a tactic for iterative full decompositions.

Definition 4.8 *Suppose* $0 \leq r$, *each* $\lambda_i (1 \leq i \leq r)$ *is a non-empty composing sequence of* ϕ*-mosaics, and* $P_0, \ldots, P_s$ $(0 \leq s)$ *are maximally propositionally consistent subsets of* $\mathrm{Cl}\phi$.

Suppose ϕ*-mosaic* $o = (A, B, C)$ *and:*

$$\begin{aligned} &m' = (A, B, P_0);\\ &y_i = (P_i, B, P_{i+1})\ (0 \leq i \leq s-1);\\ &y_s = (P_s, B, P_0);\\ &m'' = (P_0, B, C);\ \text{and}\\ &\mu = \langle y_0, \ldots, y_s\rangle. \end{aligned}$$

If $r = 0$ *suppose* $\lambda = \langle\rangle$, *the empty sequence, but otherwise, if* $r > 0$, *suppose:*

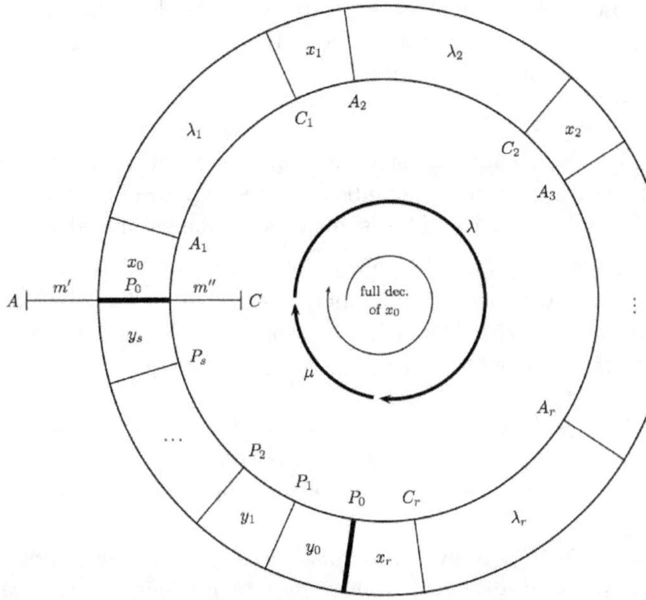

Fig. 4. Full decompositions in a shuffle

A_i is the start of the first mosaic in $\lambda_i (1 \leq i \leq r)$;
C_i is the end of the last mosaic in $\lambda_i (1 \leq i \leq r)$;
$x_0 = (P_0, B, A_1)$;
$x_i = (C_i, B, A_{i+1}), (1 \leq i \leq r-1)$;
$x_r = (C_r, B, P_0)$;
$\lambda = \langle x_0 \rangle ^\wedge \lambda_1^\wedge \langle x_1 \rangle ^\wedge \ldots ^\wedge \lambda_r^\wedge \langle x_r \rangle$.

Further suppose that m', m'', and each y_i and x_i are mosaics.

Then, we say that o is fully decomposed by the tactic shuffle $(\langle P_0, \ldots, P_s \rangle, \langle \lambda_1, \ldots, \lambda_r \rangle)$ *iff the following conditions all hold:*

F1. *o is fully decomposed by $\langle m' \rangle ^\wedge \lambda ^\wedge \mu ^\wedge \langle m'' \rangle$;*
F2. *if $r > 0$, x_0 is fully decomposed by $\lambda ^\wedge \mu ^\wedge \langle x_0 \rangle$;*
F3. *if $0 < i < r$, x_i is fully decomposed by*
$\langle x_i \rangle ^\wedge \lambda_{i+1}^\wedge \langle x_{i+1} \rangle ^\wedge \ldots ^\wedge \lambda_r^\wedge \langle x_r \rangle ^\wedge \mu ^\wedge \langle x_0 \rangle ^\wedge \lambda_1^\wedge \langle x_1 \rangle ^\wedge \ldots ^\wedge \lambda_i^\wedge \langle x_i \rangle$;
F4. *if $r > 0$, x_r is fully decomposed by $\langle x_r \rangle ^\wedge \mu ^\wedge \lambda$;*
F5. *if $0 \leq i < s$, y_i is fully decomposed by $\langle y_i, y_{i+1}, \ldots, y_s \rangle ^\wedge \lambda ^\wedge \langle y_0, \ldots, y_i \rangle$;*
F6. *y_s is fully decomposed by $\langle y_s \rangle ^\wedge \lambda ^\wedge \mu$.*

The order of the mutual full decompositions specified in F1-F6 in the definition is illustrated in the circular diagram in figure 4 (borrowed from [15]).

Note that as $s \geq 0$ there is at least one P_i involved in the shuffle. In a general linear order setting we could define a shuffle with no P_is (provided that then $r > 0$) but over the reals it turns out to be crucial to require at least one

P_i. This is because, as it is not too hard to see, a shuffle of only non-singleton closed intervals of the reals cannot be both Dedekind complete and separable (i.e. having a countable dense suborder).

4.4 Real Mosaic Systems

Finally (in [15]) we are able to introduce the notion of a real mosaic system (RMS): one in which each mosaic can be fully decomposed in terms of the three tactics and of simpler mosaics. This is our version of a saturated set of mosaics.

Definition 4.9 *For $\phi \in L(U, S)$, suppose S is a set of ϕ-mosaics and $n \geq 0$.*

A ϕ-mosaic m is a level n^+ member of S *iff m is the composition of a sequence of mosaics, each of them being either a level n member of S or fully decomposed by the tactics* lead σ *or* trail σ *with each mosaic in σ being a level n member of S.*

A ϕ-mosaic m is a level $(n+1)^-$ member of S *iff m is the composition of a sequence of mosaics, each of them being either a level n^+ member of S or fully decomposed by the tactics* lead σ *or* trail σ *with each mosaic in σ being a level n^+ member of S.*

A ϕ-mosaic $m \in S$ is a level n member of S *iff m is the composition of a sequence of mosaics with each of them being either a level n^- member of S or a mosaic which is fully decomposed by the tactic* shuff $\langle P_0, \ldots, P_s \rangle, \langle \sigma_1, \ldots, \sigma_r \rangle$ *with each mosaic in each σ_i being a level n^- member of S.*

Note that it is generally possible for mosaics to be level 0 members of some S provided that they are compositions of mosaics which can be fully decomposed by shuffles in which there are no sequences (i.e. , $r = 0$).

Definition 4.10 *For $\phi \in L(U, S)$, a* real mosaic system *of ϕ-mosaics is a set S of ϕ-mosaics such that for every $m \in S$ there exists some n such that m is a level n member of S. For any n we say that S is a real mosaic system of depth n iff every $m \in S$ is a level n member of S.*

4.5 Relativization

We will now relate the satisfiability of a formula ϕ to that of certain mosaics. Obviously, a formula will be satisfiable over the reals iff it is satisfiable over the $]0, 1[$ flow. Furthermore, this happens iff a relativized version of the formula is satisfiable somewhere in the interior of a model over $[0, 1]$. To define this relativization we need to use a new atom to indicate points in the interior. Hence the next few definitions.

Definition 4.11 *Given ϕ and an atom q which does not appear in ϕ, we define a map $* = *_q^\phi$ on formulas in* Clϕ *recursively:*

1. $*p = p \wedge q$,
2. $*\neg\alpha = \neg(*\alpha) \wedge q$,
3. $*(\alpha \wedge \beta) = *(\alpha) \wedge *(\beta) \wedge q$,
4. $*U(\alpha, \beta) = U(*\alpha, *\beta) \wedge q$, *and*
5. $*S(\alpha, \beta) = S(*\alpha, *\beta) \wedge q$.

So $*_q^\phi(\phi)$ will be a formula using only q and atoms from ϕ.

Lemma 4.12 *$*_q^\phi(\phi)$ is at most 3 times as long as ϕ.*

With the relativization machinery we can then define a relativized mosaic to be one which could correspond to the whole of a $[0,1]$ structure in which q is true of exactly the interior $]0,1[$ and the interior is a model of ϕ.

Definition 4.13 *We say that a $*_q^\phi(\phi)$-mosaic (A,B,C) is (ϕ,q)-*relativized *iff*

1. *$\neg q$ is in A and no $S(\alpha,\beta)$ is in A;*
2. *$q \in B$ and $\neg *_q^\phi(\phi) \notin B$; and*
3. *$\neg q \in C$ and no $U(\alpha,\beta)$ is in C.*

Here we confirm that ϕ is satisfiable over the reals exactly when we can find such a relativized mosaic.

Lemma 4.14 *Suppose that ϕ is a formula of $L(U,S)$ and q is an atom not appearing in ϕ. Then ϕ is $\mathbb{R}$-satisfiable iff there is some fully $[0,1]$-satisfiable (ϕ,q)-relativized $*_q^\phi(\phi)$-mosaic.*

Proof. Let $* = *_q^\phi$ and let $\zeta :]0,1[\to \mathbb{R}$ be any order preserving bijection.

Suppose that ϕ is $\mathbb{R}$-satisfiable. Say that $\mathbf{S} = (\mathbb{R},<,g)$, $s_0 \in \mathbb{R}$ and $\mathbf{S}, s_0 \models \phi$. Let $\mathbf{T} = ([0,1],<,h)$ where for atom $p \neq q$, $h(p) = \{t \in]0,1[| \zeta(t) \in g(p)\}$; and $h(q) =]0,1[$. An easy induction on the construction of formulas in $\mathrm{Cl} * \phi$ shows that $\mathbf{T}, \zeta^{-1}(s_0) \models *\phi$ and so $\mathrm{mos}_{\mathbf{T}}^{*\phi}(0,1)$ is the right mosaic.

Suppose mosaic $(A,B,C) = \mathrm{mos}(0,1)$ from structure $\mathbf{T} = ([0,1],<,h)$ is a (ϕ,q)-relativized $*(\phi)$-mosaic. Thus $q \in B$ and $\neg q \in A \cap C$. Define $\mathbf{S} = (\mathbb{R},<,g)$ via $s \in g(p)$ iff $\zeta^{-1}(s) \in h(p)$ for any atom p (including $p = q$). As $\neg * \phi \notin B$, there is some z such that $0 < z < 1$ and $\mathbf{T}, z \models *\phi$. It is easy to show that $\mathbf{S}, \zeta(z) \models \phi$. □

Our satisfiability procedure in [15] is to guess a relativized mosaic (A,B,C) and then check that (A,B,C) is fully $[0,1]$-satisfiable. Thus we now turn to the question of deciding whether a relativized mosaic is satisfiable.

4.6 Satisfiability of Mosaics

For us, in this paper, the important result from [15] is that which shows the equivalence of satisfiability of a formula to the fact of its negation not being in the cover of some mosaic in some RMS. The negation is not in the cover of a mosaic if the mosaic represents a pair of points in a model with a witness to ϕ in between. There is actually a slight further complication in that we need to deal with mosaics (relativized ones) which (by virtue of the formulas contained within each end) must be covering the whole of the structure of which they are part. This idea of relativization allows us to take care of formulas such as $U(\alpha,\beta)$ lying in the end of a mosaic: this cannot happen in a relativized mosaic.

Theorem 4.15 *([15] Theorem 72) Suppose ϕ is a formula of $L(U,S)$ and q is an atom not appearing in ϕ. Suppose $\psi = *_q^\phi(\phi)$ has length N.*

Then the following are equivalent:

1. *ϕ is $\mathbb{R}$-satisfiable;*
2. *there is a (ϕ, q)-relativized ψ-mosaic which is a level N member of some RMS.*

5 Expressiveness

In this section we use mosaics to show that if a formula is satisfiable over the reals then it is satisfiable in a compositional real model. This shows that the compositional model building method has adequate expressiveness for describing models of temporal specifications. Despite this first result not being surprising given earlier precedents [6], the new technique here allows us to to establish the new synthesis theorem later in the section.

In order to work with open intervals of the reals and match mosaics to such structures, we need to rework some definitions and several of the lemmas from [15]. The first definition gives us something like a Hintikka structure to witness a mosaic, at least as far as possible internal structure is concerned.

Definition 5.1 *Suppose ϕ is a formula of $L(U, S)$ and $m = (A, B, C)$ is a ϕ-mosaic.*

Suppose $x, y \in \mathbb{R}$ with $x < y$.

Say that structure $(]x, y[, <, h)$ supports *m iff there is a map $\mu :]x, y[\to \wp(\mathrm{Cl}\phi)$ satisfying:*

R0. for each $z \in]x, y[$, $p \in \mu(z)$ iff $z \in h(p) \cap \mathrm{Cl}\phi$;
R1. for each $z \in]x, y[$, $\mu(z)$ is a maximally propositionally consistent subset of $\mathrm{Cl}\phi$;
R2. Suppose $z \in]x, y[$. Then $U(\alpha, \beta) \in \mu(z)$ iff either
R2.1, there is u such that $z < u < y$ and $\alpha \in \mu(u)$ and for all v, if $z < v < u$ then $\beta \in \mu(v)$ or
R2.2, $\alpha \in C$ and for all v, if $z < v < y$ then $\beta \in \mu(v)$ or
R2.3, $\beta \in C$, $U(\alpha, \beta) \in C$ and for all v, if $z < v < y$, then $\beta \in \mu(v)$;
R3. the mirror image of R2 for $S(\alpha, \beta)$;
R4. $U(\alpha, \beta) \in A$ iff either
R4.1, there is u such that $x < u < y$ and $\alpha \in \mu(u)$ and for all v, if $x < v < u$ then $\beta \in \mu(v)$ or
R4.2, $\alpha \in C$ and for all v, if $x < v < y$ then $\beta \in \mu(v)$ or
R4.3, $\beta \in C$, $U(\alpha, \beta) \in C$ and for all v, if $x < v < y$, then $\beta \in \mu(v)$;
R5. the mirror image of R4 for $S(\alpha, \beta)$; and
R6. for each $\beta \in \mathrm{Cl}\phi$, β is in the cover of m iff for all u, if $x < u < y$, $\beta \in \mu(u)$.

(Also say that m is supported by $(]x, y[, <, h)$, via μ.)

A straightforward induction on the construction of formulas tells us the following.

Lemma 5.2 *If m is supported by* $\mathbf{S}$ *and that is isomorphic to* $\mathbf{T}$ *then m is also supported by* $\mathbf{T}$.

As we have seen, the class of compositional real structures is closed under appropriate semantic sum, trail, lead and shuffle operations. Thus we can proceed through an induction on mosaics in an RMS, relating them to successively more complex compositional structures.

For the full proofs see [2].

Lemma 5.3 *If m is the composition of m' and m'' with each of m' and m'' being supported by a compositional real structure then m is supported by a compositional real structure.*

Lemma 5.4 *If m is fully decomposed by the tactic* lead (σ) *with each mosaic in σ being supported by a compositional real structure, then m is supported by a compositional real structure. There is a mirror image result for* trail (σ).

Lemma 5.5 *If m is fully decomposed by the tactic*

$$\text{shuff}\,(\langle P_0, \ldots, P_s\rangle, \langle \lambda_1, \ldots, \lambda_r\rangle)$$

with each mosaic in each λ_i being supported by a compositional real structure, then m is supported by a compositional real structure.

We have shown that each of the operations of composition and using lead, trail or shuffle tactics preserves compositional supportedness of mosaics. A simple induction allows us to conclude that any mosaic in a real mosaic system is supported by a compositional real structure.

Lemma 5.6 *Suppose ϕ is a formula of $L(U, S)$ and m is a ϕ-mosaic.*

If m appears in an RMS then m is supported by a compositional real structure.

Proof. Given the real mosaic system S of ϕ-mosaics, we can easily proceed by induction on k to show that any level k member $m' \in S$ is supported by a compositional real structure. Each step of the induction is just a use of one or two of the preceding lemmas 5.3, 5.4, its mirror image and 5.5.

Suppose this is true for $k \geq -1$: it is true for $k = -1$ because there are no level -1 members of S. Lemma 5.4 tells us that any mosaic which can be fully decomposed by lead (σ) is supported if each of the mosaics in σ are. By lemma 5.3 this (and its mirror image) means that any level k^+ member of S is supported. By lemma 5.3 and lemma 5.4 and its mirror image we have that any level $(k+1)^-$ member of S is supported. By lemmas 5.3 and 5.5 it follows that any level $k+1$ member of S is supported as required. □

5.1 Relativized and Supported means Modeled

Lemma 5.7 *Suppose ϕ is a formula of $L(U, S)$ and q is an atom not appearing in ϕ. Suppose $\psi = *^{\phi}_{q}(\phi)$ has length N.*

Suppose that ψ-mosaic m is (ϕ, q)-relativized and is supported by a real structure $\mathbf{T} = (\mathbb{R}, <, h)$.

Then there is $r \in \mathbb{R}$ such that $\mathbf{T}, r \models \phi$.

Proof. Suppose ϕ is a formula of $L(U, S)$ and q is an atom not appearing in ϕ. Suppose $\psi = *^{\phi}_{q}(\phi)$ has length N.

Suppose that ψ-mosaic $m = (A, B, C)$ is (ϕ, q)-relativized and is supported by a real structure $(\mathbb{R}, <, h)$.

By lemma 5.2, m is also supported by $\mathcal{T} = (]0,1[, <, h')$ for some h'. Let $f : \mathbb{R} \to]0,1[$ be the bijection such that for all $p \in P$, for all $t \in]0,1[$, $t \in h'(p)$ iff $f(t) \in h(p)$.

Let $\mu :]0,1[\to \wp(\mathrm{Cl}\phi)$ satisfy R1-R6 in the definition of supporting.

First, we claim that for all $\alpha \leq \phi$, for all $x \in]0,1[$, $\mathbf{T}, x \models \alpha$ iff $*\alpha \in \mu(x)$. The proof is by induction on construction of α, and has a series of cases (each with forward and converse arguments) but is straightforward.

Now as m is (ϕ, q)-relativized, $\neg * \phi \notin B$.

By R6, there must be some $u \in]0,1[$ such that $*\phi \in \mu(u)$. By our induction result, $\mathbf{T}, u \models \phi$.

By lemma 2.5, $(\mathbb{R}, <, h), f(u) \models \phi$ as required. □

5.2 Finding a model of a formula

The main result is that we can use such an RMS to describe a model in our new notation.

Theorem 5.8 *A formula ϕ from $L(U, S)$ is $\mathbb{R}$-satisfiable iff there is a compositional real model of ϕ. There is some c such that, for $\mathbb{R}$-satisfiable ϕ, a satisfying model can be described by an expression of shuffles, leads, trails and sums of length $< 2^{c|\phi|^2}$ (this bound is best possible).*

Furthermore, there is an EXPTIME procedure for finding such an expression.

Proof. Suppose ϕ is a formula of $L(U, S)$ and q is an atom not appearing in ϕ. Suppose $\psi = *^{\phi}_{q}(\phi)$ has length N.

One direction of the theorem is immediate: by Definition 3.6, any compositional real model of ϕ is a real model of ϕ.

For the other direction assume ϕ is $\mathbb{R}$-satisfiable.

By theorem 4.15, there is a (ϕ, q)-relativized ψ-mosaic m which is a level N member of some RMS.

By lemma 5.6, m is supported by a compositional real structure $\mathbf{T}$, say, corresponding to compositional real expression $\mathcal{I}$.

By lemma 5.7, there is $r \in \mathbb{R}$ such that $\mathbf{T}, r \models \phi$.

The bound follows from consideration of the level of the relativized mosaic in the RMS, a level which from consideration of arguments in [15] can be shown to be at most six times the length of ϕ. We also show that the length of decomposition sequences at each level is bounded by an exponential in $|\phi|$.

We can also show that bound is best possible by considering a formula describing a binary counter. Given n, use n atoms to describe a counter which increases at discrete intervals. A formula of quadratic length (in n) can be used to specify such a model but a description of the model in our construction notation needs to be of exponential length.

Suppose that the atom p marks discrete points (so $\delta = G(p \to U(p, \neg p)$ is true), $\overline{b} = b_1, \ldots, b_n$ are the n bits of a counter that counts modulo 2^n and:

$$\phi_i = (p \wedge \bigwedge_{j=1}^{i-1} b_j) \to (b_i \to U(\neg b_i, \neg p) \wedge \neg b_i \to U(b_i, \neg p))$$

$$\psi_i = (p \wedge \neg \bigwedge_{j=1}^{i-1} b_j) \to (b_i \to U(b_i, \neg p) \wedge \neg b_i \to U(\neg b_i, \neg p))$$

The formula ϕ_i specifies if all bits less significant than b_i are true, then b_i will invert its valuation the next time p is true, and ψ_i specifies that if any less significant bit is not true, b_i will retain its current valuation at the next point that p is true.

Clearly $\delta \wedge G \bigwedge_{i=1}^{n} (\phi_i \wedge \psi_i)$ is satisfiable, and is of length $O(n^2)$. However, any real model expression describing a structure that satisfies this formula would have to contain at least 2^n distinct letters.

□

Note that thanks to the expressive completeness result in [5], we know that any satisfiable sentence of the first-order monadic logic of the reals also has a compositional real model. To find a description of a model from the sentence must be a hard problem as deciding validity in this logic is non-elementarily complex [17].

Theorem 5.9 *There is an EXPTIME procedure which given a formula ϕ from $L(U, S)$ will decide whether ϕ is $\mathbb{R}$-satisfiable or not and if so will provide an expression for a compositional model of ϕ.*

Proof. Finally, we can give an EXPTIME procedure for finding and printing out a model of any satisfiable RTL formula. The set of all ϕ-mosaics is of size exponential in the length of ϕ. There is a fairly straightforward procedure (in the style of [9]) for going through the set repeatedly and removing mosaics which can not be fully decomposed in terms of other simpler ones in the set. If ϕ is satisfiable we will eventually end up with an RMS and another straightforward EXPTIME procedure reads out the description of a model of ϕ. By repeatedly decomposing mosaics as specified in the RMS we can produce the expression in a top-down manner. □

Other results from [15] allow us to conclude that if we find all possible starting points (i.e. relativized mosaics in the RMS) and follow all possible ways of decomposing the mosaics (as given in the RMS) then we will eventually output a list of possible models of the formula which is in a certain sense exhaustive. Any real model of ϕ will be back-and-forth equivalent to one of the compositional models which is listed.

6 Conclusion

We have investigated a compositional approach to building linear temporal structures as a way of working with models on a continuous (real numbers) flow of time. Structures are built by putting together smaller structures in a recursive way, with copies of the smaller ones occupying successive intervals of time. We have formalised the approach so that such models can be described clearly and efficiently.

We have identified a sub-language of the formal compositional model building language which can be used (in a slightly modified way) to build real-flowed structures. Any RTL formula satisfiable in the reals is satisfiable in such a compositional real-flowed model. We presented an efficient method for building a real-flowed model of any given a satisfiable formula.

The approaches here may generalise to general linear models of time [2].

References

[1] Burgess, J. P. and Y. Gurevich, *The decision problem for linear temporal logic*, Notre Dame J. Formal Logic **26** (1985), pp. 115–128.

[2] French, T., J. McCabe-Dansted and M. Reynolds, *Synthesis and model checking for continuous time: Long version*, Technical report, CSSE, UWA (April 2012), "http://www.csse.uwa.edu.au/~mark/research/Online/sctm.htm".

[3] Gabbay, D., I. Hodkinson and M. Reynolds, "Temporal Logic: Mathematical Foundations and Computational Aspects, Volume 1," Oxford University Press, 1994.

[4] Gabbay, D. M. and I. M. Hodkinson, *An axiomatisation of the temporal logic with until and since over the real numbers*, J. Logic and Computation **1** (1990), pp. 229 – 260.

[5] Kamp, H., "Tense logic and the theory of linear order," Ph.D. thesis, U. Calif. (1968).

[6] Läuchli, H. and J. Leonard, *On the elementary theory of linear order*, Fundamenta Mathematicae **59** (1966), pp. 109–116.

[7] Németi, I., *Decidable versions of first order logic and cylindric-relativized set algebras*, in: *Logic Colloquium '92* (1995), pp. 171–241.

[8] Pnueli, A., *The temporal logic of programs*, in: *Proceedings of the Eighteenth Symposium on Foundations of Computer Science*, 1977, pp. 46–57.

[9] Pratt, V. R., *Models of program logics*, in: *Proc. 20th IEEE. Symposium on Foundations of Computer Science, San Juan*, 1979, pp. 115–122.

[10] Rabin, M. O., *Decidability of second order theories and automata on infinite trees*, American Mathematical Society Transactions **141** (1969), pp. 1–35.

[11] Reynolds, M., *An axiomatization for Until and Since over the reals without the IRR rule*, Studia Logica **51** (1992), pp. 165–193.

[12] Reynolds, M., *Continuous temporal models*, in: M. Stumptner, D. Corbett and M. J. Brooks, editors, *Australian Joint Conference on Artificial Intelligence*, Lecture Notes in Computer Science **2256** (2001), pp. 414–425.

[13] Reynolds, M., *The complexity of the temporal logic with "until" over general linear time*, J. Comput. Syst. Sci. **66** (2003), pp. 393–426.

[14] Reynolds, M., *Dense time reasoning via mosaics*, in: *Proceedings of the 2009 16th International Symposium on Temporal Representation and Reasoning* (2009), pp. 3–10.

[15] Reynolds, M., *The complexity of the temporal logic over the reals*, Annals of Pure and Applied Logic **161** (2010), pp. 1063–1096, online at doi:10.1016/j.apal.2010.01.002.

[16] Rosenstein, J. G., "Linear Orderings," Academic Press, New York, 1982.

[17] Stockmeyer, L., "The complexity of decision problems in automata and logic," Ph.D. thesis, M.I.T. (1974).

General Dynamic Dynamic Logic

Patrick Girard Jeremy Seligman

Department of Philosophy
University of Auckland

Fenrong Liu

Department of Philosophy
Tsinghua University

Abstract

Dynamic epistemic logic (DEL) extends purely modal epistemic logic (S5) by adding dynamic operators that change the model structure. Propositional dynamic logic (PDL) extends basic modal logic with programs that allow the definition of complex modalities. We provide a common generalisation: a logic that is 'dynamic' in both senses, and one that is not limited to S5 as its modal base. It also incorporates, and significantly generalises, all the features of existing extensions of DEL such as BMS [3] and LCC [21]. Our dynamic operators work in two steps. First, they provide a multiplicity of transformations of the original model, one for each 'action' in a purely syntactic 'action structure' (in the style of BMS). Second, they specify how to combine these multiple copies to produce a new model. In each step, we use the generality of PDL to specify the transformations. The main technical contribution of the paper is to provide an axiomatisation of this 'general dynamic dynamic logic' (GDDL). This is done by providing a computable translation of GDDL formulas to equivalent PDL formulas, thus reducing the logic to PDL, which is decidable. The proof involves switching between representing programs as terms and as automata. We also show that both BMS and LCC are special cases of GDDL, and that there are interesting applications that require the additional generality of GDDL, namely the modelling of private belief update. More recent extensions and variations of BMS and LCC are also discussed.

Keywords: Dynamic logic, BMS, LCC, PDL, belief change.

Recent research in epistemic logic extends the classical S5-analysis of knowledge with dynamic operators that model the epistemically relevant changes brought about by various acts of communication. These are represented as extensions of the basic epistemic language with expression of the form $[a]\varphi$ interpreted as 'after action a is performed, φ is the case'. The primary example of such an action is the 'public announcement' of a proposition ψ, written $!\psi$, which achieves the right effect by simply removing the $\neg\varphi$-states (those states of the model in which ψ is false), so that everyone subsequently knows that these possibilities are no longer open. [1] A rich array of dynamic operators have been introduced to deal with private communications of various sorts, and also actions that affect more than just the epistemic states of agents, the so-called 'real world changes'. [2]

Meanwhile, interest has grown in applying similar techniques to other branches of modal logic, such as doxastic logic (the logic of belief) [4,18] and preference logic [7,11,12,20]. A significant difference from the epistemic setting is the need to describe dynamic operators that change the relational structure of the underlying model, not just the size of its domain (announcement) or the propositional valuations (real-world change). For example, if one models the doxastic state of an agent by a plausibility relation between epistemically possible state, 'upgrading' a proposition φ, so that it is believed, may be modelled by an operator that transforms the plausibility relation by removing links from φ-states to $\neg\varphi$-states and adding links from $\neg\varphi$-states to φ-states. This ensures that every possible state in which φ is true becomes more plausible (for the given agent) than every possibility in which φ is false. Currently, however, there is no way of adapting the technology of BMS to model the doxastic effect on a multiplicity of agents of one or more of those agents *privately* upgrading their beliefs as a response to a less-than-public communication.

We solve this problem by providing a more general framework, inspired by Theorem 4.11 in [12], first noted in [20], which states that any dynamic operator whose effect on a model can be described in PDL (without Kleene's iteration operator $*$) can be reduced to the underlying modal logic using essentially only the standard axioms of PDL. We show how this idea can be used to extend BMS (and LCC), so that a vast range of dynamic operators can be modelled in a way that allows for private changes and real-world changes in epistemic logic, doxastic logic, preference logic, and any other normal modal logic. [3] We also extend it by adding the Kleene star, so that certain desirable frame conditions (such as transitivity) can be imposed. [4]

Section 1 introduces the concept of a 'PDL-transformation', which is a general

[1] Notoriously, it is not guaranteed that the announced proposition is subsequently known because its very announcement may change its truth value, e.g. announcing 'the sun is shining but you don't know it' results in your knowing that the sun is shining and so making the announcement false.

[2] See for example, the textbooks [23] and [19]. The initial paper on public announcement was [15] and the most significant advance came with the eponymously acronymed BMS [3]. A recent extension of BMS, incorporating real-world changes, is LCC [21], the 'Logic of Communication and Change'.

[3] In particular, we can have dynamic operators over epistemic logics weaker than S5, so catering for those who wish to avoid the controversial properties of positive or negative introspection.

[4] There are various ways in which the Kleene star can be used in dynamics. Our sense will become clear but, for example, it is *not* the sense of [13], which is known to give an undecidable logic.

way of changing models using PDL terms. These transformation are used extensively in Section 2, in which GDDL is defined semantically and then axiomatised, using a technique that exploits the possibility of representing programs both by PDL-terms and by finite state automata. We illustrate GDDL by showing how it can be used to model private belief change. Finally, Section 3 shows how BMS and LCC are special cases of GDDL, and then goes on to discuss other recent variations and extensions such as [2,9,10,22,24].

1 Preliminaries

A *Kripke signature* is a pair $\langle P, R\rangle$ of sets of symbols. The elements of P are *propositional variables* and those of R are *relation symbols.* A model of this signature, $M = \langle W, V\rangle$ consists of a set W (of *states*) and a *valuation* function V mapping each $p \in P$ to $V(p) \subseteq W$ and each $r \in R$ to $V(r) \subseteq W^2$. To describe such structures, we define $T(P, R)$ to be the set of *programs* π and $L(P, R)$ to be the set of *formulas* φ in the usual way:

$$\begin{aligned}\pi &::= r \mid \mathsf{E} \mid \varphi? \mid (\pi;\pi) \mid (\pi \cup \pi) \mid \pi^* \\ \varphi &::= p \mid \neg\varphi \mid (\varphi \vee \varphi) \mid \langle\pi\rangle\varphi\end{aligned}$$

where $r \in R$ and $p \in P$. Here, E, is the universal program that jumps between arbitrary states. In each model M, semantic values $[\![\varphi]\!]^M \subseteq W$ and $[\![\pi]\!]^M \subseteq W^2$ are given by:

$$\begin{aligned}
[\![p]\!]^M &= V(p)\\
[\![\neg\varphi]\!]^M &= W \setminus [\![\varphi]\!]^M\\
[\![(\varphi \wedge \psi)]\!]^M &= [\![\varphi]\!]^M \cap [\![\psi]\!]^M\\
[\![\langle\pi\rangle\varphi]\!]^M &= \{u \in W \mid u[\![\pi]\!]^M v \text{ and } v \in [\![\varphi]\!]^M, \text{ for some } v \in W\}
\end{aligned}$$

$$\begin{aligned}
[\![r]\!]^M &= V(r)\\
[\![\mathsf{E}]\!]^M &= W^2\\
[\![\varphi?]\!]^M &= \{\langle u,u\rangle \mid u \in [\![\varphi]\!]^M\}\\
[\![\pi_1;\pi_2]\!]^M &= \{\langle u,v\rangle \mid u[\![\pi_1]\!]^M w \text{ and } w[\![\pi_2]\!]^M v, \text{ for some } w \in W\}\\
[\![\pi_1 \cup \pi_2]\!]^M &= [\![\pi_1]\!]^M \cup [\![\pi_2]\!]^M\\
[\![\pi^*]\!]^M &= \begin{array}{l}\{\langle u,v\rangle | u = v \text{ or } u_i[\![\pi]\!]^M u_{i+1} \text{ for some } n \geq 0, u_0, \ldots, u_n \in W\\ \text{such that } u_0 = u \text{ and } u_n = v\}\end{array}
\end{aligned}$$

As usual, we also write $u[\![\pi]\!]^M v$ for $\langle u, v\rangle \in [\![\pi]\!]^M$ and $M, u \models \varphi$ for $u \in [\![\varphi]\!]^M$.

PDL-transformations We will use expressions from our language to describe changes to models. Given a model M of signature $\langle P, R\rangle$, we will show how to obtain a model ΛM of a possibly different signature $\langle Q, S\rangle$ in such a way that we retain some control over which formulas are satisfied in the new model. Specifically, we will also define a (computable) translation φ^Λ of each formula $\varphi \in L(Q, S)$ such that

$$M, u \models \varphi^\Lambda \text{ iff } \Lambda M, u \models \varphi$$

This is the content of Lemma 1.1, below. Specifically, we say that a PDL-*transformation* Λ from signature $\langle P, R\rangle$ to signature $\langle Q, S\rangle$ consists of

(i) a formula $|\Lambda| \in L(P, R)$,
(ii) an algorithm[5] for calculating $\Lambda(q) \in L(P, R)$ for each $q \in Q$ and
(iii) an algorithm for calculating a term $\Lambda(s) \in T(P, R)$ for each $s \in S$.

Now, given a model $M = \langle W, R\rangle$ of signature $\langle P, R\rangle$, we define the model ΛM of signature $\langle Q, S\rangle$ to be $\langle \Lambda W, \Lambda V\rangle$, where

$$\begin{array}{lll} \Lambda W & = & [\![|\Lambda|]\!]^M \\ \Lambda V(q) & = & [\![\Lambda(q)]\!]^M \cap \Lambda W \text{ for each } q \in Q \\ \Lambda V(s) & = & [\![\Lambda(s)]\!]^M \cap \Lambda W^2 \text{ for each } s \in S \end{array}$$

In other words, the domain of the new model is simply a restriction of the domain of the old model (defined by $|\Lambda|$) and the interpretation of the symbols of $\langle Q, S\rangle$ are given by evaluating the corresponding PDL-expressions provided by Λ, and then restricting them to the new domain.

For the translation, we can inductively compute formulas φ^Λ and terms π^Λ of signature $\langle P, R\rangle$ from each formula φ and term π of signature $\langle Q, S\rangle$, as follows:

$$\begin{array}{lllllll} q^\Lambda & = & \Lambda(q) & \quad & s^\Lambda & = & \Lambda(s); |\Lambda|? \\ (\neg\varphi)^\Lambda & = & \neg\varphi^\Lambda & & \mathsf{E}^\Lambda & = & \mathsf{E}; |\Lambda|? \\ (\varphi \wedge \psi)^\Lambda & = & (\varphi^\Lambda \wedge \psi^\Lambda) & & (\varphi?)^\Lambda & = & (\varphi^\Lambda)? \\ (\langle\pi\rangle\varphi)^\Lambda & = & \langle\pi^\Lambda\rangle\varphi^\Lambda & & (\pi_1; \pi_2)^\Lambda & = & \pi_1^\Lambda; \pi_2^\Lambda \\ & & & & (\pi_1 \cup \pi_2)^\Lambda & = & \pi_1^\Lambda \cup \pi_2^\Lambda \\ & & & & (\pi^*)^\Lambda & = & (\pi^\Lambda)^* \end{array}$$

Most of the clauses in this definition are fairly obviously what is required. Note, however, the role of formula $|\Lambda|$, which acts as a restriction on the quantifier $\langle\pi\rangle$ in s^Λ, as can be seen by expanding the semantic definition of $\langle s^\Lambda\rangle\varphi$, since

$$M, u \models \langle s^\Lambda\rangle\varphi \quad \text{iff} \quad \exists v : M, v \models |\Lambda|, u[\![s^\Lambda]\!]^M v \,\&\, M, v \models \varphi.$$

As remarked above, the definition is designed precisely so that the following result holds:

Lemma 1.1 *For each state u of ΛM and v of M, and for each formula $\varphi \in L(Q, S)$,*

$$\begin{array}{lll} M, u \models \varphi^\Lambda & \textit{iff} & \Lambda M, u \models \varphi, \textit{ and} \\ u[\![\pi^\Lambda]\!]^M v & \textit{iff} & v \in \Lambda W \textit{ and } u[\![\pi]\!]^{\Lambda M} v. \end{array}$$

[5] We refer to 'algorithms' here in an informal way, which could be made precise, but doing so would require us to be boringly pedantic about the way the symbols of the signature, for example, are presented, and to choose arbitrarily between many equally good ways of representing these algorithms. Besides, in most cases of interest, the signature $\langle Q, S\rangle$ is finite, and in this case, it is enough merely to list the various components of Λ.

Proof: See Appendix. ☺

Example Let $P = \{p_1, p_2\}$, $R = \{r_1, r_2\}$, $Q = \{q_1, q_2\}$, and $S = \{s\}$. Let Λ be the transformation from $\langle P, R\rangle$ to $\langle Q, S\rangle$ given by

$$\begin{array}{l} |\Lambda| = \langle r_1\rangle p_1 \vee \langle r_2\rangle p_2 \\ \Lambda(q_1) = \langle r_2\rangle\neg p_1 \quad \text{and} \quad \Lambda(q_2) = \langle p_2?; r_1\rangle\neg p_2 \\ \Lambda(s) = (r_1; r_2) \cup (p_1?; r_2) \end{array}$$

Then with model M as shown below, we get ΛM as follows:

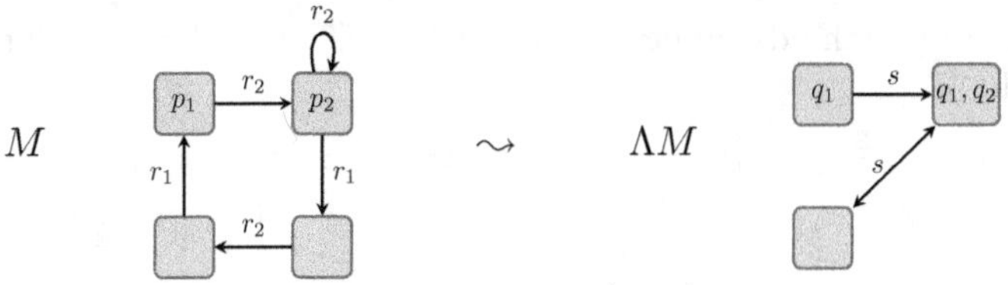

As a simple example of Lemma 1.1 in action, let φ be the formula $\langle s\rangle(q_1 \wedge q_2)$. Then φ^Λ is $\langle((r_1; r_2) \cup (p_1?; r_2)); (\langle r_1\rangle p_1 \vee \langle r_2\rangle p_2)?\rangle(\langle r_2\rangle\neg p_1 \wedge \langle p_2?; r_1\rangle\neg p_2)$. A bit of checking will confirm that $[\![\varphi]\!]^{\Lambda M}$ and $[\![\varphi^\Lambda]\!]^M$ are both equal to the set of states depicted in the left columns of these diagrams.

In what follows we will use a concise notation for PDL transformations, which we illustrate by rewriting Λ, from the above example, as

$$\langle|\langle r_1\rangle p_1 \vee \langle r_2\rangle p_2|, q_1 := \langle r_2\rangle\neg p_1, q_2 := \langle p_2?; r_1\rangle\neg p_2, s := (r_1; r_2) \cup (p_1?; r_2)\rangle$$

To simplify notation further, we will omit trivial domain restriction ($\top$) and those parts of a PDL-transformation that do not change anything. For example, the action of 'public announcement of φ' is just $\langle|\varphi|\rangle$. Finally, the identity transformation is written as I. Some PDL-transformations found in the literature [12,18,20,26] are listed below:

Add $\neg p \to p$ links	=	$\langle r := (r \cup (\neg p?; \mathsf{E}; p?))^*\rangle$ (* used to preserve transitivity)
Delete $p \to \neg p$ links ('suggestion')	=	$\langle r := ((\neg p?; r; \neg p?) \cup (p?; r; p?) \cup (\neg p?; r; p?)\rangle$
Delete $p \to \neg p$ links Add $\neg p \to p$ links ('radical upgrade')	=	$\langle r := ((\neg p?; r; \neg p?) \cup (p?; r; p?) \cup (\neg p?; \mathsf{E}; p?)\rangle$

2 General Dynamic Dynamics

Given a signature $\langle P, R\rangle$, we will define a class of dynamic operators to add to PDL to produce our dynamic dynamic logic, GDDL. Just as with the 'action structures' of BMS, we think of these operators as syntactic objects, albeit somewhat complex ones. A GDDL *dynamic operator* $[A, G, H, a]$ consists of four components:

(i) a finite structure $A = \langle D, U \rangle$ of some finite signature $\langle Q, S \rangle$ (distinct from $\langle P, R \rangle$), whose nodes $d \in D$ are called *program nodes*,

(ii) a PDL-transformation G_d from $\langle P, R \rangle$ to $\langle P, R \rangle$ for each $d \in D$,

(iii) a PDL-transformation H from $\langle P \cup Q, R \cup S \rangle$ to $\langle P, R \rangle$, and

(iv) a distinguished element $a \in D$.

We can represent the dynamic operator $[A, G, H, a]$ in the following diagrammatic style:

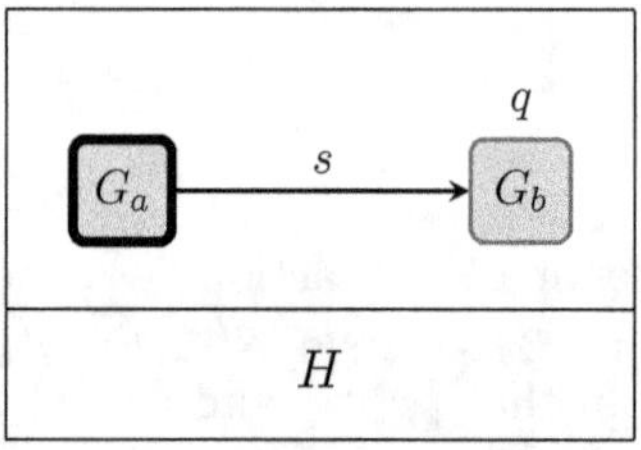

The upper section contains a representation of the structure A, with each node containing its associated PDL-transformation. Relations $U(s)$ are indicated by arrows labelled with 's', a symbol of S, and $U(q)$ is shown by labelling nodes with 'q' from Q. (The above labels and arrows are merely illustrative.) The distinguished node a is highlighted with darker edges. The lower section contains the PDL-transformation H.

The language of GDDL is given by:

$$\begin{aligned} \pi &::= r \mid \mathsf{E} \mid \varphi? \mid (\pi;\pi) \mid (\pi \cup \pi) \mid \pi^* \\ \varphi &::= p \mid \neg\varphi \mid (\varphi \vee \varphi) \mid \langle\pi\rangle\varphi \mid [A, G, H, a]\varphi \end{aligned}$$

where, again, $r \in R$, $p \in P$, and $[A, G, H, a]$ is a GDDL dynamic operator. Let $T^+(P, R)$ and $L^+(P, R)$ be the set of GDDL terms and formulas so defined. Notice, in particular, how the two senses in which the language is 'dynamic' are captured by $\langle\pi\rangle$ and $[A, G, H, a]$. $L(P, R)$ is already dynamic in the first sense but not in the second.

We think of each element d of D as representing a possible action whose effect on M is to transform it to $G_d M$. This could be an announcement, a belief or preference change, or something far more complex, depending on the application. The particular element a is the one that actually occurs. The only restriction is that the transformation is definable by PDL expressions.[6]

[6] In BMS elements of action structures are associated with formulas, called 'pre-conditions' which act to restrict the domain but which have no effect on the relational structure of the

We represent the interaction between $A = \langle D, U\rangle$ and $M = \langle W, V\rangle$ by constructing the model $GM = \langle GW, GV\rangle$ of combined signature $\langle P \cup Q, R \cup S\rangle$, as follows:

$$GW = \{\langle u, d\rangle \mid u \in [\![|G_d|]\!]^M\}$$

Then, for $\langle u, d\rangle$ and $\langle v, e\rangle$ in GW:

$$\begin{array}{lll} \langle u,d\rangle \in GV(p) & \text{iff } u \in [\![p]\!]^{G_d M} & \text{for each } p \in P \\ \langle u,d\rangle \in GV(q) & \text{iff } d \in U(q) & \text{for each } q \in Q \\ \langle u,d\rangle\, GV(r)\, \langle v,e\rangle & \text{iff } d = e \text{ and } u\, [\![r]\!]^{G_d M}\, v & \text{for each } r \in R \\ \langle u,d\rangle\, GV(s)\, \langle v,e\rangle & \text{iff } u = v \text{ and } d\, U(s)\, e & \text{for each } s \in S \end{array}$$

We can think of GM as resulting from the process of replacing each program node $d \in D$ by the transformed model $G_d M$ that results from applying the PDL-transformation G_d to M. The structure of A remains, linking these transformed models together. [7]

Finally, we use the transformation H to recover a model of signature $\langle P, R\rangle$ from GM, defining

$$[A, G, H]M \quad = \quad HGM$$

H encodes the way in which the structure of A coordinates the different programs represented by the elements of D. Again, there is great generality here. All that is required is that this means of coordination is, in some sense, PDL-definable. [8] Then, we can specify the semantics for our new dynamic operators in a standard way: [9]

$$M, u \models [A, G, H, a]\varphi \quad \text{iff} \quad [A, G, H]M, \langle u, a\rangle \models \varphi$$

Because of the generality of the approach, it is useful to consider the special case in which a GDDL-operator $[A, G, H, a]$ is defined by a single PDL-transformation Λ, defining

model. See Section 3 for details.

[7] Another useful metaphor for visualising GM is that it is a two-dimensional model in which the S links run in a horizontal direction and the R links run in a vertical direction. Whereas the S links are merely copies of their projection on to D, the R links vary. In the dth place in the horizontal direction the vertical R links form a copy of those in $G_d M$.

[8] In BMS, the coordination is built into the details of the model construction, not a parameter of the dynamic operators. Also, the signature used for A can be taken to be the same as that for M, since both are simply families of equivalence relations. Again, see Section 3 for details.

[9] Although a is not relevant to computing $[A, G, H]M$, we define $[A, G, H, a]M = [A, G, H]M$, for uniformity of notation when we consider arbitrary operators.

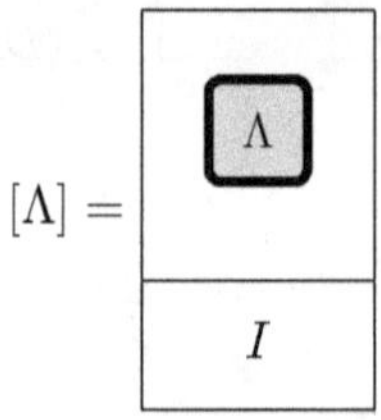

Example 2: For a purely abstract example, illustrating various aspects of the definition, take Λ to be the PDL-transformation

$$\langle |\langle r_1\rangle p_1 \vee \langle r_2\rangle p_2|, p_1 := \langle r_2\rangle\neg p_1, p_2 := \langle p_2?; r_1\rangle\neg p_2, r_2 := (r_1; r_2) \cup (p_1?; r_2)\rangle$$

Take the model M of example 1 (shown left) and the dynamic operator $[A, G, H, a]$ (shown right):

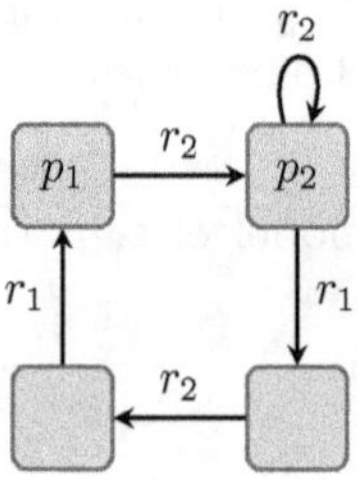

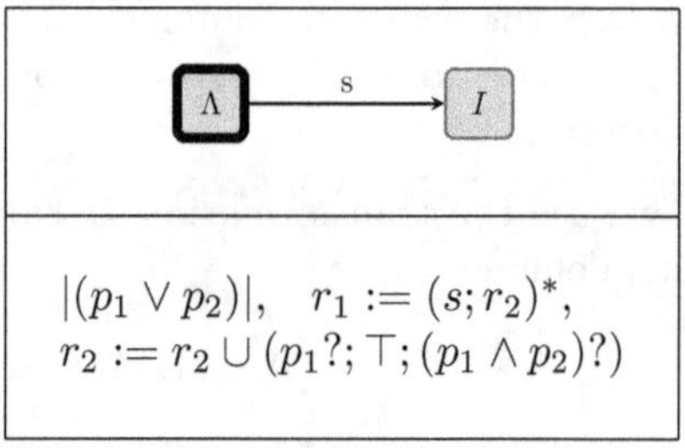

We first compute GM, then HGM:

GM

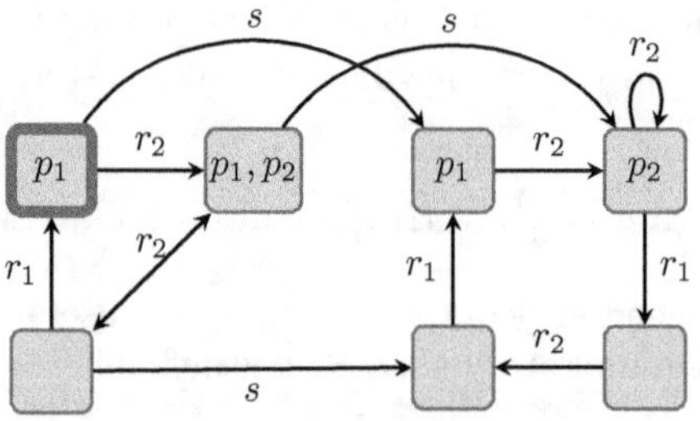

HGM

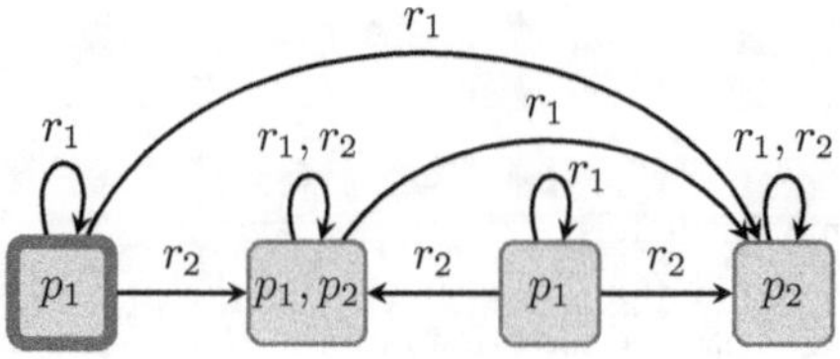

2.1 Private Belief Change

As an example of what can be done with GDDL, and a concrete illustration of the definitions in action, we will consider an application to doxastic logic. For the basic logic of belief change, we follow van Benthem's [18] account, in which the effect of a belief update is to change an agent's judgements about the relative plausibility ($\leq$) of different epistemic possibilities. The basic idea is that when updating with p, an agent should judge any p-state to be strictly more plausible than any $\neg p$-state, while retaining her earlier judgements about relative plausibility among p-states and among $\neg p$-states. Belief is defined as truth in maximally plausible states. We extend this to a multi-agent context in which belief changes are private. There are many options here, but we choose to add a separate relation ($\sim$) modelling knowledge (epistemic indistinguishability). In this example, we will assume that it is an equivalence relation, as is standard, but it should be clear that many variations are possible within the present framework.

Given a finite sets of agents I, consider a Kripke signature $\langle P, R\rangle$ with $R = \{\sim_i, \leq_i, \beta_i \mid i \in I\}$, the class $\mathcal{D}$ of models $M = \langle W, V\rangle$ of this signature for which, for each agent i, $V(\sim_i)$ is a equivalence relation, $V(\leq_i) \subseteq V(\sim_i)$ is a preorder, and

$$u\beta_i v \text{ iff } u \sim_i v \text{ and } w \leq_i v \text{ for each } w \in W \text{ such that } v \leq_i w.$$

In other words, a state v is β_i-accessible from u iff it is maximally plausible among those states that are epistemically indistinguishable from u. In this way, the formula $[\beta_i]\varphi$ states that φ is believed by i, according to the analysis of [18].[10] Updating agent i's beliefs with φ results in a new plausibility relation defined by[11] $(\varphi?; \leq_i; \varphi?) \cup (\neg\varphi?; \leq_i; \neg\varphi?) \cup (\neg\varphi?; \sim_i; \varphi?)$.

The PDL-transformation $\Uparrow^i \varphi$ ('update i's beliefs with φ') maps $\leq_i$ to this PDL term, keeping everything else the same. The fact that agent i believes ψ after upgrading her beliefs with φ should then expressed in GDDL as $[\Uparrow^i \varphi][\beta_i]\psi$.

But there is a problem. Not only is $[\Uparrow^i p][\beta_i]p$ valid (as expected), but so too is $[\Uparrow^i p][\beta_j][\beta_i]p$, for any other agent j. In other words, it is logically true that after i doxastically updates with p, not only does she believe that p but everyone else believes that she believes p. In this way, $[\Uparrow^i \varphi]$ is not really *private* at all. This situation is familiar from BMS, and we adopt a similar solution, but one that exploits the relation-change potential of PDL-transformations, defining the operator $(\uparrow^i \varphi) = [A, G, H, a]$ as follows:

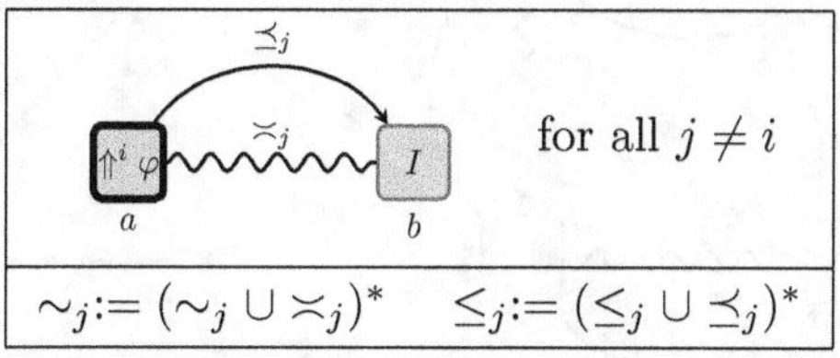

[10] Note that although β_i is in some sense redundant, in that it is fully determined by $\sim_i$ and $\leq_i$, the operator $[\beta_i]$ is not modally definable from the operators $[\sim_i]$ and $[\leq_i]$; nonetheless, the relationship between the three operators can easily be acclimatised.

[11] This differs slightly from [18] because of the restriction to the $\sim_i$-equivalence class discussed above.

Here, A is a two-state structure with domain $D = \{a, b\}$. It has signature $\langle Q, S\rangle$, where Q is empty and $S = \{\asymp_j, \preceq_j \mid j \in I\}$, and interprets $\asymp_j$ as an equivalence relation and $\preceq_j$ as a preorder, although not all of the links are shown in the diagram. The distinguished node a contains the PDL-transformation $\Uparrow^i \varphi$, representing the action of updating i's belief's with φ, and the other node, b, contains the identity transformation I, representing the action of doing nothing. Only agent i knows which of the two possible actions were performed, so $a \asymp_j b$ for all $j \neq i$. Likewise, we will assume (although there is room for more subtlety here) that each of these other agents regards it as more plausible that i's beliefs have not changed. This is captured by making $a \preceq_j b$ and not $b \preceq_j a$ for all those j. Finally, the integrating transformation H is defined by taking the (reflexive, transitive closure of) the unions of the two epistemic and the two doxastic relations.

To see how this works, we will consider an application of $(\uparrow^i \varphi)$ to the model M displayed below:[12]

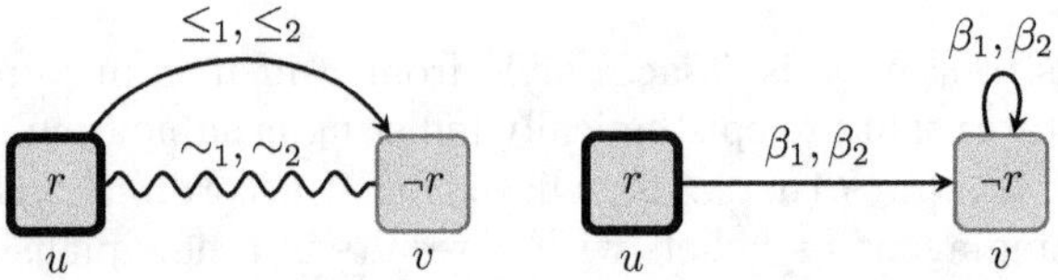

We will think of M as representing a scenario with two agents, called 1 and 2, who both falsely believe $\neg r$: 'it hasn't rained today'. Now, the operator $\uparrow^1 r$ should represent the action of agent 1 privately upgrading her belief in r, in a way that is unobserved by agent 2; perhaps she takes a furtive glance out of the window and sees someone closing an umbrella. The resulting model $(\uparrow^1 r)$ is shown below (with β_i displayed separately):

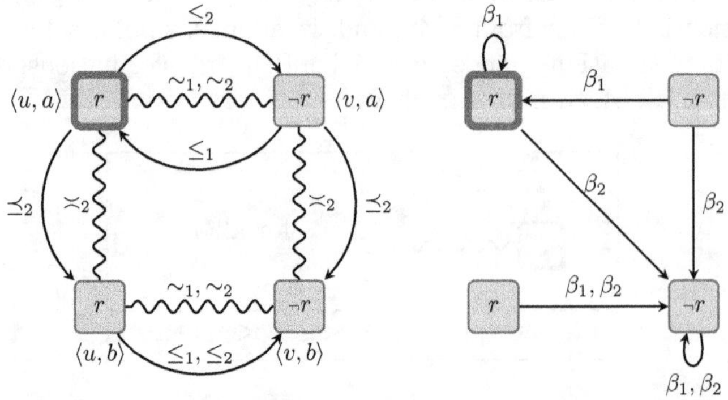

[12] Throughout this section reflexive loops and compositions of arrows for $\sim$ and $\leq$ are suppressed in diagrams, and the relations β_1 and β_2, which are defined in terms of the others, are shown separately.

Here, the designated state satisfies $[\beta_1]r$ (agent 1 believes it to be raining), $[\beta_2]\neg r$ (agent 2 still believes it is not raining) and $[\beta_2][\beta_1]\neg r$ (agent 2 also believes, falsely, that agent 1 still believes it not to be raining).[13]

2.2 Axiomatisation

The key to understanding the logic of GDDL is to find a computable translation $\varphi^{[A,G,H,a]}$ of each formula φ in $L(P,R)$ such that

$$[A,G,H,a]\varphi \leftrightarrow \varphi^{[A,G,H,a]}$$

is valid as a formula of GDDL. From this (and the replacement of logical equivalents) it follows that every formula of $L^+(P,R)$ is equivalent to a formula of $L(P,R)$, and can be proved to be so using these equivalences as axioms. We can reduce the two senses of 'dynamic' in dynamic dynamic logic to one.

How, then, to define $\varphi^{[A,G,H,a]}$? Our approach will be to define a formula $\varphi^{[A,G,d]}$ of $L(P,R)$ for each φ of $L(P \cup Q, R \cup S)$ and a program $\pi^{[A,G,d,e]}$ of $T(P,R)$ for each π of $T(P \cup Q, R \cup S)$ such that for any model M of signature $\langle P,R\rangle$, the following result holds:

Lemma 2.1 *For each $\langle u,d\rangle, \langle v,e\rangle \in GW$,*

(i) $GM, \langle u,d\rangle \models \varphi$ *iff* $M, u \models \varphi^{[A,G,d]}$, *and*
(ii) $\langle u,d\rangle [\![\pi]\!]^{GM} \langle v,e\rangle$ *iff* $u[\![\pi^{[A,G,d,e]}]\!]^M v$

This will be proved below. We can then define $\varphi^{[A,G,H,d]} = \varphi^{H[A,G,d]}$ so that

Lemma 2.2 *For each operator $[A,G,H,a]$ of* GDDL *and each formula φ of $L(P,R)$, the following is valid:*

$$[A,G,H,a]\varphi \leftrightarrow \varphi^{[A,G,H,a]}$$

Proof: From Lemmas 2.1 and 1.1. (Note that, since GM is of signature $\langle P \cup Q, R \cup S\rangle$ and $[A,G,H,d]M = HGM$ is of signature $\langle P,R\rangle$, the formula φ^H is in $L(P \cup Q, R \cup S)$, as required by Lemma 2.1.) ☺

To define $\pi^{[A,G,d,e]}$ we need a small excursion into automata theory.[14] For

[13] Although the definition of a private belief update operator is only an example to show what can be done in GDDL, it illustrates the need for some constraints in getting sensible results for epistemic logic. In particular, it is important that the epistemic relation $\asymp_i$ in the operator constrains the definition of the various transformations G_d so that each agent knows that actions affecting her own psychological state have occurred, namely, that $d \asymp_i e$ implies both $G_d(\asymp_i) = G_e(\asymp_i)$ and $G_d(\leq_i) = G_e(\leq_i)$.

[14] This excursion into automata theory is solely for the purpose of producing the reduction axioms, in a recursive way. Once the axioms are produced, however, no essential use of automata remains in the logic. We find the technique useful and illuminating but recognise that there may be an alternative approach that provides reduction axioms in a more direct way. We leave this as an open problem.

each signature, say that σ is a *basic program* of that signature if it is either a relation symbol, E, or a test. Then for each model N and a program π, define $\Sigma(N, u, v, \pi)$ to be the set of strings $\sigma_1 \dots \sigma_n$ of basic subprograms of π such that $u_i [\![\sigma_i]\!]^N u_{i+1}$ for some sequence $u_0, \dots, u_n$ of states in N with $u_0 = u$ and $u_n = v$. Say that a finite state automaton A over an alphabet of basic subprograms of a program π *represents* π iff for all models N and states u and v,

$$u[\![\pi]\!]^N v \text{ iff some word of } \Sigma(N, u, v, \pi) \text{ is accepted by } A.$$

It has been well known since [16] that every PDL program is represented by some automaton and every automaton represents some PDL program.[15] Moreover, each can be computed from the other.[16] Now, given φ in $L(P \cup Q, R \cup S)$ and π in $T(P \cup Q, R \cup S)$ and states $d, e \in D$, we will define $\varphi^{[A,G,d]}$ and $\pi^{[A,G,d,e]}$ by mutual induction.

The definition of $\varphi^{[A,G,d]}$ is straightforwardly inductive:

$$\begin{array}{ll} p^{[A,G,d]} & = G_d(p) \\ q^{[A,G,d]} & = \begin{cases} \top \text{ if } d \in U(q) \\ \bot \text{ otherwise} \end{cases} \\ (\neg\varphi)^{[A,G,d]} & = \neg\varphi^{[A,G,d]} \\ (\varphi \wedge \psi)^{[A,G,d]} & = (\varphi^{[A,G,d]} \wedge \psi^{[A,G,d]}) \\ (\langle\pi\rangle\varphi)^{[A,G,d]} & = \bigvee_{e \in D} \langle \pi^{[A,G,d,e]} \rangle (|G_e| \wedge \varphi^{[A,G,e]}) \end{array}$$

The program $\pi^{[A,G,d,e]}$ is obtained by constructing a corresponding automaton, which will refer to $\psi^{[A,G,d]}$ for subprograms $\psi?$ of π, which is inductively legitimate. This will take the next couple of paragraphs.

First, consider an automaton A_π which represents π. Let A_π have states X, of which $X_0 \subseteq X$ are initial states, $X_1 \subseteq X$ are accepting states, and for each σ (a basic subprogram of π), $T(\sigma) \subseteq X^2$ is such that there is a transition from x_1 to x_2 labelled by σ iff $\langle x_1, x_2 \rangle \in T(\sigma)$.

Now for each symbol σ in the alphabet of A_π (a basic subprogram of π) and each $c_1, c_2 \in D$, define σ^{c_1,c_2} as follows:

$$\sigma^{c_1,c_2} = \begin{cases} \psi^{[A,G,c]}? & \text{if } \sigma = \psi? \text{ and } c_1 = c_2 = c \\ G_c(\sigma) & \text{if } \sigma \in R \text{ and } c_1 = c_2 = c \\ |G_{c_1}|?; \mathsf{E}; |G_{c_2}|? & \text{if } \sigma = \mathsf{E} \\ \top? & \text{if } \sigma \in S \text{ and } \langle c_1, c_2 \rangle \in U(\sigma) \\ \bot? & \text{otherwise} \end{cases}$$

[15] The representation depends only on the compositional structure of the program not on the particular choice of basic programs, so additions to PDL such as tests and E are not a problem.

[16] The complexity of translating between the two representations has been investigated in [8].

Construct a new automaton $B_\pi^{d,e}$, whose alphabet consists of the basic programs σ^{c_1,c_2} where σ is in the alphabet of A_π, with states $X' = X \times D$, initial states $X_0' = X_0 \times \{d\}$, accepting states $X_1' = X_1 \times \{e\}$, and transition function T' defined by

$$T'(\tau) \quad = \quad \{\langle\langle x_1, c_1\rangle, \langle x_2, c_2\rangle\rangle \mid \text{for some } \sigma,\ \langle x_1, x_2\rangle \in T(\sigma) \text{ and } \sigma^{c_1,c_2} = \tau\}$$

Now, let $\pi^{[A,G,d,e]}$ be the program of $T(P,R)$ represented by $B_\pi^{d,e}$.

The two automata are designed to be synchronised in the sense given by the following technical lemma:

Lemma 2.3 *Assume that for each test $\psi?$ occurring in π, and each $\langle w, c\rangle \in GW$, $GM, \langle w, c\rangle \models \psi$ iff $M, w \models \psi^{[A,G,c]}$. Then, given $x_1, x_2 \in X$ and $\langle u, c_1\rangle, \langle v, c_2\rangle \in GW$, consider the following properties of the labels of the automata A_π and $B_\pi^{d,e}$:*

$$\gamma(\sigma): \quad \langle x_1, x_2\rangle \in T(\sigma) \text{ and } \langle u, c_1\rangle [\![\sigma]\!]^{GM} \langle v, c_2\rangle \qquad \gamma'(\tau): \ \langle\langle x_1, c_1\rangle, \langle x_2, c_2\rangle\rangle \in T'(\tau) \text{ and } u[\![\tau]\!]^M v$$

Then for each symbol τ of the alphabet of $B_\pi^{d,e}$,

$$\gamma'(\tau) \textit{ iff } \gamma(\sigma) \textit{ and } \sigma^{c_1,c_2} = \tau \textit{ for some } \sigma \textit{ in the alphabet of } A_\pi$$

Proof: See Appendix. ☺

We are now ready to prove Lemma 2.1.

Proof of Lemma 2.1: The *rank* of a formula or program is defined as follows. Formulas of rank n are not of rank $n-1$ but contain no programs of rank n, and programs of rank n not of rank $n-1$ but contain no test formulas of rank n. (In particular, formulas of rank 0 contain no programs and programs of rank 0 contain no test formulas; formulas of rank 1 contain at least one program of rank 0 but none of rank 1 and programs of rank 1 contain at least one test formula of rank 0 but none of rank 1, etc.) To prove the lemma we show, by induction on the rank n of a formula φ of $L(P \cup Q, R \cup S)$, that for $\langle u, d\rangle, \langle v, e\rangle \in W'$,

(i) $GM, u, d \models \varphi$ iff $M, u \models \varphi^{[A,G,d]}$, and

(ii) for any program π of rank $\leq n$, $\langle u, d\rangle [\![\pi]\!]^{GM} \langle v, e\rangle$ iff $u [\![\pi^{[A,G,d,e]}]\!]^M v$

We prove part 1 by induction on the structure of φ.

For propositional variables:
$GM, u, d \models p$ iff $u \in V_d(p)$ iff $u \in [\![G_d(p)]\!]^M$ iff $M, u \models G_d(p)$ iff $M, u \models p^{[A,G,d]}$
$GM, u, d \models q$ iff $d \in U(q)$ iff $q^{[A,G,d]} = \top$ iff $M, u \models q^{[A,G,d]}$ (given that $q^{[A,G,d]} \in \{\top, \bot\}$)

For Booleans ($\neg$ and $\wedge$), the proof is straightforwardly inductive.

For formulas of the form $\langle\pi\rangle\psi$, note that π must be of rank $< n$ and so for each $\langle w, f\rangle \in W'$

$$\langle u, d\rangle [\![\pi]\!]^{GM} \langle w, f\rangle \text{ iff } u[\![\pi^{[A,G,d,f]}]\!]^M v$$

and by the (inner, structural) inductive hypothesis,

$$GM, w, f \models \psi \text{ iff } M, w, \models \psi^{[A,G,f]}$$

But then the following are equivalent:

$GM, u, d \models \langle\pi\rangle\psi$
$\langle u, d\rangle [\![\pi]\!]^{GM} \langle w, f\rangle$ and $GM, w, f \models \psi$ for some $\langle w, f\rangle \in W'$
$u[\![\pi^{[A,G,d,f]}]\!]^M w$ and $M, w, \models \psi^{[A,G,f]}$ for some $\langle w, f\rangle \in W'$
$M, w \models |G_f|$, $u[\![\pi^{[A,G,d,f]}]\!]^M w$ and $M, w, \models \psi^{[A,G,f]}$ for some $f \in D, w \in W$
$M, u \models \bigvee_{f\in D}\langle\pi^{[A,G,d,f]}\rangle(|G_f| \wedge \psi^{[A,G,f]})$
$M, u \models \langle\pi\rangle\psi^{[A,G,d]}$

For part 2, we know that any test formula in π is of rank $< n$. So suppose $\langle u, d\rangle [\![\pi]\!]^{GM} \langle v, e\rangle$. Then by choice of A_π we have that some word $\sigma_1 \ldots \sigma_n$ in $\Sigma(GM, \langle u, d\rangle, \langle v, e\rangle, \pi)$ is accepted by A_π. This implies that

(i) there are $u_0, \ldots, u_n$ and $c_0, \ldots, c_n \in D$ such that $u_0 = u$, $c_0 = d$, $u_n = v$ $c_n = e$ and $\langle u_i, c_i\rangle [\![\sigma_i]\!]^{GM} \langle u_{i+1}, c_{i+1}\rangle$ for $0 \le i < n$, and

(ii) there are $x_0, \ldots, x_n \in X$ and such that $x_0 \in X_0$, $x_n \in X_1$, and $\langle x_i, x_{i+1}\rangle \in T(\sigma_i)$ for $0 \le i < n$.

Now for each i we have $\gamma_i(\sigma_i)$:

$$\langle x_i, x_{i+1}\rangle \in T(\sigma_i) \quad \text{and} \quad \langle u_i, c_i\rangle [\![\sigma_i]\!]^{GM} \langle u_{i+1}, c_{i+1}\rangle$$

So, by Lemma 2.3, for $\tau_i = \sigma_i^{c_i, c_{i+1}}$, we have $\gamma_i'(\tau_i)$:

$$\langle\langle x_i, c_i\rangle, \langle x_{i+1}, c_{i+1}\rangle\rangle \in T'(\tau_i) \quad \text{and} \quad u_i[\![\tau_i]\!]^M u_{i+1}$$

Also, since $c_0 = d$, $c_n = e$, $x_0 \in X_0$, $x_n \in X_1$, we have that $\langle x_0, c_0\rangle \in X_0'$ and $\langle x_n, c_n\rangle \in X_1'$. Thus:

(i) there are $u_0, \ldots, u_n$ such that $u_0 = u$, $u_n = v$ and $u_i[\![\tau_i]\!]^M u_{i+1}$ for $0 \le i < n$, and

(ii) there are $x_0, \ldots, x_n \in X$ and $c_0, \ldots, c_n \in D$ such that $\langle x_0, c_0\rangle \in X_0$, $\langle x_n, c_n\rangle \in X_1$, and $\langle\langle x_i, c_i\rangle, \langle x_{i+1}, c_{i+1}\rangle\rangle \in T'(\tau_i)$ for $0 \le i < n$.

This is precisely what is required for $\tau_1 \ldots \tau_n$ to be accepted by $B_\pi^{d,e}$. Then by definition of $\pi^{[A,G,d,e]}$, we have that $u[\![\pi^{[A,G,d,e]}]\!]^M v$, as required. The converse is proved similarly. ☺

Theorem 2.4 *The logic of* GDDL *is completely axiomatised by the axioms and rules of* PDL *(see Definition 4.78 in [5]) and the schema*

$$\vdash [A, G, H, a]\varphi \leftrightarrow \varphi^{[A,G,H,a]}$$

Corollary 2.5 GDDL *is decidable.*

Proof: We have a computable reduction of GDDL to its PDL fragment, which is itself decidable. ☺

3 Applications

In the remainder of the paper, we show how two well-known systems for dynamic epistemic logic (BMS and LCC) are special cases of GDDL and make further connections to more recent developments.

3.1 BMS

For BMS [3], we will be working with a signature $\langle P, R\rangle$ for which P is a (countably infinite) set of propositional variables and $R = \{K_i \mid i \in I\}$ is a set of epistemic relations, one for each agent $i \in I$, with I finite. The BMS system is not dynamic in the first (PDL) sense and so we will refer to basic modal language (in which the only terms are the atoms K_i) as $L^-(P, R)$. A model $M = \langle W, V\rangle$ of this signature is an *epistemic* model iff all the relations $V(K_i)$ are equivalence relations. So far, this is all just standard epistemic logic. The innovation was to define an *action* structure to be a structure of the form $\langle D, U, \mathsf{pre}\rangle$ for which $\langle D, U\rangle$ is an epistemic structure and $\mathsf{pre}\colon D \to L^-(P, R)$ assigns a formula to each element of D that expresses the 'precondition' of performing the action it represents. For example, if d represents the announcement of φ, then it is usually assumed that, as a precondition, the announcement must be true, and so $\mathsf{pre}(d) = \varphi$. Action structures are finite and so can be added to the syntax as dynamic operators. The full language of BMS is thus

$$\varphi \;::=\; p \mid \neg\varphi \mid (\varphi \vee \varphi) \mid \langle K_i\rangle\varphi \mid [D, U, \mathsf{pre}, a]\varphi$$

where $\langle D, U, \mathsf{pre}\rangle$ is an action structure and $a \in D$ is the designated action. Given action structure $\Delta = \langle D, U, \mathsf{pre}\rangle$ and epistemic model $M = \langle W, V\rangle$, the *product* model ΔM is defined to be $\langle \Delta W, \Delta V\rangle$, where

$$\begin{aligned}
\Delta W &= \{\langle u, d\rangle \mid M, u \models \mathsf{pre}(d)\}\\
\Delta V(p) &= \{\langle u, d\rangle \in \Delta W \mid u \in V(p)\} \text{ for each } p \in P\\
\Delta V(K_i) &= \{\langle\langle u, d\rangle, \langle v, e\rangle\rangle \in (\Delta W)^2 \mid \langle u, v\rangle \in V(K_i) \text{ and } \langle d, e\rangle \in U(K_i)\}
\end{aligned}$$

Finally, the semantics of the BMS operator $[D, U, \mathsf{pre}, a]$ is given by

$$M, u \models [D, U, \mathsf{pre}, a]\varphi \quad \text{iff} \quad \langle D, U, \mathsf{pre}\rangle M, \langle u, a\rangle \models \varphi$$

This can be seen as a special case of our general construction. First, we take a copy K_i' of each symbol K_i, because we need to keep the signature of the epistemic model distinct from that of the action structure. Then we define the structure $A = \langle D, U' \rangle$ of signature $\langle Q, S \rangle$, with $Q = \emptyset$ and $S = \{K_i' \mid i \in I\}$, such that $U'(K_i') = U(K_i)$, for each $i \in I$. For each $d \in D$ we define the transformation G_d by $\langle |G_d| = \mathsf{pre}(d) \rangle$. This captures the idea of a precondition. Finally, we take H to be the PDL-transformation given by $\langle K_i := K_i; K_i' \rangle$. To show that $[D, U, \mathsf{pre}, a]\varphi$ is logically equivalent to $[A, G, H, a]\varphi$, the following theorem is sufficient.

Theorem 3.1 *With $\Delta = \langle D, U, \mathsf{pre} \rangle$ and A, G and H defined as above,*

$$\Delta M = [A, G, H]M$$

Proof: See Appendix. ☺

It follows that every function on the class of models of a given signature that is definable by a BMS operator is also definable by a GDDL operator.

3.2 LCC

LCC [21], the Logic of Communication and Change, extends BMS in two ways: by expanding the base language to include PDL modalities, and by introducing 'real-world' change. The first extension is relatively straightforward. It just amounts to moving from $L^-(P, R)$ to $L(P, R)$, and the argument that the resulting system is a fragment of GDDL goes through as above. The second extension, to model 'real-world' change, is achieved using 'propositional substitutions', which are functions $\sigma : P \to L(P, R)$ with a finite base, meaning that σ is the identity function on all but a finite number of propositional variables. An action that changes something other than just the psychological states of agents can thus be represented by a propositional substitution σ such that, after the change, p is true of state u iff $\sigma(p)$ were true of it before the change.[17]

A LCC *action structure* $\Delta = \langle D, U, \mathsf{pre}, \mathsf{sub} \rangle$ consists of a BMS-like[18] action structure $\langle D, U, \mathsf{pre} \rangle$ and a propositional substitution function sub_d for each $d \in D$. Given an epistemic model $M = \langle W, V \rangle$, the LCC *product* model ΔM is defined to be $\langle \Delta W, \Delta V \rangle$ as for BMS, except that

$$\Delta V(p) \quad = \{\langle u, d \rangle \in \Delta W \mid u \in V(\mathsf{sub}_d(p))\} \text{ for each } p \in P$$

To extend our earlier representation of BMS operators in GDDL requires only one small change: the PDL-transformation G_d is now defined by $\langle \mathsf{pre}(d), p :=$

---

[17] The restriction to σ of finite base requires the changes to be, in some sense, local. However, the embedding of LCC in GDDL shows that what is important here is only that there is some finite representation of σ, on the basis of which σ can be recovered algorithmically.

[18] The only difference is that $\mathsf{pre}(d)$ is not restricted to $L^-(P, R)$; it may be any formula of $L(P, R)$.

$\mathsf{sub}_d(p)\rangle$. With A and H defined as for BMS, we have the required result:

Theorem 3.2 *With* $\Delta = \langle D, U, \mathsf{pre}, \mathsf{sub}\rangle$ *and* A*,* G *and* H *defined as above,*

$$\Delta M = [A, G, H]M$$

Proof: See Appendix. ☺

It follows that every function on the class of models of a given signature that is definable by a BMS operator is also definable by a GDDL operator.

3.3 Other approaches

LCC was extended in [24] by basing it on PDL with relational converse (which is still decidable). While GDDL does not include relational converse in the base language, we can think of no reason why it could not be added, in a similar manner to our addition of E over the vanilla PDL. We conjecture that other decidable extensions of PDL could be taken as the base logic. The combination of converse with nominals (which is decidable from [14,1], p.12, Theorem 3.5) should present no further difficulty. Another particularly useful decidable addition would be atomic negation on programs (see [25]) on the basis of which E and the 'window' operators are definable.

There have been several attempts to extend dynamic logic to cope with relation-changing dynamic operators. Firstly, [22] shows how, for given finite epistemic models M_1 and M_2, it is possible to define a BMS operator $[D, U, \mathsf{pre}, a]\varphi$ of BMS such that $[D, U, \mathsf{pre}, a]\varphi M_1 = M_2$ but this is far from what is needed to represent relation-changing operators that have a uniform action on all models.

Next, [2] presents an interesting extension of modal logic with 'graph-modification' operators for models of arbitrary signature (as with GDDL). There are three kinds of operator: 'label modifications' $p + \varphi$ and $p - \varphi$, which increase (and respectively decrease) $V(p)$ by $\{u \mid M, u \models \varphi\}$; 'edge label modification' operators $r + (\varphi, \psi)$ and $r - (\varphi, \psi)$, which increase (resp. decrease) $V(r)$ by the set $\{\langle u, v\rangle \mid M, u \models \varphi \text{ and } M, u \models \psi\}$; and 'new state' operators nw and $\overrightarrow{nw}$, which add a new isolated state to the domain (and shifts the point of evaluation to it, in the case of $\overrightarrow{nw}$). Of these, the label modification and edge label modification operators define PDL-transformations and so can be represented in GDDL. The new state operators are more tricky. Instead, in GDDL, we can represent the addition of a set of (isolated) states, either with or without a shift in evaluation point, but we know of no way to ensure that this set is a singleton.[19] Importantly, GDDL extends the approach of [2] by

[19] For the analogue of nw, use the operator $[A, G, H, a]$ in which A has domain $\{a, b\}$ with an empty signature, G_a is the identity, G_b maps all propositional variables to $\bot$ and all relation symbols to $\bot?$, and H is also the identity. Then if $M = \langle W, V\rangle$, $[A, G, H, a]M$ is isomorphic to $\langle W \cup W', V\rangle$ with W' disjoint from W. For $\overrightarrow{nw}$, use $[A, G, H, b]$. The size of W' can be diminished using a domain restriction component to G_b but without formulas that are

including all PDL-transformations and products. [20]

Finally, [9] and [10] propose a system of 'arrow update' operators with a novel syntax in which new transitions for a relation are introduced by specifying their pre- and post-conditions. They show that the resulting system is dynamically equivalent to BMS, and so can be seen as a way of giving an explicit syntax for the part of BMS that allows a limited form of relation-change. GDDL is a genuine extension of both, as can be seen from the example of private belief change in Section 2.1. [21]

4 Conclusion

GDDL achieves our objective of generalising existing approaches to dynamic epistemic logic to allow relational change and opens up a number of possibilities for further work. Firstly, it would be interesting to look at fragments that can be expressed in a more restricted syntax. Even the syntactically promiscuous logics of BMS and LCC exploit only a small part of the generality of GDDL operators, suggesting that other restrictions may be equally interesting in their own right. A proper study of these would require a better understanding of the class of model-transforming functions defined by GDDL. A good place to start with this investigation is the category of PDL-transformations of models of arbitrary signatures.

Secondly, we propose that many applications would profit from the ability to code appropriate GDDL operators. For example, the study of the relationship between first and higher-order psychological attitudes requires the interplay between levels available in GDDL. This arises in preference logic when trying to account for weakness of will, one analysis of which is the preference for having different preferences: I may prefer an action in which my preference for smoking is downgraded to one in which it is not. Moving from the personal setting, similar level-distinctions occur when reflecting on normative systems. Certain changes to the law, for example, may be regarded as permissible, while others are not. And with a multi-agent perspective, one could try to devise GDDL operators to model changes to conflicting normative systems, and so provide a logic for reasoning about the effect of those changes, potentially giving a new approach to reasoning about conflict resolution. When the relations in a

guaranteed to be true at only one state (such as the nominals of hybrid logic), we cannot ensure that it is a singleton.

[20] [2] also discusses 'local' graph modification operators, which significantly extend the expressive power of their system, sufficient to express the hybrid binder $\downarrow x$, and so leads to undecidability ([6]). Augmentation of GDDL with a distinguished nominal for the point of evaluation should have a similar effect.

[21] [9] only allows a uniform update for all agents (the 'common update policy') but hints at the possibility of providing distinct updates for each agent 'privately'. This is done in [10], but these updates are not really 'private', just as our $\Uparrow_i$ operator (Section 2.1) is not private. To achieve the genuine privacy of $\uparrow_i$, some (as yet unformulated) hybrid of BMS and arrow update logic would be needed.

model are understood as state-transitions of some process (as in the standard computational interpretation of PDL), GDDL operators encode operations for changing what is possible to do, and so give a basis for reasoning about design.

Thirdly, from a technical point of view, the interaction between levels raises interesting questions about the framing of general constraints on those interactions so as to ensure sensible results. We saw an example of this in our brief exploration of private belief change (Footnote 13) but as yet we have no idea about how to frame a general theory of such constraints.

These suggestions are of course very speculative but they give a sense of how GDDL could be used to open up a new area for applications. Our motivations for developing the system arose from technical considerations in an ongoing project called 'logic in the community' [17], which aims at studying the consequences of social relationships for our understanding of rational procedures. We expect there to be many further uses for GDDL in that project.

References

[1] Areces, C., P. Blackburn and M. Marx, *The computational complexity of hybrid temporal logics*, Logic Journal of the IGPL **8** (2000), pp. 653–679.

[2] Aucher, G., P. Balbiani, L. F. del Cerro and A. Herzig, *Global and local graph modifiers*, in: *Proceedings of the 5th workshop on methods for modalities (MAM5 2007)*, Electronic notes in theoretical computer science **231** (2009), pp. 293–307.

[3] Baltag, A., L. S. Moss and S. Solecki, *The logic of public announcements, common knowledge and private suspicious*, Technical Report SEN-R9922, CWI, Amsterdam (1999).

[4] Baltag, A. and S. Smets, *The logic of conditional actions*, in: R. van Rooij and K. Apt, editors, *New perspective on games and interaction*, Texts in logic and games **4**, Amsterdam University Press, 2008 pp. 9–31.

[5] Blackburn, P., M. de Rijke and Y. Venema, "Modal Logic," Cambridge University Press, Cambridge, Mass., 2001.

[6] Blackburn, P. and J. Seligman, *What are hybrid languages?*, in: M. Kracht, M. de Rijke, H. Wansing and M. Zakharyaschev, editors, *Advances in Modal Logic*, **1**, CSLI Publications, Stanford University, 1998 pp. 41–62.

[7] Girard, P., "Modal Logic for Belief and Preference Change," Ph.D. thesis, Stanford University (2008).

[8] Harel, D. and R. Sherman, *Propositional dynamic logic of flowcharts*, Information and Control **64** (1985), pp. 119–135.

[9] Kooi, B. and B. Renne, *Arrow update logic*, Review of Symbolic Logic **4** (2011), pp. 536–559.

[10] Kooi, B. and B. Renne, *Generalized arrow update logic*, in: *Proceedings of the 13th Conference on Theoretical Aspects of Rationality and Knowledge*, TARK XIII (2011), pp. 205–211.

[11] Liu, F., "Changing for the Better: Preference Dynamics and Agent Diversity," Ph.D. thesis, Institute for logic, Language and Computation, Universiteit van Amsterdam, Amsterdam, The Netherlands (2008), ILLC Dissertation series DS-2008-02.

[12] Liu, F., "Reasoning about Preference Dynamics," Synthese Library **354**, Springer, 2011.

[13] Miller, J. S. and L. S. Moss, *The undecidability of iterated modal relativization*, Studia Logica **68** (2001), pp. 1–37.

[14] Passy, S. and T. Tinchev, *An essay in combinatory dynamic logic*, Information and Computation **93** (1991), pp. 263–332.
[15] Plaza, J., *Logics of public communications*, Synthese **158** (2007), pp. 165–179.
[16] Pratt, V. R., *Using graphs to understand* PDL, in: *Logic of Programs, Workshop* (1982), pp. 387–396.
[17] Seligman, J., F. Liu and P. Girard, *Logic in the community*, in: M. Banerjee and A. Seth, editors, *ICLA*, Lecture Notes in Computer Science **6521**, 2011, pp. 178–188.
[18] van Benthem, J., *Dynamic logic for belief revision*, Journal of Applied Non-classical Logic **17** (2007), pp. 129–155.
[19] Van Benthem, J., "Modal Logic for Open Minds," CSLI lecture notes, Center for the Study of Language and Information, 2010.
[20] van Benthem, J. and F. Liu, *The dynamics of preference upgrade*, Journal of Applied Non-Classical Logics **17** (2007), pp. 157–182.
[21] van Benthem, J., J. van Eijck and B. Kooi, *Logics of communication and change*, Information and computation **204** (2006), pp. 1620–1662.
[22] van Ditmarsch, H. and B. Kooi, *Semantic results for ontic and epistemic change*, in: G. Bonanno, W. van der Hoek and M. Wooldridge, editors, *Logic and the foundations of game and decision theory (LOFT 7)*, Texts in logic and games (2008), pp. 87–117.
[23] van Ditmarsch, H., W. van der Hoek and B. Kooi, "Dynamic Epistemic Logic," Berlin: Springer, 2007.
[24] van Eijck, J. and Y. Wang, *Propositional dynamic logic as a logic of belief revision*, in: W. Hodges and R. de Queiroz, editors, *WoLLIC 2008* (2008), pp. 136–148.
[25] Walther, D., "Propositional Dynamic Logic with Negation on Atomic Programs," Master's thesis, Dresden University of Technology (2004).
[26] Zhen, L. and J. Seligman, *A logical model of the dynamics of peer pressure*, Electronic Notes in Theoretical Computer Science **278** (2011), pp. 275–288.

Appendix

Proof of Lemma 1.1: We prove the two claims simultaneously by induction. Assume that $u \in \Lambda W$.

$$\begin{array}{llll} M,u \models q^\Lambda & \text{iff} & M,u \models \Lambda(q) & \text{(definition of } q^\Lambda) \\ & \text{iff} & u \in [\![\Lambda(q)]\!]^M \cap \Lambda W & (u \in \Lambda W) \\ & \text{iff} & u \in \Lambda V(q) & \text{(definition } \Lambda V(q)) \\ & \text{iff} & \Lambda M, u \models q & \text{(definition } \Lambda M) \end{array}$$

Negation and conjunction are straightforward.

$$\begin{array}{llll} M,u \models (\langle\pi\rangle\psi)^\Lambda & \text{iff} & M,u \models \langle\pi^\Lambda\rangle\psi^\Lambda & \text{(def. } (\langle\pi\rangle\psi)^\Lambda) \\ & \text{iff} & u[\![\pi^\Lambda]\!]^M v \text{ and } M,v \models \psi^\Lambda \text{ for some } v \in W & \text{(def.)} \\ & \text{iff} & u[\![\pi]\!]^{\Lambda M} v \text{ and } \Lambda M,v \models \psi \text{ for some } v \in \Lambda W & \text{(IH)} \\ & \text{iff} & \Lambda M, u \models \langle\pi\rangle\psi & \text{(def.)} \end{array}$$

$$\begin{array}{llll} u[\![s^\Lambda]\!]^M v & \text{iff} & u[\![\Lambda(s);|\Lambda|?]\!]^M v & \text{(definition of } s^\Lambda) \\ & \text{iff} & \exists w : u[\![\Lambda(s)]\!]^M w \,\&\, w[\![|\Lambda|?]\!]^M v & \text{(definition of } [\![\Lambda(s);|\Lambda|?]\!]^M) \\ & \text{iff} & u[\![\Lambda(s)]\!]^M v \,\&\, v[\![|\Lambda|?]\!]^M v & (w[\![|\Lambda|?]\!]^M v \Rightarrow w = v) \\ & \text{iff} & u[\![\Lambda(s)]\!]^M v \,\&\, v \in [\![|\Lambda|]\!]^M & \text{(definition of } [\![|\Lambda|?]\!]^M) \\ & \text{iff} & u[\![\Lambda(s)]\!]^M v \,\&\, v \in \Lambda W & \text{(definition of } \Lambda W) \\ & \text{iff} & u[\![s]\!]^{\Lambda M} v \,\&\, v \in \Lambda W & \text{(definition of } [\![\Lambda(s)]\!]^M) \end{array}$$

$$
\begin{array}{llll}
u[\![(\psi?)^{\Lambda}]\!]^{M}v & \text{iff} & u=v \text{ and } v[\![(\psi^{\Lambda})?]\!]^{M}v & (u[\![(\psi?)^{\Lambda}]\!]^{M}v \Rightarrow u=v)\\
& \text{iff} & u=v, v\in[\![\psi^{\Lambda}]\!]^{M} & (\text{definition of } [\![(\psi^{\Lambda})?]\!]^{M})\\
& \text{iff} & u=v, v\in\Lambda W \text{ and } v\in[\![\psi]\!]^{\Lambda M} & (\text{IH, and } u\in\Lambda W)\\
& \text{iff} & u=v, v\in\Lambda W \text{ and } v[\![\psi?]\!]^{\Lambda M}v & (\text{definition of } [\![\psi?]\!]^{\Lambda M})\\
& \text{iff} & v\in\Lambda W \,\&\, u[\![\psi?]\!]^{\Lambda M}v & (u[\![\psi?]\!]^{\Lambda M}v \Rightarrow u=v)
\end{array}
$$

The remaining cases are straightforward.

☺

Proof of Lemma 2.3: In the first direction, assume that $\langle\langle x_1, c_1\rangle, \langle x_2, c_2\rangle\rangle \in T'(\tau)$ and u$[\![\tau]\!]^M v$. Then, for some σ, $\langle x_1, x_2\rangle \in T(\sigma)$ and $\sigma^{c_1,c_2} = \tau$. Since σ is a basic program, it is either a relation symbol, E, or a test.

- If $\sigma = \psi?$ and $c_1 = c_2 = c$, then $\sigma^{c_1,c_2} = \psi^{[A,G,c]}?$. So $u[\![\psi^{[A,G,c]}?]\!]^M v$ implies that $u = v$ and $M, u \models \psi^{[A,G,c]}$. Hence, $GM, \langle u, c\rangle \models \psi$, by assumption, so $\langle u, c_1\rangle[\![\sigma]\!]^{GM}\langle v, c_2\rangle$.
- If $\sigma = r \in R, c_1 = c_2 = c$, then $\sigma^{c_1,c_2} = G_c(\sigma) = r^{G_c}$. By Lemma 1.1, $u[\![r^{G_c}]\!]^v$ implies that $u[\![\sigma]\!]^{G_cM}v$. Hence, $\langle u, c_1\rangle[\![\sigma]\!]^{GM}\langle v, c_2\rangle$, by definition.
- If $\sigma = \mathsf{E}$, then $\sigma^{c_1,c_2} = |G_{c_1}|?;\mathsf{E};|G_{c_2}|?$. So $u[\![|G_{c_1}|?;\mathsf{E};|G_{c_2}|?]\!]^M v$ implies that $u \in [\![|G_{c_1}|]\!]^M$ and $v \in [\![|G_{c_2}|]\!]^M$, so $\langle u, c_1\rangle \in GW$ and $\langle v, c_2\rangle \in GW$. Therefore, $\langle u, c_1\rangle[\![\mathsf{E}]\!]^{GM}\langle v, c_2\rangle$.
- If $\sigma \in S$ and $\langle c_1, c_2\rangle \in U(\sigma)$, then $\sigma^{c_1,c_2} = \top?$, so $u[\![\top?]\!]^M v$ implies that $u = v$. Thus, $\langle u, c_1\rangle[\![\sigma]\!]^{GM}\langle v, c_2\rangle$, by definition.
- In any other case, $\sigma^{c_1,c_2} = \bot?$. But this contradicts our assumption that $u[\![\sigma^{c_1,c_2}]\!]^M v$.

Therefore, $\langle x_1, x_2\rangle \in T(\sigma)$ and $\langle u, c_1\rangle[\![\sigma]\!]^{GM}\langle v, c_2\rangle$.

In the other direction, assume that $\gamma(\sigma)$ and $\sigma^{c_1,c_2} = \tau$ for some σ. But $\langle x_1, x_2\rangle \in T(\sigma)$ and $\sigma^{c_1,c_2} = \tau$ implies that $\langle\langle x_1, c_1\rangle\rangle\langle x_2, c_2\rangle \in T'(\tau)$ by definition, so we only need to show that $u[\![\sigma^{c_1,c_2}]\!]^M v$. Again, since σ is a basic program, it is either a relation symbol, E, or a test.

- If $\sigma = \psi?, c_1 = c_2 = c$, then $\sigma^{c_1,c_2} = \psi^{[A,G,c]}?$. Now, $\langle u, c\rangle[\![\psi?]\!]^{GM}\langle v, c\rangle$ implies that $u = v$ and $GM, \langle u, c\rangle \models \psi$. By assumption, $M, u \models \psi^{[A,G,c]}$, so $u[\![\sigma^{c_1,c_2}]\!]^M v$, by definition.
- If $\sigma = r \in R$, $c_1 = c_2 = c$, then $\sigma^{c_1,c_2} = G_c(\sigma)$. But $\langle u, c\rangle[\![r]\!]^{GM}\langle v, c\rangle$ implies that $\langle u, v\rangle \in V_c(r)$, by definition, so $u[\![G_c(r)]\!]^M v$. Hence, $u[\![\sigma^{c_1,c_2}]\!]^M v$.
- If $\sigma = \mathsf{E}$, then $\sigma^{c_1,c_2} = |G_{c_1}|?;\mathsf{E};|G_{c_2}|?$. Now, $\langle u, c_1\rangle[\![\mathsf{E}]\!]^{GM}\langle v, c_2\rangle$ implies that $\langle u, c_1\rangle \in GW$ and $\langle v, c_2\rangle \in GW$, so $u \in [\![|G_{c_1}|]\!]^M$ and $v \in [\![|G_{c_2}|]\!]^M$. But $u[\![\mathsf{E}]\!]^M v$, so $u[\![|G_{c_1}|?;\mathsf{E};|G_{c_2}|?]\!]^M v$.
- If $\sigma = s \in S$ and $\langle c_1, c_2\rangle \in U(\sigma)$, then $\sigma^{c_1,c_2} = \top?$. Thus, $\langle u, c_1\rangle[\![s]\!]^{GM}\langle v, c_2\rangle$ implies that $u = v$. But $M, u \models \top$, so $u[\![\top?]\!]^M u$. Hence, $u[\![\sigma^{c_1,c_2}]\!]^M v$.

- In any other case, $\sigma^{c_1,c_2} = \bot?$. But this contradicts our assumption that $\langle u, c_1\rangle [\![\sigma]\!]^{GM} \langle v, c_2\rangle$.

Therefore, $u[\![\sigma^{c_1,c_2}]\!]^M v$.

☺

Proof of Theorem 3.1: Let $\Delta M = \langle W', V''\rangle$ and $[A, G, M]M = \langle \Lambda GW, \Lambda GV\rangle$. Then,

$$\begin{array}{llll} \Lambda GW & = & GW & (|\Lambda| = \top) \\ & = & \{\langle u, d\rangle \mid u \in [\![|G_d|]\!]^{\}} & \text{(definition)} \\ & = & \{\langle u, d\rangle \mid u \in [\![|\mathsf{pre}(d)|]\!]\} & \text{(assumption)} \\ & = & \Delta W & \end{array}$$

$$\begin{array}{llll} \Lambda GV(p) & = & [\![\Lambda G(p)]\!]^M \cap \Lambda GW & \text{(definition)} \\ & = & [\![p]\!]^M \cap \Delta W & (\Lambda GW = \Delta W) \\ & = & \Delta V(p) & \text{(definition)} \end{array}$$

$$\begin{array}{llll} [\![\Lambda G_d(r_i)]\!]^{\Lambda GM} & = & [\![\Lambda(r_i)]\!]^{\Lambda GM} & (G_d(r_i) = r_i) \\ & = & [\![r_i; s_i]\!]^{GM} & \text{(assumption)} \end{array}$$

But $\langle u, d\rangle [\![r_i; s_i]\!]^{GM} \langle v, e\rangle$ iff there exists $\langle w, f\rangle \in GW$ such that $\langle u, d\rangle [\![r_i]\!]^{GM} \langle w, f\rangle$ and $\langle w, f\rangle [\![s_i]\!]^{GM} \langle v, e\rangle$, iff $d = f$ and $w = v$, by definition. Hence, $\langle u, d\rangle [\![r_i; s_i]\!]^{GM} \langle v, e\rangle$ iff $\langle u, d\rangle [\![r_i]\!]^{GM} \langle v, d\rangle$ and $\langle v, d\rangle [\![s_i]\!]^{GM} \langle v, e\rangle$ iff $u[\![G_d(r_i)]\!]^M v$ and $d[\![s_i]\!]^A e$, by definition, iff $u[\![r_i]\!]^M v$ and $d[\![s_i]\!]^A e$. Therefore, $[\![\Lambda G_d(r_i)]\!]^{\Lambda GM} = \Delta V(r)$. ☺

Proof of Theorem 3.2: The proof is the same as the previous theorem but for the propositional case:

$$\begin{array}{llll} \Lambda G_d V(p) & = & [\![\Lambda G_d(p)]\!]^M \cap \Lambda GW & \text{(definition)} \\ & = & [\![\mathsf{sub}_d(p)]\!]^M \cap \Delta W & (\Lambda GW = \Delta W) \\ & = & \Delta V(p) & \text{(definition)} \end{array}$$

☺

Acknowledgements: We would like to thank the anonymous referees for their valuable comments. Fenrong Liu is supported by the Major Program of the National Social Science Foundation of China (NO. 11&ZD088) and the National Natural Science Foundation of China (NO.81173464). Jeremy Seligman was supported by the National Social Science Foundation Key Project of China (NO.11AZD57). He is also grateful to Auckland University for allowing a period of absence during which time the research was completed.

The Complexity of Monotone Hybrid Logics over Linear Frames and the Natural Numbers

Stefan Göller

Department of Computer Science, Universität Bremen, Germany

Arne Meier

Institute of Theoretical Computer Science, Leibniz Universität Hannover, Germany

Martin Mundhenk

Institute of Computer Science, Friedrich-Schiller-Universität Jena, Germany

Thomas Schneider

Department of Computer Science, Universität Bremen, Germany

Michael Thomas

TWT GmbH, Germany

Felix Weiß

Institute of Computer Science, Friedrich-Schiller-Universität Jena, Germany

Abstract

Hybrid logic with binders is an expressive specification language. Its satisfiability problem is undecidable in general. If frames are restricted to $\mathbb{N}$ or general linear orders, then satisfiability is known to be decidable, but of non-elementary complexity. In this paper, we consider monotone hybrid logics (i.e., the Boolean connectives are conjunction and disjunction only) over $\mathbb{N}$ and general linear orders. We show that the satisfiability problem remains non-elementary over linear orders, but its complexity drops to PSPACE-completeness over $\mathbb{N}$. We categorize the strict fragments arising from different combinations of modal and hybrid operators into NP-complete and tractable (i.e. complete for NC^1 or LOGSPACE). Interestingly, NP-completeness depends only on the fragment and not on the frame. For the cases above NP, satisfiability over linear orders is harder than over $\mathbb{N}$, while below NP it is at most as hard. In addition we examine model-theoretic properties of the fragments in question.

Keywords: satisfiability, modal logic, complexity, hybrid logic

1 Introduction

Hybrid logic is an extension of modal logic with nominals, satisfaction operators and binders. The downarrow binder $\downarrow$, which is related to the freeze operator in temporal logic [12], provides high expressivity. The price paid is the undecidability of the satisfiability problem for the hybrid language with the downarrow binder $\downarrow$ [4,11,1]. In contrast, modal logic, and its extension with nominals and the satisfaction operator, is PSPACE-complete [13,1].

In order to regain decidability, syntactic and semantic restrictions have been considered. It has been shown in [22] that the absence of certain combinations of universal operators ($\Box$, $\wedge$) with $\downarrow$ brings back decidability, and that the hybrid language with $\downarrow$ is decidable over frames of bounded width. Furthermore, this language is decidable over transitive and complete frames [17], and over frames with an equivalence relation (ER frames) [16]. Adding the at-operator @—which allows to jump to states named by nominals—leads to undecidability over transitive frames [17], but not over ER frames [16]. Over linear frames and transitive trees, $\downarrow$ on its own does not add expressivity, but combinations with @ or the global modality—an additional $\Diamond$ interpreted over the universal relation—do. These languages are decidable and of non-elementary complexity [9,17]; if the number of state variables is bounded, then they are of elementary complexity [19,24,5].

We aim for a more fine-grained distinction between fragments of different complexities by systematically restricting the set of Boolean connectives and combining this with restrictions to the modal/hybrid operators and to the underlying frames. In [15], we have focussed on four frame classes that allow cycles, and studied the complexity of satisfiability for fragments obtained by arbitrary combinations of Boolean connectives and four modal/hybrid operators. The main open question in [15] is the one for tight upper bounds for monotone fragments including the $\Box$-operator. Even though there are many logics for which the restriction to monotone Boolean connectives leads to a significant decrease in complexity, it is not straightforward, and therefore interesting to find out, where this happens for hybrid logics.

In this study, we classify the computational complexity of satisfiability for monotone fragments of hybrid logic with arbitrary combinations of the operators $\Diamond$, $\Box$, $\downarrow$ and @ over linear orders and the natural numbers. Whereas the full logic is non-elementary and decidable [17] for both frame classes, we show that in the monotone case this high complexity is gained only over linear orders and drops to PSPACE-completeness over the natural numbers. Informally speaking, the reason is that linearly ordered frames may consist of arbitrarily many dense parts that can be distinguished using the expressive power of all four operators. These dense parts and their distances are used to store information that cannot be stored in a frame without dense parts as, e.g., the natural numbers. For all other monotone fragments that contain the $\Diamond$-operator, we show NP-completeness independent on the frame class, for linear orders, all remaining fragments (i.e. the fragments without $\Diamond$) can be shown to be NC^1-complete. The reason is, informally speaking, that all (sub-)formulas of the form

$\Box\alpha$ are easily satisfied in a state without successor, which can essentially be used to reduce this problem to the satisfiability problem for monotone propositional formulae. This argument does not go through over the natural numbers, a total frame where every state has a successor. Over this frame class, we give a decision procedure that runs in logarithmic space for the fragment with all operators except $\Diamond$ (and prove a matching lower bound), and in NC^1 for all other fragments.

These results give rise to two interesting observations. First, the NP-completeness results are independent on the frame class. Second, for the fragment whose satisfiability problem is above NP, linear orders make the problem harder than the natural numbers, and for the richest fragment below NP, it is the opposite way round—the natural numbers make the problem harder than linear orders. Notice also that, in the case where Boolean operators are not restricted to monotone ones, all fragments are NP-hard.

Our results are shown in Figure 1.

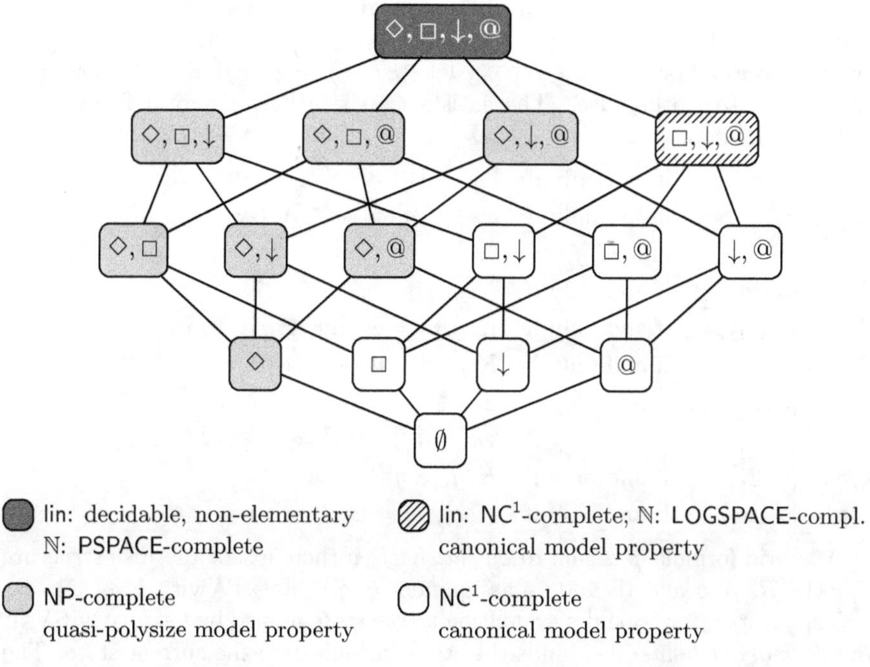

Fig. 1. Our complexity results for satisfiability over linear frames (lin) and the natural numbers ($\mathbb{N}$) for hybrid logic with monotone Boolean operators and different combinations of modal/hybrid operators

2 Preliminaries

Hybrid Logic. In the following, we introduce the notions and definitions of hybrid logic. The terminology is largely taken from [2].

Let PROP be a countable set of *atomic propositions*, NOM be a countable set of *nominals*, SVAR be a countable set of *variables* and ATOM = PROP ∪ NOM ∪ SVAR. We adhere to the common practice of denoting atomic propositions by $p, q, \ldots$, nominals by $i, j, \ldots$, and variables by $x, y, \ldots$ We define the language of *hybrid (modal) logic* $\mathcal{HL}$ as the set of well-formed formulae of the form

$$\varphi ::= a \mid \top \mid \bot \mid \neg\varphi \mid \varphi \wedge \varphi \mid \varphi \vee \varphi \mid \Diamond\varphi \mid \Box\varphi \mid {\downarrow}x.\varphi \mid @_t\varphi$$

where $a \in$ ATOM, $x \in$ SVAR and $t \in$ NOM ∪ SVAR.

We define the usual Kripke semantics only to be able to refer to already existing results. We will then simplify the standard semantics for monotone formulae. Formulae of $\mathcal{HL}$ are interpreted on *(hybrid) Kripke structures* $K = (W, R, \eta)$, consisting of a set of *states* W, a *transition relation* $R\colon W \times W$, and a *labeling function* $\eta\colon$ PROP ∪ NOM $\to \wp(W)$ that maps PROP and NOM to subsets of W with $|\eta(i)| = 1$ for all $i \in$ NOM. The relational structure (W, R) is the *Kripke frame* underlying K. In order to evaluate ↓-formulae, an assignment $g\colon$ SVAR $\to W$ is necessary. Given an assignment g, a state variable x and a state w, an *x-variant g^x_w of g* is defined by $g^x_w(x) = w$ and $g^x_w(x') = g(x')$ for all $x \neq x'$. For any $a \in$ ATOM, let $[\eta, g](a) = \{g(a)\}$ if $a \in$ SVAR and $[\eta, g](a) = \eta(a)$, otherwise. The satisfaction relation of hybrid formulae is defined as follows.

$K, g, w \models \varphi \wedge \psi$	if and only if	$\exists w' \in W(wRw' \;\&\; K, g, w' \models \varphi)$
$K, g, w \models a$	if and only if	$w \in [\eta, g](a)$, $a \in$ ATOM,
$K, g, w \models \top$,	and $K, g, w \not\models \bot$,	
$K, g, w \models \neg\varphi$	if and only if	$K, g, w \not\models \varphi$,
$K, g, w \models \varphi \wedge \psi$	if and only if	$K, g, w \models \varphi$ and $K, g, w \models \psi$,
$K, g, w \models \varphi \vee \psi$	if and only if	$K, g, w \models \varphi$ or $K, g, w \models \psi$,
$K, g, w \models \Diamond\varphi$	if and only if	$\exists w' \in W(wRw' \;\&\; K, g, w' \models \varphi)$,
$K, g, w \models \Box\varphi$	if and only if	$\forall w' \in W(wRw' \Rightarrow K, g, w' \models \varphi)$,
$K, g, w \models @_t\varphi$	if and only if	$K, g, [\eta, g](t) \models \varphi$,
$K, g, w \models {\downarrow}x.\varphi$	if and only if	$K, g^x_w, w \models \varphi$.

A hybrid formula φ is said to be *satisfiable* if there exists a Kripke structure $K = (W, R, \eta)$, a $w \in W$ and an assignment $g\colon$ SVAR $\to W$ with $K, g, w \models \varphi$.

The *at* operator $@_t$ shifts evaluation to the state named by $t \in$ NOM ∪ SVAR. The *downarrow binder* ${\downarrow}x.$ binds the state variable x to the current state. The symbols $@_x$, ${\downarrow}x.$ are called *hybrid operators* whereas the symbols $\Diamond$ and $\Box$ are called *modal operators*.

The scope of an occurrence of the binder ↓ is defined as usual. For a state variable x, an occurrence of x or $@_x$ in a formula φ is called *bound* if this occurrence is in the scope of some ↓ in φ, *free* otherwise. φ is said to contain a free state variable if some x or $@_x$ occurs free in φ.

Given two formulae φ, α and a subformula ψ of φ, we use $\varphi[\psi/\alpha]$ to denote the result of replacing each occurrence of ψ in φ with α. For considering fragments of hybrid logics, we define subsets of the language $\mathcal{HL}$ as follows. Let

O be a set of hybrid and modal operators, i.e., a subset of $\{\Diamond, \Box, \downarrow, @\}$. We define $\mathcal{HL}(O)$ to denote the set of well-formed hybrid formulae using only the operators in O, and $\mathcal{MHL}(O)$ to be the set of all formulae in $\mathcal{HL}(O)$ that do not use $\neg$.

Properties of Frames. A *frame* F is a pair (W, R), where W is a set of states and $R \subseteq W \times W$ a transition relation. A frame $F = (W, R)$ is called

- transitive if R is transitive (for all $u, v, w \in W$: $uRv \wedge vRw \rightarrow uRw$),
- linear if R is transitive, irreflexive and trichotomous ($\forall u, v \in W$: uRv or $u = v$ or vRu),

In this paper we consider the class of all linear frames, denoted by lin, and the singleton frame class $\{(\mathbb{N}, <)\}$, denoted by $\mathbb{N}$. Obviously, $\mathbb{N} \subseteq \mathsf{lin}$.

Notational convenience. We can make some simplifying assumptions about syntax and semantics, of $\mathcal{HL}(O)$ and $\mathcal{MHL}(O)$, which do not restrict generality. (1) If $\downarrow\, \in O$, then formulae do not contain any nominals. Those can be simulated by free state variables. (2) Free state variables are never bound later in the formula, and every state variable is bound at most once. The latter is no significant restriction because variables bound multiple times can be named apart, which is a well-established and computationally easy procedure. (3) Monotone formulae do not contain any atomic propositions. This restriction is correct because every monotone formula φ is satisfiable if and only if φ with all atomic propositions replaced by $\top$ is satisfiable. This justifies the following restrictions. (4) For binder-free fragments, the domain of the labelling function η is restricted to nominals, and we re-define η: $\textsc{Nom} \rightarrow W$. Furthermore, the absence of $\downarrow$ makes assignments superfluous: we write $F, w \models \varphi$ instead of $F, g, w \models \varphi$. (5) For binder fragments, the satisfaction relation $\models$ is restricted to Kripke *frames* $F = (W, <)$, where $<$ is a linear order, and assignments $g : \textsc{SVar} \rightarrow W$, i.e., we write $F, g, w \models \varphi$. (6) Over $\mathbb{N}$, we omit the single Kripke frame, i.e., we write $\eta, i \models \varphi$ with $\eta : \textsc{Nom} \rightarrow \mathbb{N}$ and $i \in \mathbb{N}$ for binder-free fragments, and $g, i \models \varphi$ with $g : \textsc{SVar} \rightarrow \mathbb{N}$ for binder fragments.

Satisfiability Problems. The *satisfiability problem* for $\mathcal{HL}(O)$ over the frame class $\mathfrak{F}$ is defined as follows:

Problem: $\mathfrak{F}$-$\mathsf{SAT}(O)$
Input: an $\mathcal{HL}(O)$-formula φ (without nominals, see above)
Output: Is there a Kripke structure K based on a frame $(W, R) \in \mathfrak{F}$, an assignment g: $\textsc{SVar} \rightarrow W$ and a $w \in W$ such that $K, g, w \models \varphi$?

The *monotone satisfiability problem* for $\mathcal{MHL}(O)$ over the frame class $\mathfrak{F}$ is defined as follows:

Problem: $\mathfrak{F}$-$\mathsf{MSAT}(O)$
Input: an $\mathcal{MHL}(O)$-formula φ without nominals and atomic propositions
Output: Is there a Kripke frame $(W, R) \in \mathfrak{F}$, an assignment g: $\textsc{SVar} \rightarrow W$ and a $w \in W$ such that $F, g, w \models \varphi$?

If $\mathfrak{F}$ is the class of all frames, we simply write $\mathsf{SAT}(O)$ or $\mathsf{MSAT}(O)$. Furthermore, we often omit the set parentheses when giving O explicitly, e.g., $\mathsf{SAT}(\Diamond, \Box, \downarrow, @)$.

Complexity Theory. We assume familiarity with the standard notions of complexity theory as, e. g., defined in [18]. In particular, we make use of the classes LOGSPACE, NLOGSPACE, NP, PSPACE, and coRE. The complexity class NONELEMENTARY is the set of all languages A that are decidable and for which there exists no $k \in \mathbb{N}$ such that A can be decided using an algorithm whose running time is bounded by $\exp_k(n)$, where $\exp_k(n)$ is the k-th iteration of the exponential function (e.g., $\exp_3(n) = 2^{2^{2^n}}$).

Furthermore, we need two non-standard complexity classes whose definition relies on circuit complexity and formal languages, see for instance [23,14]. The class NC^1 is defined as the set of languages recognizable by a logtime-uniform family of Boolean circuits of logarithmic depth and polynomial size over $\{\wedge, \vee, \neg\}$, where the fan-in of $\wedge$ and $\vee$ gates is fixed to 2. The class LOGDCFL is defined as the set of languages reducible in logarithmic space to some deterministic context-free language.

The following relations between the considered complexity classes are known.

$$\mathsf{NC}^1 \subseteq \mathsf{LOGSPACE} \subseteq \mathsf{LOGDCFL} \subseteq \mathsf{NP} \subseteq \mathsf{PSPACE} \subset \mathsf{coRE}.$$

It is unknown whether LOGDCFL contains NLOGSPACE or vice versa.

A language A is *constant-depth reducible* to D, $A \leqslant_{\mathrm{cd}} D$, if there is a logtime-uniform AC^0-circuit family with oracle gates for D that decides membership in A. Unless otherwise stated, all reductions in this paper are $\leqslant_{\mathrm{cd}}$-reductions.

Known results. The following theorem summarizes results for hybrid languages with Boolean operators $\wedge, \vee, \neg$ that are known from the literature. Since $\Box\varphi \equiv \neg\Diamond\neg\varphi$, the $\Box$-operator is implicitly present in all fragments containing $\Diamond$ and negation.

Theorem 2.1 ([1,2,3,9,17])

(1) $\mathsf{SAT}(\Diamond, \downarrow, @)$ *and* $\mathsf{SAT}(\Diamond, \downarrow)$ *are* coRE*-complete. [1]*

(2) $\mathsf{MSAT}(\Diamond, \Box)$ *is* PSPACE*-hard. [3]*

(3) $\mathfrak{F}$-$\mathsf{SAT}(\Diamond, \downarrow, @)$*, for* $\mathfrak{F} \in \{\mathsf{lin}, \mathbb{N}\}$*, are in* NONELEMENTARY*. [9,17]*

(4) $\mathfrak{F}$-$\mathsf{SAT}(\Diamond, \downarrow)$*,* $\mathfrak{F}$-$\mathsf{SAT}(\Diamond, @)$ *and* $\mathfrak{F}$-$\mathsf{SAT}(\Diamond)$*, with* $\mathfrak{F} \in \{\mathsf{lin}, \mathbb{N}\}$*, are* NP*-complete. [2,9]*

Our contribution. In this paper, we consider the monotone satisfiability problems $\mathfrak{F}$-$\mathsf{MSAT}(O)$ for $\mathfrak{F} \in \{\mathsf{lin}, \mathbb{N}\}$ and all $O \subseteq \{\Diamond, \Box, \downarrow, @\}$.

3 The hard cases: Non-elementary and PSPACE results

The hardest cases are those with the complete set of operators. In the non-monotone case, both satisfiability problems are non-elementary and decidable [17]. We show that in the monotone case even this hardness is reached, but only on linear frames, i.e. lin-MSAT($\Diamond, \Box, \downarrow, @$) is non-elementary and decidable. In contrast, on the natural numbers the complexity decreases, i.e. we show that $\mathbb{N}$-MSAT($\Diamond, \Box, \downarrow, @$) is PSPACE-complete.

Our proofs use reductions to and from fragments of first-order logic on the natural numbers. Let $\mathcal{FOL}(<, P)$ be the set of all first-order formulae that use $<$ as the unique binary relation symbol, and P as the unique unary relation symbol.[1] Let $\mathbb{N}\text{-SAT}_{\mathcal{FOL}}(<, P)$ denote the set of formulae from $\mathcal{FOL}(<, P)$ which are satisfied by a model that has $\mathbb{N}$ as its universe, interprets $<$ as the less-than relation on $\mathbb{N} \times \mathbb{N}$, and has an arbitrary interpretation for the predicate symbol P. It was shown by Stockmeyer [21] that $\mathbb{N}\text{-SAT}_{\mathcal{FOL}}(<, P)$ is non-elementary.

Let $\mathcal{FOL}(<)$ be the fragment of $\mathcal{FOL}(<, P)$ in which the predicate symbol P is not used. Accordingly, $\mathbb{N}\text{-SAT}_{\mathcal{FOL}}(<)$ denotes the set of formulae that are satisfiable over $\mathbb{N}$ and the natural interpretation of $<$. It was shown by Ferrante and Rackoff [8] that $\mathbb{N}\text{-SAT}_{\mathcal{FOL}}(<)$ is in PSPACE.

Notice that in both fragments $x = y$ can be expressed as $\neg(x < y \vee y < x)$. Moreover, every $n \in \mathbb{N}$ can be expressed by x_n in the formula $\exists x_0 \cdots \exists x_{n-1}[(\bigwedge_{i=0,1,\dots,n-1} x_i < x_{i+1}) \wedge \forall y(x_n < y \vee \bigvee_{i=0,1,\dots,n} y = x_i)]$.

Theorem 3.1 lin-MSAT($\Diamond, \Box, \downarrow, @$) *is non-elementary and decidable.*

Proof. Decidability follows from Theorem 2.1 (3). To establish non-elementary complexity, we give a reduction from $\mathbb{N}\text{-SAT}_{\mathcal{FOL}}(<, P)$.

We first show how to encode the intepretation of a predicate symbol, represented by a set $P \subseteq \mathbb{N}$, in a linear frame $F = (W, <)$ – without using atomic propositions and nominals as agreed in Section 2. Using free state variables, we can only distinguish linearly many states at any given time. We therefore use finite intervals (finite subchains of $(W, <)$) to encode whether $n \in P$. Such an interval—we call it a *marker*—has length 2 (resp. 3) if for the corresponding n holds $n \notin P$ (resp. $n \in P$). Accordingly, we call a marker of length 2 (resp. 3) *negative* (resp. *positive*). These finite intervals are separated by dense intervals—those are intervals wherein every two states have an intermediate state, e.g., $[0, 1]_{\mathbb{Q}} = \{q \in \mathbb{Q} \mid 0 \leqslant q \leqslant 1\}$. For example, the set P with $0, 2 \notin P$ and $1 \in P$ is represented by the chain in Figure 2. In our fragment, it is possible to distinguish between dense and finite intervals. We now show how to achieve this. In order to encode the alternating sequence of finite and dense intervals that represents a subset $P \subseteq \mathbb{N}$, we use the free state variable a to mark a state in a dense interval that is directly followed by the first marker. We furthermore use the following macros, where x and y are state variables that are already bound before the use of the macro, and r, s, t, u are fresh state variables.

- *The state named y is a* direct *successor of the state named x.* It suffices to

[1] I.e. $\mathcal{FOL}(<, P)$ is defined as set of all formulae φ as follows.

$$\varphi ::= \top \mid x < y \mid P(x) \mid \neg\varphi \mid \varphi \wedge \varphi \mid \varphi \vee \varphi \mid \exists x\, \varphi \mid \forall x\, \varphi$$

for variable symbols $x, y \in \text{SVAR}$.

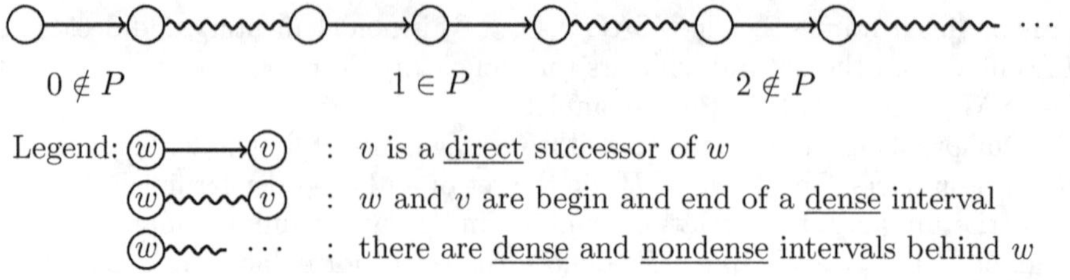

Fig. 2. An example with $0, 2 \notin P$ and $1 \in P$.

say that all successors of x are equal to, or occur after, y.

$$\mathsf{dirSuc}(x, y) := @_x \Box {\downarrow} z.(@_y z \vee @_y \Diamond z)$$

- *The state named x has no direct predecessor.* It suffices to say that, for all states r equal to, or after, the left bound a: if r is before x, then there is a state between r and x. We work around the implication by saying that one of the following three cases occurs: r is after x, or r equals x, or r is before x with a state in between.
$$\mathsf{noDirPred}(x) := @_a \Box {\downarrow} r.(@_x \Diamond r \vee @_x r \vee @_r \Diamond\Diamond x)$$
- *The state named x has a direct predecessor.* It suffices to say that there is a state r after a of which x is a direct successor.
$$\mathsf{dirPred}(x) := @_a \Diamond {\downarrow} r.\mathsf{dirSuc}(r, x)$$
- *The interval between states x, y is dense.* We say that, for all r with $x < r$: r is after y, or r has no direct predecessor.
$$\mathsf{dense}(x, y) := @_x \Box {\downarrow} r.(@_y \Diamond r \vee \mathsf{noDirPred}(r))$$
- *The state x is in a separator.* This macro says that, for some successor r of x, the interval between x and r is dense.
$$\mathsf{sep}(x) := @_x \Diamond {\downarrow} r.\mathsf{dense}(x, r)$$
- *The state x is the begin of a negative marker.* This macro says that x has a direct successor that is the begin of a separator, and x has no direct predecessor. The latter is necessary to avoid that, in the above example, the middle state of a positive marker is mistaken for the begin of a negative marker.
$$\mathsf{neg}(x) := @_x \Diamond {\downarrow} r.(\mathsf{dirSuc}(x, r) \wedge \mathsf{sep}(r)) \wedge \mathsf{noDirPred}(x)$$
- *The state x is the begin of a positive marker.* Similarly to the above macro, we express that x has a direct-successor sequence r, s with s being the begin of a separator, and x has no direct predecessor.
$$\mathsf{pos}(x) := @_x \Diamond {\downarrow} r.(\mathsf{dirSuc}(x, r) \wedge \Diamond {\downarrow} s.(\mathsf{dirSuc}(r, s) \wedge \mathsf{sep}(s))) \wedge \mathsf{noDirPred}(x)$$
- *The state x is in a separator whose end is a marker.* This macro says that, for some successor r of x, the interval between x and r is dense and r is the begin of a marker.
$$\mathsf{sepM}(x) := @_x \Diamond {\downarrow} r.(\mathsf{dense}(x, r) \wedge (\mathsf{neg}(r) \vee \mathsf{pos}(r)))$$

We now need the following two conjuncts to express that the part of the model starting at a represents a sequence of infinitely many markers.

- *a is in a separator that ends with a marker.* $\psi_1 := \mathsf{sepM}(a)$
- *Every marker has a direct successor marker.* We say that every state r after a satisfies one of the following conditions.
  - r is in a separator—this also includes that r is the end of a marker—that is followed by a marker.
  - r is the begin of a negative marker and its direct successor is the begin of a separator whose end is a marker.
  - r is the begin of a positive marker and its direct 2-step successor is the begin of a separator whose end is a marker.
  - r in the middle of a positive marker, i.e., r has a direct predecessor which is the begin of a positive marker, and r's direct successor is in a separator whose end is a marker.

$$\begin{aligned}\psi_2 := @_a\Box{\downarrow}r.\Big(&\mathsf{sepM}(r)\\ &\vee\Big(\mathsf{neg}(r)\wedge\Diamond{\downarrow}s.(\mathsf{dirSuc}(r,s)\wedge\mathsf{sepM}(s))\Big)\\ &\vee\Big(\mathsf{pos}(r)\wedge\Diamond{\downarrow}s.(\mathsf{dirSuc}(r,s)\wedge\Diamond{\downarrow}t.(\mathsf{dirSuc}(s,t)\wedge\mathsf{sepM}(t)))\Big)\\ &\vee\Big((@_a\Diamond{\downarrow}s.\mathsf{dirSuc}(s,r)\wedge\mathsf{pos}(s))\wedge\Diamond{\downarrow}t.(\mathsf{dirSuc}(r,t)\wedge\mathsf{sepM}(t))\Big)\end{aligned}$$

Finally, we encode formulae φ from $\mathcal{FOL}(<,P)$. We assume w.l.o.g. that such formulae have the shape $\varphi := Q_1x_1\ldots Q_nx_n.\beta(x_1,\ldots,x_n)$, where $Q_i\in\{\exists,\forall\}$ and β is quantifier-free with atoms $P(x)$ and $x<y$ for variables x,y, such that negations appear only directly before atoms. The transformation of φ reuses the x_i as state variables and proceeds inductively as follows.

$$\begin{aligned}f(P(x_i)) &:= \mathsf{pos}(x_i)\\ f(\neg P(x_i)) &:= \mathsf{neg}(x_i)\\ f(x_i<x_j) &:= @_{x_i}\Diamond x_j\\ f(\neg(x_i<x_j)) &:= @_{x_i}x_j\vee @_{x_j}\Diamond x_i\\ f(\alpha\wedge\beta) &:= f(\alpha)\wedge f(\beta)\\ f(\alpha\vee\beta) &:= f(\alpha)\vee f(\beta)\\ f(\exists x_i.\alpha) &:= @_a\Diamond{\downarrow}x_i.\Big((\mathsf{neg}(x_i)\vee\mathsf{pos}(x_i))\wedge f(\alpha)\Big)\\ f(\forall x_i.\alpha) &:= @_a\Box{\downarrow}x_i.\Big(\mathsf{sep}(x_i)\vee\mathsf{dirPred}(x_i)\vee f(\alpha)\Big)\end{aligned}$$

The transformation of φ into $\mathcal{MHL}(\Diamond,\Box,\downarrow,@)$ is now achieved by the function g defined as follows.

$$g(\varphi) := \psi_1\wedge\psi_2\wedge f(\varphi)$$

It is clear that the reduction function g can be computed in polynomial time. The correctness of the reduction is expressed by the following claim.

Claim 3.2 *For every formula φ from $\mathcal{FOL}(<, P)$ holds:*
$\varphi \in \mathbb{N}\text{-}\mathsf{SAT}_{\mathcal{FOL}}(<, P)$ *if and only if* $g(\varphi) \in \text{lin-}\mathsf{MSAT}(\Diamond, \Box, \downarrow, @)$.

The proof of the claim should be clear. Since $\mathbb{N}\text{-}\mathsf{SAT}_{\mathcal{FOL}}(<, P)$ is non-elementary [21], it follows that lin-MSAT$(\Diamond, \Box, \downarrow, @)$ is non-elementary, too.

Finally, we note that our reduction uses a single free state variable a, which could as well be bound to the first state of evaluation. □

The high complexity of lin-MSAT$(\Diamond, \Box, \downarrow, @)$ relies on the possibility that the linear frame alternatingly has dense and non-dense parts. If we have the natural numbers as frame for a hybrid language, we lose this possibility. As a consequence, the satisfiability problem for monotone hybrid logics over the natural numbers has a lower complexity than that over linear frames.

Theorem 3.3 $\mathbb{N}$-MSAT$(\Diamond, \Box, \downarrow, @)$ *is* PSPACE*-complete.*

Proof sketch. Let QBFSAT be the problem to decide whether a given quantified Boolean formula is valid. We show PSPACE-hardness by a polynomial-time reduction from the PSPACE-complete QBFSAT to $\mathbb{N}$-MSAT$(\Diamond, \Box, \downarrow, @)$. Let φ be an instance of QBFSAT and assume w.l.o.g. that negations occur only directly in front of atomic propositions. We define the transformation as $f \colon \varphi \mapsto \downarrow r.\Diamond\downarrow s.\Diamond h(\varphi)$ where h is given as follows: let ψ, χ be quantified Boolean formulae and let x_k be a variable in φ, then

$$h(\exists x_k \psi) := @_r \Diamond \downarrow x_k.h(\psi), \qquad h(\forall x_k \psi) := @_r \Box \downarrow x_k.h(\psi),$$
$$h(\psi \wedge \chi) := h(\psi) \wedge h(\chi), \qquad h(\psi \vee \chi) := h(\psi) \vee h(\chi),$$
$$h(\neg x_k) := @_s \Diamond x_k, \qquad h(x_k) := @_s x_k.$$

For example, the QBF $\psi = \forall x \exists y (x \wedge y) \vee (\neg x \wedge \neg y)$ is mapped to
$f(\varphi) = \downarrow r.\Diamond\downarrow s.\Diamond @_r \Box \downarrow x_0.@_r \Diamond \downarrow x_1.(@_s x_0 \wedge @_s x_1) \vee (@_s \Diamond x_0 \wedge @_s \Diamond x_1)$.

Intuitively, this construction requires the existence of an initial state named r, a successor state s that represents the truth value $\top$, and one or more successor states of s which together represent $\bot$. The quantifiers $\exists, \forall$ are replaced by the modal operators $\Diamond, \Box$ which range over s and its successor states. Finally, positive literals are enforced to be true at s, negative literals strictly after s.

For every model of $f(\varphi)$, it holds that r is situated at the first state of the model and that state has a successor labelled by s. By virtue of the function h, positive literals have to be mapped to s, whereas negative literals have to be mapped to some state other than s. An easy induction on the structure of formulae shows that $\varphi \in$ QBFSAT iff $f(\varphi) \in \mathbb{N}$-MSAT$(\Diamond, \Box, \downarrow, @)$.

We obtain PSPACE-membership via a polynomial-time reduction from $\mathbb{N}$-MSAT$(\Diamond, \Box, \downarrow, @)$ to the satisfiability problem $\mathbb{N}\text{-}\mathsf{SAT}_{\mathcal{FOL}}(<)$ for the fragment of first-order logic with the relation "$<$" interpreted over the natural numbers. Let the first order language contain all members of SVAR as variables and all members of NOM as constants. Based on the standard translation from hybrid to first-order logic [22], we devise a reduction H that maps hybrid formulae φ

and variables or constants z to first-order formulae.

$$H(p,z) := \top \text{ for } p \in \textsc{Prop} \qquad H(v,z) := v = z \quad \text{for } v \in \textsc{SVar} \cup \textsc{Nom}$$
$$H(\alpha \wedge \beta, z) := H(\alpha, z) \wedge H(\beta, z) \qquad H(\alpha \vee \beta, z) := H(\alpha, z) \vee H(\beta, z)$$
$$H(\Diamond\alpha, z) := \exists t(z < t \wedge H(\alpha, t)) \qquad H(\Box\alpha, z) := \forall t(z < t \rightarrow H(\alpha, t))$$
$$H(\downarrow x.\alpha, z) := \exists x(x = z \wedge H(\alpha, z)) \quad H(@_x\alpha, z) := H(\alpha, x)$$

In the $\Diamond$, $\Box$ and @-cases we deviate from the usual definition of the standard translation because we do not insist on using only two variables in addition to SVar—therefore it suffices to require that t is a fresh variable—and we allow constants in the second argument.

For a first-order formula ψ with variables in SVar and an assignment $g : \textsc{SVar} \rightarrow \mathbb{N}$, let $\psi[g]$ denote the first-order formula that is obtained from ψ by substituting every free occurrence of $x \in \textsc{SVar}$ by the first-order term that describes $g(x)$.

Claim 3.4 *For every instance φ of $\mathbb{N}$-MSAT$(\Diamond, \Box, \downarrow, @)$, every assignment $g : \textsc{SVar} \rightarrow \mathbb{N}$ and every $n \in \mathbb{N}$, it holds that: $g, n \models \varphi$ if and only if $(\mathbb{N}, <) \models H(\varphi, z)[g^z_n]$, where z is a new variable that does not occur in φ.*

Now, $\varphi \in \mathbb{N}$-MSAT$(\Diamond, \Box, \downarrow, @)$ if and only if $g, 0 \models \varphi \vee \Diamond\varphi$ for some assignment g. By the above claim, this is equivalent to $(\mathbb{N}, <) \models H(\varphi \vee \Diamond\varphi, z)[g^z_0]$ for some g and a new variable z, which can also be expressed as $(\mathbb{N}, <) \models \forall x(\neg(x < z) \wedge H(\varphi \vee \Diamond\varphi, z))$. This shows that $\mathbb{N}$-MSAT$(\Diamond, \Box, \downarrow, @)$ is polynomial-time reducible to $\mathbb{N}$-SAT$_{\mathcal{FOL}}(<)$, which was shown to be in PSPACE in [8]. Therefore, $\mathbb{N}$-MSAT$(\Diamond, \Box, \downarrow, @)$ is in PSPACE. □

4 The easy cases: NC^1 and LOGSPACE results

In this section, we show that the fragments without the $\Diamond$-operator have an easy satisfiability problem. Our results can be structured into four groups. First, we consider fragments without modal operators. For these fragments we obtain NC^1-completeness. Simply said, without negation and $\Diamond$ we cannot express that two nominals or state variables are not bound to the same state. Therefore, the model that binds all variables to the first state satisfies every satisfiable formula in this fragment.

Lemma 4.1 *Let $F_0 = (\{0\}, \emptyset)$ and $g_0(y) = 0$ for every $y \in \textsc{SVar}$. Then $\varphi \in$ lin-MSAT$(\downarrow, @)$ (resp. $\varphi \in \mathbb{N}$-MSAT$(\downarrow, @)$) if and only if $F_0, g_0, 0 \models \varphi$.*

Proof. The implication direction from left to right follows from the monotonicity of the considered formulas. For the other direction, notice that $F_0 \in$ lin. For frame class $\mathbb{N}$, note that if $F_0, g_0, 0 \models \varphi$ and φ has no modal operators, then $g_0, 0 \models \varphi$. □

Theorem 4.2 *Let $O \subseteq \{\downarrow, @\}$. Then* lin-MSAT$(O)$ *and* $\mathbb{N}$-MSAT(O) *are* NC^1-*complete.*

Proof. NC^1-hardness of $\mathfrak{F}$-MSAT$(\emptyset)$ follows immediately from the NC^1-completeness of the Formula Value Problem for propositional formulae [6]. It remains to show that $\mathsf{lin\text{-}MSAT}(\downarrow, @)$ and $\mathbb{N}\text{-}\mathsf{MSAT}(\downarrow, @)$ are in NC^1. In order to decide whether φ is in $\mathsf{lin\text{-}MSAT}(\downarrow, @)$, according to Lemma 4.1 it suffices to check whether the propositional formula obtained from φ deleting all occurrences of $\downarrow x.$ and $@_x$, is satisfied by the assignment that sets all atoms to *true*. According to [6] this can be done in NC^1. Since $\mathsf{lin\text{-}MSAT}(\downarrow, @) = \mathbb{N}\text{-}\mathsf{MSAT}(\downarrow, @)$ by Lemma 4.1, we obtain the same for $\mathbb{N}\text{-}\mathsf{MSAT}(\downarrow, @)$. □

Second, we consider fragments with the □-operator over linear frames. We can show NC^1-completeness here, too. The main reason is that (sub-)formulas that begin with a □ are satisfied in a state that has no successor. Therefore similar as above, every formula of this fragment that is satisfiable over linear frames is satisfied by a model with only one state.

Theorem 4.3 $\mathsf{lin\text{-}MSAT}(\Box, \downarrow, @)$ *is* NC^1*-complete.*

Proof. NC^1-hardness follows from Theorem 4.2. It remains to show that $\mathsf{lin\text{-}MSAT}(\Box, \downarrow, @) \in \mathsf{NC}^1$. We show that essentially the □-operators can be ignored.

Claim 4.4 $\mathsf{lin\text{-}MSAT}(\Box, \downarrow, @) \leqslant_{\mathrm{cd}} \mathsf{lin\text{-}MSAT}(\downarrow, @)$.

Proof of Claim. For an instance φ of $\mathsf{lin\text{-}MSAT}(\Box, \downarrow, @)$, let φ'' be the formula obtained from φ by replacing every subformula $\Box\psi$ of φ with the constant $\top$. Then φ'' is an instance of $\mathsf{lin\text{-}MSAT}(\downarrow, @)$. If $\varphi \in \mathsf{lin\text{-}MSAT}(\Box, \downarrow, @)$, then $\varphi'' \in \mathsf{lin\text{-}MSAT}(\downarrow, @)$ due to the monotonicity of φ. On the other hand, if $\varphi'' \in \mathsf{lin\text{-}MSAT}(\downarrow, @)$, then $K_0, g, 0 \models \varphi''$ (Lemma 4.1). Since $K_0, g, 0 \models \Box\alpha$ for every α, we obtain $K_0, g, 0 \models \varphi$, hence $\varphi \in \mathsf{lin\text{-}MSAT}(\Box, \downarrow, @)$. As such simple substitutions can be realized using an AC^0-circuit, the stated reduction is indeed a valid $\leqslant_{\mathrm{cd}}$-reduction from $\mathsf{lin\text{-}MSAT}(\Box, \downarrow, @)$ to $\mathsf{lin\text{-}MSAT}(\downarrow, @)$. ◇

Since $\mathsf{lin\text{-}MSAT}(\downarrow, @) \in \mathsf{NC}^1$ (Theorem 4.2) and NC^1 is closed downwards under $\leqslant_{\mathrm{cd}}$, it follows from the Claim that $\mathsf{lin\text{-}MSAT}(\Box, \downarrow, @) \in \mathsf{NC}^1$. □

It is clear that this argument does not apply to the natural numbers.

Third, we show NC^1-completeness for the fragments with □ and one of ↓ and @ over $\mathbb{N}$. They receive separate treatment because, in $(\mathbb{N}, <)$, every state has a successor, and therefore □-subformulas cannot be satisfied as easily as above. It turns out that the complexity of the satisfiability problem increases only if both hybrid operators can be used.

Theorem 4.5 $\mathbb{N}\text{-}\mathsf{MSAT}(\Box, @)$ *is* NC^1*-complete.*

Proof sketch. NC^1-hardness follows from Theorem 4.2.

We first consider $\mathbb{N}\text{-}\mathsf{MSAT}(\Box, @)$. We distinguish occurrences of nominals that are either *free*, or that are *bound* by a □, or that are bound by an @. Simply said, a free occurrence of i in α is bound by □ in $\Box\alpha$ and bound by @ in $@_x\alpha$ (even if $x \neq i$). Since the assignment g is not relevant for the considered fragment, we write $K, w \models \alpha$ for short instead of $K, g, w \models \alpha$.

Claim 4.6 *Let α' be the formula obtained from α by replacing every occurrence* *of a nominal that is bound by* $\Box$ *with* $\bot$, *and let* η *be a valuation. If* $\eta, k \models \alpha$, *then* $\eta, k \models \alpha'$.

Moreover, it turns out that binding every nominal to the initial state suffices to obtain a satisfying model.

Claim 4.7 $\varphi \in \mathbb{N}\text{-}\mathsf{MSAT}(\Box, @)$ *if and only if* $\eta_0, 0 \models \varphi$ *with* $\eta_0(x) = \{0\}$ *for every* $x \in \text{NOM}$.

Both claims together yield that, in order to decide $\varphi \in \mathbb{N}\text{-}\mathsf{MSAT}(\Box, @)$, it suffices to check whether $\eta_0, 0 \models \varphi'$. No nominal in φ' occurs bound by a $\Box$-operator. Therefore for every subformula $\Box\alpha$ of φ' and for every k holds: $\eta_0, k \models \alpha$ if and only if $\eta_0, 0 \models \alpha$. All nominals that occur free or bound by an @ evaluate to *true* in state 0 via η_0. Therefore, in order to decide $\eta_0, 0 \models \varphi'$, it suffices to ignore all $\Box$ and @-operators of φ' and evaluate it as a propositional formula under assignment η_0 that sets all atoms of φ' to *true*. This can be done in NC^1 [6]. The complete proof can be found in the Technical Report version of this paper [10]. $\Box$

Next, we consider $\mathbb{N}\text{-}\mathsf{MSAT}(\Box, \downarrow)$. According to our remarks in Section 2 about notational convenience, we assume that there are no nominals in $\mathcal{MHL}(\Box, \downarrow)$.

Theorem 4.8 $\mathbb{N}\text{-}\mathsf{MSAT}(\Box, \downarrow)$ *is* NC^1*-complete.*

Proof sketch. Now, we distinguish occurrences of state variables as the occurrences in the proof sketch above. They are either *free*, or they are *bound* by a $\Box$, or they are *bound* by $\downarrow$. Note that this phrasing differs from the standard usage of the terms 'free' and 'bound' in the context of state variables. A free occurrence of i in α is bound by $\Box$ in $\Box\alpha$, as above. It is bound by $\downarrow$ in $\downarrow i.\alpha$ only. Notice that y occurs free in $\downarrow x.y$ (for $x \neq y$).

Claim 4.9 *Let* α' *be the formula obtained from* α *by replacing every occurrence of a state variable that is bound by* $\Box$ *with* $\bot$, *and let* g *be an assignment. If* $g, k \models \alpha$, *then* $g, k \models \alpha'$.

Claim 4.10 $\varphi \in \mathbb{N}\text{-}\mathsf{MSAT}(\Box, \downarrow)$ *if and only if* $g_0, 0 \models \varphi$, *for* $g_0(x) = 0$ *for every* $x \in \text{SVAR}$.

Both claims together yield that, in order to decide $\varphi \in \mathbb{N}\text{-}\mathsf{MSAT}(\Box, \downarrow)$, it suffices to check whether $g_0, 0 \models \varphi'$. No state variable in φ' occurs bound by a $\Box$-operator. Therefore for every subformula $\Box\alpha$ of φ' and for every k holds: $g_0, k \models \alpha$ if and only if $g_0, 0 \models \alpha$. All occurrences of state variables in φ' that are bound by $\downarrow$ evaluate to *true*, because no $\Box$ occurs "between" the binding $\downarrow i$ and the occurrence of i, which means that the state where the variable is bound is the same as where the variable is used. All free occurrences of state variables evaluate to *true* in state 0 due to g_0. Therefore, in order to decide $g_0, 0 \models \varphi'$, it suffices to ignore all $\Box$ and $\downarrow$-operators of φ' and evaluate it as a propositional formula under an assignment that sets all atoms to *true*. This can be done in NC^1 [6]. The complete proof can be found in the Technical Report version of this paper [10]. $\Box$

The fourth part deals with the fragment with $\Box$ and both $\downarrow$ and @ over the natural numbers.

Lemma 4.11 $\mathbb{N}$-MSAT($\Box, \downarrow, @$) *is* LOGSPACE-*hard.*

Proof. This proof is very similar to the proof of Theorem 3.3. in [15]. We give a reduction from the problem *Order between Vertices* (ORD) which is known to be LOGSPACE-complete [7] and defined as follows.

Problem: ORD

Input: A finite set of vertices V, a successor-relation S on V, and two vertices $s, t \in V$.

Output: Is $s \leqslant_S t$, where $\leqslant_S$ denotes the unique total order induced by S on V?

Notice that (V, S) is a directed line-graph. Let (V, S, s, t) be an instance of ORD. We construct an $\mathcal{MHL}(\Box, \downarrow, @)$-formula φ that is satisfiable if and only if $s \leqslant_S t$. We use $V = \{v_0, v_1, \dots, v_n\}$ as state variables. The formula φ consists of three parts. The first part binds all variables except s to one state and the variable s to a successor of this state. The second part of φ binds a state variable v_l to the state labeled by s iff $s \leqslant_S v_l$. Let α denote the concatenation of all $@_{v_k}{\downarrow}v_l$ with $(v_k, v_l) \in S$ and $v_l \neq s$, and α^n denotes the n-fold concatenation of α. Essentially, α^n uses the assignment to collect eventually all v_i with $s \leqslant_S v_i$ in the state labeled s. The last part of φ checks whether s and t are bound to the same state after this procedure. That is, $\varphi = {\downarrow}v_0.{\downarrow}v_1.{\downarrow}v_2.\cdots{\downarrow}v_n.\Box{\downarrow}s.\ \alpha^n\ @_s t$. To prove the correctness of our reduction, we show that φ is satisfiable if and only if $s \leqslant_S t$.

Assume $s \leqslant_S t$. For an arbitrary assignment g, one can show inductively that $g, 0 \models {\downarrow}v_0.{\downarrow}v_1.\cdots{\downarrow}v_n.\Box{\downarrow}s.\ \alpha^i\ @_s r$ for $i = 0, 1, \dots, n$ and for all r that have distance i from s. Therefore it eventually holds that $g, 0 \models \varphi$. For $s \nleqslant_S t$ we show that $g, n \not\models \varphi$ for any assignment g and natural number n. Let g_0 be the assignment obtained from g after the bindings in the prefix ${\downarrow}v_0.{\downarrow}v_1.\cdots{\downarrow}v_n.\Box{\downarrow}s$ of φ, and let g_i be the assignment obtained from g_0 after evaluating the prefix of φ up to and including α^i. It holds that $g_i(s) \neq g_i(t) = 0$ for all $i = 0, 1, \dots, n$. This leads to $g_n, 0 \not\models @_s t$ and therefore $g, 0 \not\models \varphi$. $\Box$

For the upper bound, we establish a characterisation of the satisfaction relation that assigns a *unique* assignment and state of evaluation to every subformula of a given formula φ. Using this new characterisation, we devise a decision procedure that runs in logarithmic space and consists of two steps: it replaces every occurrence of any state variable x in φ with 1 if its state of evaluation agrees with that of its ${\downarrow}x$-superformula, and with 0 otherwise; it then removes all $\Box$-, $\downarrow$- and @-operators from the formula and tests whether the resulting Boolean formula is valid.

Theorem 4.12 $\mathbb{N}$-MSAT($\Box, \downarrow, @$) *is in* LOGSPACE .

The proof is technically involved and can be found in the Technical Report version of this paper [10].

5 The intermediate cases: NP results

After we have seen that all fragments without $\Diamond$ have an easy satisfiability problem, we show that $\Diamond$ together with the use of nominals makes the satisfiability problem NP-hard. Recall that, owing to the presence of nominals, $\mathcal{MHL}(\Diamond)$ is not just modal logic with the $\Diamond$-operator. The absence of $\downarrow$ makes assignments superfluous: we write $K, w \models \varphi$ instead of $K, g, w \models \varphi$.

Lemma 5.1 lin-MSAT($\Diamond$) *and* $\mathbb{N}$-MSAT($\Diamond$) *both are* NP-*hard.*

Proof sketch. We reduce from 3SAT. Let $\varphi = c_1 \wedge \ldots \wedge c_n$ be an instance of 3SAT with clauses $c_1, \ldots, c_n$ (where $c_i = (l_1^i \vee l_2^i \vee l_3^i)$ for literals l_j^i) and variables $x_1, \ldots, x_m$. We define the transformation as

$$f\colon \varphi \mapsto \Diamond(i_0 \wedge \Diamond i_1) \;\wedge\; \left(\bigwedge_{\ell=1}^{m} \Diamond(i_0 \wedge x_\ell) \vee \Diamond(i_1 \wedge x_\ell) \right) \wedge h(\varphi),$$

where i_0, i_1 and all x_ℓ are nominals, and the function h is defined as follows: let l_k^j be a literal in clause c_j, then

$$h(l_k^j) := \begin{cases} (i_1 \wedge x), & \text{if } l_k^j = x \\ (i_0 \wedge x), & \text{if } l_k^j = \neg x \end{cases}$$

$$h(c_j) := \Diamond(h(l_1^j) \vee h(l_2^j) \vee h(l_3^j)), \quad \text{where } c_j = (l_1^j \vee l_2^j \vee l_3^j);$$

$$h(c_1 \wedge \cdots \wedge c_n) := h(c_1) \wedge \cdots \wedge h(c_n).$$

Notice that f turns variables in the 3SAT instance into *nominals* in the lin-MSAT($\Diamond$) instance. The part $\Diamond(i_0 \wedge \Diamond i_1)$ enforces the existence of two successors w_1 and w_2 of the state satisfying $f(\varphi)$. The part $\bigwedge_{\ell=1}^{m} \Diamond(i_0 \wedge x_\ell) \vee \Diamond(i_1 \wedge x_\ell)$ simulates the assignment of the variables in φ, enforcing that each x_ℓ is true in either w_1 or w_2. The part $h(\varphi)$ then simulates the evaluation of φ on the assignment determined by the previous parts. With the following claim NP-hardness of lin-MSAT($\Diamond$) follows.

Claim 5.2 $\varphi \in$ 3SAT *if and only if* $h(\varphi) \in$ lin-MSAT($\Diamond$).

Using this claim, NP-hardness of lin-MSAT($\Diamond$) follows. It is straightforward to show that 3SAT reduces to $\mathbb{N}$-MSAT($\Diamond$) using the same reduction. □

We will now establish NP-membership of the problems $\mathfrak{F}$-MSAT($\Diamond, \Box, \downarrow$), $\mathfrak{F}$-MSAT($\Diamond, \Box, @$), and $\mathfrak{F}$-MSAT($\Diamond, \downarrow, @$) for $\mathfrak{F} \in \{\text{lin}, \mathbb{N}\}$. For the first two, this follows from the literature, see Theorem 2.1 (4). For the third, we observe that all modal and hybrid operators in a formula φ from the fragment $\mathcal{MHL}(\Diamond, \downarrow, @)$ are translatable into FOL by the standard translation using no universal quantifiers. The existential quantifiers introduced by the binder can be skolemised away, which corresponds to removing all binding from φ and replacing each state variable with a fresh nominal. The correctness of this translation is proven in [22]. Hence, $\mathfrak{F}$-MSAT($\Diamond, \downarrow, @$) polynomial-time reduces to $\mathfrak{F}$-MSAT($\Diamond, @$).

Lemma 5.3 $\mathsf{lin}\text{-}\mathsf{MSAT}(\Diamond, \downarrow, @)$ *and* $\mathbb{N}\text{-}\mathsf{MSAT}(\Diamond, \downarrow, @)$ *are in* NP.

From the lower bounds in Lemma 5.1 and the upper bounds in Theorem 2.1 (4) and Lemma 5.3, we obtain the following theorem.

Theorem 5.4 *Let* $\{\Diamond\} \subseteq O$, *and* $O \subsetneq \{\Diamond, \Box, \downarrow, @\}$. *Then* $\mathsf{lin}\text{-}\mathsf{MSAT}(O)$ *and* $\mathbb{N}\text{-}\mathsf{MSAT}(O)$ *are* NP-complete.

In addition to the NP-membership of the fragments captured by Theorem 5.4, we are interested in their model-theoretic properties. We show that these logics enjoy a kind of linear-size model property, precisely a quasi-quadratic size model property: over the natural numbers, every satisfiable formula has a model where two successive nominal states have at most linearly many intermediary states, and the states behind the last such state are indistinguishable. This property allows for an alternative worst-case decision procedure for satisfiability that consists of guessing a linear representation of a model of the described form and symbolically model-checking the input formula on that model. Over general linear frames, which may have dense intervals, we formulate the model property in a more general way and prove it using additional technical machinery to deal with density. However, the result then carries over to the rationals, where we are not aware of any upper complexity bound in the literature.

In [20], Sistla and Clarke showed a variation of the linear-size model property for LTL(F), which corresponds to $\mathcal{HL}(\Diamond, \Box)$ over $\mathbb{N}$: whenever $\varphi \in \mathcal{HL}(\Diamond, \Box)$ is satisfiable over $\mathbb{N}$, then it is satisfiable in the initial state of a model over $\mathbb{N}$ which has a linear-sized prefix init and a remainder final such that final is maximal with respect to the property that every type (set of all atomic propositions true in a state) occurs infinitely often, and final contains only linearly many types. Such a structure can be guessed in polynomial time, represented in polynomial space and model-checked in polynomial time. While it is straightforward to extend Sistla and Clarke's proof to cover nominals and the @ operator, it will not go through if density is allowed (frame class lin).

We establish that $\mathcal{MHL}(\Diamond, \Box, @)$ over lin has a quadratic size model property, and we subsequently show how to extend the result to the other fragments from Theorem 5.4 and how to restrict them to $\mathbb{N}$.

Theorem 5.5 $\mathcal{MHL}(\Diamond, \Box, @)$ *has the quasi-quadratic size model property with respect to* lin *and* $\mathbb{N}$.

The proof can be found in the Technical Report version of this paper [10].

As an immediate consequence, the model property in Theorem 5.5 carries over to the subfragments $\mathcal{MHL}(\Diamond, \Box)$, $\mathcal{MHL}(\Diamond, @)$, $\mathcal{MHL}(\Box, @)$, $\mathcal{MHL}(\Diamond)$, $\mathcal{MHL}(\Box)$, $\mathcal{MHL}(@)$, and $\mathcal{MHL}(\emptyset)$. Moreover, our arguments in the proofs of Theorems 4.3 and 4.12 can be used to transfer it to $\mathcal{MHL}(\Box, \downarrow, @)$. Together with the observations that

- $\mathcal{MHL}(\Diamond, \downarrow, @)$ is no more expressive than $\mathcal{MHL}(\Diamond, @)$ (see the explanation before Lemma 5.3), and
- $\mathcal{MHL}(\Diamond, \Box, \downarrow)$ is no more expressive than $\mathcal{MHL}(\Diamond, \Box)$ (because, without @, one cannot jump to named states),

we obtain the following generalisation of Theorem 5.5.

Corollary 5.6 *Let* $O \subsetneq \{\Diamond, \Box, \downarrow, @\}$. *Then* $\mathcal{MHL}(O)$ *has the quasi-quadratic size model property with respect to* lin *and* $\mathbb{N}$.

6 Conclusion

We have completely classified the complexity of all fragments of hybrid logic with monotone Boolean operators obtained from arbitrary combinations of four modal and hybrid operators, over linear frames and the natural numbers. Except for the largest such fragment over linear frames, all fragments are of elementary complexity. We have classified their complexity into PSPACE-complete, NP-complete and tractable and shown that the tractable cases are complete for either NC^1 or LOGSPACE . Surprisingly, while the largest fragment is harder over linear frames than over $(\mathbb{N}, <)$, the largest $\Diamond$-free fragment is easier over linear frames than over $(\mathbb{N}, <)$.

The question remains whether the PSPACE-complete largest fragment over $(\mathbb{N}, <)$ admits some quasi-polynomial size model property. Furthermore, this study can be extended in several possible ways: by allowing negation on atomic propositions, by considering frame classes that consist only of dense frames, such as $(\mathbb{Q}, <)$, or by considering arbitrary sets of Boolean operators in the same spirit as in [15]. For atomic negation, it follows quite easily that the largest fragment is of non-elementary complexity over $(\mathbb{N}, <)$, too, and that all fragments except $O = (\Box, \downarrow, @)$ are NP-complete. However, our proof of the quasi-quadratic model property does not immediately go through in the presence of atomic propositions. Over $(\mathbb{Q}, <)$, we conjecture that all fragments, except possibly for the largest one, have the same complexity and model properties as over $(\mathbb{N}, <)$.

References

[1] Areces, C., P. Blackburn and M. Marx, *A road-map on complexity for hybrid logics*, in: *Proc. CSL-99*, LNCS **1683**, 1999, pp. 307–321.

[2] Areces, C., P. Blackburn and M. Marx, *The computational complexity of hybrid temporal logics*, Logic Journal of the IGPL **8** (2000), pp. 653–679.

[3] Bauland, M., E. Hemaspaandra, H. Schnoor and I. Schnoor, *Generalized modal satisfiability.*, in: *Proc. STACS*, 2006, pp. 500–511.

[4] Blackburn, P. and J. Seligman, *Hybrid languages*, JoLLI **4** (1995), pp. 41–62.

[5] Bozzelli, L. and R. Lanotte, *Complexity and succinctness issues for linear-time hybrid logics*, in: *Proc. of 11th JELIA*, LNCS **5293**, 2008, pp. 48–61.

[6] Buss, S. R., *The Boolean formula value problem is in ALOGTIME*, in: *Proceedings 19th Symposium on Theory of Computing* (1987), pp. 123–131.

[7] Etessami, K., *Counting quantifiers, successor relations, and logarithmic space*, J. of Comp. and Sys. Sci. **54** (1997), pp. 400–411.

[8] Ferrante, J. and C. W. Rackoff, "The Computational Complexity of Logical Theories," Springer-Verlag, 1979.

[9] Franceschet, M., M. de Rijke and B. Schlingloff, *Hybrid logics on linear structures: Expressivity and complexity*, in: *Proc. 10th TIME*, 2003, pp. 166–173.

[10] Göller, S., A. Meier, M. Mundhenk, T. Schneider, M. Thomas and F. Weiß, *The complexity of monotone hybrid logics over linear frames and the natural numbers* (2011). URL `http://arxiv.org/abs/1204.1196`
[11] Goranko, V., *Hierarchies of modal and temporal logics with reference pointers*, Journal of Logic, Language and Information **5** (1996), pp. 1–24.
[12] Henzinger, T., *Half-order modal logic: How to prove real-time properties*, in: *Proc. PODC*, 1990, pp. 281–296.
[13] Ladner, R., *The computational complexity of provability in systems of modal propositional logic*, SIAM Journal on Computing **6** (1977), pp. 467–480.
[14] Mahajan, M., *Polynomial size log depth circuits: between* NC^1 *and* AC^1, Bulletin of the EATCS **91** (2007).
[15] Meier, A., M. Mundhenk, T. Schneider, M. Thomas, V. Weber and F. Weiss, *The complexity of satisfiability for fragments of hybrid logic - Part I*, J. Applied Logic **8** (2010), pp. 409–421.
[16] Mundhenk, M. and T. Schneider, *The complexity of hybrid logics over equivalence relations*, JoLLI **18** (2009), pp. 433–624.
[17] Mundhenk, M., T. Schneider, T. Schwentick and V. Weber, *Complexity of hybrid logics over transitive frames*, J. Applied Logic **8** (2010), pp. 422–440.
[18] Papadimitriou, C. H., "Computational Complexity," Addison-Wesley, 1994.
[19] Schwentick, T. and V. Weber, *Bounded-variable fragments of hybrid logics*, in: *Proc. 24th STACS*, LNCS **4393** (2007), pp. 561–572.
[20] Sistla, A. and E. Clarke, *The complexity of propositional linear temporal logics*, Journal of the ACM **32** (1985), pp. 733–749.
[21] Stockmeyer, L. J., "The complexity of decision problems in automata theory and logic," Ph.D. thesis, Mass.Inst.of Technology (1974).
[22] ten Cate, B. and M. Franceschet, *On the complexity of hybrid logics with binders*, in: *Proc. 19th CSL, 2005*, LNCS **3634** (2005), pp. 339–354.
[23] Vollmer, H., "Introduction to Circuit Complexity," Springer, 1999.
[24] Weber, V., *Branching-time logics repeatedly referring to states*, J. of Logic, Language and Information **18** (2009), pp. 593–624.

Labelled Tree Sequents, Tree Hypersequents and Nested (Deep) Sequents

Rajeev Goré [1]

Logic and Computation Group
Research School of Computer Science
The Australian National University
Canberra, ACT 0200, Australia

Revantha Ramanayake [2]

CNRS
LIX, École Polytechnique
91128 Palaiseau, France

Abstract

We identify a subclass of labelled sequents called "labelled tree sequents" and show that these are notational variants of tree-hypersequents in the sense that a sequent of one type can be represented naturally as a sequent of the other type. This relationship can be extended to nested (deep) sequents using the relationship between tree-hypersequents and nested (deep) sequents, which we also show.
We apply this result to transfer proof-theoretic results such as syntactic cut-admissibility between the tree-hypersequent calculus $CSGL$ and the labelled sequent calculus $G3GL$ for provability logic GL. This answers in full a question posed by Poggiolesi about the exact relationship between these calculi.
Our results pave the way to obtain cut-free tree-hypersequent and nested (deep) sequent calculi for large classes of logics using the known calculi for labelled sequents, and also to obtain a large class of labelled sequent calculi for bi-intuitionistic tense logics from the known nested (deep) sequent calculi for these logics. Importing proof-theoretic results between notational variant systems in this manner alleviates the need for independent proofs in each system. Identifying which labelled systems can be rewritten as labelled tree sequent systems may provide a method for determining the expressive limits of the nested sequent formalism.

Keywords: labelled tree sequents, notational variants, cut-elimination, proof theory.

[1] rajeev.gore@anu.edu.au

[2] revantha@lix.polytechnique.fr

1 Introduction

Gentzen [6] introduced the *sequent calculus* as a tool for studying proof systems for classical and intuitionistic logics. Gentzen sequent calculi are built from *traditional sequents* of the form $X \Rightarrow Y$ where X and Y are formula multisets. The main result is the cut-elimination theorem, which shows how to eliminate the cut-rule from these calculi. The resulting sequent calculi are said to be *cut-free.* A significant drawback of the Gentzen sequent calculus is the difficulty of adapting the calculus to new logics. For example, although there is a traditional cut-free Gentzen sequent calculus for $S4$, there is no known traditional cut-free Gentzen calculus for $S5$ despite the fact that the logic $S5$ can be directly obtained from a Hilbert calculus [3] for $S4$ by the addition of a single axiom corresponding to symmetry. This has lead to various generalisations of the Gentzen sequent calculus in an attempt to present many different (modal) logics in a single modular proof-theoretic framework with nice properties.

Hypersequent calculi [18,1] generalise Gentzen sequent calculi by using a /-separated list of traditional sequents (a *hypersequent*) rather than just a single one. Usually, the order of the sequents is not important so a multiset can be used instead of a list. In this case the hypersequent $X \Rightarrow Y/U \Rightarrow V$ is the same as $U \Rightarrow V/X \Rightarrow Y$, for example.

Tree-hypersequents [16] are defined from hypersequents through the addition of the symbols ; and ( ) to the syntax, and by attaching importance to the *order* of the traditional sequents. Furthermore, the placement of the / and ; symbols play a crucial role in the semantic meaning of a tree-hypersequent, enabling each tree-hypersequent to be associated with a tree-like frame. For example, the tree-hypersequents $-/(-/(-;-));-$ and $-/-/-/(-;-)$, where the dashes stand for sequents, correspond to the (tree) frame figures below left and below right respectively:

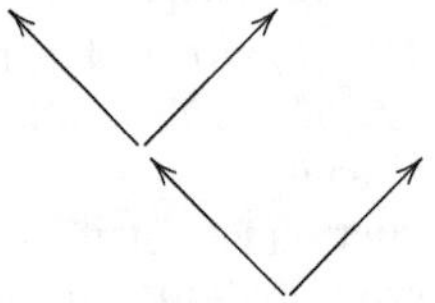

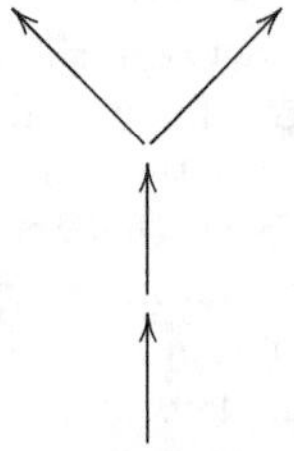

Nested (Deep) Sequents [11,2] generalise traditional sequents through the addition of the symbol [ ] to the syntax, giving us unordered but nested expressions of the form $\Rightarrow [\cdot, [\cdot], [\cdot]], [\cdot]$ and $\Rightarrow [\cdot, [\cdot, [\cdot], [\cdot]]]$ to capture the same frames as above. Further connections with display sequents are also known [7].

Labelled sequents [10,4,13] generalise the traditional sequent by the prefixing of indices or labels to formulae occurring in the sequent. A labelled sequent can be viewed as a directed graph with sequents at each node [20].

Negri [14] has presented a method for generating cut-free labelled sequent calculi for a large family of modal logics. These labelled sequent calculi incorporate the frame accessibility relation into the syntax of the calculi.

All of these extended sequent formalism are modular, since a new logic can be presented by the inclusion of extra rules corresponding to the properties of its accessibility relation or to the appropriate modal axioms. The modal logic $S5$, for example, can be given a cut-free presentation by adding rules for reflexivity, transitivity and symmetry to the base calculus, or by adding rules that capture the axioms T, 4 and 5, as appropriate to the formalism.

A *labelled tree sequent* is a special instance of a labelled sequent where the underlying graph structure is restricted to a tree. Labelled tree sequents have appeared in various guises in the literature, where they have been used to construct calculi for non-classical logics (for example, see [9]). We observe that restricting the underlying graph structure of a labelled tree sequent to a tree (forcing irreflexivity, for example) does not limit the logics that we can handle to simply K-like or GL-like logics. This is because the formulae used to construct the inference rules may of course contain modalities, consequently enriching the expressiveness of the framework. As a trivial example, a standard Gentzen sequent calculus $\mathcal{C}_{S4}$ for the reflexive and transitive logic $S4$ induces a labelled tree sequent calculus C'_{S4} for $S4$, obtained by replacing each traditional sequent $\Gamma \Rightarrow \Delta$ in each inference rule in $\mathcal{C}_{S4}$ with the labelled tree sequent $x : \Gamma \Rightarrow x : \Delta$ where $x : \Gamma$ is obtained by prefixing each formula in Γ with the label x.

Here we establish mappings between tree-hypersequent calculi and labelled tree sequent calculi. This result shows that these systems are notational variants. Using this result it becomes possible to transfer proof-theoretic results between these systems, alleviating the need for independent proofs in each system. As an application of this work, we answer in full a question posed by Poggiolesi [17] regarding the relationship between tree-hypersequent and labelled sequent calculi for provability logic. We envisage that the existing results in the general labelled sequent framework may be coerced under suitable restrictions to provide new proof systems for tree-hypersequent and nested sequent calculi.

Related Work. This paper is based on work appearing in Ramanayake's [19] PhD thesis. The independent but contemporaneous work of Fitting [5] shows that there is a to-and-fro correspondence between prefixed tableaux and Brünnler's deep sequents [2], and coincidentally, uses exactly the same term "notational variants" as does Ramanayake.

Fitting does not prove syntactic cut-admissibility for his prefixed tableaux, but uses the standard semantic completeness proof to establish "cut-free completeness". On the other hand, in this paper, we obtain syntactic transformations between the various calculi that we study, which lead to syntactic proofs of cut-admissibility.

Fitting notes that prefixed tableaux are subsumed by Negri's [14] labelled systems, and states that clarifying the relationship between labelled systems and Bruennler's deep sequent systems is an interesting question. We answer this question as follows.

Since deep sequents are an independent re-invention of Kashima's much older notion of nested sequents [11], we use the term nested (deep) sequents uniformly. In Section 5, we present to-and-fro maps between labelled tree sequent and nested (deep) sequent systems, thus providing an answer to Fitting's question. Finally, since the labelled tree sequent is a proper subclass of the labelled sequent, by ascertaining which labelled sequents cannot be written as labelled tree sequents, it may be possible to determine the expressive limits of the nested sequent formalism. This is a primary motivation for studying labelled tree sequents, rather than just working with the more expressive labelled sequents of Negri.

2 Preliminaries

The *basic modal language* $\mathcal{ML}$ is defined using a countably infinite set $Atms = \{p_i\}_{i \in \mathbb{N}}$ of propositional variables, the usual propositional connectives $\neg$, $\vee$, $\wedge$ and $\supset$, the unary modal operators $\Box$ and $\Diamond$, and the parenthesis symbols ( ).

A *formula* is an expression generated by the following grammar

$$A \ ::= \ p \in Atms \mid \neg A \mid (A \vee B) \mid (A \wedge B) \mid (A \supset B) \mid \Box A \mid \Diamond A$$

In this paper we work exclusively with classical modal logics. In this context we have the freedom of working with certain proper subsets of $\{\neg, \vee, \wedge, \supset, \Box, \Diamond\}$, as the missing language elements can be defined in terms of the remaining ones.

A *traditional sequent* (denoted $X \Rightarrow Y$) is an ordered pair (X, Y) where X (the 'antecedent') and Y (the 'succedent') are finite multisets of formulae.

Definition 2.1 (Gentzen sequent calculus) *The* Gentzen sequent calculus *consists of some set of traditional sequents (the* initial sequents*) and some set of* inference rules*, each of the form*

$$\frac{\mathcal{S}_1 \ldots \mathcal{S}_n}{\mathcal{S}}$$

where the traditional sequents $\mathcal{S}_1, \ldots, \mathcal{S}_n$ are called the premises *of the rule, and $\mathcal{S}$ is called the* conclusion *sequent.*

2.1 Tree-hypersequent calculi

A tree-hypersequent is built from traditional sequents using the symbols '/' and ';', and the parenthesis symbols ( ).

Definition 2.2 *A* tree-hypersequent *is defined inductively as follows:*

(i) if $\mathcal{S}$ is a traditional sequent, then $\mathcal{S}$ is a tree-hypersequent,

(ii) if $\mathcal{S}$ is a traditional sequent and $G_1, G_2, \ldots, G_n$ are tree-hypersequents, then $\mathcal{S}/(G_1; G_2; \ldots; G_n)$ is a tree-hypersequent.

We write THS as an abbreviation for the term 'tree-hypersequent(s)'.

As usual, we often introduce or delete parentheses for the sake of clarity. Following are some examples of THS:

$$p \Rightarrow p \vee q \qquad\qquad p \Rightarrow q/((q \Rightarrow r/\Box r \Rightarrow \neg s); l \Rightarrow \Box(p \supset s))$$

We write $G\{\overset{\pi}{H}\}$ to mean that π is an occurrence of the THS H in the THS G. Within the context of any discussion within this paper, we will be concerned with at most *one fixed occurrence* of a THS H in G. Thus, following standard practice, we will refer only implicitly to the specific occurrence, by dropping the occurrence name π and writing $G\{H\}$ to mean that H occurs at a 'distinguised position' in G. This will not cause ambiguity in practice. For a THS H', we write $G\{H'\}$ to mean the THS obtained from $G\{H\}$ by replacing *that* fixed occurrence of H with H'.

Throughout this paper, we will use an underlined capitalised letter of the form $\underline{X}$ (possibly with subscripts) to denote a ;-separated sequence $G_1; \ldots ; G_n$ of THS. We define the notion of *equivalent position* between occurrences of the traditional sequents $\mathcal{S}_1$ and $\mathcal{S}_2$ at distinguished positions in the THS $G_1\{\mathcal{S}_1\}$ and $G_2\{\mathcal{S}_2\}$ respectively (denoted $G_1\{\mathcal{S}_1\} \sim G_2\{\mathcal{S}_2\}$) as follows:

(i) if G is $\mathcal{S}_1$ and H is $\mathcal{S}_2$, then $G\{\mathcal{S}_1\} \sim H\{\mathcal{S}_2\}$;

(ii) if G is $\mathcal{S}_1/\underline{X}_1$ and H is $\mathcal{S}_2/\underline{X}_2$, where the distinguished positions of $\mathcal{S}_1$ and $\mathcal{S}_2$ are the pictured ones, then $G\{\mathcal{S}_1\} \sim H\{\mathcal{S}_2\}$;

(iii) if $H_1\{\mathcal{S}_1\} \sim H_2\{\mathcal{S}_2\}$ then $\mathcal{T}_1/(H_1\{\mathcal{S}_1\}; \underline{X}_1) \sim \mathcal{T}_2/(H_2\{\mathcal{S}_2\}; \underline{X}_2)$ where $\mathcal{T}_1$ and $\mathcal{T}_2$ are traditional sequents.

The intended interpretation $\mathcal{I}$ of a THS as a formula is:

$$(X \Rightarrow Y)^{\mathcal{I}} = \wedge X \supset \vee Y \qquad (\mathcal{S}/(G_1; \ldots G_n))^{\mathcal{I}} = \mathcal{S}^{\mathcal{I}} \vee \Box G_1^{\mathcal{I}} \vee \ldots \vee \Box G_n^{\mathcal{I}}$$

Definition 2.3 (tree-hypersequent calculus) *Obtained from Definition 2.1 with the phrase 'traditional sequent' replaced with 'tree-hypersequent'.*

Define $X \Rightarrow Y \otimes U \Rightarrow V$ as $X, U \Rightarrow Y, V$. Let $H\{\mathcal{S}\}$ and $H'\{\mathcal{S}'\}$ be THS such that $H\{\mathcal{S}\} \sim H'\{\mathcal{S}'\}$. Then define $H\{\mathcal{S}\} \star H'\{\mathcal{S}'\}$ inductively as follows:

(i) $\mathcal{S} \star \mathcal{S}' = \mathcal{S} \otimes \mathcal{S}'$

(ii) $(\mathcal{S}/\underline{X}) \star (\mathcal{S}'/\underline{Y}) = (\mathcal{S} \otimes \mathcal{S}'/\underline{X}; \underline{Y})$

(iii) $(\mathcal{T}/H\{\mathcal{S}\}; \underline{X}) \star (\mathcal{T}'/H'\{\mathcal{S}'\}; \underline{Y}) = \mathcal{T} \otimes \mathcal{T}'/(H\{\mathcal{S}\} \star H'\{\mathcal{S}'\}; \underline{X}; \underline{Y})$ where $\mathcal{T}$ and $\mathcal{T}'$ are traditional sequents:

We define the cut-rule as follows: for THS $G\{X \Rightarrow Y, A\}$ and $G'\{A, U \Rightarrow V\}$ such that $G\{X \Rightarrow Y, A\} \sim G'\{A, U \Rightarrow V\}$,

$$\frac{G\{X \Rightarrow Y, A\} \qquad G'\{A, U \Rightarrow V\}}{G\{X \Rightarrow Y\} \star G'\{U \Rightarrow V\}} \; cut$$

The $\star$ operation can be viewed as a merge operation on trees, and it ensures that the conclusion sequent of the cut-rule is indeed a THS.

2.2 Labelled sequent calculi

Fitting [4] has described the incorporation of frame semantics into tableau proof systems for the purpose of obtaining tableau systems for certain logics. Approaches to internalise the frame semantics into the Gentzen sequent calculus via the labelling of formulae appear in Mints [13], Vigano [21] and Kushida and

Okada [12]. Both approaches originate from Kanger's "spotted formulae" [10]. Here, we use the labelled systems for modal logic presented in Negri [14].

Assume that we have at our disposal an infinite set $\mathbb{SV}$ of ('state') variables disjoint from the set of propositional variables. We will use the letters $x, y, z \ldots$ to denote state variables. A *labelled formula* has the form $x : A$ where x is a state variable and A is a formula. If $X = \{A_1, \ldots A_n\}$ is a formula multiset, then $x : X$ denotes the multiset $\{x : A_1, \ldots, x : A_n\}$ of labelled formulae. Notice that if the formula multiset X is empty, then the labelled formula multiset $x : X$ is also empty. A *relation term* is a term of the form Rxy where x and y are variables. A (possibly empty) set of relations terms is called a *relation set*. A *labelled sequent* (denoted $\mathcal{R}, X \Rightarrow Y$) is the ordered triple $(\mathcal{R}, X, Y)$ where $\mathcal{R}$ is a relation set and X ('antecedent') and Y ('succedent') are multisets of labelled formulae.

Definition 2.4 (labelled sequent calculus) *Obtained from Definition 2.1 with the phrase 'traditional sequent' replaced with 'labelled sequent'. Moreover, each inference rule may include a* standard variable restriction *of the form "z does not appear in the conclusion sequent of the rule" for some state variable z.*

Observe that the standard variable restriction is a specific type of side condition on an inference rule.

2.3 Labelled tree sequent calculi

We begin by introducing some terminology and notation.

A *frame* is a pair $F = (W, R)$ where W is a non-empty set of states and R is a binary relation on W. For $x \in W$, define the subframe $x\uparrow$ in the usual way [3] as (W', R') where W' is the minimal upward closed set of $\{x\}$ wrt R, and R' is the restriction of R to W'. When $F = x\uparrow$ we say that F is *generated* by x and x is said to be a root of F. If a frame has a root it is said to be rooted. A rooted frame whose underlying *undirected* graph does not contain a path from any state back to itself (ie. no cycles) is called a *tree*. For example, a frame containing a reflexive state is not a tree. Although a rooted frame may have multiple roots, due to the prohibition of cycles, a tree has exactly one root.

If S is a set of states and Γ is a multiset of labelled formulae, then Γ_S is the multiset $\{x : A \mid x \in S \text{ and } x : A \in \Gamma\}$. So $\Gamma_{\{x\}}$ is the multiset of labelled formulae in Γ that are labelled with the state x. With a slight abuse of notation, we write this multiset as Γ_x.

A (possibly empty) set of relations terms (ie. terms of the form Rxy) is called a *relation set*. For a relation set $\mathcal{R}$, the *frame Fr*($\mathcal{R}$) *defined by* $\mathcal{R}$ is given by $(|\mathcal{R}|, \mathcal{R})$ where $|\mathcal{R}| = \{x \mid Rxv \in \mathcal{R} \text{ or } Rvx \in \mathcal{R} \text{ for some state } v\}$. In the reverse direction, given a frame $F = (W, R)$, write $\text{Rel}(F)$ for the relation set corresponding to R, and let $|F| = W$.

Definition 2.5 (treelike) *A relation set $\mathcal{R}$ is treelike if the frame defined by $\mathcal{R}$ is a tree or $\mathcal{R}$ is empty.*

For a non-empty relation set $\mathcal{R}$ that is treelike, let root($\mathcal{R}$) denote the root of this tree.

To illustrate this definition, consider the relation sets $\{Rxx\}$, $\{Rxy, Ruv\}$, $\{Rxy, Rzy\}$, and $\{Rxy, Rxz, Ryu, Rzu\}$. The frames defined by these sets are, respectively,

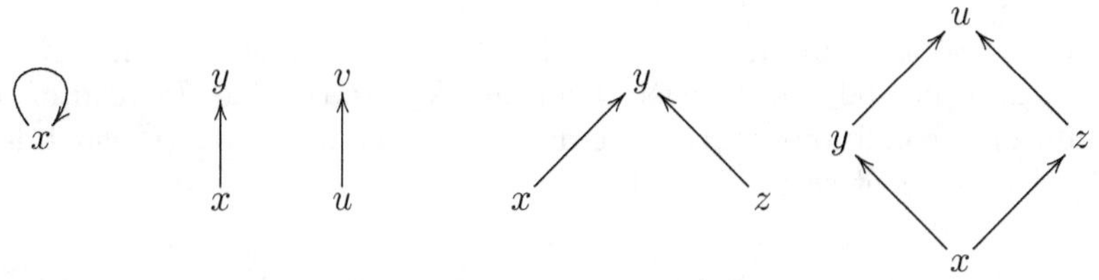

None of the above relation sets are treelike because the frames defined by their relation sets are not trees. In the above frames from left-to-right, frame 1 contains a reflexive state (and hence a cycle); frame 2 and frame 3 are not rooted. Finally, frame 4 is not a tree because the underlying undirected graph contains a cycle.

Definition 2.6 (labelled tree sequent) *A* labelled tree-sequent *is a labelled sequent of the form* $\mathcal{R}, X \Rightarrow Y$ *where*

(i) $\mathcal{R}$ *is treelike, and*

(ii) if $\mathcal{R} = \emptyset$ *then* X *has the form* $\{x : A_1, \ldots, x : A_n\}$ *and* Y *has the form* $\{x : B_1, \ldots, x : B_m\}$ *for some state variable* x *(ie. all labelled formulae in* X *and* Y *have the same label), and*

(iii) if $\mathcal{R} \neq \emptyset$ *then every state variable* x *that occurs in either* X *or* Y *(in some labelled formula* $x : A$ *for some formula* A*) also occurs in* $\mathcal{R}$ *(ie as a term* Rxu *or* Rux *for some state* u*).*

We write LTS *as an abbreviation for the term "labelled tree sequent(s)"*

Each of the following is a LTS:

$$x : A \Rightarrow x : A \qquad \Rightarrow y : A \qquad Rxy, Rxz, x : A \Rightarrow y : A$$

Notice that it is possible for a state variable to occur in the relation set and not in the X, Y multisets (this is what happens with the state variable z in the example above far right). The following are *not* LTS:

$$x : A \Rightarrow x : A, z : A \qquad Rxy, x : A \Rightarrow z : A \qquad Rxy, Ryz, Rxz \Rightarrow$$

From left-to-right above, the first labelled sequent is not a LTS because the relation set is empty and yet two distinct state variables occur in the sequent, violating condition (ii). The next sequent violates condition (iii) because the state variable z appears in the succedent as $z : A$ but it does not appear in the relation set. The final sequent violates condition (i) because the relation set is not treelike.

Definition 2.7 (labelled tree sequent calculus) *A* labelled tree sequent calculus *is a labelled sequent calculus whose initial sequents and inference rules are constructed from* LTS.

Negri [14] uses the following cut-rule for labelled sequent calculi:

$$\frac{\mathcal{R}_1, X \Rightarrow Y, x : A \qquad \mathcal{R}_2, x : A, U \Rightarrow V}{\mathcal{R}_1 \cup \mathcal{R}_2, X, U \Rightarrow Y, V}\ cut$$

We cannot use this rule directly in a labelled tree sequent calculus because $\mathcal{R}_1 \cup \mathcal{R}_2$ need not be treelike even if $\mathcal{R}_1$ and $\mathcal{R}_2$ are treelike. Instead of placing additional conditions on the cut-rule, we define an 'additive' cut-rule for labelled tree sequent calculi as follows:

$$\frac{\mathcal{R}, X \Rightarrow Y, x : A \qquad \mathcal{R}, x : A, X \Rightarrow Y}{\mathcal{R}, X \Rightarrow Y}\ cut_{LTS}$$

We close this section by revising some standard terminology. The terminology is applicable to each of the sequents and calculi we have defined in this section.

An *initial sequent instance* in the calculus $\mathcal{C}$ is a substitution instance of propositional variables/formulae (and state variables, if applicable) of an initial sequent from $\mathcal{C}$. A *rule instance* in the calculus $\mathcal{C}$ is a substitution instance of propositional variables/formulae (and state variables, if applicable) of one of the inference rules from $\mathcal{C}$. A *derivation* in the calculus $\mathcal{C}$ is defined in the usual way, as either an initial sequent instance, or an application of a rule instance to derivations of the premises of the rule. If there is a derivation of some sequent $\mathcal{S}$ in $\mathcal{C}$, then we say that $\mathcal{S}$ is derivable in $\mathcal{C}$. The *height* of a derivation is defined in the usual way as the maximum depth of the derivation tree. We write $\vdash^{\delta}_{\mathcal{C}} \mathcal{S}$ to mean that there is a derivation δ of the sequent $\mathcal{S}$ in $\mathcal{C}$. To avoid having to name the derivation we simply write $\vdash_{\mathcal{C}} \mathcal{S}$.

We say that an inference rule ρ is *admissible* in $\mathcal{C}$ if whenever premises of any rule instance of ρ is derivable in $\mathcal{C}$, then so is the conclusion of the rule instance. Following standard terminology, we say that a calculus $\mathcal{C}$ is *sound and complete* for the logic L if it derives exactly the theorems of L. Formally, for every formula A:

$$\vdash_{\mathcal{C}} \ \Rightarrow A \text{ (or } \vdash_{\mathcal{C}} \ \Rightarrow x : A \text{ if } \mathcal{C} \text{ is a labelled sequent calculus) iff } A \in L$$

The results we present in this paper are syntactic in the sense that there is an underlying algorithm witnessing each result.

3 Maps between THS and LTS

If $X \Rightarrow Y$ and $U \Rightarrow V$ are traditional sequents, recall that we defined $X \Rightarrow Y \otimes U \Rightarrow V$ to be the traditional sequent $X, U \Rightarrow Y, V$. Overloading the operator, if $\mathcal{R}_1, X \Rightarrow Y$ and $\mathcal{R}_2, U \Rightarrow V$ are two labelled sequents, then define $(\mathcal{R}_1, X \Rightarrow Y) \otimes (\mathcal{R}_2, U \Rightarrow V)$ to be the labelled sequent $\mathcal{R}_1 \cup \mathcal{R}_2, X, U \Rightarrow Y, V$. Because the order of elements in a multiset is irrelevant, in each case $\otimes$ is associative and commutative.

Definition 3.1 (*THS* **to** *LTS*) *For a state variable x, define the mapping $\mathbb{TL}_x$ from a THS to a LTS as follows, where the variable $x_{\bar{s}}$ ($\bar{s} \in \mathbb{N}^*$) passed* *to a recursive call of* $\mathbb{TL}$ *is never reused:*

$$\mathbb{TL}_x(X \Rightarrow Y) = x : X \Rightarrow x : Y$$
$$\mathbb{TL}_x(X \Rightarrow Y/G_1; \ldots; G_n) = (\otimes_{j=1}^{n} \mathbb{TL}_{x_j}(G_j)) \otimes (Rxx_1, \ldots, Rxx_n, x : X \Rightarrow x : Y)$$

Of course, we need to verify for arbitrary *THS* H that $\mathbb{TL}_x(H)$ is indeed a *LTS*. This is a straightforward induction on the structure of H. In the base case (H is $X \Rightarrow Y$) the image $x : X \Rightarrow x : Y$ is a *LTS*. In the inductive case (H is $X \Rightarrow Y/G_1; \ldots; G_n$), by the induction hypothesis we have that $\mathbb{TL}_{x_i}(G_i)$ $(1 \leq i \leq n)$ is a *LTS*. Then it is easy to see that the relation set $\mathcal{R}$ of $\mathbb{TL}_x(H)$ is non-empty and treelike — in particular, $Rxx_1, \ldots, Rxx_n \in \mathcal{R}$ and the frame defined by $\mathcal{R}$ has root x. It remains to check condition (iii) of Definition 2.6. The state variable of a labelled formula in $\mathbb{TL}_x(H)$ either is x or occurs in the relation set $\mathcal{R}_j$ of the *LTS* $\mathbb{TL}_{x_j}(G_j)$ for some j. Since x certainly occurs in $\mathcal{R}$ and $\mathcal{R}_j \subseteq \mathcal{R}$, the condition is satisfied in each case.

Example 3.2 If H is the *THS* $p \Rightarrow q/((q \Rightarrow r/\Box r \Rightarrow \neg s); l \Rightarrow \Box(p \supset s))$,

$$\begin{aligned}\mathbb{TL}_x(H) &= \mathbb{TL}_{x_1}(q \Rightarrow r/\Box r \Rightarrow \neg s) \otimes \mathbb{TL}_{x_2}(l \Rightarrow \Box(p \supset s)) \\ &\qquad \otimes (Rxx_1, Rxx_2, x : p \Rightarrow x : q) \\ &= \mathbb{TL}_{x_{11}}(\Box r \Rightarrow \neg s) \otimes (Rx_1x_{11}, x_1 : q \Rightarrow x_1 : r) \\ &\qquad \otimes (x_2 : l \Rightarrow x_2 : \Box(p \supset s)) \otimes (Rxx_1, Rxx_2, x : p \Rightarrow x : q) \\ &= (x_{11} : \Box r \Rightarrow x_{11} : \neg s) \otimes (Rx_1x_{11}, x_1 : q \Rightarrow x_1 : r) \\ &\qquad \otimes (x_2 : l \Rightarrow x_2 : \Box(p \supset s)) \otimes (Rxx_1, Rxx_2, x : p \Rightarrow x : q)\end{aligned}$$

The last equality simplifies to the *LTS*

$$Rx_1x_{11}, Rxx_1, Rxx_2, x_{11} : \Box r, x_1 : q, x_2 : l, x : p \Rightarrow x_{11} : \neg s, x_1 : r, x_2 : \Box(p \supset s), x : q$$

We sometimes suppress the subscript, writing $\mathbb{TL}$ instead of $\mathbb{TL}_x$ for the sake of clarity when the state variable that is used is not important. Observe that $\mathbb{TL}G$ assigns a unique state variable to each traditional sequent $\mathcal{S}$ appearing in the *THS* G. Moreover, given $G_1\{\mathcal{S}_1\} \sim G_2\{\mathcal{S}_2\}$, without loss of generality we may assume that the state variable assigned to $\mathcal{S}_1$ in $\mathbb{TL}(G_1\{\mathcal{S}_1\})$ and $\mathcal{S}_2$ in $\mathbb{TL}(G_2\{\mathcal{S}_2\})$ is identical.

Definition 3.3 *Define the function* $\mathbb{LT}$ *from an LTS* $\mathcal{R}, X \Rightarrow Y$ *to a THS as:*

$\mathcal{R} = \emptyset$: *then* $\mathcal{R}, X \Rightarrow Y$ *must have the form* $x : U \Rightarrow x : V$ *for some state variable* x, *so let* $\mathbb{LT}(x : U \Rightarrow x : V) = (U \Rightarrow V)$

$\mathcal{R} \neq \emptyset$: *then suppose that* $x = root(\mathcal{R})$ *and* $\mathcal{R}_x = \{Rxy_1, \ldots, Rxy_n\}$, *and let* $\Delta_i = |y_i\uparrow|$, *and let*

$$\mathbb{LT}(\mathcal{R}, X \Rightarrow Y) = X_x \Rightarrow Y_x/(\mathbb{LT}(Rel(y_1\uparrow), X_{\Delta_1} \Rightarrow Y_{\Delta_1}); \ldots; \mathbb{LT}(Rel(y_n\uparrow), X_{\Delta_n} \Rightarrow Y_{\Delta_n})).$$

Recall that $\mathbb{SV}$ denotes the set of state variables. Let $\mathrm{Var}(\mathcal{S}) \subset \mathbb{SV}$ denote the finite set of state variables occurring in the labelled sequent S. A *renaming of* $\mathcal{S}$ is a one-to-one function $f_{\mathcal{S}} : \mathrm{Var}(\mathcal{S}) \mapsto \mathbb{SV}$ (by one-to-one we mean that if $f_{\mathcal{S}}(x) = f_{\mathcal{S}}(y)$ then $x = y$). We write $Dom(f_{\mathcal{S}})$ and $Im(f_{\mathcal{S}})$ to denote the domain and image of $f_{\mathcal{S}}$ respectively.

For any labelled sequent $\mathcal{S}'$ and renaming $f_{\mathcal{S}}$ of the labelled sequent $\mathcal{S}$, let $\mathcal{S}'_{f_{\mathcal{S}}}$ be the labelled sequent obtained from $\mathcal{S}'$ by the simultaneous and uniform substitution $x \mapsto f_{\mathcal{S}}(x)$ for all $x \in Dom(f_{\mathcal{S}}) \cap \mathrm{Var}(\mathcal{S}')$. Notice that $\mathcal{S}'_{f_{\mathcal{S}}}$ need not be an LTS even if $\mathcal{S}'$ is a LTS.

Example 3.4 Consider the following LTS $\mathcal{S}$ (below left) and $\mathcal{S}'$ (below right):

$$x : A \Rightarrow x : A \qquad\qquad Rxy, x : A \Rightarrow y : B$$

Let the renaming $f_{\mathcal{S}}$ of $\mathcal{S}$ be the function mapping $x \mapsto y$. Then $\mathcal{S}'_{f_{\mathcal{S}}}$ is the labelled sequent $Ryy, y : A \Rightarrow y : B$. Clearly this sequent is not a LTS.

However, if $\mathcal{S}$ is a LTS, then for any renaming $f_{\mathcal{S}}$ of $\mathcal{S}$, it is easy to see that $\mathcal{S}_{f_{\mathcal{S}}}$ must be a LTS.

Lemma 3.5 *Let G denote a THS and let $\mathcal{S}$ denote a LTS. Then*

(i) $\mathbb{TL}G$ is a labelled tree-sequent.

(ii) $\mathbb{LT}\mathcal{S}$ is a THS.

(iii) $\mathbb{LT}(\mathbb{TL}G)$ is G, and $\mathbb{TL}(\mathbb{LT}\mathcal{S})$ is $\mathcal{S}_{f_{\mathcal{S}}}$ for some renaming $f_{\mathcal{S}}$ of $\mathcal{S}$.

Proof. The proofs of (i) and (ii) are straightforward, following from an inspection of the functions $\mathbb{TL}$ and $\mathbb{LT}$. In the case of (iii), observe that Definition 2.6(iii) ensures that no labelled formulae are 'lost' when passing from $\mathcal{S}$ to $\mathbb{LT}\mathcal{S}$. However, since the $\mathbb{TL}$ function *assigns* state variables, it may be necessary to 'swap' label names in order to obtain equality of the LTS $\mathbb{TL}(\mathbb{LT}\mathcal{S})$ and $\mathcal{S}$. That is, there is some renaming $f_{\mathcal{S}}$ of $\mathcal{S}$ such that $\mathbb{TL}(\mathbb{LT}\mathcal{S})$ is $\mathcal{S}_{f_{\mathcal{S}}}$. ⊣

Lemma 3.6 (substitution lemma) *Suppose that $\mathcal{C}$ is a LTS calculus and $\mathcal{S}$ is a LTS. Let $f_{\mathcal{S}}$ be an arbitrary renaming of $\mathcal{S}$. If $\vdash^{\delta}_{\mathcal{C}} \mathcal{S}$ then there is an effective transformation to a derivation δ' such that $\vdash^{\delta'}_{\mathcal{C}} \mathcal{S}_{f_{\mathcal{S}}}$.*

Proof. Induction on the height of δ. If the height is one, then δ must be an initial sequent. It is easy to see that $\mathcal{S}_{f_{\mathcal{S}}}$ is also an initial sequent.

Now suppose that the last rule in δ is the LTS inference rule ρ, with premises $\mathcal{S}_1, \ldots, \mathcal{S}_n$. The proof is not completely trivial since f $\cup_i \mathrm{Var}(\mathcal{S}_i)$ contains state variables not in $\mathcal{S}$ (for example, due to a standard variable restriction on ρ), then it is possible that $(\mathcal{S}_i)_{f_{\mathcal{S}}}$ is not a LTS even though $\mathcal{S}_i$ is a LTS. For example, suppose that f is the renaming $x \mapsto y$ of the $LTS \Rightarrow x : \Box A$, and consider the following rule instance of ρ:

$$\frac{\mathcal{S}_1 = Rxy, y : \Box A \Rightarrow y : A}{\mathcal{S} = \qquad\qquad \Rightarrow x : \Box A}$$

Then $(\mathcal{S}_1)_f$ is the labelled sequent $Ryy, y : \Box A \Rightarrow y : A$ which is not a LTS (because the relation set contains the cycle Ryy) although $\mathcal{S}_1$ is a LTS.

Returning to the proof, the solution is to define first a one-to-one function g from $\cup_i \mathrm{Var}(\mathcal{S}_i) \setminus Dom(f_S)$ to fresh state variables (in particular, to variables outside $Im(f_S)$). Then $((\mathcal{S}_i)_g)_{f_S}$ is a LTS. Then $((\cdot)_g)_{f_S}$ implicitly defines a renaming $\mathrm{Var}(\mathcal{S}_i) \mapsto \mathbb{SV}$ for $\mathcal{S}_i$ for each i. So $((\mathcal{S}_i)_g)_{f_S}$ is a LTS.

Continuing the example above, set g as the map $y \mapsto z$, so

$$\frac{((\mathcal{S}_1)_g)_f = Ryz, z : \Box A \Rightarrow z : A}{(\mathcal{S}_g)_f = \qquad\qquad \Rightarrow y : \Box A}$$

and this is a legal rule instance of ρ.

Once again, returning to the proof, by the induction hypothesis we can obtain derivations of $((\mathcal{S}_i)_g)_{f_S}$ in $\mathcal{C}$. Moreover, observe that

$$\frac{((\mathcal{S}_1)_g)_{f_S} \ldots ((\mathcal{S}_n)_g)_{f_S}}{(\mathcal{S}_g)_{f_S}}$$

is a rule instance of ρ. Hence we have a derivation of $(\mathcal{S}_g)_f$. Since $\mathrm{Var}(\mathcal{S}) \cap Dom(g) = \emptyset$, it follows that $(\mathcal{S}_g)_{f_S} = \mathcal{S}_{f_S}$. ⊣

We remind the reader that this substitution lemma pertains to LTS calculi as given in Definition 2.7. In particular, this lemma may not apply to calculi containing pathological rules that are not invariant under renaming, such as the following rule:

$$\frac{x \neq a}{x : A \Rightarrow x : B} \ (a \text{ is some fixed state variable})$$

3.1 Inference rules induced by $\mathbb{TL}$ and $\mathbb{LT}$

It is straightforward to construct an inference rule for THS from an inference rule for LTS and *vice versa* under the maps $\mathbb{LT}$ and $\mathbb{TL}$. We illustrate with some detailed examples. Following standard practice, the active formulae in the conclusion (resp. premise) of an inference rule are called *principal* (*auxiliary*) formulae.

Example 3.7 Consider the following inference rule $R\Box$:

$$\frac{\mathcal{R}, \overbrace{Rxy, y : \Box A}^{\text{principal}}, \Gamma \Rightarrow \Delta, \overbrace{y : A}^{\text{principal}}}{\mathcal{R}, \Gamma \Rightarrow \Delta, \underbrace{x : \Box A}_{\text{principal}}} R\Box$$

where y does not appear in the conclusion of the rule. We can write the premise and conclusion, respectively, as

$$(\mathcal{R}, \Gamma \Rightarrow \Delta) \otimes (Rxy, y : \Box A \Rightarrow y : A)$$
$$(\mathcal{R}, \Gamma \Rightarrow \Delta) \otimes (\Rightarrow x : \Box A)$$

The sequent $\mathcal{R}, \Gamma \Rightarrow \Delta$ is an arbitrary LTS except it does not contain y, hence it follows that

$$\mathbb{LT}(\mathcal{R}, \Gamma \Rightarrow \Delta) = G\{\overbrace{X \Rightarrow Y}^{x} / (\underline{X}; \overbrace{\emptyset}^{y})\}$$

for arbitrary $G, X, Y, \underline{X}$ — the braces indicate the locations corresponding to the state variables, and we use $\emptyset$ as a metalevel symbol to explicate that there is no position in the THS corresponding to y. Meanwhile we have

$$\mathbb{LT}(Rxy, y : \Box A \Rightarrow y : A) = \overbrace{\ \Rightarrow\ }^{x} / \overbrace{\Box A \Rightarrow A}^{y}$$
$$\mathbb{LT}(\Rightarrow x : \Box A) = \underbrace{\ \Rightarrow \Box A}_{x}$$

Thus, the image $\mathbb{LT}(R\Box)$ of $R\Box$ under $\mathbb{LT}$ is the THS inference rule:

$$\frac{G\{X \Rightarrow Y/\Box A \Rightarrow A\}}{G\{X \Rightarrow Y, \Box A\}}\ \mathbb{LT}(R\Box)$$

Example 3.8 For the other direction, consider the THS rule $\Box K_{gl}$.

$$\frac{G\{\overbrace{X \Rightarrow Y}^{x} / \overbrace{\Box A \Rightarrow A}^{y}\}}{G\{\underbrace{X \Rightarrow Y, \Box A}_{x} / \underbrace{\ \emptyset\ }_{y}\}}\ \Box K_{gl}$$

As before, we have used braces to identify location in the THS with state variables, and $\emptyset$ as a metalevel symbol to explicate that there is no position in the conclusion THS corresponding to y. Notice that the sequent $\Box A \Rightarrow A$ in the premise disappears in the conclusion. Equivalently, the location correponding to the variable y is not populated. Applying $\mathbb{TL}_x$ to the premise of $\Box K_{gl}$ we get the LTS

$$\mathcal{R}, Rxy, y : \Box A, \Gamma \Rightarrow \Delta, y : A$$

where $\mathcal{R}, \Gamma$ and Δ are arbitrary. Applying $\mathbb{TL}_x$ to the conclusion of $\Box K_{gl}$ we get the LTS

$$\mathcal{R}, \Gamma \Rightarrow \Delta, x : \Box A$$

where y does not appear in the conclusion. We thus obtain the LTS rule $\mathbb{TL}\Box K_{gl}$.

$$\frac{\mathcal{R}, Rxy, y : \Box A, \Gamma \Rightarrow \Delta, y : A}{\mathcal{R}, \Gamma \Rightarrow \Delta, x : \Box A}\ \mathbb{TL}\Box K_{gl}$$

with the standard variable restriction "y does not occur in the the conclusion of $\mathbb{TL}\Box K_{gl}$".

3.2 Calculi induced by $\mathbb{TL}$ and $\mathbb{LT}$

We can now construct a THS calculus from a LTS calculus and *vice versa*. If $\mathcal{C}$ is a THS calculus, then let $\mathbb{TL}\mathcal{C}$ denote the calculus consisting of the image of every initial sequent and inference rule in $\mathcal{C}$ under $\mathbb{TL}$. Similarly, if $\mathcal{C}$ is a LTS calculus, then let $\mathbb{LT}\mathcal{C}$ denote the calculus consisting of the image of every initial sequent and inference rule in $\mathcal{C}$ under $\mathbb{LT}$.

Lemma 3.9 *Let $\mathcal{C}$ be a THS calculus. Then,*

(i) for any THS G, we have $\vdash_{\mathcal{C}} G$ *iff* $\vdash_{\mathbb{TL}\mathcal{C}} \mathbb{TL}G$

(ii) for any LTS $\mathcal{S}$*, we have* $\vdash_{\mathbb{TL}\mathcal{C}} \mathcal{S}$ *iff* $\vdash_{\mathcal{C}} \mathbb{LT}\mathcal{S}$.

In each case, the respective derivations in $\mathcal{C}$ *and* $\mathbb{TL}\mathcal{C}$ *have identical height.*

Proof. Proof of (i). Suppose that $\vdash^{\delta}_{\mathcal{C}} G$. We need to show that $\mathbb{TL}G$ is derivable in $\mathbb{TL}\mathcal{C}$. We can obtain a derivation δ' of $\mathbb{TL}G$ from δ by replacing every *THS* G' appearing in δ with $\mathbb{TL}G'$, and every rule ρ with $\mathbb{TL}\rho$ — by the definition of $\mathbb{TL}\mathcal{C}$, the resulting object is a derivation in the calculus $\mathbb{TL}\mathcal{C}$ with endsequent $\mathbb{TL}G$. In particular, notice that if ρ is a legal rule instance in $\mathcal{C}$, then $\mathbb{TL}\rho$ will obey any relevant standard variable restrictions in $\mathbb{TL}\mathcal{C}$. Moreover, by construction, δ and δ' have identical height.

Proof of (ii) is analogous to the above. ⊣

Corollary 3.10 *For any THS calculus* $\mathcal{C}$ *and formula A we have* $\vdash_{\mathcal{C}} \Rightarrow A$ *iff* $\vdash_{\mathbb{TL}\mathcal{C}} \Rightarrow x : A$.

Proof. Immediate from Lemma 3.9. ⊣

4 Poggiolesi's *CSGL* and Negri's *G3GL*

Negri [14] has given a labelled sequent calculus *G3GL* for provability logic *GL* as part of a systematic program to present labelled sequent calculi for modal logics. Subsequently Poggiolesi [17] presented the *THS* calculus *CSGL* for *GL* and proved syntactic cut-admissibility. In that work, Poggiolesi [17] states:

> *"As it has probably already emerged in the previous sections, CSGL is quite similar to Negris calculus G3GL [see [14]]: indeed, except for the rule* 4 *that only characterizes CSGL, the propositional and modal rules of the two calculi seem to be based on a same intuition. Given this situation, a question naturally arises: what is the exact relation between the two calculi? Is it possible to find a translation from the THS calculi to the labeled calculi and vice versa?"*

Here we establish the following.

(i) We answer in full the question raised by Poggiolesi. In particular, we give a translation between *CSGL* and *G3GL*; and

(ii) Show that *CSGL* is sound and complete for provability logic *GL* and prove syntactic cut-admissibility utilising the *existing* proofs of these results for *G3GL*. In contrast, Poggiolesi [17] has to provide a new proof for each result, in particular, dealing with the many cases that arise in the proof of syntactic cut-admissibility. Since many proof-theoretical properties (invertibility of the inference rules, for example) are preserved under the notational variants translation, we get these results directly, once again alleviating the need for independent proofs.

A key aspect of our work is the coercion of results from the labelled sequent calculus *G3GL* into the *LTS* calculus $\mathbb{TL}CSGL$ (Theorem 4.4).

Initial THS: $G\{p, X \Rightarrow Y, p\} \quad G\{\Box A, X \Rightarrow Y, \Box A\}$

Propositional rules:

$$\frac{G\{X \Rightarrow Y, A\}}{G\{\neg A, X \Rightarrow Y\}}\neg A \qquad \frac{G\{A, X \Rightarrow Y\}}{G\{X \Rightarrow Y, \neg A\}}\neg K$$

$$\frac{G\{A, B, X \Rightarrow Y\}}{G\{A \wedge B, X \Rightarrow Y\}}\wedge A \qquad \frac{G\{X \Rightarrow Y, A\} \qquad G\{X \Rightarrow Y, B\}}{G\{X \Rightarrow Y, A \wedge B\}}\wedge K$$

Modal rules:

$$\frac{G\{\Box A, X \Rightarrow Y/(U \Rightarrow V, \Box A/\underline{X})\} \qquad G\{\Box A, X \Rightarrow Y/(A, U \Rightarrow V/\underline{X})\}}{G\{\Box A, X \Rightarrow Y/(U \Rightarrow V/\underline{X})\}}\Box A$$

$$\frac{G\{X \Rightarrow Y/\Box A \Rightarrow A\}}{G\{X \Rightarrow Y, \Box A\}}\Box K$$

Special logical rule:

$$\frac{G\{\Box A, X \Rightarrow Y/(\Box A, U \Rightarrow V/\underline{X})\}}{G\{\Box A, X \Rightarrow Y/(U \Rightarrow V/\underline{X})\}}4$$

Table 1
$CSGL$: the THS calculus of Poggiolesi [17].

Initial LTS: $\mathcal{R}, x : p, \Gamma \Rightarrow \Delta, x : p \quad \mathcal{R}, x : \Box A, \Gamma \Rightarrow \Delta, x : \Box A$

Propositional rules:

$$\frac{\mathcal{R}, \Gamma \Rightarrow \Delta, x : A}{\mathcal{R}, x : \neg A, \Gamma \Rightarrow \Delta}\neg A \qquad \frac{\mathcal{R}, x : A, \Gamma \Rightarrow \Delta}{\mathcal{R}, \Gamma \Rightarrow \Delta, x : \neg A}\neg K$$

$$\frac{\mathcal{R}, x : A, x : B, \Gamma \Rightarrow \Delta}{\mathcal{R}, x : A \wedge B, \Gamma \Rightarrow \Delta}\wedge A \qquad \frac{\mathcal{R}, \Gamma \Rightarrow \Delta, x : A \qquad \mathcal{R}, \Gamma \Rightarrow \Delta, x : B}{\mathcal{R}, \Gamma \Rightarrow \Delta, x : A \wedge B}\wedge K$$

Modal rules:

$$\frac{\mathcal{R}, Rxy, x : \Box A, \Gamma \Rightarrow \Delta, y : \Box A \qquad \mathcal{R}, Rxy, x : \Box A, y : A, \Gamma \Rightarrow \Delta}{\mathcal{R}, Rxy, x : \Box A, \Gamma \Rightarrow \Delta}\mathbb{TL}\Box A$$

$$\frac{\mathcal{R}, Rxy, y : \Box A, \Gamma \Rightarrow \Delta, y : A}{\mathcal{R}, \Gamma \Rightarrow \Delta, x : \Box A}\mathbb{TL}\Box K$$

Special logical rule:

$$\frac{\mathcal{R}, Rxy, x : \Box A, y : \Box A, \Gamma \Rightarrow \Delta}{\mathcal{R}, Rxy, x : \Box A, \Gamma \Rightarrow \Delta}\mathbb{TL}4$$

Table 2
$\mathbb{TL}CSGL$: the LTS calculus obtained from $CSGL$ under the mapping $\mathbb{TL}$. Rule $\mathbb{TL}\Box K$ has the standard restriction that y does not appear in the conclusion.

4.1 The calculus $CSGL$ and $\mathbb{TL}CSGL$

Poggiolesi's THS calculus $CSGL$ [17] is given in Table 1. From this calculus we construct the LTS calculus $\mathbb{TL}CSGL$ (Table 2) following the procedure given in the previous section.

For a relation term or labelled formula α, define the left and right weakening rules as follows:

$$\frac{\mathcal{R}, \Gamma \Rightarrow \Delta}{\mathcal{R}, \alpha, \Gamma \Rightarrow \Delta} LW \quad \frac{\mathcal{R}, \Gamma \Rightarrow \Delta}{\mathcal{R}, \Gamma \Rightarrow \Delta, \alpha} RW$$

We remind the reader that each of the above rules when viewed as a LTS inference rule (as opposed to a labelled sequent inference rule) has the restriction that the premise and conclusion is a LTS. The left and right contraction rules are defined as follows:

$$\frac{\mathcal{R}, x : A, x : A, \Gamma \Rightarrow \Delta}{\mathcal{R}, x : A, \Gamma \Rightarrow \Delta} LC \quad \frac{\mathcal{R}, \Gamma \Rightarrow \Delta, x : A, x : A}{\mathcal{R}, \Gamma \Rightarrow \Delta, x : A} RC$$

Lemma 4.1 *The rules LW and RW for weakening and the rules LC and RC for contraction are height-preserving syntactically admissible in* $\mathbb{TL}CSGL$.

Proof. Poggiolesi [17] shows that the corresponding THS rules (ie. the weakening and contraction rules under $\mathbb{LT}$) are height-preserving syntactically admissible in $CSGL$. By Theorem 3.9, the mapping between derivations in $CSGL$ and $\mathbb{TL}CSGL$ is height-preserving. Hence the analogous results apply to $\mathbb{TL}CSGL$ too, so we are done. ⊣

4.2 Negri's calculus *G3GL*

If we compare the $\mathbb{TL}CSGL$ calculus with Negri's labelled sequent calculus $G3GL$, the only differences are that

(i) in $G3GL$, the treelike condition on the relation set of every labelled sequent is removed, and

(ii) $G3GL$ does not contain the inference rule $\mathbb{TL}4$, and

(iii) $G3GL$ contains the initial sequent (*Irrefl*) and inference rule (*Trans*):

$$\mathcal{R}, Rxx, \Gamma \Rightarrow \Delta \ (Irref) \qquad \frac{\mathcal{R}, Rxz, Rxy, Ryz, \Gamma \Rightarrow \Delta}{\mathcal{R}, Rxy, Ryz, \Gamma \Rightarrow \Delta} \ (Trans)$$

For those rules in $G3GL$ that also occur in $\mathbb{TL}CSGL$, we will use the rule labelling of $\mathbb{TL}CSGL$. For example, we write $\mathbb{TL}\Box K$ instead of the label $R\Box{-}L$ used in [14]. Strictly speaking, the calculus $G3GL$ also contains rules for the disjunction and implication connectives. Since these connectives can be written in terms of negation and conjunction, for our purposes there is no harm in this omission. The rules LW and RW as well as the contraction rules LC and RC are height-preserving admissible in $G3GL$ [14].

Theorem 4.2 (Negri) *The labelled sequent calculus G3GL (i) has syntactic cut-admissibility, and (ii) is sound and complete for the logic GL.*

Proof. See Negri [14]. ⊣

4.3 Results

Let $G3GL + \mathbb{TL}4$ be the calculus obtained by the addition of the rule $\mathbb{TL}4$ to $G3GL$, where the rule $\mathbb{TL}4$ will no longer be subject to the restriction that its premise and conclusion are LTS. Suppose that ρ is the following instance of (*Trans*) in a derivation in $G3GL + \mathbb{TL}4$:

$$\frac{\mathcal{R}, Rxy, Ryz, Rxz, \Gamma \Rightarrow \Delta}{\mathcal{R}, Rxy, Ryz, \Gamma \Rightarrow \Delta}\ (Trans)$$

Define the *width* of ρ to be the number of rule occurrences above ρ that make the term Rxz principal. Observe that Rxz can be principal in this way only due to a rule from $\{\mathbb{TL}\Box A, \mathbb{TL}4, (Trans), (Irref)\}$.

Lemma 4.3 *Let δ be a derivation in G3GL+$\mathbb{TL}4$ not containing (Irref). Then the (Trans) rule is eliminable from δ.*

Proof. The non-trivial case is when there is a positive number $s+1$ of occurrences of (*Trans*) in δ. Let ρ be an arbitrary topmost occurrence of (*Trans*) in δ:

$$\begin{array}{c} \text{no } (Trans) \text{ rules} \\ \dfrac{Rxy, Ryz, Rxz, \mathcal{R}, \Gamma \Rightarrow \Delta}{Rxy, Ryz, \mathcal{R}, \Gamma \Rightarrow \Delta}\ \rho \\ \vdots \\ \mathcal{R}_0, X \Rightarrow Y \end{array}$$

Let γ denote the subderivation of δ deriving $Rxy, Ryz, \mathcal{R}, \Gamma \Rightarrow \Delta$. We claim that ρ is eliminable. Proof by induction on the width n of ρ.

If $n = 0$ then there is no rule above ρ that makes Rxz principal. It is clear that we can transform γ by deleting the Rxz term from every sequent above ρ to obtain directly a derivation γ' of $Rxy, Ryz, \mathcal{R}, \Gamma \Rightarrow \Delta$. Replacing the subderivation γ in δ with γ' we have eliminated ρ and thus the new derivation contains only s occurrences of (*Trans*).

Now suppose that $n = k+1$. Since δ does not contain (*Irref*) and because ρ is a topmost occurrence of (*Trans*), the Rxz term must be principal due to either a $\mathbb{TL}\Box A$ rule or a $\mathbb{TL}4$ rule.

Case I (Rxz principal by $\mathbb{TL}\Box A$). Then δ has the following form, where we have written the two premises of the $\mathbb{TL}\Box A$ rule one above the other:

$$\begin{array}{c} \dfrac{\begin{array}{c} Rxy, Ryz, Rxz, \mathcal{R}', x : \Box B, \Gamma' \Rightarrow \Delta', z : \Box B \\ Rxy, Ryz, Rxz, \mathcal{R}', x : \Box B, z : B, \Gamma' \Rightarrow \Delta' \end{array}}{Rxy, Ryz, Rxz, \mathcal{R}', x : \Box B, \Gamma' \Rightarrow \Delta'}\ \mathbb{TL}\Box A \\ \vdots \\ \dfrac{Rxy, Ryz, Rxz, \mathcal{R}, \Gamma \Rightarrow \Delta}{Rxy, Ryz, \mathcal{R}, \Gamma \Rightarrow \Delta}\ \rho \\ \vdots \\ \mathcal{R}_0, X \Rightarrow Y \end{array}$$

Apply the admissible rule LW with $y : \Box B$ to each of the premises of $\mathbb{TL}\Box A$. Then apply $\mathbb{TL}\Box A$ to these sequents and proceed as follows (once again we write the two premises of the $\mathbb{TL}\Box A$ rule one above the other):

$$\cfrac{\cfrac{\begin{array}{c}Rxy, Ryz, Rxz, \mathcal{R}', x : \Box B, y : \Box B, \Gamma' \Rightarrow \Delta', z : \Box B \\ Rxy, Ryz, Rxz, \mathcal{R}', x : \Box B, z : B, y : \Box B, \Gamma' \Rightarrow \Delta'\end{array}}{Rxy, Ryz, Rxz, \mathcal{R}', x : \Box B, y : \Box B, \Gamma' \Rightarrow \Delta'}\ \mathbb{TL}\Box A}{Rxy, Ryz, Rxz, \mathcal{R}', x : \Box B, \Gamma' \Rightarrow \Delta'}\ \mathbb{TL}4$$

$$\vdots$$

$$\cfrac{Rxy, Ryz, Rxz, \mathcal{R}, \Gamma \Rightarrow \Delta}{Rxy, Ryz, \mathcal{R}, \Gamma \Rightarrow \Delta}\ \rho$$

$$\vdots$$

$$\mathcal{R}_0, X \Rightarrow Y$$

Notice that in the $\mathbb{TL}\Box A$ and $\mathbb{TL}4$ rules in the above proof diagram, it is the Ryz and Rxy term respectively that is principal (and not the Rxz term). As a result the width of ρ is reduced to k. Eliminate ρ using the induction hypothesis.

Case II (Rxz principal by $\mathbb{TL}4$). Then δ has the following form:

$$\cfrac{Rxy, Ryz, Rxz, \mathcal{R}', x : \Box B, z : \Box B, \Gamma' \Rightarrow \Delta'}{Rxy, Ryz, Rxz, \mathcal{R}', x : \Box B, \Gamma' \Rightarrow \Delta'}\ \mathbb{TL}4$$

$$\vdots$$

$$\cfrac{Rxy, Ryz, Rxz, \mathcal{R}, \Gamma \Rightarrow \Delta}{Rxy, Ryz, \mathcal{R}, \Gamma \Rightarrow \Delta}\ \rho$$

$$\vdots$$

$$\mathcal{R}_0, X \Rightarrow Y$$

Apply the admissible rule LW with $y : \Box B$ to the premise of $\mathbb{TL}4$. Then apply $\mathbb{TL}4$ to this sequent and proceed as follows:

$$\cfrac{\cfrac{\dots}{Rxy, Ryz, Rxz, \mathcal{R}', x : \Box B, y : \Box B, \Gamma' \Rightarrow \Delta'}\ \mathbb{TL}4}{Rxy, Ryz, Rxz, \mathcal{R}', x : \Box B, \Gamma' \Rightarrow \Delta'}\ \mathbb{TL}4$$

$$\vdots$$

$$\cfrac{Rxy, Ryz, Rxz, \mathcal{R}, \Gamma \Rightarrow \Delta}{Rxy, Ryz, \mathcal{R}, \Gamma \Rightarrow \Delta}\ \rho$$

$$\vdots$$

$$\mathcal{R}_0, X \Rightarrow Y$$

Notice that in the two $\mathbb{TL}4$ rules in the above proof diagram, it is the Ryz and Rxy term respectively that is principal (and not the Rxz term). As a result the width of ρ is reduced to k. Eliminate ρ using the induction hypothesis.

We have shown how to reduce the number of occurrences of (*Trans*) from $s + 1$ to s. As ρ was an arbitrary topmost occurrence, the result follows from an induction argument. $\dashv$

The following result connects Negri's labelled sequent calculus $G3GL$ and the LTS calculus $\mathbb{TL}CSGL$. Together with Corollary 3.10, this completely answers the question posed in Poggiolesi [17].

Theorem 4.4 *For any formula A, $\vdash_{\mathbb{TL}CSGL} \Rightarrow x : A$ iff $\vdash_{G3GL} \Rightarrow x : A$. Moreover the translation between the corresponding derivations is effective.*

Proof. For the left-to-right direction it suffices to show that $\mathbb{TL}4$ is syntactically admissible in $G3GL$. First, working in $G3GL$ (because we are working in a labelled sequent calculus, the relation sets that occur in the derivation need not be treelike), observe that:

$$
\dfrac{\dfrac{\dfrac{\dfrac{\dfrac{z:\Box A \Rightarrow z:\Box A}{Rxz, x:\Box A, z:\Box A \Rightarrow z:A, z:\Box A} \qquad \dfrac{z:A \Rightarrow z:A}{Rxz, z:A, x:\Box A, z:\Box A \Rightarrow z:A}}{Rxz, x:\Box A, z:\Box A \Rightarrow z:A}\ \mathbb{TL}\Box A}{Rxy, Ryz, Rxz, x:\Box A, z:\Box A \Rightarrow z:A}\ LW}{Rxy, Ryz, x:\Box A, z:\Box A \Rightarrow z:A}\ (\mathit{Trans})}{Rxy, x:\Box A \Rightarrow y:\Box A}\ \mathbb{TL}\Box K
$$

Suppose that we are given a derivation of the premise $\mathcal{R}, Rxy, y : \Box A, x : \Box A, X \Rightarrow Y$ of $\mathbb{TL}4$. From the cut-rule and the above derivation we get a derivation of $\mathcal{R}, Rxy, x : \Box A, x : \Box A, X \Rightarrow Y$. By Theorem 4.2 we can obtain a cut-free derivation of this sequent. Since the left contraction rule LC is admissible in $G3GL$ [14], we get $\mathcal{R}, Rxy, x : \Box A, X \Rightarrow Y$ and thus $\mathbb{TL}4$ is syntactically admissible in $G3GL$.

Now for the right-to-left direction. First observe that the derivation of $\Rightarrow x : A$ does not contain any occurrences of the initial sequent (*Irref*). To see this, observe that in any $G3GL$ derivation, viewed downwards, a state variable occurrence y can disappear from premise sequent to conclusion sequent only via the $\mathbb{TL}\Box K$ rule — all the other rules preserve the set of state variables in the relation set. Moreover, for this to occur, the variable y must occur exactly *once* in the relation set of the premise of $\mathbb{TL}\Box K$ (in a term of the form Rxy for some variable x distinct from y). Now, if the given derivation contains the initial sequent (*Irref*) $\mathcal{R}, Ryy, X \Rightarrow Y$, then the relation set of the initial sequent contains at least two occurrences of y. It follows that the relation set of every sequent below this initial sequent in δ will contain these two occurrences of y, contradicting the fact that the endsequent has the form $\Rightarrow x : A$.

Suppose that we are given a derivation δ of the $LTS \Rightarrow x : A$ in $G3GL$. We need to obtain a derivation of $\Rightarrow x : A$ in $\mathbb{TL}CSGL$. By Lemma 4.3, there is a derivation δ' of $\Rightarrow x : A$ in $G3GL + \mathbb{TL}4$ containing no occurrences of (*Trans*). To complete the proof, we will show that δ' is a derivation in $\mathbb{TL}CSGL$. It suffices to show that every labelled sequent in δ' is a LTS. By inspection, every rule in $G3GL + \mathbb{TL}4$ with the exception of the (*Trans*) rule has the property that if the conclusion is a LTS, then so are the premise(s). Since $\Rightarrow x : A$ is a LTS by assumption, every sequent in δ' must be a LTS so we are done. $\dashv$

A comment regarding the initial sequent $\mathcal{R}, Rxx, \Gamma \Rightarrow \Delta$ (*Irref*). Negri uses this initial sequent in the proof of cut-admissibility for $G3GL$ to argue that there cannot be a labelled sequent with a relation set (in our terminology) containing $\{Rxx_1, Rx_1x_2, \ldots Rx_nx\}$ (a 'loop'). We saw above that (*Irref*) cannot occur in any *G3GL* derivation of a sequent of the form $\Rightarrow x : A$. By definition, the relation set of a *LTS* can never contain such a loop so there is no initial *LTS* in $\mathbb{TLCSGL}$ corresponding to (*Irref*) in *G3GL*.

Theorem 4.5 *The calculus CSGL (i) is sound and complete for the logic GL, and (ii) has syntactic cut-admissibility.*

Proof. Follows from Theorem 4.2 using Corollary 3.10 and Theorem 4.4. ⊣

Note that although the above proofs make use of the results for *G3GL* [14], these results are syntactic because the proofs for *G3GL* are syntactic.

5 Conclusion

We have shown that *THS* and *LTS* are notational variants, allowing us to transfer proof-theoretic results including syntactic cut-admissibility between these formalisms, thus alleviating the need for independent proofs in each system. We have answered in full Poggiolesi's question regarding the relationship between the *THS* calculus *CSGL* and the labelled sequent calculus *G3GL*.

It is straightforward to construct mappings between *THS* and the nested (deep) sequents that Fitting [5] refers to in his work (adapting to Brünnler's [2] formulation of nested sequents is analogous). Define a nested sequent to be of the form $\Gamma, [\mathcal{N}_1], \ldots, [\mathcal{N}_k]$ where Γ is a formula multiset and $\mathcal{N}_1, \ldots, \mathcal{N}_k$ $(k \geq 0)$ are nested sequents. Nested sequent calculi consist of initial sequents and inference rules built from nested sequents. Following the formulations used by Brünnler and Fitting, nested sequent inference rules are permitted to operate at any level of nesting. That is, the auxiliary formulae — the 'active' formulae in the premise — of a rule instance are permitted to occur inside the scope of [ ]. For example, here is a rule instance of Fitting's $\wedge$-introduction rule:

$$\frac{p, [q, [r]] \qquad p, [q, [s]]}{p, [q, [r \wedge s]]}$$

However, a nested sequent inference rule is not permitted to operate *inside* a formula — ie. a proper subformula cannot be the auxiliary formula of a rule instance. For example, the following is *forbidden* as the auxiliary formulae are the proper subformulae r (of $q \vee r$) and s (of $q \vee s$):

$$\frac{p, [q \vee r] \qquad p, [q \vee s]}{p, [q \vee (r \wedge s)]}$$

Informally speaking, this means that the deep inference applies to the nesting but not to subformulae. This restriction leads to straightforward maps between *THS* and nested sequents as shown below.

Given a traditional sequent $X \Rightarrow Y$, we will assume that we have at hand a suitable concrete representation $X \Rightarrow_{1S} Y$ of $X \Rightarrow Y$ as a formula multiset — think of $X \Rightarrow_{1S} Y$ as a one-sided sequent. Also asssume that given a formula multiset Γ, we have at hand a suitable concrete representation Γ^s of Γ

as a traditional sequent. Then the maps $\mathbb{TN}$ and $\mathbb{NT}$ respectively map THS to nested sequents and *vice versa.* In the following: $X \Rightarrow Y$ is a traditional sequent; $G_1, \ldots, G_k$ are THS; Γ is a formula multiset; and $\mathcal{N}_1, \ldots, \mathcal{N}_k$ are nested sequents.

$$\mathbb{TN}(X \Rightarrow Y/(G_1; \ldots; G_k)) = (X \Rightarrow_{1S} Y, [\mathbb{TN}G_1], \ldots, [\mathbb{TN}G_k])$$
$$\mathbb{NT}(\Gamma, [\mathcal{N}_1], \ldots, [\mathcal{N}_k]) = (\Gamma^s/(\mathbb{NT}\mathcal{N}_1; \ldots; \mathbb{NT}\mathcal{N}_k))$$

In this way we can show that THS and nested sequents are notational variants. By computing the calculi induced by these mappings our results can be extended to these systems. Since THS and LTS are notational variants, this provides an answer to the question posed by Fitting concerning how labelled systems and nested (deep) sequent systems relate.

Negri [14,15] has identified a large class of modal and intermediate logics that can be presented using cut-free labelled sequent calculi. We would like to identify the subclass of such labelled sequent calculi that can be coerced into the LTS framework. In this way we could directly obtain proofs of syntactic cut-admissibility for LTS and THS calculi for suitable logics from the existing proofs for labelled sequent calculi. Theorem 4.4 in this paper is an example of such a result. Hein [8] has conjectured that such a coercion is possible for modal logics axiomatised by 3/4 Lemmon-Scott formulae $\{\Diamond^h \Box^i p \supset \Box^j p | h, i, j \geq 0\}$. This investigation is the subject of future work.

References

[1] Avron, A., *The method of hypersequents in the proof theory of propositional non-classical logics*, in: *Logic: from foundations to applications (Staffordshire, 1993)*, Oxford Sci. Publ., Oxford Univ. Press, New York, 1996 pp. 1–32.

[2] Brünnler, K., *Deep sequent systems for modal logic*, in: *Advances in modal logic. Vol. 6*, Coll. Publ., London, 2006 pp. 107–119.

[3] Chagrov, A. and M. Zakharyaschev, "Modal logic," Oxford Logic Guides **35**, The Clarendon Press Oxford University Press, New York, 1997, xvi+605 pp., oxford Science Publications.

[4] Fitting, M., "Proof methods for modal and intuitionistic logics," Synthese Library **169**, D. Reidel Publishing Co., Dordrecht, 1983, viii+555 pp.

[5] Fitting, M., *Prefixed tableaus and nested sequents*, Ann. Pure Appl. Logic **163** (2012), pp. 291–313.

[6] Gentzen, G., "The collected papers of Gerhard Gentzen," Edited by M. E. Szabo. Studies in Logic and the Foundations of Mathematics, North-Holland Publishing Co., Amsterdam, 1969, xii+338 pp. (2 plates) pp.

[7] Goré, R., L. Postniece and A. Tiu, *Taming displayed tense logics using nested sequents with deep inference*, in: M. Giese and A. Waaler, editors, *TABLEAUX*, Lecture Notes in Computer Science **5607** (2009), pp. 189–204.
URL `http://dx.doi.org/10.1007/978-3-642-02716-1`

[8] Hein, R., "Geometric Theories and Modal Logic in the Calculus of Structures," Master's thesis, Technische Universität Dresden (2005).

[9] Ishigaki, R. and K. Kikuchi, *Tree-sequent methods for subintuitionistic predicate logics*, in: *Automated reasoning with analytic tableaux and related methods*, Lecture Notes in Comput. Sci. **4548**, Springer, Berlin, 2007 pp. 149–164.

[10] Kanger, S., "Provability in logic," Stockholm Studies in Philosophy 1, Almqvist and Wiksell, Stockholm, 1957.
[11] Kashima, R., *Cut-free sequent calculi for some tense logics*, Studia Logica **53** (1994), pp. 119–135.
[12] Kushida, H. and M. Okada, *A proof-theoretic study of the correspondence of classical logic and modal logic*, J. Symbolic Logic **68** (2003), pp. 1403–1414.
[13] Mints, G., *Indexed systems of sequents and cut-elimination*, J. Philos. Logic **26** (1997), pp. 671–696.
[14] Negri, S., *Proof analysis in modal logic*, J. Philos. Logic **34** (2005), pp. 507–544.
[15] Negri, S., *Proof analysis in non-classical logics*, in: *Logic Colloquium 2005*, Lect. Notes Log. **28**, Assoc. Symbol. Logic, Urbana, IL, 2008 pp. 107–128.
[16] Poggiolesi, F., *The method of tree-hypersequents for modal propositional logic*, in: *Towards mathematical philosophy*, Trends Log. Stud. Log. Libr. **28**, Springer, Dordrecht, 2009 pp. 31–51.
[17] Poggiolesi, F., *A purely syntactic and cut-free sequent calculus for the modal logic of provability*, The Review of Symbolic Logic **2** (2009), pp. 593–611.
[18] Pottinger, G., *Uniform, cut-free formulations of T, S4 and S5*, Abstract in JSL **48** (1983), pp. 900–901.
[19] Ramanayake, R., "Cut-elimination for provability logics and some results in display logic," Ph.D. thesis, Research School of Computer Science, The Australian National University, Canberra. (2011).
URL `http://users.cecs.anu.edu.au/~rpg/Revantha.Ramanayake/thesis.pdf`
[20] Restall, G., *Comparing modal sequent systems*, `http://consequently.org/papers/comparingmodal.pdf`.
[21] Viganò, L., "Labelled non-classical logics," Kluwer Academic Publishers, Dordrecht, 2000, xiv+291 pp., with a foreword by Dov M. Gabbay.

Extending $\mathcal{ALCQ}$ with Bounded Self-Reference

Daniel Gorín

Department of Computer Science, Friedrich-Alexander-Universität Erlangen-Nürnberg

Lutz Schröder [1]

Department of Computer Science, Friedrich-Alexander-Universität Erlangen-Nürnberg

Abstract

Self-reference has been recognized as a useful feature in description logics but is known to cause substantial problems with decidability. We have shown in previous work that the basic description logic $\mathcal{ALC}$ remains decidable, and in fact retains its low complexity, when extended with a bounded form of self-reference where only one variable (denoted me following previous work by Marx) is allowed, and no more than two relational steps are allowed to intercede between binding and use of me (this result is optimal in the sense that already allowing three steps leads to undecidability). Here, we extend these results to $\mathcal{ALCQ}$, i.e. $\mathcal{ALC}$ extended with qualified number restrictions, and analyse the expressivity of the arising logic, $\mathcal{ALCQ}\mathsf{me}_2$. In fact it turns out the expressive power of $\mathcal{ALCQ}\mathsf{me}_2$ is identical to that of $\mathcal{ALCHIQ}be$, the extension of $\mathcal{ALCQ}$ with role inverses, role hierarchies, safe Boolean combinations of roles, and a simple self-loop construct. However, while there is a straightforwardly defined polynomial translation from $\mathcal{ALCHIQ}be$ to $\mathcal{ALCQ}\mathsf{me}_2$, the translation from $\mathcal{ALCQ}\mathsf{me}_2$ to $\mathcal{ALCHIQ}be$ has an exponential blowup in the formula size. To establish the desired complexity bounds, we therefore provide a polynomial satisfiability-preserving encoding of $\mathcal{ALCQ}\mathsf{me}_2$ into $\mathcal{ALCHIQ}be$ and prove that the latter is decidable in EXPTIME.

Keywords: Description logic, hybrid logic, binding constructs, self-reference, complexity, expressivity, qualified number restrictions

Introduction

In a very broad sense, hybrid logics extend modal logics with special symbols, called *nominals*, that name individual states of models. The assignment of nominals to states can be either *static*, akin to that of *constant symbols* in first-order logic, or *dynamic*. The latter is typically achieved through the $\downarrow$-binder: if x is a nominal, then $\downarrow x.\phi$ is true at a state a if ϕ is true in a under the

[1] The research reported here forms part of the DFG project *Generic Algorithmic Methods in Modal and Hybrid Logics* (SCHR 1115-2)

assumption that x names a. The difference can be dramatic: while extending a modal logic with static hybrid machinery often results in only a moderate increase of computational complexity (if at all), adding $\downarrow$ to the basic modal logic immediately leads to undecidability. This has proven to be hard to repair: it persists when only one nominal is allowed to occur [12], also if one restricts to typically well-behaved classes of models [14] and even if one weakens the semantics of $\downarrow$ [2].

Decidability is regained in the uni-modal case on certain classes of models [14] and when restricting to the rather weak fragment that avoids the so-called $\Box\downarrow\Box$-pattern where the $\downarrow$ occurs under the scope of a $\Box$ and the bound nominal occurs under another $\Box$ [16]. In recent work [9], we have isolated another decidable fragment; it is obtained by allowing only one nominal to be bound by $\downarrow$ and ensuring that between every $\downarrow$ and the usage points of the bound nominal, no more than two modalities occur (when three modalities are allowed to occur, decidability is again lost); we refer to this restriction as limiting the *depth* of occurrences of the bound nominal. In [12], Marx proposes the usage of pronouns I and me as suggestive notation for the hybrid language were only one nominal can be bound (with I implicitly binding the nominal me); using this notation, an example of a depth 2 formula is given by

$$[\mathsf{hasParent}]\mathsf{I}.\neg\langle\mathsf{likes}\rangle(\mathsf{Club} \wedge \langle\mathsf{accepts}\rangle\mathsf{me}), \tag{1}$$

which, intuitively, describes those whose parents share Groucho's taste for clubs (observe, incidentally, that (1) falls inside the $\Box\downarrow\Box$-pattern, so it is outside the decidable fragment of [16]).

Despite its innocent look, the depth-2 fragment of the I-me hybrid logic turns out to be expressively interesting. In particular, the nominal-free language (that is, free of nominals other than me) with the universal modality is capable of expressing all formulas of the guarded fragment over the correspondence language; and this containment is strict: the latter has the finite model property while the former does not [9]. Nevertheless, the nominal-free fragment has the same complexity as basic $\mathcal{ALC}$: PSPACE-complete over acyclic TBoxes (even when the satisfaction operator $@_{\mathsf{me}}$ is added), and EXPTIME-complete over general TBoxes, equivalently in presence of the universal modality.

Given these results, it is interesting to understand the effect of *hybridizing* other modal logics in this restricted fashion. Here, we consider the case of *graded modal logic*, or in description logic parlance, the logic $\mathcal{ALCQ}$, whose extension with the I-me construct limited to depth 2 we denote $\mathcal{ALCQ}\mathsf{me}_2$. To avoid unnecessary duplication of concepts and notations, we will work using the language and terminology of description logics (DLs) throughout [5].

To get a taste of $\mathcal{ALCQ}\mathsf{me}_2$, consider an asymmetric social-networking platform such as Twitter as the domain of knowledge representation. In this context we can express the concept of a *celebrity* as

$$\mathsf{I}.{\geq_N}\mathsf{isFollowedBy}.\neg\exists\mathsf{isFollowedBy}.\mathsf{me}$$

for some suitable choice of $N \gg 0$. In words, a celebrity in Twitter is someone with a large base of followers that are not followed back.

It turns out that $\mathcal{ALCQ}\mathsf{me}_2$ is expressive enough to encode inverses (aka *past* modalities), union, intersection, and relative complementation of roles (i.e. relations), as well as role inclusions. This puts it, prima facie, close to the DL known as $\mathcal{ALCHIQ}b$. It is the main object of this work to make this comparison systematic.

In fact it is clear that $\mathcal{ALCQ}\mathsf{me}_2$ contains, via a straightforward semantics-preserving translation that has only a polynomial blowup in the formula size, a slight extension of $\mathcal{ALCHIQ}b$, namely by the self-loop construct $\exists R.\mathsf{Self}$ also found, e.g., in the DL $s\mathcal{ROIQ}$ [10]; we denote this extension by $\mathcal{ALCHIQ}be$. Conversely, we show that indeed there exists also a semantics-preserving translation from $\mathcal{ALCQ}\mathsf{me}_2$ to $\mathcal{ALCHIQ}be$, so that we can say in a precise sense that $\mathcal{ALCQ}\mathsf{me}_2$ and $\mathcal{ALCHIQ}be$ have the same expressive power; however, this converse translation has an exponential blowup in the formula size. In order to fix the computational complexity of $\mathcal{ALCQ}\mathsf{me}_2$, we therefore define a second translation into $\mathcal{ALCHIQ}be$. This translation has only a polynomial blowup; it preserves satisfiability but is not semantics-preserving in the proper sense, as it introduces new underdefined relation symbols and satisfaction of the original formula does not imply satisfaction of its translation (only *satisfiability* of the latter). We then show by reduction to $\mathcal{ALCIQ}b$ [18] that $\mathcal{ALCHIQ}be$ is decidable in PSPACE over the empty TBox, and in EXPTIME over general TBoxes, thus implying that the same bounds hold for $\mathcal{ALCQ}\mathsf{me}_2$, which is hence no harder than $\mathcal{ALCQ}$ (or, for that matter, basic $\mathcal{ALC}$). Our proof can be easily modified to show that, in general, adding the self-looping construct to a description logic is harmless in terms of computational complexity.

The material is organized as follows. We first introduce the description logics we will use as well as some basic terminology (for a deeper introduction, refer to [5]). In particular, we define the logics $\mathcal{ALCHIQ}be$ and $\mathcal{ALCQ}\mathsf{me}_2$. In Section 2 we discuss in a general fashion various ways in which one can compare the expressiveness of two description logics, in order to pave the ground for a comparison of the various translations that we define. We discuss the embedding of $\mathcal{ALCHIQ}be$ into $\mathcal{ALCQ}\mathsf{me}_2$ in Section 3, and the converse semantics-preserving translation in Section 4. Moreover, we establish a polynomial satisfiability-preserving translation of $\mathcal{ALCQ}\mathsf{me}_2$ into $\mathcal{ALCHIQ}be$ (Proposition 4.5). Finally, we present our complexity results in Section 5.

1 Description Logics and Self-Referential Constructs

Description logics (DLs) are typically described as arising from a combination of a number of more or less standard language features. The features we will be interested in are qualified number restrictions, role-hierarchies, inverses, safe Boolean combination of roles and the "self-looping" concept constructor. We now introduce a DL called $\mathcal{ALCHIQ}be$ that contains all of these features, and later consider some of its better known fragments.

Assume a vocabulary $\langle \mathsf{N_C}, \mathsf{N_R}\rangle$ composed of two disjoint and countably infinite sets $\mathsf{N_C} = \{A_1, A_2, \ldots\}$ and $\mathsf{N_R} = \{R_1, R_2, \ldots\}$ of *atomic* concepts and roles, respectively. The set of (complex) concepts C, D and roles R, S of $\mathcal{ALCHIQ}be$ are respectively given by

$$C, D := A_i \mid \neg C \mid C \sqcap D \mid \exists R.\mathsf{Self} \mid \geq_n R.C$$
$$R, S := R_i \mid R^- \mid R \sqcap S \mid R \sqcup S \mid R - S$$

where $A_i \in \mathsf{N_C}$; $R_i \in \mathsf{N_R}$ and n is a positive integer. The concept $\exists R.\mathsf{Self}$ is meant to denote the individuals related via R to themselves (but notice that Self in itself is not a concept).

For brevity, we will employ standard notation, such as $\bot \equiv A \sqcap \neg A$, for some $A \in \mathsf{N_C}$; $\top \equiv \neg\bot$; $\leq_n R.C \equiv \neg \geq_{n+1}.C$, for every $n \geq 0$; $\exists R.C \equiv \geq_1 R.C$ and $\forall R.C \equiv \neg\exists R.\neg C$. One can also define $\forall R.\mathsf{Self} \equiv \leq_1 R.\top \sqcap (\forall R.\bot \sqcup \exists R.\mathsf{Self})$. We will sometimes write $\geq_{n-1} R.C$ in cases where $n \geq 1$, and then assume that $\geq_0 R.C$ is a synonym for $\top$.

When measuring formula size, we assume numbers expressed in *binary*. We shall use *rank*(C) to mean the maximal number of nested qualified number restrictions (i.e., concepts of the form $\geq_n R.D$) occurring in C (this is also known as the *modal depth* of C).

An *interpretation* or *model* is a structure $\mathcal{I} = \langle \Delta^\mathcal{I}, \cdot^\mathcal{I} \rangle$ where $\Delta^\mathcal{I}$ is a non-empty set (the *domain* of $\mathcal{I}$); $A^\mathcal{I} \subseteq \Delta^\mathcal{I}$ for each $A \in \mathsf{N_C}$; and, for $R_i \in \mathsf{N_R}$, $R_i^\mathcal{I} \subseteq \Delta^\mathcal{I} \times \Delta^\mathcal{I}$. We write $R^\mathcal{I}(a)$ for the set $\{b : (a, b) \in R^\mathcal{I}\}$. Let $\mathcal{I} = \langle \Delta^\mathcal{I}, \cdot^\mathcal{I} \rangle$ be an interpretation. The extension of complex concepts and roles under $\mathcal{I}$ is defined as:

$$
\begin{aligned}
(\neg C)^\mathcal{I} &= \Delta^\mathcal{I} - C^\mathcal{I} & (R - S)^\mathcal{I} &= R^\mathcal{I} - S^\mathcal{I}\\
(C \sqcap D)^\mathcal{I} &= C^\mathcal{I} \cap D^\mathcal{I} & (R \sqcap S)^\mathcal{I} &= R^\mathcal{I} \cap S^\mathcal{I}\\
\exists R.\mathsf{Self} &= \{a : (a, a) \in R^\mathcal{I}\} & (R \sqcup S)^\mathcal{I} &= R^\mathcal{I} \cup S^\mathcal{I}\\
(\geq_n R.C)^\mathcal{I} &= \{a : |R^\mathcal{I}(a) \cap C^\mathcal{I}| \geq n\} & (R^-)^\mathcal{I} &= \{(b, a) : (a, b) \in R^\mathcal{I}\}
\end{aligned}
$$

In description logics, a *TBox* is a set of "concept inclusion axioms" of the form $C \sqsubseteq D$. Sometimes the term "general TBox" is used to stress the fact that concepts occurring in it are arbitrary (in opposition to, say, *acyclic TBoxes*, where the dependency graph induced by the atomic concepts in the TBox must be acyclic). Apart from a TBox, in $\mathcal{ALCHIQ}be$, one has an *RBox* that contains "role inclusion axioms" of the form $R \sqsubseteq S$. Let $\mathcal{T}$ and $\mathcal{H}$ be a TBox and a RBox, respectively. An interpretation $\mathcal{I}$ *satisfies* $(\mathcal{T}, \mathcal{H})$ whenever $C^\mathcal{I} \subseteq D^\mathcal{I}$ for all $C \sqsubseteq D \in \mathcal{T}$ and $R^\mathcal{I} \subseteq S^\mathcal{I}$ for all $R \sqsubseteq S \in \mathcal{H}$. A concept C is *satisfiable over* $(\mathcal{T}, \mathcal{H})$ if $C^\mathcal{I} \neq \emptyset$ for some interpretation $\mathcal{I}$ that satisfies $(\mathcal{T}, \mathcal{H})$. Finally, C is *satisfiable* if it is satisfiable over the empty TBox and the empty RBox.

Many well-known description logics are obtained as fragments of the logic $\mathcal{ALCHIQ}be$ just introduced. E.g., $\mathcal{ALCHIQ}b$ results from dropping the $\exists R.\mathsf{Self}$ concept-constructor (which is part of the very expressive description logic $s\mathcal{ROIQ}$ [10]) and $\mathcal{ALCHIQ}$ is obtained if, additionally, one removes role-constructors $\sqcap$, $\sqcup$ and $-$. If, in any of these cases, attention is restricted to empty RBoxes, the logics obtained are respectively called $\mathcal{ALCIQ}be$, $\mathcal{ALCIQ}b$

and $\mathcal{ALCIQ}$; note that $\mathcal{ALCIQ}b$ played a key role in the proof of the EXPTIME upper bound for $\mathcal{SHIQ}$ [18]. We obtain the logic $\mathcal{ALCQ}$ by removing the inverse role constructor $\cdot^-$ from $\mathcal{ALCIQ}$. If we furthermore restrict n to 1 in qualified-number restrictions $\geq_n R$ (and hence to $n = 0$ in $\leq_n R$), the basic description logic $\mathcal{ALC}$ is obtained.

It is worth noticing that in [18], safe Boolean roles are defined as containing *role negation* $\neg R$, instead of *relativized negation* $R - S$ as defined above. However, the *safety* condition for role expressions used there is that they must contain at least one non-negated role in each clause of the disjunctive normal form. It is straightforward to verify that this is equivalent to the definition we gave. Moreover, it is interesting to observe that this restriction is not casual, as unsafe roles can be used, for instance, to express global cardinality constraints as in $\leq_5 R \sqcup \neg R.\, C$, which are known to increase complexity; e.g. $\mathcal{ALCIQ}$ with cardinality restrictions is NEXPTIME-complete [17].

Regarding computational complexity, the concept satisfiability problem for $\mathcal{ALCIQ}b$ is known to be PSPACE-complete over the empty TBox, and EXPTIME-complete over general TBoxes [18]. The latter problem remains EXPTIME-complete for $\mathcal{ALCHIQ}b$ [11].

Meet $\mathcal{ALCQ}\mathsf{me}_2$: Bounded self-referentiality for $\mathcal{ALCQ}$

In [12], Marx introduced the I-me construct as a convenient notation for the single-variable fragment of hybrid logic with the $\downarrow$-binder (cf. [3]). In this notation, I plays the role of a $\downarrow$-binder and me is the bound nominal. We want to extend the logic $\mathcal{ALCQ}$ with the I-me construct under the restriction that me is never separated from its binding I by more than two qualified number restrictions. We call the resulting logic $\mathcal{ALCQ}\mathsf{me}_2$.

At a syntactic level, we want to add to the concept language of $\mathcal{ALCQ}$ a new concept me and a concept constructor $\mathsf{I}.C$, where C must satisfy certain requirements with respect to the occurrences of me. Again, we assume two sets $\mathsf{N_C}$ and $\mathsf{N_R}$ of atomic concepts and roles; the concept language of $\mathcal{ALCQ}\mathsf{me}_2$ then corresponds to $\mathcal{F}_1$ in the following grammar:

$$\begin{aligned} \mathcal{F}_1 \ni C, C' &:= A_i \mid \mathsf{me} \mid \neg C \mid C \sqcap C' \mid \mathsf{I}.{\geq_n} R_i.\, C \mid {\geq_n} R_i.\, D \\ \mathcal{F}_2 \ni D, D' &:= A_i \mid \mathsf{me} \mid \neg D \mid D \sqcap D' \mid \mathsf{I}.{\geq_n} R_i.\, C \end{aligned} \tag{2}$$

Here also, $A_i \in \mathsf{N_C}$ and $R_i \in \mathsf{N_R}$, and n is a positive integer. We will say that me occurs *free* in a concept C if it is not under the scope of an I. Similarly, we say that a concept C occurs in a *relational context* in D whenever C occurs under the scope of a $\geq_n R_i$ or $\mathsf{I}.{\geq_n} R_i$ in D.

One can easily get an idea of why (2) works by verifying that, for instance, $\mathsf{I}.{\geq_n} R.{\geq_m} S.{\geq_l} T.\mathsf{me}$ is not well-formed. More generally, one can check that $\mathcal{F}_2 \subsetneq \mathcal{F}_1$ and that each free occurrence of me in C can occur in at most one relational context, if $C \in \mathcal{F}_1$, and in no relational context if $C \in \mathcal{F}_2$. This means that, in fact, a concept such as ${\geq_n} R.{\geq_m} S.\mathsf{me}$ is not well-formed, although $\mathsf{I}.{\geq_n} R.{\geq_m} S.\mathsf{me}$ is so.

While according to (2), I can only occur in front of a qualified number restriction, we shall liberally use it in other positions, by letting it commute through Boolean operations and putting $\mathsf{l}.A \equiv A$ for $A \in \mathsf{N_C}$, $\mathsf{l}.\mathsf{l}.C \equiv \mathsf{l}.C$, and $\mathsf{l}.\mathsf{me} \equiv \top$. That said, for definitions and proofs we will stick to the language defined by (2).

Let $\langle \Delta^{\mathcal{I}}, \cdot^{\mathcal{I}} \rangle$ be an interpretation. For every $a \in \Delta^{\mathcal{I}}$, we use $C_a^{\mathcal{I}}$ to denote the extension of an $\mathcal{ALCQ}\mathsf{me}_2$-concept C under the assumption that me stands for a. This is defined as follows:

$$\begin{aligned} A_a^{\mathcal{I}} &= A^{\mathcal{I}} & \mathsf{me}_a^{\mathcal{I}} &= \{a\} \\ (\neg C)_a^{\mathcal{I}} &= \Delta^{\mathcal{I}} - C_a^{\mathcal{I}} & (C \sqcap D)_a^{\mathcal{I}} &= C_a^{\mathcal{I}} \cap D_a^{\mathcal{I}} \\ (\mathsf{l}.{\geq_n} R.C)_a^{\mathcal{I}} &= \{b : b \in ({\geq_n} R.C)_b^{\mathcal{I}}\} & ({\geq_n} R.C)_a^{\mathcal{I}} &= \{b : |R^{\mathcal{I}}(b) \cap C_a^{\mathcal{I}}| \geq n\} \end{aligned}$$

We say that C is a *closed* concept if it has no free occurrences of me. One can easily show the following:

Proposition 1.1 *If C is a closed concept, then $C_a^{\mathcal{I}} = C_b^{\mathcal{I}}$ for all $a, b \in \Delta^{\mathcal{I}}$.*

We thus denote the extension of a closed concept C just by $C^{\mathcal{I}}$. From Proposition 1.1 it is clear that $\mathcal{ALCQ}\mathsf{me}_2$ is a conservative extension of $\mathcal{ALCQ}$.

In the context of $\mathcal{ALCQ}\mathsf{me}_2$, a TBox is a collection of axioms of the form $C \sqsubseteq D$, where C and D are closed (but otherwise arbitrary) concepts. An interpretation $\mathcal{I}$ *satisfies* a TBox $\mathcal{T}$ whenever $C^{\mathcal{I}} \subseteq D^{\mathcal{I}}$ for all $C \sqsubseteq D \in \mathcal{T}$. A *closed* concept C is *satisfiable over* $\mathcal{T}$ if $C^{\mathcal{I}} \neq \emptyset$ for some interpretation $\mathcal{I}$ that satisfies $\mathcal{T}$. Finally, C is *satisfiable* if it is satisfiable over the empty TBox.

The object of the current work is to investigate the properties of $\mathcal{ALCQ}\mathsf{me}_2$. In the end, one wants to see how $\mathcal{ALCQ}\mathsf{me}_2$ stands with respect to its better-known fellow logics. In the following sections, we investigate then what is the *expressive power* and the *computational complexity* of $\mathcal{ALCQ}\mathsf{me}_2$.

2 Prerequisites for a Discussion of Relative Expressivity

The term "expressivity" or "expressive power" has come to mean different, although related, things. This has the unfortunate effect that it may lead to seemingly contradictory claims regarding the relative expressivity of two logics, both being correct once one pinpoints the precise notion of expressivity being used in each case. To prevent this type of misunderstandings, we will give a short account of the main notions of expressivity that can be usually found in the literature.

When comparing the expressive power of two logics, it certainly helps if they have a perfect match on the objects used as interpretations. In practice, though, this would be too strong a restriction: for instance, it would exclude all so-called *TBox-internalization results*, where a fresh atomic role is needed to behave as a universal modality (so the models of the logic doing the internalization need to interpret more symbols). If we want some liberty in this respect, we need to have the means to compare logics whose languages and/or models do not perfectly coincide. Having said that, we will avoid an overly abstract presentation by focusing on the case of "description logics" (a more abstract, very general presentation would require, for instance, moving to the framework of *institutions* [7]).

Let us make more precise what we mean by "description logics". These can be characterized by a *concept language*, built from atomic concepts and roles; and the possibility of building *theories*, in the form of TBoxes (sets of concept-related axioms), RBoxes (sets of role-related axioms), ABoxes (sets of axioms regarding named individuals), etc. Theories can be assumed to form a monoid, so one can talk about the *empty theory* or take the *union* of two theories. A model for a description logic is a non-empty domain plus an interpretation for each atomic concept, role, named-individual, etc.

In the various notions we introduce next, we will typically say that "a logic $\mathcal{L}_2$ has at least the same power (for some task) as a logic $\mathcal{L}_1$", and by this it should be understood, intuitively, that a certain logical task in $\mathcal{L}_1$ can be *uniformly reduced to the same task* on $\mathcal{L}_2$, perhaps *over an extended language*. This means that models for $\mathcal{L}_2$ will be *expansions* of models for $\mathcal{L}_1$, which leads to a natural way of mapping $\mathcal{L}_2$-models to $\mathcal{L}_1$-models (this can be seen as an specialization of the notion of *logic comorphism* [8]).

Definition 2.1 *Let $\mathcal{S}_1 = \langle \mathsf{N_{C1}}; \mathsf{N_{R1}} \rangle$ and $\mathcal{S}_2 = \langle \mathsf{N_{C2}}; \mathsf{N_{R2}} \rangle$ be two collections of atomic concepts and roles such that $\mathsf{N_{C1}} \subseteq \mathsf{N_{C2}}$ and $\mathsf{N_{R1}} \subseteq \mathsf{N_{R2}}$ hold (we write $\mathcal{S}_1 \subseteq \mathcal{S}_2$). The* forgetful mapping *β from $\mathcal{S}_2$-interpretations to $\mathcal{S}_1$-interpretations is defined by i) $\Delta^{\beta(\mathcal{I})} = \Delta^{\mathcal{I}}$; ii) $A^{\beta(\mathcal{I})} = A^{\mathcal{I}}$, for every $A \in \mathsf{N_{C1}}$ and iii) $R^{\beta(\mathcal{I})} = R^{\mathcal{I}}$, for every $R \in \mathsf{N_{R1}}$. Moreover, given a consistent $\mathcal{L}_2$-theory $\mathcal{T}$ over $\mathcal{S}_2$, we say that $\mathcal{T}$ is* oblivious *to $\mathcal{S}_1$ if for every model $\mathcal{I}_1$ for $\mathcal{S}_1$ there exists a model $\mathcal{I}_2$ satisfying $\mathcal{T}$ such that $\mathcal{I}_1 = \beta(\mathcal{I}_2)$.*

Intuitively, a theory is oblivious to $\mathcal{S}_1$ if it only refers to symbols that are not in $\mathcal{S}_1$. We are now ready to look at the notions of expressivity that will concern us.

Querying power. Let C be a concept of a description logic $\mathcal{L}_1$ over vocabulary $\mathcal{S}_1$. One can view C as a *query* over $\mathcal{S}_1$-interpretations by taking $C^{\mathcal{I}}$ to be the result of the query over $\mathcal{I}$. To a first approximation, we can say that $\mathcal{L}_2$ has at least the same querying power as $\mathcal{L}_1$ if there is an effective mapping α from $\mathcal{L}_1$-concepts to $\mathcal{L}_2$-concepts such that $C^{\mathcal{I}} = \alpha(C)^{\mathcal{I}}$. This assumes that concepts in the domain and image of α are built over $\mathcal{S}_1$. To lift this restriction, we shall say that $\mathcal{L}_2$ *has at least the same querying power as* $\mathcal{L}_1$, notation $\mathcal{L}_1 \leq_Q \mathcal{L}_2$, if there exist a mapping δ taking a vocabulary $\mathcal{S}_1$ to an $\mathcal{L}_2$-theory $\mathcal{T}$ oblivious to $\mathcal{S}_1$; and an effective mapping α such that, for all models $\mathcal{I}_2$ satisfying $\mathcal{T}$, $C^{\beta(\mathcal{I}_2)} = \alpha(C)^{\mathcal{I}_2}$. We write $\mathcal{L}_1 \leq_{Q,\delta} \mathcal{L}_2$ when we want to make explicit the mapping δ employed.

This definition essentially says that given a query C and a model $\mathcal{I}_1$ to be queried, we can find some appropriate $\mathcal{L}_2$- model $\mathcal{I}_2$ satisfying $\mathcal{T}$ (e.g., by an exhaustive search), and query it using $\alpha(C)$, instead.

Classification power. Let $\mathcal{T}$ be a theory for a description logic $\mathcal{L}_1$ over $\mathcal{S}_1$. Let us write $\mathcal{I}_1 \models_{\mathcal{L}_1} \mathcal{T}$ to denote that model $\mathcal{I}_1$ satisfies all the axioms of $\mathcal{T}$. We can then say that $\mathcal{T}$ splits the class of $\mathcal{S}_1$-models in two, implicitly defining a property on models. We then say that $\mathcal{L}_2$ *has at least the same classification power as* $\mathcal{L}_1$, notation $\mathcal{L}_1 \leq_C \mathcal{L}_2$, if there exists a mapping δ taking a vocabulary $\mathcal{S}_1$ to an $\mathcal{L}_2$-theory $\mathcal{T}_2$ oblivious to $\mathcal{S}_1$; and an effective mapping γ of $\mathcal{L}_1$-theories to $\mathcal{L}_2$-theories such that for every $\mathcal{L}_1$-theory $\mathcal{T}_1$ and every $\mathcal{L}_2$-model $\mathcal{I}_2$ of $\mathcal{T}_2$, it holds that $\beta(\mathcal{I}_2) \models_{\mathcal{L}_1} \mathcal{T}_1$ iff $\mathcal{I}_2 \models_{\mathcal{L}_2} \gamma(\mathcal{T}_1)$. In all cases, we assume γ to be defined axiom-wise, i.e., $\gamma(\mathcal{T}) = \bigcup_{a \in \mathcal{T}} \gamma'(a)$, where γ' maps axioms to (finite) theories; moreover, we identify γ with γ'. We write $\mathcal{L}_1 \leq_{C,\delta} \mathcal{L}_2$ when we want to make explicit the mapping δ employed.

It is possible to slightly weaken this notion as follows. We say that $\mathcal{L}_2$ *has at least weakly the same classification power as* $\mathcal{L}_1$, notation $\mathcal{L}_1 \leq_C^{\exists} \mathcal{L}_2$, if there exists a mapping δ taking a vocabulary $\mathcal{S}_1$ to an $\mathcal{L}_2$-theory $\mathcal{T}_2$ oblivious to $\mathcal{S}_1$; and an effective mapping γ of $\mathcal{L}_1$-theories to $\mathcal{L}_2$-theories such that for every $\mathcal{L}_1$-theory $\mathcal{T}_1$ and every $\mathcal{L}_1$-model $\mathcal{I}_1$, $\mathcal{I}_1 \models_{\mathcal{L}_1} \mathcal{T}_1$ iff *there exists* a model $\mathcal{I}_2$ of $\mathcal{T}_2$ such that $\beta(\mathcal{I}_2) = \mathcal{I}_1$ and $\mathcal{I}_2 \models_{\mathcal{L}_2} \gamma(\mathcal{T}_1)$. Observe that $\leq_C^{\exists}$ differs from $\leq_C$ in the quantification pattern of $\mathcal{L}_2$-models.

While $\mathcal{L}_1 \leq_{C,\delta} \mathcal{L}_2$ means that we can reduce the problem of deciding if an $\mathcal{L}_1$-model satisfies an $\mathcal{L}_1$-theory $\mathcal{T}_1$ to that of deciding if a certain $\mathcal{L}_2$-model (e.g., any model we find that satisfies the theory given by δ) satisfies the $\mathcal{L}_2$-theory $g(\mathcal{T}_1)$, if we have that $\mathcal{L}_1 \leq_C^{\exists} \mathcal{L}_2$, then the same problem is reduced to (potentially) *many* instances of the decision problem for $\mathcal{L}_2$ (i.e., one for each model $\mathcal{I}_2$ that satisfies the theory given by δ). One can regard the latter notion as a strong form of *satisfiability preserving* translation (see below).

Local and global reasoning power. The two criteria above could be encompassed under the term *descriptive power*, since they refer to the capacity of a logic to describe models or individuals in a model. The following criteria, on the other hand, refer to the *inferences* that can be made.

Consider first the set $V_{\mathcal{L}}$ of *valid concepts of* $\mathcal{L}$. We say that $\mathcal{L}_2$ *has at least the same local reasoning power as* $\mathcal{L}_1$, notation $\mathcal{L}_1 \leq_{R^l} \mathcal{L}_2$, if there exists an effective mapping α from $\mathcal{L}_1$-concepts to $\mathcal{L}_2$-concepts such that, for every $\mathcal{L}_1$-concept C, $C \in V_{\mathcal{L}_1}$ iff $\alpha(C) \in V_{\mathcal{L}_2}$. Similarly, for a theory $\mathcal{T}$, let $V_{\mathcal{L}}(\mathcal{T})$ denote the set of $\mathcal{L}$-consequences of $\mathcal{T}$, that is the set $\{C : \mathcal{I} \models_{\mathcal{L}} \mathcal{T} \Rightarrow C^{\mathcal{I}} = \Delta^{\mathcal{I}}, \forall \mathcal{I}\}$. We then say that $\mathcal{L}_2$ *has at least the same global reasoning power as* $\mathcal{L}_1$, notation $\mathcal{L}_1 \leq_{R_g} \mathcal{L}_2$, if there exist two effective mappings α and γ such that, for every theory $\mathcal{T}$ of $\mathcal{L}_1$, and every $\mathcal{L}_1$-concept C, $C \in V_{\mathcal{L}_1}(\mathcal{T})$ iff $\alpha(C) \in V_{\mathcal{L}_2}(\gamma(\mathcal{T}))$.

It is worth noticing that $\mathcal{L}_2$ having at least the same local reasoning power as $\mathcal{L}_1$ intuitively means that local reasoning $\mathcal{L}_1$ can be reduced to local reasoning in $\mathcal{L}_2$.

Proposition 2.2 *The following hold (where δ_0 maps a signature to the empty theory):*

(i) *If $\mathcal{L}_1 \leq_{Q,\delta_0} \mathcal{L}_2$, then $\mathcal{L}_1 \leq_{R^l} \mathcal{L}_2$,*

(ii) *If $\mathcal{L}_1 \leq_{Q,\delta_0} \mathcal{L}_2$ and $\mathcal{L}_1 \leq_{C,\delta_0} \mathcal{L}_2$, then $\mathcal{L}_1 \leq_{R^g} \mathcal{L}_2$.*

(iii) *If $\mathcal{L}_1 \leq_{R^g} \mathcal{L}_2$, then $\mathcal{L}_1 \leq_{R^l} \mathcal{L}_2$.*

Notice that for the last case, we use that γ preserves empty theories.

We say that $\mathcal{L}_1$ and $\mathcal{L}_2$ have *equal querying power* if each one has at least the same expressive power as the other. This extends to the other notions in the obvious way.

All the above notions are qualitative in nature. A quantitative comparison can be done by looking at the blowup incurred by the mapping on concepts and/or theories. So-called *succinctness* results (see, e.g., [1]) typically involve lower-bounds (at least exponential, usually) for the blowup incurred by the mapping on concepts for the notion of classification power. An analogous notion of succinctness for querying power was investigated in [6]. The mappings used for defining local and global reasoning power are usually called *satisfiability preserving translations*, and are used as a tool for proving complexity results; the blowup in this case impacts on the final complexity.

Proposition 2.3 *The following relations hold:*

(i) $\mathcal{ALCIQ}be \leq_Q \mathcal{ALCHIQ}be$ *and* $\mathcal{ALCHIQ}be \leq_Q \mathcal{ALCIQ}be$.

(ii) $\mathcal{ALCIQ}be \leq_C \mathcal{ALCHIQ}be$.

(iii) $\mathcal{ALCIQ}be \leq_{R^g} \mathcal{ALCHIQ}be$ *and* $\mathcal{ALCHIQ}e \leq_{R^g} \mathcal{ALCIQ}be$.

Proof Clearly, role-hierarchies add no querying power (since the model is fixed in this case), so it is no surprise that $\mathcal{ALCHIQ}be \leq_Q \mathcal{ALCIQ}be$. It is proved in [18] that $\mathcal{ALCHIQ} \leq_{R^g} \mathcal{ALCIQ}b$; essentially, every role S in a concept C is replaced by the conjunction $R_1 \sqcap R_2 \ldots \sqcap R_n \sqcap S$ of all roles $R_i \sqsubseteq S$. This is trivially extended to the case with $\exists R.\mathsf{Self}$. □

3 Lower Bounds for Expressivity

We want to show that $\mathcal{ALCQ}\mathsf{me}_2$ has enough expressive power to accommodate $\mathcal{ALCHIQ}be$ for all the tasks outlined in the previous section. Moreover, the blowup in all cases will be shown to be polynomial. We shall assume, throughout this section, that $\mathcal{ALCHIQ}be$ formulas are defined over a vocabulary $\mathcal{S}_1 = \langle \mathsf{N_{C1}}, \mathsf{N_{R1}} \rangle$; $\mathcal{ALCQ}\mathsf{me}_2$ concepts, on the other hand, will be built over $\mathcal{S}_2 = \langle \mathsf{N_{C2}}, \mathsf{N_{R2}} \rangle$, with $\mathcal{S}_1 \subseteq \mathcal{S}_2$. Further assumptions regarding $\mathcal{S}_2$ will be made as needed.

Let us start by considering the querying power of $\mathcal{ALCQ}\mathsf{me}_2$. Clearly, role-hierarchies can be disregarded here, so we concentrate on the features of $\mathcal{ALCIQ}be$. The first thing to observe is that the $\exists R.\mathsf{Self}$ concept corresponds to $\mathsf{I}.\exists R.\mathsf{me}$. To account for the inverse role of R_i we assume $\mathsf{N_{R2}}$ to contain a (fresh) role $\tilde{S}_{R_i^-}$ for each $R_i \in \mathsf{N_{R1}}$; its meaning is then defined by way of the TBox axiom:

$$(\mathsf{I}.\forall R_i.\exists \tilde{S}_{R_i^-}.\mathsf{me} \quad) \sqcap (\mathsf{I}.\forall \tilde{S}_{R_i^-}.\exists R_i.\mathsf{me}). \tag{3}$$

In any model that satisfies axiom (3), $\tilde{S}_{R_i^-}$ will be the inverse role of R_i. In order to deal, in addition, with safe Boolean combination of roles, we further assume that $\mathsf{N_{R2}}$ contains a fresh role $\tilde{S}_R$ for each complex role expression R built over atomic roles in $\mathsf{N_{R1}}$. We give them meaning via the following TBox axioms (where we identify $\tilde{S}_{R_i}$ with R_i, for $R_i \in \mathsf{N_{R1}}$):

$$(\mathsf{I}.\forall\tilde{S}_{R^-}.\exists\tilde{S}_R.\mathsf{me}) \sqcap (\mathsf{I}.\forall\tilde{S}_R.\exists\tilde{S}_{R^-}.\mathsf{me}) \tag{4}$$

$$\mathsf{I}.\forall\tilde{S}_{R\sqcup S}.(\exists\tilde{S}_{R^-}.\mathsf{me} \sqcup \exists\tilde{S}_{S^-}.\mathsf{me}) \tag{5}$$

$$\mathsf{I}.\forall\tilde{S}_{R\sqcap S}.(\exists\tilde{S}_{R^-}.\mathsf{me} \sqcap \exists\tilde{S}_{S^-}.\mathsf{me}) \tag{6}$$

$$\mathsf{I}.\forall\tilde{S}_{R-S}.(\exists\tilde{S}_{R^-}.\mathsf{me} \sqcap \neg\exists\tilde{S}_{S^-}.\mathsf{me}) \tag{7}$$

$$\mathsf{I}.\forall\tilde{S}_R.\exists\tilde{S}_{(R\sqcup S)^-}.\mathsf{me} \tag{8}$$

$$\mathsf{I}.\forall\tilde{S}_R.((\exists\tilde{S}_{S^-}.\mathsf{me} \sqcap \exists\tilde{S}_{(R\sqcap S)^-}.\mathsf{me}) \sqcup (\neg\exists\tilde{S}_{S^-}.\mathsf{me} \sqcap \exists\tilde{S}_{(R-S)^-}.\mathsf{me})) \tag{9}$$

Clearly, (4) makes inverses behave correctly for non-atomic roles as well; taking advantage of this, we have a sound definition for the symbols in (5)–(7) while (8)–(9) make it complete (one can add axioms for commutativity of $\sqcap$ and $\sqcup$, if needed). All this together immediately gives us:

Proposition 3.1 $\mathcal{ALCHIQ}be \leq_Q \mathcal{ALCQ}\mathsf{me}_2$.

We can move now to classification power. Because of the analysis above, it suffices to show how to encode role hierarchies in $\mathcal{ALCQ}\mathsf{me}_2$. Using the symbols $\tilde{S}_R$ already defined (again, identifying $\tilde{S}_{R_i}$ with $R_i \in \mathsf{N_{C1}}$), it is clear that a role inclusion axiom of the form $R \sqsubseteq S$ can be defined with the TBox axiom:

$$\mathsf{I}.\forall\tilde{S}_{R^-}.\exists\tilde{S}_S.\mathsf{me} \tag{10}$$

We then have

Proposition 3.2 $\mathcal{ALCHIQ}be \leq_C \mathcal{ALCQ}\mathsf{me}_2$.

Notice, though, that it is not possible to immediately conclude, from all these results, that $\mathcal{ALCHIQ}be \leq_{R^l} \mathcal{ALCQ}\mathsf{me}_2$. In essence, the mapping α induced in the proof of Proposition 3.1 requires the support of a non-empty TBox. This, in turn, means that neither can we conclude yet that $\mathcal{ALCHIQ}be \leq_{R^g} \mathcal{ALCQ}\mathsf{me}_2$. We turn, then, to proving that this reduction holds as well. For this, one needs a standard auxiliary result. For a relation $T \subseteq X \times X$, define $T^0 = Id_X$ and $T^{n+1} = T^n \cup (T \circ T^n)$; one then has the following:

Lemma 3.3 *Let C be a closed $\mathcal{ALCQ}\mathsf{me}_2$-concept over a vocabulary $\mathcal{S} = \langle \mathsf{N_C}, \mathsf{N_R}\rangle$ and let $a \in C^\mathcal{I}$ for some $\mathcal{I}$. Moreover, let $\mathcal{I}_{k,a}$ denote the restriction of $\mathcal{I}$ to the domain $\Delta^{\mathcal{I}_{k,a}} = \bigcup_{R_i \in \mathsf{N_R}} \{b : aR_i^k b\}$. Then we have that $a \in C^{\mathcal{I}_{k,a}}$ whenever $k \geq rank(C)$.*

With this lemma we can prove,

Theorem 3.4 $\mathcal{ALCHIQ}be \leq_{R^g} \mathcal{ALCQ}\mathsf{me}_2$.

Proof We need to define a mapping γ of axioms to finite $\mathcal{ALCQ}\mathsf{me}_2$-TBoxes for the $\mathcal{ALCHIQ}be$-TBox, and a mapping α from $\mathcal{ALCHIQ}be$-concepts to $\mathcal{ALCQ}\mathsf{me}_2$-concepts. For the former, we can just reuse the mapping from the proof of Proposition 3.2, with the proviso that each mapped axiom introduces also the finitely many definitional axioms (from the mapping δ of the vocabulary) for the symbols that occur in it. Of course, we cannot do this for α and the problem is how to locally define the symbols that are not mentioned in the $\mathcal{ALCHIQ}be$-TBox. We can then assume, without loss of generality, that we are working with an empty TBox. Let C be an $\mathcal{ALCHIQ}be$-concept, and let C' be the $\mathcal{ALCQ}\mathsf{me}_2$-concept obtained from the translation in Proposition 3.1. From Lemma 3.3, we know that it is enough to define the $\tilde{S}_R$ symbols occurring in C' until depth $rank(C') = m$. For this, we resort to a fresh role U, which we force to behave as a "universal role up to depth m" by means of the concept

$$D = \bigsqcap_{i=1}^{k} \bigsqcap_{j=0}^{m} \forall^j U.\mathsf{I}.\forall R_i.(\exists U.\mathsf{me} \sqcap \mathsf{I}.\forall U.\exists U.\mathsf{me}) \tag{11}$$

where $R_1 \ldots R_k$ are the atomic roles occurring in C' and $\forall^n R.D$ denotes D if $n = 0$, and $\forall R.\forall^{n-1} R.D$ otherwise. Having defined U this way, we can give the desired meaning, up to depth m, to every role $\tilde{S}_R$ occurring in C' using suitable concepts $E(R)$; e.g.,

$$E(R_i^-) = \bigsqcap_{j=0}^{m} \forall^j U.(\mathsf{I}.\forall R_i.\exists \tilde{S}_{R_i^-}.\mathsf{me} \sqcap \mathsf{I}.\forall \tilde{S}_{R_i^-}.\exists R_i.\mathsf{me}). \tag{12}$$

The required concept then has the shape $C' \sqcap D \sqcap \bigwedge_R E(R)$ where R ranges over all complex roles occurring in C. □

4 A Tight Upper Bound for Expressive Power

We have seen in the previous section that $\mathcal{ALCQ}\mathsf{me}_2$ is at least as expressive as $\mathcal{ALCHIQ}be$. We now make this bound tight by showing that $\mathcal{ALCIQ}be$ is also as expressive as $\mathcal{ALCQ}\mathsf{me}_2$. It must be observed, though, that in this case we will incur an exponential blowup. We will later show that a weaker result can be obtained with a polynomial blowup.

We begin, then, by showing that $\mathcal{ALCQ}\mathsf{me}_2 \leq_{Q,\delta_0} \mathcal{ALCIQ}be$, where δ_0 maps a vocabulary to the empty TBox. From this it will be straightforward to derive the remaining bounds. The proof of the first result will go in two steps: we show that every $\mathcal{ALCQ}\mathsf{me}_2$-concept can be taken to a normal form from which an equivalent $\mathcal{ALCIQ}be$ concept can be derived. Both translation steps will cause an exponential blowup, but no additional symbols will be introduced.

Definition 4.1 *Assume a fixed, finite set of roles $\mathcal{R} = \{R_1 \ldots R_n\}$. We let $P(\mathcal{R})$ denote the set of all maximal satisfiable conjunctive clauses over the set $\{\exists R_1.\mathsf{me}, \ldots \exists R_n.\mathsf{me}\}$. Moreover, we say that $\mathsf{I}.{\geq_n} R.D$ is an* $\mathcal{R}$-covered concept *if D is of the form $\bigsqcup_{C_i \in P(\mathcal{R})}(C_i \sqcap D_i)$; in this case, the C_i are called* selectors. *Finally, a concept C is in* expanded form *if i) every $\mathsf{I}.{\geq_n} R.D$ occurring in C is an $\mathcal{R}$-covered concept, for some $\mathcal{R}$, and ii)* me *occurs in C only in the selectors of covered concepts and in concepts of the form $\mathsf{I}.\exists R.\mathsf{me}$.*

E.g., an $\{S\}$-covered concept is of the form $\mathsf{I}.{\geq_n} R.(((\exists S.\mathsf{me}) \sqcap D_1) \sqcup \neg((\exists S.\mathsf{me}) \sqcap D_2))$. The intuition behind this definition is that if $\mathsf{I}.{\geq_n} R.D$ is an $\mathcal{R}$-covered concept, then D describes a property D_i for the R-successors of the point of evaluation that is determined by the way they link back to it with respect to some set of roles $\mathcal{R}$. Notice that every R-successor of an element satisfying an $\mathcal{R}$-covered concept will satisfy one and only one of its selectors.

Lemma 4.2 *Every closed concept is equivalent to a closed concept in expanded form.*

Proof We define two mappings γ_1 and γ_2 such that for all closed $\mathcal{ALCQ}\mathsf{me}_2$-concept C, $\gamma_2(\gamma_1(C))$ is in expanded form and equivalent to C. Intuitively, γ_1 deals with those me that occur at depth 1 while γ_2 does so for those at depth 2. To achieve this, γ_2 turns every $(\mathsf{l}.)\geq_n R.D$ in $\gamma_1(C)$ into a covered concept. In what follows, we will use $C[x/y]$ to denote the substitution of all the *top-level* occurrences of x in C by y; here "top-level" means "not under the scope of any qualified number restriction" (e.g. $(\mathsf{me} \sqcap \geq_n R.\mathsf{me})[\mathsf{me}/A] = A \sqcap \geq_n R.\mathsf{me}$). For $x \in 2 = \{0,1\}$ and a concept C, we let C^x denote C, if $x = 1$ and $\neg C$ otherwise.

We start by defining γ_1, whose only non-trivial case is $\gamma_1(\mathsf{l}.\geq_n R.C)$, given by

$$\bigsqcup_{x,y,z\in 2} (\mathsf{l}.\exists R.\mathsf{me}^x \sqcap \gamma_1(\mathsf{l}.C[\mathsf{me}/\top])^y \sqcap \gamma_1(\mathsf{l}.C[\mathsf{me}/\bot])^z \sqcap \mathsf{l}.\geq_{n-xy+xz} R.\gamma_1(C[\mathsf{me}/\bot])) \tag{13}$$

where $\mathsf{l}.C[\mathsf{me}/\top]$ (meant to be read $\mathsf{l}.(C[\mathsf{me}/\top])$) distributes l among Booleans as expected and, for $n - xy = 0$, we have $\geq_{n-xy+xz} R.D \equiv \top$ (even when $n - xy + xz = 1$). In essence, $\gamma_1(\mathsf{l}.\geq_n R.C)$ eliminates the top-level me in C, replacing it with $\bot$. For this to be valid, some adjustments need to be made: if $\exists R.\mathsf{me}$ and $\mathsf{l}.C[\mathsf{me}/\top]$ hold (at me), then we need one witness less (zero witnesses for the case $n = 1$), but if $\exists R.\mathsf{me}$ and $\mathsf{l}.C[\mathsf{me}/\bot]$ hold (at me) we need to account for an extra "false positive" (there is a corner case when $n, x, y, z = 1$, where zero witnesses are needed as well).

For γ_2 we require some additional notation. We use $C[x \in X/f(x)]$ to denote the uniform substitution in C of every concept $x \in X$ by $f(x)$. For $f, g \in 2^X$, we define the function $fg \in 2^X$ as $fg(x) = f(x) \cdot g(x)$. In addition, we let $F(C)$ be the set of all concepts of the form $\geq_m S.D$ occurring top-level in C without an $\mathsf{l}.$ in front; e.g., $F((\mathsf{l}.\geq_5 R.\mathsf{me}) \sqcap \neg\geq_3 R.A) = \{\geq_3 R.A\}$. Given $f \in 2^{F(C)}$, we say that f is $\mathcal{R}_C$-consistent if there are no distinct $\geq_n R.D$ and $\geq_m R.E$ in $F(C)$ such that $f(\geq_n R.D) \neq f(\geq_m R.E)$, and denote by $2_c^{F(C)}$ the set of all $\mathcal{R}_C$-consistent functions in $2^{F(C)}$. Now, each $f \in 2_c^{F(C)}$ induces these conjunctive clauses:

$$Cl_*(f) = \textstyle\bigsqcap_{\geq_m S.D\in F(C)} D[\mathsf{me}/*]^{f(\geq_m S.D)} \quad (* \in \{\top, \bot\}) \tag{14}$$

$$Cl_\exists(f) = \textstyle\bigsqcap_{\geq_m S.D\in F(C)} \exists S.\mathsf{me}^{f(\geq_m S.D)} \tag{15}$$

Moreover, given $f, g, h \in 2_c^{F(C)}$ we define $C[f, g, h]$, as:

$$C[f, g, h] = C[x = \geq_m S.D \in F(C)/\geq_{m-fh(x)+gh(x)} S.D[\mathsf{me}/\bot]] \tag{16}$$

Again, for $m - fh(x) = 0$, we assume $\geq_{m-fh(x)+gh(x)} S.D \equiv \top$. We can now define γ_2, the only non-trivial case being that for $\gamma_2(\mathsf{l}.\geq_n R.C)$, given by

$$\bigsqcup_{f,g\in 2_c^{F(C)}} [\gamma_2(Cl_\top(f) \sqcap Cl_\bot(g)) \sqcap \mathsf{l}.\geq_n R.\bigsqcup_{h\in 2_c^{F(C)}} (Cl_\exists(h) \sqcap \gamma_2(C[f, g, h]))] \tag{17}$$

The principle behind (17) is analogous to that used for γ_1. In this case we need an equivalent of x, y and z in (13) for each concept in $F(C)$, and we use functions f, g and h for this. Observe that $\gamma_2(\mathsf{I}.\geq_n R.C)$ is an $\mathcal{R}$-covered concept, where $\mathcal{R}$ is the set $\{S : \geq_m S.D \in F(C)\}$. Using this insights, it is not hard to verify that $\gamma_2(\gamma_1(C))$ is equivalent to C and in expanded form. □

Proposition 4.3 *$\mathcal{ALCQ}\mathsf{me}_2 \leq_Q \mathcal{ALCIQ}be$.*

Proof By Lemma 4.2 we can assume that closed $\mathcal{ALCQ}\mathsf{me}_2$-concepts are in expanded form. We can therefore define a translation δ, mapping concepts in this form to $\mathcal{ALCIQ}be$-concepts, by making it commute with the Booleans, stipulating $\delta(\mathsf{I}.\exists R.\mathsf{me}) = \exists R.\mathsf{Self}$ and $\delta(\geq_n R.\mathsf{me}) = \geq_n R.\delta(\mathsf{me})$; and taking $\delta(\mathsf{I}.\geq_n R.((C_1 \sqcap D_1) \sqcup \ldots \sqcup (C_m \sqcap D_m))$, where the C_i are the selectors of the $\mathcal{R}$-covered concept, to be

$$\bigsqcup_{n=\sum_{j=1}^{m} k_j} \bigsqcap_{1 \leq i \leq m} \geq_{k_j} \theta(R, C_i).\delta(D_j) \tag{18}$$

where θ maps a role and selector to a safe role expression as follows:

$$\begin{aligned}\theta(R, \exists S_1.\mathsf{me} \sqcap \cdots \sqcap \exists S_l.\mathsf{me} \sqcap \neg\exists S_{l+1}.\mathsf{me} \sqcap \cdots \sqcap \neg\exists S_l.\mathsf{me}) = \\ (R \sqcap S_1^- \sqcap \cdots \sqcap S_l^-) - (S_{l+1}^- \sqcap \cdots \sqcap S_m^-)\end{aligned} \tag{19}$$

What $\delta(\mathsf{I}.\geq_n R.((C_1 \sqcap D_1) \sqcup \ldots \sqcup (C_m \sqcap D_m))$ does is to consider all the possible ways in which we can distribute the n required successors among the m equivalence classes given by the selectors. The number of cases is bounded by the number of *partitions of* n, which is exponential on n [4]. □

It is not hard to verify that from this proof one also gets the following:

Corollary 4.4 *$\mathcal{ALCQ}\mathsf{me}_2 \leq_C \mathcal{ALCIQ}be$ and $\mathcal{ALCQ}\mathsf{me}_2 \leq_{R^g} \mathcal{ALCIQ}be$.*

All the intermediate transformations in the above proofs incur exponential blowups. It is not hard, however, to verify that the overall blowup in formula size is still only exponential. In order to obtain a polynomial blowup, we need to relax the properties we require of the translation from preservation of *satisfaction* to preservation of *satisfiability*:

Proposition 4.5 *$\mathcal{ALCQ}\mathsf{me}_2 \leq_C^{\exists} \mathcal{ALCIQ}be$, with only a polynomial blowup.*

Proof We introduce a satisfiability preserving translation that adds fresh atomic concept and role symbols, and ensures that a model for the translated formula is turned into a model for the $\mathcal{ALCQ}\mathsf{me}_2$-formula simply by ignoring the interpretation of the extra symbols. We can view the translation as a two-step process. First, we map an $\mathcal{ALCQ}\mathsf{me}_2$-concept to an $\mathcal{ALCQ}b\mathsf{me}_2$-concept (i.e., a concept of $\mathcal{ALCQ}\mathsf{me}_2$ enriched with safe Boolean combinations of roles) in such a way that if $\geq_n R.C$ occurs in the resulting concept, then $C \neq \top$ implies $n = 1$. This is straightforward to achieve: a concept $(\mathsf{I}.)\geq_n R.C$, with $n > 1$, occurring in a positive context (i.e., under an even number of negations) is mapped to $\geq_n (R \sqcap R').\top \sqcap (\mathsf{I}.)\forall R'.C$, for a fresh R'; while the same concept in a negative context is mapped to $\geq_n (R \sqcap R').\top \sqcup \neg(\mathsf{I}.)\forall(R - R').\neg C$, with R' fresh (these transformations need to be performed top-down).

We therefore need only eliminate all the occurrences of $\mathsf{I}.\forall R.C$ in the resulting concept. We can deal with the top-level occurrences of me in C replacing $\mathsf{I}.\forall R.C$ by

$$\mathsf{I}.\forall R.C[\mathsf{me}/D] \sqcap ((\exists R.\mathsf{Self} \sqcap D \sqcap \leq_1 R.D) \sqcup (\neg\exists R.\mathsf{Self} \sqcap \neg D \sqcap \forall R.\neg D)) \quad (20)$$

where D is fresh and $C[\mathsf{me}/D]$ is as in the proof of Lemma 4.2. Let $\mathsf{I}.\forall R.C$ occur in the concept resulting after this transformation; if me happens to occur free in C, then some $\forall S.D$ must occur in C with me free (top-level) in D. We can therefore substitute $\mathsf{I}.\forall R.C$ by

$$\mathsf{I}.\forall R.C[\forall S.D/\forall S.D[\mathsf{me}/E]] \sqcap E \sqcap (\forall (R \sqcap S^-).\leq_1 S.E) \sqcap (\forall (R - S^-).\forall S.\neg E) \quad (21)$$

where E is fresh. After all the occurrences of me are eliminated, the $\mathsf{I}.$ can be removed as well. It is not hard to see that these transformations incur in a polynomial blowup and that any model for the translated formula is trivially turned into a model for the original one by applying the forgetful mapping of Definition 2.1. □

5 Decidability and Complexity

It is time to discuss the decidability and complexity of the local satisfiability problem for $\mathcal{ALCQ}\mathsf{me}_2$, both over empty and over general TBoxes. Because Proposition 4.5 gives us an effective, polynomial, satisfiability preserving reduction of $\mathcal{ALCQ}\mathsf{me}_2$-concepts to $\mathcal{ALCIQ}be$-concepts, and because Proposition 3.1 gives us a reduction in the opposite direction, the complexities of $\mathcal{ALCQ}\mathsf{me}_2$ and $\mathcal{ALCIQ}be$ must match.

While the logic $\mathcal{ALCIQ}b$ is known to be PSPACE-complete for satisfiability over empty TBoxes and EXPTIME-complete for satisfiability over general TBoxes [18], not much appears to be known about the complexity of $\mathcal{ALCIQ}be$. We can conclude that it is decidable and at most in N2EXPTIME (for general TBoxes) from the fact that the extension of $s\mathcal{ROIQ}$ with Boolean combination of *simple roles* is N2EXPTIME-complete [13] and that in $\mathcal{ALCIQ}be$ all atomic roles would be "simple" in $s\mathcal{ROIQ}$ terminology. In this section we will then show that $\mathcal{ALCIQ}be$ is, in fact, not harder than $\mathcal{ALCIQ}b$.

The technique we will employ is a variation of the *internalized tableaux* of [9]. Although we will give a specialized proof for the case of $\mathcal{ALCIQ}be$, it should be clear that our construction can be used to show that adding the $\exists R.\mathsf{Self}$ concept to a description logic in general does not raise its computational complexity. The relevant property to have is the *self-loop free model property*, i.e. if a concept (resp. theory) is satisfiable, then it is satisfiable in a model without self-loops.

Lemma 5.1 *$\mathcal{ALCIQ}b$ has the self-loop free model property.*

Proof Given an interpretation $\mathcal{I}$, one can build an equivalent interpretation $\mathcal{J}$ that is self-loop free as follows. Take as domain of $\mathcal{J}$ two copies of the domain of $\mathcal{I}$, i.e., $\Delta^{\mathcal{J}} = 2 \times \Delta^{\mathcal{I}}$. Concepts and roles are preserved, except for self-loops, which are replaced by links to the corresponding element on the other copy; formally we have that $A^{\mathcal{J}} = 2 \times A^{\mathcal{I}}$ and $R^{\mathcal{J}}$ is given by

$$\begin{aligned} &\{((x,a),(x,b)) : x \in 2, (a,b) \in R^{\mathcal{I}}, a \neq b\} \\ \cup &\{((x,a),(y,a)) : x,y \in 2, x \neq y, (a,a) \in R^{\mathcal{I}}\} \end{aligned} \tag{22}$$

The proof that $\mathcal{I}$ and $\mathcal{J}$ are equivalent is left to the reader. □

Lemma 5.2 *TBox satisfiability in $\mathcal{ALCIQ}be$ can be polynomially reduced to TBox satisfiability in $\mathcal{ALCIQ}b$.*

Proof Let $\mathcal{T}$ be an $\mathcal{ALCIQ}be$ TBox over a vocabulary $\langle \mathsf{N_C}, \mathsf{N_R} \rangle$, and let Σ be the smallest set that contains every concept occurring in $\mathcal{T}$, is closed under subconcepts and single negations, and contains the concept $\exists R.\mathsf{Self}$ for every role expression R occurring in $\mathcal{T}$. Let $\mathsf{N'_C} = \mathsf{N_C} \cup \{B_C : C \in \Sigma\}$; we then define the $\mathcal{ALCIQ}b$ TBox $\mathcal{T}'$, over the vocabulary $\langle \mathsf{N'_C}, \mathsf{N_R} \rangle$. It contains an axiom $B_C \sqsubseteq B_D$ for each $C \sqsubseteq D$ in $\mathcal{T}$, plus the following definitions:

$$\begin{aligned} B_A &\equiv A \ (A \in \mathsf{N_C}) & B_{\neg C} &\equiv \neg B_C \\ B_{C \sqcap D} &\equiv B_C \sqcap B_D & B_{\exists R.\mathsf{Self}} &\equiv B_{\exists R^-.\mathsf{Self}} \\ B_{\exists (R \sqcap S).\mathsf{Self}} &\equiv B_{\exists R.\mathsf{Self}} \sqcap B_{\exists S.\mathsf{Self}} & B_{\exists (R \sqcup S).\mathsf{Self}} &\equiv B_{\exists R.\mathsf{Self}} \sqcup B_{\exists S.\mathsf{Self}} \\ B_{\exists (R - S).\mathsf{Self}} &\equiv B_{\exists R.\mathsf{Self}} \sqcap \neg B_{\exists S.\mathsf{Self}} \end{aligned}$$

$$B_{\geq_n R.C} \equiv ((B_C \sqcap B_{\exists R.\mathsf{Self}}) \sqcap \geq_{n-1} R.B_C) \sqcup (\neg(B_C \sqcap B_{\exists R.\mathsf{Self}}) \sqcap \geq_n R.B_C).$$

Recall that for $n = 1$, we identify $\geq_{n-1} R.C$ with $\top$. Clearly, $\mathcal{T}$ has size linear in the size of $(\mathcal{T}, \Sigma)$, which in turn is linear in $\mathcal{T}$. It only remains to see that $\mathcal{T}$ and $\mathcal{T}'$ are equisatisfiable. It is clear that any model $\mathcal{I}$ for $\mathcal{T}$ is expanded to a model $\mathcal{I}'$ for $\mathcal{T}'$ by setting $B_C^{\mathcal{I}'} = C^{\mathcal{I}}$, for all $C \in \Sigma$. For the other direction, let $\mathcal{I}'$ be a model for $\mathcal{T}'$ and notice that by Lemma 5.1, we can assume $\mathcal{I}'$ to contain no self-loops. We then obtain a model $\mathcal{I}$ by restricting $\mathcal{I}'$ to $\langle \mathsf{N_C}, \mathsf{N_R} \rangle$ and adding the necessary self-loops, i.e., setting $R^{\mathcal{I}} = R^{\mathcal{I}'} \cup \{(a,a) : a \in B_{\exists R.\mathsf{Self}}^{\mathcal{I}'}\}$. It is not hard to verify that $\mathcal{I}$ is a model for $\mathcal{T}$. □

Theorem 5.3 *The problem of concept satisfiability over general TBoxes for the logic $\mathcal{ALCIQ}be$ is EXPTIME-complete.*

Proof This follows directly from Lemma 5.2 (observing that satisfiability over general TBoxes can be reduced to TBox satisfiability) and EXPTIME-completeness of TBox satisfiability for $\mathcal{ALCIQ}b$ [18]. □

Theorem 5.4 *The concept satisfiability problem (over empty TBoxes) for the logic $\mathcal{ALCIQ}be$ is PSPACE-complete.*

Proof Given that satisfiability for $\mathcal{ALCIQ}be$ is PSPACE-complete [18], we only need to give a polynomial, satisfiability preserving translation from $\mathcal{ALCIQ}be$ to $\mathcal{ALCIQ}b$. This is done following the idea of the proof of Lemma 5.2: given an $\mathcal{ALCIQ}be$-concept C containing atomic roles $R_1 \ldots R_m$, one builds the $\mathcal{ALCHIQ}b$-concept $B_C \sqcap \forall^n (R_1 \sqcup \ldots \sqcup R_m).\mathit{Def}(B)$, with $n = \mathit{rank}(C)$, $\forall^0 R.D \equiv D$, $\forall^{n+1} R.D = \forall R.\forall^n R.D$, and $\mathit{Def}(B)$ is the conjunction of the definitional axioms for the fresh concepts B_D (with $D \in \Sigma$)

given in the proof of Lemma 5.2. Equisatisfiability follows by a standard argument. □

Corollary 5.5 *Satisfiability over general (resp. empty) TBoxes for $\mathcal{ALCQ}\mathsf{me}_2$ is EXPTIME-complete (resp. PSPACE-complete).*

6 Conclusions

Although it is known that decidability breaks easily when self-referential concepts are added to description logics, we have shown that, under controlled conditions, it is indeed feasible to extend $\mathcal{ALCQ}$ (i.e. $\mathcal{ALC}$ with qualified number restrictions, equivalently graded multi-modal logic) with a form of self-referential reasoning without affecting the complexity. Specifically, we have defined the logic $\mathcal{ALCQ}\mathsf{me}_2$, which includes the I-me construct that allows naming one state at a time for future reference but limits occurrences of me to (modal) depth at most 2 from the binding I. We have shown that $\mathcal{ALCQ}\mathsf{me}_2$ is expressively equivalent to the DL $\mathcal{ALCHIQ}be$, which includes role inverses and hierarchies, safe Boolean combinations of roles, and the self-loop construct $\exists R.\,\mathsf{Self}$. The translation of $\mathcal{ALCQ}\mathsf{me}_2$ into $\mathcal{ALCHIQ}be$, however, has an exponential blowup (it remains an open question whether this can be avoided). We have therefore given a second reduction of $\mathcal{ALCQ}\mathsf{me}_2$ to $\mathcal{ALCHIQ}be$ that is only satisfiability-preserving but has polynomial blowup. After subsequent analysis of the complexity of $\mathcal{ALCHIQ}be$, this has allowed us to prove that $\mathcal{ALCQ}\mathsf{me}_2$ is decidable in PSPACE over the empty TBox, and in EXPTIME over general TBoxes — the same bounds as for $\mathcal{ALCQ}$ or indeed basic $\mathcal{ALC}$.

In future research, we will study the addition of controlled binding constructions to richer DLs. Because $\mathcal{ALCHIQ}$ is embedded in $\mathcal{ALCQ}\mathsf{me}_2$, it is clear that adding nominals to the mix must put us at least in NEXPTIME, and we conjecture NEXPTIME-completeness for this language. It would also be interesting to know in which cases the interaction with transitive roles is safe. A further branch of investigation is the generalization of the results at depth 2 to extended description logics featuring, e.g., uncertainty or defaults, modelled generically in the framework of coalgebraic logic (see, e.g., [15]), thus improving a result at depth 1 obtained in our previous work [9].

Acknowledgements The authors wish to thank the referees of an earlier version of this paper for suggesting the possibility of proving Proposition 4.5. Moreover, we acknowledge discussions with Erwin R. Catesbeiana on the merits of consistency.

References

[1] Adler, M. and N. Immerman, *An n! lower bound on formula size*, ACM Trans. Comput. Logic **4** (2003), pp. 296–314.

[2] Areces, C., D. Figueira, S. Figueira and S. Mera, *The expressive power of memory logics*, Rev. Symb. Log. **2** (2011), pp. 290–318.

[3] Areces, C. and B. ten Cate, *Hybrid logics*, in: P. Blackburn, J. van Benthem and F. Wolter, editors, *Handbook of Modal Logic*, Elsevier, 2007 pp. 821–868.

[4] Ayoub, R., "An Introduction to the Analytic Theory of Numbers," Number 10 in Mathematical Surveys, American Mathematical Society, 1963.

[5] Baader, F., D. Calvanese, D. L. McGuinness, D. Nardi and P. F. Patel-Schneider, editors, "The Description Logic Handbook," Cambridge University Press, 2003.

[6] Figueira, S. and D. Gorín, *On the size of shortest modal descriptions*, in: *Advances in Modal Logic, vol. 8* (2010), pp. 120–139.

[7] Goguen, J. A. and R. M. Burstall, *Institutions: Abstract model theory for specification and programming*, J. ACM **39** (1992), pp. 95–146.

[8] Goguen, J. A. and G. Rosu, *Institution morphisms*, Formal Asp. Comput. **13** (2002), pp. 274–307.

[9] Gorín, D. and L. Schröder, *Narcissists are easy, stepmothers are hard*, in: *Foundations of Software Science and Computation Structures, FoSSaCS 2012*, LNCS **7213** (2012), pp. 240–254.

[10] Horrocks, I., O. Kutz and U. Sattler, *The even more irresistible* $\mathcal{SROIQ}$, in: *Proc. of the 10th Int. Conf. on Principles of Knowledge Represe ntation and Reasoning (KR2006)* (2006), pp. 57–67.

[11] Hustadt, U., B. Motik and U. Sattler, *A decomposition rule for decision procedures by resolution-based calculi*, in: F. Baader and A. Voronkov, editors, *Proc. of the 11th Int. Conference on Logic for Programming Artificial Intelligence and Reasoning (LPAR 2004)*, LNAI **3452** (2005), pp. 21–35.

[12] Marx, M., *Narcissists, stepmothers and spies*, in: *Proc. of the 2002 International Workshop on Description Logics (DL'02)*, CEUR **53**, 2002.

[13] Rudolph, S., M. Krötzsch and P. Hitzler, *Cheap boolean role constructors for description logics*, in: *Proc. of 11th European Conf. on Logics in Artificial Intelligence (JELIA)*, LNAI (2008), pp. 362–374.

[14] Schneider, T., "The Complexity of Hybrid Logics over Restricted Classes of Frames," Ph.D. thesis, Univ. of Jena (2007).

[15] Schröder, L. and D. Pattinson, *PSPACE bounds for rank-1 modal logics*, ACM Trans. Comput. Log. **10** (2009), pp. 13:1–13:33.

[16] ten Cate, B. and M. Franceschet, *On the complexity of hybrid logics with binders*, in: *Computer Science Logic, CSL 2005*, LNCS **3634**, 2005, pp. 339–354.

[17] Tobies, S., *The complexity of reasoning with cardinality restrictions in expressive description logics*, J. Artif. Intell. Res. **12** (2000), pp. 199–217.

[18] Tobies, S., "Complexity results and practical algorithms for logics in Knowledge Representation," Ph.D. thesis, RWTH Aachen (2001).

Refinement Quantified Logics of Knowledge and Belief for Multiple Agents

James Hales [1] Tim French Rowan Davies

Computer Science and Software Engineering
The University of Western Australia
Perth, Australia

Abstract

Given the "possible worlds" interpretation of modal logic, a refinement of a Kripke model is another Kripke model in which an agent has ruled out some possible worlds to be consistent with some new information. The refinements of a finite Kripke model have been shown to correspond to the results of applying arbitrary action models to the Kripke model [10]. Refinement modal logics add quantifiers over such refinements to existing modal logics. Work by van Ditmarsch, French and Pinchinat [11] gave an axiomatisation for the refinement modal logic over the class of unrestricted Kripke models, for a single agent. Recent work by Hales, French and Davies [13] extended these results, restricting the quantification to the class of doxastic and epistemic models for a single agent. Here we extend these results further, to the classes of doxastic and epistemic models for multiple agents. The generalisation to multiple agents for doxastic and epistemic models is not straightforward and requires novel techniques, particularly for the epistemic case. We provide sound and complete axiomatisations for the considered logics, and a provably correct translations to their underlying modal logics, corollaries of which are expressivity and decidability results.

Keywords: Modal logic, Epistemic logic, Doxastic logic, Bisimulation quantifier, Refinement quantifier, Temporal epistemic logic, Multi-agent system, Action models

1 Introduction

This paper examines the extension of multi-agent doxastic and epistemic logics by refinement quantifiers. Refinement quantifiers were introduced by van Ditmarsch and French [10] to capture a general notion of informative updates in the context of epistemic logic. Informative updates, such as public announcements correspond to an agent receiving new information and incorporating this into their knowledge state. These are discussed in great detail by van Ditmarsch, van der Hoek and Kooi [12].

When we move from explicit updates to arbitrary updates we move from the question "Is ϕ true after the agent learns ψ?", to the question, "Is it possible

[1] Acknowledges the support of the Prescott Postgraduate Scholarship.

that the agent can learn *something* in such a way that ϕ is true?". When the new information is constrained to be be expressible as an epistemic formula we have Arbitrary Public Announcement Logic which was shown to be undecidable by van Ditmarsch and French [9]. A refinement quantifier is a weaker operator than an arbitrary announcement. Refinement quantified logics have been shown to be decidable, and axiomatized for the logics K and the modal μ-calculus by van Ditmarsch, French and Pinchinat [11], and for single-agent *KD45* and *S5* by Hales, French and Davies [13]. This paper goes to the original motivation for refinement quantifiers as informative updates in multi-agent systems. We present sound and complete axiomatizations for refinement quantified multi-agent *KD45* and *S5*, and derive expressivity and decidability results for each logic.

Refinements may be thought of as the result of an iterated process of duplicating and removing successor worlds in a Kripke model. This process will preserve an agent's positive knowledge (things they know) but may not preserve their general knowledge (things they merely suspected to be true are not guaranteed to be true in a refinement). We define the refinements of a Kripke model using a relationship between Kripke models, also called a refinement, which is simply the reverse direction of a simulation, which are a generalisation of bisimulations [5]. As we are in a multi-agent setting, we specify refinement relations with respect to sets of agents, so that the knowledge of all other agents is preserved, except possibly what they may know of the refined agents knowledge state.

Refinements have an important role in multi-agent epistemic logic. Refinements preserve the positive knowledge of all agents [10], so they naturally generalise anything that may be considered an informative update. Examples include public announcements [3], group announcements [1] and action models [12]. A refinement quantifier over some epistemic property, ϕ, corresponds to the question of whether we can provide information to the set of agents in such a way that ϕ will be true. Dynamics systems of knowledge have applications in reasoning about games, autonomous agent negotiation, and communication systems, and in these contexts the refinement operation corresponds to whether one knowledge knowledge state is reachable from another.

Our strategy for proving completeness follows the approaches of D'Agostino and Lenzi [6], (and subsequently [11,13]) of giving a provably correct translation into a sublanguage with a known completeness result.

2 Technical Preliminaries

We recall the definitions given by van Ditmarsch, French, and Pinchinsat [11] in describing the refinement modal logic, and adapt those definitions to be based on doxastic logic, *KD45*, and epistemic logic, *S5*. Specifically, we restrict the Kripke models under discussion to those in the class of $\mathcal{KD}45$ models when we are discussing the extension of the refinement modal logic to *KD45*, which we call the refinement doxastic logic, or $KD45_\forall$, and to those in the class of $\mathcal{S}5$ models when we are discussing the extension to *S5*, which we call the refinement epistemic logic, or $S5_{\forall}$.

Let A be a non-empty, finite set of agents, and let P be a non-empty, countable set of propositional atoms.

Definition 2.1 (Kripke model) *A Kripke model $M = (S, R, V)$ consists of a domain S, which is a set of states (or worlds), accessibility $R : A \to \mathcal{P}(S \times S)$, and a valuation $V : P \to \mathcal{P}(S)$. The class of all Kripke models is called $\mathcal{K}$. We write $M \in \mathcal{K}$ to denote that M is a Kripke model.*

For $R(a)$ we write R_a. Given two states $s, s' \in S$, we write $R_a(s, s')$ to denote that $(s, s') \in R_a$. We write sR_a for $\{t|(s,t) \in R_a\}$. As we will be required to discuss several models at once, we will use the convention that $M = (S, R, V)$, $M' = (S', R', V')$, $M^\gamma = (S^\gamma, R^\gamma, V^\gamma)$, etc. For $s \in S$ we will let M_s refer to the pair (M, s), or the pointed Kripke model M at state s.

Definition 2.2 (Doxastic model) *A doxastic model is a Kripke model $M = (S, R, V)$ such that the relation R_a is serial, transitive, and Euclidean for all $a \in A$. The class of all doxastic models is called $\mathcal{KD}45$. We write $M \in \mathcal{KD}45$ to denote that M is a doxastic model.*

Definition 2.3 (Epistemic model) *An epistemic model is a Kripke model $M = (S, R, V)$ such that the relation R_a is an equivalence relation for all $a \in A$. The class of all epistemic models is called $\mathcal{S}5$. We write $M \in \mathcal{S}5$ to denote that M is an epistemic model.*

This paper covers results in both $\mathcal{KD}45$ and $\mathcal{S}5$; as such we will assume that all models are doxastic models when discussing *KD45* or $KD45_{\forall}$, and that all models are epistemic models when discussing *S5* or $S5_{\forall}$.

Definition 2.4 (Bisimulation) *Let $M = (S, R, V)$ and $M' = (S', R', V')$ be Kripke models. A non-empty relation $\mathcal{R} \subseteq S \times S'$ is a bisimulation if and only if for all $s \in S$ and $s' \in S'$, with $(s, s') \in \mathcal{R}$, for all $a \in A$:*

atoms *$s \in V(p)$ if and only if $s' \in V'(p)$ for all $p \in P$*

***forth-**a* *for all $t \in S$, if $R_a(s,t)$, then there is a $t' \in S'$ such that $R'_a(s', t')$ and $(t, t') \in \mathcal{R}$*

***back-**a* *for all $t' \in S'$, if $R'_a(s', t')$, then there is a $t \in S$ such that $R_a(s,t)$ and $(t, t') \in \mathcal{R}$.*

We call M_s and $M'_{s'}$ bisimilar, and write $M_s \leftrightarrows M'_{s'}$ to denote that there is a bisimulation between M_s and $M'_{s'}$.

Definition 2.5 (Simulation and refinement) *Let M and M' be Kripke models and let $B \subseteq A$ be a set of agents. A non-empty relation $\mathcal{R} \subseteq S \times S'$ is a B-simulation if and only if it satisfies* **atoms**, **forth-**a *for every $a \in A$ and* **back-**a *for every $a \in A \setminus B$.*

If $s \in S$ and $s' \in S'$ such that $(s, s') \in \mathcal{R}$, we call $M'_{s'}$ a B-simulation of M_s and call M_s a B-refinement of $M'_{s'}$. We write $M'_{s'} \succeq_B M_s$, or equivalently, $M_s \preceq_B M'_{s'}$ to denote this.

In the case where $B = A$ we use the terms simulation and refinement in place of A-simulation and A-refinement, and we write $M'_{s'} \succeq M_s$, or equivalently, $M_s \preceq M'_{s'}$. In the case where $B = \{a\}$ for some $a \in A$ we simply use the terms a-simulation and a-refinement, and we write $M'_{s'} \succeq_a M_s$ or $M_s \preceq_a M'_{s'}$.

The refinements of a Kripke model correspond to the results of executing arbitrary action models on the Kripke model [10]. In an epistemic setting, a refinement can therefore be considered as the result of an informative update in which each agent's positive knowledge is preserved, and other knowledge may potentially vary [10]. A B-refinement corresponds to a more restricted informative update in which only the agents in the set B are directly provided with information, and the knowledge of agents not in B is preserved, except for where it concerns non-positive knowledge of agents in B. An example of this is given in the following section, in Example 3.5. How the notion of B-refinements relates to action models is a question left for future work.

3 Syntax and semantics

Here we define the syntax and semantics of the logics $KD45_\forall$ and $S5_\forall$, which restrict the logic $K_\forall$, defined by van Ditmarsch, French, and Pinchinat to models and refinements of models that are in $\mathcal{KD}45$ or $\mathcal{S}5$ respectively.

The same syntax used for $K_\forall$ is used for $KD45_\forall$ and $S5_\forall$, and so we will define it only once, as $\mathcal{L}_\forall$.

Definition 3.1 (Language of $\mathcal{L}_\forall$) *Given a finite set of agents A and a set of propositional atoms P, the language of $\mathcal{L}_\forall$ is inductively defined as*

$$\phi ::= p \mid \neg\phi \mid (\phi \wedge \phi) \mid \Box_a\phi \mid \forall_B\phi$$

where $a \in A$, $B \subseteq A$ and $p \in P$.

We use all of the standard abbreviations for modal logics, in addition to the abbreviation $\exists_B\phi ::= \neg\forall_B\neg\phi$. We abbreviate $\exists_{\{a\}}$ and $\forall_{\{a\}}$ as $\exists_a$ and $\forall_a$ respectively. Similarly we abbreviate $\exists_A$ and $\forall_A$ as $\exists$ and $\forall$ respectively.

We refer to the language $\mathcal{L}$, of modal formulae, which is simply $\mathcal{L}_\forall$ without the $\forall_B$ operator, and the language $\mathcal{L}_0$, of propositional formulae, which is $\mathcal{L}$ without the $\Box_a$ operator. We use the notation $\phi \leq \psi$ to mean that ϕ is a (non-strict) subformula of ψ.

We also use the cover operator, following the definitions given by Bílková, Palmigiano, and Venema [4]. The cover operator, $\nabla_a\Gamma$ is an abbreviation defined by $\nabla_a\Gamma ::= \Box_a \bigvee_{\gamma\in\Gamma}\gamma \wedge \bigwedge_{\gamma\in\Gamma}\Diamond_a\gamma$, where Γ is a finite set of formulae. We note that the modal operators $\Box_a$, $\Diamond_a$ and ∇_a are interdefineable, as $\Box_a\phi \leftrightarrow \nabla_a\{\phi\} \vee \nabla_a\emptyset$ and $\Diamond_a\phi \leftrightarrow \nabla_a\{\phi, \top\}$. This is the basis of our axiomatisations, as it is for the axiomatisation of the single-agent logic $K_\forall^{(1)}$, presented by van Ditmarsch, French and Pinchinat [11], and the axiomatisations for the single-agent logics $KD45_\forall^{(1)}$ and $S5_\forall^{(1)}$ presented by Hales, French and Davies [13]. The cover operator allows us to define normal forms for modal logics that allow us to only consider conjunctions of modalities in specific situations for our axiomatisations and provably correct translations.

The semantics for $K_\forall$, $KD45_\forall$ and $S5_\forall$ are very similar, and so we will introduce a generalised semantics that can be applied to all three.

Definition 3.2 (Semantics of $C_\forall$) *Let $\mathcal{C}$ be a class of Kripke models, and let $M = (S, R, V)$ be a Kripke model taken from the class $\mathcal{C}$. The interpretation of $\phi \in \mathcal{L}_\forall$ is defined by induction.*

$$\begin{array}{ll} M_s \models p & \textit{iff } s \in V_p \\ M_s \models \neg\phi & \textit{iff } M_s \not\models \phi \\ M_s \models \phi \wedge \psi & \textit{iff } M_s \models \phi \textit{ and } M_s \models \psi \\ M_s \models \Box_a\phi & \textit{iff for all } t \in S : (s,t) \in R_a \textit{ implies } M_t \models \phi \\ M_s \models \forall_B\phi & \textit{iff for all } M'_{s'} \in \mathcal{C} : M_s \succeq_B M'_{s'} \textit{ implies } M'_{s'} \models \phi \end{array}$$

The logics $K_\forall$, $KD45_\forall$ and $S5_\forall$ are instances of $C_\forall$ with the classes $\mathcal{K}$, $\mathcal{KD}45$ and $\mathcal{S}5$ respectively. The difference between these logics are the class of models that formulae are interpreted over, and the class of models that the refinement quantifier, $\forall_B$ quantifies over. It should be emphasised that the interpretation of the refinement operator, $\forall_B$, varies for each logic, as the refinements considered in the interpretation of each logic must be taken from the appropriate class of models. It is for this reason that $KD45_\forall$ and $S5_\forall$ are not conservative extensions of $K_\forall$. For example, $\exists_a\Box_a\bot$ is valid in $K_\forall$, but not in $KD45_\forall$ or $S5_\forall$. This is because given any pointed model in $\mathcal{K}$, one can construct an a-refinement from that model by deleting the a-edges starting at the designated state; in this resulting a-refinement, $\Box_a\bot$ is satisfied, and hence $\exists_a\Box_a\bot$ is satisfied in the original model. However because of the seriality of $\mathcal{KD}45$ and $\mathcal{S}5$ models, $\Box_a\bot$ is not even satisfiable in $KD45_\forall$ or $S5_\forall$, and as $\exists_a$ quantifies over $\mathcal{KD}45$ or $\mathcal{S}5$ models in these cases, therefore $\exists_a\Box_a\bot$ is not satisfiable either.

Lemma 3.3 *The logics $K_\forall$, $KD45_\forall$ and $S5_\forall$ are bisimulation invariant.*

The proof for bisimulation invariance in $K_\forall$, given by van Ditmarsch, French and Pinchinat [11] applies to $KD45_\forall$ and $S5_\forall$.

Example 3.4 Imagine a scenario where an agent is presented with three cards face down, and asked to identify which is the ace of spades (let's suppose it is the *left* card). As an agent's knowledge is only ever based on reliable evidence, it follows that given any informative update, there is always a further informative update after which the agent knows the location of the ace:

$$left \rightarrow \forall\exists\Box left. \tag{1}$$

This scenario is represented in Figure 1. We can also imagine a corresponding scenario in terms of the agent's *belief* rather than *knowledge*. Here an agent may believe that the ace is in fact the *centre* card, despite this not being the case. In this setting the formula (1) does not hold. We also note that once an agent holds a belief, no informative update will cause the agent to revise that belief.

$$\Box(right \vee centre) \rightarrow \forall\Box(right \vee centre). \tag{2}$$

That is, we do not consider belief *revision* in the sense of [2] but rather belief *refinement*. This situation is depicted in Figure 2. This allows incorrect information, but requires that the information provided is consistent because of the requirement that $\mathcal{KD}45$ models are serial.

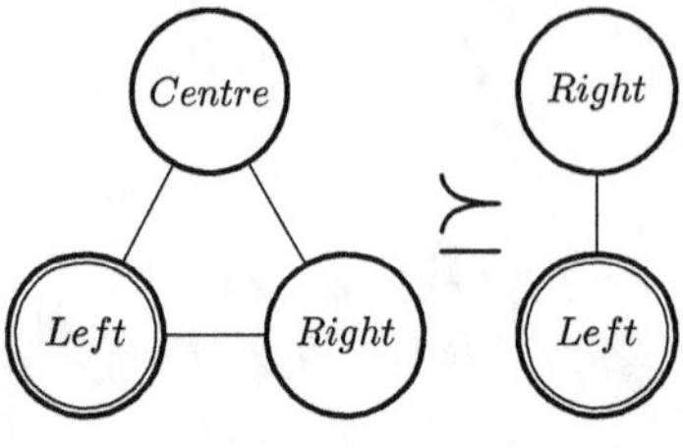

Fig. 1. The initial state of an agent's uncertainty from Example 3.4, with an example refinement, both in $\mathcal{S}5$.

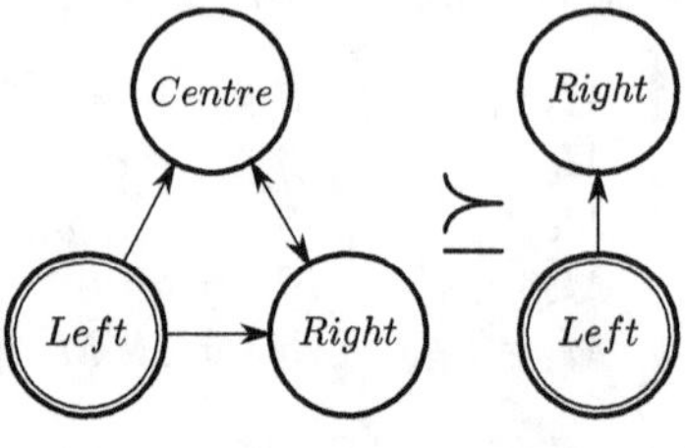

Fig. 2. The initial state of an agent's uncertainty from Example 3.4, with an example refinement, this time in $\mathcal{KD}45$.

Example 3.5 Now consider a situation where we have three agents: James, Rowan and Tim, and three coloured cards: Red, Green Blue. The cards are dealt one to each player, and it becomes universally known that Tim does not have the blue card. If James has a red card, then James is able to deduce that Tim must have the green card and hence Rowan has the blue card. However, if James has the blue card, then he remains uncertain about which card Tim has. This situation is represented in Figure 3. Here we use the triple $[b, g, r]$ to indicate that in the corresponding world, James has the blue card, Rowan has the green card, and Tim has the red card.

Suppose now that Tim is able to request information from an oracle. He asks whether Rowan has the red card, and the oracle answers. James hears Tim ask the question, but he does not hear the answer. Rowan hears Tim ask the oracle a question, but was not sure if Tim asked whether Rowan has the red card or the green card. The new knowledge state is represented in Figure 4

Note that while only Tim queried the oracle (so the informative update was a refinement over the agent set $\{Tim\}$), Rowan and James were able to learn about *how* Tim's knowledge changed.

These examples show how refinements captured a very general notion of informative update. Quantifying over refinements therefore allows us to determine whether certain knowledge states are achievable among a set of agents, and has applications in designing and verifying security protocols [7] and reasoning about bidding strategies in games [8] .

4 Refinement doxastic logic

In this section we consider the refinement doxastic logic, $KD45_\forall$. We provide a sound and complete axiomatisation of the multi-agent refinement doxastic logic, and provide expressivity and decidability results.

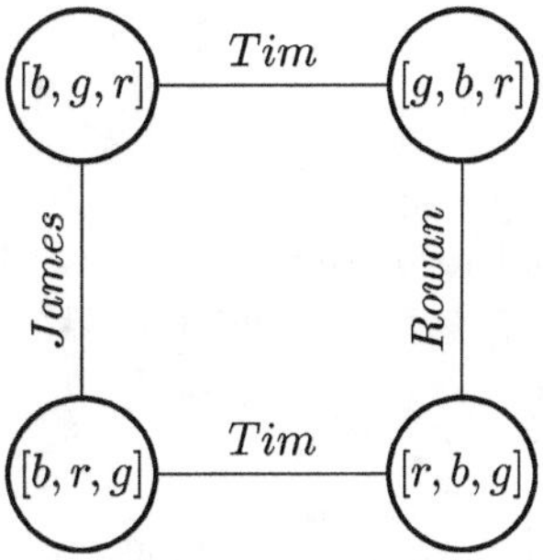

Fig. 3. The initial knowledge state in Example 3.5, in $\mathcal{S}5$. Reflexive, transitive and Euclidean edges are omitted.

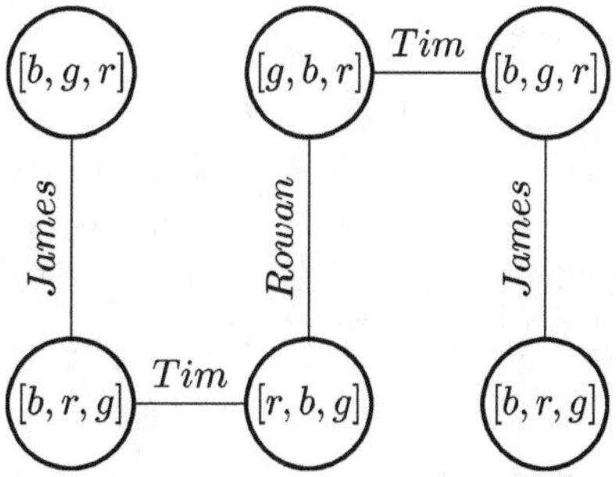

Fig. 4. The final knowledge state in Example 3.5. It can be seen to be a $\{Tim\}$-refinement of the scenario in Figure 3 by relating worlds with equal assignments.

As with previous work by van Ditmarsch, French and Pinchinat [11], and Hales, French and Davies [13], completeness, expressivity and decidability results are shown by a provably correct translation from the language of refinement modal formulae to the language of modal formulae. The translation and axiomatisation rely on a special normal form for doxastic logic, in terms of the cover operator. We first introduce this normal form in the technical preliminaries of this section, and in the following subsection we introduce the axiomatisation, prove its soundness, and provide the provably correct translation from which our other results follow.

4.1 Technical preliminaries

The axiomatisation of the single-agent refinement modal logic by van Ditmarsch, French and Pinchinat [11] relied on a cover logic disjunctive normal form, formulated in terms of the cover operator. The following axiomatisation of the single-agent refinement doxastic logic by Hales, French and Davies [13] relied on a restricted version of this normal form, the cover logic prenex normal form. The cover logic prenex normal form restricts the formulae inside cover operators to be only propositional formulae. In the single-agent doxastic logic, all formulae may be expressed in this prenex normal form, however this is not true in the multi-agent setting. We introduce a generalisation of the prenex normal form to the multi-agent setting, which we call the cover logic alternating disjunctive normal form.

We first introduce the (non-cover logic) alternating disjunctive normal form, and then introduce its cover logic version. These are analogous to the prenex normal form and the corresponding cover logic version used by Hales, French and Davies [13].

Definition 4.1 (Alternating disjunctive normal form (ADNF)) *A formula in a-*alternating disjunctive normal form *(abbreviated as a-ADNF) is defined by the following abstract syntax, where α is a formula in a-ADNF:*

$$\alpha ::= \delta \mid \alpha \vee \alpha$$
$$\delta ::= \pi \mid \Box_b \gamma_b \mid \Diamond_b \gamma_b \mid \delta \wedge \delta$$

where $\pi \in \mathcal{L}_0$, $b \in A \setminus \{a\}$, and γ_b stands for a formula in b-ADNF.

A formula in alternating disjunctive normal form *(abbreviated as ADNF) is defined by the following abstract syntax, where α is a formula in ADNF:*

$$\alpha ::= \delta \mid \alpha \vee \alpha$$
$$\delta ::= \pi \mid \Box_a \gamma_a \mid \Diamond_a \gamma_a \mid \delta \wedge \delta$$

where $\pi \in \mathcal{L}_0$, $a \in A$, and γ_a stands for a formula in a-ADNF.

The alternating disjunctive normal form essentially prohibits direct nestings of modal operators of a particular agent inside modal operators of the same agent. For example, the formula $\Box_a \Diamond_a p$ *is not* in ADNF, because the $\Diamond_a$ operator is nested directly within the $\Box_a$ operator, but the formula $\Box_a \Box_b \Diamond_a p$ *is* in ADNF because although $\Diamond_a$ is nested within the $\Box_a$ operator, there is a $\Box_b$ operator inbetween. In the case where there is only one agent in the language, this is the same as the prenex normal form of Hales, French and Davies [13], where modal operators may only contain propositional formulae. We will now show that every formula of $\mathcal{L}$ is equivalent to a formula in ADNF, under the semantics of *KD45*.

Lemma 4.2 *We have the following equivalences in KD45:*

$$\Box_a(\pi \vee (\alpha \wedge \Box_a \beta)) \leftrightarrow (\Box_a(\pi \vee \alpha) \wedge \Box_a \beta) \vee (\Box_a \pi \wedge \neg \Box_a \beta)$$
$$\Box_a(\pi \vee (\alpha \wedge \Diamond_a \beta)) \leftrightarrow (\Box_a(\pi \vee \alpha) \wedge \Diamond_a \beta) \vee (\Box_a \pi \wedge \neg \Diamond_a \beta)$$

This is proven by Meyer and van der Hoek [14] for the single-agent epistemic logic, $S5^{(1)}$, however the same proof also applies to *KD45*.

Meyer and van der Hoek remarked that the only use of the reflexivity axiom of *S5*, **T**, in the proof, is in the form of the theorems $\vdash \Box\Box\phi \to \Box\phi$, and $\vdash \Box\neg\Box\phi \to \neg\Box\phi$. Therefore the proof holds for any logic which replaces **T** with axioms entailing both of these properties. Both of these properties are valid in *KD45*, and therefore the proof by Meyer and van der Hoek [14] applies to this result.

Lemma 4.3 *Every formula of $\mathcal{L}$ is equivalent to a formula in ADNF, under the semantics of KD45.*

Proof. We use a similar reasoning to the proof for prenex normal form, given by Meyer and van der Hoek [14]. If we proceed by induction, assuming that all formulae within a-modal operators are already in the appropriate a-ADNF, then we can iteratively apply the equivalences from Lemma 4.2 in order to replace subformulae where modal operators belonging to a particular agent appear directly within a modal operator of the same agent. □

The axiomatisation of $KD45_\forall$ that we present in the next section is described in terms of the cover operator, ∇, which we introduced previously. As a cover logic version of the prenex normal form was used in the axiomatisation of the single-agent logic $KD45_{\forall}^{(1)}$, we use a cover logic version of the ADNF for the multi-agent $KD45_{\forall}$ logic.

We first introduce the (non-alternating) cover logic disjunctive normal form.

Definition 4.4 (Cover logic disjunctive normal form (CDNF)) *A formula in* cover logic disjunctive normal form *(abbreviated as CDNF) is defined by the following abstract syntax:*

$$\alpha ::= \pi \wedge \bigwedge_{b \in B} \nabla_b \Gamma_b \mid \alpha \vee \alpha$$

where $\pi \in \mathcal{L}_0$, $B \subseteq A$ and Γ_b stands for a finite set of formulae in CDNF.

Lemma 4.5 *Every formula of $\mathcal{L}$ is equivalent to a formula in cover logic disjunctive normal form, under the semantics of K.*

This is shown by Janin and Walukiewicz [15] in the context of the modal μ-calculus, where the CDNF also contains μ and ν operators, but we note that the result and proof holds if the μ and ν operators are removed. We also note that as this result is in K, it also holds for *KD45* and *S5*.

The CADNF is essentially a cover logic version of the ADNF. It prohibits cover operators of a particular agent from appearing directly within the scope of cover operators of the same agent. In the case where there is only one agent in the language, this is the same as the cover logic prenex normal form of Hales, French and Davies [13].

Definition 4.6 (Cover logic alternating disj. normal form (CADNF)) *A formula in a-*cover logic alternating disjunctive normal form *(abbreviated as a-CADNF) is defined by the following abstract syntax:*

$$\alpha ::= \pi \wedge \bigwedge_{b \in B} \nabla_b \Gamma_b \mid \alpha \vee \alpha$$

where $\pi \in \mathcal{L}_0$, $B \subseteq A \setminus \{a\}$, and Γ_b stands for a finite, non-empty set of formulae in b-CADNF.

A formula in cover logic alternating disjunctive normal form *(abbreviated as CADNF) is defined by the following abstract syntax:*

$$\alpha ::= \pi \wedge \bigwedge_{a \in B} \nabla_a \Gamma_a \mid \alpha \vee \alpha$$

where $\pi \in \mathcal{L}_0$, $B \subseteq A$, and Γ_a stands for a finite, non-empty set of formulae in a-CADNF.

Lemma 4.7 *Every formula of $\mathcal{L}$ is equivalent to a formula in cover logic alternating disjunctive normal form, under the semantics of KD45.*

Proof. By Lemma 4.3 we can write any formula in ADNF. By Lemma 4.5, we can rewrite this formula in ADNF into CDNF. We note that as the results for Lemma 4.5 are for K-equivalence, the algorithm for this conversion preserves the property that modal operators of a particular agent are not nested inside modal operators of the same agent (as otherwise the resulting formula would not be K-equivalent). Therefore the result is in CADNF. □

The CADNF is used in the formulation of our axiomatisation, and is relied upon for our soundness proofs, and as the basis of our provably correct translation used for the completeness, expressivity and decidability results.

4.2 Axiomatisation

We provide an axiomatisation of the multi-agent refinement quantified doxastic logic, $KD45_\forall$, and prove its soundness and completeness. Expressivity and decidability results are given as corollaries.

Definition 4.8 (Axiomatisation $\mathbf{RML_{KD45}}$) *The axiomatisation* $\mathbf{RML_{KD45}}$ *is a substitution schema consisting of the following axioms:*

$$
\begin{array}{rl}
\mathbf{P} & \textit{All propositional tautologies} \\
\mathbf{K} & \Box_a(\phi \to \psi) \to (\Box_a\phi \to \Box_a\psi) \\
\mathbf{D} & \Box_a\phi \to \Diamond_a\phi \\
\mathbf{4} & \Box_a\phi \to \Box_a\Box_a\phi \\
\mathbf{5} & \Diamond_a\phi \to \Box_a\Diamond_a\phi \\
\mathbf{R} & \forall_B(\phi \to \psi) \to (\forall_B\phi \to \forall_B\psi) \\
\mathbf{RP} & \forall_B\pi \leftrightarrow \pi \textit{ where } \pi \textit{ is a propositional formula} \\
\mathbf{RKD45} & \exists_B\nabla_a\Gamma_a \leftrightarrow \nabla_a\left(\{\exists_B\gamma \mid \gamma \in \Gamma_a\} \cup \{\top\}\right) \textit{ where } a \in B \\
\mathbf{RComm} & \exists_B\nabla_a\Gamma_a \leftrightarrow \nabla_a\{\exists_B\gamma \mid \gamma \in \Gamma_a\} \textit{ where } a \notin B \\
\mathbf{RDist} & \exists_B\bigwedge_{c\in C}\nabla_c\Gamma_c \leftrightarrow \bigwedge_{c\in C}\exists_B\nabla_c\Gamma_c \textit{ where } C \subseteq A
\end{array}
$$

where for every $a \in A$, the set Γ_a is a non-empty set of formulae in a-ADNF.

Along with the rules:

$$
\begin{array}{rl}
\mathbf{MP} & \textit{From } \vdash \phi \to \psi \textit{ and } \vdash \phi, \textit{ infer } \vdash \psi \\
\mathbf{NecK} & \textit{From } \vdash \phi \textit{ infer } \vdash \Box_a\phi \\
\mathbf{NecR} & \textit{From } \vdash \phi \textit{ infer } \vdash \forall_B\phi
\end{array}
$$

The axiomatisation $\mathbf{RML_{KD45}}$ is similar to the axiomatisation for the single-agent case considered previously[13] in many respects. Notable differences are that $\mathbf{RML_{KD45}}$ relies on the formulae in cover operators being in a-ADNF instead of being propositional formulae, and that $\mathbf{RML_{KD45}}$ also introduces the **RComm** and **RDist** axioms, which are required to handle the interactions between different agents.

Each of the **RKD45**, **RComm** axioms serve to push $\exists_B$ operators inside a modality, specifically a cover operator, so that it is applied to each formula inside the cover operator. The axiom **RDist** allows us to push the $\exists_B$ operator inside conjunctions in a very specific case. The axiom **R** allows us to push $\exists_B$ operators inside disjunctions, and the **RP** axiom allows us to eliminate $\exists_B$ operators applied to propositional atoms. This forms the basis of our provably correct translation from $\mathcal{L}_\forall$ to $\mathcal{L}$.

The restriction to a-ADNF for the **RKD45**, **RComm** and **RDist** axioms is necessary to resolve inconsistencies that may be caused by positive and negative introspection of belief that is present in *KD45*. For example, we will look at a counter-example to the **RKD45** axiom if we relax the restriction. Let $M = (S, R, V)$ where $S = \{1, 2, 3\}$, $R_a = \{(1, 2), (1, 3), (2, 3), (3, 2)\}$ and $V(p) = \{2\}$. Then $M_1 \models \nabla_a\{\exists_a\Box_a p, \exists_a\Box_a\neg p, \top\}$. However the formula $\nabla_a\{\Box_a p, \Box_a\neg p\}$ is not satisfiable in $\mathcal{KD}45$ and so $M_1 \not\models \exists_a\nabla_a\{\Box_a p, \Box_a\neg p\}$. This problem is not present in $K_\forall$, but exists in $KD45_\forall$ because of positive and negative introspection. The same problem is present in $KD45_\forall^{(1)}$ (the counter-example used here involves only one agent), and is avoided by use of the cover logic prenex normal form.

Lemma 4.9 *The axiomatisation* $\mathbf{RML_{KD45}}$ *is sound with respect to the semantic class* $\mathcal{KD}45$.

Proof. The soundness of the axioms **P** and **K**, **D**, **4**, **5** and the rules **MP** and **NecK** can be shown by the same reasoning used to show that they are sound in *KD45*. The soundness of the axioms **RP** and **R**, and the rule **NecR** can be shown by the same reasoning used to show that they are sound in the single-agent refinement quantified modal logic, as shown by van Ditmarsch, French and Pinchinat [11].

All that remains to be shown is the soundness of **RKD45**, **RComm**, and **RDist**.

RKD45 ($\Longrightarrow$) Trivial. The refinements at successors that we require to satisfy the right-hand side of the equivalence are the successors of the refinement that is entailed by the left-hand side of the equivalence.

($\Longleftarrow$) Let $M_s \in \mathcal{KD}45$ be a doxastic model such that $M_s \models \nabla_a(\{\exists_B\gamma \mid \gamma \in \Gamma_a\} \cup \{\top\})$, where $a \in B$ and γ is an a-ADNF for every $\gamma \in \Gamma_a$. Then for every $\gamma \in \Gamma_a$ there exists some $t^\gamma \in sR_a$ and $M^\gamma_{t^\gamma} \in \mathcal{KD}45$ such that $M^\gamma_{t^\gamma} \preceq_B M_{t^\gamma}$ (via a B-simulation $\mathfrak{R}^\gamma$) and $M^\gamma_{t^\gamma} \models \gamma$. Without loss of generality, we assume that each of the M^γ are disjoint.

We need to show that $M_s \models \exists_B\nabla_a\Gamma_a$. To do this we will construct a model $M'_{s'} \in \mathcal{KD}45$, such that $M'_{s'} \preceq_B M_s$ and show that $M'_{s'} \models \nabla_a\Gamma_a$.

We begin by constructing the model $M' = (S', R', V')$ where:

$$S' = \{s'\} \cup \{t^{\gamma\prime} \mid \gamma \in \Gamma_a\} \cup S \cup \bigcup_{\gamma\in\Gamma_a} S^\gamma$$

$$R'_a = \{(s', t^{\gamma\prime}) \mid \gamma \in \Gamma_a\} \cup \{t^{\gamma\prime} \mid \gamma \in \Gamma_a\}^2 \cup R_a \cup \bigcup_{\gamma\in\Gamma_a} R^\gamma_a$$

$$R'_b = \{(s', t) \mid t \in sR_b\} \cup \{(t^{\gamma\prime}, u) \mid u \in t^\gamma R^\gamma_b\} \cup R_b \cup \bigcup_{\gamma\in\Gamma_a} R^\gamma_b$$

$$V'(p) = \{s' \mid s \in V(p)\} \cup \{t^{\gamma\prime} \mid \gamma \in \Gamma_a, t^\gamma \in V^\gamma(p)\} \cup V(p) \cup \bigcup_{\gamma\in\Gamma_a} V^\gamma(p)$$

where $b \in A \setminus \{a\}$, $p \in P$, and s' and each of the $t^{\gamma\prime}$ are new states that do not appear in S or any of the S^γ. This construction is shown in Figure 5.

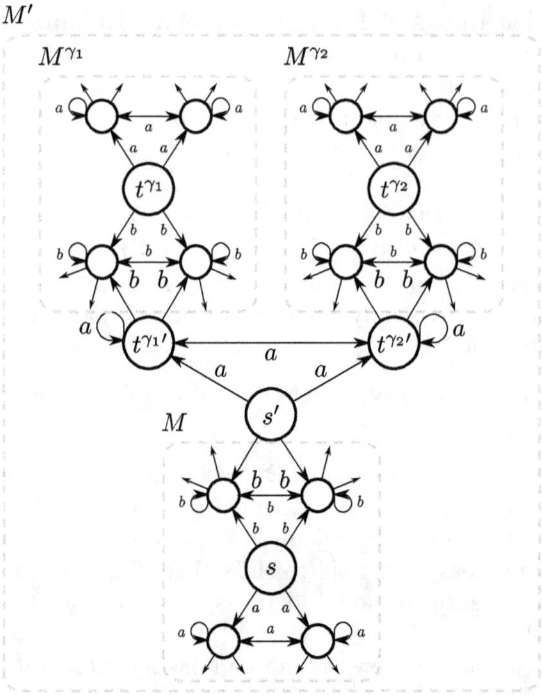

Fig. 5. Construction of refinement $M'_{s'} \preceq_B M_s$.

We note that by construction $M' \in \mathcal{KD}45$ and $M'_{s'} \preceq_B M_s$, via the B-simulation $\mathfrak{R} = \{s', s\} \cup \{(t^{\gamma\prime}, t^{\gamma}) \mid \gamma \in \Gamma_a\} \cup \{(t,t) \mid t \in S\} \cup \bigcup_{\gamma \in \Gamma_a} \mathfrak{R}^{\gamma}$.

We must show that $M'_{s'} \models \nabla_a \Gamma_a$. To do this we show for every $\gamma \in \Gamma_a$ that $M'_{t^{\gamma\prime}} \models \gamma$. Let $\gamma \in \Gamma_a$. As γ is an a-ADNF, it is a disjunction of conjunctions of propositional formulae and formulae of the form $\Box_b \phi$ and $\Diamond_b \phi$, where $b \neq a$. By construction, the valuation of $M'_{t^{\gamma\prime}}$ is identical to the valuation of $M^{\gamma}_{t^{\gamma}}$, therefore for any $\pi \in \mathcal{L}_0$ we have that $M'_{t^{\gamma\prime}} \models \pi$ if and only if $M^{\gamma}_{t^{\gamma}} \models \pi$. Furthermore we note for every $u \in S^{\gamma}$ that $M'_u \leftrightarroweq M^{\gamma}_u$, and so for any $\psi \in \mathcal{L}$ we have that $M'_{t^{\gamma}} \models \psi$ if and only if $M^{\gamma}_{t^{\gamma}} \models \psi$; in particular this is the case when $\psi = \Box_b \phi$ or $\psi = \Diamond_b \phi$. By construction, $t^{\gamma\prime} R'_b = t^{\gamma} R'_b$, and so $M'_{t^{\gamma\prime}} \models \Box_b \phi$ if and only if $M'_{t^{\gamma}} \models \Box_b \phi$ if and only if $M^{\gamma}_{t^{\gamma}} \models \Box_b \phi$. As by hypothesis we have $M^{\gamma}_{t^{\gamma}} \models \gamma$ we therefore have that $M'_{t^{\gamma\prime}} \models \gamma$. As this holds for every $\gamma \in \Gamma_a$, it follows that $M'_{s'} \models \nabla_a \Gamma_a$.

As $M'_{s'} \preceq_B M_s$, and $M'_{s'} \models \nabla_a \Gamma_a$ we therefore have that $M_s \models \exists_B \nabla_a \Gamma_a$. Therefore **RKD45** is sound.

RComm ($\Longrightarrow$) Trivial.

($\Longleftarrow$) The proof is similar to the proof for **RKD45**. The construction for **RComm**, and the B-simulation used to show that it is a B-refinement are identical. Slightly different reasoning must be used to show that this construction is a B-refinement, reflecting the fact that in this case $a \notin B$, but this is straightforward. The reasoning used to show that in the resulting B-refinement that $M'_{s'} \models \nabla_a \Gamma_a$ is identical to the reasoning used in **RKD45**.

RDist ($\Longrightarrow$) Trivial.

($\Longleftarrow$) Let $M_s \in \mathcal{KD}45$ be a doxastic model such that $M_s \models \bigwedge_{c \in C} \exists_B \nabla_c \Gamma_c$, where $B, C \subseteq A$ and Γ_c is a set of c-ADNF for every $c \in C$. Then for every $c \in C$ there exists some $M^c_{s^c} \in \mathcal{KD}45$ such that $M^c_{s^c} \preceq_B M_s$ (via a B-simulation $\mathfrak{R}^c$) and $M^c_{s^c} \models \nabla_c \Gamma_c$. Without loss of generality, we assume that each of the M^c are disjoint.

We need to show that $M_s \models \exists_B \bigwedge_{c \in C} \nabla_c \Gamma_c$. To do this we will construct a model $M'_{s'} \in \mathcal{KD}45$ such that $M'_{s'} \preceq_B M_s$ and show that $M'_{s'} \models \bigwedge_{c \in C} \nabla_c \Gamma_c$.

We begin by constructing the model $M' = (S', R', V')$ where:

$$\begin{aligned}
S' &= \{s'\} \cup S \cup \bigcup_{c \in C} S^c \\
R'_c &= \{(s', t) \mid t \in s^c R^c_c\} \cup R_c \cup \bigcup_{d \in C} R^d_c \text{ for } c \in C \\
R'_c &= \{(s', t) \mid t \in s R_c\} \cup R_c \cup \bigcup_{d \in C} R^d_c \text{ for } c \notin C \\
V'(p) &= \{s' \mid s \in V(p)\} \cup V(p) \cup \bigcup_{c \in C} V^c(p)
\end{aligned}$$

where $p \in P$ and s' is a new state that does not appear in S or any of the S^c.

We note that by construction $M' \in \mathcal{KD}45$ and $M'_{s'} \preceq_B M_s$, via the B-simulation $\mathfrak{R} = \{(s', s)\} \cup \{(t, t) \mid t \in S\} \cup \bigcup_{c \in C} \mathfrak{R}^c$.

We use a similar argument to that used in **RKD45** to show that $M'_{s'} \models \bigwedge_{c \in C} \nabla_c \Gamma_c$. We note for every $c \in C$ and $u \in S^c$ that $M'_u \leftrightarroweq M^c_u$. By construction we have that $s' R'_c = s^c R'_c$, and therefore from $M^c_{s^c} \models \nabla_c \Gamma_c$ we have that $M'_{s'} \models \nabla_c \Gamma_c$.

As $M'_{s'} \preceq_B M_s$, and $M'_{s'} \models \nabla_c \Gamma_c$ we therefore have that $M_s \models \exists_B \bigwedge_{c \in C} \nabla_c \Gamma_c$. Therefore **RDist** is sound.

Therefore the axiomatisation $\mathbf{RML_{KD45}}$ is sound. □

The construction we use to show the soundness of **RKD45** is similar to the constructions used to show the soundness of the axioms **GK** for the axiomatisation of $K^{(1)}_\forall$ [11] and **GKD45** for the axiomatisation of $KD45^{(1)}_\forall$ [13]. In each construction we begin by assuming for each $\gamma \in \Gamma$ the existence of initial refinements $M^\gamma_{t^\gamma}$ at some successor of M_s, where $M^\gamma_{t^\gamma} \models \gamma$. The construction for **GK**, shown in Figure 6 simply combines each initial refinement M^γ with a new state s' into a combined refinement $M'_{s'}$. As the only new edges on any of the initial refinements are in-bound edges, the interpretation of formulae in these initial refinements are preserved by this construction. In particular this means for each $\gamma \in \Gamma$ that $M'_{t^\gamma} \models \gamma$, and so it is then a simple matter to show that $M'_{s'} \models \nabla \Gamma$. The construction used for **GKD45** on the other hand, shown in Figure 7, does not use the whole of each initial refinement, but rather only uses a duplicate $t^{\gamma\prime}$ of the root state t^γ that shares the same valuation, but not the same successors. This works because the axiom **GKD45** assumes that each

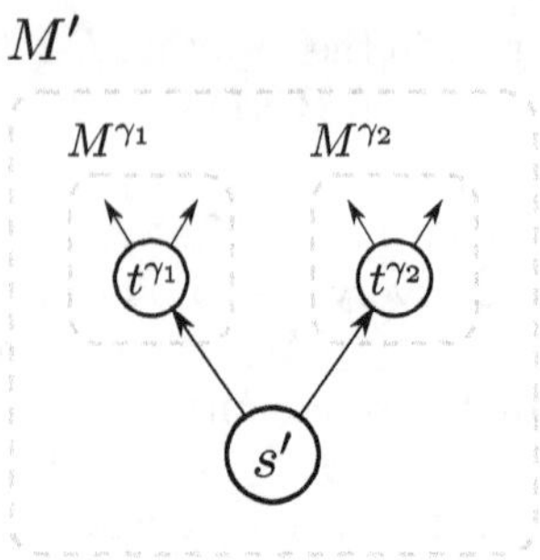

Fig. 6. Construction of refinement $M'_{s'} \preceq M_s$ to show the soundness of **GK** in $K_\forall^{(1)}$

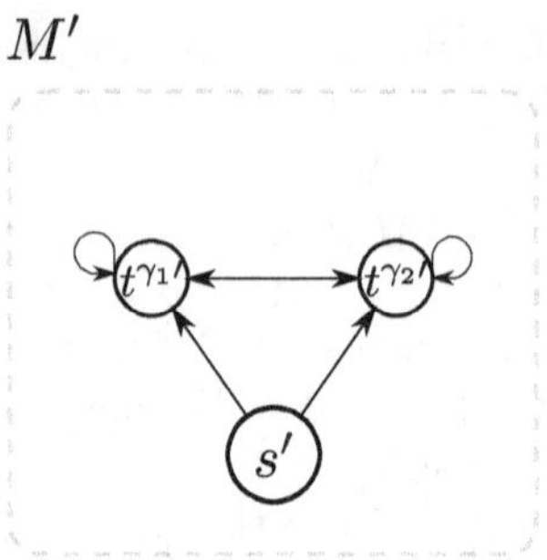

Fig. 7. Construction of refinement $M'_{s'} \preceq M_s$ to show the soundness of **GKD45** in $KD45_\forall^{(1)}$

$\gamma \in \Gamma$ is a propositional formula, so only duplicating the valuation is required to preserve the interpretation of these formulae. Moreover, using only the root state is necessary because using the whole of each initial refinement, as in the construction for **GK**, would be problematic in the setting of $\mathcal{KD}45$; the edges from s' to each state t^γ may necessitate the addition of additional edges from s' to other states in each M^γ, in order to satisfy the transitive property of $\mathcal{KD}45$, and this would mean that we may not have $M'_{s'} \models \nabla\Gamma$.

The construction used for **RKD45**, shown in Figure 5, uses elements from both the **GK** and **GKD45** constructions. As in **GK**, we include the initial refinements into the combined refinement, and we only add in-bound edges to each initial refinement, so that we get the same bisimilarity property relied upon for the **GK** construction. However rather than adding edges from the new root state s' to each state t^γ, we instead introduce *proxy states* $t^{\gamma\prime}$, corresponding to each t^γ. The proxy states duplicate the propositional valuations and b-successors, where $b \neq a$, of each root state t^γ. In this resulting combined refinement, each proxy state $t^{\gamma\prime}$ preserves the truth of every a-ADNF formula from the corresponding state t^γ. The proxy states avoid difficulties due to the additional transitive and Euclidean edges that would otherwise be required in order to satisfy the properties of $\mathcal{KD}45$, similar to the difficulties avoided by the construction used for **GKD45**. In the case where we have only a single agent, the construction for **RKD45** is essentially the same as the construction used for **GKD45**.

We now show the completeness of the axiomatisation $\mathbf{RML_{KD45}}$ by a provably correct translation from $\mathcal{L}_\forall$ to $\mathcal{L}$. Completeness then follows from the completeness of $KD45$.

We first introduce some equivalences used by our translation.

Lemma 4.10 *The following are provable equivalences using* $\mathbf{RML_{KD45}}$*:*

(i) $\exists_B(\phi \vee \psi) \leftrightarrow \exists_B\phi \vee \exists_B\psi$

(ii) $\exists_B(\pi \wedge \bigwedge_{c\in C} \nabla_c\Gamma_c) \leftrightarrow \pi \wedge \bigwedge_{c\in C\cap B} \nabla_c(\{\exists_B\gamma \mid \gamma \in \Gamma_c\} \cup \{\top\}) \wedge \bigwedge_{c\in C\setminus B} \nabla_c\{\exists_B\gamma \mid$

$\gamma \in \Gamma_c\}$

where $\pi \in \mathcal{L}_0$, $B, C \subseteq A$, *and for every* $c \in C$ *the set* Γ_c *is a non-empty set of c-ADNF formulae.*

Proof. (1) is derivable using **P** and **R**. (2) is derivable using **P**, **R**, **RP**, **RKD45**, **RComm** and **RDist**. (3) is derivable using **P**, **R**, **RP**, **RComm** and **RDist**. □

Lemma 4.11 *Every formula of* $\mathcal{L}_\forall$ *is provably equivalent to a formula of* $\mathcal{L}$ *via the axiomatisation of* $\mathbf{RML_{KD45}}$.

Proof. We proceed by iteratively removing $\exists_B$-operators from our formula via provable equivalences. Take any subformula of the form $\exists_B \phi$, where $\phi \in \mathcal{L}$, and rewrite ϕ in CADNF. This is a provable equivalence in $\mathbf{RML_{KD45}}$ as the axioms of *KD45* appear in $\mathbf{RML_{KD45}}$. We then iteratively apply the equivalences from Lemma 4.10, pushing the $\exists_B$ inside disjunctions and cover operators, until the only $\exists_B$ operators are applied to propositional formulae. We can then use the axiom **RP** to remove these $\exists_B$ operators. □

Theorem 4.12 *The axiomatisation* $\mathbf{RML_{KD45}}$ *is sound and complete with respect to the semantic class* $\mathcal{KD}45$.

Proof. Soundness is proven in Lemma 4.9. As in [11,13], completeness is via the provably correct translation from Lemma 4.11 to a sublanguage $\mathcal{L}$ of $\mathcal{L}_\forall$, which has completeness for the semantic class $\mathcal{KD}45$ via the axioms of *KD45* that appear in $\mathbf{RML_{KD45}}$. □

Corollary 4.13 *The logic* $KD45_\forall$ *is expressively equivalent to KD45.*

Corollary 4.14 *The logic* $KD45_\forall$ *is decidable.*

We note that the decision procedure for $KD45_\forall$ is via the translation to *KD45*, and that this translation has a non-elementary complexity. Better decision procedures are left to future work.

5 Refinement epistemic logic

In this section we consider the refinement epistemic logic, $S5_\forall$. We provide an axiomatisation of the multi-agent refinement epistemic logic, and provide expressivity and decidability results. Our axiomatisation, soundness and completeness results follow the same general format of those we have seen previously, but using a different technique in place of a disjunctive normal form.

5.1 Technical preliminaries

Although the alternating disjunctive normal form is a valid normal form for epistemic logic, it is not sufficiently restricted to give a sound axiomatisation of a similar form to the axiomatisation $\mathbf{RML_{KD45}}$. We instead introduce the notion of an *explicit* formula, and formulate our axiomatisation in terms of formulae in this form.

Definition 5.1 (Explicit formulae) *Let* $\pi \in \mathcal{L}_0$ *be a propositional formula,* $B \subseteq A$ *be a finite set of agents and for every* $b \in B$ *let* $\Gamma_b \subseteq \mathcal{L}$ *be a finite* *set of formulae. Let $\gamma^0 \in \mathcal{L}$ be a formula such that for every $b \in B$ we have $\gamma^0 \in \Gamma_b$. Let $\Phi = \{\phi \mid b \in B, \gamma \in \Gamma_b, \phi \leq \gamma\}$ be a set of subformulae of all of the formulae in each set Γ_b. Finally, let α be a formula of the form*

$$\alpha = \pi \wedge \gamma^0 \wedge \bigwedge_{b \in B} \nabla_b \Gamma_b$$

Then α is an explicit formula *if the following conditions hold:*

(i) *For every $b \in B$, $\gamma \in \Gamma_b$, $\phi \in \Phi$: either $\vdash_{S5} \gamma \to \phi$ or $\vdash_{S5} \gamma \to \neg\phi$.*

(ii) *For every $b \in B$, $\gamma \in \Gamma_b$, $\Box_b\phi \in \Phi$: $\vdash_{S5} \gamma \to \Box_b\phi$ if and only if for every $\gamma' \in \Gamma_b$ we have $\vdash_{S5} \gamma' \to \phi$.*

Explicit formulae essentially remove a number of elements of choice from the interpretation of a formula: formulae appearing within the cover operators at the top level must explicitly specify which of their subformulae (and subformulae of other formulae appearing in cover operators) are true or false; the formula must explicitly specify which formula in the cover operator is true at the current state; and each cover operator $\nabla_b\Gamma_b$ in the formula must also explicitly agree on which $\Box_b\phi$ subformulae are true.

We will now show that every formula of $\mathcal{L}$ is equivalent to a disjunction of explicit formulae while preserving its *S5* interpretation.

Lemma 5.2 *Every formula of $\mathcal{L}$ is equivalent to a disjunction of explicit formulae, under the semantics of S5.*

Proof. Lemma 4.5 gives equivalence to a disjunction of formulae of the form $\pi \wedge \bigwedge_{b \in B} \nabla_b\Gamma_b$, so we only need show the lemma for such a formula, say δ.

Let $\Phi = \{\phi \mid b \in B, \gamma \in \Gamma_b, \phi \leq \gamma\}$. Then replace each γ in δ with a disjunction over the truth of the subformulae ϕ, giving an equivalent δ':

$$\delta' = \pi \wedge \bigwedge_{b \in B} \nabla_b \left\{ \bigvee_{\Psi \subseteq \Phi} \left(\gamma \wedge \bigwedge_{\phi \in \Psi} \phi \wedge \bigwedge_{\phi \in \Phi \setminus \Psi} \neg\phi \right) \mid \gamma \in \Gamma_b \right\}$$

We move these disjunctions outwards by iteratively applying the equivalence

$$\nabla_b(\{\rho \vee \sigma\} \cup \Gamma) \equiv_{S5} \nabla_b(\{\rho\} \cup \Gamma) \vee \nabla_b(\{\sigma\} \cup \Gamma) \vee \nabla_b(\{\rho, \sigma\} \cup \Gamma)$$

Then distributivity of conjunction over disjunction gives a disjunction of formulae of the form

$$\pi \wedge \bigwedge_{b \in B} \nabla_b\Gamma'_b$$

where each Γ'_b has only elements of the form $\gamma \wedge \bigwedge_{\phi \in \Psi} \phi \wedge \bigwedge_{\phi \in \Phi \setminus \Psi} \neg\phi$. Then, by reflexivity this is equivalent to

$$\pi \wedge \bigwedge_{b \in B} \left(\bigvee_{\gamma' \in \Gamma'_b} \gamma' \wedge \nabla_b\Gamma'_b \right)$$

and again distributivity gives an equivalent disjunction of formulae of the form

$$\pi \wedge \bigwedge_{b \in B} (\gamma'_b \wedge \nabla_b \Gamma'_b)$$

where we omit inconsistent disjuncts, and the Γ'_b are as previously.

Then for every $b, c \in B$ we must have that $\gamma'_b \to \gamma'_c$ or $\gamma'_b \to \neg\gamma'_c$, since γ'_c is a conjunction of elements of $\phi \in \Phi$ and their negations, and γ'_b contains either ϕ or $\neg\phi$ for each. But, $\gamma'_b \to \neg\gamma'_c$ leads to the disjunct being inconsistent, so $\gamma'_b \to \gamma'_c$, and by the same reasoning $\gamma'_c \to \gamma'_b$, thus all γ'_b are equivalent.

So for any $b \in B$ we can let $\gamma'_0 = \gamma'_b$, and rewrite our disjunct as

$$\pi \wedge \gamma'_0 \wedge \bigwedge_{b \in B} (\nabla_b \Gamma'_b)$$

This is of the appropriate form for an explicit formula, and it satisfies the two conditions, as follows. The relevant set of subformulae is

$$\Phi' = \{\phi' \mid b \in B, \gamma' \in \Gamma'_b, \phi' \leq \gamma'\}$$

and each γ' still has the form

$$\gamma \wedge \bigwedge_{\phi \in \Psi} \phi \wedge \bigwedge_{\phi \in \Phi \setminus \Psi} \neg\phi$$

(i) Always either $\gamma' \to \phi'$ or $\gamma' \to \neg\phi'$ because ϕ' is a conjunction of formulae $\phi \in \Phi$ and their negations, and for each $\gamma' \to \phi$ or $\gamma' \to \neg\phi$.

(ii) Let $b \in B$, $\gamma' \in \Gamma'_b$ and $\Box_b\phi' \in \Phi'$. Suppose that $\vdash_{S5} \gamma' \to \Box_b\phi'$ and there exists some $\gamma'' \in \Gamma'_b$ such that $\nvdash_{S5} \gamma'' \to \phi'$. Then from positive and negative introspection we have that $\vdash_{S5} \nabla_b\Gamma'_b \to \Box_b\phi'$. Furthermore, from the first property of an explicit formula we have that $\vdash_{S5} \gamma'' \to \neg\phi'$, and so $\vdash_{S5} \nabla_b\Gamma'_b \to \Diamond_b\neg\phi' \to \neg\Box_b\phi$. But this is a contradiction.

Suppose instead that $\vdash_{S5} \gamma' \to \neg\Box_b\phi'$ and for every $\gamma'' \in \Gamma'_b$ we have that $\vdash_{S5} \gamma'' \to \phi'$. Then from the first hypothesis we have that $\vdash_{S5} \Diamond_b\neg\Box_b\phi' \to \Diamond_b\neg\phi'$, but from the second hypothesis we have that $\vdash_{S5} \nabla_b\Gamma'_b \to \Box_b\phi'$, a contradiction.

Therefore every formula is equivalent to a disjunction of explicit formulae.□

We also note that if we take an explicit formula and remove some of the cover operators from the conjunction, that the result is still an explicit formula.

Lemma 5.3 *Let* $\alpha = \pi \wedge \gamma^0 \wedge \bigwedge_{b \in B} \nabla_b\Gamma_b$ *be an explicit formula, let* $C \subseteq B$ *and let* $\beta = \pi \wedge \gamma^0 \wedge \bigwedge_{c \in C} \nabla_c\Gamma_c$. *Then* β *is also an explicit formula.*

This result follows directly from the definition of explicit formulae once we realise that the set Φ of subformulae is the same for α and β.

Explicit formulae will be used to formulate our axiomatisation, and the equivalence of epistemic formulae to disjunctions of explicit formulae will be used as part of the provably correct translation that we use in the completeness proof. The requirement for explicit formulae will be discussed briefly after introducing the axiomatisation in the next section.

5.2 Axiomatisation

We provide an axiomatisation of multi-agent refinement epistemic logic, $S5_\forall$, and prove its soundness and completeness. Expressivity and decidability results are given as corollaries.

Definition 5.4 (Axiomatisation $\mathbf{RML_{S5}}$) *The axiomatisation* $\mathbf{RML_{S5}}$ *is a substitution schema consisting of the following axioms:*

P *All propositional tautologies*
K $\Box_a(\phi \to \psi) \to (\Box_a\phi \to \Box_a\psi)$
T $\Box_a\phi \to \phi$
5 $\Diamond_a\phi \to \Box_a\Diamond_a\phi$
R $\forall_B(\phi \to \psi) \to (\forall_B\phi \to \forall_B\psi)$
RP $\forall_B\pi \leftrightarrow \pi$ *where* π *is a propositional formula*
RS5 $\exists_B(\gamma_0 \wedge \nabla_a\Gamma_a) \leftrightarrow \exists_B\gamma_0 \wedge \nabla_a(\{\exists_B\gamma \mid \gamma \in \Gamma_a\} \cup \{\top\})$ *where* $a \in B$
RComm $\exists_B(\gamma_0 \wedge \nabla_a\Gamma_a) \leftrightarrow \exists_B\gamma_0 \wedge \nabla_a\{\exists_B\gamma \mid \gamma \in \Gamma_a\}$ *where* $a \notin B$
RDist $\exists_B \bigwedge_{c\in C}(\gamma_0 \wedge \nabla_c\Gamma_c) \leftrightarrow \bigwedge_{c\in C}\exists_B(\gamma_0 \wedge \nabla_c\Gamma_c)$ *where* $C \subseteq A$

where $\gamma_0 \wedge \nabla_a\Gamma_a$ *and* $\gamma_0 \wedge \bigwedge_{c\in C}\nabla_c\Gamma_c$ *are explicit formulae.*
Along with the rules:

MP *From* $\vdash \phi \to \psi$ *and* $\vdash \phi$, *infer* $\vdash \psi$
NecK *From* $\vdash \phi$ *infer* $\vdash \Box_a\phi$
NecR *From* $\vdash \phi$ *infer* $\vdash \forall_B\phi$

The axiomatisation $\mathbf{RML_{S5}}$ is similar to the axiomatisation $\mathbf{RML_{KD45}}$, except that it contains the axioms for *S5*, and that the axioms **RS5**, **RComm** and **RDist** rely on the notion of explicit formulae, and make specific reference to which formula from within the cover operator is true at the current state. As in $\mathbf{RML_{KD45}}$, the **RS5**, **RComm**, **RDist**, **R** and **RP** axioms together form the basis of our provably correct translation to epistemic logic.

The restriction to cover logic alternating disjunctive normal form that $\mathbf{RML_{KD45}}$ uses is not sufficient for the axiomatisation $\mathbf{RML_{S5}}$. For example, we consider a counter-example to **RS5** if we replace the restriction to explicit formulae with a restriction to alternating disjunctive normal formulae. Let $M = (S, R, V)$ where $S = \{1,2\}$, $R_a = \{(1,1),(1,2),(2,1),(2,2)\}$, $R_b = \{(1,1),(2,2)\}$ and $V(p) = \{1\}$. Then $M_1 \models \exists_a\Box_b\Box_a p \wedge \Diamond_a\exists_a\Box_b\Box_a p \wedge \Diamond_a\exists_a\Box_b\Box_a\neg p$. However the formula $\nabla_a\{\Box_b\Box_a p, \Box_b\Box_a\neg p\}$ is not satisfiable in $\mathcal{S}5$, and so $M_1 \not\models \exists_a(\Box_b\Box_a p \wedge \nabla_a\{\Box_b\Box_a p, \Box_b\Box_a\neg p\})$. In $KD45_\forall$ this is not a problem, but the problem arises in $\mathcal{S}5$ because of the addition of reflexivity; in $S5_\forall$ we have that $\vdash_{S5} \Box_b\Box_a p \to \Box_a p$ and $\vdash_{S5} \Box_b\Box_a\neg p \to \Box_a\neg p$, and so we get the same contradiction as in our previous counter-example in $KD45_\forall$.

The construction that we use to show the soundness of **RS5** is similar to the construction that we used to show the soundness of **RKD45** in Lemma 4.9. The main differences are that in the construction for **RS5** all relations must be reflexive, transitive and Euclidean, and that the root state of the combined refinement is just another $t^{\gamma'}$ state, and is not treated differently as the root state was for the **RKD45** construction. In the construction for **RKD45**, the fact that the interpretation of a-ADNF formulae is preserved comes from the fact that the the states of the initial refinements that are included in the combined refinement are bisimilar to the corresponding states from the uncombined initial refinements. This bisimilarity is because the only new edges added in the construction on states in the initial refinements are in-bound edges. In $\mathcal{S}5$ we cannot *only* have in-bound edges into the states from the initial refinements, as in $\mathcal{S}5$ every edge must be symmetric. We must instead use a different strategy to ensure that combining the initial refinements does not vary the interpretation of each formula $\gamma \in \Gamma_a$ that we are interested in. The use of explicit formulae provides us with additional restrictions on the properties of each initial refinement M^γ, which are sufficient to show that combining the initial refinements does not vary the interpretation of the formulae $\gamma \in \Gamma_a$.

Lemma 5.5 *The axiomatisation* $\mathbf{RML_{S5}}$ *is sound with respect to the semantic class* $\mathcal{S}5$.

Proof. The soundness of the axioms **P** and **K**, **T**, **5** and the rules **MP** and **NecK** can be shown by the same reasoning used to show that they are sound in *S5*. As the axioms **RP** and **R**, and the rule **NecR** involve only a single agent, their soundness can be shown by the same reasoning used to show that they are sound in the single-agent refinement quantified modal logic, as shown by van Ditmarsch, French and Pinchinat [11].

All that remains to be shown is the soundness of **RS5**, **RComm** and **RDist**.

RS5 ($\Longrightarrow$) Trivial.

($\Longleftarrow$) Let $M_s \in \mathcal{S}5$ be an epistemic model such that $M_s \models \exists_B \gamma_0 \wedge \nabla_a (\{\exists_B \gamma \mid \gamma \in \Gamma_a\} \cup \{\top\})$, where $a \in B$ and $\gamma_0 \wedge \nabla_a \Gamma_a$ is an explicit formula. Then for every $\gamma \in \Gamma_a$ there exists some $t^\gamma \in sR_a$ and $M^\gamma_{t^\gamma} \preceq_B M_{t^\gamma}$ (via a B-simulation $\mathfrak{R}^\gamma$) such that $M^\gamma_{t^\gamma} \models \gamma$. We also have that $M_s \models \exists_B \gamma_0$, where $\gamma_0 \in \Gamma_a$, and so we may assume that $M^{\gamma_0}_{t^{\gamma_0}} \preceq_B M_s$. Without loss of generality, assume that each of the M^γ are disjoint.

We need to show that $M_s \models \exists_B (\gamma_0 \wedge \nabla_a \Gamma_a)$. To do this we will construct a model $M'_{t^{\gamma_0 \prime}} \in \mathcal{S}5$ such that $M'_{t^{\gamma_0 \prime}} \preceq_B M_s$ and show that $M'_{t^{\gamma_0 \prime}} \models \gamma_0 \wedge \nabla_a \Gamma_a$.

We begin by constructing the model $M' = (S', R', V')$ where:

$$S' = \{t^{\gamma'} \mid \gamma \in \Gamma_a\} \cup \bigcup_{\gamma \in \Gamma_a} S^\gamma$$

$$R'_a = (\{t^{\gamma'} \mid \gamma \in \Gamma_a\})^2 \cup \bigcup_{\gamma \in \Gamma_a} R^\gamma_a$$

$$R'_b = \bigcup_{\gamma \in \Gamma_a} \left((\{t^{\gamma'}\} \cup t^\gamma R^\gamma_b)^2 \cup R^\gamma_b \right)$$

$$V'(p) = \{t^{\gamma'} \mid \gamma \in \Gamma_a, t^\gamma \in V(p)\} \cup \bigcup_{\gamma \in \Gamma_a} V^\gamma(p)$$

where $b \in A \setminus \{a\}$, $p \in P$ and each $t^{\gamma\prime}$ is a new state that does not appear in S or any of the S^γ.

We note that by construction $M' \in \mathcal{S}5$ and $M'_{t^{\gamma_0\prime}} \preceq_B M_s$, via the B-simulation $\mathfrak{R} = \{(t^{\gamma\prime}, t^\gamma) \mid \gamma \in \Gamma_a\} \cup \bigcup_{\gamma \in \Gamma_a} \mathfrak{R}^\gamma$.

For every $\gamma \in \Gamma_a$, we can view $t^{\gamma\prime}$ as initially being a bisimilar copy of t^γ in M^γ, but with its a-successors pruned. The result would be an B-refinement of $M^\gamma_{t^\gamma}$. M' is formed by joining each $t^{\gamma\prime}$ state with a-edges, the result being an B-refinement of our original model M.

We must show that $M'_{t^{\gamma_0\prime}} \models \gamma_0 \wedge \nabla_a \Gamma_a$. To do this we show by induction on the structure of formulae in Φ that for every $\phi \in \Phi$, $\gamma \in \Gamma_a$ that: $M'_{t^{\gamma\prime}} \models \phi$ if and only if $M^\gamma_{t^\gamma} \models \phi$, and for every $u \in S^\gamma$ that $M'_u \models \phi$ if and only if $M^\gamma_u \models \phi$.

The base case, where $\phi = p$ for some $p \in P$ follows by construction. The case where $\phi = \neg\alpha$ or $\phi = \alpha \wedge \beta$ follows directly from the induction hypothesis.

Let $b \in A$, $\phi = \Box_b \alpha$, $\gamma \in \Gamma_b$ and $u \in S^\gamma$. Then $M'_u \models \Box_b \alpha$ if and only if for every $v \in uR'_b$ we have that $M'_v \models \alpha$. By construction either $uR'_b = uR^\gamma_b$ or $uR'_b = uR^\gamma_b \cup \{t^{\gamma\prime}\}$. Suppose that $uR'_b = uR^\gamma_b$. Then by the induction hypothesis, $M'_v \models \alpha$ for every $v \in uR'_b$ if and only if $M^\gamma_v \models \alpha$ for every $v \in uR'_b = uR^\gamma_b$ if and only if $M^\gamma_u \models \Box_b \alpha$. Suppose instead that $uR'_b = uR^\gamma_b \cup \{t^{\gamma\prime}\}$. By the induction hypothesis, $M'_v \models \alpha$ for every $v \in uR^\gamma_b$ if and only if $M^\gamma_v \models \alpha$ for every $v \in uR^\gamma_b$. We note that $t^{\gamma\prime} \in uR'_b$ if and only if $t^\gamma \in uR'_b$, and that by the induction hypothesis $M'_{t^{\gamma\prime}} \models \alpha$ if and only if $M^\gamma_{t^\gamma} \models \alpha$. Therefore $M'_v \models \alpha$ for every $v \in uR'_b$ if and only if $M^\gamma_v \models \alpha$ for every $v \in uR^\gamma_b$ if and only if $M^\gamma_u \models \Box_b \alpha$. Similar reasoning shows that $M'_{t^{\gamma\prime}} \models \Box_b \alpha$ if and only if $M^\gamma_{t^\gamma} \models \Box_b \alpha$; we note that to show this in the case where $b = a$ that we must rely on property (ii) in the definition of explicit formulae, and the fact that the a-successors of each state $t^{\gamma\prime}$ for $\gamma \in \Gamma$ are only the other states $t^{\gamma'\prime}$ for $\gamma' \in \Gamma$.

Therefore by induction we have for every $\gamma \in \Gamma_a$ that $M'_{t^{\gamma\prime}} \models \gamma$ if and only if $M^\gamma_{t^\gamma} \models \gamma$. As for every $\gamma \in \Gamma_a$ we have that $M^\gamma_{t^\gamma} \models \gamma$, this gives us $M'_{t^{\gamma\prime}} \models \gamma$. This also gives us $M'_{t^{\gamma_0\prime}} \models \gamma_0 \wedge \nabla_a \Gamma_a$. As we have shown that $M'_{t^{\gamma_0\prime}} \preceq_B M_s$ this gives us $M_s \models \exists_B (\gamma_0 \wedge \nabla_a \Gamma_a)$. Therefore **RS5** is sound.

RComm ($\Longrightarrow$) Trivial.

($\Longleftarrow$) As in the case for **RComm** in $\mathbf{RML_{KD45}}$, the proof is similar to the proof for **RS5**. The inductive proof to show that the constructed refinement satisfies $\gamma_b \wedge \nabla_b \Gamma_b$ is similar to the proof for **RS5**, except that where we treat the agent a specially, we instead treat b specially.

RDist ($\Longrightarrow$) Trivial.

($\Longleftarrow$) Let $M_s \in \mathcal{S}5$ be an epistemic model such that $M_s \models \bigwedge_{c \in C} \exists_B (\gamma_0 \wedge \nabla_c \Gamma_c)$. Then for every $c \in C$ there exists some $M^c_{s^c} \preceq_B M_s$ (via a B-simulation $\mathfrak{R}^c$) such that $M^c_{s^c} \models \gamma_0 \wedge \nabla_c \Gamma_c$. Without loss of generality, we assume that each of the M^c are disjoint.

We need to show that $M_s \models \exists_B \bigwedge_{c \in C} (\gamma_0 \wedge \nabla_c \Gamma_c)$. To do this we will construct a model $M'_{s'} \in \mathcal{S}5$, show that $M'_{s'} \preceq_B M_s$ and show that $M'_{s'} \models \bigwedge_{c \in C} (\gamma_0 \wedge \nabla_c \Gamma_c)$.

We begin by constructing the model $M' = (S', R', V')$ where:

$$S' = \{s'\} \cup \bigcup_{c \in C} S^c$$

$$R'_c = (\{s'\} \cup s^c R^c_c)^2 \cup R_c \cup \bigcup_{d \in C} R^d_c \text{ for } c \in C$$

$$R'_c = (\{s'\} \cup sR_c)^2 \cup R_c \cup \bigcup_{d \in C} R^d_c \text{ for } c \notin C$$

$$V'(p) = \{s' \mid s \in V(p)\} \cup V(p) \cup \bigcup_{c \in C} V^c(p)$$

where $p \in P$ and s' is a new state that does not appear in S or any of the S^c.

We note that by construction $M' \in \mathcal{S}5$ and $M'_{s'} \preceq_a M_s$, via the a-simulation $\mathfrak{R}$, where $\mathfrak{R} = (s', s) \cup \bigcup_{b \in A} \mathfrak{R}^b$.

We must show that $M'_{s'} \models \bigwedge_{c \in C}(\gamma_0 \cup \nabla_c \Gamma_c)$. To do this we show by induction on the structure of formulae in Φ that for every $\phi \in \Phi$, $c \in C$, that: $M'_{s'} \models \phi$ if and only if $M^c_{s^c} \models \phi$, and for every $u \in S^c$ that $M'_u \models \phi$ if and only if $M^c_u \models \phi$. We use similar reasoning as used in the proof for **RS5**.

Therefore we have that $M'_{s'} \preceq_B M_s$ and $M'_{s'} \models \bigwedge_{c \in C}(\gamma_0 \cup \nabla_c \Gamma_c)$. Therefore **RDist** is sound.

Therefore $\mathbf{RML_{S5}}$ is sound with respect to the semantic class $\mathcal{S}5$. □

We show the completeness of the axiomatisation $\mathbf{RML_{S5}}$ in a similar fashion to the completeness proof of $\mathbf{RML_{KD45}}$, by a provably correct translation from $\mathcal{L}_\forall$ to $\mathcal{L}$. Completeness then follows from the completeness of *S5*.

As for the completeness proof for $\mathbf{RML_{KD45}}$, we introduce some similar equivalences that will be used by our translation.

Lemma 5.6 *The following are provable equivalences using* $\mathbf{RML_{S5}}$*:*

(i) $\exists_a(\phi \vee \psi) \leftrightarrow \exists_a \phi \vee \exists_a \psi$

(ii) $\exists_a(\pi \wedge \gamma_0 \wedge \bigwedge_{b \in A} \nabla_b \Gamma_b) \leftrightarrow \pi \wedge \exists_a \gamma_0 \wedge \exists_a \bigwedge_{\gamma \in \Gamma_a} \Diamond_a \exists_a \gamma \wedge \bigwedge_{b \in A} \nabla_b \{\exists_a \gamma \mid \gamma \in \Gamma_b\}$

where $\pi \wedge \gamma_0 \wedge \bigwedge_{b \in A} \nabla_b \Gamma_b$ *is an explicit formula.*

This can be shown by following similar reasoning as used for the proof of Lemma 4.10, but by substituting $\mathbf{RML_{S5}}$ axioms for $\mathbf{RML_{KD45}}$ axioms. We also rely on Lemma 5.3 to ensure that the result of applying **RDist** is a conjunction of $\exists_a$ operators applied to explicit formulae.

Lemma 5.7 *Every formula of* $S5_\forall$ *is provably equivalent to a formula of S5.*

This can be shown using similar reasoning to Lemma 4.11, but instead of rewriting modal subformulae into cover logic alernating disjunctive normal form, we use Lemma 5.2 to rewrite subformulae as explicit formulae, and instead of using the equivalences from Lemma 4.10, we use the equivalences from Lemma 5.6.

Theorem 5.8 *The axiomatisation* $\mathbf{RML_{S5}}$ *is sound and complete with respect to the semantic class* $\mathcal{S}5$.

Corollary 5.9 *The logic $S5_\forall$ is expressively equivalent to S5.*

Corollary 5.10 *The logic $S5_\forall$ is decidable.*

As in the case for $KD45_\forall$, the decision procedure for $S5_\forall$ is via the translation to *S5*, and this translation has a non-elementary complexity. Better decision procedures are left to future work.

References

[1] Ågotnes, T., P. Balbiani, H. van Ditmarsch and P. Seban, *Group announcement logic*, Journal of Applied Logic **8** (2010), pp. 62–81.

[2] Alchourrón, C., P. Gärdenfors and D. Makinson, *On the logic of theory change: Partial meet contraction and revision functions*, Journal of symbolic logic (1985), pp. 510–530.

[3] Balbiani, P., A. Baltag, H. van Ditmarsch, A. Herzig, T. Hoshi and T. de Lima, *What can we achieve by arbitrary announcements?: A dynamic take on fitch's knowability*, in: *Proceedings of the 11th conference on Theoretical aspects of rationality and knowledge*, ACM, 2007, pp. 42–51.

[4] Bílková, M., A. Palmigiano and Y. Venema, *Proof systems for the coalgebraic cover modality*, Advances in Modal Logic **7** (2008), pp. 1–21.

[5] Blackburn, P., M. de Rijke and Y. Venema, "Modal logic," Cambridge Univ Pr, 2002, 341–343 pp.

[6] D'Agostino, G. and G. Lenzi, *An axiomatization of bisimulation quantifiers via the mu-calculus*, Theor. Comput. Sci. **338** (2005), pp. 64–95.

[7] Dechesne, F. and Y. Wang, *To know or not to know: epistemic approaches to security protocol verification*, Synthese **177** (2010), pp. 51–76.

[8] van Ditmarsch, H., *The logic of pit*, Synthese **149** (2006), pp. 343–374.

[9] van Ditmarsch, H. and T. French, *Undecidability for arbitrary public announcement logic*, Advances in Modal Logic **7** (2008), pp. 23–42.

[10] van Ditmarsch, H. and T. French, *Simulation and information: Quantifying over epistemic events*, Knowledge Representation for Agents and Multi-Agent Systems (2009), pp. 51–65.

[11] van Ditmarsch, H., T. French and S. Pinchinat, *Future event logic: axioms and complexity*, in: *Advances in Modal Logic*, 2010, pp. 24–27.

[12] van Ditmarsch, H., W. van der Hoek and B. Kooi, "Dynamic epistemic logic," Springer Verlag, 2007.

[13] Hales, J., T. French and R. Davies, *Refinement quantified logics of knowledge*, Electronic Notes in Theoretical Computer Science **278** (2011), pp. 85–98.

[14] van der Hoek, W. and J. Meyer, "Epistemic logic for AI and computer science," Cambridge Univ Pr, 2004, 35–38 pp.

[15] Janin, D. and I. Walukiewicz, *Automata for the modal μ-calculus and related results*, Mathematical Foundations of Computer Science 1995 (1995), pp. 552–562.

On Modal Products with the Logic of 'Elsewhere'

Christopher Hampson Agi Kurucz

Department of Informatics
King's College London
Strand, London, WC2R 2LS, U.K.

Abstract

The finitely axiomatisable and decidable modal logic **Diff** of 'elsewhere' (or 'difference operator') is known to be quite similar to **S5**. Their validity problems have the same coNP complexity, and their Kripke frames have similar structures: equivalence relations for **S5**, and 'almost' equivalence relations, with the possibility of some irreflexive points, for **Diff**. However, their behaviour may differ dramatically as components of two-dimensional logics. Here we consider the decision problems of modal product logics of the form $L \times$ **Diff**. We present some cases where the transition from $L \times$ **S5** to $L \times$ **Diff** not only increases the complexity of the validity problem, but in fact introduces undecidability, sometimes even non-recursive enumerability.

Keywords: difference operator, products of modal logics, decision problems

1 Introduction

Von Wright's 'logic of elsewhere' [24] is the set **Diff** of propositional modal formulas that are valid in all *difference frames*, that is to say, relational structures $\mathfrak{F} = (W, R)$ where for all $u, v \in W$, uRv iff $u \neq v$. Segerberg [20] gives a complete axiomatisation of **Diff**: He shows that **Diff** is the smallest set of modal formulas (having $\Box$ and $\Diamond$ as modal operators) that is closed under the rules of Substitution, Modus Ponens and Necessitation $\varphi/\Box\varphi$, and contains all propositional tautologies and the formulas

$$\Box(p \to q) \to (\Box p \to \Box q)$$
$$p \to \Box\Diamond p$$
$$\Diamond\Diamond p \to (p \vee \Diamond p)$$

So an arbitrary Kripke frame for **Diff** may contain both reflexive and irreflexive points, but it is always symmetric and *pseudo-transitive*:

$$\forall x, y, z \, \big(R(x,y) \wedge R(y,z) \to (x = z \vee R(x,z))\big). \tag{1}$$

Note that it is not hard to see [3] that

every rooted frame for **Diff** is a p-morphic image of a difference frame. (2)

One can express the *universal modality* and the '*precisely one*' modality with the help of a difference diamond:

$$\forall\psi = \psi \wedge \neg\Diamond\neg\psi, \qquad \Diamond^{=1}\psi = (\psi \vee \Diamond\psi) \wedge \neg\Diamond(\psi \wedge \Diamond\psi).$$

Then, for any model $\mathfrak{M}$ over some difference frame (W, R), and any $x \in W$,

$$\begin{aligned} \mathfrak{M}, x \models \forall\psi &\quad \text{iff} \quad \mathfrak{M}, y \models \psi, \text{ for all } y \in W, \\ \mathfrak{M}, x \models \Diamond^{=1}\psi &\quad \text{iff} \quad |\{y \in W : \mathfrak{M}, y \models \psi\}| = 1. \end{aligned}$$

In this paper we take the first steps in investigating decision problems of two-dimensional product logics with **Diff**. We find them intriguing because of the following reason. It is known that in general the existence of a polynomial reduction of a logic L_1 to a logic L_2 does not imply that $L_1 \times L$ is polynomially reducible to $L_2 \times L$ (see e.g. [5, Remark 6.19]). However, in cases when there exist so called 'model level' reductions between L_1 and L_2, such reductions may be 'lifted' to the products (see Sections 2.8, 6.3 and 6.5 in [5]). Both **Diff** and the well-known modal logic **S5** of equivalence relations not only share the same coNP-complete validity problems [3,13], but in fact their frames closely resemble one another. So one might have hoped for such a 'liftable' reduction. However, here we present some cases where the transition from $L \times$ **S5** to $L \times$ **Diff** not only increases the complexity of the validity problem, but in fact introduces undecidability, sometimes even non-recursive enumerability.

The product construction as a combination method on modal logics was introduced in [19,21,6], and has been extensively studied ever since. Modal products are connected to several other multi-dimensional logical formalisms, see [5,12] for surveys and references. Here we discuss the following special case of the general construction: Given a bimodal frame $\mathfrak{F}_h = (W_h, R_h^1, R_h^2)$ and a unimodal frame $\mathfrak{F}_v = (W_v, R_v)$, their *product* is defined to be the 3-modal frame

$$\mathfrak{F}_h \times \mathfrak{F}_v = (W_h \times W_v, \bar{R}_h^1, \bar{R}_h^2, \bar{R}_v),$$

where $W_h \times W_v$ is the Cartesian product of W_h and W_v and, for all $x, x' \in W_h$, $y, y' \in W_v$, $i = 1, 2$,

$$\begin{aligned} (x, y)\bar{R}_h^i(x', y') &\quad \text{iff} \quad xR_h^i x' \text{ and } y = y', \\ (x, y)\bar{R}_v(x', y') &\quad \text{iff} \quad yR_v y' \text{ and } x = x'. \end{aligned}$$

Frames of this form will be called *product frames* throughout. Now let L_h be a Kripke complete bimodal logic in the language with boxes $\Box_h^1$, $\Box_h^2$ and diamonds $\Diamond_h^1$, $\Diamond_h^2$. Let L_v be a Kripke complete unimodal logic in the language with box $\Box_v$ and diamond $\Diamond_v$. Their product $L_h \times L_v$ is then the set of all 3-modal formulas, in the language having $\Box_h^1$, $\Box_h^2$, $\Box_v$ and $\Diamond_h^1$, $\Diamond_h^2$, $\Diamond_v$, that are valid in all product frames $\mathfrak{F}_h \times \mathfrak{F}_v$, where $\mathfrak{F}_h$ is a frame for L_h, and $\mathfrak{F}_v$ is a frame for L_v. (Here we assume that $\Box_h^i$ and $\Diamond_h^i$ are interpreted by $\bar{R}_h^i$, while $\Box_v$ and $\Diamond_v$ are interpreted by $\bar{R}_v$.) It is easy to see that in fact it is enough to consider *rooted* frames for both component logics [5, Prop.3.7]:

$$L_h \times L_v = \{\varphi : \varphi \text{ is valid in every } \mathfrak{F}_h \times \mathfrak{F}_v, \text{ where } \mathfrak{F}_h \text{ is a rooted frame for } L_h, \text{ and } \mathfrak{F}_v \text{ is a rooted frame for } L_v.\} \quad (3)$$

Our notation and terminology is mostly standard. However, we assume that the reader is familiar with basic notions of propositional multi-modal logic and its possible world (or relational) semantics, and we use these without explicit references. For concepts and statements not defined or proved here, consult, for example, [1,2].

2 Results

In this section we illustrate that the transition from $L \times \mathbf{S5}$ to $L \times \mathbf{Diff}$ might introduce undecidability, and, in some cases, even non-recursively enumerability.

Our first example of such an L is the logic $\mathbf{K}_u$, the bimodal logic of all Kripke frames of the form $\mathfrak{F} = (W, R, W \times W)$, that is, the first relation is arbitrary, and the second is the universal relation on W. By a standard unravelling argument, it can be shown that an arbitrary rooted frame for $\mathbf{K}_u$ is always a p-morphic image of some frame $(W, R, W \times W)$, where (W, R) is a disjoint union of irreflexive, intransitive trees. As the product construction on frames commutes with taking p-morphic images [5, Prop.3.10], by (2) and (3) we obtain that

$\mathbf{K}_u \times \mathbf{Diff}$ is determined by product frames $\mathfrak{F}_h \times \mathfrak{F}_v$, where $\mathfrak{F}_h = (W_h, R_h, W_h \times W_h)$ is such that (W_h, R_h) is a disjoint union of irreflexive, intransitive trees, and $\mathfrak{F}_v = (W_v, R_v)$ is a difference frame. (4)

The validity problem of $\mathbf{K}_u$ is EXPTIME-complete [23,10]. The following theorem is in contrast with the decidability of $\mathbf{K}_u \times \mathbf{S5}$ (see [25] and [5, Thm.6.58]):

Theorem 2.1 $\mathbf{K}_u \times \mathbf{Diff}$ *is undecidable.*

Proof. We reduce the undecidable *non-halting problem* for two-counter Minsky machines [15] to the $\mathbf{K}_u \times \mathbf{Diff}$-satisfiability problem.

A *two-counter Minsky machine* is a finite sequence of instructions $M = (I_0, \dots, I_T)$, where each I_t, for $t < T$, is from the set

$$\{\mathsf{zero}_i, \mathsf{inc}_i, \mathsf{dec}_i(j) : i = 0, 1,\ j \leq T\},$$

and $I_T = \mathsf{halt}$. A *configuration* of M is a triple (k, ℓ, m) of natural numbers, with k being the index of the current instruction, and ℓ, m the current contents of the two registers. The (unique) *computation* of M is the function $f_M : \omega \to (\omega \times \omega \times \omega)$ defined by taking $f_M(0) = (0, 0, 0)$, and if $f_M(n) = (k, \ell, m)$ then

$k \leq T$ and

$$f_M(n+1) = \begin{cases} (k+1, 0, m), & \text{if } I_k = \mathsf{zero}_0, \\ (k+1, \ell, 0), & \text{if } I_k = \mathsf{zero}_1, \\ (k+1, \ell+1, m), & \text{if } I_k = \mathsf{inc}_0, \\ (k+1, \ell, m+1), & \text{if } I_k = \mathsf{inc}_1, \\ (k+1, \ell-1, m), & \text{if } I_k = \mathsf{dec}_0(j) \text{ and } \ell > 0, \\ (k+1, \ell, m-1), & \text{if } I_k = \mathsf{dec}_1(j) \text{ and } m > 0, \\ (j, 0, m), & \text{if } I_k = \mathsf{dec}_0(j) \text{ and } \ell = 0, \\ (j, \ell, 0), & \text{if } I_k = \mathsf{dec}_1(j) \text{ and } m = 0, \\ (k, \ell, m) & \text{if } I_k = \mathsf{halt}. \end{cases}$$

We write $f_M(n) = (\mathsf{i}^M(n), \mathsf{c}_0^M(n), \mathsf{c}_1^M(n))$ to indicate the role of the numbers in the configurations. We define the halting number H_M of M as

$$H_M = \begin{cases} n+1, & \text{if } n \text{ is the smallest number with } I_{\mathsf{i}^M(n)} = \mathsf{halt}, \\ \omega, & \text{if there is no } n < \omega \text{ with } I_{\mathsf{i}^M(n)} = \mathsf{halt}. \end{cases}$$

We say that M *halts* if and only if $H_M < \omega$.

Now, given a Minsky machine M as above, we will define a 3-modal formula φ_M, whose length is recursive (in fact, linear) in T. We will use the language having $\Diamond_h, \Box_h$ for the 'horizontal' **K**-modalities, $\forall_h$ for the 'horizontal' universal modality, and $\Diamond_v$ for the 'vertical' difference operator. We will also use the following 'vertical' abbreviations:

$$\exists_v \psi = \psi \vee \Diamond_v \psi,$$
$$\forall_v \psi = \neg \exists_v \neg \psi,$$
$$\Diamond_v^{=1} \psi = (\psi \vee \Diamond_v \psi) \wedge \neg \Diamond_v (\psi \wedge \Diamond_v \psi).$$

The idea is to encode the configuration of M as M evolves over time; with each **K**-succession representing one time-step in the computation of M. We take two propositional variables c_0 and c_1 that will emulate the counters in each of the two registers of M: the number of points in each vertical **Diff**-cluster satisfying c_i will represent the contents of the ith register.

We introduce the following abbreviations, for $i = 0, 1$, that will dictate how the counters in each register are manipulated:

$$\psi_{\mathsf{inc}}(i) = \Diamond_v^{=1}(\neg c_i \wedge \Diamond_h c_i) \wedge \forall_v (c_i \to \Box_h c_i)$$
$$\psi_{\mathsf{dec}}(i) = \Diamond_v^{=1}(c_i \wedge \Diamond_h \neg c_i) \wedge \forall_v (\neg c_i \to \Box_h \neg c_i)$$
$$\psi_{\mathsf{fix}}(i) = \forall_v (c_i \leftrightarrow \Diamond_h c_i)$$
$$\psi_{\mathsf{zero}}(i) = \forall_v \Box_h \neg c_i$$

For example, $\psi_{\mathsf{inc}}(i)$ stipulates that there is *exactly one* vertically accessible point that evolves from satisfying $\neg c_i$ to satisfying c_i, while every vertically accessible point that satisfies c_i remains satisfying c_i in its **K**-successors; hence the number of points satisfying c_i is incremented by exactly one from one vertical **Diff**-cluster to the next.

We take a propositional variable s_t, for each $t \leq T$, to encode the internal state of M as M evolves over time, and define φ_M to be the conjunction of the following formulas:

$$s_0 \wedge \forall_v \neg c_0 \wedge \forall_v \neg c_1 \tag{5}$$

$$\forall_h \bigvee_{t \leq T} s_t \wedge \bigwedge_{t \neq t' \leq T} \forall_h \neg (s_t \wedge s_{t'}) \tag{6}$$

$$\bigwedge_{t \leq T} \forall_h (\Diamond_h s_t \to \Box_h s_t) \tag{7}$$

$$\bigwedge_{i=0,1} \forall_h \forall_v (\Diamond_h c_i \to \Box_h c_i) \tag{8}$$

$$\bigwedge_{\substack{t<T,\, i=0,1 \\ I_t = \mathsf{zero}_i}} \forall_h \big(s_t \to \Diamond_h s_{t+1} \wedge \psi_{\mathsf{zero}}(i) \wedge \psi_{\mathsf{fix}}(1-i)\big) \tag{9}$$

$$\bigwedge_{\substack{t<T,\, i=0,1 \\ I_t = \mathsf{inc}_i}} \forall_h \big(s_t \to \Diamond_h s_{t+1} \wedge \psi_{\mathsf{inc}}(i) \wedge \psi_{\mathsf{fix}}(1-i)\big) \tag{10}$$

$$\bigwedge_{\substack{t<T,\, i=0,1 \\ I_t = \mathsf{dec}_i(j)}} \forall_h \big((s_t \wedge \exists_v c_i) \to \Diamond_h s_{t+1} \wedge \psi_{\mathsf{dec}}(i) \wedge \psi_{\mathsf{fix}}(1-i)\big) \tag{11}$$

$$\bigwedge_{\substack{t<T,\, i=0,1 \\ I_t = \mathsf{dec}_i(j)}} \forall_h \big((s_t \wedge \forall_v \neg c_i) \to \Diamond_h s_j \wedge \psi_{\mathsf{fix}}(0) \wedge \psi_{\mathsf{fix}}(1)\big) \tag{12}$$

The first formula (5) encodes the initial configuration of M, while (6) stipulates that every point horizontally accessible from the root must always satisfy exactly one state variable s_t, for $t \leq T$. Formulas (7) and (8) ensure that any two distinct **K**-chains encode the same sequence of configurations. The remaining formulas (9)–(12) specify the behaviour of the machine, depending on the sequence of instructions set down by M.

Now suppose that φ_M is $\mathbf{K}_u \times \mathbf{Diff}$-satisfiable. By (4), we may assume that $\mathfrak{M}, (x_0, y_0) \models \varphi_M$ for some model $\mathfrak{M}$ based on a product frame $\mathfrak{F}_h \times \mathfrak{F}_v$, where $\mathfrak{F}_h = (W_h, R_h, W_h \times W_h)$ is such that (W_h, R_h) is a disjoint union of irreflexive, intransitive trees, and $\mathfrak{F}_v = (W_v, R_v)$ is a difference frame. The following notion is then well-defined, for every $x \in W_h$:

$$d(x_0, x) = \begin{cases} 0, & \text{if } x = x_0, \\ n, & \text{if there exist } w_0, \ldots, w_n \text{ with } x_0 = w_0 R_h \ldots R_h w_n = x, \\ \omega, & \text{otherwise.} \end{cases}$$

Now we have the following claim (where we use $(x, y) \models \psi$ as a shorthand for $\mathfrak{M}, (x, y) \models \psi$):

CLAIM **2.1.1** *If $(x_0, y_0) \models \varphi_M$, then for all $n < H_M$, for all $x \in W_h$ with $d(x_0, x) = n$, and for all $t \leq T$, the following hold:*

(i) $(x, y_0) \models s_t$ if and only if $t = \mathsf{i}^M(n)$,

(ii) $|\{y \in W_v : (x, y) \models c_0\}| = \mathsf{c}_0^M(n)$,

(iii) $|\{y \in W_v : (x, y) \models c_1\}| = \mathsf{c}_1^M(n)$.

(Here $|U|$ denotes the cardinality of set U.)

Proof. We prove this by induction on n. If $n = 0$, then the statements hold by (5) and (6). Now suppose for induction that the statements (i), (ii) and (iii) hold for all $x' \in W_h$ with $d(x_0, x') = n$, and let $x \in W_h$ be such that $d(x_0, x) = n + 1$. Then there is a unique $x' \in W_h$ such that $d(x_0, x') = n$ and $x' R_h x$. Let $t = \mathsf{i}^M(n)$. Then, by the IH, $(x', y_0) \models s_t$. There are six cases, depending on the form of I_t. Let us consider two examples:

- $I_t = \mathsf{zero}_0$:
  Then $\mathsf{i}^M(n+1) = \mathsf{i}^M(n) + 1 = t + 1$, $\mathsf{c}_0^M(n+1) = 0$, and $\mathsf{c}_1^M(n+1) = \mathsf{c}_1^M(n)$. By (9), we have $(x', y_0) \models \Diamond_h s_{t+1}$, and so by (6) and (7), (i) holds. Also by (9), we have $(x', y_0) \models \psi_{\mathsf{zero}}(0)$, and so $(x, y) \models \neg c_0$, for all $y \in W_v$, as required in (ii). Further, by (9), we have $(x', y_0) \models \psi_{\mathsf{fix}}(1)$, and so by (8), for all $y \in W_v$, $(x', y) \models c_1$ if and only if $(x, y) \models c_1$, as required in (iii).
- $I_t = \mathsf{dec}_1(j)$:
  Suppose first that $\mathsf{c}_1^M(n) > 0$. Then $(x', y_0) \models \exists_v c_1$, by the IH, and we have $\mathsf{i}^M(n+1) = \mathsf{i}^M(n) + 1 = t + 1$, $\mathsf{c}_0^M(n+1) = \mathsf{c}_0^M(n)$, and $\mathsf{c}_1^M(n+1) = \mathsf{c}_1^M(n) - 1$. By (11), we have $(x', y_0) \models \Diamond_h s_{t+1}$, and so by (6) and (7), (i) holds. Also by (11), we have $(x', y_0) \models \psi_{\mathsf{dec}}(1)$. Therefore, there is $y^* \in W_v$ with $(x', y^*) \models c_1 \wedge \Diamond_h \neg c_1$, and for all $y \in W_v$, $y \neq y^*$, $(x', y) \models c_1$ if and only if $(x, y) \models c_1$. By (8), we also have $(x, y^*) \models \neg c_1$, as required in (iii). Further, by (11), we have $(x', y_0) \models \psi_{\mathsf{fix}}(0)$, and so by (8), for all $y \in W_v$, $(x', y) \models c_0$ if and only if $(x, y) \models c_0$, as required in (ii).

  Now suppose that $\mathsf{c}_1^M(n) = 0$. Then $(x', y_0) \models \forall_v \neg c_1$, by the IH, and we have $\mathsf{i}^M(n+1) = j$, $\mathsf{c}_0^M(n+1) = \mathsf{c}_0^M(n)$, and $\mathsf{c}_1^M(n+1) = \mathsf{c}_1^M(n) = 0$. By (12), we have $(x', y_0) \models \Diamond_h s_j$, and so by (6) and (7), (i) holds. Also by (12), we have $(x', y_0) \models \psi_{\mathsf{fix}}(i)$, for $i = 0, 1$, and so by (8), for all $y \in W_v$, $(x', y) \models c_i$ if and only if $(x, y) \models c_i$, as required in (ii) and (iii).

The other cases are similar and are left to the reader. □

As a consequence of Claim 2.1.1, we obtain the following:

$$\text{if } \varphi_M \wedge \forall_h \neg s_T \text{ is } \mathbf{K}_u \times \mathbf{Diff}\text{-satisfiable, then } M \text{ does not halt.} \tag{13}$$

On the other hand, we also have that

$$\text{if } M \text{ does not halt, then } \varphi_M \wedge \forall_h \neg s_T \text{ is } \mathbf{K}_u \times \mathbf{Diff}\text{-satisfiable.} \tag{14}$$

Indeed, suppose that M does not halt. Let $\mathfrak{F}_h = (\omega, +1, \omega \times \omega)$, and let $\mathfrak{F}_v$ be the difference frame on ω. We define a model $\mathfrak{M}$ on $\mathfrak{F}_h \times \mathfrak{F}_v$ by taking, for all

$n, m < \omega$, $t \leq T$, and $i = 0, 1$,

$$\mathfrak{M}, (n, m) \models s_t \quad \text{iff} \quad \mathsf{i}^M(n) = t \text{ and } m = 0,$$
$$\mathfrak{M}, (n, m) \models c_i \quad \text{iff} \quad m < \mathsf{c}_i^M(n).$$

It is then straightforward to check that $\mathfrak{M}, (0, 0) \models \varphi_M \wedge \forall_h \neg s_T$. The theorem now follows from (13) and (14). □

Note that the class of all frames, for each of $\mathbf{K}_u$ and **Diff**, can be defined by a recursive set of first-order sentences in the frame-correspondence language. Therefore, the product logic $\mathbf{K}_u \times \mathbf{Diff}$ is *recursively enumerable* [6]. So Theorem 2.1 implies that $\mathbf{K}_u \times \mathbf{Diff}$ *lacks the effective* (or *bounded*) *finite model property:* The size of a frame necessary to falsify any given formula φ that does not belong to $\mathbf{K}_u \times \mathbf{Diff}$ cannot be bound by a function recursive in the length of φ. However, as $\mathbf{K}_u \times \mathbf{Diff}$ is not finitely axiomatisable [9], in principle it can happen that we cannot enumerate the finite frames for $\mathbf{K}_u \times \mathbf{Diff}$, and so $\mathbf{K}_u \times \mathbf{Diff}$ might have the (abstract) finite model property. It is easy to see that it does not have the finite model property w.r.t. product frames: For example, take the formula φ_M defined in the proof above for the two-counter Minsky machine $M = (\mathsf{inc}_0, \mathsf{dec}_1(0), \mathsf{halt})$.

Next, instead of frames with a universal modality as first components, we consider frames of the form (W, R, R^*), where (W, R) is an irreflexive, intransitive tree, and R^* is the *reflexive and transitive closure* of R. The modal operator corresponding to R^* is sometimes called *master modality*, or *common knowledge operator* in epistemic logics. Examples of logics determined by frames of this kind are

- $\mathbf{K}_C$, the bimodal logic of all frames of the form (W, R, R^*),
- $\mathbf{PTL}_{X\Box}$, the 'next-time, future' fragment of Propositional Temporal Logic over $(\omega, +1, \leq)$ as time-line, and
- $\mathbf{PDL}_1^-$, test-free Propositional Dynamic Logic with just one atomic program and its Kleene star closure.

Both $\mathbf{K}_C$ and $\mathbf{PDL}_1^-$ have the same EXPTIME-completeness as $\mathbf{K}_u$ [8,4,16], while $\mathbf{PTL}_{X\Box}$ is PSPACE-complete [22]. Furthermore, each of these logics (and, indeed, $\mathbf{K}_u$) are polynomially reducible to full **PDL** using 'model level' reductions that can be 'lifted' to products, see [5, Sections 6.3,6.5]. As is shown in [5, Thm.6.49], the validity problem of $\mathbf{PDL} \times \mathbf{S5}$ is decidable in coN2EXPTIME, and so $\mathbf{K}_C \times \mathbf{S5}$, $\mathbf{PTL}_{X\Box} \times \mathbf{S5}$, and $\mathbf{PDL}_1^- \times \mathbf{S5}$ are also decidable. (Note that all these logics are EXPSPACE-hard, and $\mathbf{PTL}_{X\Box} \times \mathbf{S5}$ is in fact EXPSPACE-complete, see [5, Thms.6.65,6.66].)

Theorem 2.2 *Let $\mathcal{C}$ be any class of frames such that*

- *every frame in $\mathcal{C}$ is of the form $\mathfrak{F}_h \times \mathfrak{F}_v$, where $\mathfrak{F}_h = (W, R, R^*)$ with (W, R) being an irreflexive, intransitive tree, and $\mathfrak{F}_v$ is a difference frame;*
- $(\omega, +1, \leq) \in \mathcal{C}$.

Let L be the set of all 3-modal formulas that are valid in all frames in $\mathcal{C}$. Then

L is not recursively enumerable.

Proof. We reduce the undecidable but recursively enumerable *halting problem* for two-counter Minsky machines to the L-satisfiability problem.

Let M be a Minsky machine, and let φ_M be the formula defined in the proof of Theorem 2.1. It is straightforward to see that $\varphi_M \wedge \neg\forall_h\neg s_T$ is satisfiable in a frame in $\mathcal{C}$ if and only if M halts. (Here the notation $\forall_h$ is a bit misleading, as the corresponding relation is not 'horizontally' universal any more, but the reflexive and transitive closure of the relation corresponding to $\Box_h$.) □

Corollary 2.3 *The logics* $\mathbf{K}_C \times \mathbf{Diff}$, $\mathbf{PTL}_{X\Box} \times \mathbf{Diff}$, *and* $\mathbf{PDL}_1^- \times \mathbf{Diff}$ *are not recursively enumerable.*

3 Discussion

We conclude the paper with a few remarks on related formalisms and further research.

- As **Diff** can be regarded as a fragment of hybrid logic, our Theorems 2.1 and 2.2 imply the undecidability of some hybrid product logics (see [18]).
- We gave examples of bimodal logics L where $L \times \mathbf{Diff}$ is undecidable. Products of standard decidable unimodal logics and **Diff** have not been investigated. Is there an example for undecidability among them? In particular, is any of $\mathbf{K} \times \mathbf{Diff}$, $\mathbf{K4} \times \mathbf{Diff}$, $\mathbf{S5} \times \mathbf{Diff}$, or $\mathbf{Diff} \times \mathbf{Diff}$ decidable? Each of these products with **S5** in place of **Diff** is known to be decidable, logics like $\mathbf{K} \times \mathbf{S5}$ and $\mathbf{S5} \times \mathbf{S5}$ are even coNEXPTIME-complete. Even if these kinds of products with **Diff** turn out to be decidable, we cannot always hope for proofs that are completely analogous to the **S5**-cases: in contrast to the product finite model property of, say, $\mathbf{K} \times \mathbf{S5}$ or $\mathbf{S5} \times \mathbf{S5}$, it turns out that $\mathbf{Diff} \times \mathbf{Diff}$ has no (abstract) finite model property [9]. Concerning attempts at filtration arguments, note that no logic of the form $L \times \mathbf{Diff}$ is finitely axiomatisable, whenever L is between **K** and **S5** [9].
- We proved Theorems 2.1 and 2.2 using reductions of the halting problem for two-counter Minsky machines. It appears that this technique is slightly different to other proofs of undecidability results about product logics, which use reductions of the halting problem for Turing machines or $\omega \times \omega$-tilings. One might think that these latter undecidable problems are tailor-made for product logics: product frames are by definition grid-like, so it should not be hard to encode these 'grid-based' problems into them. Indeed, if we have both next-time and universal or master modalities in both dimensions, then this is rather straightforward (see, for example, the case of $\mathbf{K}_u \times \mathbf{K}_u$ in [5, Thm.5.37]). However, if some of this machinery is missing, then often the grid needs to be encoded by 'diagonal' points, and some other tricks may be needed [7,14,17,11]. We failed to apply these kinds of tricks in the undecidability proofs given here. In order to understand the boundaries of each technique, it would be interesting to know whether there is a natural way to fully encode the $\omega \times \omega$-grid in frames for $L \times \mathbf{Diff}$-logics.

References

[1] Blackburn, P., M. de Rijke and Y. Venema, "Modal Logic," Cambridge University Press, 2001.

[2] Chagrov, A. and M. Zakharyaschev, "Modal Logic," Oxford Logic Guides **35**, Clarendon Press, Oxford, 1997.

[3] de Rijke, M., *The modal logic of inequality*, Journal of Symbolic Logic **57** (1992), pp. 566–584.

[4] Fischer, M. and R. Ladner, *Propositional dynamic logic of regular programs*, Journal of Computer and System Sciences **18** (1979), pp. 194–211.

[5] Gabbay, D., A. Kurucz, F. Wolter and M. Zakharyaschev, "Many-Dimensional Modal Logics: Theory and Applications," Studies in Logic and the Foundations of Mathematics **148**, Elsevier, 2003.

[6] Gabbay, D. and V. Shehtman, *Products of modal logics. Part I*, Journal of the IGPL **6** (1998), pp. 73–146.

[7] Gabelaia, D., A. Kurucz, F. Wolter and M. Zakharyaschev, *Products of 'transitive' modal logics*, Journal of Symbolic Logic **70** (2005), pp. 993–1021.

[8] Halpern, J. and Y. Moses, *A guide to completeness and complexity for modal logics of knowledge and belief*, Artificial Intelligence **54** (1992), pp. 319–379.

[9] Hampson, C. and A. Kurucz, *Axiomatisation and decision problems of modal product logics with the difference operator* (2012), (manuscript).

[10] Hemaspaandra, E., *The price of universality*, Notre Dame Journal of Formal Logic **37** (1996), pp. 174–203.

[11] Kikot, S. and A. Kurucz, *Undecidable two dimensional modal product logics with diagonal constant* (2012), (submitted).

[12] Kurucz, A., *Combining modal logics*, in: P. Blackburn, J. van Benthem and F. Wolter, editors, *Handbook of Modal Logic*, Studies in Logic and Practical Reasoning **3**, Elsevier, 2007 pp. 869–924.

[13] Ladner, R., *The computational complexity of provability in systems of modal logic*, SIAM Journal on Computing **6** (1977), pp. 467–480.

[14] Marx, M. and M. Reynolds, *Undecidability of compass logic*, Journal of Logic and Computation **9** (1999), pp. 897–914.

[15] Minsky, M., "Finite and infinite machines," Prentice-Hall, 1967.

[16] Pratt, V., *Models of program logics*, in: *Proceedings of the 20th IEEE Symposium on Foundations of Computer Science*, 1979, pp. 115–122.

[17] Reynolds, M. and M. Zakharyaschev, *On the products of linear modal logics*, Journal of Logic and Computation **11** (2001), pp. 909–931.

[18] Sano, K., *Axiomatizing hybrid products: How can we reason many-dimensionally in hybrid logic?*, Journal of Applied Logic **8** (2010), pp. 459–474.

[19] Segerberg, K., *Two-dimensional modal logic*, Journal of Philosophical Logic **2** (1973), pp. 77–96.

[20] Segerberg, K., *A note on the logic of elsewhere*, Theoria **46** (1980), pp. 183–187.

[21] Shehtman, V., *Two-dimensional modal logics*, Mathematical Notices of the USSR Academy of Sciences **23** (1978), pp. 417–424, (Translated from Russian).

[22] Sistla, A. and E. Clarke, *The complexity of propositional linear temporal logics*, Journal of the Association for Computing Machinery **32** (1985), pp. 733–749.

[23] Spaan, E., "Complexity of Modal Logics," Ph.D. thesis, Department of Mathematics and Computer Science, University of Amsterdam (1993).

[24] von Wright, G., *A modal logic of place*, in: E. Sosa, editor, *The philosophy of Nicolas Rescher*, Dordrecht, 1979 pp. 65–73.

[25] Wolter, F. and M. Zakharyaschev, *Modal description logics: modalizing roles*, Fundamenta Informaticae **39** (1999), pp. 411–438.

A Uniform Logic of Information Dynamics

Wesley H. Holliday Tomohiro Hoshi Thomas F. Icard, III

Logical Dynamics Lab, Center for the Study of Language and Information
Cordura Hall, 210 Panama Street, Stanford, CA 94305
Department of Philosophy, Building 90, Stanford University, Stanford, CA 94305

Abstract

Unlike standard modal logics, many dynamic epistemic logics are not closed under uniform substitution. A distinction therefore arises between the logic and its *substitution core*, the set of formulas all of whose substitution instances are valid. The classic example of a non-uniform dynamic epistemic logic is Public Announcement Logic (PAL), and a well-known open problem is to axiomatize the substitution core of PAL. In this paper we solve this problem for PAL over the class of all relational models with infinitely many agents, PAL-$\mathbf{K}_\omega$, as well as standard extensions thereof, e.g., PAL-$\mathbf{T}_\omega$, PAL-$\mathbf{S4}_\omega$, and PAL-$\mathbf{S5}_\omega$. We introduce a new Uniform Public Announcement Logic (UPAL), prove completeness of a deductive system with respect to UPAL semantics, and show that this system axiomatizes the substitution core of PAL.

Keywords: dynamic epistemic logic, Public Announcement Logic, schematic validity, substitution core, uniform substitution

1 Introduction

One of the striking features of many of the *dynamic epistemic logics* [28,19,13,9,4] studied in the last twenty years is the failure of closure under *uniform substitution* in these systems. Given a valid principle of information dynamics in such a system, uniformly substituting complex epistemic formulas for atomic sentences in the principle may result in an *invalid* instance. Such failures of closure under uniform substitution turn out to reveal insights into the nature of information change [1,7,11,24,8]. They also raise the question: what are the more robust principles of information dynamics that are valid in all instances, that are *schematically* valid? Even for the simplest system of dynamic epistemic logic, Public Announcement Logic (PAL) [28], the answer has been unknown. In van Benthem's "Open Problems in Logical Dynamics" [3], Question 1 is whether the set of schematic validities of PAL is axiomatizable. [1]

[1] Dynamic epistemic logics are not the only non-uniform modal logics to have been studied. Other examples include Buss's [16] modal logic of "pure provability," Åqvist's [10] two-dimensional modal logic (see [31]), Carnap's [17] modal system for logical necessity (see [12,30]), an epistemic-doxastic logic proposed by Halpern [21], and the full computation tree logic CTL* (see [29]). Among propositional logics, inquisitive logic [27,18] is a non-uniform

In this paper, we give an axiomatization of the set of schematic validities—or *substitution core*—of PAL over the class of all relational models with infinitely many agents, PAL-$\mathbf{K}_\omega$, as well as standard extensions thereof, e.g., PAL-$\mathbf{T}_\omega$, PAL-$\mathbf{S4}_\omega$, and PAL-$\mathbf{S5}_\omega$. After reviewing the basics of PAL in §1.1, we introduce the idea of Uniform Public Announcement Logic (UPAL) in §1.2, prove completeness of a UPAL deductive system in §3 with respect to alternative semantics introduced in §2, and show that it axiomatizes the substitution core of PAL in §4. In §5, we demonstrate our techniques with examples, and in §6 we conclude by discussing extensions of these techniques to other logics.

Although much could be said about the conceptual significance of UPAL as a uniform logic of information dynamics, here we only present the formal results. For conceptual discussion of PAL, we refer the reader to the textbooks [9,4]. Our work here supports a theme of other recent work in dynamic epistemic logic: despite its apparent simplicity, PAL and its variants prove to be a rich source for mathematical investigation (see, e.g., [3,2,25,24,22,32,26,5,23,33]).

1.1 Review of PAL

We begin our review of PAL with the language we will use throughout.

Definition 1.1 *For a set* At *of atomic sentences and a set* Agt *of agent symbols with* $|\mathsf{Agt}| = \kappa$*, the language* $\mathcal{L}^{\kappa}_{\mathsf{PAL}}$ *is generated by the following grammar:*

$$\varphi ::= \top \mid p \mid \neg\varphi \mid \varphi \wedge \varphi \mid \Diamond_a \varphi \mid \langle\varphi\rangle\varphi,$$

where $p \in \mathsf{At}$ *and* $a \in \mathsf{Agt}$. *We define* $\Box_a\varphi$ *as* $\neg\Diamond_a\neg\varphi$ *and* $[\varphi]\psi$ *as* $\neg\langle\varphi\rangle\neg\psi$.

- $\mathsf{Sub}(\varphi)$ *is the set of subformulas of* φ;
- $\mathsf{At}(\varphi) = \mathsf{At} \cap \mathsf{Sub}(\varphi)$;
- $\mathsf{Agt}(\varphi) = \{a \in \mathsf{Agt} \mid \Diamond_a\psi \in \mathsf{Sub}(\varphi)$ *for some* $\psi \in \mathcal{L}^{\kappa}_{\mathsf{PAL}}\}$;
- $\mathsf{An}(\varphi) = \{\chi \in \mathcal{L}^{\kappa}_{\mathsf{PAL}} \mid \langle\chi\rangle\psi \in \mathsf{Sub}(\varphi)$ *for some* $\psi \in \mathcal{L}^{\kappa}_{\mathsf{PAL}}\}$.

We will be primarily concerned with the language $\mathcal{L}^{\omega}_{\mathsf{PAL}}$ with infinitely many agents, which leads to a more elegant treatment than $\mathcal{L}^{n}_{\mathsf{PAL}}$ for some arbitrary finite n. In §6 we will briefly discuss the single-agent and finite-agent cases.

We will consider two interpretations of $\mathcal{L}^{\kappa}_{\mathsf{PAL}}$, one now and one in §2. The standard interpretation uses the following models and truth definition.

Definition 1.2 *Models for PAL are tuples of the form* $\mathcal{M} = \langle W, \{R_a\}_{a\in\mathsf{Agt}}, V\rangle$*, where* W *is a non-empty set,* R_a *is a binary relation on* W*, and* $V\colon \mathsf{At} \to \mathcal{P}(W)$.

Definition 1.3 *Given a PAL model* $\mathcal{M} = \langle W, \{R_a\}_{a\in\mathsf{Agt}}, V\rangle$ *with* $w \in W$,

---

example. In some of these cases, the schematically valid fragment—or *substitution core*—turns out to be another known system. For example, the substitution core of Carnap's system **C** is **S5** [30], and the substitution core of inquisitive logic is Medvedev Logic [18, §3.4].

$\varphi, \psi \in \mathcal{L}^{\kappa}_{\mathsf{PAL}}$, and $p \in \mathsf{At}$, we define $\mathcal{M}, w \vDash \varphi$ as follows:

$$
\begin{array}{lll}
\mathcal{M}, w \vDash \top; & & \\
\mathcal{M}, w \vDash p & \text{iff} & w \in V(p); \\
\mathcal{M}, w \vDash \neg\varphi & \text{iff} & \mathcal{M}, w \nvDash \varphi; \\
\mathcal{M}, w \vDash \varphi \wedge \psi & \text{iff} & \mathcal{M}, w \vDash \varphi \text{ and } \mathcal{M}, w \vDash \psi; \\
\mathcal{M}, w \vDash \Diamond_a \varphi & \text{iff} & \exists v \in W \colon wR_a v \text{ and } \mathcal{M}, v \vDash \varphi; \\
\mathcal{M}, w \vDash \langle\varphi\rangle\psi & \text{iff} & \mathcal{M}, w \vDash \varphi \text{ and } \mathcal{M}_{|\varphi}, w \vDash \psi,
\end{array}
$$

where $\mathcal{M}_{|\varphi} = \langle W_{|\varphi}, \{R_{a_{|\varphi}}\}_{a \in \mathsf{Agt}}, V_{|\varphi}\rangle$ is the model such that

$$
\begin{array}{l}
W_{|\varphi} = \{v \in W \mid \mathcal{M}, v \vDash \varphi\}; \\
\forall a \in \mathsf{Agt} \colon R_{a_{|\varphi}} = R_a \cap (W_{|\varphi} \times W_{|\varphi}); \\
\forall p \in \mathsf{At} \colon V_{|\varphi}(p) = V(p) \cap W_{|\varphi}.
\end{array}
$$

We use the notation $[\![\varphi]\!]^{\mathcal{M}} = \{v \in W \mid \mathcal{M}, v \vDash \varphi\}$. For a class of models C, $Th_{\mathcal{L}^{\kappa}_{\mathsf{PAL}}}(\mathsf{C})$ is the set of formulas of $\mathcal{L}^{\kappa}_{\mathsf{PAL}}$ that are valid over C.

For the following statements, we use the standard nomenclature for normal modal logics, e.g., **K**, **T**, **S4**, and **S5** for the unimodal logics and $\mathbf{K}_\kappa$, $\mathbf{T}_\kappa$, $\mathbf{S4}_\kappa$, and $\mathbf{S5}_\kappa$ for their multimodal versions with $|\mathsf{Agt}| = \kappa$ (assume κ countable). Let $\mathrm{Mod}(\mathbf{L}_\kappa)$ be the class of all models of the logic $\mathbf{L}_\kappa$, so $\mathrm{Mod}(\mathbf{K}_\kappa)$ is the class of all models, $\mathrm{Mod}(\mathbf{T}_\kappa)$ is the class of models with reflexive R_a relations, etc. We write $\mathtt{L}_\kappa$ for the Hilbert-style deductive system whose set of theorems is $\mathbf{L}_\kappa$, and for any deductive system $\mathtt{S}$, we write $\vdash_{\mathtt{S}} \varphi$ when φ is a theorem of $\mathtt{S}$.

Theorem 1.4 (PAL Axiomatization [28]) *Let $\mathtt{PAL\text{-}L}_\kappa$ be the system extending $\mathtt{L}_\kappa$ with the following rule and axioms:*[2]

i.	*(replacement)*	$\dfrac{\psi \leftrightarrow \chi}{\varphi(\psi/p) \leftrightarrow \varphi(\chi/p)}$
ii.	*(atomic reduction)*	$\langle\varphi\rangle p \leftrightarrow (\varphi \wedge p)$
iii.	*(negation reduction)*	$\langle\varphi\rangle\neg\psi \leftrightarrow (\varphi \wedge \neg\langle\varphi\rangle\psi)$
iv.	*(conjunction reduction)*	$\langle\varphi\rangle(\psi \wedge \chi) \leftrightarrow (\langle\varphi\rangle\psi \wedge \langle\varphi\rangle\chi)$
v.	*(diamond reduction)*	$\langle\varphi\rangle\Diamond_a\psi \leftrightarrow (\varphi \wedge \Diamond_a\langle\varphi\rangle\psi)$.

For all $\varphi \in \mathcal{L}^{\kappa}_{\mathsf{PAL}}$,

$$\vdash_{\mathtt{PAL\text{-}K}_\kappa} \varphi \textit{ iff } \varphi \in Th_{\mathcal{L}^{\kappa}_{\mathsf{PAL}}}(Mod(\mathbf{K}_\kappa)).$$

The same result holds for $\mathtt{T}_\kappa/\boldsymbol{T}_\kappa$, $\mathtt{S4}_\kappa/\boldsymbol{S4}_\kappa$, and $\mathtt{S5}_\kappa/\boldsymbol{S5}_\kappa$ in place of $\mathtt{K}_\kappa/\boldsymbol{K}_\kappa$.

---

[2] If $\mathtt{L}_\kappa$ contains the rule of uniform substitution, then we must either restrict this rule so that in $\mathtt{PAL\text{-}L}_\kappa$ we can only substitute into formulas φ with $\mathsf{An}(\varphi) = \emptyset$, or remove the rule and add for each axiom of $\mathtt{L}_\kappa$ all substitution instances of that axiom with formulas in $\mathcal{L}^{\kappa}_{\mathsf{PAL}}$. Either way, we take the rules of modus ponens and $\Box_a$-necessitation from $\mathtt{L}_\kappa$ to apply in $\mathtt{PAL\text{-}L}_\kappa$ to all formulas. Finally, for $\varphi, \psi \in \mathcal{L}^{\kappa}_{\mathsf{PAL}}$ and $p \in \mathsf{At}(\varphi)$, $\varphi(\psi/p)$ is the formula obtained by replacing all occurrences of p in φ by ψ. For alternative axiomatizations of PAL, see [32,33].

Although we have taken diamond operators as primitive for convenience in later sections, typically the PAL axiomatization is stated in terms of box operators by replacing axiom schemas ii - v by the following: $[\varphi]p \leftrightarrow (\varphi \to p)$; $[\varphi]\neg\psi \leftrightarrow (\varphi \to \neg[\varphi]\psi)$; $[\varphi](\psi \wedge \chi) \leftrightarrow ([\varphi]\psi \wedge [\varphi]\chi)$; $[\varphi]\Box_a\psi \leftrightarrow (\varphi \to \Box_a[\varphi]\psi)$.

1.2 Introduction to UPAL

As noted above, one of the striking features of PAL is that it is not closed under uniform substitution. In the terminology of Goldblatt [20], PAL is not a *uniform* modal logic. For example, the valid atomic reduction axiom has invalid substitution instances, e.g., $\langle p\rangle\Box_a p \leftrightarrow (p \wedge \Box_a p)$. Given this observation, a distinction arises between PAL and its *substitution core*, defined as follows.

Definition 1.5 *A substitution is any* $\sigma\colon \mathsf{At} \to \mathcal{L}^{\kappa}_{\mathsf{PAL}}$*; and* $(\cdot)^{\sigma}\colon \mathcal{L}^{\kappa}_{\mathsf{PAL}} \to \mathcal{L}^{\kappa}_{\mathsf{PAL}}$ *is the extension such that* $(\varphi)^{\sigma}$ *is obtained from* φ *by replacing each* $p \in \mathsf{At}(\varphi)$ *by* $\sigma(p)$ *[14, Def. 1.18]. The* substitution core *of* $Th_{\mathcal{L}^{\kappa}_{\mathsf{PAL}}}(\mathsf{C})$ *is the set*

$$\{\varphi \in \mathcal{L}^{\kappa}_{\mathsf{PAL}}\colon (\varphi)^{\sigma} \in Th_{\mathcal{L}^{\kappa}_{\mathsf{PAL}}}(\mathsf{C}) \text{ for all substitutions } \sigma\}.$$

Formulas in the substitution core of $Th_{\mathcal{L}^{\kappa}_{\mathsf{PAL}}}(\mathsf{C})$ *are* schematically valid *over* C.

Examples of formulas that are in $\mathrm{Th}_{\mathcal{L}^{\kappa}_{\mathsf{PAL}}}(\mathrm{Mod}(\mathbf{K}_{\kappa}))$ but are not in the substitution core of $\mathrm{Th}_{\mathcal{L}^{\kappa}_{\mathsf{PAL}}}(\mathrm{Mod}(\mathbf{K}_{\kappa}))$ include the following (for $\kappa \geq 1$): [3]

$[p]p$	$\Box_a p \to [p]\Box_a p$
$[p]\Box_a p$	$\Box_a p \to [p](p \to \Box_a p)$
$[p](p \to \Box_a p)$	$\Box_a(p \to q) \to (\langle q\rangle\Box_a r \to \langle p\rangle\Box_a r)$
$[p \wedge \neg\Box_a p]\neg(p \wedge \neg\Box_a p)$	$(\langle p\rangle\Box_a r \wedge \langle q\rangle\Box_a r) \to \langle p \vee q\rangle\Box_a r.$

We discuss the epistemic significance of such failures of uniformity in [23]. Burgess [15] explains the logical significance of uniformity as follows:

> The standard aim of logicians at least from Russell onward has been to characterize the class [of] all formulas *all of whose instantiations are true.* Thus, though Russell was a logical atomist, when he endorsed $p \vee {\sim} p$ as [a] law of logic, he did not mean to be committing himself only to the view that the disjunction of any *logically atomic* statement with its negation is true, but rather to be committing himself to the view that the disjunction of *any statement whatsoever* with its negation is true .... This has remained the standard employment of statement letters ever since, not only among Russell's successors in the classical tradition, but also among the great majority of formal logicians who have thought classical logic to be in need of additions and/or amendments, including C. I. Lewis, the founder of modern modal logic. With such an understanding of the role of statement letters, it is clear that if A is a law of logic, and B is any substitution in A, then B also is a law of logic .... Thus it is that the rule of substitution applies not

[3] The first two principles in the second column are schematically valid over transitive single-agent models, but not over all single-agent models or over transitive multi-agent models.

> only in classical logic, but in standard, Lewis-style modal logics (as well as in intuitionistic, temporal, relevance, quantum, and other logics). None of this is meant to deny that there may be circumstances where it is legitimate to adopt some other understanding of the role of statement letters. If one does so, however, it is indispensable to note the conceptual distinction, and highly advisable to make a notational and terminological distinction. (147-148)

In PAL, an atomic sentence p has the same truth value at any pointed models $\mathcal{M}, w$ and $\mathcal{M}_{|\varphi}, w$, whereas a formula containing a modal operator may have different truth values at $\mathcal{M}, w$ and $\mathcal{M}_{|\varphi}, w$, which is why uniform substitution does not preserve PAL-validity. Hence in PAL an atomic sentence cannot be thought of as a *propositional variable* in the ordinary sense of something that stands in for any proposition. By contrast, if we consider the substitution core of PAL as a logic in its own right, for which semantics will be given in §2, then we can think of the atomic sentences as genuine propositional variables.

The distinction between PAL and its substitution core leads to Question 1 in van Benthem's list of "Open Problems in Logical Dynamics" [3]:

Question 1 ([2,3,4]) *Is the substitution core of PAL axiomatizable?*

To answer this question, we will introduce a new framework of Uniform Public Announcement Logic (UPAL), which we use to prove the following result.

Theorem 1.6 (Axiomatization of the PAL Substitution Core) *Let* $\mathtt{UPAL\text{-}L}_{\mathcal{K}}$ *be the system extending* $\mathtt{L}_{\mathcal{K}}$ *with the following rules and axioms:*[4]

1. (*uniformity*) $\dfrac{\varphi}{(\varphi)^{\sigma}}$ *for any substitution* σ
2. (*necessitation*) $\dfrac{\varphi}{[p]\varphi}$
3. (*extensionality*) $\dfrac{\varphi \leftrightarrow \psi}{\langle\varphi\rangle p \leftrightarrow \langle\psi\rangle p}$
4. (*distribution*) $[p](q \to r) \to ([p]q \to [p]r)$
5. (*p-seriality*) $p \to \langle p\rangle\top$
6. (*truthfulness*) $\langle p\rangle\top \to p$
7. ($\top$*-reflexivity*) $p \to \langle\top\rangle p$
8. (*functionality*) $\langle p\rangle q \to [p]q$
9. (*pa-commutativity*) $\langle p\rangle\Diamond_a q \to \Diamond_a\langle p\rangle q$
10. (*ap-commutativity*) $\Diamond_a\langle p\rangle q \to [p]\Diamond_a q$
11. (*composition*) $\langle p\rangle\langle q\rangle r \leftrightarrow \langle\langle p\rangle q\rangle r$.

[4] As in $\mathtt{PAL\text{-}L}_{\mathcal{K}}$, in $\mathtt{UPAL\text{-}L}_{\mathcal{K}}$ we take the rules of modus ponens and $\Box_a$-necessitation from $\mathtt{L}_{\mathcal{K}}$ to apply to all formulas in $\mathcal{L}^{\mathcal{K}}_{\mathsf{PAL}}$.

For all $\varphi \in \mathcal{L}^{\omega}_{\mathsf{PAL}}$,

$$\vdash_{\mathtt{UPAL\text{-}K}_\omega} \varphi \textit{ iff } \varphi \textit{ is in the substitution core of } Th_{\mathcal{L}^{\omega}_{\mathsf{PAL}}}(Mod(\mathbf{K}_\omega)).$$

The same result holds for $\mathtt{T}_\omega/\mathbf{T}_\omega$, $\mathtt{S4}_\omega/\mathbf{S4}_\omega$, and $\mathtt{S5}_\omega/\mathbf{S5}_\omega$ in place of $\mathtt{K}_\omega/\mathbf{K}_\omega$, with only minor adjustments to the proof (see note 5).

Theorem 1.7 (Axiomatization of the PAL Substitution Core cont.)

1. $\vdash_{\mathtt{UPAL\text{-}T}_\omega} \varphi$ *iff* φ *is in the substitution core of* $Th_{\mathcal{L}^{\omega}_{\mathsf{PAL}}}(Mod(\mathbf{T}_\omega))$;
2. $\vdash_{\mathtt{UPAL\text{-}S4}_\omega} \varphi$ *iff* φ *is in the substitution core of* $Th_{\mathcal{L}^{\omega}_{\mathsf{PAL}}}(Mod(\mathbf{S4}_\omega))$;
3. $\vdash_{\mathtt{UPAL\text{-}S5}_\omega} \varphi$ *iff* φ *is in the substitution core of* $Th_{\mathcal{L}^{\omega}_{\mathsf{PAL}}}(Mod(\mathbf{S5}_\omega))$.

Unless the specific base system $\mathtt{L}_\kappa$ matters, we simply write 'UPAL' and 'PAL'. It is easy to check that all the axioms of PAL except atomic reduction are derivable in UPAL, and the rule of replacement is an admissible rule in UPAL. Another system with the same theorems as UPAL, but presented in a format closer to that of the typical box version of PAL, is the following (with $\bot := \neg\top$):

I.	(uniformity)	$\dfrac{\varphi}{(\varphi)^\sigma}$ for any substitution σ
II.	(RE)	$\dfrac{\varphi \leftrightarrow \psi}{[p]\varphi \leftrightarrow [p]\psi}$
III.	([ ]-extensionality)	$\dfrac{\varphi \leftrightarrow \psi}{[\varphi]p \leftrightarrow [\psi]p}$
IV.	(N)	$[p]\top$
V.	($\top$-reflexivity)	$[\top]p \to p$
VI.	($\bot$-reduction)	$[p]\bot \leftrightarrow \neg p$
VII.	($\neg$-reduction)	$[p]\neg q \leftrightarrow (p \to \neg[p]q)$
VIII.	($\wedge$-reduction)	$[p](q \wedge r) \leftrightarrow ([p]q \wedge [p]r)$
IX.	($\Box_a$-reduction)	$[p]\Box_a q \leftrightarrow (p \to \Box_a[p]q)$
X.	([ ]-composition)	$[p][q]r \leftrightarrow [p \wedge [p]q]r$.

We have formulated UPAL as in Theorem 1.6 to make clear the correspondence between axioms and the semantic conditions in Definition 2.3 below, as well as to make clear the specific properties used in the steps of our main proof.

2 Semantics for UPAL

In this section we introduce semantics for Uniform Public Announcement Logic, for which the system of UPAL is shown to be sound and complete in §3.

Definition 2.1 *Models for UPAL are tuples* $\mathfrak{M}$ *of the form* $\langle M, \{\mathcal{R}_a\}_{a\in\mathsf{Agt}}, \{\mathcal{R}_\varphi\}_{\varphi\in\mathcal{L}^{\kappa}_{\mathsf{PAL}}}, \mathcal{V}\rangle$, *where* M *is a non-empty set,* $\mathcal{R}_a$ *and* $\mathcal{R}_\varphi$ *are binary relations on* M, *and* $\mathcal{V}\colon \mathsf{At} \to \mathcal{P}(M)$.

Unlike in the PAL truth definition, in the UPAL truth definition we treat $\langle\varphi\rangle$ like any other modal operator.

Definition 2.2 *Given a UPAL model* $\mathfrak{M} = \langle M, \{\mathcal{R}_a\}_{a\in\mathsf{Agt}}, \{\mathcal{R}_\varphi\}_{\varphi\in\mathcal{L}^\kappa_{\mathsf{PAL}}}, \mathcal{V}\rangle$ *with* $w \in M$, $\varphi, \psi \in \mathcal{L}^\kappa_{\mathsf{PAL}}$, *and* $p \in \mathsf{At}$, *we define* $\mathfrak{M}, w \Vdash \varphi$ *as follows:*

$$
\begin{array}{lll}
\mathfrak{M}, w \Vdash \top; & & \\
\mathfrak{M}, w \Vdash p & \text{iff} & w \in \mathcal{V}(p); \\
\mathfrak{M}, w \Vdash \neg\varphi & \text{iff} & \mathfrak{M}, w \nVdash \varphi; \\
\mathfrak{M}, w \Vdash \varphi \wedge \psi & \text{iff} & \mathfrak{M}, w \Vdash \varphi \text{ and } \mathfrak{M}, w \Vdash \psi; \\
\mathfrak{M}, w \Vdash \Diamond_a \varphi & \textit{iff} & \exists v \in M \colon w\mathcal{R}_a v \textit{ and } \mathfrak{M}, v \Vdash \varphi; \\
\mathfrak{M}, w \Vdash \langle\varphi\rangle\psi & \textit{iff} & \exists v \in M \colon w\mathcal{R}_\varphi v \textit{ and } \mathfrak{M}, v \Vdash \psi.
\end{array}
$$

We use the notation $\|\varphi\|^{\mathfrak{M}} = \{v \in M \mid \mathfrak{M}, v \Vdash \varphi\}$.

Instead of giving the $\langle\varphi\rangle$ operators a special truth clause, we ensure that they behave in a PAL-like way by imposing constraints on the $\mathcal{R}_\varphi$ relations in Definition 2.3 below. Wang and Cao [33] have independently proposed a semantics for PAL in this style, with respect to which they prove that PAL is complete. The difference comes in the specific constraints for UPAL vs. PAL.

Definition 2.3 *A UPAL model* $\mathfrak{M} = \langle M, \{\mathcal{R}_a\}_{a\in\mathsf{Agt}}, \{\mathcal{R}_\varphi\}_{\varphi\in\mathcal{L}^\kappa_{\mathsf{PAL}}}, \mathcal{V}\rangle$ *is* legal *iff the following conditions hold for all* $\psi, \chi \in \mathcal{L}^\kappa_{\mathsf{PAL}}$, $w, v \in M$, *and* $a \in \mathsf{Agt}$*:*

(extensionality)	*if* $\|\psi\|^{\mathfrak{M}} = \|\chi\|^{\mathfrak{M}}$, *then* $\mathcal{R}_\psi = \mathcal{R}_\chi$;
(ψ-seriality)	*if* $w \in \|\psi\|^{\mathfrak{M}}$, *then* $\exists v\colon w\mathcal{R}_\psi v$;
(truthfulness)	*if* $w\mathcal{R}_\psi v$, *then* $w \in \|\psi\|^{\mathfrak{M}}$;
($\top$-reflexivity)	$w\mathcal{R}_\top w$;
(functionality)	*if* $w\mathcal{R}_\psi v$, *then for all* $u \in M$, $w\mathcal{R}_\psi u$ *implies* $u = v$;
(ψa-commutativity)	*if* $w\mathcal{R}_\psi v$ *and* $v\mathcal{R}_a u$, *then* $\exists z\colon w\mathcal{R}_a z$ *and* $z\mathcal{R}_\psi u$;
($a\psi$-commutativity)	*if* $w\mathcal{R}_a v$, $v\mathcal{R}_\psi u$ *and* $w \in \|\psi\|^{\mathfrak{M}}$, *then* $\exists z\colon w\mathcal{R}_\psi z$ *and* $z\mathcal{R}_a u$;
(composition)	$\mathcal{R}_{\langle\psi\rangle\chi} = \mathcal{R}_\psi \circ \mathcal{R}_\chi$.

In §4, we will also refer to weaker versions of the first and third conditions:

(extensionality for φ)	*if* $\psi, \chi \in \mathsf{An}(\varphi) \cup \{\top\}$ *and* $\|\psi\|^{\mathfrak{M}} = \|\chi\|^{\mathfrak{M}}$, *then* $\mathcal{R}_\psi = \mathcal{R}_\chi$;
(truthfulness for φ)	*if* $\psi \in \mathsf{An}(\varphi) \cup \{\top\}$ *and* $w\mathcal{R}_\psi v$, *then* $w \in \|\psi\|^{\mathfrak{M}}$.

It is easy to see that each of the axioms of UPAL in Theorem 1.6 corresponds to the condition of the same name written in boldface in Definition 2.3.

3 Completeness of UPAL

In this section, we take our first step toward proving Theorem 1.6 by proving:

Theorem 3.1 (Soundness and Completeness) *The system of* UPAL-K_ω *given in Theorem 1.6 is sound and complete for the class of legal UPAL models.*

Soundness is straightforward. To prove completeness, we use the standard canonical model argument.

Definition 3.2 *The canonical model* $\mathfrak{M}^c = \langle M^c, \{\mathcal{R}^c_a\}_{a\in\mathsf{Agt}}, \{\mathcal{R}^c_\varphi\}_{\varphi\in\mathcal{L}^\kappa_{\mathsf{PAL}}}, \mathcal{V}^c\rangle$ *is defined as follows:*

1. $M^c = \{\Gamma \mid \Gamma$ *is a maximally UPAL-K_ω-consistent set*$\}$;
2. $\Gamma\mathcal{R}^c_a\Delta$ *iff* $\psi \in \Delta$ *implies* $\Diamond_a\psi \in \Gamma$;
3. $\Gamma\mathcal{R}^c_\varphi\Delta$ *iff* $\psi \in \Delta$ *implies* $\langle\varphi\rangle\psi \in \Gamma$;
4. $\mathcal{V}^c(p) = \{\Gamma \in M^c \mid p \in \Gamma\}$.

The following fact, easily shown, will be used in the proof of Lemma 3.5.

Fact 3.3 *For all* $\Gamma \in M^c$, $\varphi \in \mathcal{L}^\kappa_{\mathsf{PAL}}$, *if* $\langle\varphi\rangle\top \in \Gamma$, *then* $\{\psi \mid \langle\varphi\rangle\psi \in \Gamma\} \in M^c$.

The proof of the truth lemma is completely standard [14, §4.2].

Lemma 3.4 (Truth) *For all* $\Gamma \in M^c$ *and* $\varphi \in \mathcal{L}^\kappa_{\mathsf{PAL}}$,

$$\mathfrak{M}^c, \Gamma \Vdash \varphi \text{ iff } \varphi \in \Gamma.$$

To complete the proof of Theorem 3.1, we need only check the following.

Lemma 3.5 (Legality) $\mathfrak{M}^c$ *is a legal model.*

Proof. Suppose $\|\varphi\|^{\mathfrak{M}^c} = \|\psi\|^{\mathfrak{M}^c}$, so by Lemma 3.4 and the properties of maximally consistent sets, $\varphi \leftrightarrow \psi \in \Gamma$ for all $\Gamma \in M^c$. Hence $\vdash_{\mathsf{UPAL}\text{-}\mathsf{K}_\omega} \varphi \leftrightarrow \psi$, for if $\neg(\varphi \leftrightarrow \psi)$ is UPAL-K_ω-consistent, then $\neg(\varphi \leftrightarrow \psi) \in \Delta$ for some $\Delta \in M^c$, contrary to what was just shown. It follows that for any $\alpha \in \mathcal{L}^\kappa_{\mathsf{PAL}}$, $\vdash_{\mathsf{UPAL}\text{-}\mathsf{K}_\omega} \langle\varphi\rangle\alpha \leftrightarrow \langle\psi\rangle\alpha$, given the extensionality and uniformity rules of UPAL-K_ω. Hence if $\Gamma_1\mathcal{R}^c_\varphi\Gamma_2$, then for all $\alpha \in \Gamma_2$, $\langle\varphi\rangle\alpha \in \Gamma_1$ and $\langle\psi\rangle\alpha \in \Gamma_1$ by the consistency of Γ_1, which means $\Gamma_1\mathcal{R}^c_\psi\Gamma_2$. The argument in the other direction is the same, whence $\mathcal{R}^c_\varphi = \mathcal{R}^c_\psi$. $\mathfrak{M}^c$ satisfies **extensionality**.

Suppose $\Gamma_1\mathcal{R}^c_{\langle\varphi\rangle\psi}\Gamma_2$, so for all $\alpha \in \Gamma_2$, $\langle\langle\varphi\rangle\psi\rangle\alpha \in \Gamma_1$. Hence $\langle\varphi\rangle\langle\psi\rangle\alpha \in \Gamma_1$ given the composition axiom and uniformity rule of UPAL-K_ω, so $\langle\varphi\rangle\top \in \Gamma_1$ by normal modal reasoning with the distribution axiom. It follows by Fact 3.3 and Definition 3.2.3 that there is some Σ_1 such that $\Gamma_1\mathcal{R}_\varphi\Sigma_1$ and $\langle\psi\rangle\alpha \in \Sigma_1$, and by similar reasoning that there is some Σ_2 such that $\Sigma_1\mathcal{R}_\psi\Sigma_2$ and $\alpha \in \Sigma_2$. Hence $\Gamma_2 \subseteq \Sigma_2$, so $\Gamma_2 = \Sigma_2$ given that Γ_2 is maximal. Therefore, $\mathcal{R}^c_{\langle\varphi\rangle\psi} \subseteq \mathcal{R}^c_\varphi \circ \mathcal{R}^c_\psi$. The argument in the other direction is similar. $\mathfrak{M}^c$ satisfies **composition**.

We leave the other legality conditions to the reader. □

4 Bridging UPAL and PAL

In this section, we show that UPAL axiomatizes the substitution core of PAL. It is easy to check that all of the axioms of UPAL are PAL schematic validities, and all of the rules of UPAL preserve schematic validity, so UPAL derives only PAL schematic validities. To prove that UPAL derives all PAL schematic validities, we show that if φ is not derivable from UPAL, so by Theorem 3.1 there is a legal UPAL model falsifying φ, then there is a substitution τ and a PAL model falsifying $(\varphi)^\tau$, in which case φ is not schematically valid over PAL models.

Proposition 4.1 *For any $\varphi \in \mathcal{L}^{\omega}_{\mathsf{PAL}}$, if there is a legal UPAL model $\mathfrak{M} = \langle M, \{\mathcal{R}_a\}_{a\in\mathsf{Agt}}, \{\mathcal{R}_\psi\}_{\psi\in\mathcal{L}^{\omega}_{\mathsf{PAL}}}, \mathcal{V}\rangle$ with $w_0 \in M$ such that $\mathfrak{M}, w_0 \nVdash \varphi$, then there is a PAL model $\mathcal{N} = \langle N_0, \{S_a\}_{a\in\mathsf{Agt}}, U\rangle$ with $w_0 \in N_0$ and a substitution τ such that $\mathcal{N}, w_0 \nvDash (\varphi)^\tau$.*

Our first step in proving Proposition 4.1 is to show that we can reduce φ to a certain simple form, which will help us in constructing the substitution τ.

Definition 4.2 *The set of* simple *formulas is generated by the grammar*

$$\varphi ::= \top \mid p \mid \neg\varphi \mid \varphi \wedge \varphi \mid \Diamond_a \varphi \mid \langle\varphi\rangle p,$$

where $p \in \mathsf{At}$ and $a \in \mathsf{Agt}$.

Proposition 4.3 *For every $\varphi \in \mathcal{L}^{\kappa}_{\mathsf{PAL}}$, there is a simple formula $\varphi' \in \mathcal{L}^{\kappa}_{\mathsf{PAL}}$ that is equivalent to φ over legal UPAL models (and all PAL models).*

Proof. The proof is similar to the standard PAL reduction argument [9, §7.4], only we do not perform atomic reduction steps, and we use the composition axiom of UPAL to eliminate consecutive occurrences of dynamic operators. □

By Proposition 4.3, given that $\mathfrak{M}$ is legal, we can assume that φ is simple. Before constructing $\mathcal{N}$ and τ, we show that our initial model $\mathfrak{M}$ can be transformed into an intermediate model $\mathfrak{N}$ that satisfies a property (part 2 of Lemma 4.4) that we will take advantage of in our proofs below. We will return to the role of this property in relating UPAL to PAL in Example 5.2 and §6.

For what follows, we need some new notation. First, let

$$\mathcal{R}_{\mathsf{Agt}} = \bigcup_{a\in\mathsf{Agt}} \mathcal{R}_a;$$

$\mathcal{R}^*$ is the reflexive transitive closure of $\mathcal{R}$; and $\mathcal{R}(w) = \{v \in M \mid w\mathcal{R}v\}$.

Lemma 4.4 *For any legal model $\mathfrak{M} = \langle M, \{\mathcal{R}_a\}_{a\in\mathsf{Agt}}, \{\mathcal{R}_\varphi\}_{\varphi\in\mathcal{L}^{\omega}_{\mathsf{PAL}}}, \mathcal{V}\rangle$ with $w_0 \in M$ such that $\mathfrak{M}, w_0 \Vdash \varphi$, there is a model $\mathfrak{N} = \langle N, \{\mathcal{S}_a\}_{a\in\mathsf{Agt}}, \{\mathcal{S}_\varphi\}_{\varphi\in\mathcal{L}^{\omega}_{\mathsf{PAL}}}, \mathcal{U}\rangle$ with $w_0 \in N$ such that*

1. *$\mathfrak{N}, w_0 \Vdash \varphi$;*
2. *if $\alpha, \beta \in \mathsf{An}(\varphi) \cup \{\top\}$ and $\|\alpha\|^{\mathfrak{N}} \neq \|\beta\|^{\mathfrak{N}}$, then*

$$\|\alpha\|^{\mathfrak{N}} \cap \mathcal{S}^*_{\mathsf{Agt}}(w_0) \neq \|\beta\|^{\mathfrak{N}} \cap \mathcal{S}^*_{\mathsf{Agt}}(w_0).$$

3. *$\mathfrak{N}$ satisfies $\top$-**reflexivity, functionality, extensionality for** φ and **truthfulness for** φ.*

Proof. Consider some $\alpha, \beta \in \mathsf{An}(\varphi) \cup \{\top\}$ such that $\|\alpha\|^{\mathfrak{M}} \neq \|\beta\|^{\mathfrak{M}}$. Hence there is some $v \in M$ such that $\mathfrak{M}, v \nVdash \alpha \leftrightarrow \beta$. Let $\mathfrak{M}'$ be exactly like $\mathfrak{M}$

except that for some $x \notin \mathsf{Agt}(\varphi)$, $w_0 \mathcal{R}'_x v$.[5] Then it is easy to show that for all $\psi \in \mathsf{Sub}(\varphi)$ and $u \in M$,

$$\mathfrak{M}', u \Vdash \psi \text{ iff } \mathfrak{M}, u \Vdash \psi.$$

Hence $\mathfrak{M}', w_0 \Vdash \varphi$ and $\mathfrak{M}', v \nVdash \alpha \leftrightarrow \beta$. Then given $w_0 \mathcal{R}'_x v$, we have

$$\|\alpha\|^{\mathfrak{M}'} \cap \mathcal{R}'^{*}_{\mathsf{Agt}}(w_0) \neq \|\beta\|^{\mathfrak{M}'} \cap \mathcal{R}'^{*}_{\mathsf{Agt}}(w_0).$$

Finally, one can check that $\mathfrak{M}'$ satisfies $\top$**-reflexivity**, **functionality**, **extensionality for** φ and **truthfulness for** φ by the construction. By repeating this procedure, starting now with $\mathfrak{M}'$, for each of the finitely many α and β as described above, one obtains a model $\mathfrak{N}$ as described in Lemma 4.4 □

Obtaining $\mathfrak{N}$ from $\mathfrak{M}$ as in Lemma 4.4, we now define our PAL model $\mathcal{N} = \langle N_0, \{S_a\}_{a\in\mathsf{Agt}}, U\rangle$. Let $N_0 = \mathcal{S}^*_{\mathsf{Agt}}(w_0)$; for some $z \notin \mathsf{Agt}(\varphi)$, let S_z be the universal relation on N_0; and for each $a \in \mathsf{Agt}$ with $a \neq z$, let S_a be the restriction of $\mathcal{S}_a$ to N_0. We will define the valuation U after constructing the substitution τ. The following facts will be used in the proof of Lemma 4.8.

Fact 4.5

1. *For all* $a \in \mathsf{Agt}$ *and* $w \in N_0$, $S_a(w) = \mathcal{S}_a(w)$.
2. *if* $\|\alpha\|^{\mathfrak{N}} \cap N_0 = \|\beta\|^{\mathfrak{N}} \cap N_0$, *then for all* $u \in N_0$,

$$\mathfrak{N}, u \Vdash \langle\alpha\rangle\chi \text{ iff } \mathfrak{N}, u \Vdash \langle\beta\rangle\chi.$$

Proof. Part 1 is obvious. For part 2, if $\|\alpha\|^{\mathfrak{N}} \cap N_0 = \|\beta\|^{\mathfrak{N}} \cap N_0$, then $\|\alpha\|^{\mathfrak{N}} = \|\beta\|^{\mathfrak{N}}$ by Lemma 4.4.2, so $\mathcal{S}_\alpha = \mathcal{S}_\beta$ by Lemma 4.4.3 (**extensionality for** φ).□

Remark 4.6 There is another way of transforming the UPAL model $\mathfrak{M} = \langle M, \{\mathcal{R}_a\}_{a\in\mathsf{Agt}}, \{\mathcal{R}_\varphi\}_{\varphi\in\mathcal{L}^\omega_{\mathsf{PAL}}}, \mathcal{V}\rangle$ into a PAL model $\mathcal{N}$ sufficient for our purposes. First, let $\mathfrak{N} = \langle N, \{\mathcal{S}_a\}_{a\in\mathsf{Agt}}, \{\mathcal{S}_\varphi\}_{\varphi\in\mathcal{L}^\omega_{\mathsf{PAL}}}, \mathcal{U}\rangle$ be exactly like $\mathfrak{M}$ except that for some $z \notin \mathsf{Agt}(\varphi)$, $\mathcal{S}_z$ is the universal relation on N, and observe that $\mathfrak{N}$ satisfies the conditions of Lemma 4.4. Second, take $\mathcal{N} = \langle N_0, \{S_a\}_{a\in\mathsf{Agt}}, U\rangle$ such that $N_0 = N$, $S_a = \mathcal{S}_a$, and U is defined as below, and observe that Fact 4.5 holds. Then the proof can proceed as below. The difference is that this approach takes the domain of the PAL model to be the entire domain of the UPAL model $\mathfrak{N}$, with S_z as the universal relation on this entire domain, whereas our approach takes the domain of the PAL model to be just that of the "epistemic submodel" *generated by* w_0 *in* $\mathfrak{N}$, $\mathcal{S}^*_{\mathsf{Agt}}(w_0)$, with S_z as the universal relation on this set. We prefer the latter approach because it allows us to work with smaller PAL models when we carry out the construction with concrete examples as in §5.

[5] As noted after Theorem 1.6, we can modify our proof for other models classes. For example, for the class of models with equivalence relations, in this step we can define $\mathcal{R}'_x$ to be the smallest equivalence relation extending $\mathcal{R}_x$ such that $w_0 \mathcal{R}'_x v$. Note that since $\alpha, \beta \in \mathsf{An}(\varphi) \cup \{\top\}$ and $x \notin \mathsf{Agt}(\varphi)$, no matter how we define $\mathcal{R}'_x$, the following claim in the text still holds.

To construct $\tau(p)$ for $p \in \mathsf{At}(\varphi)$, let $B_1, \ldots, B_m$ be the sequence of all B_i such that $\langle B_i \rangle p \in \mathsf{Sub}(\varphi)$, and let $B_0 := \top$. For $0 \leq i, j \leq m$, if $\|B_i\|^{\mathfrak{N}} \cap N_0 = \|B_j\|^{\mathfrak{N}} \cap N_0$, delete one of B_i or B_j from the list (but never B_0), until there is no such pair. Call the resulting sequence $A_0, \ldots, A_n$, and define

$$s(i) = \{j \mid 0 \leq j \leq n \text{ and } \|A_j\|^{\mathfrak{N}} \cap N_0 \subsetneq \|A_i\|^{\mathfrak{N}} \cap N_0\}.$$

Extend the language with new variables $p_0, \ldots, p_n$ and $a_0, \ldots, a_n$, and define $\tau(p) = \gamma_0 \wedge \cdots \wedge \gamma_n$ such that

$$\gamma_i := (\Box_z a_i \wedge \bigwedge_{j \in s(i)} \neg \Box_z a_j) \to p_i.$$

Having extended the language for each $p \in \mathsf{At}(\varphi)$, define the valuation U for N_0 such that for each $p \in \mathsf{At}(\varphi)$, $U(p) = \mathcal{U}(p) \cap N_0$, and for the new variables:

(a) $U(p_i) = \{w \in N_0 \mid \exists u\colon w\mathcal{S}_{A_i} u \text{ and } u \in \mathcal{U}(p)\}$;

(b) $U(a_i) = \|A_i\|^{\mathfrak{N}} \cap N_0$.

Hence:

(**a**) $[\![p_i]\!]^{\mathcal{N}} = \{w \in N_0 \mid \exists u\colon w\mathcal{S}_{A_i} u \text{ and } u \in \mathcal{U}(p)\}$;

(**b**) $[\![a_i]\!]^{\mathcal{N}} = \|A_i\|^{\mathfrak{N}} \cap N_0$.

Note that it follows from (**a**) and the UPAL truth definition that

(**c**) $[\![p_i]\!]^{\mathcal{N}} = \|\langle A_i \rangle p\|^{\mathfrak{N}} \cap N_0$.

Using these facts, we will show that $\mathfrak{N}, w_0 \nVdash \varphi$ implies $\mathcal{N}, w_0 \nvDash \tau(\varphi)$.

Lemma 4.7 *For all* $0 \leq i \leq n$,

$$[\![\tau(p)]\!]^{\mathcal{N}_{|a_i}} = [\![p_i]\!]^{\mathcal{N}}.$$

Proof. We first show that for $0 \leq i, j \leq n$, $i \neq j$:

(i) $[\![\gamma_i]\!]^{\mathcal{N}_{|a_i}} = [\![p_i]\!]^{\mathcal{N}_{|a_i}}$;

(ii) $[\![\gamma_j]\!]^{\mathcal{N}_{|a_i}} = [\![a_i]\!]^{\mathcal{N}_{|a_i}}$ $(= N_{0|a_i})$.

For (i), we claim that

$$[\![\Box_z a_i \wedge \bigwedge_{k \in s(i)} \neg \Box_z a_k]\!]^{\mathcal{N}_{|a_i}} = N_{0|a_i}.$$

Since a_i is atomic, $[\![\Box_z a_i]\!]^{\mathcal{N}_{|a_i}} = N_{0|a_i}$. By definition of the s function and (**b**), for all $k \in s(i)$, $[\![a_k]\!]^{\mathcal{N}} \subsetneq [\![a_i]\!]^{\mathcal{N}}$, so $[\![\neg \Box_z a_k]\!]^{\mathcal{N}_{|a_i}} = N_{0|a_i}$. Hence the claimed equation holds, so $[\![\gamma_i]\!]^{\mathcal{N}_{|a_i}} = [\![p_i]\!]^{\mathcal{N}_{|a_i}}$ given the structure of γ_i.

For (ii), we claim that for $j \neq i$,

$$[\![\Box_z a_j \wedge \bigwedge_{k \in s(j)} \neg \Box_z a_k]\!]^{\mathcal{N}_{|a_i}} = \emptyset.$$

By construction of the sequence $A_0, \ldots, A_n$ for p and (**b**), $[\![a_j]\!]^{\mathcal{N}} \neq [\![a_i]\!]^{\mathcal{N}}$. Hence if not $[\![a_i]\!]^{\mathcal{N}} \subsetneq [\![a_j]\!]^{\mathcal{N}}$, then $[\![a_i]\!]^{\mathcal{N}} \not\subseteq [\![a_j]\!]^{\mathcal{N}}$, so $[\![\Box_z a_j]\!]^{\mathcal{N}|a_i} = \emptyset$ because S_z is the universal relation on N_0. If $[\![a_i]\!]^{\mathcal{N}} \subsetneq [\![a_j]\!]^{\mathcal{N}}$, then by (**b**) and the definition of s, $i \in s(j)$; since a_i is atomic, $[\![\neg\Box_z a_i]\!]^{\mathcal{N}|a_i} = \emptyset$. In either case the claimed equation holds, so $[\![\gamma_j]\!]^{\mathcal{N}|a_i} = N_{0|a_i}$ given the structure of γ_j.

Given the construction of τ, (i) and (ii) imply:

$$[\![\tau(p)]\!]^{\mathcal{N}|a_i} = [\![\gamma_i]\!]^{\mathcal{N}|a_i} \cap \bigcap_{j \neq i} [\![\gamma_j]\!]^{\mathcal{N}|a_i} = [\![p_i]\!]^{\mathcal{N}|a_i} \cap [\![a_i]\!]^{\mathcal{N}|a_i} = [\![p_i]\!]^{\mathcal{N}},$$

where the last equality holds because $[\![p_i]\!]^{\mathcal{N}} \subseteq [\![a_i]\!]^{\mathcal{N}}$, which follows from (**a**), (**b**), and the fact that $\mathfrak{N}$ satisfies **truthfulness for** φ. □

We now establish the connection between the UPAL model $\mathfrak{N}$ on the one hand and the PAL model $\mathcal{N}$ and substitution τ on the other.

Lemma 4.8 *For all simple subformulas χ of φ,*

$$[\![(\chi)^\tau]\!]^{\mathcal{N}} = \|\chi\|^{\mathfrak{N}} \cap N_0.$$

Proof. By induction on χ. For the base case, we must show $[\![(p)^\tau]\!]^{\mathcal{N}} = \|p\|^{\mathfrak{N}} \cap N_0$ for $p \in \mathsf{At}(\varphi)$. By construction of the sequence $A_0, \ldots, A_n$ for p, $A_0 = \top$, so $\|A_0\|^{\mathfrak{N}} \cap N_0 = N_0$. Then by (**b**), $[\![a_0]\!]^{\mathcal{N}} = N_0$, and hence

$$\begin{aligned}
[\![(p)^\tau]\!]^{\mathcal{N}} &= [\![(p)^\tau]\!]^{\mathcal{N}|a_0} \\
&= [\![p_0]\!]^{\mathcal{N}} && \text{by Lemma 4.7} \\
&= \{w \in N_0 \mid \exists u : w\mathcal{S}_{A_0}u \text{ and } u \in \mathcal{U}(p)\} && \text{by (a)} \\
&= \{w \in N_0 \mid w \in \mathcal{U}(p)\} && \text{by } \top\textbf{-reflexivity} \\
& && \text{and } \textbf{functionality} \\
&= \|p\|^{\mathfrak{N}} \cap N_0.
\end{aligned}$$

The boolean cases are straightforward. Next, we must show $[\![(\Box_a\varphi)^\tau]\!]^{\mathcal{N}} = \|\Box_a\varphi\|^{\mathfrak{N}} \cap N_0$. For the inductive hypothesis, $[\![(\varphi)^\tau]\!]^{\mathcal{N}} = \|\varphi\|^{\mathfrak{N}} \cap N_0$, so

$$\begin{aligned}
[\![(\Box_a\varphi)^\tau]\!]^{\mathcal{N}} &= [\![\Box_a(\varphi)^\tau]\!]^{\mathcal{N}} \\
&= \{w \in N_0 \mid \; S_a(w) \subseteq [\![(\varphi)^\tau]\!]^{\mathcal{N}}\} \\
&= \{w \in N_0 \mid \; S_a(w) \subseteq \|\varphi\|^{\mathfrak{N}} \cap N_0\} \\
&= \{w \in N_0 \mid \; S_a(w) \subseteq \|\varphi\|^{\mathfrak{N}}\} && \text{given } S_a \subseteq N_0 \times N_0 \\
&= \{w \in N_0 \mid \; \mathcal{S}_a(w) \subseteq \|\varphi\|^{\mathfrak{N}}\} && \text{by Fact 4.5.1} \\
&= \|\Box_a\varphi\|^{\mathfrak{N}} \cap N_0.
\end{aligned}$$

Finally, we must show $[\![(\langle B_i\rangle p)^\tau]\!]^{\mathcal{N}} = \|\langle B_i\rangle p\|^{\mathfrak{N}} \cap N_0$. For the inductive hypothesis, $[\![(B_i)^\tau]\!]^{\mathcal{N}} = \|B_i\|^{\mathfrak{N}} \cap N_0$. By construction of the sequence $A_0, \ldots, A_n$ for $p \in \mathsf{At}(\varphi)$, there is some A_j such that

$$(\star) \quad \|B_i\|^{\mathfrak{N}} \cap N_0 = \|A_j\|^{\mathfrak{N}} \cap N_0.$$

Therefore,

$$\begin{aligned}
[\![(B_i)^\tau]\!]^{\mathcal{N}} &= \|A_j\|^{\mathfrak{N}} \cap N_0 \\
&= [\![a_j]\!]^{\mathcal{N}} && \text{by (b)},
\end{aligned}$$

and hence

$$\begin{aligned}
[\![(\langle B_i\rangle p)^\tau]\!]^{\mathcal{N}} &= [\![\langle (B_i)^\tau\rangle (p)^\tau]\!]^{\mathcal{N}} \\
&= [\![\langle a_j\rangle (p)^\tau]\!]^{\mathcal{N}} \\
&= [\![(p)^\tau]\!]^{\mathcal{N}|a_j} \\
&= [\![p_j]\!]^{\mathcal{N}} && \text{by Lemma 4.7} \\
&= \|\langle A_j\rangle p\|^{\mathfrak{N}} \cap N_0 && \text{by } (\mathbf{c}) \\
&= \|\langle B_i\rangle p\|^{\mathfrak{N}} \cap N_0 && \text{given } (\star) \text{ and Fact 4.5.2.}
\end{aligned}$$

The proof by induction is complete. □

With the following fact, we complete the proof of Proposition 4.1.

Fact 4.9 $\mathcal{N}, w_0 \nvDash (\varphi)^\tau$.

Proof. Immediate from Lemma 4.8 given $\mathfrak{N}, w_0 \nVdash \varphi$. □

5 Examples

In this section, we work out two examples illustrating how the techniques of §4 allow us to find, for any formula φ that is valid but not schematically valid in PAL, a PAL model that falsifies a substitution instance of φ. The proof in §4 shows that all we need to do is find a legal UPAL model falsifying φ. However, since legal UPAL models are generally large, we would like to instead find a small UPAL model falsifying φ, from which we can read off a PAL model that falsifies a substitution instance of φ. In fact, we can always do so provided that the model satisfies a weaker condition than legality. For a given $\varphi \in \mathcal{L}^{\kappa}_{\mathsf{PAL}}$, we say that a UPAL model $\mathfrak{M}$ is *φ-legal* iff it satisfies all of the legality conditions of Definition 2.3 when we replace **ψ-seriality** with:

(**ψ-seriality for** φ) if $\psi \in \mathsf{An}(\varphi) \cup \{\top\}$ and $w \in \|\psi\|^{\mathfrak{M}}$, then $\exists v$: $w\mathcal{R}_\psi v$.

Hence in a φ-legal model, we can let all of the infinitely many $\mathcal{R}_\psi$ relations irrelevant to φ be empty, which makes constructing φ-legal models easier. With this new notion, we can state a simple method for finding a PAL model that falsifies a substitution instance of the non-schematically valid φ:

Step 1. Transform φ into an equivalent simple formula φ'.

Step 2. Find a φ'-legal pointed UPAL model $\mathfrak{M}, w_0$ such that $\mathfrak{M}, w_0 \nVdash \varphi'$.

Step 3. Obtain $\mathcal{N}$ and τ from $\mathfrak{M}, w_0$ as in §4 so that $\mathcal{N}, w_0 \nvDash (\varphi')^\tau$.

Since $\varphi \leftrightarrow \varphi'$ is schematically valid in PAL, we have $\mathcal{N}, w \nvDash (\varphi)^\tau$, as desired. The key to this method is that the construction in §4 also establishes the following variant of Proposition 4.1:

Proposition 5.1 *For any* simple $\varphi \in \mathcal{L}^{\omega}_{\mathsf{PAL}}$, *if there is a* φ-legal *UPAL model* $\mathfrak{M} = \langle M, \{\mathcal{R}_a\}_{a\in\mathsf{Agt}}, \{\mathcal{R}_\psi\}_{\psi\in\mathcal{L}^{\omega}_{\mathsf{PAL}}}, \mathcal{V}\rangle$ *with* $w_0 \in M$ *such that* $\mathfrak{M}, w_0 \nVdash \varphi$, *then there is a PAL model* $\mathcal{N} = \langle N_0, \{S_a\}_{a\in\mathsf{Agt}}, U\rangle$ *with* $w_0 \in N_0$ *and a substitution* τ *such that* $\mathcal{N}, w_0 \nvDash (\varphi)^\tau$.

This proposition holds because if φ is already simple, then the only properties of $\mathfrak{M}$ used in the proof of Fact 4.9 are $\top$**-reflexivity**, **functionality**, **extensionality for** φ and **truthfulness for** φ, which are part of φ-legality.

Finally, if φ does not contain any occurrence of a dynamic operator in the scope of any other, then we can simply skip Step 1 and do Steps 2 and 3 for φ itself. One can check that the construction in §4 works not only with a simple formula, but more generally with any formula with the scope restriction.

Example 5.2 Consider the PAL-valid formula $\varphi := [p]p$, which is already simple. Let us try to falsify φ in a φ-legal UPAL model. The obvious first try is $\mathfrak{M}$ in Fig. 1, which is indeed a φ-legal UPAL model, in which all $\mathcal{R}_a$ relations are empty. (We simplify the diagrams by omitting all reflexive $\mathcal{R}_\top$ loops.) However, $\mathfrak{M}$ has an un-PAL-like property: although $\|\top\|^{\mathfrak{M}} \cap \mathcal{R}^*_{\mathsf{Agt}}(w_0) = \|p\|^{\mathfrak{M}} \cap \mathcal{R}^*_{\mathsf{Agt}}(w_0)$, we have $w_0 \mathcal{R}_\top w_0$ but not $w_0 \mathcal{R}_p w_0$. (See §6 for why this is un-PAL-like.) To eliminate this property, we modify $\mathfrak{M}$ to $\mathfrak{N} = \langle N, \{\mathcal{S}_a\}_{a \in \mathsf{Agt}}, \{\mathcal{S}_\psi\}_{\psi \in \mathcal{L}^\omega_{\mathsf{PAL}}}, \mathcal{U}\rangle$ in Fig. 1 as in Lemma 4.4.[6] Next, following the procedure in §4, we obtain the PAL model $\mathcal{N} = \langle N_0, \{S_a\}_{a \in \mathsf{Agt}}, U\rangle$ in Fig. 1 and the substitution τ given below.

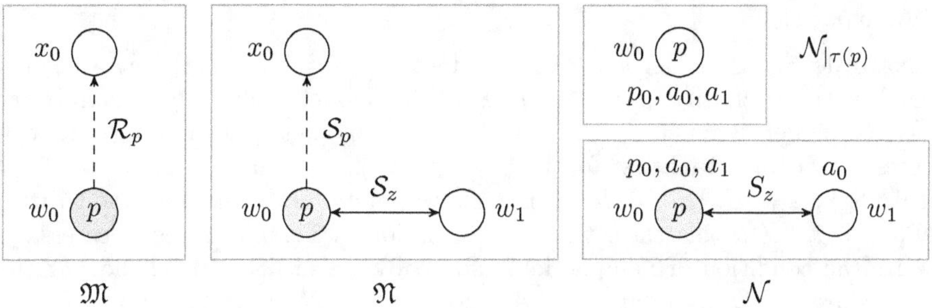

Fig. 1. UPAL and PAL Models for Example 5.2

Where $A_0 := \top$, $A_1 := p$, and a_0, a_1, p_0, and p_1 are the new atoms, we define the valuation U in $\mathcal{N}$ such that:

$U(a_0) = \|A_0\|^{\mathfrak{N}} \cap N_0 = \{w_0, w_1\}$;
$U(a_1) = \|A_1\|^{\mathfrak{N}} \cap N_0 = \{w_0\}$;
$U(p_0) = \{w \in N_0 \mid \exists u\colon w\mathcal{S}_{A_0}u \text{ and } u \in \mathcal{U}(p)\} = \{w_0\}$;
$U(p_1) = \{w \in N_0 \mid \exists u\colon w\mathcal{S}_{A_1}u \text{ and } u \in \mathcal{U}(p)\} = \emptyset$.

Defining the function s such that

$$s(i) = \{j \mid 0 \le j \le n \text{ and } \|A_j\|^{\mathfrak{N}} \cap N_0 \subsetneq \|A_i\|^{\mathfrak{N}} \cap N_0\},$$

[6] In fact, the construction of Lemma 4.4 would connect w_0 to x_0 by $\mathcal{R}_z$, but note that we can always connect w_0 to a new point falsifying $\alpha \leftrightarrow \beta$ (in this case, $\top \leftrightarrow p$) instead.

we have $s(0) = \{1\}$ and $s(1) = \emptyset$. Defining $\tau(p) = \gamma_0 \wedge \cdots \wedge \gamma_n$ such that

$$\gamma_i := (\Box_z a_i \wedge \bigwedge_{j \in s(i)} \neg\Box_z a_j) \to p_i,$$

we have

$$\tau(p) = ((\Box_z a_0 \wedge \neg\Box_z a_1) \to p_0) \wedge (\Box_z a_1 \to p_1).$$

Observe:

$$[\![(\Box_z a_0 \wedge \neg\Box_z a_1) \to p_0]\!]^{\mathcal{N}} = \{w_0\};$$
$$[\![\Box_z a_1 \to p_1]\!]^{\mathcal{N}} = \{w_0, w_1\};$$
$$[\![\tau(p)]\!]^{\mathcal{N}} = \{w_0\}.$$

Hence $\mathcal{N}_{|\tau(p)}$ is the model displayed in the upper-right in Fig. 1. Observe:

$$[\![(\Box_z a_0 \wedge \neg\Box_z a_1) \to p_0]\!]^{\mathcal{N}_{|\tau(p)}} = \{w_0\};$$
$$[\![\Box_z a_1 \to p_1]\!]^{\mathcal{N}_{|\tau(p)}} = \emptyset;$$
$$[\![\tau(p)]\!]^{\mathcal{N}_{|\tau(p)}} = \emptyset.$$

Hence $\mathcal{N}, w_0 \nvDash ([p]p)^{\tau}$, so our starting formula φ is not schematically valid over PAL models.

Example 5.3 Consider the PAL-valid formula $\varphi := [p \wedge \neg\Box_b p]\neg(p \wedge \neg\Box_b p)$.[7] Let us try to falsify φ in a φ-legal UPAL model. The obvious first try is the model $\mathfrak{A}$ in Fig. 2. However, $\mathfrak{A}$ is not φ-legal, since it violates **ψz-commutativity** for $\psi := p \wedge \neg\Box_b p$. By modifying $\mathfrak{A}$ to $\mathfrak{N} = \langle N, \{\mathcal{S}_a\}_{a\in\mathsf{Agt}}, \{\mathcal{S}_\psi\}_{\psi\in\mathcal{L}^{\omega}_{\mathsf{PAL}}}, \mathcal{U}\rangle$ in Fig. 2, we obtain a φ-legal UPAL model with $\mathfrak{N}, w_0 \nVdash \varphi$. (In this case, the transformation of Lemma 4.4 is uncecessary, since the condition of Lemma 4.4.2 is already satisfied by $\mathfrak{N}$.) Following the procedure of §4, we obtain the PAL model $\mathcal{N} = \langle N_0, \{S_a\}_{a\in\mathsf{Agt}}, U\rangle$ in Fig. 3 and the substitution τ given below.

Where $A_0 := \top$, $A_1 := p \wedge \neg\Box_b p$, and a_0, a_1, p_0, and p_1 are the new atoms, we define the valuation U in $\mathcal{N}$ such that:

$$U(a_0) = \|A_0\|^{\mathfrak{N}} \cap N_0 = \{w_0, w_1, w_2\};$$
$$U(a_1) = \|A_1\|^{\mathfrak{N}} \cap N_0 = \{w_0, w_1\};$$
$$U(p_0) = \{w \in N_0 \mid \exists u\colon w\mathcal{S}_{A_0}u \text{ and } u \in \mathcal{U}(p)\} = \{w_0, w_1\};$$
$$U(p_1) = \{w \in N_0 \mid \exists u\colon w\mathcal{S}_{A_1}u \text{ and } u \in \mathcal{U}(p)\} = \{w_0\}.$$

Defining the function s as before, we have $s(0) = \{1\}$ and $s(1) = \emptyset$. Since this is the same s as in Example 5.2, the substitution is also the same:

$$\tau(p) = ((\Box_z a_0 \wedge \neg\Box_z a_1) \to p_0) \wedge (\Box_z a_1 \to p_1).$$

[7] Here we could transform $\varphi := [p \wedge \neg\Box_b p]\neg(p \wedge \neg\Box_b p)$ into the simple

$$\varphi' := (p \wedge \neg\Box_b p) \to \neg([p \wedge \neg\Box_b p]p \wedge ((p \wedge \neg\Box_b p) \to \neg((p \wedge \neg\Box_b p) \to \Box_b[p \wedge \neg\Box_b p]p))),$$

but as noted before Example 5.2, if φ does not contain any occurrence of a dynamic operator in the scope of any other, then we can skip Step 1 and do Steps 2 and 3 for φ itself.

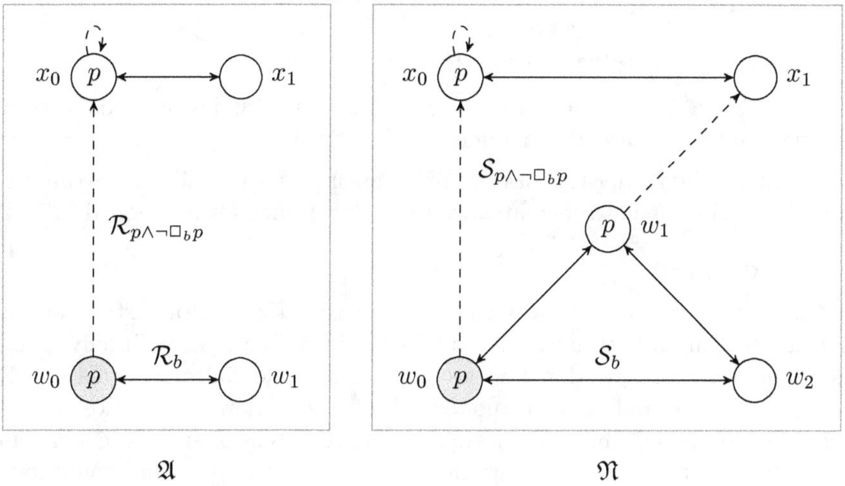

Fig. 2. UPAL Models for Example 5.3

Note that since the construction of $\mathcal{N}$ from $\mathfrak{N}$ is such that $S_z = S_b$, we can simply take $\Box_z$ to be $\Box_b$ in $\tau(p)$, so that $\mathsf{Agt}((\varphi)^\tau) = \mathsf{Agt}(\varphi) = \{b\}$.

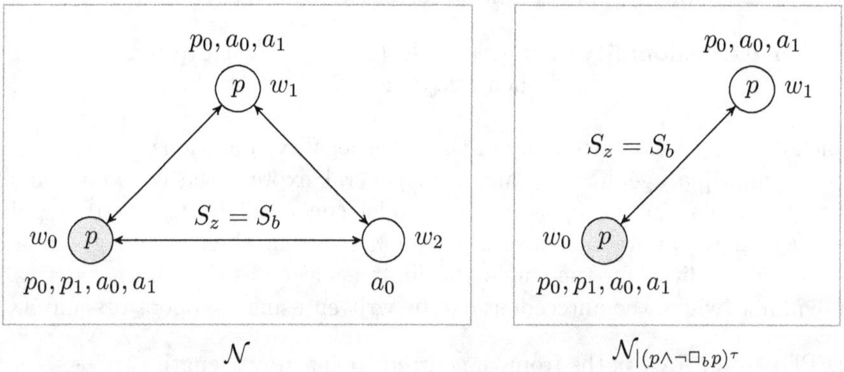

Fig. 3. PAL Models for Example 5.3

Observe:

$[\![(\Box_z a_0 \wedge \neg\Box_z a_1) \to p_0]\!]^{\mathcal{N}} = \{w_0, w_1\}$;

$[\![\Box_z a_1 \to p_1]\!]^{\mathcal{N}} = \{w_0, w_1, w_2\}$;

$[\![\tau(p)]\!]^{\mathcal{N}} = \{w_0, w_1\}$;

$[\![\tau(p) \wedge \neg\Box_b \tau(p)]\!]^{\mathcal{N}} = \{w_0, w_1\}$.

Hence $\mathcal{N}_{|(p\wedge\neg\Box_b p)^\tau}$ is the model displayed on the right in Fig. 3. Observe:

$[\![(\Box_z a_0 \wedge \neg\Box_z a_1) \to p_0]\!]^{\mathcal{N}_{|(p\wedge\neg\Box_b p)^\tau}} = \{w_0, w_1\}$;

$[\![\Box_z a_1 \to p_1]\!]^{\mathcal{N}_{|(p\wedge\neg\Box_b p)^\tau}} = \{w_0\}$;

$$[\![\tau(p)]\!]^{\mathcal{N}|(p\wedge\neg\Box_b p)^\tau} = \{w_0\};$$
$$[\![\tau(p)\wedge\neg\Box_b\tau(p)]\!]^{\mathcal{N}|(p\wedge\neg\Box_b p)^\tau} = \{w_0\}.$$

Hence $\mathcal{N}, w_0 \nvDash ([p\wedge\neg\Box_b p]\neg(p\wedge\neg\Box_b p))^\tau$, so our starting formula φ is not schematically valid over PAL models.

We invite the reader to work out other examples using UPAL, starting from the other valid but not schematically valid PAL principles mentioned in §1.2.

6 Discussion

In this paper, we have shown that UPAL axiomatizes the substitution core of PAL with infinitely many agents. In this final section, we briefly discuss the axiomatization question for the single-agent and finite-agent cases. For a given language and class of models, the key question is how close we can come to expressing that two formulas are co-extensional in the epistemic submodel generated by the current point. For example, this condition is expressed by the formula $\Box_a^+(\varphi\leftrightarrow\psi)$ (where $\Box_a^+\alpha := \alpha\wedge\Box_a\alpha$) in *single*-agent PAL over *transitive* models. In this case, we get a new schematic validity in PAL:

$$\text{(inner extensionality)}\quad \Box_a^+(\varphi\leftrightarrow\psi)\rightarrow(\langle\varphi\rangle\alpha\leftrightarrow\langle\psi\rangle\alpha).$$

The corresponding legality condition for UPAL models is:

(inner extensionality) if $\|\varphi\|^{\mathfrak{M}}\cap\mathcal{R}_a(w) = \|\psi\|^{\mathfrak{M}}\cap\mathcal{R}_a(w)$, then $w\mathcal{R}_\varphi v$ iff $w\mathcal{R}_\psi v$,

which does not follow from any of the other legality conditions.

For multiple agents, we cannot in general express the co-extensionality of two formulas in the epistemic submodel generated by the current point; however, if we allow our models to be *non-serial*, then we do get related schematic validities for the single and finite-agent cases that are not derivable in UPAL-K$_n$ (where the antecedent can be written using $\Box_a$ operators and $\bot$): [8]

(FPE) "all $\mathcal{R}_{\mathsf{Agt}}$-paths from the current point are of length $\leq n$" $\rightarrow$ $(E^n(\varphi\leftrightarrow\psi)\rightarrow(\langle\varphi\rangle\alpha\leftrightarrow\langle\psi\rangle\alpha))$,

where

$$E^0\alpha := \alpha\wedge\bigwedge_{a\in\mathsf{Agt}}\Box_a\alpha \text{ and } E^n\alpha := \alpha\wedge E^0E^{n-1}\alpha.$$

The corresponding legality condition for UPAL is:

(FPE) if $\mathcal{R}^*_{\mathsf{Agt}}(w)$ is path-finite and $\|\varphi\|^{\mathfrak{M}}\cap\mathcal{R}^*_{\mathsf{Agt}}(w) = \|\psi\|^{\mathfrak{M}}\cap\mathcal{R}^*_{\mathsf{Agt}}(w)$, then $w\mathcal{R}_\varphi v$ iff $w\mathcal{R}_\psi v$,

[8] The (FPE) axioms are also schematically valid over serial models, because the antecedent is always false, but then they are also derivable using the seriality axiom $\Diamond_a\top$.

where $\mathcal{R}^*_{\mathsf{Agt}}(w)$ is path-finite just in case every $\mathcal{R}_{\mathsf{Agt}}$-path from w ends in a dead-end point in finitely many steps. This shows why the axiomatization of the substitution core of PAL-$\mathbf{K}_\omega$ is more elegant than that of PAL-$\mathbf{K}_n$: with infinitely many agents we cannot express the "everybody knows" modality E, so we do not need to add to UPAL the infinitely many FPE axioms.

Finally, if we consider PAL with the standard *common knowledge* operator C, then we can express co-extensionality in the generated epistemic submodel using the formula $C(\varphi \leftrightarrow \psi)$, in which case we get the new schematic validity

(common extensionality) $\quad C(\varphi \leftrightarrow \psi) \rightarrow (\langle\varphi\rangle\alpha \leftrightarrow \langle\psi\rangle\alpha).$

The corresponding legality condition in UPAL is:

(**common extensionality**) $\quad$ if $\|\varphi\|^{\mathfrak{M}} \cap \mathcal{R}^*_{\mathsf{Agt}}(w) = \|\psi\|^{\mathfrak{M}} \cap \mathcal{R}^*_{\mathsf{Agt}}(w)$, then $w\mathcal{R}_\varphi v$ iff $w\mathcal{R}_\psi v$.

We leave it to future work to give analyses for the above languages analogous to the analysis we have given here for $\mathcal{L}^{\omega}_{\mathsf{PAL}}$. A natural next step is to axiomatize the substitution core of the system of PAL-RC [6] with relativized common knowledge. Relativized common knowledge $C(\varphi, \psi)$ is interpreted in UPAL models exactly as in PAL models. We conjecture that UPAL together with the relativized common knowledge reduction axiom $\langle p\rangle C(q,r) \leftrightarrow C_A(\langle p\rangle q, \langle p\rangle r)$, the common extensionality axiom above, and the appropriate base logic (see [6]) axiomatizes the substitution core of PAL-RC with finitely or infinitely many agents over any of the model classes we have discussed. Indeed, it can be shown using arguments similar to those of §4 that the set of formulas in the language $\mathcal{L}^{\kappa}_{\mathsf{PAL\text{-}RC}}$ that are valid over legal UPAL models with **common extensionality** is exactly the substitution core of PAL-RC. Hence it only remains to prove that the extended system just described—call it UPAL-RC—is sound and complete for this model class. Such a proof requires a finite canonical model construction to deal with common knowledge, and we cannot go into the details here.

Another natural step is to attempt to apply the strategies of this paper to axiomatize the substitution cores of other dynamics epistemic logics, including the full system of DEL [4, Ch. 4]. One may imagine a general program of "uniformizing" dynamic epistemic logics, of which UPAL is only the beginning.

Acknowledgement

We wish to thank Johan van Benthem for stimulating our interest in the topic of this paper and the anonymous referees for their very helpful comments.

References

[1] van Benthem, J., *What One May Come to Know*, Analysis **64** (2004), pp. 95–105.
[2] van Benthem, J., *One is a Lonely Number: Logic and Communication*, in: Z. Chatzidakis, P. Koepke and W. Pohlers, editors, *Logic Colloquium '02*, ASL & A.K. Peters, 2006 pp. 96–129.

[3] van Benthem, J., *Open Problems in Logical Dynamics*, in: D. Gabbay, S. Goncharov and M. Zakharyashev, editors, *Mathematical Problems from Applied Logic I*, Springer, 2006 pp. 137–192.
[4] van Benthem, J., "Logical Dynamics of Information and Interaction," Cambridge University Press, 2011.
[5] van Benthem, J., *Two Logical Faces of Belief Revision* (2011), manuscript.
[6] van Benthem, J., J. van Eijck and B. Kooi, *Logics of communication and change*, Information and Computation **204** (2006), pp. 1620–1662.
[7] van Ditmarsch, H. and B. Kooi, *The Secret of My Success*, Synthese **151** (2006), pp. 201–232.
[8] van Ditmarsch, H., W. van der Hoek and P. Iliev, *Everything is Knowable – How to Get to Know Whether a Proposition is True*, Theoria **78** (2011), pp. 93–114.
[9] van Ditmarsch, H., W. van der Hoek and B. Kooi, "Dynamic Epistemic Logic," Springer, 2008.
[10] Aqvist, L., *Modal Logic with Subjunctive Conditionals and Dispositional Predicates*, Journal of Philosophical Logic **2** (1973), pp. 1–76.
[11] Balbiani, P., A. Baltag, H. van Ditmarsch, A. Herzig, T. Hoshi and T. de Lima, *'Knowable' as 'known after an announcement'*, The Review of Symbolic Logic **1** (2008), pp. 305–334.
[12] Ballarin, R., *Validity and Necessity*, Journal of Philosophical Logic **34** (2005), pp. 275–303.
[13] Baltag, A., L. Moss and S. Solecki, *The Logic of Public Announcements, Common Knowledge and Private Suspicions*, in: I. Gilboa, editor, *Proceedings of the 7th Conference on Theoretical Aspects of Rationality and Knowledge (TARK 98)*, Morgan Kaufmann, 1998 pp. 43–56.
[14] Blackburn, P., M. de Rijke and Y. Venema, "Modal Logic," Cambridge University Press, 2001.
[15] Burgess, J. P., *Which Modal Models are the Right Ones (for Logical Necessity)?*, Theoria **18** (2003), pp. 145–158.
[16] Buss, S. R., *The Modal Logic of Pure Provability*, Notre Dame Journal of Formal Logic **31** (1990), pp. 225–231.
[17] Carnap, R., *Modalities and Quantification*, The Journal of Symbolic Logic **11** (1946), pp. 33–64.
[18] Ciardelli, I. A., "Inquisitive Semantics and Intermediate Logics," Master's thesis, University of Amsterdam (2009), ILLC Master of Logic Thesis Series MoL-2009-11.
[19] Gerbrandy, J. and W. Groenevelt, *Reasoning about Information Change*, Journal of Logic, Language and Information **6** (1997), pp. 147–169.
[20] Goldblatt, R., "Logics of Time and Computation," CSLI Press, 1992.
[21] Halpern, J. Y., *Should Knowledge Entail Belief?*, Journal of Philosophical Logic **25** (1996), pp. 483–494.
[22] Holliday, W. H., T. Hoshi and T. F. Icard, III, *Schematic Validity in Dynamic Epistemic Logic: Decidability*, in: H. van Ditmarsch, J. Lang and S. Ju, editors, *Proceedings of the Third International Workshop on Logic, Rationality and Interaction (LORI-III)*, Lecture Notes in Artificial Intelligence **6953**, Springer, 2011 pp. 87–96.
[23] Holliday, W. H., T. Hoshi and T. F. Icard, III, *Information Dynamics and Uniform Substitution* (2012), manuscript.
[24] Holliday, W. H. and T. F. Icard, III, *Moorean Phenomena in Epistemic Logic*, in: L. Beklemishev, V. Goranko and V. Shehtman, editors, *Advances in Modal Logic*, **8**, College Publications, 2010 pp. 178–199.
[25] Kooi, B., *Expressivity and completeness for public update logics via reduction axioms*, Journal of Applied Non-Classical Logics **17** (2007), pp. 231–253.
[26] Ma, M., *Mathematics of Public Announcements*, in: H. van Ditmarsch, J. Lang and S. Ju, editors, *Proceedings of the Third International Workshop on Logic, Rationality and Interaction (LORI-III)*, Lecture Notes in Artificial Intelligence **6953**, Springer, 2011 pp. 193–205.
[27] Mascarenhas, S., "Inquisitive Semantics and Logic," Master's thesis, University of Amsterdam (2009), ILLC Master of Logic Thesis Series MoL-2009-18.

[28] Plaza, J., *Logics of public communications*, in: M. Emrich, M. Pfeifer, M. Hadzikadic and Z. Ras, editors, *Proceedings of the 4th International Symposium on Methodologies for Intelligent Systems*, Oak Ridge National Laboratory, 1989 pp. 201–216.
[29] Reynolds, M., *An Axiomatization of Full Computation Tree Logic*, The Journal of Symbolic Logic **66** (2001), pp. 1011–1057.
[30] Schurz, G., *Logic, matter of form, and closure under substitution*, in: M. Bilkova and L. Behounek, editors, *The Logica Yearbook*, Filosofia, 2005 pp. 33–46.
[31] Segerberg, K., *Two-dimensional modal logic*, Journal of Philosophical Logic **2** (1973), pp. 77–96.
[32] Wang, Y., *On Axiomatizations of PAL*, in: H. van Ditmarsch, J. Lang and S. Ju, editors, *Proceedings of the Third International Workshop on Logic, Rationality and Interaction (LORI-III)*, Lecture Notes in Artificial Intelligence **6953**, Springer, 2011 pp. 314–327.
[33] Wang, Y. and Q. Cao, *On Axiomatizations of Public Announcement Logic*, manuscript.

Expressiveness of Positive Coalgebraic Logic

Krzysztof Kapulkin [1]

Department of Mathematics
University of Pittsburgh, Pittsburgh PA, USA

Alexander Kurz [2]

Department of Computer Science
University of Leicester, United Kingdom

Jiří Velebil [3]

Department of Mathematics
Faculty of Electrical Engineering
Czech Technical University in Prague, Czech Republic

Abstract

From the point of view of modal logic, coalgebraic logic over posets is the natural coalgebraic generalisation of positive modal logic. From the point of view of coalgebra, posets arise if one is interested in simulations as opposed to bisimulations. From a categorical point of view, one moves from ordinary categories to enriched categories. We show that the basic setup of coalgebraic logic extends to this more general setting and that every finitary functor on posets has a logic that is expressive, that is, has the Hennessy-Milner property.

Keywords: Coalgebra, Modal Logic, Poset

1 Introduction

We study the logic of coalgebras over posets and show that to any functor $T : \mathsf{Pos} \to \mathsf{Pos}$ one can associate a positive modal logic $\mathcal{L}_T$, that is, a modal logic without negation. Moreover, this logic has the Hennessy-Milner property (= is expressive) if T is finitary. For example, this extends to posets the familiar result that the modal logic **K** distinguishes non-bisimilar states of finitely

[1] `kkapulki@andrew.cmu.edu`

[2] `kurz@mcs.le.ac.uk` The author acknowledges the support of the EPSRC, EP/G041296/1.

[3] `velebil@math.feld.cvut.cz` The author acknowledges the support of the grant No. P202/11/1632 of the Czech Science Foundation.

branching Kripke models. We will also show expressiveness for some non-finitary functors such as the one giving rise to image-finite labelled transition systems with infinitely many labels.

As in the classical set-based situation the notion of (bi)similarity of interest is classified by the final T-coalgebra. But over Pos the final coalgebra carries a partial order, thus classifying similarity [34,40,17,21,29,9]. Accordingly, $\mathcal{L}_T$ will always be invariant under T-similarity and characterise it at least if T is finitary.

In the set-based situation, expressiveness follows if there is an *injection* mapping elements of the final coalgebra to their theories. Over Pos, if we want the logic not only to separate points but also to characterise the order on the final coalgebra, we need to consider order-reflecting injections (=*embeddings*). Consequently, whereas any set-functor T preserves injections (with non-empty domain), over Pos we need to explicitely require that T *preserves embeddings*.

Moreover, we want a strong expressiveness result stating that expressiveness can be achieved by *monotone* modal operators. As usual in the coalgebraic setting, we obtain modal operators from T via predicate liftings [31,36] and monotonicy of the modal operators in the usual sense coincides with monotonicity of the predicate liftings. As can be seen from (13), monotonicity of the predicate liftings requires the collection $[X, Y]$ of monotone maps from X to Y to be considered as a poset. Accordingly, also T will need to preserve the order between maps, that is, we need to require that T is *locally monotone*. Technically speaking this means that we are working in the setting of categories enriched over Pos.

To summarise then, from a technical point of view, we transfer to the setting of coalgebras enriched over Pos Schröder's theorem [36] stating that for any finitary set-functor the logic of all predicate liftings is expressive, which now becomes that for any finitary, locally-monotone, and embedding preserving poset-functor the logic of all monotone predicate liftings is expressive.

From a category theoretic point of view, one may ask whether instead of just treating Pos it would be more appropriate to immediatly treat locally presentable categories [6] in general. Whereas this seems entirely natural from the coalgebraic point of view, it is problematic from the logical point of view: In the spirit of Stone duality, both Set and Pos are in a dual adjunction with Boolean algebras (BA) and distributive lattices (DL), respectively. This allows us to systematically associate a logic $\mathcal{L}_T$ for T-coalgebras of any functor T on Set or Pos. The main idea here, going back to Domain Theory in Logical Form [1] and to the duality for modal algebras and Kripke frames [16], is to obtain the logic $\mathcal{L}_T$ from the functor $L : \mathsf{BA} \to \mathsf{BA}$ 'dual' to $T : \mathsf{Set} \to \mathsf{Set}$. That this is possible for arbitrary functors T on Set was shown in [26] and it is one contribution of this paper to show that this carries over from Set to Pos, as long as we are willing to work in the setting of categories enriched over Pos.

An important aspect of Stone duality is that, although we start with a general functor T, we obtain on the algebraic side a logic given concretely by a set of modal operators of finite arity and a set of equations. Furthermore, equational logic provides us with a proof system. To come back to the question of how to generalise beyond posets, it is not clear what then should replace distributive lattices and equational logic. We expect that future developments will take the lead from the observation that Pos itself is enriched over a two-element category of "truth-values", suggesting to replace Pos by a category of $\mathcal{V}$-categories [20] (rather than by a locally finitely presentable category), thus generalising to many-valued modal logics.

Acknowledgements We are grateful to numerous anonymous referees whose criticism helped to improve the paper since its main result was first presented in [19].

2 Preliminaries

We review some basic material on the Stone duality approach to coalgebraic logic and on posets. More will be introduced later where the need arises.

2.1 Logical Connections

The basic ingredient of set-based coalgebraic logic is the adjunction

$$\mathsf{Set}^{op} \underset{Pred}{\overset{Stone}{\underset{\longrightarrow}{\overset{\longleftarrow}{\bot}}}} \mathsf{BA} \tag{1}$$

where *Pred* and *Stone* are the "predicate" and "Stone" functors, respectively. The functor *Pred* endows the powerset with the natural structure of a Boolean algebra and *Stone* takes the set of ultrafilters on a given Boolean algebra. Many nice properties of the above adjunction follow from the fact that it is given by a two-element set 2, that acts as a *schizophrenic object* in the sense of [33]. We will refer to the above adjunction as (an instance of) a *logical connection*.

Stone's representation theorem states that the unit $\eta_A : A \to \mathit{Pred}\,\mathit{Stone}\,A$ of the above adjunction is injective, which is a way of proving the completeness theorem of classical propositional logic.

By choosing the categories Set and BA we also have made a choice of over which category we will consider the coalgebras (here over Set), and where we will compute with the formulas of the relevant logic (here in BA).

Recall that, given a functor $T : \mathsf{Set} \to \mathsf{Set}$, a T-*coalgebra* (notation: (X, ξ) or just ξ) is a map $\xi : X \to TX$. A morphism $f : \xi \to \xi'$ is a map $f : X \to X'$ such that $Tf \cdot \xi = \xi' \cdot f$.

The rest of the set-based coalgebraic logic is therefore determined by a choice of a "behaviour" functor $T : \mathsf{Set} \to \mathsf{Set}$ and a functor $L : \mathsf{BA} \to \mathsf{BA}$ that captures the "logic" of coalgebras for T. The choice of T is made first and the functor L is subsequently computed to encode the modal operators and axioms describing T.

Thus, the full picture of set-based coalgebraic modal logic can be conveniently described by the following diagram [24]

$$T^{op} \circlearrowright \mathsf{Set}^{op} \underset{Pred}{\overset{Stone}{\underset{\longrightarrow}{\overset{\longleftarrow}{\bot}}}} \mathsf{BA} \circlearrowleft L \tag{2}$$

The syntax and proof system of the induced modal logic are given by (a presentation) of the initial L-algebra[4] in BA and the semantics by a natural transformation

$$\delta_X : LPredX \to PredT^{op}X \tag{3}$$

Explicitly, δ associates to any coalgebra (X,ξ) the L-algebra $(PredX, Pred\xi \cdot \delta)$ and the map from the initial L-algebra to $PredX$ gives the semantics of the formulas of the logic.

Example 2.1 We recover Kripke frames and modal algebras by taking TX to be the powerset $\mathcal{P}X$ of X and LA to be the free Boolean algebra generated by $\{\Box a \mid a \in A\}$ modulo the equations stating that $\Box$ preserves finite meets. δ is defined by $\delta(\Box a) = \{Y \subseteq X \mid Y \subseteq a\}$.

That the category of T-coalgebras in the example above is isomorphic to the category of Kripke frames and bounded morphisms appears in [3], see also [35]. That the category of L-algebras is isomorphic to the category of modal algebras or Boolean algebras with operators is due to [2]. The generalisation of this classic correspondence [16] to general T is due to [26]. Let us also remark already that this example gives the logic of all predicate liftings of $\mathcal{P}$ since all predicate liftings can be obtained from $\Box$ and Boolean operations.

2.2 Posets

We are interested in coalgebras over the category Pos of posets and monotone maps. We denote by $V : \mathsf{Pos} \to \mathsf{Set}$ the forgetful functor and by $D : \mathsf{Set} \to \mathsf{Pos}$ its left-adjoint, which sends a set to the corresponding discrete poset. D has a further left-adjoint $C : \mathsf{Pos} \to \mathsf{Set}$ sending a poset to the set of its connected components. Consequently, D preserves limits and colimits. Note that $VD = \mathrm{Id}$.

Definition 2.2 *An* embedding $f : X \rightarrowtail Y$ *in* Pos *is a map that is monotone and order-reflecting, ie* $x \leq y \Leftrightarrow f(x) \leq f(y)$.

Proposition 2.3 *A morphism* $f : X \rightarrowtail Y$ *is an embedding in* Pos *if and only if it is a regular mono, that is, an equalizer.*

Notation. 2 denotes the linear order $0 < 1$. Given posets X,Y we write $[X, Y]$ for the poset of monotone maps, ordered pointwise.

Assumption. In order to be able to use the (enriched) Yoneda lemma, we assume that all functors $T : \mathsf{Pos} \to \mathsf{Pos}$ are *locally monotone*, that is, $f \leq g$ implies $Tf \leq Tg$.

2.3 Coalgebras over posets

Given a locally monotone $T : \mathsf{Pos} \to \mathsf{Pos}$, we will study the category $\mathsf{Coalg}(T)$ of *T-coalgebras*

$$\xi : X \to TX.$$

[4] An *L-algebra* (notation: (A, α) or just α) is an arrow $\alpha : LA \to A$ in BA. A morphism $f : \alpha \to \alpha'$ is an arrow $f : A \to A'$ in BA such that $f \cdot \alpha = \alpha' \cdot Lf$.

A *coalgebra morphism* $f : \xi \to \xi'$ is a monotone map $f : X \to X'$ such that $Tf \cdot \xi = \xi' \cdot f$ holds. We consider coalgebra homomorphisms to be ordered pointwise, i.e., $\mathsf{Coalg}(T)$ as *enriched* over Pos.

Coalgebras over posets have recently been studied by Levy [29]. Given a set-functor H and a so-called H-relator Γ, and following earlier work by, e.g., [38,17], he defines the notion of Γ-simulation between two H-coalgebras. Further he associates a functor $T : \mathsf{Pos} \to \mathsf{Pos}$ to Γ and shows that that the final T-coalgebra is fully abstract w.r.t. Γ-simulation. For our purposes we can summarise [29] as follows. Say that $R : X \nrightarrow Y$ is a *monotone relation* from X to Y if $R \subseteq X \times Y$ and $R = \leq_X ; R ; \leq_Y$ where ; denotes relational composition. A monotone relation $R : X \nrightarrow X'$ is a *simulation* from $\xi : X \to TX$ to $\xi' : X' \to TX'$ if

$$R \subseteq (\xi \times \xi')^{-1}((\mathsf{Rel}(T))(R)).$$

Here, $\mathsf{Rel}(T)$ is the relation lifting of T, that is, see [11],

$$\mathsf{Rel}(T)(R) = \{(a, a') \in TX \times TX' \mid \exists w \in TR \,.\, a = T\pi_X(w), a' = T\pi_{X'}(w)\}.$$

Alternatively, *similarity* can be defined via the final coalgebra: x is simulated by y if

$$!_\xi(x) \leq !_{\xi'}(x')$$

where ! denotes arrows into the final T-coalgebra.

The two definitions of similarity are equivalent under reasonable assumptions on the functor T by the Rutten-Worrell coinduction theorem [34, Thm 4.1], [40, Thm 5.10].

Example 2.4 We obtain syntax and (a slightly generalised) semantics of positive modal logic [15] by taking TX to be set of convex subsets of X and LA to be the free distributive lattice generated by $\{\Box a, \Diamond a \mid a \in A\}$ modulo the equations stating that $\Box$ preserves finite meets, $\Diamond$ preserves finite joins and the equations (1) of [15]. δ is defined by $\delta(\Box a)$ as in Example 2.1 and $\delta(\Diamond a) = \{Y \subseteq X \mid Y \cap a \neq \emptyset\}$.

In this example similarity agrees with bisimilarity due to the special nature of convex sets. The usual notions of similarity are obtained by taking upsets or downsets, see Example 3.4.

3 Functors on posets

The relationship between the modal logic **K** and positive modal logic [15] can be explained via the observation that the convex powerset functor is the extension of the powerset functor, see Definition 3.1.

Any finitary set functor H arises as a coequaliser

$$\textstyle\coprod_{n,m<\omega} Hn \times \mathsf{Set}(n,m) \times X^m \rightrightarrows \coprod_{n<\omega} Hn \times X^n \longrightarrow HX \qquad (4)$$

where the upper map takes $(\sigma \in Hn, f : n \to m, v : m \to X)$ to $(\sigma, v \cdot f)$ and the lower map to $(Hf(\sigma), v)$. In more familiar notation, the coequaliser amounts to imposing the equations

$$\sigma(x_{f(1)}, \ldots x_{f(n)}) = Hf(\sigma)(x_1, \ldots x_m)$$

where n, m range over non-negative integers, f over maps $n \to m$, and σ over Hn.

Using the inclusion $D : \mathsf{Set} \to \mathsf{Pos}$, we can calculate this coequaliser not only for sets X but also for posets X, as follows.

Definition 3.1 *Let H be a finitary set functor. Define $\bar{H} : \mathsf{Pos} \to \mathsf{Pos}$ via the coequaliser in* Pos

$$\coprod_{n,m<\omega} DHn \times [Dn, Dm] \times [Dm, X] \rightrightarrows \coprod_{n<\omega} DHn \times [Dn, X] \longrightarrow \bar{H}X$$

Remark 3.2 (i) One reason for defining $\bar{H}$ via the coequaliser is that then $\bar{H}$ is locally monotone. The more immediate DHV (with $V : \mathsf{Pos} \to \mathsf{Set}$ the forgetful functor) is not locally monotone in general.

(ii) Note that $\bar{H}$ extends H in the sense that we have $\bar{H}D = DH$. The reason is that $D : \mathsf{Set} \to \mathsf{Pos}$ is a full co-reflective subcategory and, therefore, colimits of diagrams $\mathcal{C} \to \mathsf{Pos}$ factoring through D are already in Set, see the proof of Proposition 3.5.3 in Borceux [12, Vol 1].

(iii) D extends to a functor $\mathsf{Coalg}(H) \to \mathsf{Coalg}(\bar{H})$, due to $\bar{H}D = DH$.

Extensions of functors from Set to Pos are investigated in [9]. For example, we know that the final $\bar{H}$-coalgebra, if it exists, is discrete.

Example 3.3 (i) A *polynomial endofunctor* $H\colon \mathsf{Set} \to \mathsf{Set}$ is a functor given by

$$H(X) = \coprod_{n<\omega} \Sigma_n \times X^n.$$

The order on $\bar{H}(X, \leq)$ is the point-wise order induced by $(X, \leq)$.

(ii) The finite powerset functor $\mathcal{P}_\omega : \mathsf{Set} \to \mathsf{Set}$ extends to $\bar{\mathcal{P}}_\omega : \mathsf{Pos} \to \mathsf{Pos}$ mapping a poset X to the set of finitely generated convex subsets of X. The order on $\bar{\mathcal{P}}_\omega(X)$ is known as the Egli-Milner order, explicitly, for $A, B \in \bar{\mathcal{P}}_\omega(X)$ we have $A \leq B$ iff

$$\forall x \in A \,.\, \exists y \in B \,.\, x \leq y \ \wedge\ \forall y \in B \,.\, \exists x \in A \,.\, x \leq y,$$

With the exception of the first one, the following examples do not extend set-functors as they do not map discrete sets to discrete sets. Accordingly, interesting (non-symmetric) notions of similarity are obtained.

Example 3.4 (i) An example of a functor that does not preserve embeddings is the one which maps a poset to the discrete poset of its connected components.

(ii) $\mathcal{U}p_\omega : \mathsf{Pos} \to \mathsf{Pos}$ is the covariant functor which maps a poset to the set of all finitely generated up-sets ordered by reverse inclusion. Spelling out the definition of simulation from Section 2.3, we obtain that R is a simulation if $xRx' \Rightarrow \xi(x)\mathsf{Rel}(\mathcal{U}p_\omega)(R)\xi'(x')$, which is equivalent to

$$\forall y' \in \xi'(x') \,.\, \exists y \in \xi(x) \,.\, yRy'$$

(iii) $\mathcal{D}own_\omega : \mathsf{Pos} \to \mathsf{Pos}$ is the covariant functor which maps a poset to the set of all finitely generated down-sets ordered by inclusion. Here we have that R is a simulation if xRx' implies that

$$\forall y \in \xi(x) \,.\, \exists y' \in \xi'(x') \,.\, yRy'$$

(iv) Let A be a poset and $T : \mathsf{Pos} \to \mathsf{Pos}$, $TX = A \ltimes X$, where $\ltimes$ refers to the lexicographic ordering: $A \ltimes X$ has carrier $A \times X$ and the order is given by $(a, x) < (a', x') \Leftrightarrow (a < a' \vee (a = a' \wedge x < x'))$. In its second argument $\ltimes$ is functorial and locally monotone. Pavlović and Pratt [32] showed that the final $\mathbb{N} \ltimes \mathrm{Id}$-coalgebra is isomorphic to the non-negative real numbers.

(v) Consider $T : \mathsf{Pos} \to \mathsf{Pos}$, $TX = X^2$, that is, X is mapped to the poset of pairs (x_1, x_2) with $x_1 \leq x_2, x_1, x_2 \in X$.

(vi) Write $X \lhd X$ for the functor that makes two disjoint copies of X with everything on the left being smaller than anything on the right.

(vii) Allwein and Harrison [8] advertise the use of partially-ordered modalities. If A is an ordered set and $T : \mathsf{Pos} \to \mathsf{Pos}$ a functor, then $[A, T]$ is a functor which has A-indexed T-modalities. This generalises the approach of [8] to the situation where not only modalities, but also carriers of coalgebras may be partially ordered.

4 Logic for Coalgebras over Posets

Technically, in this paper, we replace the adjunction between Set and BA of diagram (1) by an adjunction

$$\mathsf{Pos}^{op} \underset{Pred}{\overset{Stone}{\underset{\bot}{\leftrightarrows}}} \mathsf{DL} \tag{5}$$

between Pos and DL, the category of distributive lattices. The above adjunction is to be considered as an adjunction in the *enriched sense*. This means that both the predicate functor *Pred* and the Stone functor *Stone* are locally monotone and that there is an isomorphism

$$\mathsf{Pos}(X, \mathit{Stone} A) = \mathsf{Pos}^{op}(\mathit{Stone} A, X) \cong \mathsf{DL}(A, \mathit{Pred} X)$$

of *posets*, natural in X and A.

The predicate functor *Pred* assigns to a poset X the poset $[X, 2]$ of monotone maps $X \to 2$ endowed with the distributive lattice structure induced by 2. Observe that it means that the following diagram

$$\begin{array}{ccc} \mathsf{Pos}^{op} & \xrightarrow{\;Pred\;} & \mathsf{DL} \\ & {\scriptstyle [-,2]}\searrow \quad \swarrow{\scriptstyle U} & \\ & \mathsf{Pos} & \end{array} \tag{6}$$

commutes up to isomorphism, where U denotes the obvious (locally monotone!) forgetful functor from DL to Pos. Observe that, in elementary terms, $\mathit{Pred}X$ is the distributive lattice of upper-sets in X. Also note that there is an adjunction

$$F \dashv U : \mathsf{DL} \to \mathsf{Pos}.$$

The Stone functor $\mathit{Stone} : \mathsf{DL} \to \mathsf{Pos}^{op}$ assigns the poset $\mathsf{DL}(A, \mathbb{2})$ of prime filters on A to the distributive lattice A.

Notice that the adjunction of diagram (5) is built in the same way as the one of diagram (1), with $\mathbb{2}$ instead of 2. This "sameness" can be stated precisely by introducing schizophrenic objects and adjunctions they generate in the enriched setting. See [27] for the development of the theory along the lines of [33]. Let us just comment that basic ideas of [33] carry over to enrichment over Pos without any difficulties.

The whole picture of poset-based coalgebraic logic will therefore be given by the diagram

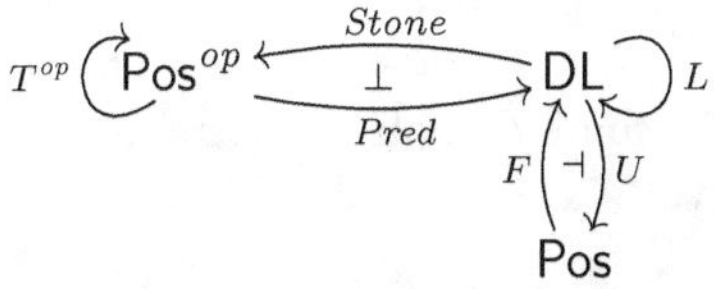

(7)

As before the syntax and axioms of the logic will be given by a functor L and the semantics will be given by a natural transformation

$$\delta_X : L\mathit{Pred}X \to \mathit{Pred}T^{op}X \tag{8}$$

Definition 4.1 *We call a pair* (L, δ) *as in* (7) *and* (8) *a* logic for T.

Remark 4.2 The adjunction $F \dashv U$ is important in order to be able to present the functor L concretely by operations and equations. As any finitary variety, DL is a completion $\mathrm{inc} : \mathsf{DL}_0 \to \mathsf{DL}$ of the full subcategory DL_0 of finitely generated free algebras, see [7] for details. Consequently, any functor $L' : \mathsf{DL}_0 \to \mathsf{DL}$ can be extended continuously to a functor $L : \mathsf{DL} \to \mathsf{DL}$. Technically, this can be expressed by saying that L is the left-Kan extension [28] of L' along inc. Conversely, if a functor L arises as such a left-Kan extension, we say that L *is determined by finitely generated free algebras*, i.e., in the terminology of [39], L is *finitely based* w.r.t. $F \dashv U$. The important fact for us is that such a functor can be presented by operations and equations [26]. We do not have the space for a full account on this, but we will see more details in Section 4.2.

We think of δ as the one-step semantics of the logic and may write for $y \in TX$ and $b \in L\mathit{Pred}X$

$$y \Vdash b \iff y \in \delta(b) \tag{9}$$

To go from the one-step semantics to the 'global' semantics, we have to iterate the one-step logic-constructor L and form the initial L-algebra as the colimit of the initial L-sequence, see, e.g., [24]. This colimit exists if L is finitary, which will be the case in the examples we will look at later. For now we assume that the initial L-algebra $L\mathcal{L} \to \mathcal{L}$ exists and we consider $\mathcal{L}$ as the set of formulas of the logic. The semantics of a formula $\varphi \in \mathcal{L}$ w.r.t. a coalgebra $\xi : X \to TX$ is then given by the unique L-algebra morphism $[\![\cdot]\!]_\xi$

$$\begin{array}{ccc} L\mathcal{L} & \longrightarrow & \mathcal{L} \\ {\scriptstyle L[\![\cdot]\!]_\xi}\downarrow & & \downarrow{\scriptstyle [\![\cdot]\!]_\xi} \\ LPredX \xrightarrow{\delta_X} PredTX & \xrightarrow{Pred\xi} & PredX \end{array} \tag{10}$$

If L has a presentation by operations and equation as in Remark 4.2, a formula can be represented as $\boxdot(\varphi_1, \dots \varphi_n)$ where $\boxdot$ is an n-ary 'modal' operation symbol and $\varphi_i \in \mathcal{L}$. Then (10) can be written as the inductive clause

$$x \in [\![\boxdot(\varphi_1, \dots \varphi_n)]\!]_\xi \;\Leftrightarrow\; \xi(x) \in \delta_X(\boxdot([\![\varphi_1]\!]_\xi, \dots [\![\varphi_n]\!]_\xi))$$

4.1 Expressiveness

We discuss the expressiveness of coalgebraic logics following Klin [23, Theorem 4.2] (see [18, Theorem 4]). Let us say that a logic for T-coalgebras is *expressive* if all elements x, y of the final T-coalgebra *can be separated by a formula*, or, more precisely, if we have

$$x \not\leq y \;\Rightarrow\; \exists \varphi \in \mathcal{L} \,.\, x \Vdash \varphi \;\&\; y \not\Vdash \varphi \tag{11}$$

This means that the logic not only separates elements, but also characterises the order, namely, $x \leq y$ iff $\forall \varphi \in \mathcal{L} \,.\, x \Vdash \varphi \Rightarrow y \Vdash \varphi$.

The formal treatment will follow an idea analogous to the one introduced by Pattinson [31] for completeness arguments: We first define what it means to be *one-step* expressive and then show that one-step expressiveness extends to expressiveness. One advantage of this one-step approach is that it often gives rise to modularity [13,37]: If we are interested in an inductively defined class of functors and we can show that all constructions preserve a certain property P one-step-wise, then it follows that all functors in that class have property P.

Coming back to one-step expressiveness, first note that the natural transformation δ induces its *mate* $\tau : T^{op} Stone \to Stone L$ given explicitly by the pasting [5]

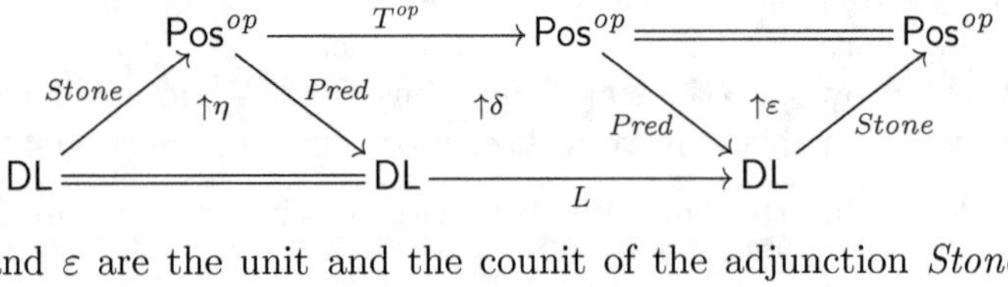
Pos^op
T^op
Pos^op
Pos^op
Stone
↑η
Pred
↑δ
Pred
↑ε
Stone
DL
DL
L
DL
(12)

where η and ε are the unit and the counit of the adjunction $Stone \dashv Pred$, see (7). Explicitly, using the notation from (9), we have

$$\begin{aligned} \tau_A : T^{op} Stone A &\to Stone LA \\ y &\mapsto \{b \in LA \mid y \Vdash L(\eta)(b).\} \end{aligned}$$

[5] That is, in Pos^{op} we have that τ is given by $\varepsilon T^{op} Stone \cdot Stone \delta Stone \cdot Stone L \eta$.

Thus, if τ_A is injective, then all elements of $T^{op}\,Stone A$ can be separated by some 'one-step formula' $b \in LA$ and if, moreover, τ_A is an embedding then the logic even characterises the order on $T^{op}\,Stone A$. Thus we see that τ being an embedding is the one-step version of (11).

Alternatively, one could try to define one-step expressiveness by the surjectivity of δ, but this is too strong: It means that every subset is the extension of some formula.

Definition 4.3 *We call* (L, δ) *expressive if* (11) *holds and one-step expressive if the mate* τ *of* δ *is an embedding.*

Remark 4.4 These definitions work over Set as well as over Pos. Note that basic modal logic is not one-step expressive for $\mathcal{P} : \mathsf{Set} \to \mathsf{Set}$. Indeed, if $\alpha : LA \to A$ is the Lindenbaum-algebra of the logic, then α is an isomorphism and we have $\tau_A : \mathcal{P}^{op}\,Stone A \to Stone LA \cong Stone A$, which is not injective. On the other hand, basic modal logic is one-step expressive for the finite powerset $\mathcal{P}_\omega$.

Theorem 4.5 *Let* $T : \mathsf{Pos} \to \mathsf{Pos}$ *be finitary and embedding-preserving and let* (L, δ) *be a logic for* T. *If the logic is one-step expressive, then it is expressive.*

Proof. We have the following diagram in Pos:

$$\begin{array}{ccccccccc}
Stone2 = 1 & \longrightarrow & Stone L2 & \longrightarrow & Stone L^2 2 & \longrightarrow \ldots \longrightarrow & Stone L^k 2 & \longrightarrow \\
\uparrow \theta_0 & & \uparrow \theta_1 & & \uparrow \theta_2 & & \uparrow \theta_k & \\
1 & \longrightarrow & T1 & \longrightarrow & T^2 1 & \longrightarrow \ldots \longrightarrow & T^k 1 & \longrightarrow
\end{array}$$

We define θ_0 to be the identity on 1 and $\theta_{k+1} = \tau_{L^k 2} \cdot T(\theta_k)$. We need an auxiliary claim: There is an ordinal α such that $T^\alpha 1$ is the final coalgebra and $T^\alpha 1 \to T^\omega 1$ is an embedding. Indeed, generalising a well-known result of Worrell [41], Adámek [4] shows in the proof of his Theorem 4.6 the following: (1a) $T^{\omega+1}1 \to T^\omega 1$ is an embedding; (1b) this property is preserved at successor ordinals; (1c) this property is preserved at limit ordinals. [6] Consequently, beyond ω, the final sequence consists only of embeddings and thus must converge to the final coalgebra.

Now suppose $x \not\leq y$ in the final coalgebra $\zeta : T^\alpha 1 \to T^{\alpha+1}1$. Due to the claim, we have $x \not\leq y$ in $T^\omega 1$. But this means that there is $n < \omega$ such that $x \not\leq y$ in $T^n 1$. Since T preserves embeddings, θ_n is an embedding, hence $\theta_n(x) \not\subseteq \theta_n(y)$. Thus there must be $\varphi \in \theta_n(x)$ with $\varphi \notin \theta_n(y)$. It is routine, if lengthy, to verify that this implies that $x \in [\![\varphi]\!]_\zeta$ and $y \notin [\![\varphi]\!]_\zeta$. □

Remark 4.6 The proof can be seen as a final sequence version of the proof of Klin [23, Theorem4.2], which in turn can be seen as a category theoretic analysis of Schröder [36].

---

[6] This part of the proof of Theorem 4.6 in [4] does not need its assumption that T preserves epis; and the argument for (1c) is not specific to the ordinal 2ω.

A benefit of the final sequence approach is that we also get a result for functors that are not necessarily finitary. Adapting [25], we say about two states x, x' of two coalgebras $\xi : X \to TX, \xi' : X' \to TX'$ that x is *ω-simulated* by x' if $!_n(x) \leq !_n(x')$ for all $n < \omega$, where $!_n(x) : X \to T^n 1$ is the map induced by ξ. A logic is called *ω-expressive* if whenever x is not ω-simulated by x', then $\exists \varphi \in \mathcal{L} . x \Vdash \varphi \ \& \ y \nVdash \varphi$. The proof of the following corollary then repeats the last three sentences of the proof of the theorem.

Corollary 4.7 *Let T be a (not necessarily finitary) endofunctor on* Pos *preserving embeddings and let (L, δ) be a one-step expressive logic for T.*[7] *Then (L, δ) is ω-expressive.*

4.2 Predicate Liftings

As in the Set-based case L and δ can be described more explicitly by predicate liftings, introduced by Pattinson [31]. To transport this approach to our setting, we only need to generalise the n-th power (n is a finite set) of a set S to the "n-th power of a poset S", where n now a finite poset. It turns out that the universal property (natural in X)

$$\mathsf{Set}(X, S^n) \cong \mathsf{Set}(n, [X, S])$$

of the n-fold power of a set S can be taken verbatim to define the desired (enriched) notion for posets.

Definition 4.8 *Suppose S is a poset and n is a finite poset. The* n-fold cotensor *of S is a poset $n \pitchfork S$ together with an isomorphism*

$$\mathsf{Pos}(X, n \pitchfork S) \cong \mathsf{Pos}(n, [X, S])$$

natural in X, where $[X, Y]$ denotes the poset of monotone maps from X to Y.

It is easy to verify that $n \pitchfork S$ is the poset $[n, S]$ of all monotone maps from n to S, ordered pointwise. The reason to introduce the $\pitchfork$ notation is that it carries over if one wants to replace Pos in (5) by, for example, Priestley spaces [14] (which are dually equivalent to DL): then X, S are Priestley spaces but n remains a poset, so we could not write $[n, S]$ anymore.

Definition 4.9 *An n-ary* predicate lifting *for a functor T is a natural transformation λ given by components*

$$\lambda_X : [X, n \pitchfork 2] \to [TX, 2] \qquad (13)$$

where n can be any finite poset.

As opposed to the Set-based case, all predicate liftings are monotone since each λ_X is an arrow in Pos and hence a monotone map.

---

[7] One-step expressiveness can be weakened. It is enough to require that the τ_A are embeddings for those $A = L^k 2$ appearing in the final sequence construction. An example is given by the (not-necessarily finitely generated) convex subsets functor.

Remark 4.10 It follows from the (enriched) Yoneda lemma that the poset of all predicate liftings is order-isomorphic to

$$[T(n \pitchfork 2), 2]. \tag{14}$$

Recall also that, for every finite poset n, the poset $[T(n \pitchfork 2), 2]$ is an underlying poset of a distributive lattice that we denote by Λn (where Λ should remind of lifting). The nature of the adjunction (5) and diagram (6) allows us to infer the isomorphism

$$\Lambda n \cong PredT^{op} StoneFn \tag{15}$$

where Fn denotes the free DL over a finite poset n. The assignment $n \mapsto \Lambda n$ can be viewed as a finitary signature in the enriched sense: arities are finite posets and n-ary operations (i.e., n-ary predicate liftings) form a distributive lattice.

The following remark explains how predicate liftings fit into diagram (7). Recall that $U : \mathsf{DL} \to \mathsf{Pos}$ denotes the forgetful functor.

Remark 4.11 Predicate liftings induce a functor $L : \mathsf{DL} \to \mathsf{DL}$. The idea is that LA is the distributive lattice of "modal formulas of depth one, labelled in elements of a distributive lattice A". Hence L is the polynomial functor corresponding to the above signature Λ and the precise formula is given by

$$LA = \coprod_{n \text{ finite poset}} [n, UA] \otimes \Lambda n \tag{16}$$

where by $\otimes$ we denote the $[n, UA]$-fold *tensor* of the distributive lattice Λn. In general, P-fold tensor of a distributive lattice A is a distributive lattice $P \otimes A$ together with an isomorphism

$$\mathsf{DL}(P \otimes A, B) \cong \mathsf{Pos}(P, \mathsf{DL}(A, B))$$

natural in B. Since (16) is a left Kan extension formula (in the appropriate enriched sense), L comes equipped for canonical reasons with a natural transformation given by:

$$\delta_X : LPredX \to PredT^{op}X \tag{17}$$

$$\lambda(a_1, \ldots, a_n) \mapsto \lambda \cdot Ta' : TX \to T(n \pitchfork 2) \to 2, \tag{18}$$

where $a' : X \to n \pitchfork 2$ is the transpose of $a : n \to UPredX = [X, 2]$, see Definition 4.8.

Remark 4.12 Instead of taking all predicate liftings as in (15), we can consider any collection $\Lambda'_n \subseteq U\Lambda n$ of predicate liftings. It only means that in (16) we replace Λn by the DL freely generated from Λ'_n. The corresponding logic (L, δ) is then still given by (18) where now $\lambda \in \Lambda'_n$.

Proposition 4.13 *The functor L preserves surjective homomorphisms.*

Proof. As L is determined by its values on finitely generated free algebras, see Remark 4.2, it preserves sifted colimits and, therefore, reflexive coequalisers [7]. But every surjective homomorphism is the reflexive coequaliser of its kernel pair. □

4.3 The Logic of all Predicate Liftings is Expressive

The next theorem shows that the logic of all predicate liftings, or also the logic of all predicate liftings with discrete arities, is expressive. The key observations are contained in the following lemma.

Lemma 4.14 *Every finite poset X can be embedded into a poset $n \pitchfork 2$ for some finite discrete poset n.*

Proof. Write n for the discrete poset of elements of X. The embedding is given by $X \to [X^{\mathrm{op}}, 2] \to [n, 2]$ where the first map takes principal down-sets and the second is the embedding from monotone maps $X^{\mathrm{op}} \to 2$ into all not-necessarily monotone maps. □

The next lemma requires embeddings as opposed to only injections. It is also the place where it comes in that DL has finite meets and joins. Finally, we note for future generalisations that the lemma makes use of the following: (i) every poset is a filtered colimit of finite posets where we can take the cone to consist of embeddings (reg monos); (ii) if X is finite and $X \to \mathit{Stone}A$ is reg mono then the transpose $A \to \mathit{Pred}X$ is reg epi; (iii) L preserves reg epis.

Lemma 4.15 *Consider a finite poset X and an embedding $c : X \to \mathit{Stone}A$. Denote by $c^\sharp : A \to \mathit{Pred}X$ the transpose of c. Then $\mathit{Stone}Lc^\sharp : \mathit{Stone}L\mathit{Pred}X \to \mathit{Stone}LA$ is an embedding in* Pos.

Proof. We first prove that $c^\sharp$ is a surjection. To this end, it is convenient to identify $\mathit{Pred}X$ with the set of upsets of X (ordered by inclusion), $\mathit{Stone}A$ with the set of prime filters on A (ordered by inclusion), and to abbreviate $a \in c(x)$ by $x \Vdash a$. In this notation $c(x) = \{a \in A \mid x \Vdash a\}$ and $c^\sharp(a) = \{x \in X \mid x \Vdash a\}$. Moreover, that c is an embedding means that $x \not\leq y$ iff there is $a_{xy} \in A$ such that $x \Vdash a_{xy}$ and $y \nVdash a_{xy}$. Now consider a 'principal' upset $\uparrow x \in \mathit{Pred}X$. Since X is finite we find $a_x = \bigwedge\{a_{xy} \mid y \in X \text{ and } x \not\leq y\}$ in A. It follows $c^\sharp(a_x) = \uparrow x$. Since every upset in $\mathit{Pred}X$ is a finite join of principal upsets, $c^\sharp$ is onto.

Now that $c^\sharp$ is a surjection, by Proposition 4.13, we have that $Lc^\sharp$ is a surjection as well. But in all equational classes of algebras all epimorphisms are regular, so $Lc^\sharp$ is a regular epi in DL. Finally, $\mathit{Stone}Lc^\sharp$ is a regular epi in Pos^{op} since it is an image of a regular epi under a left adjoint. □

Theorem 4.16 *If $T : \mathsf{Pos} \to \mathsf{Pos}$ is finitary, locally monotone, and embedding-preserving, and Λ consists of all predicate liftings with finite arities, then τ is an embedding.*

Proof. (For reasons of type setting, we write $S = \mathit{Stone}$ and $P = \mathit{Pred}$ inside this proof.) By the definition of τ, we need to show that the composite

$$SL\varepsilon_A \cdot S\delta_{SA} \cdot \eta_{TSA} : TSA \to SPTSA \to SLPSA \to SLA$$

is an embedding. Consider $c : X \to SA$ and its transpose $c^\sharp : A \to PX$. Now

$$\begin{array}{ccccccc} TSA & \xrightarrow{\eta_{TSA}} & SPTSA & \xrightarrow{S\delta_{SA}} & SLPSA & \xrightarrow{SL\varepsilon_A} & SLA \\ {\scriptstyle Tc}\uparrow & & & & & \nearrow{\scriptstyle SLc^\sharp} & \\ TX & \xrightarrow{\eta_{TX}} & SPTX & \xrightarrow{S\delta_X} & SLPX & & \end{array} \tag{19}$$

commutes because of the naturality of η and δ. Further, Pos being a locally finitely presentable category [6], SA is a filtered colimit $c_i : X_i \to SA$ where we can take the X_i to be finite and the c_i to be embeddings. Therefore, since T preserves filtered colimits, we only need that the lower composite of (19)

$$SLc^\sharp \cdot S\delta_X \cdot \eta_{TX} : TX \to SPTX \to SLPX \to SLA$$

is an embedding for finite X and embeddings $c : X \to SA$. By Lemma 4.15 we need to show that

$$\alpha = S\delta_X \cdot \eta_{TX} : TX \to SPTX \to SLPX$$

is an embedding. According to (18), α maps a point $t \in TX$ together with a formula $\lambda(a)$ to its truth value via

$$\lambda \cdot Ta : TX \to T(n \pitchfork 2) \to 2. \tag{20}$$

To show that α is order-reflecting, consider $t_1, t_2 \in TX$, $t_1 \not\leq t_2$. Due to Lemma 4.14 and T preserving embeddings, there is $a : X \to n \pitchfork 2$ such that $Ta(t_1) \not\leq Ta(t_2)$. Define $\lambda : T(n \pitchfork 2) \to 2$ as $\lambda(x) = 1 \Leftrightarrow Ta(t_1) \leq x$, which is monotone. We now have $\lambda(Ta(t_1)) \not\leq \lambda(Ta(t_2))$, hence α is order-reflecting. □

Corollary 4.17 *Let* $T : \mathsf{Pos} \to \mathsf{Pos}$ *be finitary, locally monotone, and embedding-preserving. The logic of all predicate liftings (with discrete arities) is expressive.*

Remark 4.18 This theorem is not a consequence of [23, Thm 4.4] since we need strong monos instead of monos and also because we want that discrete arities suffice for expressiveness. For a precise comparison of the two theorems, we give below a common generalisation of both, but the only instances we know are Theorem 4.16 and [23, Thm 4.4].

Theorem: Let $\mathcal{X}$ be a locally finitely presentable category [6], $F \dashv U : \mathcal{A} \to \mathsf{Set}$, $\mathcal{A}_0 \hookrightarrow \mathcal{A}$ a full, dense subcategory, and $P \dashv S : \mathcal{A}^{\mathrm{op}} \to \mathcal{X}$. Let $LA = PTSA$ for $A \in \mathcal{A}_0$ and extend via colimits to $L : \mathcal{A} \to \mathcal{A}$. Let $(\mathcal{E}, \mathcal{M})$ be either the (StrongEpi,Mono) or a (Epi,RegMono) factorisation system [6, 1.61] on $\mathcal{X}$. Moreover we require: T preserves arrows in $\mathcal{M}$; the unit $X \to SPX$ is pointwise in $\mathcal{M}$; (*) given X finitely presentable, $X \to \bar{X}$ in $\mathcal{E}$, and $m : \bar{X} \to SA$ in $\mathcal{M}$, there is $A' \in \mathcal{A}$ and $\varphi : A' \to A$ such that $S\varphi \cdot m$ in $\mathcal{M}$. Then the mate $\tau : TS \to ST$ of $\delta : LP \to PT$ is in $\mathcal{M}$.

The technical condition (*) can be established in the case of (Epi,StrongMono)-factorisation if $\mathcal{X}$ is, as in [23, Thm 4.4], strongly locally finitely presentable. In our case, (*) follows from Lemma 4.15.

An interesting example of a non-finitary functor that has an expressive logic is $\mathcal{P}_\omega^{Act}$ where *Act* is an infinite set of 'actions' and $\mathcal{P}_\omega$ is finite powerset. Its coalgebras are image-finite transition systems. We can extend our results to these in the manner of [41].

Theorem 4.19 *If T_i is a family of finitary, locally monotone functors preserving embeddings, then $\prod_i T_i$ has an expressive logic.*

Proof. By Theorem 4.16 each T_i has a logic (L_i, δ_i) with the mates $\tau_i : T_i S \to SL_i$ being embeddings. Hence $\tau : \prod_i T_i S \to \prod_i SL_i \cong S \coprod_i L_i$ is an embedding and therefore $(\coprod L_i, \delta)$ with now δ being the mate of τ is one-step expressive. It follows from Corollary 4.7 that $\coprod_i L_i$ is ω-expressive and from (the poset-version of) [41, Theorem 13] that the final T-coalgebra is embedded in the ω-limit of the final sequence, hence ω-expressiveness implying expressiveness.□

Remark 4.20 The same proofs, with Pos replaced by Set also give proofs of the expressivity of coalgebraic logic over Set. The set-analogue of Theorem 4.19 is also of interest. We also strengthen the results of [36,22] in that monotone predicate liftings with negation-free propositional logic are expressive for finitary set-functors.

4.4 Separating Sets of Predicate Liftings

In concrete examples, we are interested in generating logics from small sets Λ of predicate liftings. Going back to Remark 4.12 and (18) and (20), we see that for expressivity, it is enough to require that the set of predicate liftings is separating. We adapt this notion from [31,36] to the ordered setting.

Definition 4.21 *A collection Λ' of predicate liftings is* separating *if the family*

$$(\widehat{\lambda}_X : TX \to [[X, n \pitchfork 2], 2])_{\lambda \in \Lambda n}$$

is jointly order-reflecting, that is, for all finite X and all $t_1, t_2 \in TX$ with $t_1 \not\leq t_2$ there are a finite poset n and $a : X \to n \pitchfork 2$ and $\lambda \in \Lambda'_n$ such that $\lambda \cdot Ta(t_1) = 1$ and $\lambda \cdot Ta(t_2) = 0$.

An inspection of the proof of Theorem 4.16 shows

Proposition 4.22 *A collection Λ' of predicate liftings is separating only if the mate τ of δ is an embedding.*

Corollary 4.23 *If Λ' is separating then the logic given by Λ' is expressive.*

Let us look at a couple of the Examples 3.4

Example 4.24 (i) For $T = \mathcal{U}p_\omega : \mathsf{Pos} \to \mathsf{Pos}$ we take one unary predicate lifting: LA is generated by $\Box a, a \in A$. (8) is given by

$$\Box a \;\mapsto\; \texttt{lambda}\ b.\texttt{if}\ b \subseteq a\ \texttt{then}\ 1\ \texttt{else}\ 0 : \mathcal{U}pX \to 2$$

which we can also read as defining a predicate lifting in the form of (13) (with $a \in \mathsf{Pos}(X, 2)$ and $b \in \mathcal{U}p(X)$). In the form of (14) it is a function $\mathcal{U}p(2) \to 2$ mapping $\{0, 1\}$ to 0 and $\{1\}$ and $\emptyset$ to 1.

(ii) For the functor $X \mapsto X \triangleleft X$ we take two unary modal operators $[l]$ and $[r]$ and, in addition to usual axioms for disjoint union saying that $[l]$ and $[r]$ preserve the DL-operations, we also have $[l]a \leq [r]a$, reflecting that formulas denote up-sets.

5 Conclusion

We have developed the predicate lifting approach to coalgebras over posets. Let us note that this also includes coalgebras over the category Pre of preorders: The adjunction $\mathsf{Pre}^{\mathrm{op}} \rightleftarrows \mathsf{Pos}^{\mathrm{op}} \rightleftarrows \mathsf{DL}$ retains all the necessary properties from (5). The purpose of this observation is the importance of Pre as a base category for coalgebras in the study of simulation: Since two-way simulation is in general different from bisimulation, one needs to work with preorders if one wants to classify bisimulation and simulation in the same final coalgebra. It should be of interest to use this to study in a systematic way logics of simulation and how they relate to logics of bisimulation.

The logic of all monotone predicate liftings is also of interest for set-functors, for example if one wants to extend coalgebraic logics by fixed points. The logic of all monotone predicate liftings of a functor $H : \mathsf{Set} \to \mathsf{Set}$ can be investigated via the logic of all (necessarily monotone) predicate liftings $\bar{H} : \mathsf{Pos} \to \mathsf{Pos}$ as in Definition 3.1, leading to a systematic investigation into the relationship between the BA-logic for T-coalgebras and their positive DL-fragments [10].

Further work on which we embarked already includes an account of Moss's [30] cover modality ∇ over Pos [11]. Investigations into many-valued logics based on replacing 2 by a commutative quantale are ongoing.

References

[1] Abramsky, S., *Domain theory in logical form*, Ann. Pure Appl. Logic **51** (1991).

[2] Abramsky, S., *A Cook's tour of the finitary non-well-founded sets*, in: *We Will Show Them: Essays in Honour of Dov Gabbay*, College Publications, 2005 Presented at BCTCS 1988.

[3] Aczel, P., "Non-Well-Founded Sets," CSLI, Stanford, 1988.

[4] Adámek, J., *On final coalgebras of continuous functors*, Theor. Comp. Sci. **294** (2003).

[5] Adámek, J., H. Herrlich and G. E. Strecker, "Abstract and concrete categories," John Wiley & Sons, 1990.

[6] Adámek, J. and J. Rosický, "Locally presentable and accessible categories," Cambridge University Press, 1994.

[7] Adámek, J., J. Rosický and E. Vitale, "Algebraic theories," Cambridge University Press, 2011.

[8] Allwein, G. and W. Harrison, *Partially-ordered modalities*, in: *AiML'10*, 2010.

[9] Balan, A. and A. Kurz, *Finitary functors: From Set to Preord and Poset*, in: *CALCO'11*, 2011.

[10] Balan, A., A. Kurz and J. Velebil, *On coalgebraic logic over posets*, in: *Short Contributions of CMCS'2012*.

[11] Bílková, M., D. Petrişan, A. Kurz and J. Velebil, *Relation liftings on preorders and posets*, in: *CALCO'11*, 2011.

[12] Borceux, F., "Handbook of categorical algebra," Cambridge University Press, 1994.

[13] Cîrstea, C. and D. Pattinson, *Modular proof systems for coalgebraic logics*, Theor. Comp. Sci. **338** (2007).

[14] Davey, B. A. and H. A. Priestley, "Introduction to lattices and order," Cambridge University Press, 1990.
[15] Dunn, J., *Positive modal logic*, Studia Logica **55** (1995).
[16] Goldblatt, R., *Metamathematics of modal logic*, in: *Mathematics of Modality*, CSLI Lecture Notes **43**, CSLI, 1993 .
[17] Hughes, J. and B. Jacobs, *Simulations in coalgebra*, Theor. Comp. Sci. **327** (2004).
[18] Jacobs, B. and A. Sokolova, *Exemplaric expressivity of modal logics*, J. Log. Comp. **20** (2010).
[19] Kapulkin, K., A. Kurz and J. Velebil, *Expressivity of coalgebraic logic over posets*, in: *CMCS'10, Short Contributions*, 2010.
[20] Kelly, G., "Basic concepts of enriched category theory," Cambridge University Press, 1982.
[21] Klin, B., "An Abstract Coalgebraic Approach to Process Equivalence for Well-Behaved Operational Semantics," Ph.D. thesis, University of Aarhus (2004).
[22] Klin, B., *The least fibred lifting and the expressivity of coalgebraic modal logic*, in: *CALCO'05*, 2005.
[23] Klin, B., *Coalgebraic modal logic beyond sets*, in: *MFPS'07*, 2007.
[24] Kurz, A., *Coalgebras and their logics*, SIGACT News **37** (2006).
[25] Kurz, A. and D. Pattinson, *Coalgebraic modal logic of finite rank*, Math. Structures Comp. Sci. **15** (2005).
[26] Kurz, A. and J. Rosický, *Strongly complete logics for coalgebras* (2006), manuscript.
[27] Kurz, A. and J. Velebil, *Enriched logical connections*, Appl. Categ. Structures (2011), online first.
[28] Mac Lane, S. "Categories for the working mathematician," Springer, 1971.
[29] Levy, P., *Similarity quotients as final coalgebras*, in: *FoSSaCS'11*, 2011.
[30] Moss, L., *Coalgebraic logic*, Ann. Pure Appl. Logic **96** (1999).
[31] Pattinson, D., *Coalgebraic modal logic: Soundness, completeness and decidability of local consequence*, Theor. Comp. Sci. **309** (2003).
[32] Pavlović, D. and V. Pratt, *The continuum as a final coalgebra*, Theor. Comp. Sci. **280** (2002).
[33] Porst, H. E. and W. Tholen, *Concrete dualities*, in: "Category theory at work", Heldermann Verlag, (1991).
[34] Rutten, J., *Relators and metric bisimulations*, in: *CMCS'98*, 1998.
[35] Rutten, J., *Universal coalgebra: A theory of systems*, Theor. Comp. Sci. **249** (2000).
[36] Schröder, L., *Expressivity of coalgebraic modal logic: The limits and beyond*, in: *FOSSACS'05*, 2005.
[37] Schröder, L. and D. Pattinson, *Modular algorithms for heterogeneous modal logics*, in: *International Colloquium on Automata, Languages and Programming, ICALP 2007*, 2007, pp. 459–471.
[38] Thijs, A., "Simulation and Fixpoint Semantics," Ph.D. thesis, Groningen (1996).
[39] Velebil, J. and A. Kurz, *Equational presentations of functors and monads*, Math. Structures Comp. Sci. **21** (2011).
[40] Worrell, J., *Coinduction for recursive data types: partial order, metric spaces and ω-categories*, in: *CMCS'00*, 2000.
[41] Worrell, J., *On the final sequence of an finitary set functor*, Theor. Comp. Sci. **338** (2005).

A Proof of Remark 4.18

Theorem: Let $\mathcal{X}$ be a locally finitely presentable category [6], $F \dashv U : \mathcal{A} \to \mathsf{Set}$, $\mathcal{A}_0 \hookrightarrow \mathcal{A}$ a full, dense subcategory, and $P \dashv S : \mathcal{A}^{\mathrm{op}} \to \mathcal{X}$. Let $LA = PTSA$ for $A \in \mathcal{A}_0$ and extend via colimits to $L : \mathcal{A} \to \mathcal{A}$. Let $(\mathcal{E}, \mathcal{M})$ be either the (StrongEpi,Mono) or a (Epi,RegMono) factorisation system [6, 1.61] on $\mathcal{X}$. Moreover we require: T preserves arrows in $\mathcal{M}$; the unit $X \to SPX$ is pointwise in $\mathcal{M}$; (*) given X finitely presentable, $X \to \bar{X}$ in $\mathcal{E}$, and $m : \bar{X} \to$

SA in $\mathcal{M}$, there is $A' \in \mathcal{A}$ and $\varphi : A' \to A$ such that $S\varphi \cdot m$ in $\mathcal{M}$. Then the mate $\tau : TS \to ST$ of $\delta : LP \to PT$ is in $\mathcal{M}$.

Remark. $\mathcal{X}$ always has the (StrongEpi,Mono) factorisation system [6, 1.61] and it has the (Epi,RegMono) factorisation system iff regular monos are closed under composition [5, 14.22]. In the latter case, we have RegMono=StrongMono [5, 14.14]. The proof of the theorem will be the same in both cases as it only uses the following properties shared by both factorisation systems: If $X_j \to X$ is a filtered colimit and all arrows of a cone $X_j \to Y$ are in $\mathcal{M}$ then the induced arrow $X \to Y$ is in $\mathcal{M}$, see [6, 1.60]. If $f \cdot g \in \mathcal{M}$, then $g \in \mathcal{M}$, see [5, 14.11].

Proof. There is a colimit $A_i \to A$ with $A_i \in \mathcal{A}_0$ and hence a colimit $PTSA_i \to LA$. To show $TSA \to SLA$ in $\mathcal{M}$, it suffices to show that $TSA \to SLA \to \prod_i SPTSA_i$ is in $\mathcal{M}$. Since η is natural and pointwise in $\mathcal{M}$ and since $\mathcal{M}$ is closed under products [5, 14.15], it suffices to show that $TSA \to \prod_i TSA_i$ is in $\mathcal{M}$. Observe that we have a filtered colimit $X_j \to SA$ and factoring $X_j \to SA$ as $X_j \to \bar{X}_j \to SA$ gives a filtered colimit $\bar{X}_j \to SA$ of arrows in $\mathcal{M}$ and hence also a filtered colimit $T\bar{X}_j \to TSA$ arrows in $\mathcal{M}$; due to [6, 1.62] it suffices to show that $T\bar{X}_j \to \prod_i TSA_i$ in $\mathcal{M}$. But this follows from (*).

Modal Logic of Some Products of Neighborhood Frames

Andrey Kudinov [1]

Institute for Information Transmission Problem
Bolshoy Karetny per. 19,
Moscow, Russia

Abstract

We consider modal logics of products of neighborhood frames and prove that for any pair L and L' of logics from set $\{\mathsf{S4}, \mathsf{D4}, \mathsf{D}, \mathsf{T}\}$ modal logic of products of L-neighborhood frames and L'-neighborhood frames is the fusion of L and L'.

Keywords: Modal logic, neighborhood frame, topological semantics, product of neighborhood frames, fusion of logics.

1 Introduction

Neighborhood frames as a generalization of Kripke semantics for modal logic were invented independently by Dana Scott [9] and Richard Montague [7]. Neighborhood semantics is more general than Kripke semantics and in case of normal reflexive and transitive logics coincide with topological semantics. In this paper we consider product of neighborhood frames, which was introduced by Sano in [8]. It is a generalization of product of topological spaces [2] presented in [1].

The product of neighborhood frames is defined in the vein of the product of Kripke frames (see [11] and [12]). But, there are some differences. In any product of Kripke frames axioms of commutativity and Church-Rosser property are valid. Nonetheless, as it was shown in [1], the logic of the products of all topological spaces is the fusion of logics $\mathsf{S4} \otimes \mathsf{S4}$.

In his recent work [13] Uridia considers derivational semantics for products of topological spaces. He proves that the logic of all topological spaces is the fusion of logics $\mathsf{D4} \otimes \mathsf{D4}$. And in fact $\mathsf{D4} \otimes \mathsf{D4}$ is complete w.r.t. the product of the rational numbers $\mathbb{Q}$. Derivational and topological semantics can be considered as a special case of neighborhood semantics. So the result of [13] and corresponding result for $\mathsf{S4}$ from [1] can be obtained as corollaries from the main result of this paper.

[1] This work was supported by RFBR grants 11-01-00958-a and 11-01-93107-a

[2] "Product of topological spaces" is a well-known notion in Topology but it is different from what we use here (for details see [1])

Neighborhood frames are usually considered in the context of non-normal logics since they are usually complete w.r.t. non-normal logics and Kripke frames are not. In this paper, however, we will consider only monotone neighborhood frames, that correspond to normal modal logics. In some sense the results of this paper (and of [1], [13]) shows that neighborhood semantics in some sense is more natural for products of normal modal logics, since there are no need to add extra axioms (at least in some cases).

2 Language and logics

In this paper we study propositional modal logic with modal operators. A formula is defined recursively as follows:

$$\phi ::= p \mid \bot \mid \phi \to \phi \mid \Box_i \phi,$$

where $p \in \mathrm{PROP}$ is a propositional letter and $\Box_i$ is a modal operator. Other connectives are introduced as abbreviations: classical connectives are expressed through $\bot$ and $\to$, dual modal operators $\Diamond_i$ are expressed as follows $\Diamond_i = \neg\Box_i\neg$.

Definition 2.1 *A* normal logic *(or* a logic, *for short) is a set of modal formulas closed under Substitution* $\left(\frac{A(p_i)}{A(B)}\right)$, *Modus Ponens* $\left(\frac{A,\, A\to B}{B}\right)$ *and two Generalization rules* $\left(\frac{A}{\Box_i A}\right)$; *containing all classic tautologies and the following axioms*

$$\Box_i(p \to q) \to (\Box_i p \to \Box_i q).$$

$\mathsf{K_n}$ *denotes* the minimal normal modal logic with n modalities *and* $\mathsf{K} = \mathsf{K_1}$.

Let L be a logic and let Γ be a set of formulas, then $\mathsf{L} + \Gamma$ denotes the minimal logic containing L and Γ. If $\Gamma = \{A\}$, then we write $\mathsf{L} + A$ rather than $\mathsf{L} + \{A\}$

Definition 2.2 *Let* $\mathsf{L_1}$ *and* $\mathsf{L_2}$ *be two modal logics with one modality* $\Box$ *then* fusion *of these logics is*

$$\mathsf{L_1} \otimes \mathsf{L_2} = K_2 + \mathsf{L}_{1(\Box\to\Box_1)} + \mathsf{L}_{2(\Box\to\Box_2)};$$

where $\mathsf{L}_{i(\Box\to\Box_i)}$ *is the set of all formulas from* $\mathsf{L_i}$ *where all* $\Box$ *replaced by* $\Box_i$.

In this paper we consider the following four well-known logics:

$$\mathsf{D} = \mathsf{K} + \Box p \to \Diamond p;$$
$$\mathsf{T} = \mathsf{K} + \Box p \to p;$$
$$\mathsf{D4} = \mathsf{D} + \Box p \to \Box\Box p;$$
$$\mathsf{S4} = \mathsf{T} + \Box p \to \Box\Box p.$$

3 Kripke frames

The notion of Kripke frames and Kripke models is well known (see [2]), so we only define special kind of frames that we are using in this paper. We can call them *fractal* frames because their basic property is that any cone is isomorphic to the whole frame. In particular, we consider four types of infinite trees with fixed branching: irreflexive and transitive, reflexive and transitive, irreflexive and non-transitive (any point sees only next level) and reflexive and non-transitive.

Definition 3.1 *Let A be a nonempty set.*

$$A^* = \{a_1 \dots a_k \mid a_i \in A\}$$

be the set of all finite sequences of elements from A, including the empty sequence Λ. Elements from A^* we will denote by letters with an arrow (e.g. $\vec{a} \in A^*$) The length of sequence $\vec{a} = a_1 \dots a_k$ is k (Notation: $l(\vec{a}) = k$) the length of the empty sequence equals 0 ($l(\Lambda) = 0$). Concatenation is denoted by "$\cdot$": $(a_1 \dots a_k) \cdot (b_1 \dots b_l) = \vec{a} \cdot \vec{b} = a_1 \dots a_k b_1 \dots b_l$.

Definition 3.2 *Let A be a nonempty set. We define an infinite frame $F_{in}[A] = (A^*, R)$, such that for $\vec{a}, \vec{b} \in A^*$*

$$\vec{a} R \vec{b} \iff \exists x \in A \left(\vec{b} = \vec{a} \cdot x \right).$$

We also defined

$$F_{rn}[A] = (A^*, R^r), \text{ where } R^r = R \cup Id \text{ — reflexive closure;}$$
$$F_{it}[A] = (A^*, R^*), \text{ where } R^* = \bigcup_{i=1}^{\infty} R^i \text{ — transitive closure;}$$
$$F_{rt}[A] = (A^*, R^{r*}).$$

So "t" stands for transitive, "n" — for non-transitive "r" for reflexive and "i" for irreflexive.

The following easy-to-prove proposition shows that frames $F_{\xi\eta}[A]$ (where $\xi \in \{i, r\}$ and $\eta \in \{t, n\}$)) are indeed fractal.

Proposition 3.3 *Let $F = F_{\xi\eta}[A] = (A^*, R)$ then*

$$\vec{a} R (\vec{a} \cdot \vec{c}) \iff \Lambda R \vec{c}.$$

Definition 3.4 *Let $F_1 = F_{\xi_1\eta_1}[A] = (A^*, R_1)$ be and $F_2 = F_{\xi_2\eta_2}[B] = (B^*, R_2)$, where $\xi_1, \xi_2 \in \{i, r\}$ and $\eta_1, \eta_2, \in \{t, n\}$), $A \cap B = \varnothing$, $A = \{a_1, a_2, \dots\}$ and $B = \{b_1, b_2, \dots\}$ then we define frame $F_1 \otimes F_2 = (W, R_1', R_2')$, as follows*

$$W = (A \sqcup B)^*$$
$$\vec{x} R_1' \vec{y} \iff \vec{y} = \vec{x} \cdot \vec{z} \text{ for some } \vec{z} \in A^* \text{ such that } \Lambda R_1 \vec{z}$$
$$\vec{x} R_2' \vec{y} \iff \vec{y} = \vec{x} \cdot \vec{z} \text{ for some } \vec{z} \in B^* \text{ such that } \Lambda R_2 \vec{z}$$

Proposition 3.5 ([6], [4]) *Let F_1 and F_2 be as in Definition 3.4 then*

$$Log(F_1 \otimes F_2) = Log(F_1) \otimes Log(F_2). \qquad (1)$$

Let as define four frames: $F_{in} = F_{in}[\omega]$, $F_{rn} = F_{rn}[\omega]$, $F_{it} = F_{it}[\omega]$ and $F_{rt} = F_{rt}[\omega]$

Proposition 3.6 *For just defined frames*

(i) $Log(F_{in}) = \mathsf{D}$;

(ii) $Log(F_{rn}) = \mathsf{T}$;

(iii) $Log(F_{it}) = \mathsf{D4}$;

(iv) $Log(F_{rt}) = \mathsf{S4}$.

4 Neighborhood frames

In this section we consider neighborhood frames. All definitions and lemmas of this section are well-known and can be found in [10] and [3].

Definition 4.1 *A* (monotone) neighborhood frame *(or an n-frame) is a pair $\mathfrak{X} = (X, \tau)$, where X is a nonempty set and $\tau : X \to 2^{2^X}$ such that $\tau(x)$ is a filter on X for any x. We call function τ the* neighborhood function *of $\mathfrak{X}$ and sets from $\tau(x)$ we call* neighborhoods of x. The neighborhood model *(n-model) is a pair $(\mathfrak{X}, V)$, where $\mathfrak{X} = (X, \tau)$ is a n-frame and $V : PV \to 2^X$ is* a valuation. *In a similar way we define* neighborhood 2-frame *(n-2-frame) as (X, τ_1, τ_2) such that $\tau_i(x)$ is a filter on X for any x, and a* n-2-model.

Definition 4.2 The valuation of a formula φ *at a point of a n-model $M = (\mathfrak{X}, V)$ is defined by induction as usual for boolean connectives and for modalities as follows*

$$M, x \models \Box_i \psi \iff \exists V \in \tau_i(x) \forall y \in V (M, y \models \psi).$$

Formula is valid in a n-model M if it is valid at all points of M (notation $M \models \varphi$). Formula is valid in a n-frame $\mathfrak{X}$ if it is valid in all models based on $\mathfrak{X}$ (notation $\mathfrak{X} \models \varphi$). We write $\mathfrak{X} \models L$ if for any $\varphi \in L$, $\mathfrak{X} \models \varphi$. Logic of a class of n-frames $\mathcal{C}$ as $Log(\mathcal{C}) = \{\varphi \mid \mathfrak{X} \models \varphi$ for some $\mathfrak{X} \in \mathcal{C}\}$. For logic L we also define $nV(L) = \{\mathfrak{X} \mid \mathfrak{X}$ is an n-frame and $\mathfrak{X} \models L\}$.

Definition 4.3 *Let $F = (W, R)$ be a Kripke frame. We define n-frame $\mathcal{N}(F) = (W, \tau)$ as follows. For any $w \in W$*

$$\tau(w) = \{U \mid R(w) \subseteq U \subseteq W\}.$$

Lemma 4.4 *Let $F = (W, R)$ be a Kripke frame. Then*

$$Log(\mathcal{N}(F)) = Log(F).$$

The proof is straightforward.

Definition 4.5 *Let $\mathfrak{X} = (X, \tau_1, \ldots)$ and $\mathcal{Y} = (Y, \sigma_1, \ldots)$ be n-frames. Then function $f : X \to Y$ is a* bounded morphism *if*

(i) *f is surjective;*

(ii) *for any $x \in X$ and $U \in \tau_i(x)$ $f(U) \in \sigma_i(f(x))$;*

(iii) *for any $x \in X$ and $V \in \sigma_i(f(x))$ there exists $U \in \tau_i(x)$, such that $f(U) \subseteq V$.*

In notation $f : \mathfrak{X} \twoheadrightarrow \mathcal{Y}$.

Lemma 4.6 *Let $\mathfrak{X} = (X, \tau_1, \ldots)$, $\mathcal{Y} = (Y, \sigma_1, \ldots)$ be n-frames, $f : \mathfrak{X} \twoheadrightarrow \mathcal{Y}$ V' is a valuation on $\mathcal{Y}$. We define $V(p) = f^{-1}(V'(p))$. Then*

$$\mathfrak{X}, V, x \models \varphi \iff \mathcal{Y}, V', f(x) \models \varphi.$$

The proof is by standard induction on length of formula.

Corollary 4.7 *If $f : \mathfrak{X} \twoheadrightarrow \mathcal{Y}$ then $Log(\mathcal{Y}) \subseteq Log(\mathfrak{X})$.*

Definition 4.8 *Let $\mathfrak{X}_1 = (X_1, \tau_1)$ and $\mathfrak{X}_2 = (X_2, \tau_2)$ be two n-frames. Then* the product *of these n-frames is an n-2-frame defined as follows*

$$\begin{aligned}
&\mathfrak{X}_1 \times \mathfrak{X}_2 = (X_1 \times X_2, \tau'_1, \tau'_2),\\
&\tau'_1(x_1, x_2) = \{U \subseteq X_1 \times X_2 \mid \exists V(V \in \tau_1(x_1) \;\&\; V \times \{x_2\} \subseteq U)\},\\
&\tau'_2(x_1, x_2) = \{U \subseteq X_1 \times X_2 \mid \exists V(V \in \tau_2(x_2) \;\&\; \{x_1\} \times V \subseteq U)\}.
\end{aligned}$$

Definition 4.9 *For two unimodal logics L_1 and L_2 we define* n-product *of them as follows*

$$\mathsf{L}_1 \times_n \mathsf{L}_2 = Log(\{\mathfrak{X}_1 \times \mathfrak{X}_2 \mid \mathfrak{X}_1 \in nV(L_1) \;\&\; \mathfrak{X}_2 \in nV(L_2)\})$$

Note that $\mathfrak{X}_1 \times \mathfrak{X}_2$ if we forget about one of its neighborhood functions say τ'_2 then $\mathfrak{X}_1 \times \mathfrak{X}_2$ will be a disjoint union of L_1 n-frames. Hence

Proposition 4.10 ([8]) *For two unimodal logics L_1 and L_2*

$$\mathsf{L}_1 \otimes \mathsf{L}_2 \subseteq \mathsf{L}_1 \times_n \mathsf{L}_2.$$

5 Main construction

The construction in this section was inspired by [1], but it is not a straightforward generalization. In case of $\mathsf{S4} \times_\mathsf{n} \mathsf{S4}$ it is, in essence, very similar to the construction in [1]. However, here we operate only with words (finite or infinite), and not with numbers and fractions. It makes proofs shorter and allows us to generalize the results to non-transitive cases.

Let $F = (A^*, R) = F_{\xi\eta}[A]$ and $0 \notin A$. We define set of "pseudo-infinite" sequences

$$X = \{a_1 a_2 \ldots \mid a_i \in A \cup \{0\} \;\&\; \exists N \forall k \geq N(a_k = 0)\}.$$

Define $f_F : X \to A^*$ which "fogets" all zeros. For $\alpha \in X$ such that $\alpha = a_1 a_2 \ldots$ we define

$$\begin{aligned}
st(\alpha) &= \min\{N \mid \forall k \geq N(a_k = 0)\};\\
\alpha|_k &= a_1 \ldots a_k;\\
U_k(\alpha) &= \{\beta \in X \mid \alpha|_m = \beta|_m \;\&\; f_F(\alpha) R f_F(\beta), \text{ where } m = \max(k, st(\alpha))\}.
\end{aligned}$$

Lemma 5.1 *$U_k(\alpha) \subseteq U_m(\alpha)$ whenever $k \geq m$.*

Proof. Let $\beta \in U_k(\alpha)$. Since $\alpha|_k = \beta|_k$ and $k \geq m$ then $\alpha|_m = \beta|_m$. Hence, $\beta \in U_m(\alpha)$. □

Definition 5.2 *Due to Lemma 5.1 sets $U_n(\alpha)$ forms a filter base. So we can define*

$$\tau(\alpha) - \textit{the filter with base } \{U_n(\alpha) \mid n \in \omega\};$$
$$\mathcal{N}_\omega(F) = (X, \tau) \text{ — \textit{is} the n-frame based on } F.$$

Lemma 5.3 *Let $F = (A^*, R) = F_{\xi,\eta}[A]$ then*

$$f_F : \mathcal{N}_\omega(F) \twoheadrightarrow \mathcal{N}(F).$$

Proof. From now on in this proof we will omit the subindex in f_F. Let $\mathcal{N}_\omega(F) = (X, \tau)$. Since for any $\overrightarrow{x} \in A^*$ sequence $\overrightarrow{x} \cdot 0^\omega \in X$ and $f(\overrightarrow{x} \cdot 0^\omega) = \overrightarrow{x}$ then f is surjective.

Assume, that $x \in X$ and $U \in \tau(x)$. We need to prove that $R(f(x)) \subseteq f(U)$. There is m such that $U_m(x) \subseteq U$ and since $f(U_m(x)) = R(f(x))$ then

$$R(f(x)) = f(U_m(x)) \subseteq f(U).$$

Assume that $x \in X$ and V is a neighborhood of x, i.e. $R(f(x)) \subseteq V$. We need to prove that there exists $U \in \tau(x)$, such that $f(U) \subseteq V$. As U we take $U_m(x)$ for some $m \geq st(x)$, then

$$f(U_m(x)) = R(f(x)) \subseteq V.$$

□

Corollary 5.4 *For frame $F = F_{\xi\eta}[A]$ $Log(\mathcal{N}_\omega(F)) \subseteq Log(F)$.*

Proof. It follows from Lemmas 4.4, 4.5 and 5.3

$$Log(\mathcal{N}_\omega(F)) \subseteq Log(\mathcal{N}(F)) = Log(F).$$

□

Proposition 5.5 *Let $F_{in} = F_{in}[\omega]$, $F_{rn} = F_{rn}[\omega]$, $F_{it} = F_{it}[\omega]$ and $F_{rt} = F_{rt}[\omega]$ then*

(i) *$Log(\mathcal{N}_\omega(F_{in})) = \mathsf{D}$;*

(ii) *$Log(\mathcal{N}_\omega(F_{rn})) = \mathsf{T}$;*

(iii) *$Log(\mathcal{N}_\omega(F_{it})) = \mathsf{D4}$;*

(iv) *$Log(\mathcal{N}_\omega(F_{rt})) = \mathsf{S4}$.*

Proof. In all these cases the inclusion from left to right is covered by Corollary 5.4 and Proposition 3.6.

Let us check the inclusion in converse direction.

(i). It is easy to check that n-frame $\mathfrak{X} = (X, \tau) \models \mathsf{D}$ iff for each $x \in X$ $\varnothing \notin \tau(x)$. For $\mathcal{N}_\omega(F_{in})$ and $\mathcal{N}_\omega(F_{it})$ it obviously true.

(ii). It is easy to check that n-frame $\mathfrak{X} = (X, \tau) \models \mathsf{T}$ iff $x \in U \in \tau(x)$ for each x and U. For $\mathcal{N}_\omega(F_{rn})$ and $\mathcal{N}_\omega(F_{rt})$ it is obviously true.

(iii) and (iv). It is well-known (see e.g. [5]) that $\mathfrak{X} = (X, \tau) \models \Box p \to \Box\Box p$ iff for each $U \in \tau(x)$ $\{y \,|\, U \in \tau(y)\} \in \tau(x)$. Indeed, it follows from the fact that for any $y \in U_m(x)$ and any k $U_k(y) \subseteq U_m(x)$. □

Let $F_1 = (A^*, R_1) = F_{\xi_1\eta_1}[A]$ and $F_2 = (B^*, R_2) = F_{\xi_2\eta_2}[B]$ we assume that $A \cap B = \varnothing$, $A = \{a_1, a_2, \ldots\}$ and $B = \{b_1, b_2, \ldots\}$. Consider the product of n-frames $\mathfrak{X}_1 = (X_1, \tau_1) = \mathcal{N}_\omega(F_1)$ and $\mathfrak{X}_2 = (X_2, \tau_2) = \mathcal{N}_\omega(F_2)$

$$\mathfrak{X} = (X_1 \times X_2, \tau_1', \tau_2') = \mathcal{N}_\omega(F_1) \times_n \mathcal{N}_\omega(F_2).$$

We define function $g : \mathfrak{X}_1 \times \mathfrak{X}_2 \to (A \cup B)^*$ as follows. For $(\alpha, \beta) \in \mathfrak{X}_1 \times \mathfrak{X}_2$, such that $\alpha = x_1 x_2 \ldots$ and $\beta = y_1 y_2 \ldots$, $x_i \in A \cup \{0\}$, $y_j \in B \cup \{0\}$, we define $g(\alpha, \beta)$ to be the finite sequence which we get after eliminating all zeros from the infinite sequence $x_1 y_1 x_2 y_2 \ldots$.

Lemma 5.6 *Function g defined above is a bounded morphism: $g : \mathfrak{X} \twoheadrightarrow \mathcal{N}(F_1 \otimes F_2)$.*

Proof. Let $\vec{z} = z_1 z_2 \ldots z_n \in (A \cup B)^*$. Define for $i \le n$

$$x_i = \begin{cases} z_i, \text{ if } z_i \in A; \\ 0, \text{ if } z_i \notin A; \end{cases} \qquad y_i = \begin{cases} z_i, \text{ if } z_i \in B; \\ 0, \text{ if } z_i \notin B. \end{cases}$$

Let $\alpha = x_1 x_2 \ldots x_n 0^\omega$ and $\beta = y_1 y_2 \ldots y_n 0^\omega$ then $g(\alpha, \beta) = \vec{z}$. Hence g is surjective.

The next two conditions we check only for τ_1 and for τ_2 it is similar. Assume, that $(\alpha, \beta) \in X_1 \times X_2$ and $U \in \tau_1(\alpha, \beta)$. We need to prove that $R_1'(g(\alpha, \beta)) \subseteq g(U)$. There is $m > \max\{st(\alpha), st(\beta)\}$ such that $U_m'(\alpha) \times \{\beta\} \subseteq U$ and since $g(U_m'(\alpha) \times \{\beta\}) = R_1'(g(\alpha, \beta))$ then

$$R_1'(g(\alpha, \beta)) = g(U_m'(\alpha) \times \{\beta\}) \subseteq g(U);$$

where $U_m'(\alpha)$ is the corresponding neighborhood from $\mathfrak{X}_1$.

Assume that $(\alpha, \beta) \in X_1 \times X_2$ and $R_1'(g(\alpha, \beta)) \subseteq V$. We need to prove that there exists $U \in \tau_1'(\alpha, \beta)$, such that $g(U) \subseteq V$. As U we take $U_m'(\alpha) \times \{\beta\}$ for some $m > \max\{st(\alpha), st(\beta)\}$, then

$$g(U_m'(\alpha) \times \{\beta\}) = R_1'(g(\alpha, \beta)) \subseteq V.$$

□

Corollary 5.7 *Let $F_1 = (A^*, R_1) = F_{\xi_1\eta_1}[A]$ and $F_2 = (B^*, R_2) = F_{\xi_2\eta_2}[B]$ then $Log(\mathcal{N}_\omega(F_1) \times_n \mathcal{N}_\omega(F_2)) \subseteq Log(F_1) \otimes Log(F_2)$.*

It immediately follows from Lemmas 5.6, 4.6 and Proposition 4.10.

Corollary 5.8 *Let $F_1, F_2 \in \{F_{in}, F_{rn}, F_{it}, F_{rt}\}$ then $Log(\mathcal{N}_\omega(F_1) \times_n \mathcal{N}_\omega(F_2)) = Log(F_1) \otimes Log(F_2)$.*

Proof. The left-to-right inclusion follows from Corollary 5.7.

To prove right-to-left inclusion we notice that due to Proposition 5.5 $Log(\mathcal{N}_\omega(F_i)) = Log(F_i)$ $(i = 1, 2)$ and due to Proposition 4.10

$$Log(\mathcal{N}_\omega(F_1)) \otimes Log(\mathcal{N}_\omega(F_2)) \subseteq Log(\mathcal{N}_\omega(F_1)) \times_n Log(\mathcal{N}_\omega(F_2)).$$

□

6 Completeness results

Theorem 6.1 *Let* $\mathsf{L}_1, \mathsf{L}_2 \in \{\mathsf{S4}, \mathsf{D4}, \mathsf{D}, \mathsf{T}\}$ *then*

$$\mathsf{L}_1 \times_n \mathsf{L}_2 = \mathsf{L}_1 \otimes \mathsf{L}_2.$$

Proof. Logics $\mathsf{L}_1 = Log(F_1)$ and $\mathsf{L}_2 = Log(F_2)$ for some $F_1, F_2 \in \{F_{in}, F_{rn}, F_{it}, F_{rt}\}$. By Corollary 5.8

$$\mathsf{L}_1 \times_n \mathsf{L}_2 = Log(\mathcal{N}_\omega(F_1) \times_n \mathcal{N}_\omega(F_2)) = Log(F_1) \otimes Log(F_2) = \mathsf{L}_1 \otimes \mathsf{L}_2.$$

□

The following fact was proved in [1].

Corollary 6.2 *Let* $\mathfrak{X} = (\mathbb{Q}, \tau)$ *where* $\mathbb{Q}$ *is the set of rational numbers and* τ *is based on the standard topology on* $\mathbb{Q}$, *i.e.* $\tau(x) = \{U \mid \exists V(x \in V \text{ is open and } V \subseteq U)\}$. *Then*

$$Log(\mathfrak{X} \times_n \mathfrak{X}) = \mathsf{S4} \otimes \mathsf{S4}.$$

Proof. Let $\mathcal{N}_\omega(F_{rt}) = (X, \tau)$. We can assume that $X = \{\overrightarrow{x} \cdot 0^\omega \mid \overrightarrow{x} \in \mathbb{Z}\}$. Note that X is a countable set and neighborhood function τ is based on topology generated by the lexicographical order $<_l$ on X. According to the classical result of Cantor, since the lexicographical order on X is dense, $(X, <_l)$ isomorphic to $(\mathbb{Q}, <)$ (see) and corresponding topological spaces are homeomorphic. Hence,

$$Log(\mathfrak{X} \times_n \mathfrak{X}) = Log(\mathcal{N}_\omega(F_{rt}) \times \mathcal{N}_\omega(F_{rt})) = \mathsf{S4} \otimes \mathsf{S4}.$$

□

The following fact was announced[3] in [13].

Corollary 6.3 *Let* $\mathfrak{X} = (\mathbb{Q}, \tau)$ *where* $\mathbb{Q}$ *is the set of rational numbers and* τ *is based on the standard topology on* $\mathbb{Q}$, *i.e.* $\tau(x) = \{U \mid \exists V(x \in V \text{ is open and } V \setminus \{x\} \subseteq U)\}$. *Then*

$$Log(\mathfrak{X} \times_n \mathfrak{X}) = \mathsf{D4} \otimes \mathsf{D4}.$$

Proof. Let $\mathfrak{X} = (\mathbb{Q}, \tau)$, where τ is based on derivation operator in $\mathbb{Q}$. Since $(X, <_l)$ isomorphic to $(\mathbb{Q}, <)$ then

$$Log(\mathfrak{X} \times_n \mathfrak{X}) = Log(\mathcal{N}_\omega(F_{it}) \times \mathcal{N}_\omega(F_{it})) = \mathsf{D4} \otimes \mathsf{D4}.$$

□

[3] The talk at the conference was very detailed, but to my knowledge, the full proof has not been published yet.

7 Conclusion

There are several ways to continue research. One of them is to try and extend the technique to other logics (e.g. K). The other way is to add the third modality which corresponds to the following neighborhood function $\tau'(x, y) = \{U \mid \exists V_1 \in \tau_1(x) \,\&\, \exists V_2 \in \tau_2(y)\, (V_1 \times V_2 \subseteq U)\}$. Similar construction was considered in [1] for the topological semantic.

References

[1] Benthem, J., G. Bezhanishvili, B. Cate and D. Sarenac, *Multimodal logics of products of topologies*, Studia Logica **84** (2006), pp. 369–392.

[2] Blackburn, P., M. de Rijke and Y. Venema, "Modal Logic," Cambridge University Press, 2002.

[3] Chellas, B., "Modal Logic: An Introduction." Cambridge University Press, Cambridge, 1980.

[4] Fine, K. and G. Schurz, *Transfer theorems for multimodal logics*, in: J. Copeland, editor, *Logic and Reality: Proceedings of the Arthur Prior Memorial Conference* (1996), pp. 169–213.

[5] Hansen, H., "Monotonic modal logic," Master's thesis, ILLC, University of Amsterdam (2003).

[6] Kracht, M. and F. Wolter, *Properties of independently axiomatizable bimodal logics.*, Journal Symbolic Logic **56** (1991), pp. 1469–1485.

[7] Montague, R., *Universal grammar*, Theoria **36** (1970), pp. 373–398.

[8] Sano, K., *Axiomatizing hybrid products of monotone neighborhood frames*, Electr. Notes Theor. Comput. Sci. **273** (2011), pp. 51–67.

[9] Scott, D., *Advice on modal logic*, in: *Philosophical Problems in Logic: Some Recent Developments*, D. Reidel, 1970 pp. 143–173.

[10] Segerberg, K., "An essay in classical modal logic," Filosofiska föreningen och Filosofiska institutionen vid Uppsala universitet (Uppsala), 1971.

[11] Segerberg, K., *Two-dimensional modal logic*, Journal of Philosophical Logic **2** (1973), pp. 77–96.

[12] Shehtman, V., *Two-dimensional modal logic*, Mathematical Notices of USSR Academy of Science **23** (1978), pp. 417–424, (Translated from Russian).

[13] Uridia, L., *The modal logic of the bi-topological rational plane*, in: *Topology, Algebra, and Categories in Logic (TACL'11)*, 2011.

On Modal Logics of Hamming Spaces

Andrey Kudinov Ilya Shapirovsky Valentin Shehtman

Institute for Information Transmission Problems, Russian Academy of Sciences
B. Karetny 19, 127994 Moscow, Russia

National Research University Higher School of Economics, Moscow, Russia

Abstract

With a set S of words in an alphabet A we associate the frame (S, H), where sHt iff s and t are words of the same length and $h(s,t) = 1$ for the Hamming distance h. We investigate some unimodal logics of these frames. We show that if the length of words n is fixed and finite, the logics are closely related to many-dimensional products $\mathbf{S5}^n$, so in many cases they are undecidable and not finitely axiomatizable.
The relation H can be extended to infinite sequences. In this case we prove some completeness theorems characterizing the well-known modal logics **DB** and **TB** in terms of the Hamming distance.

Keywords: Decidability, undecidability, finite axiomatizability, Hamming distance, product of modal logics.

1 Introduction

According to a rather old viewpoint on modalities, a proposition is possible in a world if it is true in a similar world. An abstract 'similarity' relation should be reflexive and symmetric, thus the corresponding modal logic is **TB**.

Similarity can be treated more specifically in different ways — e.g. one of them was developed by Z. Pawlak and E. Orlowska and others within the framework of information systems.

In this paper we study another, rather modest, but mathematically explicit approach to similarity of possible worlds — every world is presented by a sequence of properties (coded by symbols of a certain alphabet), and two sequences are similar if they coincide at all positions but one. So we investigate unimodal logics of the following relation on words: sHt iff s and t are words of the same length and $h(s,t) = 1$, where h is the Hamming distance. Recall that the Hamming distance between two words of equal length is the number of positions, where they differ.

Unimodal logics of this kind are closely related to many-dimensional modal logics. If we consider the n-th power (cf. [3]) of an inequality frame $(A, \neq)$

$$(A, \neq)^n = (A^n, \neq_0, \ldots, \neq_{n-1}),$$

we can see that $(A^n, H) = (A^n, \neq_0 \cup \cdots \cup \neq_{n-1})$. Thus the $\Box$-operator in the logic of (A^n, H) can be interpreted as the conjunction of all $\Box$-operators in the logic of $(A, \neq)^n$.

If $n = 2$, we obtain a particular case of the $\curlyvee$-product studied in the recent paper [11]. There it was shown that under certain conditions the product modalities can be expressed in terms of the "Hamming-like" $\Box$-operator. In this paper we extend the corresponding construction to higher dimensions.

If the length of words (the dimension of the product) is fixed, the modal operator corresponding to H is rather expressive and as we will show, it can be used to describe various geometric objects in A^n (like spheres or hyperplanes). As usual in many-dimensional logic, there is a price for this richness – logics of these frames for dimensions higher than two are undecidable; the Hamming modal operator allows us to express all modalities of the n-dimensional product of **S5** (up to permutation). Hence we obtain undecidability for logics of the relation H on words of fixed length greater than two.

However, for "long" words (ω-sequences) there are some positive results. We show that the relation H on binary sequences yields the well-known logic **DB** (the minimal symmetric serial logic). This result can also be formulated in terms of a certain relation on sets — 'the symmetric difference of two sets is a singleton'. For $\overline{H}$, the reflexive closure of H, the results are extended to sequences over any infinite I, and A with $|A| > 1$ — the corresponding logic is always **TB**, the minimal symmetric reflexive logic.

2 Preliminaries

2.1 Syntax and semantics

We skip standard definitions concerning propositional modal logics (all logics are supposed normal). $ML(\Diamond_0, \ldots, \Diamond_{n-1})$ denotes the set of all modal formulas constructed from the countable set of propositional variables PV using the classical connectives $\wedge$, $\neg$ and the unary connectives $\Diamond_0, \ldots, \Diamond_{n-1}$. Other connectives are defined in the usual way, in particular, $\Box_i\varphi = \neg\Diamond_i\neg\varphi$. $\Diamond$ and $\Box$ abbreviate $\Diamond_0$ and $\Box_0$, respectively.

$PV(\varphi)$ denotes the set of all variables occurring in a formula φ. The *modal depth* of a formula φ is denoted by $md(\varphi)$.

K denotes the minimal unimodal logic. For a logic L and a set of formulas Γ, $L + \Gamma$ is the minimal logic containing L and Γ. $L + \varphi$ abbreviates $L + \{\varphi\}$. Our basic logics are

$$\mathbf{KB} := \mathbf{K} + \Diamond\Box p \to p, \quad \mathbf{TB} := \mathbf{KB} + p \to \Diamond p, \quad \mathbf{DB} := \mathbf{KB} + \Diamond\top.$$

The notions of a frame, a (Kripke) model, validity and the truth are standard, cf. [1]. The truth of a formula φ at a world w in a model M is denoted by $\mathsf{M}, w \vDash \varphi$; the validity of φ in a frame F by $\mathsf{F} \vDash \varphi$. For a set of formulas Ψ, $\mathsf{F} \vDash \Psi$ means $\mathsf{F} \vDash \varphi$ for all $\varphi \in \Psi$. F is called an *L-frame* for a logic L if $\mathsf{F} \vDash L$.

Log$(\mathfrak{F})$ denotes the logic of the class $\mathfrak{F}$ (i.e., the set of all valid formulas). For a frame $\mathsf{F} = (W, R)$, *Log*(W, R) and *Log*(F) abbreviate *Log*$(\{\mathsf{F}\})$.

A formula φ is *satisfiable in a frame* F *at a point* w (or briefly, *satisfiable at* F, w) if $(\mathsf{F}, \theta), w \vDash \varphi$ for some valuation θ. For a class of frames $\mathfrak{F}$, φ is *satisfiable in* $\mathfrak{F}$ (or $\mathfrak{F}$*-satisfiable*), if φ is satisfiable in F for some $\mathsf{F} \in \mathfrak{F}$. A formula φ is L*-satisfiable* if φ is satisfiable in some L-frame.

For a relation R on a set W, R^n denotes its n-th iteration. So R^0 is the identity relation on W and $R^{n+1} = R \circ R^n$ for $n \geq 0$, where $\circ$ is the composition of relations. Put $R^{\leq n} := R^0 \cup \dots \cup R^n$.

R^{-1} is the converse to R, and $R^{\pm} := R \cup R^{-1}$ is the symmetric closure of R. For $x \in W$, $R(x) := \{y \mid xRy\}$.

The cardinality of a set W is denoted by $|W|$. The restriction of a function (frame, model) f to a set S is denoted by $f|S$. If (W, R) is a frame and $W \supseteq V \neq \varnothing$, the restriction $(W, R)|V$ $(= (V, R|V) = (V, R \cap (V \times V)))$ is usually denoted by (V, R).

A *point-generated subframe* with a root u of a frame $\mathsf{F} = (W, R)$ is

$$\mathsf{F} \uparrow u := \mathsf{F}| \bigcup_{n \geq 0} R^n(u).$$

If $\mathsf{F} = \mathsf{F} \uparrow u$, u is called a *root* of F. Recall that

$$Log(\mathsf{F}) = \bigcap_{u \in W} Log(\mathsf{F} \uparrow u),$$

by the generation lemma.

A frame (W, R) with a root u is called a *tree* if $R^{-1}(u) = \varnothing$ and $R^{-1}(x)$ is a singleton for any $x \neq u$.

$f : \mathsf{F} \twoheadrightarrow \mathsf{G}$ denotes that f is a p-morphism from F onto G; the existence of a p-morphism is denoted by $\mathsf{F} \twoheadrightarrow \mathsf{G}$.

Recall that $\mathsf{F} \twoheadrightarrow \mathsf{G}$ implies $Log(\mathsf{F}) \subseteq Log(\mathsf{G})$, by the p-morhism lemma.

The next proposition is easily proved by induction on k (cf. e.g. [1]).

Proposition 2.1 *Let* M *be a Kripke model over a frame* (W, R). *Then for any* x *in* M, *for any* φ *of modal depth* $\leq k$

$$\mathsf{M}, x \vDash \varphi \iff \mathsf{M}|R^{\leq k}(x), x \vDash \varphi.$$

2.2 Words and trees

An *alphabet* is an arbitrary nonempty set. $A^* := \{\lambda\} \cup A \cup A^2 \cup \dots$ is the set of all *words* in an alphabet A, where λ denotes the empty word. st denotes the concatenation of words s and t; $|s|$ is the length of a word s.

For $s, t \in A^*$, put [1]

$$\begin{aligned} s \lhd t &\iff t = as \text{ for some } a \in A, \\ s \unlhd t &\iff s \lhd t \text{ or } s = t, \\ s \sqsubseteq t &\iff t = rs \text{ for some } r \in A^*. \end{aligned}$$

[1] Note that in this paper words are ordered 'from the end'. This is done for representing numbers written in numeral systems, see the proof of Theorem 5.3.

We regard natural numbers as ordinals. For $n, k > 0$, $T_{n,k}$ is the set of all words of length at most n in the alphabet $k = \{0, \ldots, k-1\}$.

A nonempty set of words W (or a frame $(W, \lhd)$) is a *standard tree* if $(W, \sqsubseteq)$ has the least element u_0 and W is downwards closed:

$$\forall u \forall v \, (u \in W \;\&\; u \neq u_0 \;\&\; v \lhd u \Longrightarrow v \in W).$$

Proposition 2.2

(i) *If* φ *is* **DB**-*satisfiable, then there exists* $n \geq 2$ *such that* φ *is satisfiable at* $(T_{n,n}, \lhd^{\pm}), \lambda$.

(ii) *If* φ *is* **TB**-*satisfiable, then there exists* $n \geq 2$ *such that* φ *is satisfiable at* $(T_{n,n}, \unlhd^{\pm}), \lambda$.

Proof. (1) Suppose φ is **DB**-satisfiable. It is well-known that **DB**-frames are serial symmetric and **DB** has the finite model property. So by the generation lemma we obtain a finite **DB**-frame $\mathsf{F} = (W, R)$ with a root u such that φ is satisfiable at F, u.

Now we use unravelling (cf. [3]). Given a finite **DB**-frame $\mathsf{F} = (W, R)$ with a root u such that φ is satisfiable at F, u , we construct a tree $\mathsf{F}^{\sharp} = (W^{\sharp}, R^{\sharp})$, which consists of all R-paths in F starting at u; $\alpha R^{\sharp} \beta$ iff β is obtained by adding a world at the end of α. There is p-morphism $\mathsf{F}^{\sharp} \twoheadrightarrow \mathsf{F}$ sending every path to its end.

Put $n := \max(|W|, md(\varphi))$. Since F is serial, it follows that $1 \leq |R^{\sharp}(\alpha)| \leq n$ for any $\alpha \in W^{\sharp}$.

We may further assume that $|R^{\sharp}(\alpha)| = n$ for all α in $\mathsf{F}^{\sharp}$ (in fact, $1 \leq |R^{\sharp}(\alpha)| \leq n$, but $R^{\sharp}(\alpha)$ can be extended by adding virtual copies of one of its elements; we skip the routine details here). Therefore $\mathsf{F}^{\sharp}$ is isomorphic to $(n^*, \lhd)$.

Since F is symmetric, it readily follows that $(n^*, \lhd^{\pm}) \twoheadrightarrow \mathsf{F}$, and a p-morphism sends the root to the root. So φ is satisfiable at $(n^*, \lhd^{\pm}), \lambda$. Since $md(\varphi) \leq n$, by Proposition 2.1, it is satisfiable at $(n^*, \lhd^{\pm}) | (\lhd^{\pm})^{\leq n}(\lambda), \lambda$. The latter frame is exactly $(T_{n,n}, \lhd^{\pm})$.

(2) The argument for **TB** is almost the same — take the reflexive symmetric closure instead of the symmetric closure. □

2.3 Distances, products, and powersets

For a frame $\mathsf{F} = (W, R)$ we define *the distance function*

$$d(u, v) := \min\{k \geq 0 \mid u(R^{\pm})^k v\}$$

if the latter set is nonempty and ∞ otherwise. If F is connected (i.e., $(W, R^{\pm})$ is rooted), $d(u, v)$ is always finite.

Proposition 2.3 *Let* T *be a standard tree,* d *the distance in* $(T, \lhd)$, $V \subseteq T$, $|V| > 1$, *and* $d(v_1, v_2) = 2$ *for all distinct* $v_1, v_2 \in V$. *Then there exists a unique* $u \in T$ *such that* $u \lhd^{\pm} v$ *for all* $v \in V$.

Proof. If $v_1, v_2 \in V$, $v_1 \neq v_2$, then there exists a unique point u such that $u \lhd^{\pm} v_1$ and $u \lhd^{\pm} v_2$. Also, for any $v \in V$, if $d(v, v_2) = d(v, v_1) = 2$, then $u \lhd^{\pm} v$. □

For $\mathbf{x}, \mathbf{y} \in A^n$, by $h(\mathbf{x}, \mathbf{y})$ we denote the Hamming distance between $\mathbf{x}$ and $\mathbf{y}$:

$$h(\mathbf{x}, \mathbf{y}) = |\{i \mid x_i \neq y_i\}|.$$

For $\mathbf{x}, \mathbf{y} \in A^*$, put

$$\mathbf{x} H \mathbf{y} \iff |\mathbf{x}| = |\mathbf{y}| \ \& \ h(\mathbf{x}, \mathbf{y}) = 1.$$

So h is the distance in the frame (A^*, H).

$\mathcal{P}(U)$ denotes the power set of a set U; $U \triangle V$ denotes the symmetric difference of sets U, V. The frame $(\{0,1\}^n, H)$ is obviously isomorphic to $(\mathcal{P}(n), \triangle_1)$, where

$$U \triangle_1 V \quad \iff \quad |U \triangle V| = 1.$$

Consider an inequality frame [2] $(A, \neq)$ and its power

$$(A, \neq)^n = (A^n, \neq_0, \ldots, \neq_{n-1});$$

so

$$\mathbf{x} \neq_i \mathbf{y} \iff x_i \neq y_i \text{ and } x_j = y_j \text{ for all } j \neq i.$$

Clearly, $(A^n, H) = (A^n, \neq_0 \cup \cdots \cup \neq_{n-1})$.

Consider the translation $t^{(n)} : ML(\Diamond) \to ML(\Diamond_0, \ldots, \Diamond_{n-1})$ preserving the atoms and the boolean connectives and such that

$$t^{(n)}(\Diamond\varphi) = \Diamond_0 t^{(n)}(\varphi) \vee \cdots \vee \Diamond_{n-1} t^{(n)}(\varphi).$$

The following is trivial:

Proposition 2.4 *For any $A \neq \varnothing$, $n > 0$, $\theta : PV \to A^n$, φ, $\mathbf{x} \in A^n$,*

$$((A, \neq)^n, \theta), \mathbf{x} \vDash t^{(n)}(\varphi) \iff ((A^n, H), \theta), \mathbf{x} \vDash \varphi.$$

Theorem 2.5 (i) *For any infinite A, $n > 0$*

$$Log(A^n, H) = Log(\omega^n, H).$$

(ii) *$Log(\omega^n, H)$ is recursively axiomatizable.*

Proof. We apply Theorem 4.3 from [4].

Let $\mathcal{C}$ be the class $\{(A, \neq) \mid A \text{ is infinite }\}$, which is obviously axiomatizable by a recursive set Σ of classical pure equality formulas. Let

$$\mathcal{C}^{[n]} = \{\mathsf{F}^n \mid \mathsf{F} \in \mathcal{C}\}.$$

---

[2] A more precise notation is $(A, \neq_A)$.

Recall that for any n-modal formula φ there is a 'cubic translation' $\varphi^n(x_0, \ldots, x_{n-1})$; it translates a propositional variable p_j as an atom $P_j(x_0, \ldots, x_{n-1})$, and

$$(\Box_i \varphi)^n(x_0, \ldots, x_{n-1}) = \forall y(y \neq x_i \to [y/x_i]\varphi^n(x_0, \ldots, x_{n-1})).$$

Then we have

$$\mathcal{C}^{[n]} \models \varphi \Leftrightarrow \mathcal{C} \models \overline{\forall}\varphi^n \Leftrightarrow \Sigma \vdash \varphi^n,$$

where $\overline{\forall}\varphi^n$ denotes the universal second-order sentence obtained by quantifying over all parameters and predicate symbols occurring in φ^n. This implies that $Log(\mathcal{C}^{[n]})$ is RE.

On the other hand, it follows that $Log(\mathcal{C}^{[n]}) = Log((A, \neq)^n)$ for any infinite A. In fact, the above equivalence also holds for $\mathcal{C} = \{(A, \neq)\}$, by the Löwenheim - Skolem theorem.

Since the H-modality is recursively encoded in this language, the result follows. □

3 Non-finitely axiomatizable logics

In this section we recall some known results on logics of Hamming frames: for an infinite alphabet A, the logics $Log(A^n, H)$ are non-finitely axiomatizable for all $n > 0$ (in particular, this holds for the inequality relation on an infinite set). As the paper [8] is published only in Russian, we reproduce its relevant parts here.

Theorems on non-finite axiomatizability for many-dimensional modal logics usually require an intricate technique (see e.g. [5]). In our case we prove a stronger result (non-axiomatizability in finitely many variables) by using a rather simple construction. We base on the approach from [10], as far as we know the first result of this kind in the context of modal logic.

Definition 3.1 *For a modal logic L we define its m-*fragments*:*

$$L\lceil m := \{\varphi \in L \mid PV(\varphi) \subseteq \{p_1, \ldots, p_m\}\} \text{ for } m \geq 1,$$
$$L\lceil 0 := \{\varphi \in L \mid PV(\varphi) = \varnothing\}.$$

L is called m-variable axiomatizable *if $\boldsymbol{K} + L\lceil m = L$. A logic is called* finite-variable axiomatizable *if it is m-variable axiomatizable for some m.*

Frames F, G are called (modally) m-equivalent *(in symbols, $\mathsf{F} \sim_m \mathsf{G}$) if $Log(\mathsf{F})\lceil m = Log(\mathsf{G})\lceil m$.*

Lemma 3.2 *[10],[12] Let Λ be a logic, $m \geq 0$, and suppose there exist frames G, G' such that $G \sim_m G'$, $\Lambda \subseteq Log(G)$, $\Lambda \not\subseteq Log(G')$. Then Λ is not m-variable axiomatizable.*

The next fact is a slight modification of the Jankov-Fine lemma (cf. [2], [12]).

Lemma 3.3 *(see [2], [12]) Let $\mathsf{F} = (W, R)$ be a frame such that $R^{\leq k}$ is transitive for some $k > 0$ and let G be a finite frame. Then $Log(\mathsf{F}) \subseteq Log(\mathsf{G})$ iff there exists a point w in F such that $\mathsf{F} \uparrow w$ $(= \mathsf{F}|R^{\leq k}(w)) \twoheadrightarrow \mathsf{G}$.*

Proposition 3.4 *Let f be a monotonic*[3] *map from a frame (W, R) to a frame (V, S). If there exists $W_0 \subseteq W$ such that $|W_0| > |V|$ and $w_1 R w_2$ for all different w_1, w_2 from W_0, then (V, S) contains a reflexive point.*

Proof. Since $|W_0| > |V|$, there exist $w_1, w_2 \in W_0$ such that $w_1 \neq w_2$ and $f(w_1) = f(w_2)$. Then $w_1 R w_2$ by assumption, so $f(w_1) S f(w_2) = f(w_1)$ by the monotonicity of f. □

For $m > 0$ consider the frames $\mathsf{K}_m := (m, \neq_m)$, $\mathsf{K}'_m := (m, R_m)$, where $\neq_m$ is the inequality relation on m; $R_m := \{(m-1, m-1)\} \cup \neq_m$. Thus K_m is an irreflexive clique with m points, and K'_m is obtained from K_{m+1} by sticking two points into one reflexive point.

Lemma 3.5 *[12]* $\mathsf{K}_{2^m+1} \sim_m \mathsf{K}'_{2^m}$ *for any $m \geq 0$.*

Theorem 3.6 *[8] If A is infinite, then for any $n > 0$ the logic $Log(A^n, H)$ is not finite-variable axiomatizable.*

Proof. By Theorem 2.5, we may assume that A is the set of integers.

By Lemmas 3.2, 3.5, and 3.3, it is sufficient to show that for any $m > 1$
(i) $(A^n, H) \not\twoheadrightarrow \mathsf{K}_m$,
(ii) $(A^n, H) \twoheadrightarrow \mathsf{K}'_m$.

Note that (i) follows from Proposition 3.4, since A is infinite.

Let $a_0, \ldots, a_{m-2}$ be different elements of A.

If $n = 1$, then we define $f(a_i) = i$ for $m-1$ different elements $a_0, \ldots, a_{m-2}$ of A and $f(w) = m-1$ for all other elements from A. It is easy to check that $f : (A, \neq) \twoheadrightarrow \mathsf{K}'_m$.

Suppose $n > 1$. Let

$$V_i := \{\mathbf{x} \mid x_0 + \cdots + x_{n-1} = a_i\} \text{ for } 0 \leq i < m-1,$$
$$V_{m-1} := A^n - \bigcup_{0 \leq i < m-1} V_i.$$

Clearly, A^n is the disjoint union of these sets. Moreover, for all $i < m-1$, we have

$$\text{if } \mathbf{x} \in V_i, \text{ then } H(\mathbf{x}) \cap V_i = \varnothing; \tag{1}$$
$$\text{if } \mathbf{x} \notin V_i, \text{ then } H(\mathbf{x}) \cap V_i \neq \varnothing; \tag{2}$$
$$\text{for each } \mathbf{x} \in A^n,\ H(\mathbf{x}) \cap V_{m-1} \neq \varnothing. \tag{3}$$

Let us check (1). Let $i < m$, $\mathbf{x} \in V_i$, $\mathbf{x} H \mathbf{y}$. If follows that for some l we have $x_l \neq y_l$, and $x_j = y_j$ for all $j \neq l$. Then $\sum_{0 \leq j < n} x_j \neq \sum_{0 \leq j < n} y_j$, so $\mathbf{y} \notin V_i$.

To obtain (2), take $\mathbf{y}$ such that $x_i = y_i$ for $i > 0$ and $y_0 = a_i - \sum_{1 \leq j < n} x_j$. Then $\mathbf{y} \in V_i$ and $\mathbf{x} H \mathbf{y}$.

Let us check (3). Consider the set

$$U := \{\mathbf{y} \mid x_0 \neq y_0 \ \&\ \forall j > 0\, (y_j = x_j)\}.$$

[3] I.e., f is a homomorphism.

Then $U \subseteq H(\mathbf{x})$. Since U is infinite, and for each $i < m-1$, $U \cap V_i$ is a singleton, we have $U \cap V_{m-1} \neq \varnothing$.

Now we define a map $f : A^n \to m$ by putting $f(\mathbf{x}) := i$ iff $\mathbf{x} \in V_i$. Since every V_i is nonempty, f is surjective. If $\mathbf{x}H\mathbf{y}$ and $\mathbf{x} \notin V_{m-1}$, then using (1) we get $f(\mathbf{x}) \neq f(\mathbf{y})$; hence, f is monotonic. From (2) and (3) it follows that f satisfies the 'lift property', i.e.

$$f(x)R_m i \Rightarrow \exists y \in H(x)\ f(y) = i.$$

Thus f is a p-morphism. □

Note that if $n = 1$ then $(A, H) = (A, \neq)$ and Theorem 3.6 yields

Corollary 3.7 *If A is an infinite set, then $Log(A, \neq)$ is not finite-variable axiomatizable.*

Remark 3.8 This gives us a simple example of a modal logic which is not finitely axiomatizable, but has a finitely axiomatizable conservative extension. Namely, by Corollary 3.7 the logic $Log(\mathbb{R}^2, \neq)$ is not f.a., whereas the topological modal logic with the difference modality of $\mathbb{R}^2$ has a finite axiomatization [7].

Let C_0 and C_1 be a reflexive and an irreflexive singleton, respectively. From Lemma 3.5 it follows that for any $m \geq 0$

$$\mathsf{K}_{2^m+1} \times \mathsf{C}_1 \sim_m \mathsf{K}'_{2^m} \times C_1, \quad \mathsf{K}_{2^m+1} \times \mathsf{C}_0 \sim_m \mathsf{K}'_{2^m} \times \mathsf{C}_0.$$

Let A be an infinite set. If $|B| > 1$, then $(B, \neq) \twoheadrightarrow \mathsf{C}_1$, so

$$\begin{array}{ll} (A,\neq) \times (B,\neq) \twoheadrightarrow \mathsf{K}'_m \times \mathsf{C}_1, & (A,\neq) \times (B,\neq) \not\twoheadrightarrow \mathsf{K}_m \times \mathsf{C}_1, \\ (A,\neq) \times \mathsf{C}_0 \twoheadrightarrow \mathsf{K}'_m \times \mathsf{C}_0, & (A,\neq) \times \mathsf{C}_0 \not\twoheadrightarrow \mathsf{K}_m \times \mathsf{C}_0, \end{array}$$

which leads to

Corollary 3.9 *[8] Suppose B is nonempty, A is infinite. Then $Log((A, \neq) \times (B, \neq))$ is not finite-variable axiomatizable.*

Recently [6] a similar result was obtained for products with the minimal difference logic **DL**: all logics in the interval between **K** × **DL** and **S5** × **DL** are not finite-variable axiomatizable. However, the logics described by the previous corollary are not in this interval.

4 Undecidability

In this section we prove undecidability results on logics of Hamming spaces.

Fix $n \geq 2$. Our aim is to define unimodal operators (formulas) which will emulate $\mathbf{S5}^n$-modalities in frames (A^n, H). For this purpose we use a formula $sets^{(n)}$ encoding the product structure on (A^n, H) up to permutation of coordinates.

Put $\Box^0\varphi = \varphi$, $\Box^{l+1}\varphi = \Box\Box^l\varphi$, $\Box^{\leq l}\varphi = \bigwedge_{0 \leq i \leq l} \Box^i\varphi$, $\Diamond^l\varphi = \neg\Box^l\neg\varphi$, $\Diamond^{\leq l}\varphi = \neg\Box^{\leq l}\neg\varphi$. Note that the operator $\Box^{\leq n}$ acts like the universal modality on (A^n, H): $((A^n, H), \theta), \mathbf{x} \vDash \Box^{\leq n}\varphi$ for some $\mathbf{x}$ iff $((A^n, H), \theta), \mathbf{x} \vDash \varphi$ for all $\mathbf{x} \in A^n$.

For each set $U \subseteq n$ we fix a variable p_U. Let $sets^{(n)}$ be the conjunction of the following formulas:

$$p_\varnothing \wedge \neg\Diamond p_\varnothing \tag{4}$$

$$\Box^{\leq n}\left(\bigvee_{U\subseteq n} p_U \wedge \bigwedge_{U,V\subseteq n,\ U\neq V} (p_U \to \neg p_V)\right) \tag{5}$$

$$\Box^{\leq n}\left(\bigwedge_{U,V\subseteq n,\ |U\triangle V|=1} (p_U \to \Diamond p_V)\right) \tag{6}$$

$$\Box^{\leq n}\left(\bigwedge_{U,V\subseteq n,\ |U\triangle V|>1} (p_U \to \neg\Diamond p_V)\right) \tag{7}$$

(Note that if we also add the conjuncts $p_U \to \neg\Diamond p_U$ for all nonempty $U \subseteq n$, then we obtain the frame formula for the frame $(\mathcal{P}(n), \triangle_1)$ at the point $\varnothing$.)

For $\mathbf{x}, \mathbf{y} \in A^n$ and a permutation $\sigma : n \to n$, let

$$D_\sigma(\mathbf{x}, \mathbf{y}) = \{i \mid x_{\sigma(i)} \neq y_{\sigma(i)}\}.$$

The meaning of the formula $sets^{(n)}$ is explained by the following key fact.

Lemma 4.1 *Let* $|A| > 1$, $((A^n, H), \theta), \mathbf{r} \vDash sets^{(n)}$. *Then there exists a unique permutation* $\sigma : n \to n$ *such that for any* $\mathbf{x} \in A^n$ *and* $V \subseteq n$,

$$D_\sigma(\mathbf{r}, \mathbf{x}) = V \iff ((A^n, H), \theta), \mathbf{x} \vDash p_V. \tag{8}$$

Proof. For any $\mathbf{x}$ in A^n we define sets $A_0(\mathbf{x}), \ldots, A_{n-1}(\mathbf{x})$ ($\mathbf{x}$-*axis*):

$$A_i(\mathbf{x}) = \{\mathbf{y} \mid \mathbf{y}H\mathbf{x} \ \&\ y_i \neq x_i\}.$$

Clearly, $H(\mathbf{x})$ is the disjoint union of the sets $A_i(\mathbf{x})$, and for $0 \leq i \neq j < n$,

$$\mathbf{y}H\mathbf{z} \text{ for no } \mathbf{y} \in A_i,\ \mathbf{z} \in A_j. \tag{9}$$

Put $\mathsf{M} = ((A^n, H), \theta)$. Let σ be the binary relation on n such that

$$(i, j) \in \sigma \iff \mathsf{M}, \mathbf{y} \vDash p_{\{i\}} \text{ for some } \mathbf{y} \in A_j(\mathbf{r}).$$

First, let us show that σ is a permutation $n \to n$.

By (6), $\mathsf{M}, \mathbf{r} \vDash \Diamond p_{\{0\}} \wedge \ldots \wedge \Diamond p_{\{n-1\}}$. Let $i < n$. Then for some $j < n$, $\mathbf{y} \in A_j(\mathbf{r})$, we have $\mathsf{M}, \mathbf{y} \vDash p_{\{i\}}$. Suppose $k \neq i$, $k < n$; by (7), $\mathsf{M}, \mathbf{y} \vDash \neg\Diamond p_{\{k\}}$; by (5), $\mathsf{M}, \mathbf{y} \vDash \neg p_{\{k\}}$; so we have $\mathsf{M}, \mathbf{z} \nvDash p_{\{k\}}$ for all $\mathbf{z} \in A_j(\mathbf{r})$. It means that σ is a function from n to n. If $\mathbf{z} \in H(\mathbf{r})$ then $\mathsf{M}, \mathbf{z} \vDash \neg p_\varnothing$ by (4), and if $|V| > 1$ then $\mathsf{M}, \mathbf{z} \vDash \neg p_V$ by (7). By (5), $\mathsf{M}, \mathbf{z} \vDash p_V$ for some V, and it follows that V is a singleton. It means that σ is surjective, i.e., it is a permutation.

Since σ is a bijection, then $h(\mathbf{r},\mathbf{x}) = |D_\sigma(\mathbf{r},\mathbf{x})|$ for any $\mathbf{x}$ in M.

Let us check (8). By (5), for every $\mathbf{x}$ in M there exists a unique $U \subseteq n$ such that $\mathbf{x} \vDash p_U$. It follows that we only have to prove the left-to-right direction.

We proceed by induction on $h(\mathbf{r},\mathbf{x})$.

If $h(\mathbf{r},\mathbf{x}) = 0$, then $\mathbf{x} = \mathbf{r}$, $D_\sigma(\mathbf{r},\mathbf{x}) = \varnothing$; by (4), $\mathsf{M},\mathbf{r} \vDash p_\varnothing$.

For the induction step, let $0 \le l < n-1$ and suppose that (8) holds for all $\mathbf{y} \in H^{\le l}(\mathbf{r})$. Let

$$h(\mathbf{x},\mathbf{r}) = l+1,\ \mathsf{M},\mathbf{x} \vDash p_U,\ D_\sigma(\mathbf{r},\mathbf{x}) = V.$$

We have to show that $U = V$.

If $l = 0$, then $\mathbf{x} \in H(\mathbf{r})$ and, as shown above, $U = \{i\}$ for some i; then, by the definition of σ, $\mathbf{x} \in A_{\sigma(i)}(\mathbf{r})$, which means $V = \{i\}$.

Suppose $l > 0$. Then $|V| = h(\mathbf{r},\mathbf{x}) = l+1$. Let $V = \{i_0,\dots,i_l\}$. Take points $\mathbf{y}^{(0)},\dots,\mathbf{y}^{(l)}$ such that for $j \le l$

$$\mathbf{y}^{(j)} \in A_{\sigma(i_j)}(\mathbf{x}) \text{ and } D_\sigma(\mathbf{r},\mathbf{y}^{(j)}) = V - \{i_j\},$$

that is $y_k^{(j)} = x_k$ for $k \neq \sigma(i_j)$, $y_k^{(j)} = r_k$ for $k = \sigma(i_j)$.

Then $h(\mathbf{r},\mathbf{y}^{(j)}) = l$, and by the induction hypothesis,

$$\mathsf{M},\mathbf{y}^{(j)} \vDash p_{V-\{i_j\}} \tag{10}$$

Since $\mathbf{y}^{(j)} \in H(\mathbf{x})$, then by (7) we have

$$|U \mathbin{\triangle} (V - \{i_j\})| \le 1 \text{ for all } j \le l. \tag{11}$$

If $U = V - \{i_j\}$ for some $j \le l$, then $\mathsf{M},\mathbf{y}^{(k)} \vDash \Diamond p_{V-\{i_j\}}$ for some $k \neq j$ (note that $l > 0$), which contradicts (7). Thus $U \mathbin{\triangle}_1 (V - \{i_j\})$ for all $j \le l$, and $|U| = l-1$ or $|U| = l+1$.

In the first case, by (11) we have $U \subseteq V - \{i_j\}$ for all j, so $U \subseteq \bigcap\limits_{0\le j\le l} (V - \{i_j\}) = \varnothing$. It follows that $l = 1$. Let $\mathbf{y} = \mathbf{y}^{(0)}$, $\{i\} = D_\sigma(\mathbf{r},\mathbf{y})$. For each $k \neq i$ we choose a point $\mathbf{z}_k$ such that $\mathbf{y}H\mathbf{z}_k$ and $\mathsf{M},\mathbf{z} \vDash p_{\{i,k\}}$ (such points exist, because (6) and (10) imply $\mathsf{M},\mathbf{y} \vDash \Diamond p_{\{i,k\}}$). Let Z be the set of these points. Since $|Z| = n-1$, by the pigeonhole principle there exist two points from the set $Z \cup \{\mathbf{r},\mathbf{x}\}$ which belong to the same $\mathbf{y}$-axis. If these points are $\mathbf{r}$ and $\mathbf{x}$, we have a contradiction with (4), in all other cases we have a contradiction with (7).

It follows that $|U| = l+1$, so $U \supseteq V - \{i_j\}$ for all j, so

$$U \supseteq \bigcup_{0\le j\le l} (V - \{i_j\}) = V.$$

Since $|U| = |V|$, we obtain $U = V$. □

If $((A^n,H),\theta),\mathbf{r} \vDash \mathit{sets}^{(n)}$, then the permutation satisfying (8) is called *the* $(\theta,\mathbf{r})$*-permutation.*

For $0 \leq i < n$, put

$$plane_i^{(n)} = \bigvee_{U \subseteq n,\ i \notin U} p_U.$$

From the above lemma we immediately obtain

Proposition 4.2 *Let* $|A| > 1$, $((A^n, H), \theta), \mathbf{r} \models sets^{(n)}$, σ *be the* $(\theta, \mathbf{r})$-*permutation. Then for any* $\mathbf{x} \in A^n$, $0 \leq i < n$,

$$((A^n, H), \theta), \mathbf{x} \models plane_i^{(n)} \iff r_{\sigma(i)} = x_{\sigma(i)}. \tag{12}$$

Finally, we define formulas $\Diamond_i^{(n)}$, $0 \leq i < n$:

$$\begin{aligned}\Diamond_i^{(n)} &= s \vee ((plane_i^{(n)} \to \Diamond(\neg plane_i^{(n)} \wedge s))\wedge \\ &\wedge(\neg plane_i^{(n)} \to \Diamond(plane_i^{(n)} \wedge (s \vee (\Diamond(\neg plane_i^{(n)} \wedge s)))))).\end{aligned}$$

For a formula φ, let $\Diamond_i^{(n)}\varphi$ denote the result of substitution of s for φ in $\Diamond_i^{(n)}$.

Let us illustrate the meaning of the above formulas. $\Diamond_i^{(n)}\varphi$ is true at a point $\mathbf{x}$ iff either φ is true at $\mathbf{x}$, or φ is true at a point $\mathbf{y} \in H(\mathbf{x})$ such that the following holds: if $\mathbf{x} \in plane_i^{(n)}$, then $\mathbf{y} \notin plane_i^{(n)}$; if $\mathbf{x} \notin plane_i^{(n)}$, then either $\mathbf{y} \in plane_i^{(n)}$, or $\mathbf{y} \notin plane_i^{(n)}$ and there exists $\mathbf{z} \in plane_i^{(n)}$ such that $\mathbf{x}H\mathbf{z}H\mathbf{y}$. In all cases we "move" along $\sigma(i)$-direction; on the other hand, any point in $A_{\sigma(i)}(\mathbf{x})$ is reachable. It means that the set of all "possible" points is

$$\{\mathbf{y} \mid x_j = y_j \text{ for all } j \neq \sigma(i)\},$$

and we have

Proposition 4.3 *Let* $|A| > 1$, $((A^n, H), \theta), \mathbf{r} \models sets^{(n)}$, σ *be the* $(\theta, \mathbf{r})$-*permutation. Then for any* $\mathbf{x} \in A^n$,

$$((A^n, H), \theta), \mathbf{x} \models \Diamond_i^{(n)} \iff \exists \mathbf{y}\, (D_\sigma(\mathbf{x}, \mathbf{y}) \subseteq \{i\}\ \&\ ((A^n, H), \theta), \mathbf{y} \models s)$$

For a formula φ in the n-modal language $ML(\Diamond_0, \ldots, \Diamond_{n-1})$, we define the unimodal formula $[\varphi]^{(n)}$:

$$[p]^{(n)} = p \text{ for } p \in PV; \quad [\phi \wedge \psi]^{(n)} = [\phi]^{(n)} \wedge [\psi]^{(n)}; \quad [\neg\phi]^{(n)} = \neg([\phi]^{(n)});$$
$$[\Diamond_i \phi]^{(n)} = \Diamond_i^{(n)} [\phi]^{(n)}.$$

Lemma 4.4 *Let* $|A| > 1$, $((A^n, H), \theta), \mathbf{r} \models sets^{(n)}$, σ *be the* $(\theta, \mathbf{r})$-*permutation,* $(A, A \times A)^n = (A^n, R_0, \ldots R_{n-1})$. *Then for any* n-*modal formula* φ *with* $PV(\varphi) \cap PV(sets^{(n)}) = \varnothing$, *for any* $\mathbf{x} \in A^n$, *we have*

$$((A^n, R_{\sigma(0)}, \ldots R_{\sigma(n-1)}), \theta), \mathbf{x} \models \varphi \iff ((A^n, H), \theta), \mathbf{x} \models [\varphi]^{(n)} \tag{13}$$

Proof. Note that $R_{\sigma(i)} = \{(\mathbf{x}, \mathbf{y}) \mid D_\sigma(\mathbf{x}, \mathbf{y}) \subseteq \{i\}\}$. Thus using Proposition 4.3, the proof can be obtained by straightforward induction on the length of φ (see e.g. [11, Lemma 3.5] for details). □

Theorem 4.5 *For $|A| > 1$, $n \geq 2$, for any n-modal formula φ that does not share variables with $sets^{(n)}$, we have:*

$$\varphi \text{ is } (A, A \times A)^n\text{-satisfiable} \iff sets^{(n)} \wedge [\varphi]^{(n)} \text{ is } (A^n, H)\text{-satisfiable.}$$

Proof. Let $\mathsf{F} = (A, A \times A)^n = (A^n, R_0, \ldots, R_{n-1})$.

($\Longleftarrow$). For any permutation $\sigma : n \to n$, the frames $(A^n, R_{\sigma(0)}, \ldots R_{\sigma(n-1)})$ and F are isomorphic, so by Lemma 4.4, φ is F-satisfiable.

($\Longrightarrow$). Suppose $(\mathsf{F}, \theta), \mathbf{r} \models \varphi$. Let σ be the identity map on n. Put $\eta(p) = \theta(p)$ for all $p \notin PV(sets^{(n)})$; for $V \subseteq n$, put

$$\eta(p_V) = \{\mathbf{x} \mid D_\sigma(\mathbf{r}, \mathbf{x}) = V\}.$$

Since $PV(\varphi) \cap PV(sets^{(n)}) = \varnothing$, $(\mathsf{F}, \eta), \mathbf{r} \models \varphi$. By a straightforward argument, $((A^n, H), \eta), \mathbf{r} \models sets^{(n)}$. By Lemma 4.4, $((A^n, H), \eta), \mathbf{r} \models [\varphi]^{(n)}$, so $sets^{(n)} \wedge [\varphi]^{(n)}$ is (A^n, H)-satisfiable. □

Since all logics $\mathbf{S5}^n$, $n > 2$, are undecidable [9], and

$$\mathbf{S5}^n = Log(\{(A, A \times A)^n \mid A \neq \varnothing\}) = Log((\omega, \omega \times \omega)^n),$$

we have the following corollaries.

Corollary 4.6 *For any $n > 2$, the logic $Log(\{(A^n, H) \mid A \neq \varnothing\})$ is undecidable.*

Corollary 4.7 *For any $n > 2$, the logic $Log(\omega^n, H)$ is undecidable.*

5 Completeness

In spite of undecidability proved in section 4, there are some positive results on logics of Hamming frames over "long" words. For functions $f, g : I \to A$, put

$$f\, H\, g \iff |\{i \mid i \in I,\ f(i) \neq g(i)\}| = 1.$$

Our first positive result is the completeness of **DB** with respect to the Hamming frame of infinite (0,1)-sequences $(2^\omega, H)$. We formulate it in terms of sets of natural numbers. Clearly, the frame $(2^\omega, H)$ is isomorphic to the frame $(\mathcal{P}(\omega), \Delta_1)$.

Definition 5.1 *Let (W, R), (V, S) be frames, $x \in W$, $f : W \to V$, $n > 0$. f is an* n-reduction at x from *(W, R) to (V, S), if the restriction of f on $R^{\leq n}(x)$ is monotonic, and for any $y \in R^{\leq n-1}(x)$, $z \in V$, if $f(y) S z$ then $y R y'$ and $f(y') = z$ for some y'.*

Lemma 5.2 *Let (W, R), (V, S) be frames, $x \in W$, f be an n-reduction at x from (W, R) to (V, S). Then for any φ with $md(\varphi) \leq n$, if φ is satisfiable at $f(x)$ in (V, S), then φ is satisfiable at x in (W, R).*

Proof. Suppose $((V, S), \theta), f(x) \models \varphi$. For $p \in PV$, put $\eta(p) = f^{-1}(\theta(p))$. Then by induction on the modal depth of a formula, it is easy to check that

$$((W, R), \eta), y \models \psi \iff ((V, S), \theta), f(y) \models \psi$$

for any ψ, l, y such that $0 \leq l \leq n - md(\psi)$ and $y \in R^{\leq l}(x)$. $\square$

For a set U, $\mathcal{P}_{fin}(U)$ denotes the set of all finite subsets of U.

Theorem 5.3 $Log(\mathcal{P}(\omega), \triangle_1)) = Log(\mathcal{P}_{fin}(\omega), \triangle_1)) = \boldsymbol{DB}$.

Proof. First, we introduce some auxiliary definitions. Fix $n \geq 2$. For $x \in \omega$, let $\overline{x}$ be the leaf $a_n \dots a_1 \in T_{n,n}$ such that $x = mn^n + \sum_{1 \leq i \leq n} a_i n^{i-1}$ for some $m \geq 0$.

Let d denote the distance in $(T_{n,n}, \lhd)$.

Let $x \in \omega$, $\overline{x} = a_n \dots a_1$. For $w = b_r \dots b_1 \in T_{n-1,n}$, let $[x\!:\!w]$ be the word $u \in T_{n,n}$ such that $d(\overline{x}, u) = d(\overline{x}, w) - 1$. Clearly, $[x\!:\!w]$ exists and unique: if $w \sqsubseteq \overline{x}$, then $[x\!:\!w] = a_{r+1} \dots a_1$ $(= b_{r+1} \dots b_1)$, otherwise (in this case $r > 0$) $[x\!:\!w] = b_{r-1} \dots b_1$. Note that $w \lhd^{\pm} [x\!:\!w]$. By a straightforward argument, for any $u, v \in T_{n-1,n}$

$$u \lhd^{\pm} v \;\&\; d(\overline{x}, u) < d(\overline{x}, v) \iff [x\!:\!v] = u, \tag{14}$$

$$u \lhd^{\pm} v \implies d(\overline{x}, u) \neq d(\overline{x}, v). \tag{15}$$

For $l \geq -1$, put $\mathcal{P}_l(\omega) = \{V \mid V \subset U,\ |V| \leq l\}$. By induction, we construct a sequence of functions $f_i : \mathcal{P}_i(\omega) \to T_{i,n}$, $0 \leq i \leq n$, such that:

(i) if $S, S' \in \mathcal{P}_i(\omega)$, $S \triangle_1 S'$, then $f_i(S) \lhd^{\pm} f_i(S')$;

(ii) if $S \in \mathcal{P}_{i-1}(\omega)$, $x > \max(S)$, then $f_i(S \cup \{x\}) = [x\!:\!f_i(S)]$;

Put $f_0(\varnothing) = \lambda$. Clearly, f_0 satisfies (i) and (ii).

Let $l < n$ and suppose f_l is already constructed and satisfies (i), (ii). We define f_{l+1} as follows.

Consider $S \subset \omega$. If $|S| \leq l$, put $f_{l+1}(S) = f_l(S)$. Suppose $|S| = l + 1$. Let $S = \{x_0, \dots, x_l\}$, $S_i = S - \{x_i\}$. For any $i, j \leq l$, if $i \neq j$, then $S_i \triangle_1 (S_i \cap S_j) \triangle_1 S_j$, so $d(f_l(S_i), f_l(S_j)) \in \{0, 2\}$ by (i).

If $f_l(S_i) \neq f_l(S_j)$ for some $i, j \leq l$, then by Proposition 2.3 there exists a unique u such that $u \lhd^{\pm} f(S_k)$ for all $k \leq l$. We put

$$f_{l+1}(S) = u.$$

Note that $|u| < |f(S_i)|$ or $|u| < |f(S_j)|$, so $f_{l+1}(S) \in T_{l-1,n}$.

If $f_l(S_i) = f_l(S_j)$ for all $i, j \leq l$, we put

$$f_{l+1}(S) = [\max(S)\!:\!f_l(S_0)].$$

Since $[\max(S)\!:\!f_l(S_0)] \lhd^{\pm} f_l(S_0)$, then $f_{l+1}(S) \in T_{l+1,n}$.

Let us check that f_{l+1} satisfies (i) and (ii).

To show (i), suppose $S \triangle_1 S'$, $S, S' \in \mathcal{P}_{l+1}(\omega)$.

If both S, S' in $\mathcal{P}_l(\omega)$, then (i) holds by the induction hypothesis. Suppose $|S| = l + 1$. If $f(S - \{x\}) \neq f(S - \{y\})$ for some $x, y \in S$, then by the above definition $f(S) \lhd^{\pm} f(S - \{x\})$ for any $x \in S$, and (i) holds, since $S' = S - \{x\}$

for some $x \in S$. Otherwise, $f(S) = [\max(S){:}f_l(S')] \lhd^{\pm} f_l(S')$, which proves (i).

The key step in our proof is to check (ii). To this end, suppose $S \in \mathcal{P}_l(\omega)$, $x > \max(S)$. Consider the following two cases.

Case 1. $f_{l+1}(S) = f_{l+1}((S - \{y\}) \cup \{x\})$ for all $y \in S$. Then, by the definition of f_{l+1},

$$f_{l+1}(S \cup \{x\}) = [\max(S \cup \{x\}){:}f_l(S)] = [x{:}f_l(S)].$$

Case 2. $f_{l+1}(S) \neq f_{l+1}(S' \cup \{x\})$, where $S' = S - \{y\}$ for some $y \in S$. Since $x > \max(S')$, by applying the induction hypothesis to $f_l(S' \cup \{x\})$, we obtain

$$f_{l+1}(S' \cup \{x\}) = f_l(S' \cup \{x\}) \stackrel{IH}{=} [x{:}f_l(S')] = [x{:}f_{l+1}(S')].$$

It follows that $f_{l+1}(S) \neq [x{:}f_{l+1}(S')]$. By (i), $f_{l+1}(S) \lhd^{\pm} f_{l+1}(S')$, so (14) implies $d(\overline{x}, f_{l+1}(S)) \geq d(\overline{x}, f_{l+1}(S'))$; by (15), $d(\overline{x}, f_{l+1}(S)) > d(\overline{x}, f_{l+1}(S'))$, and by (14) again, we obtain $[x{:}f_{l+1}(S)] = f_{l+1}(S')$. Thus

$$f_{l+1}(S) \lhd^{\pm} [x{:}f_{l+1}(S)] = f_{l+1}(S') \lhd^{\pm} [x{:}f_{l+1}(S')] = f_{l+1}(S' \cup \{x\}). \quad (16)$$

By the construction of f_{l+1}, $f_{l+1}(S \cup \{x\}) = u$, where

$$u \lhd^{\pm} f_l(S) = f_{l+1}(S), \quad u \lhd^{\pm} f_l(S' \cup \{x\}) = f_{l+1}(S' \cup \{x\}).$$

So by (16) we obtain $f_{l+1}(S \cup \{x\}) = [x{:}f_l(S)]$, q.e.d.

Let us check that f_n is an n-reduction at $\varnothing$ from $(\mathcal{P}_{fin}(\omega), \triangle_1)$ to $(T_{n,n}, \lhd^{\pm})$. f_n is monotonic by (i). Suppose $S \in \mathcal{P}_{n-1}(\omega)$, $u \in T_{n,n}$, $f_n(S) \lhd^{\pm} u$. If $f_n(S) \lhd u$, take x such that $x > \max(S)$, $u \sqsubseteq \overline{x}$; if $u \lhd f_n(S)$, then $f_n(S) \neq \lambda$ and there exists x such that $x > \max(S)$, $f_n(S) \not\sqsubseteq \overline{x}$. In both cases we obtain $f_n(S \cup \{x\}) = [x{:}f_n(S)] = u$.

By Propositions 5.2 and 2.2, we obtain that any **DB**-satisfiable formula is satisfiable in $(\mathcal{P}_{fin}(\omega), \triangle_1)$.

It follows that $Log(\mathcal{P}_{fin}(\omega), \triangle_1) \subseteq$ **DB**. Since $(\mathcal{P}_{fin}(\omega), \triangle_1)$ is a generated subframe of $(\mathcal{P}(\omega), \triangle_1)$, then $Log(\mathcal{P}(\omega), \triangle_1) \subseteq$ **DB**. The converse inclusions are trivial. □

The above proof gives us the following semantic characterization of the logic **TB**.

For sets S, S', put

$$S \triangle_{\leq 1} S' \iff |S \triangle S'| \leq 1.$$

Since an n-reduction between two frames is also an n-reduction between their reflexive closures, we have

Corollary 5.4 $Log(\mathcal{P}(\omega), \triangle_{\leq 1}) = Log(\mathcal{P}_{fin}(\omega), \triangle_{\leq 1}) =$ ***TB***.

Put $f\,\overline{H}\,g \iff f\,H\,g$ or $f = g$.

Theorem 5.5 *Let I be an infinite set, $|A| > 1$. Then $Log(A^I, \overline{H}) =$ **TB**.*

Proof. Note that the considered frame is reflexive and symmetric, so $Log(A^I, \overline{H}) \supseteq$ **TB**.

To prove completeness, assume without any loss of generality that $\{0, 1\} \subseteq A$, $\omega \subseteq I$.

We define $F : A^I \to 2^I$. For $f \in A^I$, to define $F(f) : I \to 2$, we put

$$F(f)(x) = \begin{cases} 0 & f(x) = 0 \\ 1 & f(x) \neq 0 \end{cases}$$

One can easily see that $F : (A^I, \overline{H}) \twoheadrightarrow (2^I, \overline{H})$, so $Log(A^I, \overline{H}) \subseteq Log(2^I, \overline{H})$.

For $f : I \to 2$, let $G(f)$ be the restriction of f to ω. Clearly, $G : (2^I, \overline{H}) \twoheadrightarrow (2^\omega, \overline{H})$, so $Log(2^I, \overline{H}) \subseteq Log(2^\omega, \overline{H})$. By Corollary 5.4, $Log(2^\omega, \overline{H}) =$ **TB**. It follows that

$$\mathbf{TB} \subseteq Log(A^I, \overline{H}) \subseteq Log(2^I, \overline{H}) \subseteq \mathbf{TB}.$$

□

Theorem 5.6 *For any infinite set I, $Log(\mathcal{P}(I), \triangle_{\leq 1}) =$ **TB**.*

Proof. Follows from Theorem 5.5 for $A = 2$. □

6 Some open questions

For a set I, let $\mathfrak{H}(I) = \{(A^I, H) \mid |A| > 1\}$, $\overline{\mathfrak{H}}(I) = \{(A^I, \overline{H}) \mid |A| > 1\}$.

For any $I \supseteq J$, we have $Log(A^I, \overline{H}) \subseteq Log(A^J, \overline{H})$, since the restriction (projection) operator is a p-morphism. The classes $\mathfrak{H}(n)$, $\overline{\mathfrak{H}}(n)$ are elementary, so their logics have the countable frame property. It follows that $Log(\overline{\mathfrak{H}}(n)) = Log(\omega^n, \overline{H})$ (see the proof of Theorem 5.5). So we have

$$\begin{aligned} \mathbf{S5} = Log(\omega, \omega \times \omega) = Log(\overline{\mathfrak{H}}(1)) \supsetneq Log(\overline{\mathfrak{H}}(2)) = Log(\omega^2, \overline{H}) \supsetneq \\ \supsetneq Log(\overline{\mathfrak{H}}(3)) = Log(\omega^3, \overline{H}) \supsetneq \cdots \supsetneq Log(\overline{\mathfrak{H}}(\omega)) = Log(2^\omega, \overline{H}) = \mathbf{TB}. \end{aligned}$$

(the fact that inclusions are strict is very simple: for example, these logics can be distinguished by formulas of finite width, see e.g. [2]).

The irreflexive case is even less clear. However, for both cases we have

Problem 6.1

$Log(\overline{\mathfrak{H}}(\omega)) \stackrel{?}{=} \bigcap\limits_{n<\omega} Log(\omega^n, \overline{H})$;

$Log(\mathfrak{H}(\omega)) \stackrel{?}{=} \bigcap\limits_{n<\omega} Log(\mathfrak{H}(n))$.

Since $\mathbf{S5}^2$ is decidable (see e.g. [5]), $Log(\omega^2, \overline{H})$ is also decidable.

Problem 6.2 *Is there a decidable logic $Log(\omega^n, \overline{H})$, $n = 3, 4, \ldots$?*

The logic $Log(\mathfrak{H}(2))$ is a fragment of the logic $Log(\{(A, \neq)^2 \mid |A| > 1\})$.

Problem 6.3 *Is the logic $Log(\mathfrak{H}(2))$ decidable?*

Problem 6.4 *Does there exist a finitely axiomatizable logic $Log(\omega^n, \overline{H})$, for $n = 2, 3, \ldots$?*

Finding reflexive analogues of frames K_m, K'_m (if they exist) seems nontrivial.

Acknowledgements

We would like to thank the anonymous referees for their useful comments.

The work on this paper was supported by Poncelet Laboratory (UMI 2615 of CNRS and Independent University of Moscow), and by RFBR (projects No. 11-01-00958, 11-01-92471, 11-01-93107).

References

[1] Blackburn, P., M. D. Rijke and Y. Venema, "Modal Logic," Cambridge University Press, 2002.

[2] Chagrov, A. and M. Zakharyaschev, "Modal logic," Clarendon Press (Oxford and New York), 1997.

[3] Gabbay, D. and V. Shehtman, *Products of modal logics, part 1*, Logic Journal of the IGPL **6** (1998), pp. 73–146.

[4] Gabbay, D. and V. Shehtman, *Products of modal logics, part 2: Relativised quantifiers in classical logic*, Logic Journal of the IGPL **8** (2000), pp. 165–210.

[5] Gabbay, D. M., A. Kurucz, F. Wolter and M. Zakharyaschev, "Many-dimensional modal logics: theory and applications," Studies in Logic and the Foundations of Mathematics, North-Holland, Amsterdam, 2003.

[6] Hampson, C. and A. Kurucz, *Axiomatisation and decision problems of modal product logics with the difference operator* (2012), (manuscript).

[7] Kudinov, A., *Difference modality in topological spaces*, in: *Algebraic and Topological Methods in Non-classical Logics II, Abstracts, Universitat de Barcelona*, Barcelona, 2005, pp. 50–51.

[8] Kudinov, A. and I. Shapirovsky, *Some examples of modal logics without a finite axiomatisation (in Russian)*, in: *Proceedings of the conference on Information Technologies and Systems (ITaS'10)*, 2010, pp. 258–262.

[9] Maddux, R., *The equational theory of CA_3 is undecidable*, Journal of Symbolic Logic **45** (1980), pp. 311–316.

[10] Maksimova, L., D. Skvortsov and V. Shehtman, *The impossibility of a finite axiomatization of Medvedev's logic of finite problems*, Soviet Mathematics - Doklady **20** (1979), pp. 394–398.

[11] Shapirovsky, I., *Simulation of two dimensions in unimodal logics*, in: *Advances in Modal Logic*, 2010, pp. 371–392.

[12] Shapirovsky, I. and V. Shehtman, *Modal logics of regions and Minkowski spacetime*, Journal of Logic and Computation **15** (2005), pp. 559–574.

Finite Frames for $\mathbf{K4.3} \times \mathbf{S5}$ are Decidable

Agi Kurucz

Department of Informatics
King's College London
Strand, London WC2R 2LS, U.K.

Sérgio Marcelino [1]

Dep. Matemática
Instituto Superior Técnico
Universidade Técnica de Lisboa, Portugal
and
SQIG, Instituto de Telecomunicações, Lisboa, Portugal

Abstract

If a modal logic L is finitely axiomatisable, then it is of course decidable whether a finite frame is a frame for L: one just has to check the finitely many axioms in it. If L is not finitely axiomatisable, then this might not be the case. For example, it is shown in [7] that the finite frame problem is undecidable for every L between the product logics $\mathbf{K} \times \mathbf{K} \times \mathbf{K}$ and $\mathbf{S5} \times \mathbf{S5} \times \mathbf{S5}$. Here we show that the finite frame problem for the modal product logic $\mathbf{K4.3} \times \mathbf{S5}$ is decidable. $\mathbf{K4.3} \times \mathbf{S5}$ is outside the scope of both the finite axiomatisation results of [4], and the non-finite axiomatisability results of [11]. So it is not known whether $\mathbf{K4.3} \times \mathbf{S5}$ is finitely axiomatisable. Here we also discuss whether our results bring us any closer to either proving non-finite axiomatisability of $\mathbf{K4.3} \times \mathbf{S5}$, or finding an explicit, possibly infinite, axiomatisation of it.

Keywords: products of modal logics, finite frame problem, axiomatisation

---

1 Introduction and results

The product construction as a combination method for modal logics was introduced in [13,14,4], and has been extensively studied ever since. Modal products are connected to several other multi-dimensional logical formalisms, see [3,9] for surveys and references. Here we consider only two-dimensional products, but the definitions can be generalised to higher dimensions. In what follows we assume that the reader is familiar with basic notions of propositional multi-modal logic and its possible world (or relational) semantics, and we use these

[1] Sérgio Marcelino was partially supported by FCT and EU FEDER, via the project FCT PEst-OE/EEI/LA0008/2011, the postdoc grant SFRH/BPD/76513/2011, and the PQDR initiative of SQIG.

without explicit references. For concepts and statements not defined or proved here, consult, for example, [1,2].

Given two Kripke frames $\mathfrak{F}_0 = \langle W_0, R_0 \rangle$ and $\mathfrak{F}_1 = \langle W_1, R_1 \rangle$, their *product* is defined to be the 2-frame

$$\mathfrak{F}_0 \times \mathfrak{F}_1 = \langle W_0 \times W_1, \bar{R}_0, \bar{R}_1 \rangle,$$

where $W_0 \times W_1$ is the Cartesian product of W_0 and W_1 and, for all $x, x' \in W_0$, $y, y' \in W_1$,

$$\langle x, y \rangle \bar{R}_0 \langle x', y' \rangle \quad \text{iff} \quad x R_0 x' \text{ and } y = y',$$
$$\langle x, y \rangle \bar{R}_1 \langle x', y' \rangle \quad \text{iff} \quad y R_1 y' \text{ and } x = x'.$$

Frames of this form will be called *product frames* throughout. Now let L_0 and L_1 be Kripke complete modal logics in the languages with $\Box_0$ and $\Box_1$, respectively. Their product $L_0 \times L_1$ is then the set of all bimodal formulas, in the language having both $\Box_0$ and $\Box_1$, that are valid in all product frames $\mathfrak{F}_0 \times \mathfrak{F}_1$, where $\mathfrak{F}_0$ is a frame for L_0, and $\mathfrak{F}_1$ is a frame for L_1. (Here we assume that $\Box_0$ is interpreted by $\bar{R}_0$, while $\Box_1$ is interpreted by $\bar{R}_1$.) Note that $L_0 \times L_1$ always contains the *fusion* $L_0 \oplus L_1$ of L_0 and L_1: the smallest normal bimodal logic that contains L_0 for $\Box_0$ and L_1 for $\Box_1$. Therefore, *any* product frame $\mathfrak{F}_0 \times \mathfrak{F}_1$ for $L_0 \times L_1$ is such that $\mathfrak{F}_i$ is a frame for L_i, for $i = 0, 1$.

A modal product logic $L_0 \times L_1$ is Kripke complete by definition: it is defined as a set of formulas that are valid in some class $\mathcal{C}$ of frames. However, there are frames for $L_0 \times L_1$ that are not in $\mathcal{C}$. So even if it is decidable whether a finite 2-modal frame is in $\mathcal{C}$ or not, the *finite frame problem* for $L_0 \times L_1$ is not necessarily decidable. If $L_0 \times L_1$ is finitely axiomatisable, then it is of course decidable whether a finite frame is a frame for $L_0 \times L_1$: one just has to check the finitely many axioms in it. But if $L_0 \times L_1$ is not finitely axiomatisable, then this might not be the case, even if the component logics L_0 and L_1 are both finitely axiomatisable, and so the class of product frames for $L_0 \times L_1$ is decidable. We do not know two-dimensional examples of this kind, but there are non-finitely axiomatisable higher dimensional product logics with undecidable finite frame problems (such as **K** × **K** × **K** and **S5** × **S5** × **S5**), see [7].

Below we summarise the known results on the axiomatisation problem for two-dimensional product logics:

(1) If both unimodal logics L_0 and L_1 are such that their classes of Kripke frames are definable by recursive sets of first-order sentences, then their product $L_0 \times L_1$ is a recursively enumerable bimodal logic [4].

(2) If both L_0 and L_1 are finitely axiomatisable by modal formulas having universal Horn first-order correspondents, then $L_0 \times L_1$ is finitely axiomatisable [4]. For example, if each L_i is either **K** (the logic of all frames), or **K4** (the logic of all transitive frames), or **S4** (the logic of all reflexive and transitive frames), or **S5** (the logic of all equivalence frames), then $L_0 \times L_1$ is finitely axiomatisable.

(3) The result in (2) cannot be generalised to products of logics axiomatised by formulas having universal (but not necessarily Horn) first-order components.

A counterexample is the finitely axiomatisable modal logic **K4.3**, determined by the frames $\langle W, R\rangle$, where R is transitive and *weakly connected*:

$$\forall x, y, z \in W\ \big(xRy \wedge xRz \to (y = z \vee yRz \vee zRy)\big).$$

(A rooted transitive and weakly connected relation is a *linearly ordered sequence of clusters.*) As shown in [11], there are product logics with a 'linear' first component that are not axiomatisable finitely: For example, if L_0 is any of the logics **K4.3**, **S4.3**, $\mathsf{Logic\_of}\{\langle\omega, \leq\rangle\}$, and L_1 is any of the logics **K**, **K4**, **S4**, **GL**, **Grz**, then $L_0 \times L_1$ is not axiomatisable using finitely many propositional variables.

However, there are recursively enumerable product logics that are outside the scope of both (2) and (3) above, so it is not known whether they are finitely axiomatisable or not. A notable example is **K4.3** $\times$ **S5**. In this paper we show the following:

Theorem 1.1 *It is decidable whether a finite 2-frame is a frame for* **K4.3**$\times$**S5**.

It is clearly enough to decide the frame problem for finite *rooted* 2-frames. As both being transitive and weakly connected, and being an equivalence relation are first-order definable, the respective classes of all frames for **K4.3** and **S5** are closed under ultraproducts. As **K4.3** and **S5** are modal logics, their classes of frames are also closed under point-generated subframes. So, by [10, Thm.2.10], we obtain that, for every finite rooted 2-frame $\mathfrak{F}$, $\mathfrak{F}$ is a frame for **K4.3** $\times$ **S5** iff $\mathfrak{F}$ is a p-morphic image of a product frame for **K4.3** $\times$ **S5**. So it is enough to show the following:

Theorem 1.2 *It is decidable whether a finite rooted 2-frame is a p-morphic image of a product frame for* **K4.3** $\times$ **S5**.

Note that if every finite frame for **K4.3** $\times$ **S5** were the p-morphic image of a *finite* product frame for **K4.3** $\times$ **S5**, then we could enumerate finite frames for **K4.3** $\times$ **S5**. As **K4.3** $\times$ **S5** is recursively enumerable, we can always enumerate those finite frames that are not frames for **K4.3** $\times$ **S5**. So this would provide us with a decision algorithm for the finite frame problem. However, take, say, the 2-frame $\mathfrak{F} = \langle W, \leq, W \times W\rangle$, where $W = \{x, y\}$ and $x \leq x \leq y \leq y$. Then it is easy to see that $\mathfrak{F}$ is a p-morphic image of $\langle\omega, \leq\rangle \times \langle\omega, \omega \times \omega\rangle$, but there is no finite product frame $\mathfrak{G}$ for **K4.3** $\times$ **S5** such that $\mathfrak{F}$ is a p-morphic image of $\mathfrak{G}$.

To explain our decision algorithm, now we have a closer look at some properties of 2-frames for **K4.3** $\oplus$ **S5**, that is, where the first relation is transitive and weakly connected, and the second relation is an equivalence. To emphasise these facts, the transitive and weakly connected relations in our 2-frames will always be denoted by $\leq$, and the equivalence relations by $\sim$. This will not necessarily mean that $\leq$ is reflexive: there might be 'reflexive' points in our frames with $x \leq x$, and some other 'irreflexive' ones with $y \not\leq y$. (This is a slight abuse of notation, as we will also denote by $\leq$ the usual — reflexive and antisymmetric — linear order on the natural numbers.) So from now on, let

$\mathfrak{F} = \langle W, \leq, \sim \rangle$ be a 2-frame for **K4.3** $\oplus$ **S5**. We will use the following notation:

$$C_x = \{x' : x \leq x' \text{ and } x' \leq x\},$$
$$x < y \quad \text{iff} \quad x \leq y \text{ and } y \nleq x,$$
$$x \ll y \quad \text{iff} \quad x < y \text{ and } \forall x' \, (x \leq x' < y \to x' \in C_x),$$
$$[x, y] = \{u : x \leq u \leq y\}, \qquad [x, y) = \{u : x \leq u < y\},$$
$$(x, y] = \{u : x < u \leq y\}, \qquad (x, y) = \{u : x < u < y\}.$$

Observe that if x is irreflexive, then C_x is not the '$\leq$-cluster' of x in the usual sense, but $C_x = \emptyset$. Also, the above 'intervals' are not the usual ones either, as $x \notin [x, y]$ or $x \notin [x, y)$ for irreflexive x. For any $X \subseteq W$, we let

$$\mathit{min}\, X = \{x \in X : \text{there is no } x' \in X \text{ with } x' < x\}, \text{ and}$$
$$\mathit{max}\, X = \{x \in X : \text{there is no } x' \in X \text{ with } x < x'\}.$$

Note that $\mathit{min}\, X$ and $\mathit{max}\, X$ are nonempty, whenever X is finite and nonempty. For any $n > 0$ and $X, Y \subseteq W$, we let

$$X \overset{n}{\rightsquigarrow} Y \quad \text{iff} \quad \forall x_1, \ldots, x_n \in X \, (x_1 \leq \cdots \leq x_n \to \exists y_1, \ldots, y_n \in Y \, (y_1 \leq \cdots \leq y_n \land \bigwedge_{1 \leq i \leq n} x_i \sim y_i)).$$

For $n = 1$, we omit the superscript and write $X \rightsquigarrow Y$:

$$X \rightsquigarrow Y \qquad \text{iff} \qquad \forall x \in X \, \exists y \in Y \; x \sim y.$$

If $X = \{x\}$ then we write $x \rightsquigarrow Y$ instead of $\{x\} \rightsquigarrow Y$. Clearly, as $\sim$ is transitive, $\rightsquigarrow$ is a transitive relation on the subsets of W: if $X \rightsquigarrow Y$ and $Y \rightsquigarrow Z$, then $X \rightsquigarrow Z$. Note that if $x \nleq x$ then $C_x = \emptyset$, and so $C_x \rightsquigarrow Y$ always holds. Observe that $X \rightsquigarrow Y$ does not always follow from $X \overset{2}{\rightsquigarrow} Y$, as there might exist some $x \in X$ with neither $x \leq x'$ nor $x' \leq x$, for any $x' \in X$.

Next, we introduce some important properties of our 2-frames, expressed in the first-order frame-correspondence language having binary predicate symbols $\leq$ and $\sim$. First of all, let

$$\mathsf{sq}(x, y, z, w) \qquad \text{iff} \qquad x \sim y \leq z \land x \leq w \sim z.$$

When $\mathsf{sq}(x, y, z, w)$ holds, we visualise this fact with the picture

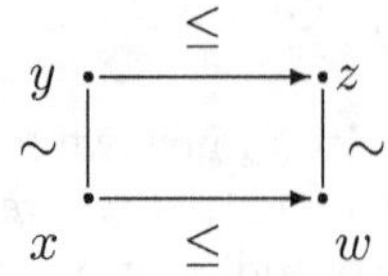

The locations of x, y, z, w in this picture motivate the notation for the remaining first-order properties of our frames (l = left, r = right, u = up, d = down):

$$\psi_u(x, y, z, w): \quad \mathsf{sq}(x, y, z, w) \land [y, z) \rightsquigarrow [x, w]$$
$$\psi_d(x, y, z, w): \quad \mathsf{sq}(x, y, z, w) \land [x, w) \rightsquigarrow [y, z]$$
$$\psi_b(x, y, z, w): \quad \psi_u(x, y, z, w) \land \psi_d(x, y, z, w)$$

$$\begin{aligned}
\psi_{u^2}(x,y,z,w): &\quad \mathsf{sq}(x,y,z,w) \wedge [y,z)\overset{2}{\rightsquigarrow}[x,w] \\
\psi_{d^2}(x,y,z,w): &\quad \mathsf{sq}(x,y,z,w) \wedge [x,w)\overset{2}{\rightsquigarrow}[y,z] \\
\psi_{(u,d^2)}(x,y,z,w): &\quad \mathsf{sq}(x,y,z,w) \wedge \\
&\qquad\qquad \forall a\,\big(a \in [y,z) \to \exists b\,(b \in [x,w] \wedge \psi_{d^2}(b,a,z,w))\big) \\
\Phi_l: &\quad \forall x,y,z\,\big(x \sim y \le z \to \exists w\, \psi_b(x,y,z,w)\big) \\
\Phi_r^+: &\quad \forall x,w,z\,\big(x \le w \sim z \to \exists y\,\big(\psi_{u^2}(x,y,z,w) \wedge \\
&\qquad\qquad \psi_{d^2}(x,y,z,w) \wedge \psi_{(u,d^2)}(x,y,z,w)\big)\big) \\
\Phi: &\quad \Phi_l \wedge \Phi_r^+
\end{aligned}$$

Observe that $\psi_u(x,y,z,w)$ follows from $\psi_{(u,d^2)}(x,y,z,w)$.

Now we are in a position to formulate our main result:

Theorem 1.3 *For every finite rooted 2-frame $\mathfrak{F} = \langle W, \le, \sim\rangle$ for* **K4.3 ⊕ S5**, *$\mathfrak{F}$ is a p-morphic image of a product frame for* **K4.3 × S5** *iff Φ holds in $\mathfrak{F}$.*

The formula Φ is quite complex (Π_3). Figure 1 shows that we cannot hope for a much simpler one: $\mathfrak{F}$ is a frame for **S4.3 ⊕ S5**, where Φ_r^+ fails (see the indicated x, w, z), but Φ_l,

$$\forall x,w,z\,\big(x \le w \sim z \to \exists y\,\big(\psi_{u^2}(x,y,z,w) \wedge \psi_{d^2}(x,y,z,w)\big)\big), \text{ and}$$
$$\forall x,w,z\,\big(x \le w \sim z \to \exists y\,\big(\psi_{u^2}(x,y,z,w) \wedge \psi_{(u,d^2)}(x,y,z,w)\big)\big)$$

all hold (the arrows and ellipses represent the reflexive, transitive and weakly connected $\le$, and the triangles and circles the $\sim$-equivalence classes).

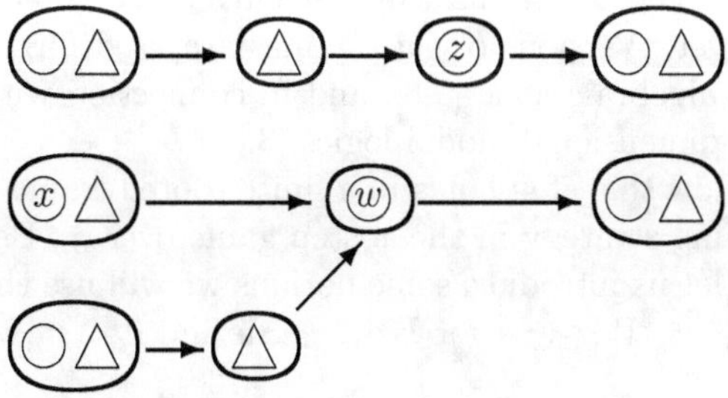

Fig. 1. A frame $\mathfrak{F}$ showing that something like Φ is needed.

The paper is organised as follows. The main steps of the proof of Theorem 1.3 are discussed in Section 2. The more technical claims and lemmas are proved in Section 3. Finally, in Section 4 we discuss some related open problems, possible extensions of our results, and also whether they bring us any closer to either proving non-finite axiomatisability of **K4.3 × S5**, or finding an explicit, possibly infinite, axiomatisation of it.

2 P-morphic images of product frames for K4.3 × S5

We begin with a general observation about p-morphic images of transitive and weakly connected frames.

Claim 2.1 *Let f be a p-morphism from some transitive and weakly connected frame $\mathfrak{F}_0 = \langle W_0, \leq_0 \rangle$ onto a frame $\mathfrak{F}_1 = \langle W_1, \leq_1 \rangle$. For all $a, b \in W_0$, $x_1, \dots, x_n \in W_1$, if $a \leq_0 b$ and $f(a) \leq_1 x_1 \leq_1 \cdots \leq_1 x_n <_1 f(b)$, then there exist $c_1, \dots, c_n \in W_0$ such that $a \leq_0 c_1 \leq_0 \cdots \leq_0 c_n <_0 b$ and $f(c_i) = x_i$, for $i = 1, \dots, n$.*

Proof. Take some $a, b \in W_0$, $x_1, \dots, x_n \in W_1$ such that $a \leq_0 b$ and $f(a) \leq_1 x_1 \leq_1 \cdots \leq_1 x_n <_1 f(b)$. By the backward condition on f, there exists $c_1, \dots, c_n \in W_0$ such that $a \leq_0 c_1 \leq_0 \cdots \leq_0 c_n$ and $f(c_i) = x_i$, for $i = 1, \dots, n$. As $\leq_0$ is transitive, we have $a \leq_0 c_n$. As $\leq_0$ is weakly connected, we have either $c_n = b$, or $b \leq_0 c_n$, or $c_n \leq_0 b$. But $f(c_n) = x_n <_1 f(b)$, so the first two cases cannot hold. Therefore, $c_n <_0 b$ follows. □

It is straightforward to check that Φ holds in every product frame for **K4.3** × **S5**. And, using Claim 2.1, it is not hard to check either that Φ is preserved under taking p-morphic images of frames for **K4.3** ⊕ **S5**. So we have:

Proposition 2.2 *If $\mathfrak{F}$ is a p-morphic image of a product frame for* **K4.3** × **S5**, *then Φ holds in $\mathfrak{F}$.*

We have to work a bit more to prove the other direction of Theorem 1.3. Given a rooted 2-frame $\mathfrak{F} = \langle W, \leq, \sim \rangle$ for **K4.3** ⊕ **S5**, we will define a 'p-morphism game' between two players ∀ (male) and ∃ (female) over $\mathfrak{F}$. In this game, ∃ constructs step-by-step, (special) homomorphisms from larger and larger **K4.3** × **S5**-product frames to $\mathfrak{F}$, and ∀ tries to challenge her by pointing out possible 'defects': reasons why her current homomorphism is not an onto p-morphism yet. Versions of such games are used for building complete representations in algebraic logic [5,6], and in connection with axiomatisation problems of multi-dimensional modal logics [3,8].

We will then show that if Φ holds in a finite rooted frame $\mathfrak{F}$ for **K4.3** ⊕ **S5**, then ∃ has a winning strategy in the ω-step game over $\mathfrak{F}$. Before defining the rules of the game, let us introduce some notions we will use throughout. Given a rooted 2-frame $\mathfrak{F} = \langle W, \leq, \sim \rangle$ for **K4.3** ⊕ **S5** and $0 < m, n < \omega$, we call an $n \times m$ matrix

$$\langle x_j^i \in W : i < m,\ j < n \rangle$$

a *perfect grid*, if either $m = 1$ and $x_i^0 \sim x_j^0$ for all $i, j < n$, or $m > 1$ and the following hold:

(pg1) $x_j^i \sim x_k^i$, for all $i < m$, $j, k < n$,

(pg2) either $x_j^i \ll x_j^{i+1}$ or $x_j^i \in C_{x_j^{i+1}}$, for all $i < m-1$, $j < n$,

(pg3) for all $i < m-1$, $j < n$, if $x_j^i \ll x_j^{i+1}$ then for all $k < n$, either $C_{x_j^i} \rightsquigarrow C_{x_k^i}$ or $C_{x_j^i} \rightsquigarrow C_{x_k^{i+1}}$.

(See Figure 2 for an example, where the arrows and ellipses represent $\leq$, and the triangles and circles the $\sim$-equivalence classes.)

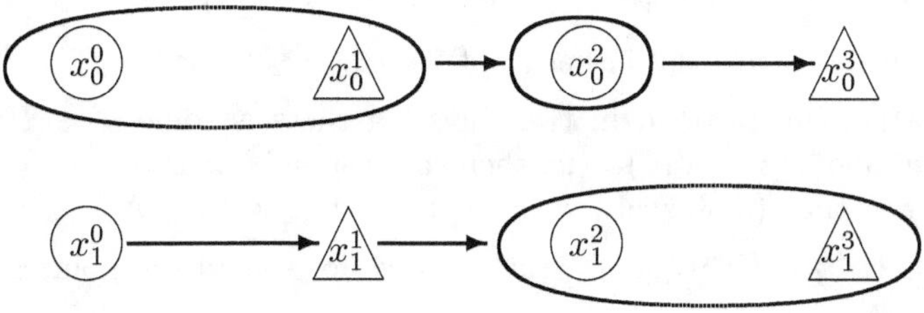

Fig. 2. A perfect grid $\langle x_j^i : i < 4,\, j < 2\rangle$.

Observe that if $\langle x_j^i : i < m,\, j < n\rangle$ is a perfect grid, then for all $k < \ell \leq m$, $\langle x_j^i : k \leq i \leq \ell,\, j < n\rangle$ is a perfect grid as well. If $m = 2$ then we call the $2n$-tuple $\langle x_0^0, \ldots, x_{n-1}^0, x_0^1, \ldots, x_{n-1}^1\rangle$ a *perfect atomic grid.* Clearly, if $m > 1$ and $\langle x_j^i : i < m,\, j < n\rangle$ is a perfect grid, then $\langle x_0^i, \ldots, x_{n-1}^i, x_0^{i+1}, \ldots, x_{n-1}^{i+1}\rangle$ is a perfect atomic grid, for each $i < m - 1$.

Given an $n \times m$ matrix $\bar{x} = \langle x_j^i : i < m,\, j < n\rangle$ and an $n \times k$ matrix $\bar{y} = \langle y_j^i : i < k,\, j < n\rangle$ such that $x_j^{m-1} = y_j^0$, for all $j < n$, their *union* $\bar{x} \sqcup \bar{y}$ is the $n \times (m + k - 1)$ matrix $\langle z_j^i : i < m + k - 1,\, j < n\rangle$, defined by taking, for all $j < n$,

$$z_j^i = \begin{cases} x_j^i, & \text{if } i < m, \\ y_j^{i-m+1}, & \text{if } m - 1 \leq i < m + k - 1. \end{cases}$$

It is easy to see the following claim:

Claim 2.3 *If $\bar{x} = \langle x_j^i : i < m,\, j < n\rangle$ and $\bar{y} = \langle y_j^i : i < k,\, j < n\rangle$ are perfect grids such that $x_j^{m-1} = y_j^0$, for all $j < n$, then $\bar{x} \sqcup \bar{y}$ is a perfect grid as well.*

Given a rooted 2-frame $\mathfrak{F} = \langle W, \leq, \sim\rangle$ for **K4.3$\oplus$S5**, we define an $\mathfrak{F}$-*network* to be a tuple $N = \langle U^N, <^N, V^N, f^N\rangle$ such that the following hold:

- $U^N = \{u_0, \ldots, u_m\}$ for some $m < \omega$,
- $<^N$ is an irreflexive linear order on U^N with $u_0 <^N \cdots <^N u_m$,
- $V^N = \{v_0, \ldots, v_n\}$ for some $n < \omega$,
- f^N is a function from $U^N \times V^N$ to W such that $\langle f^N(u_i, v_j) : i \leq m,\, j \leq n\rangle$ is a perfect grid.

It is not hard to see, using (pg1) and (pg2), that if N is an $\mathfrak{F}$-network, then f^N is a homomorphism from the product frame $\langle U^N, <^N\rangle \times \langle V^N, V^N \times V^N\rangle$ to $\mathfrak{F}$.

Now we define a *game* $\mathcal{G}_\omega(\mathfrak{F})$ between $\forall$ and $\exists$. They build a countable sequence of $\mathfrak{F}$-networks $N_0 \subseteq N_1 \subseteq \cdots \subseteq N_k \subseteq \ldots$. (Here $N_k \subseteq N_{k+1}$ means that $U^{N_k} \subseteq U^{N_{k+1}}$, $<^{N_k} \subseteq <^{N_{k+1}}$, $V^{N_k} \subseteq V^{N_{k+1}}$, and $f^{N_k} \subseteq f^{N_{k+1}}$.) In round 0, $\forall$ picks a root r of $\mathfrak{F}$, and $\exists$ responds with $U^{N_0} = \{u_0\}$, $<^{N_0} = \emptyset$, $V^{N_0} = \{v_0\}$, and $f^{N_0}(u_0, v_0) = r$.

In round k ($0 < k < \omega$), some sequence $N_0 \subseteq \cdots \subseteq N_{k-1}$ of $\mathfrak{F}$-networks has already been built. $\forall$ picks

- a pair $\langle u, v\rangle \in U^{N_{k-1}} \times V^{N_{k-1}}$, and
- a point $w \in W$ such that either (a) $f^{N_{k-1}}(u,v) \leq w$, or (b) $f^{N_{k-1}}(u,v) \sim w$.

In case (a), $\exists$ can respond in two ways. If there is some $u' \in U^{N_{k-1}}$ with $u <^{N_{k-1}} u'$ and $f^{N_{k-1}}(u',v) = w$, then she responds with $N_k = N_{k-1}$. Otherwise, she responds (if she can) with some $\mathfrak{F}$-network $N_k \supseteq N_{k-1}$ such that

- $U^{N_{k-1}} \cup \{u^+\} \subseteq U^{N_k}$ and $f^{N_k}(u^+, v) = w$, for some fresh point u^+, and
- $V^{N_k} = V^{N_{k-1}}$.

In case (b), again $\exists$ can respond in two ways. If there is some $v' \in V^{N_{k-1}}$ with $f^{N_{k-1}}(u,v') = w$, then she responds with $N_k = N_{k-1}$. Otherwise, she responds (if she can) with some $\mathfrak{F}$-network $N_k \supseteq N_{k-1}$ such that

- $V^{N_k} = V^{N_{k-1}} \cup \{v^+\}$ and $f^{N_k}(u, v^+) = w$, for some fresh point v^+.

If $\exists$ can respond in each round k for $k < \omega$ then *she wins the play.* We say that $\exists$ *has a winning strategy in* $\mathcal{G}_\omega(\mathfrak{F})$ if she can win all plays, whatever moves $\forall$ takes in the rounds.

Proposition 2.4 *Let $\mathfrak{F}$ be a countable rooted 2-frame for* **K4.3** $\oplus$ **S5**. *If $\exists$ has a winning strategy in $\mathcal{G}_\omega(\mathfrak{F})$, then $\mathfrak{F}$ is a p-morphic image of a product frame for* **K4.3** $\times$ **S5**.

Proof. Consider a play of the game $\mathcal{G}_\omega(\mathfrak{F})$ when $\forall$ eventually picks all possible pairs and corresponding $\leq$- or $\sim$-connected points in $\mathfrak{F}$ (since $\mathfrak{F}$ is countable, he can do this). If $\exists$ uses her strategy, then she succeeds to construct a countable ascending chain of $\mathfrak{F}$-networks whose union gives a p-morphism from some **K4.3** $\times$ **S5**-product frame onto $\mathfrak{F}$. □

Proposition 2.5 *Let $\mathfrak{F}$ be a finite rooted 2-frame for* **K4.3** $\oplus$ **S5** *such that Φ holds in $\mathfrak{F}$. Then $\exists$ has a winning strategy in $\mathcal{G}_\omega(\mathfrak{F})$.*

Proof. We prove that, for all $k < \omega$, $\exists$ can survive round k in every play, no matter what moves $\forall$ takes in the rounds. We prove this by induction on k. For $k = 0$ this is obvious. So assume inductively that some sequence $N_0 \subseteq \cdots \subseteq N_{k-1}$ of $\mathfrak{F}$-networks has already been built, for some $0 < k < \omega$. Suppose that $U^{N_{k-1}} = \{u_0, \ldots, u_m\}$ such that $u_0 <^{N_{k-1}} \cdots <^{N_{k-1}} u_m$, and $V^{N_{k-1}} = \{v_0, \ldots, v_n\}$. Next, $\forall$ picks some $\langle u, v\rangle \in U^{N_{k-1}} \times V^{N_{k-1}}$ and $w \in W$. There are several cases, depending on how $f^{N_{k-1}}(u,v)$ and w are related. In each case we show how $\exists$ can respond with an N_k satisfying the requirements. We omit those cases where $\exists$'s response is fully determined by the rules of the game.

<u>*Case* (a).1.</u> $f^{N_{k-1}}(u,v) \leq w$, for all $u' \in U^{N_{k-1}}$, if $u <^{N_{k-1}} u'$ then $f^{N_{k-1}}(u',v) \neq w$, but there exists $u^* \in U^{N_{k-1}}$ such that $u <^{N_{k-1}} u^*$ and $f^{N_{k-1}}(u^*,v) \not\leq w$.

By the IH, $f^{N_{k-1}}$ is a homomorphism, and so $f^{N_{k-1}}(u,v) \leq f^{N_{k-1}}(u^*,v)$ follows. Thus, by weak connectedness of $\leq$, we have $w < f^{N_{k-1}}(u^*,v)$. Therefore, as $U^{N_{k-1}}$ is finite, there are $<^{N_{k-1}}$-successor points $u', u'' \in U^{N_{k-1}}$ such that

$$f^{N_{k-1}}(u', v) \leq w < f^{N_{k-1}}(u'', v). \tag{1}$$

To simplify notation, we let $x_i = f^{N_{k-1}}(u', v_i)$, $y_i = f^{N_{k-1}}(u'', v_i)$, for all $i \leq n$. By the IH, we have that

$$\langle x_0, \ldots, x_n, y_0, \ldots, y_n \rangle \text{ is a perfect atomic grid.} \tag{2}$$

We may assume that $v = v_0$, and so we have $x_0 \ll y_0$ by (1) and (2). Therefore, by (pg3), for each $i \leq n$, we have either $C_{x_0} \rightsquigarrow C_{x_i}$ or $C_{x_0} \rightsquigarrow C_{y_i}$. We now define w_i, for each $i \leq n$ (see Figure 3). Let $w_0 = w$, so by (1) and (2), we have $w_0 \in C_{x_0}$. For every $0 < i \leq n$,

- if $C_{x_0} \rightsquigarrow C_{x_i}$, then we choose some $w_i \in C_{x_i}$ with $w_0 \sim w_i$, and
- if $C_{x_0} \not\rightsquigarrow C_{x_i}$, then $C_{x_0} \rightsquigarrow C_{y_i}$ and we choose some $w_i \in C_{y_i}$ with $w_0 \sim w_i$.

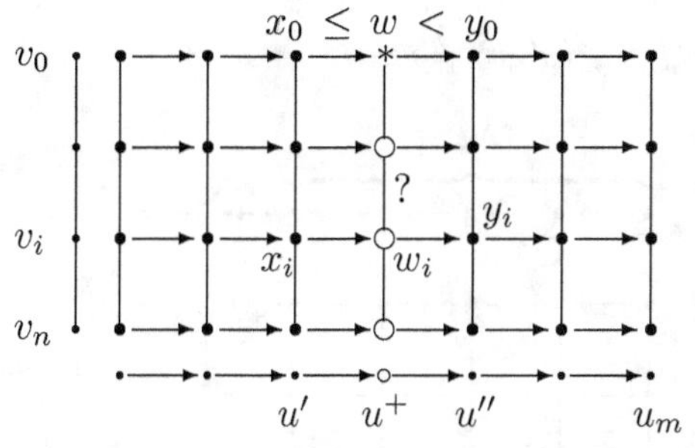

Fig. 3. Case (a).1 of the p-morphism game.

Claim **2.5.1**

(i) $\langle x_0, \ldots, x_n, w_0, \ldots, w_n \rangle$ *is a perfect atomic grid.*

(ii) $\langle w_0, \ldots, w_n, y_0, \ldots, y_n \rangle$ *is a perfect atomic grid.*

Proof. Let us prove (pg3) first. (i): Let $i \leq n$ be such that $x_i \ll w_i$. Then $w_i \notin C_{x_i}$, so by the definition of w_i, we have

$$C_{x_0} \not\rightsquigarrow C_{x_i}, \tag{3}$$

$w_i \in C_{y_i}$, and so

$$x_i \ll y_i. \tag{4}$$

Take some $j < n$. There are two cases:

- $w_j \in C_{y_j}$. Then, by (4) and (2), either $C_{x_i} \rightsquigarrow C_{x_j}$ or $C_{x_i} \rightsquigarrow C_{y_j} = C_{w_j}$.
- $w_j \notin C_{y_j}$. Then $C_{x_0} \rightsquigarrow C_{x_j}$ by the definition of w_j. Therefore, $C_{x_j} \not\rightsquigarrow C_{x_i}$ follows by (3), and so $C_{x_i} \rightsquigarrow C_{x_j}$ by Claim 3.1.

(ii): Let $i \leq n$ be such that $w_i \ll y_i$. Then $w_i \notin C_{y_i}$, so

$$C_{x_0} \rightsquigarrow C_{x_i}, \tag{5}$$
$$w_i \in C_{x_i}, \tag{6}$$

and so (4) holds. Take some $j < n$. There are two cases:

- $w_j \in C_{x_j}$. Then by (6), (4) and (2), either $C_{w_i} = C_{x_i} \rightsquigarrow C_{x_j} = C_{w_j}$ or $C_{w_i} = C_{x_i} \rightsquigarrow C_{y_j}$.
- $w_j \notin C_{x_j}$. Then $C_{x_0} \not\rightsquigarrow C_{x_j}$ by the definition of w_j. Therefore, $C_{x_i} \not\rightsquigarrow C_{x_j}$ follows by (5), and so we have $C_{w_i} = C_{x_i} \rightsquigarrow C_{y_j}$ by (6), (4) and (2).

As (pg1) and (pg2) clearly hold in both cases, the proof of Claim 2.5.1 is completed. □

Now take a fresh point u^+. Let $U^{N_k} = U^{N_{k-1}} \cup \{u^+\}$, let $<^{N_k} \supseteq <^{N_{k-1}}$ be such that $u' <^{N_k} u^+ <^{N_k} u''$, and let $f^{N_k}(u^+, v_i) = w_i$, for $i < n$. By Claim 2.5.1, the obtained N_k is an $\mathfrak{F}$-network extending N_{k-1} as required.

<u>*Case* (a).2.</u> $f^{N_{k-1}}(u, v) \leq w$, and for all $u' \in U^{N_{k-1}}$, if $u <^{N_{k-1}} u'$ then $f^{N_{k-1}}(u', v) \leq w$ and $f^{N_{k-1}}(u', v) \neq w$.

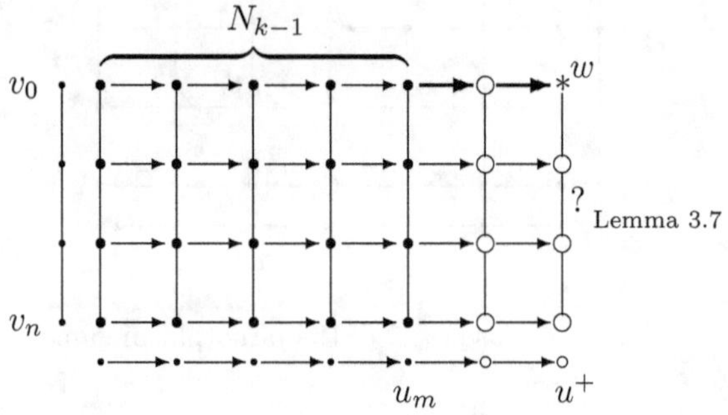

Fig. 4. Case (a).2 of the p-morphism game.

Then $f^{N_{k-1}}(u_m, v) \leq w$. We may assume that $v = v_0$ (see Figure 4). By the IH, we have $f^{N_{k-1}}(u_m, v_i) \sim f^{N_{k-1}}(u_m, v_j)$, for all $i, j \leq n$. So, by Lemma 3.7, there exists $t > 0$ and a perfect grid $\bar{z} = \langle z_j^\ell : \ell \leq t,\ j \leq n \rangle$ such that $z_j^0 = f^{N_{k-1}}(u_m, v_j)$, for $j \leq n$, and $z_0^t = w$. By the IH, $\bar{f} = \langle f^{N_{k-1}}(u_i, v_j) : i \leq m,\ j \leq n \rangle$ is a perfect grid, and so by Claim 2.3, $\bar{f} \sqcup \bar{z}$ is a perfect grid as well. Therefore, if we define

- $U^{N_k} = U^{N_{k-1}} \cup \{u_\ell^+ : 0 < \ell \leq t\}$, $u^+ = u_t^+$,
- $f^{N_k}(u_\ell^+, v_j) = z_j^\ell$, for $0 < \ell \leq t$, $j \leq n$,

then we obtain an $\mathfrak{F}$-network N_k extending N_{k-1} as required.

<u>*Case* (b).</u> $f^{N_{k-1}}(u, v) \sim w$, and $w \neq f^{N_{k-1}}(u, v')$ for all $v' \in V^{N_{k-1}}$.

Suppose $u = u_p$ for some $p \leq m$ (see Figure 5). By the IH, $\langle f^{N_{k-1}}(u_i, v_j) : i \leq p,\ j \leq n \rangle$ is a perfect grid, and $w \sim f^{N_{k-1}}(u_p, v) \sim f^{N_{k-1}}(u_p, v_n)$. So by

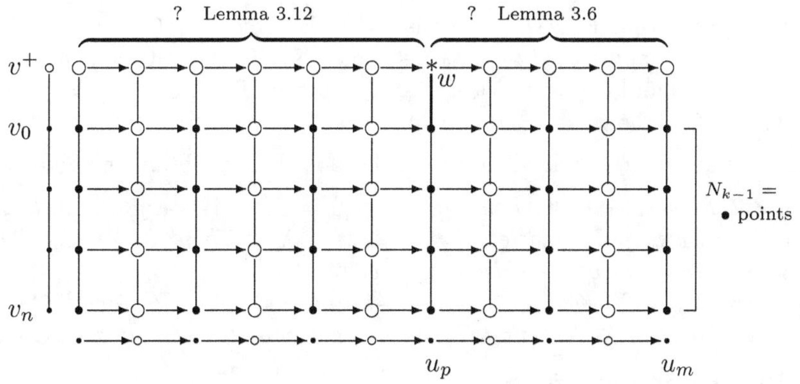

Fig. 5. Case (b) of the p-morphism game.

Lemma 3.12, there exist $s_i < \omega$ $(i \leq p)$ and a perfect grid $\bar{z} = \langle z_j^\ell : \ell \leq s_p,\ j \leq n+1\rangle$ such that $0 = s_0 < s_1 < \cdots < s_p$, $z_{n+1}^{s_p} = w$, and $z_j^{s_i} = f^{N_{k-1}}(u_i, v_j)$, for $i \leq p$, $j \leq n$.

By the IH, $\langle f^{N_{k-1}}(u_{p+i}, v_j) : i \leq m-p,\ j \leq n\rangle$ is a perfect grid as well. As we have $w \sim f^{N_{k-1}}(u_p, v) \sim f^{N_{k-1}}(u_p, v_n)$, by Lemma 3.6 there exist $t_i < \omega$ $(i \leq m-p)$ and a perfect grid $\bar{y} = \langle y_j^t : t \leq t_{m-p},\ j \leq n+1\rangle$ such that $0 = t_0 < t_1 < \cdots < t_{m-p}$, $y_{n+1}^0 = w$, and $y_j^{t_i} = f^{N_{k-1}}(u_{p+i}, v_j)$, for $i \leq m-p$, $j \leq n$.

By Claim 2.3, $\bar{z} \sqcup \bar{y} = \langle x_j^\ell : \ell \leq s_p + t_{m-p} - 1,\ j \leq n+1\rangle$ is a perfect grid, and therefore by defining

- $U^{N_k} = U^{N_{k-1}} \cup \{u_\ell^+ : \ell < s_p + t_{m-p} - 1,\ \ell \neq s_i, s_p + t_j \text{ for } i \leq p,\ j \leq m-p\}$,
- $V^{N_k} = V^{N_{k-1}} \cup \{v^+\}$,
- $f^{N_k}(u_\ell^+, v_j) = x_j^\ell$, for $u_\ell^+ \in U^{N_k}$, $j \leq n$, and
- $f^{N_k}(u_p, v^+) = w$, $f^{N_k}(u_\ell^+, v^+) = x_{n+1}^\ell$, for $u_\ell^+ \in U^{N_k}$,

we obtain an $\mathfrak{F}$-network N_k extending N_{k-1} as required. This completes the proof of Proposition 2.5. □

3 How Φ helps $\exists$ to have a winning strategy in $\mathcal{G}_\omega(\mathfrak{F})$

In this section we state and prove the claims and lemmas that are used in the proof of Proposition 2.5. The material is divided into two subsections. In Section 3.1 we discuss those statements that describe plays of the game played 'on the left', that is, when $\exists$ makes use of the the fact that the finite frame $\mathfrak{F}$ validates Φ_l. Then in Section 3.2 we describe those plays of the game that are played 'on the right', that is, when $\exists$ also needs to use the conjunct Φ_r^+ of Φ.

Throughout, $\mathfrak{F} = \langle W, \leq, \sim\rangle$ is a finite rooted 2-frame for **K4.3** $\oplus$ **S5**. We begin with two claims that are very important throughout:

Claim 3.1 *Suppose that Φ_l holds in $\mathfrak{F}$, and let $x, y \in W$ be such that $x \sim y$. Then, either $C_x \rightsquigarrow C_y$ or $C_y \rightsquigarrow C_x$.*

Proof. Suppose that $C_x \not\rightsquigarrow C_y$, that is, there is some $a \in C_x$ with $a \not\rightsquigarrow C_y$. Then $y \sim x \leq a$, and so by Φ_l, there is some b such that $\psi_d(y, x, a, b)$ holds. Therefore, $y \leq b$ and $a \sim b$, so $b \notin C_y$, and so $y < b$. Thus, $C_y \subseteq [y, b)$, and so $C_y \rightsquigarrow [x, a] = C_x$ follows by $\psi_d(y, x, a, b)$. □

As $\rightsquigarrow$ is a transitive relation on the subsets of W, we obtain the following:

Claim 3.2 *Suppose that Φ_l holds in $\mathfrak{F}$, let $\emptyset \neq X \subseteq W$ be finite such that $x \sim y$ for all $x, y \in X$, and let $\mathcal{C} = \{C_x : x \in X\}$. Then $\langle \mathcal{C}, \rightsquigarrow \rangle$ is a finite linearly ordered chain of '$\rightsquigarrow$-clusters'. In particular,*

(i) *there is $x_i \in X$ such that C_{x_i} is $\rightsquigarrow$-initial in $\mathcal{C}$: $C_{x_i} \rightsquigarrow C$ for all $C \in \mathcal{C}$;*
(ii) *there is $x_f \in X$ such that C_{x_f} is $\rightsquigarrow$-final in $\mathcal{C}$: $C \rightsquigarrow C_{x_f}$ for all $C \in \mathcal{C}$.*

3.1 Playing on the left

We start with formulating and proving a general structural property of finite frames validating Φ_l (Lemma 3.3). Then in Lemma 3.4 we show that this structural property can be generalised to extensions of perfect atomic grids. This property is then used in Lemma 3.5 to help $\exists$ maintaining a perfect grid, whenever $\forall$ challenges to extend a perfect atomic grid with a '$\leq$-move' (see Case (a).2 in the proof of Prop. 2.5). Then Lemma 3.5 is used as the base case in the inductive proof of Lemma 3.6. Finally, Lemma 3.6 is used in the inductive proof of Lemma 3.7. This last lemma states that any perfect grid can be extended by $\exists$, whenever $\forall$ plays a '$\leq$-move' of the above kind.

Given $x, y, z, w, a \in W$, we write $\mathsf{left}(x, y, z, w, a)$ if the following hold:

(le1) $\mathsf{sq}(x, y, z, w)$ and $x \leq a \leq w$,

(le2) $C_y \rightsquigarrow C_a$,

(le3) $[x, a) \rightsquigarrow C_y$,

(le4) either $a \in C_w$, or $C_a \rightsquigarrow C_y$, or $C_a \rightsquigarrow C_z$,

(le5) $(a, w) \rightsquigarrow C_z$.

Lemma 3.3 *Suppose that Φ_l holds in $\mathfrak{F}$. For all $x, y, z \in W$, if $x \sim y \leq y \ll z$ then there exist w^*, a^* such that $\mathsf{left}(x, y, z, w^*, a^*)$ holds.*

Proof. By Φ_l, there exists w with $\psi_b(x, y, z, w)$. If $w \in C_x$ then let $w^* = a^* = w$, and we clearly have $\mathsf{left}(x, y, z, w^*, a^*)$ as required.

So suppose that

$$\text{there is no } w \in C_x \text{ with } \psi_b(x, y, z, w), \tag{7}$$

and let

$$w^+ \in max\{w : x < w \text{ and } \psi_b(x, y, z, w)\} \tag{8}$$

(as $\mathfrak{F}$ is finite, there is such w^+ by Φ_l and (7)). Now there are two cases: either $[x, w^+) \rightsquigarrow C_y$, or $[x, w^+) \not\rightsquigarrow C_y$.

<u>*Case 1.*</u> $[x, w^+) \rightsquigarrow C_y$.
As $\psi_b(x, y, z, w^+)$ and $y \ll z$, we have $C_y \rightsquigarrow [x, w^+]$. As $y \leq y$, there exists $a \in [x, w^+]$ with $a \rightsquigarrow C_y$. Let

$$a^* \in max\{a \in [x, w^+] : a \rightsquigarrow C_y\} \tag{9}$$

(there is such a^* as $\mathfrak{F}$ is finite). We claim that

$$\mathsf{left}(x, y, z, w^+, a^*), \tag{10}$$

and so $w^* = w^+$ will do. Indeed, we clearly have $x \leq a^* \leq w^+$, so we have (le1) by (8). (le2): Let $b^* \in C_y$ be such that $a^* \sim b^*$. By Φ_l, there exists w' with $\psi_b(a^*, b^*, z, w')$.

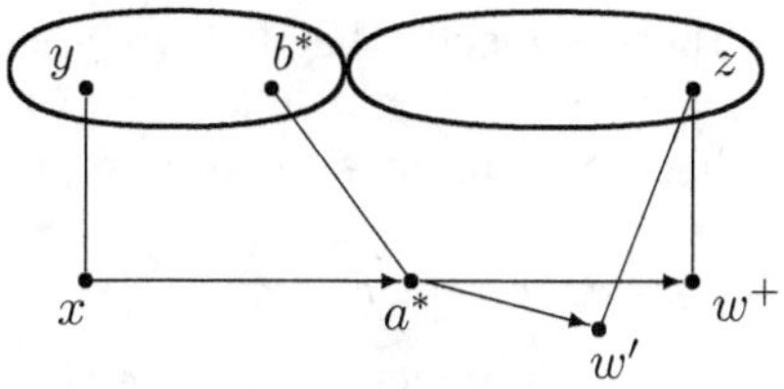

We claim that

$$\psi_b(x, y, z, w'). \tag{11}$$

Indeed, on the one hand, if $b \in [y, z)$ then $b \in [b^*, z)$, and so $b \rightsquigarrow [a^*, w']$ by $\psi_b(a^*, b^*, z, w')$. As $x \leq a^*$, this implies $b \rightsquigarrow [x, w']$. On the other hand, if $a \in [x, w')$ then there are two cases:

- $a \in [x, a^*)$. Then $a \in [x, w^+)$, and so $a \rightsquigarrow [y, z]$ by (8).
- $a = a^*$ or $a \in [a^*, w')$. Then $a \rightsquigarrow [b^*, z] = [y, z]$ by $\psi_b(a^*, b^*, z, w')$.

So in both cases we have $a \rightsquigarrow [y, z]$, and so (11) is proved.

Now (7) and (11) imply that $x < w'$. Therefore, by (11) and (8), we have $w^+ \not\leq w'$. As $x \leq w^+$ and $x \leq w'$, by the weak connectedness of $\leq$ we have

$$\text{either } w' = w^+ \text{ or } w' \leq w^+. \tag{12}$$

Now we can show (le2), that is, $C_y \rightsquigarrow C_{a^*}$. Take some $b \in C_y$. Then $b \in [b^*, z)$, and so by $\psi_b(a^*, b^*, z, w')$, we have $b \rightsquigarrow [a^*, w']$. By (12), this implies $b \rightsquigarrow [a^*, w^+]$, that is, $b \sim a$ for some $a \in [a^*, w^+]$. Thus, $a \in [x, w^+]$ and $a \rightsquigarrow C_y$, and so by (9), we have $a^* \not< a$. As we also have $a^* \leq a$, this implies $a \in C_{a^*}$, as required in (le2).

(le3): As we are in the case when $[x, w^+) \rightsquigarrow C_y$, we also have $[x, a^*) \rightsquigarrow C_y$ by $a^* \leq w^+$, and so (le3) holds.

(le4) and (le5): If $a^* \in C_{w^+}$ then (le4) holds. If $a^* < w^+$, then take any $a \in [a^*, w^+)$. As $a \in [x, w^+)$ and we are in the case when $[x, w^+) \rightsquigarrow C_y$, we have $a \rightsquigarrow C_y$, proving $C_{a^*} \rightsquigarrow C_y$, and so (le4). Moreover, by (9), we also have $a^* \not< a$, and so $a \in C_{a^*}$ follows. Therefore, $a^* \ll w^+$, and so $\emptyset = (a^*, w^+) \rightsquigarrow C_z$, as required in (le5), completing the proof of (10).

<u>*Case 2.*</u> $[x, w^+) \not\rightsquigarrow C_y$.
Then there is some $r \in [x, w^+)$ with $r \not\rightsquigarrow C_y$. Let

$$r^* \in \mathit{min}\{r \in [x, w^+) : r \not\rightsquigarrow C_y\} \tag{13}$$

(there is such r^* as $\mathfrak{F}$ is finite). As $\psi_b(x, y, z, w^+)$ by (8), we have

$$r^* \rightsquigarrow C_z. \tag{14}$$

Now let $s^* \in C_z$ be such that $r^* \sim s^*$. By Φ_l, there is w^* with $\psi_b(r^*, s^*, z, w^*)$. Thus, we have

$$[r^*, w^*) \rightsquigarrow C_z. \tag{15}$$

We also need to define a^*. To this end, we claim that

$$\{a \in [x, r^*] : a \rightsquigarrow C_y\} \text{ is not empty.} \tag{16}$$

Indeed, by Φ_l and $y \le y$, there is a such that $\psi_b(x, y, y, a)$ holds. Thus, $a \sim y$ and $[x, a) \rightsquigarrow C_y$, and so $a \neq r^*$ and $r^* \not\le a$ follow from (13). As $x \le r^*$ and $x \le a$, the weak connectedness of $\le$ implies that $a \le r^*$, proving (16). Now let

$$a^* \in max\{a \in [x, r^*] : a \rightsquigarrow C_y\} \tag{17}$$

(there is such a^* by (16) and the finiteness of $\mathfrak{F}$). We claim that

$$\mathsf{left}(x, y, z, w^*, a^*). \tag{18}$$

Indeed, we have $x \le a^* \le r^* \le w^*$, so (le1) holds.

(le2): As $a^* \rightsquigarrow C_y$ by (17), there is $b^* \in C_y$ be such that $a^* \sim b^*$. By Φ_l, there is s with $\psi_b(b^*, a^*, r^*, s)$, and so $b^* \le s$. As $r^* \sim s$ and $r^* \not\rightsquigarrow C_y$ by (13), we have $s \notin C_y = C_{b^*}$, and so $b^* < s$ follows. Now take any $b \in C_y$. Then $b \in [b^*, s)$, and so $\psi_b(b^*, a^*, r^*, s)$ implies that there is some $a \in [a^*, r^*]$ with $a \sim b$. Therefore, $a \in [x, r^*]$ and $a \rightsquigarrow C_y$, so $a^* \not< a$ by (17). But we also have $a^* \le a$, and so $a \in C_{a^*}$ follows, as required in (le2).

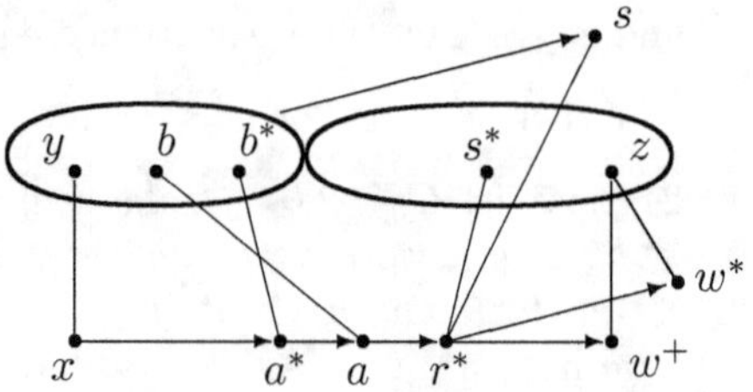

(le3): As $a^* \le r^* < w^+$ by (17), we have $[x, a^*) \rightsquigarrow C_y$ by (13).

For (le4) and (le5), first we claim that

$$\text{either } C_{a^*} = C_{r^*} \text{ or } a^* \ll r^*. \tag{19}$$

Indeed, we have $a^* \le r^*$ by (17). Suppose that $C_{a^*} \neq C_{r^*}$, and let $a \in [a^*, r^*)$. Then $a \in [x, w^+)$ and $a < r^*$, so $a \rightsquigarrow C_y$ follows by (13). As $a \in [x, r^*]$, we have $a^* \not< a$ by (17). Therefore, $a \in C_{a^*}$ follows from $a^* \le a$, as required in (19).

(le5): $(a^*, w^*) \rightsquigarrow C_z$ follows from (14), (15) and (19).

(le4): If $a^* \in C_{w^*}$, then (le4) holds. If $a^* < w^*$, then by (19) there are two cases:

- $C_{a^*} = C_{r^*}$. Then $r^* < w^*$ and $C_{a^*} \subseteq [r^*, w^*)$. So $C_{a^*} \rightsquigarrow C_z$ follows by (15).
- $a^* \ll r^*$. Then $C_{a^*} \rightsquigarrow C_y$ follows by (13).

So (le4) holds in both cases, completing the proof of (18). □

Lemma 3.4 *Suppose that Φ_l holds in $\mathfrak{F}$, and let $\langle x_0, \ldots, x_{n-1}, y_0, \ldots, y_{n-1}\rangle$ be a perfect atomic grid for some $n > 0$. For all $x \in W$, if $x \sim x_0$ then there exists y such that $y \sim y_0$ and one of the following* (I) *or* (II) *holds:*

(I) *Either $y \in C_x$ and for all $j < n$, if $x_j \ll y_j$ then $C_{x_j} \rightsquigarrow C_x = C_y$.*

(II) *Or $x < y$ and:*

(a) *For all $j < n$, if $x_j \in C_{y_j}$ or $x_j \not\leq x_j$, then $[x, y) \rightsquigarrow C_{y_j}$.*

(b) *For all $j < n$, if $x_j \leq x_j \ll y_j$, then there is a_j with $\mathsf{left}(x, x_j, y_j, y, a_j)$, that is,*

$$\mathsf{sq}(x, x_j, y_j, y) \text{ and } x \leq a_j \leq y, \tag{20}$$

$$C_{x_j} \rightsquigarrow C_{a_j}, \tag{21}$$

$$[x, a_j) \rightsquigarrow C_{x_j}, \tag{22}$$

$$\text{either } a_j \in C_y, \text{ or } C_{a_j} \rightsquigarrow C_{x_j}, \text{ or } C_{a_j} \rightsquigarrow C_{y_j}, \tag{23}$$

$$(a_j, y) \rightsquigarrow C_{y_j}. \tag{24}$$

Proof. There are two cases:

Case 1. For all $j < n$, either $x_j \in C_{y_j}$ or $x_j \not\leq x_j$.
By (pg1) and Claim 3.2, there is $i < n$ such that

$$C_{y_i} \text{ is } \rightsquigarrow\text{-initial in } \{C_{y_j} : j < n\}. \tag{25}$$

By Φ_l, there is some y such that

$$\psi_b(x, x_i, y_i, y). \tag{26}$$

There are two cases, either $y \in C_x$, or $x < y$:

- $y \in C_x$. As for all $j < n$ with $x_j \ll y_j$, we have $x_j \not\leq x_j$, it follows that $\emptyset = C_{x_j} \rightsquigarrow C_x = C_y$, as required in (I).
- $x < y$. Then $[x, y) \rightsquigarrow [x_i, y_i]$ by (26). As either $x_i \in C_{y_i}$ or $x_i \not\leq x_i$, we have $[x_i, y_i] = C_{y_i}$ by (pg2). Therefore, by (25) and the transitivity of $\rightsquigarrow$, it follows that $[x, y) \rightsquigarrow C_{y_j}$, for all $j < n$, as required in (II).

Case 2. There is some $j < n$ such that $x_j \leq x_j \ll y_j$.
By (pg1) and Claim 3.2, there exists some $f < n$ such that C_{x_f} is $\rightsquigarrow$-final in $\{C_{x_j} : j < n,\ x_j \leq x_j \ll y_j\}$. Also, there is $i < n$ such that C_{y_i} is $\rightsquigarrow$-initial in $\{C_{y_j} : j < n,\ x_j \leq x_j \ll y_j, \text{ and } C_{x_f} \rightsquigarrow C_{x_j}\}$. Observe that then

$$C_{y_i} \text{ is } \rightsquigarrow\text{-initial in } \{C_{y_j} : j < n,\ x_j \leq x_j \ll y_j, \text{ and } C_{x_i} \rightsquigarrow C_{x_j}\}, \text{ and} \tag{27}$$

$$C_{x_i} \text{ is } \rightsquigarrow\text{-final in } \{C_{x_j} : j < n,\ x_j \leq x_j \ll y_j\}. \tag{28}$$

Now, by Lemma 3.3, there exist y^*, a^* such that

$$\mathsf{left}(x, x_i, y_i, y^*, a^*). \tag{29}$$

There are two cases, either $y^* \in C_x$, or $x < y^*$. If $y^* \in C_x$, then we let $y = y^*$, and claim that (I) holds. Indeed, by (29) we have $a^* \in C_x$, and so $C_{x_i} \rightsquigarrow C_{a^*} = C_x = C_y$, again by (29). Thus by (28), $C_{x_j} \rightsquigarrow C_x = C_y$ for all $j < n$ with $x_j \leq x_j \ll y_j$. Also, if $j < n$ is such that $x_j \not\leq x_j$, then $C_{x_j} = \emptyset$, and so $C_{x_j} \rightsquigarrow C_x = C_y$, as required in (I).

So suppose that $x < y^*$. We will define some y, and show that

$$\mathsf{sq}(x, x_i, y_i, y) \text{ and } x \leq a^* \leq y, \text{ and} \tag{30}$$

$$(a^*, y) \rightsquigarrow C_{y_j}, \text{ for all } j < n. \tag{31}$$

Then

$$\mathsf{left}(x, x_i, y_i, y, a^*) \tag{32}$$

will follow from (29), as the other conjuncts in $\mathsf{left}(x, x_i, y_i, y, a^*)$ do not depend on y, but only on a^*. (Observe that (31) is more than what is required in $\mathsf{left}(x, x_i, y_i, y, a^*)$: it is for all $j < n$, not just for i.)

To this end, we consider three cases:

- $y_i \rightsquigarrow C_{a^*}$. Then we choose some $y \in C_{a^*}$ such that $y_i \sim y$, and so (30)–(31) clearly hold.
- $y_i \not\rightsquigarrow C_{a^*}$ and $(a^*, y^*) \rightsquigarrow C_{y_j}$, for all $j < n$. Then we let $y = y^*$, and (30)–(31) clearly hold.
- $y_i \not\rightsquigarrow C_{a^*}$ and $(a^*, y^*) \not\rightsquigarrow C_{y_j}$, for some $j < n$. Then let

$$u^* \in \mathit{min}\{u \in (a^*, y^*) : u \not\rightsquigarrow C_{y_j} \text{ for some } j < n\} \tag{33}$$

(there is such u^* as $\mathfrak{F}$ is finite), and let $j^* < n$ be such that $u^* \not\rightsquigarrow C_{y_{j^*}}$. As $(a^*, y^*) \rightsquigarrow C_{y_i}$ follows from (29), we then have $C_{y_i} \not\rightsquigarrow C_{y_{j^*}}$. Therefore, by (27), we have $C_{x_i} \not\rightsquigarrow C_{x_{j^*}}$, and so $C_{x_i} \rightsquigarrow C_{y_{j^*}}$ follows by $x_i \ll y_i$ and (pg3). We also have $C_{x_i} \rightsquigarrow C_{a^*}$ by (29). Therefore, there are $r \in C_{y_{j^*}}$ and $s \in C_{a^*}$ such that $r \sim s$. By Φ_l, there is v^* such that $\psi_b(r, s, u^*, v^*)$ holds. As $u^* \not\rightsquigarrow C_{y_{j^*}}$ by (33), we have $y_{j^*} < v^*$. So by $\psi_b(r, s, u^*, v^*)$, there is some $y \in [s, u^*]$ such that $y \sim y_{j^*}$. Now, as $s \in C_{a^*}$, we have $x \leq a^* \leq s \leq y$, and so (30) follows from (pg1). Also, as $y \leq u^* < y^*$, we have (31) by (33).

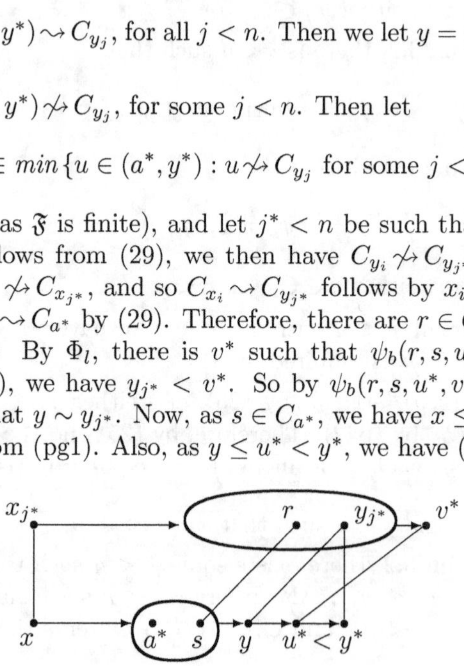

So we proved that y satisfies (30)–(32) in all three cases. Note that y is defined such that

$$\text{if } y_i \rightsquigarrow C_{a^*} \text{ then } y \in C_{a^*}. \tag{34}$$

Next, we show that (30)–(32) imply that (II) holds for y. The following claim will be used several times:

CLAIM **3.4.1** *If $a^* < y$ and $j < n$ is such that $C_{x_i} \rightsquigarrow C_{y_j}$, then $C_{a^*} \rightsquigarrow C_{y_j}$.*

Proof. By (32), we have $C_{x_i} \rightsquigarrow C_{a^*}$. If $C_{x_i} \rightsquigarrow C_{y_j}$, there exist $u \in C_{a^*}$, $v \in C_{y_j}$ with $u \sim v$. So by Claim 3.1, we have either $C_{a^*} \rightsquigarrow C_{y_j}$ or $C_{y_j} \rightsquigarrow C_{a^*}$. If $C_{y_j} \rightsquigarrow C_{a^*}$ were the case, then we would have $y_j \rightsquigarrow C_{a^*}$, and so $y_i \rightsquigarrow C_{a^*}$ would follow by (pg1). By (34), we would have $y \in C_{a^*}$, contradicting $a^* < y$. Therefore, we have $C_{a^*} \rightsquigarrow C_{y_j}$. □

Proof of (II)(a): Let $j < n$ be such that $x_j \in C_{y_j}$ or $x_j \not\leq x_j$.
By $x_i \ll y_i$ and (pg3), we have

$$C_{x_i} \rightsquigarrow C_{y_j}. \tag{35}$$

Now there are two cases: either $a^* \in C_y$, or $a^* < y$. In each case, we claim to have $[x, y) \rightsquigarrow C_{y_j}$, as required in (II)(a). Indeed,

- $a^* \in C_y$. Then $[x, y) = [x, a^*)$, and we have $[x, a^*) \rightsquigarrow C_{x_i}$ by (32). So $[x, y) \rightsquigarrow C_{y_j}$ follows by (35).
- $a^* < y$. Then we have:
  · $[x, a^*) \rightsquigarrow C_{x_i}$ by (32), and so $[x, a^*) \rightsquigarrow C_{y_j}$ by (35);
  · $C_{a^*} \rightsquigarrow C_{y_j}$ by (35) and Claim 3.4.1;
  · $(a^*, y) \rightsquigarrow C_{y_j}$ by (31).

Proof of (II)(b): Let $j < n$ be such that $x_j \leq x_j \ll y_j$.
There are two cases, either $[x, a^*) \rightsquigarrow C_{x_j}$, or $[x, a^*) \not\rightsquigarrow C_{x_j}$. In both cases, first we define a_j and then show that (20)–(24) (that is, $\mathsf{left}(x, x_j, y_j, y, a_j)$) hold.

- $[x, a^*) \rightsquigarrow C_{x_j}$. Then we let $a_j = a^*$, and we clearly have (20) and (22). By (28), we have $C_{x_j} \rightsquigarrow C_{x_i}$, and by (32), we have $C_{x_i} \rightsquigarrow C_{a_j}$. So $C_{x_j} \rightsquigarrow C_{a_j}$ follows, proving (21). We have (24) by (31). Finally, let us prove (23), that is, either $a_j \in C_y$, or $C_{a_j} \rightsquigarrow C_{x_j}$ or $C_{a_j} \rightsquigarrow C_{y_j}$: Suppose that $a_j = a^* < y$. By (32), there are two cases: either $C_{a^*} \rightsquigarrow C_{x_i}$ or $C_{a^*} \rightsquigarrow C_{y_i}$.
  · $C_{a^*} \rightsquigarrow C_{x_i}$. Then, by $x_i \ll y_i$ and (pg3), we have either $C_{x_i} \rightsquigarrow C_{x_j}$ or $C_{x_i} \rightsquigarrow C_{y_j}$, so (23) follows.
  · $C_{a^*} \rightsquigarrow C_{y_i}$. If $C_{x_i} \rightsquigarrow C_{x_j}$, then $C_{y_i} \rightsquigarrow C_{y_j}$ follows by (27), and so we have $C_{a^*} \rightsquigarrow C_{y_j}$. If $C_{x_i} \not\rightsquigarrow C_{x_j}$, then by $x_i \ll y_i$ and (pg3), we have $C_{x_i} \rightsquigarrow C_{y_j}$. So by Claim 3.4.1, we have $C_{a^*} \rightsquigarrow C_{y_j}$, as required in (23).
- $[x, a^*) \not\rightsquigarrow C_{x_j}$. By Lemma 3.3, there are a_j, y_j^* such that

$$\mathsf{left}(x, x_j, y_j, y_j^*, a_j). \tag{36}$$

  We claim that $\mathsf{left}(x, x_j, y_j, a_j)$ as well, that is, (20)–(24) hold. Indeed, by (36), we have $x \leq a_j$ and $[x, a_j) \rightsquigarrow C_{x_j}$. As $x \leq a^*$ and $[x, a^*) \not\rightsquigarrow C_{x_j}$, by the weak connectivity of $\leq$ it follows that

$$x \leq a_j < a^* \leq y, \tag{37}$$

  as required in (20). As (21) and (22) do not depend on y, they hold because of (36). Next, by (32), we have $[x, a^*) \rightsquigarrow C_{x_i}$, and so $C_{x_i} \not\rightsquigarrow C_{x_j}$ follows from

$[x, a^*) \not\rightsquigarrow C_{x_j}$. So by $x_i \ll y_i$ and (pg3), we have

$$C_{x_i} \rightsquigarrow C_{y_j}, \tag{38}$$

and so

$$[x, a^*) \rightsquigarrow C_{y_j}. \tag{39}$$

For (23): We have $C_{a_j} \rightsquigarrow C_{y_j}$ by (37) and (39). For (24): (37) and (39) imply $(a_j, a^*) \rightsquigarrow C_{y_j}$. So if $a^* \in C_y$, then $(a_j, y) \rightsquigarrow C_{y_j}$ follows. If $a^* < y$, then $C_{a^*} \rightsquigarrow C_{y_j}$ follows by (38) and Claim 3.4.1. Also, we have $(a^*, y) \rightsquigarrow C_{y_j}$ by (31). Therefore, $(a_j, y) \rightsquigarrow C_{y_j}$ holds, as required.

So we proved (II)(b), and the proof of Lemma 3.4 is completed. □

Lemma 3.5 *Suppose that Φ_l holds in $\mathfrak{F}$, and let $\langle x_0, \ldots, x_{n-1}, y_0, \ldots, y_{n-1} \rangle$ be a perfect atomic grid for some $n > 0$. For all $x \in W$, if $x \sim x_0$ then there exist $k > 0$ and a perfect grid $\langle z_j^\ell : \ell \le k, j \le n \rangle$ such that $z_j^0 = x_j$, $z_j^k = y_j$, for $j < n$, and $z_n^0 = x$.*

Proof. By Lemma 3.4, there is y such that either (I) or (II) of the lemma holds. If (I) holds, that is, $y \in C_x$, then let $k = 1$, $z_n^0 = x$, and $z_n^1 = y$. Of course, we let $z_j^0 = x_j$ and $z_j^1 = y_j$, for $j < n$. It is straightforward to show that $\langle z_0^0, \ldots, z_n^0, z_0^1, \ldots, z_n^1 \rangle$ is a perfect atomic grid.

Suppose that (II) holds, that is $x < y$, and for all $j < n$ with $x_j \le x_j \ll y_j$, we have some a_j as in (II)(b). Then let $k > 0$, and $z_n^0, \ldots z_n^k$ be such that $x = z_n^0 \ll \cdots \ll z_n^k = y$ (that is, we take a point from each $\le$-cluster between x and y). Of course, we let $z_j^0 = x_j$, $z_j^k = y_j$, for all $j < n$. Next, we define a number $\ell_j < k$, for every $j < n$ as follows:

- If $x_j \in C_{y_j}$ or $x_j \not\le x_j$, then let $\ell_j = 0$.
- If $x_j \le x_j \ll y_j$, then there are several cases, depending on the location of a_j in $[x, y]$:
  - · If $a_j \in C_y$, then let $\ell_j = k - 1$.
  - · If $a_j < y$ and $C_{a_j} \rightsquigarrow C_{x_j}$, then let ℓ_j be such that $z_n^{\ell_j} \in C_{a_j}$.
  - · If $a_j < y$, $C_{a_j} \not\rightsquigarrow C_{x_j}$, and $a_j \in C_x$, then let $\ell_j = 0$.
  - · If $a_j < y$, $C_{a_j} \not\rightsquigarrow C_{x_j}$, and $x < a_j$, then let ℓ_j be such that $z_n^{\ell_j+1} \in C_{a_j}$.

The following claim is a straightforward consequence of (II)(a) and (22)–(24) in (II)(b):

Claim **3.5.1**

(ii) *Either $C_{z_n^0} \rightsquigarrow C_{x_i}$, or ($\ell_j = 0$ and $C_{z_n^0} \rightsquigarrow C_{y_j}$).*

(ii) *$z_n^\ell \rightsquigarrow C_{x_i}$ and $C_{z_n^\ell} \rightsquigarrow C_{x_i}$, for all ℓ with $0 < \ell \le \ell_j$.*

(iii) *$z_n^\ell \rightsquigarrow C_{y_i}$ and $C_{z_n^\ell} \rightsquigarrow C_{y_i}$, for all ℓ with $\ell_j < \ell < k$.*

We use Claim 3.5.1(ii) and (iii) to define z_j^ℓ, for each $0 < \ell < k$ and $j < n$:

- If $0 < \ell \le \ell_j$, then choose $z_j^\ell \in C_{x_j}$ such that $z_n^\ell \sim z_j^\ell$.
- If $\ell_j < \ell < k$, then choose $z_j^\ell \in C_{y_j}$ such that $z_n^\ell \sim z_j^\ell$.

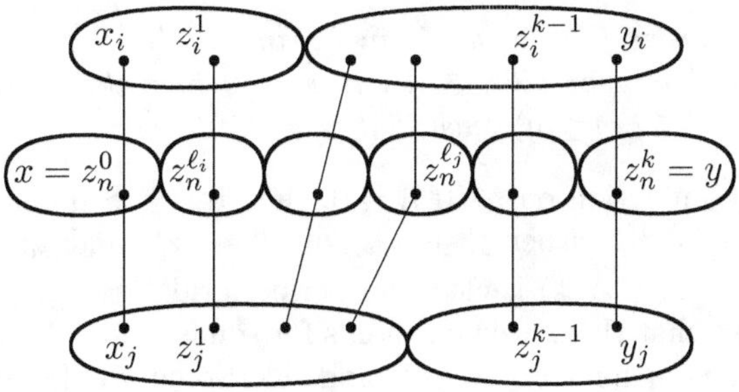

As a consequence of Claim 3.5.1, and (21), we obtain the following:

CLAIM **3.5.2** *For all $j < n$,*

(i) *either $C_{z_n^0} \rightsquigarrow C_{z_j^0}$ or $C_{z_n^0} \rightsquigarrow C_{z_j^1}$;*

(ii) *$C_{z_n^\ell} \rightsquigarrow C_{z_j^\ell}$, whenever $0 < \ell < k$;*

(iii) *if $x_j \ll y_j$ then either $C_{z_j^{\ell_j}} \rightsquigarrow C_{z_n^{\ell_j}}$ or $C_{z_j^{\ell_j}} \rightsquigarrow C_{z_n^{\ell_j+1}}$.*

Now we claim that $\langle z_j^\ell : \ell \leq k,\ j \leq n\rangle$ is a perfect grid as required. Indeed (pg1) and (pg2) clearly hold. Let us prove that (pg3) holds as well, that is, for all $\ell < k$, $i, j \leq n$,

$$\text{if } z_i^\ell \ll z_i^{\ell+1} \text{ then either } C_{z_i^\ell} \rightsquigarrow C_{z_j^\ell} \text{ or } C_{z_i^\ell} \rightsquigarrow C_{z_j^{\ell+1}}. \tag{40}$$

If $i = j$, this clearly holds. Otherwise, there are three cases:

- $i = n$, $j < n$. Then (40) holds by Claim 3.5.2(i) and (ii).
- $i < n$, $j = n$. If $z_i^\ell \ll z_i^{\ell+1}$ then $\ell = \ell_i$ and (40) holds by Claim 3.5.2(iii).
- $i, j < n$. Again, if $z_i^\ell \ll z_i^{\ell+1}$ then $\ell = \ell_i$, and so either $C_{z_i^{\ell_i}} \rightsquigarrow C_{z_n^{\ell_i}}$ or $C_{z_i^{\ell_i}} \rightsquigarrow C_{z_n^{\ell_i+1}}$, by Claim 3.5.2(iii). Now either $C_{z_i^{\ell_i}} \rightsquigarrow C_{z_j^{\ell_i}}$ or $C_{z_i^{\ell_i}} \rightsquigarrow C_{z_j^{\ell_i+1}}$ follow by Claim 3.5.2(i) and (ii),

completing the proof of Lemma 3.5. □

Lemma 3.6 *Suppose that Φ_l holds in $\mathfrak{F}$, and let $\langle x_j^i : i \leq m,\ j < n\rangle$ be a perfect grid, for some $m, n < \omega$, $n > 0$. For all $x \in W$, if $x \sim x_0^0$ then there exist $t_i < \omega$ ($i \leq m$) and a perfect grid $\langle z_j^\ell : \ell \leq t_m,\ j \leq n\rangle$ such that $0 = t_0 < t_1 < \cdots < t_m$, $z_j^{t_i} = x_j^i$, for $i \leq m$, $j < n$, and $z_n^0 = x$.*

Proof. It is by induction on m. For $m = 0$ the statement is obvious. Suppose the statement holds for some $m < \omega$. Let $\langle x_j^i : i \leq m+1,\ j < n\rangle$ be a perfect grid, and let $x \in W$ be such that $x \sim x_0^0$. Then $\langle x_j^i : i \leq m,\ j < n\rangle$ is a perfect grid, and so by the IH, there exist $t_i < \omega$, for $i \leq m$, and a perfect grid $\bar{z} = \langle z_j^\ell : \ell \leq t_m,\ j \leq n\rangle$ such that $0 = t_0 < t_1 < \cdots < t_m$, $z_j^{t_i} = x_j^i$, for $i \leq m$, $j < n$, and $z_n^0 = x$. We also have that $\langle x_0^m, \ldots, x_{n-1}^m, x_0^{m+1}, \ldots, x_{n-1}^{m+1}\rangle$ is a perfect atomic grid, and $z_n^{t_m} \sim z_0^{t_m} = x_0^m$. So by Lemma 3.5, there exist $k > 0$ and a perfect grid $\bar{y} = \langle y_j^\ell : \ell \leq k,\ j \leq n\rangle$ such that $y_j^0 = x_j^m$, for $j < n$, $y_n^0 = z_n^{t_m}$ and $y_j^k = x_j^{m+1}$, for $j < n$. By Claim 2.3, $\bar{z} \sqcup \bar{y}$ is a perfect grid as required. □

Lemma 3.7 *Suppose that Φ_l holds in $\mathfrak{F}$, and let $\langle y_j : j \leq n\rangle$ be such that $y_i \sim y_j$ for $i, j \leq n$. For all $y \in W$, if $y_0 \leq y$ then there exist $t > 0$ and a perfect grid $\langle z_j^\ell : \ell \leq t,\ j \leq n\rangle$ such that $z_0^t = y$ and $z_j^0 = y_j$, for $j \leq n$.*

Proof. It is by induction on n. If $n = 0$, then take $t > 0$ and $z_0^0, \dots, z_0^t$ such that $y_0 = z_0^0$, $y = z_0^t$, either $z_0^0 \in C_{z_0^1}$ or $z_0^0 \ll z_0^1$, and $z_0^\ell \ll z_0^{\ell+1}$, for all $1 \leq \ell < t$. Then $\langle z_0^0, \dots, z_0^t\rangle$ is clearly a perfect grid.

Now suppose that the statement holds for some $n < \omega$. Let $\langle y_j : j \leq n+1\rangle$ be such that $y_i \sim y_j$ for $i, j \leq n+1$, and take some $y \in W$ with $y_0 \leq y$. By the IH, there exist $m > 0$ and a perfect grid $\langle x_j^i : i \leq m,\ j \leq n\rangle$ such that $x_0^m = y$ and $x_j^0 = y_j$, for $j \leq n$. As $y_{n+1} \sim y_0 = x_0^0$, by Lemma 3.6 there exist $t_i < \omega$ $(i \leq m)$ and a perfect grid $\bar{z} = \langle z_j^\ell : \ell \leq t_m,\ j \leq n+1\rangle$ such that $0 = t_0 < t_1 < \cdots < t_m$, $z_j^{t_i} = x_j^i$, for $i \leq m$, $j \leq n$, and $z_{n+1}^0 = y_{n+1}$. Therefore, $z_0^{t_m} = x_0^m = y$, $z_j^0 = z_j^{t_0} = x_j^0 = y_j$, for $j \leq n$, and $z_{n+1}^0 = y_{n+1}$, showing that $\bar{z}$ is a perfect grid as required. □

3.2 Playing on the right

Similarly to Section 3.1, here we start with formulating and proving a general structural property of finite frames validating Φ (Lemma 3.8). Observe that the 'right' conjunct Φ_r^+ of Φ is kind of 'stronger' than its 'left' conjunct Φ_l. Perhaps this is why the 'right' property below is considerably simpler than the corresponding 'left' property (see Lemma 3.3 above). Then in Lemma 3.10 we show that this structural property can be generalised to extensions of perfect atomic grids. This property is then used in Lemma 3.11 to help $\exists$ maintaining a perfect grid, whenever $\forall$ challenges to extend a perfect atomic grid with a '$\sim$-move' (see Case (b) in the proof of Prop. 2.5). Finally, Lemma 3.11 is used as the base case in the inductive proof of Lemma 3.12 that, together with Lemma 3.6, show that any perfect grid can be extended by $\exists$, whenever $\forall$ plays a '$\sim$-move'.

Given $x, y, z, w \in W$, we write $\mathsf{right}(x, y, z, w)$ if the following hold:

(r1) $\mathsf{sq}(x, y, z, w)$,

(r2) either $x \in C_w$ or $C_x \rightsquigarrow C_y$,

(r3) either $y \in C_z$, or $C_y \rightsquigarrow C_x$, or $C_y \rightsquigarrow C_w$,

(r4) $(y, z) \rightsquigarrow C_w$.

Lemma 3.8 *Suppose that Φ holds in $\mathfrak{F}$. For all $x, w, z \in W$, if $x \leq x \ll w \sim z$ then there exists y^* such that $\mathsf{right}(x, y^*, z, w)$ holds.*

Proof. If $C_x \rightsquigarrow C_z$, then there is $y^* \in C_z$ with $x \sim y^*$. It is straightforward to see that $\mathsf{right}(x, y^*, z, w)$ holds. So suppose that

$$C_x \not\rightsquigarrow C_z, \tag{41}$$

and let

$$y^+ \in \mathit{min}\{y : \psi_{all}(x, y, z, w)\}, \tag{42}$$

where $\psi_{all}(x,y,z,w)$ is a shorthand for

$$\psi_{u^2}(x,y,z,w) \wedge \psi_{d^2}(x,y,z,w) \wedge \psi_{(u,d^2)}(x,y,z,w).$$

(As $\mathfrak{F}$ is finite, there is such y^+ by Φ_r^+.) Now there are two cases: either $[y^+,z) \rightsquigarrow C_w$, or $[y^+,z) \not\rightsquigarrow C_w$.

<u>*Case 1.*</u> $[y^+,z) \rightsquigarrow C_w$.
We claim that $\mathsf{right}(x,y^+,z,w)$ holds, and so $y^* = y^+$ will do. Indeed, we clearly have (r1). (r3) and (r4) hold by $[y^+,z) \rightsquigarrow C_w$. For (r2): By (41), there is some $a \in C_x$ with $a \not\rightsquigarrow C_z$. We have $\psi_{d^2}(x,y^+,z,w)$ by (42), and so $x \le x \le a < w$ implies that there are b,b' such that $y^+ \le b \le b' \le z$, $x \sim b$, and $a \sim b'$. Thus $b' \notin C_z$, and so $b \le b' < z$ follows. Now $[y^+,z) \rightsquigarrow C_w$ implies that $b \rightsquigarrow C_w$, and so $y^+ \rightsquigarrow C_w$ follows from $y^+ \sim x \sim b$. Therefore, there is some $w' \in C_w$ with $y^+ \sim w'$. By Φ_r^+, there is y' such that $\psi_{all}(x,y',y^+,w')$.

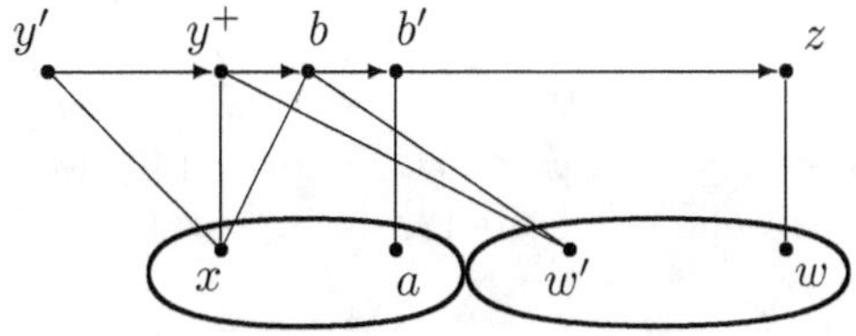

It is straightforward to check that $\psi_{all}(x,y',z,w)$ also holds. So by (42), we have $y' \in C_{y^+}$, and so $C_x \rightsquigarrow C_{y^+}$ follows by $x \le x < w$ and $\psi_{d^2}(x,y',y^+,w')$, completing the proof of (r2).

<u>*Case 2.*</u> $[y^+,z) \not\rightsquigarrow C_w$.
Then let

$$b^+ \in max\{b \in [y^+,z) : b \not\rightsquigarrow C_w\}. \tag{43}$$

(there is such b^+ as $\mathfrak{F}$ is finite). We have $\psi_{(u,d^2)}(x,y^+,z,w)$ by (42), so there is $a^+ \in [x,w]$ such that $a^+ \sim b^+$ and

$$[a^+,w) \overset{2}{\rightsquigarrow} [b^+,z]. \tag{44}$$

By (43), we have $b^+ \not\rightsquigarrow C_w$, and so $a^+ \in C_x$.

We claim that there exists b^* such that

$$b^* \in C_{b^+} \cup \{b^+\},\ b^* \not\rightsquigarrow C_w \text{ and } b^* \not\rightsquigarrow C_z. \tag{45}$$

Indeed, if $b^+ \not\rightsquigarrow C_z$ then (45) holds for $b^* = b^+$. So suppose that $b^+ \rightsquigarrow C_z$. As by (43) we also have $b^+ \not\rightsquigarrow C_w$, it follows that $C_z \not\rightsquigarrow C_w$. So by Claim 3.1, we have $C_w \rightsquigarrow C_z$, and so $C_x \not\rightsquigarrow C_w$ follows by (41). Also by (41), there is some $a^* \in C_x$ such that $a^* \not\rightsquigarrow C_z$. By $C_w \rightsquigarrow C_z$, we also have $a^* \not\rightsquigarrow C_w$. As $a^+ \le a^* \le a^* < w$, by (44) there exists $b^* \in [b^+,z]$ with $a^* \sim b^*$. As $a^* \not\rightsquigarrow C_z$, we have $b^* \not\rightsquigarrow C_z$ and $b^* \notin C_z$. Thus $b^* \in [b^+,z) \subseteq [y^+,z)$ follows. As $a^* \not\rightsquigarrow C_w$, we also have $b^* \not\rightsquigarrow C_w$. Therefore, by (43), we obtain that $b^* \in C_{b^+}$, as required in (45).

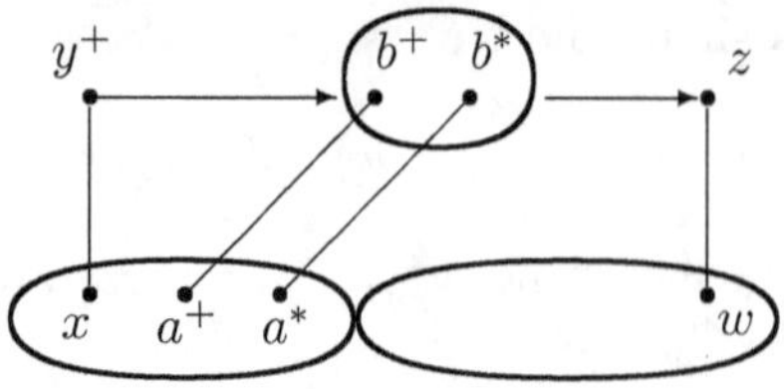

So take some b^* satisfying (45). By (43), we have

$$b^* \in max\{b \in [y^+, z) : b \not\rightsquigarrow C_w\}. \tag{46}$$

We claim that

$$C_x \rightsquigarrow C_{b^*}. \tag{47}$$

Indeed, as we have $\psi_{(u,d^2)}(x, y^+, z, w)$ by (42), there is $c' \in [x, w]$ such that $[c', w) \overset{2}{\rightsquigarrow} [b^*, z]$ and $c' \sim b^*$. As $b^* \not\rightsquigarrow C_w$, it follows that $c' \in C_x$. Now take any $c \in C_x$. Then $c' \le c \le c' < w$, and so there exist b, b' such that $b^* \le b \le b' \le z$, $c \sim b$ and $c' \sim b'$. Thus $b' \sim b^*$ and by (45) we have $b' \notin C_z$ and $b' \not\rightsquigarrow C_w$. Therefore, $y^+ \le b^* \le b \le b' < z$ follows, and by (46) we have that $b' \in C_{b^*}$. Therefore, $b \in C_{b^*}$ as well, as required in (47).

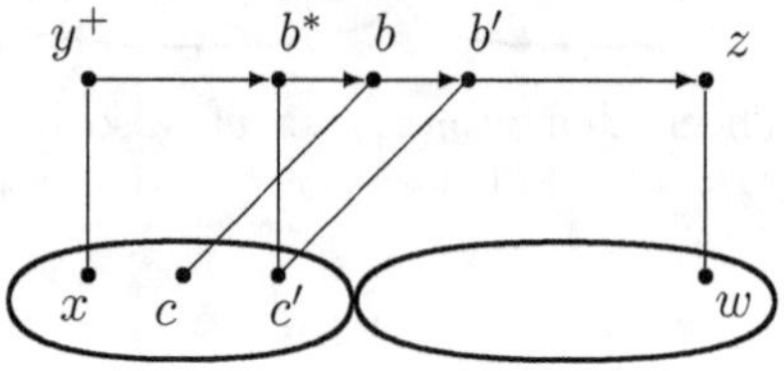

Now by (47), there is $y^* \in C_{b^*}$ such that $x \sim y^*$. We claim that $\mathsf{right}(x, y^*, z, w)$ holds. Indeed, (r1) is clear, (r2) is (47), and (r4) holds by (46). For (r3): We show that $C_{y^*} \rightsquigarrow C_x$. Take some $d \in C_{y^*} = C_{b^*}$. Then $y^+ \le d \le b^* < z$. As by (42) we have $\psi_{u^2}(x, y^+, z, w)$, this implies that there exist e, e^* such that $x \le e \le e^* \le w$, $e \sim d$ and $e^* \sim b^*$.

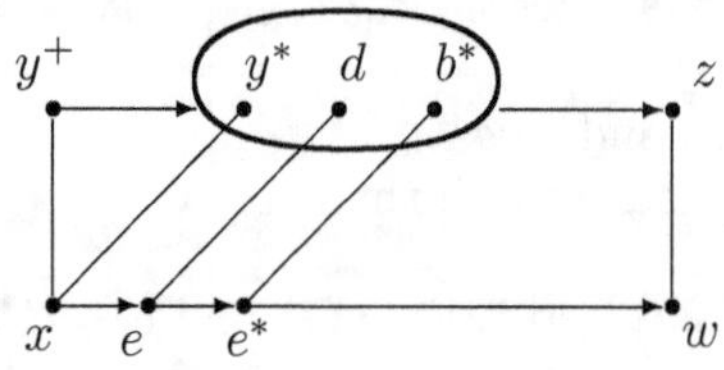

As $b^* \not\rightsquigarrow C_w$ by (46), we have $e^* \in C_x$, and so $e \in C_x$ follows, as required. □

The following claim will be useful in subsequent proofs:

Claim 3.9 *Suppose that Φ_r^+ holds in $\mathfrak{F}$. If $y^+ \in min\{y : \psi_u(x, y, z, w)\}$, then $C_x \rightsquigarrow C_{y^+}$.*

Proof. If $C_x = \emptyset$, then this holds. So take some $a \in C_x$. As $a \le x \sim y^+$, by Φ_r^+ there exists b such that $\psi_{(u,d^2)}(a, b, y^+, x)$, and so $\psi_u(a, b, y^+, x)$. As $x \le a \sim b$, by Φ_r^+ again, there exists y' such that $\psi_u(x, y', b, a)$. So we have $y' \leq b \leq y^+$, and $[y', y^+) \cup \{y^+\} \rightsquigarrow C_x$. So it is straightforward to check that $\psi_u(x, y', z, w)$ holds. Therefore, by $y^+ \in min\{y : \psi_u(x, y, z, w)\}$, we have $y' \not< y^+$, and so $y' \in C_{y^+}$. Therefore, $b \in C_{y^+}$ follows, proving $C_x \rightsquigarrow C_{y^+}$. □

Lemma 3.10 *Suppose that Φ holds in $\mathfrak{F}$, and let $\langle x_0, \ldots, x_{n-1}, y_0, \ldots, y_{n-1}\rangle$ be a perfect atomic grid for some $n > 0$. For all $y \in W$, if $y \sim y_0$ then there exists x such that, for every $j < n$, $\mathsf{right}(x_j, x, y, y_j)$ holds, that is,*

$$\mathsf{sq}(x_j, x, y, y_j), \tag{48}$$

$$\textit{either } x_j \in C_{y_j} \textit{ or } C_{x_j} \rightsquigarrow C_x, \tag{49}$$

$$\textit{either } x \in C_y, \textit{ or } C_x \rightsquigarrow C_{x_j}, \textit{ or } C_x \rightsquigarrow C_{y_j}, \tag{50}$$

$$(x, y) \rightsquigarrow C_{y_j}. \tag{51}$$

Proof. By (pg1), Φ_l and Claim 3.2, there is $i < n$ such that

$$C_{y_i} \text{ is } \rightsquigarrow\text{-initial in } \{C_{y_j} : j < n\}. \tag{52}$$

We claim that there exists x^* such that

$$\mathsf{sq}(x_i, x^*, y, y_i), \tag{53}$$

$$C_{x_i} \rightsquigarrow C_{x^*}, \tag{54}$$

$$\text{either } x^* \in C_y, \text{ or } C_{x^*} \rightsquigarrow C_{x_i}, \text{ or } C_{x^*} \rightsquigarrow C_{y_i}, \tag{55}$$

$$(x^*, y) \rightsquigarrow C_{y_i}. \tag{56}$$

Indeed, if $x_i \leq x_i \ll y_i$ then such an x^* exists by Lemma 3.8. If $x_i \in C_{y_i}$ or $x_i \not\leq x_i$, then let $x^* \in min\{x' : \psi_u(x_i, x', y, y_i)\}$ (there exists such x^* by Φ_r^+ and the finiteness of $\mathfrak{F}$). Then (53), (55), and (56) follow from $\psi_u(x_i, x^*, y, y_i)$ and $[x_i, y_i] = C_{y_i}$, and (54) follows from Claim 3.9.

Now we consider two cases:

<u>*Case 1.*</u> For all $j < n$, if $x_j \leq x_j \ll y_j$ then $C_{x_j} \rightsquigarrow C_{x_i}$.
Then we let $x = x^*$, and claim that (48)–(51) hold, for all $j < n$. Indeed, take some $j < n$. Then (48) is clear. For (49): If $x_j \in C_{y_j}$ or $x_j \not\leq x_j$, then (49) clearly holds. If $x_j \leq x_j \ll y_j$ then $C_{x_j} \rightsquigarrow C_{x_i}$, so (49) follows from (54). For (50): By (55), there are three cases:

- $x \in C_y$. Then (50) holds.
- $C_x \rightsquigarrow C_{y_i}$. Then $C_x \rightsquigarrow C_{y_j}$ by (52).
- $C_x \rightsquigarrow C_{x_i}$ and $C_x \not\rightsquigarrow C_{y_i}$. Then $x_i \leq x_i \ll y_i$, and by (pg3) we have either $C_{x_i} \rightsquigarrow C_{x_j}$ or $C_{x_i} \rightsquigarrow C_{y_j}$. So (50) follows by the transitivity of $\rightsquigarrow$.

Finally, (51) follows from (56) and (52).

<u>*Case 2.*</u> There is some $j < n$ with $x_j \leq x_j \ll y_j$ and $C_{x_j} \not\rightsquigarrow C_{x_i}$.
By (pg1), Φ_l and Claim 3.2, there is $f < n$ such that

$$C_{x_f} \text{ is } \rightsquigarrow\text{-final in } \{C_{x_j} : j < n,\ x_j \leq x_j \ll y_j\}. \tag{57}$$

We claim that

$$C_{x_f} \rightsquigarrow C_{y_i}. \tag{58}$$

Indeed, if $x_i \in C_{y_i}$ or $x_i \not\leq x_i$, then this holds by $x_f \ll y_f$ and (pg3). If $x_i \leq x_i \ll y_i$, then $C_{x_f} \not\rightsquigarrow C_{x_i}$ by our assumption on Case 2 and (57), and so $C_{x_f} \rightsquigarrow C_{y_i}$ follows again by $x_f \ll y_f$ and (pg3).

As $x_f \leq x_f \sim x^*$, by Φ_r^+ and the finiteness of $\mathfrak{F}$, there is some x such that

$$x \in min\{x' : \psi_u(x_f, x', x^*, x_f)\}. \tag{59}$$

We claim that, for all $j < n$, we have $\mathsf{right}(x_j, x, y, y_j)$, that is, (48)–(51) hold. Indeed, take some $j < n$. Then (48) is clear. For (49): By (59) and Claim 3.9, we have that $C_{x_f} \rightsquigarrow C_x$. If $x_j \notin C_{y_j}$, then $C_{x_j} \rightsquigarrow C_x$ follows by (57).

In order to show (50) and (51), we claim that

$$\text{either } x \in C_y, \text{ or } [x, y) \rightsquigarrow C_{y_i}. \tag{60}$$

Indeed, suppose that $x \notin C_y$ and take some $a \in [x, y)$. There are three cases:

- $a \in [x, x^*) \cup \{x^*\}$. Then $a \rightsquigarrow C_{x_f}$ by (59), and so $a \rightsquigarrow C_{y_i}$ follows by (58).
- $x^* \notin C_y$ and $a \in C_{x^*}$. Then by (55), either $a \rightsquigarrow C_{y_i}$, or $a \rightsquigarrow C_{x_i}$. In the latter case, either $C_{x_i} = C_{y_i}$, or $C_{x_i} \rightsquigarrow C_{x_f}$ by (57), and so $a \rightsquigarrow C_{y_i}$ follows by (58).
- $a \in (x^*, y)$. Then $a \rightsquigarrow C_{y_i}$ by (56).

Now let us show (50): If $x \notin C_y$, then we have $C_x \rightsquigarrow C_{y_i}$ by (60), and so $C_x \rightsquigarrow C_{y_j}$ follows by (52). And for (51): We have $(x, y) \rightsquigarrow C_{y_i}$ by (60), and so $(x, y) \rightsquigarrow C_{y_j}$ follows by (52). □

Lemma 3.11 *Suppose that Φ holds in $\mathfrak{F}$, and let $\langle x_0, \ldots, x_{n-1}, y_0, \ldots, y_{n-1}\rangle$ be a perfect atomic grid for some $n > 0$. For all $y \in W$, if $y \sim y_0$ then there exist $k > 0$ and a perfect grid $\langle z_j^\ell : \ell \leq k, j \leq n\rangle$ such that $z_j^0 = x_j$, $z_j^k = y_j$, for $j < n$, and $z_n^k = y$.*

Proof. By Lemma 3.10, there is x such that $\mathsf{right}(x_j, x, y, y_j)$ holds, for every $j < n$. If $x \in C_y$ then let $k = 1$, $z_n^0 = x$, $z_n^1 = y$, and $z_j^0 = x_j$, $z_j^1 = y_j$, for all $j < n$. It is straightforward to show that $\langle z_0^0, \ldots, z_n^0, z_0^1, \ldots, z_n^1\rangle$ is a perfect atomic grid.

If $x < y$, then let $k > 0$ and $z_n^0, \ldots z_n^k$ be such that $x = z_n^0 \ll \cdots \ll z_n^k = y$ (that is, we take a point from each $\leq$-cluster between x and y). Of course, we let $z_j^0 = x_j$, and $z_j^k = y_j$, for all $j < n$. Next, for each $j < n$, we have $(x, y) \rightsquigarrow C_{y_j}$ by (51). Therefore, for each $0 < \ell < k$, there exists $z_j^\ell \in C_{y_j}$ such that $z_n^\ell \sim z_j^\ell$. We claim that $\langle z_j^\ell : \ell \leq k, j \leq n\rangle$ is a perfect grid as required. Indeed (pg1) and (pg2) clearly hold. Let us prove that (pg3) holds as well, that is, for all $\ell < k$, $i, j \leq n$, if $z_i^\ell \ll z_i^{\ell+1}$ then either $C_{z_i^\ell} \rightsquigarrow C_{z_j^\ell}$ or $C_{z_i^\ell} \rightsquigarrow C_{z_j^{\ell+1}}$. Indeed, if $i = j$, this clearly holds. Otherwise, there are three cases:

- $i = n$, $j < n$. Then $C_{z_n^0} = C_x$, and we have either $C_x \rightsquigarrow C_{x_j} = C_{z_j^0}$ or $C_x \rightsquigarrow C_{y_j} = C_{z_j^1}$ by (50). Also, if $0 < \ell < k$ then $C_{z_n^\ell} \subseteq (x, y) \rightsquigarrow C_{y_j} = C_{z_j^\ell}$ by (51).

- $i < n, j = n$. If $z_i^\ell \ll z_i^{\ell+1}$, then $\ell = 0$ and $x_i \ll y_i$, and so $C_{z_i^0} = C_{x_i} \rightsquigarrow C_x = C_{z_n^0}$ by (49).
- $i, j < n$. Again, if $z_i^\ell \ll z_i^{\ell+1}$ then $\ell = 0$ and $x_i \ll y_i$. So by (pg3), either $C_{z_i^0} = C_{x_i} \rightsquigarrow C_{x_j} = C_{z_j^0}$ or $C_{z_i^0} = C_{x_i} \rightsquigarrow C_{y_j} = C_{z_j^1}$,

completing the proof of Lemma 3.11. □

Lemma 3.12 *Suppose that Φ holds in $\mathfrak{F}$, and let $\langle x_j^i : i \leq m, j < n\rangle$ be a perfect grid, for some $m, n < \omega$, $n > 0$. For all $x \in W$, if $x \sim x_0^m$ then there exist $s_i < \omega$ $(i \leq m)$ and a perfect grid $\langle z_j^\ell : \ell \leq s_m, j \leq n\rangle$ such that $0 = s_0 < s_1 < \cdots < s_m$, $z_j^{s_i} = x_j^i$, for $j < n$, $i \leq m$, and $z_n^{s_m} = x$,*

Proof. It is by induction on m. For $m = 0$ the statement is obvious. Suppose the statement holds for some $m < \omega$. Let $\langle x_j^i : i \leq m+1, j < n\rangle$ be a perfect grid, and let $x \in W$ be such that $x \sim x_0^m$. Then $\langle x_j^i : 1 \leq i \leq m+1, j < n\rangle$ is a perfect grid, and so by the IH, there exist $s_i < \omega$, for $1 \leq i \leq m+1$, and a perfect grid $\bar{z} = \langle z_j^\ell : 1 \leq \ell \leq s_{m+1}, j \leq n\rangle$ such that $1 = t_1 < t_2 < \cdots < t_{m+1}$, $z_j^{t_i} = x_j^i$, for $1 \leq i \leq m+1$, $j < n$, and $z_n^{t_{m+1}} = x$. We also have that $\langle x_0^0, \ldots, x_{n-1}^0, x_0^1, \ldots, x_{n-1}^1\rangle$ is a perfect atomic grid, and $z_n^{t_1} \sim z_0^{t_1} = x_0^1$. So by Lemma 3.11, there exist $k > 0$ and a perfect grid $\bar{y} = \langle y_j^\ell : \ell \leq k, j \leq n\rangle$ such that $y_j^0 = x_j^0$, for $j < n$, $y_j^k = x_j^1$, for $j < n$, and $y_n^k = z_n^1$. By Claim 2.3, $\bar{y} \sqcup \bar{z}$ is a perfect grid as required. □

4 Discussion

Our results can be extended to **S4.3** $\times$ **S5**, even with some simplifications to the formula Φ. Theorem 1.3 also holds for Logic_of$\{\langle\omega, <\rangle\} \times$ **S5**. However, as the class of all frames for Logic_of$\{\langle\omega, <\rangle\}$ is not closed under ultraproducts, it is not known whether Logic_of$\{\langle\omega, <\rangle\} \times$ **S5** has other finite frames as well, frames that are not p-morphic images of product frames. It would also be interesting to know whether any of the logics (such as the decidable **K4.3** $\times$ **K**, or the undecidable but recursively enumerable **K4.3** $\times$ **K4**) that are within the scope of the non-finite axiomatisability results of [11] has a decidable finite frame problem.

Are we any closer to either proving non-finite axiomatisability of **K4.3** $\times$ **S5**, or finding an explicit, possibly infinite, axiomatisation of it? On the one hand, a way of proving that a product logic L is not finitely axiomatisable is constructing a sequence $\langle\mathfrak{F}_n : n < \omega\rangle$ of *finite* frames such that no $\mathfrak{F}_n$ is a frame for L, but some countable elementary substructure $\mathfrak{G}$ of a non-trivial ultraproduct of the $\mathfrak{F}_n$ is a p-morphic image of a product frame for L. Since the formula Φ we use to decide the finite frame problem for **K4.3** $\times$ **S5** is a first-order formula in the frame-correspondence language, if it fails in every $\mathfrak{F}_n$ then, by Łos' theorem, it fails in any ultraproduct as well, and so it fails in $\mathfrak{G}$. But Φ holds in every product frame and preserved under p-morphic images. So our result implies that we cannot hope for an argument of this kind to work, and have to do something else, possibly constructing *infinite* $\mathfrak{F}_n$.

On the other hand, it can be shown that our first-order formula Φ is not reflected under ultrafilter extensions, and so not modally definable. However, there is a bimodal formula φ such that

- for every 2-frame $\mathfrak{F}$ for $\mathbf{K4.3} \oplus \mathbf{S5}$, if Φ holds in $\mathfrak{F}$, then φ is valid in $\mathfrak{F}$;
- for every finite 2-frame $\mathfrak{F}$ for $\mathbf{K4.3} \oplus \mathbf{S5}$, if φ is valid in $\mathfrak{F}$, then Φ holds in $\mathfrak{F}$.

So if L_φ is the smallest normal bimodal logic containing $\mathbf{K4.3} \oplus \mathbf{S5}$ and φ, then we have $L_\varphi \subseteq \mathbf{K4.3} \times \mathbf{S5}$. However, in order to show the converse inclusion, one would need to show that L_φ has the *finite model property*. And we have no idea about that. Note that it is not known either whether $\mathbf{K4.3} \times \mathbf{S5}$ has the finite model property w.r.t. arbitrary (not necessarily product) frames. $\mathbf{K4.3}^t \times \mathbf{S5}$ lacks the finite model property [12], where $\mathbf{K4.3}^t$ is the temporal extension of $\mathbf{K4.3}$ with a 'past box'. Note that $\mathbf{K4.3}^t \times \mathbf{S5}$ (and so $\mathbf{K4.3} \times \mathbf{S5}$) is decidable [12].

Acknowledgements. We are grateful to Ian Hodkinson for discussions.

References

[1] Blackburn, P., M. de Rijke and Y. Venema, "Modal Logic," Cambridge University Press, 2001.

[2] Chagrov, A. and M. Zakharyaschev, "Modal Logic," Oxford Logic Guides **35**, Clarendon Press, Oxford, 1997.

[3] Gabbay, D., A. Kurucz, F. Wolter and M. Zakharyaschev, "Many-Dimensional Modal Logics: Theory and Applications," Studies in Logic and the Foundations of Mathematics **148**, Elsevier, 2003.

[4] Gabbay, D. and V. Shehtman, *Products of modal logics. Part I*, Journal of the IGPL **6** (1998), pp. 73–146.

[5] Hirsch, R. and I. Hodkinson, *Complete representations in algebraic logic*, Journal of Symbolic Logic **62** (1997), pp. 816–847.

[6] Hirsch, R. and I. Hodkinson, "Relation Algebras by Games," Studies in Logic and the Foundations of Mathematics **147**, Elsevier, North Holland, 2002.

[7] Hirsch, R., I. Hodkinson and A. Kurucz, *On modal logics between* $\mathbf{K} \times \mathbf{K} \times \mathbf{K}$ *and* $\mathbf{S5} \times \mathbf{S5} \times \mathbf{S5}$, Journal of Symbolic Logic **67** (2002), pp. 221–234.

[8] Kurucz, A., *On axiomatising products of Kripke frames*, Journal of Symbolic Logic **65** (2000), pp. 923–945.

[9] Kurucz, A., *Combining modal logics*, in: P. Blackburn, J. van Benthem and F. Wolter, editors, *Handbook of Modal Logic*, Studies in Logic and Practical Reasoning **3**, Elsevier, 2007 pp. 869–924.

[10] Kurucz, A., *On the complexity of modal axiomatisations over many-dimensional structures*, in: L. Beklemishev, V. Goranko and V. Shehtman, editors, *Advances in Modal Logic, Volume 8*, College Publications, 2010 pp. 256–270.

[11] Kurucz, A. and S. Marcelino, *Non-finitely axiomatisable two-dimensional modal logics*, Journal of Symbolic Logic (to appear).

[12] Reynolds, M., *A decidable temporal logic of parallelism*, Notre Dame Journal of Formal Logic **38** (1997), pp. 419–436.

[13] Segerberg, K., *Two-dimensional modal logic*, Journal of Philosophical Logic **2** (1973), pp. 77–96.

[14] Shehtman, V., *Two-dimensional modal logics*, Mathematical Notices of the USSR Academy of Sciences **23** (1978), pp. 417–424, (Translated from Russian).

Justifications, Ontology, and Conservativity

Roman Kuznets [1] Thomas Studer

Institut für Informatik und angewandte Mathematik
Universität Bern

Abstract

An ontologically transparent semantics for justifications that interprets justifications as sets of formulas they justify has been recently presented by Artemov. However, this semantics of modular models has only been studied for the case of the basic justification logic J, corresponding to the modal logic K. It has been left open how to extend and relate modular models to the already existing symbolic and epistemic semantics for justification logics with additional axioms, in particular, for logics of knowledge with factive justifications.
We introduce modular models for extensions of J with any combination of the axioms (jd), (jt), (j4), (j5), and (jb), which are the explicit counterparts of standard modal axioms. After establishing soundness and completeness results, we examine the relationship of modular models to more traditional symbolic and epistemic models. This comparison yields several new semantics, including symbolic models for logics of belief with negative introspection (j5) and models for logics with the axiom (jb). Besides pure justification logics, we also consider logics with both justifications and a belief/knowledge modal operator of the same strength. In particular, we use modular models to study the conditions under which the addition of such an operator to a justification logic yields a conservative extension.

1 Introduction

Justification logics are epistemic logics that feature explicit justifications to evidence the agent's knowledge and/or belief. Instead of formulas $\Box A$, for *A is known*, the language of justification logic includes formulas of the form $t{:}A$ that stand for *A is known for the reason t*, where t is a so-called *justification term*. Justification logics also include operations on these terms to reflect the agent's reasoning power. For instance, if $A \to B$ is known for a reason s and A is known for a reason t, then B is known for the reason $s \cdot t$, where the binary operation $\cdot$ models the agent's ability to apply modus ponens.

The first justification logic, the Logic of Proofs LP, was originally developed by Artemov [1,2] to provide a classical provability semantics for intuitionistic logic. To this end, he introduced an arithmetic semantics for LP in which

[1] Supported by the SNSF grant PZ00P2–131706.

justification terms are interpreted as proofs in Peano arithmetic and the operations on terms correspond to computable operations on proofs in PA. Artemov established arithmetical completeness of LP with respect to this provability semantics and developed an algorithm embedding the modal logic S4 into LP, which, together with the well-known embeddings of intuitionistic logic into S4, solved the long-standing problem of finding a classical provability semantics for intuitionistic logic and S4.

The first non-arithmetic semantics for justification logics was introduced by Mkrtychev [25] in order to obtain decidability results for LP. In these models, which are now called M-models, an evaluation function $*$ assigns to each justification term t a set of formulas $*(t)$. The underlying principle is that

$$\Vdash t{:}F \qquad \Longleftrightarrow \qquad F \text{ is evidenced by } t \text{ according to } * \,. \tag{1}$$

However, if justifications are assumed to be factive, i.e., they can only support true facts, the clarity of (1) in M-models is muddled by truth thrown into the mix. Historically, it was a pragmatic choice of efficiency over philosophical transparency.

Later, Fitting [14], working independently from Mkrtychev, presented an epistemic, i.e., Kripke-style, semantics for justification logics with essentially the same machinery for handling justification terms as in M-models. In this semantics, commonly referred to as F-models, the truth of formulas also invades the causal space of justifications: a formula $t{:}F$ holds at a world w if and only if

(i) F is evidenced by t at w and

(ii) F is true at all worlds that the agent considers possible at w.

It is, therefore, standard to speak of terms as being admissible evidence that is not, however, decisive. F-models can be easily extended to the multi-agent case. Hence, they provide a powerful tool for epistemic applications of justification logics. For instance, F-models have been used for new analyses of traditional epistemic puzzles [4,5] and for investigating the evidential dynamics of public announcements [10,12] and common knowledge [3,9,11].

Despite their many applications, F-models present the same compromise as factive M-models. For the sake of efficiency, justification and truth are intertwined in that t is only evidence for F if F is true (in the same M-model or at all accessible worlds in the F-model). The philosophical objections to such a paradigm also have practical roots. In court, evidence is used to determine the truth of the matter. However, if the acceptability of the evidence were to depend on this truth, it would create a vicious circle. A clear ontological separation between justification and truth is achieved in *modular models* recently introduced by Artemov [6] (although they are less practicable than M-models or F-models, in some cases).

Similar to F-models, modular models consist of a Kripke structure together with an evaluation function $*_w$ for each world w. However, unlike in F-models, no formula is required to be true for t to be evidence for F at a world w. Additionally, modular models satisfy the condition of *justification yields be-*

lief (JYB), which provides a connection between justifications and the traditional possible world semantics for knowledge and belief. This connection principle states that *having evidence for F yields a belief that F.*

Artemov has studied modular models only for the case of the basic justification logic J. Extending them to justification logics with additional axioms, in particular to logics of knowledge with factive justifications, has been left open. We introduce modular models for extensions of J with any combination of the axioms (jd), (jt), (j4), (j5), and (jb), which are the explicit counterparts of standard modal axioms. The connection principle *justification yields belief* makes it possible to give uniform proofs for soundness and completeness for all these logics, which cannot be done in the case of F-models, where this connection principle need not hold. In particular, to obtain the soundness of a justification logic with respect to F-models, the models have to fulfill additional properties that depend on the axioms included in the logic. For modular models, however, these properties, e.g., monotonicity (for (jt)) or the strong evidence property (for (j5)), naturally follow from the fact that justification yields belief. We illustrate this point by developing F-models for justification logics with the axiom (jb) via modular models for these logics. The definition of the latter is trivially read from the axioms, whereas F-models additionally require the strong evidence property, which is not directly related to the axiom system of the logic.

This semantics for justification logics with the axiom (jb) is newly developed in this paper. This axiom was introduced by Brünnler et al. [8] as a justification counterpart of the usual axiom (b) from modal logic. The exploration of the relationship between modular models and M-models also leads to new M-models for justification logics of belief that include the axiom (j5) (some models of this kind are studied in [27]). This relationship, however, is more difficult to study mainly because we are not aware of a conceptually clear way of defining what constitutes an M-model.

Besides pure justification logics, we also consider logics with both justifications and a modal operator for knowledge/belief of the same strength. In particular, we show that the addition of such an operator to a justification logic yields a conservative extension for logics the justifications in which either have to be factive or may not be consistent. For logics with consistent but not necessarily factive justifications, conservativity requires a sufficient store of evidence, i.e., it is necessary to possess evidence for all axioms of the logic. We show that this additional requirement is essential by providing a counterexample to conservativity for the case when no evidence is present for any axiom.

2 Syntax

Justification terms are built from *constants* c_i and *variables* x_i according to the following grammar:

$$t ::= c_i \mid x_i \mid (t \cdot t) \mid (t + t) \mid\ !t \mid\ ?t \mid \bar{?}t \ .$$

We denote the *set of terms* by Tm. A term is called *ground* if it does not contain variables. The operations of $\cdot$ and $+$ are assumed to be left-associative in order to omit unnecessary parentheses. *Formulas* are built from *atomic propositions* p_i according to the following grammar:

$$F ::= p_i \mid \neg F \mid (F \to F) \mid t{:}F \ .$$

Prop and $\mathcal{L}_j$ denote the *set of atomic propositions* and the *set of formulas* respectively. We define $(A \wedge B) := \neg(A \to \neg B)$ and $\bot := p \wedge \neg p$ for a fixed atomic proposition p.

We consider a family of justification logics that differ in their axioms and in the availability of justifications for these axioms. By an *axiom* we understand a set of formulas in the language $\mathcal{L}_j$, called *axiom instances.* We consider sets of axioms that have form $\mathsf{L}(X) = \mathsf{J} \cup X$, where $\mathsf{J} = \{\text{A1}, \text{A2}, \text{A3}\}$ is the smallest set of axioms with

A1 finite complete axiomatization of classical propositional logic,

A2 $t{:}(A \to B) \to (s{:}A \to (t \cdot s){:}B)$,

A3 $t{:}A \to (t + s){:}A$ and $s{:}A \to (t + s){:}A$,

and additional axioms $X \subseteq \{(\text{jd}), (\text{jt}), (\text{j4}), (\text{j5}), (\text{jb})\}$ with

(jd) $t{:}\bot \to \bot$,

(jt) $t{:}A \to A$,

(j4) $t{:}A \to\, !t{:}t{:}A$,

(j5) $\neg t{:}A \to\, ?t{:}\neg t{:}A$,

(jb) $\neg A \to \bar{?}t{:}\neg t{:}A$.

We often write L instead of $\mathsf{L}(X)$ if the set of axioms X is not important. For a formula F and an axiom A, we write $F \in \text{A}$ to mean that F is an instance of A.

The axiom (jb) was recently introduced in [8] (it was independently proposed by Ghari in an unpublished manuscript [18]). Note that the formulation of (jb) in [8] is slightly different, namely: $A \to \bar{?}t{:}\neg t{:}\neg A$.

A *constant specification* CS for a set of axioms L is any subset

$$\mathsf{CS} \subseteq \{c{:}F \mid c \text{ is a constant and there is an axiom A} \in \mathsf{L} \text{ such that } F \in \text{A}\}.$$

Constant specifications determine axiom instances for which the logic provides justifications. A constant specification CS for a set of axioms L is called *axiomatically appropriate* (for L) if for each axiom $\text{A} \in \mathsf{L}$ and for each axiom instance $F \in \text{A}$, there is a constant c such that $c{:}F \in \mathsf{CS}$.

A *justification logic* L_{CS} is determined by its set of axioms L and its constant specification CS (for L). Whenever L_{CS} is used, it is assumed that CS is a constant specification for L. The *deductive system* L_{CS} is the Hilbert system given by the axioms L and by the rules modus ponens

$$\frac{A \quad A \to B}{B}\ (\text{MP})$$

and axiom necessitation

$$\overline{!\cdots!c{:}\cdots{:}!!c{:}!c{:}c{:}F}\ (\text{AN!}), \quad \text{where } c{:}F \in \mathsf{CS}\ .$$

When (j4) $\in \mathsf{L}$, a simplified axiom necessitation rule can be used:

$$\overline{c{:}F}\ (\text{AN}), \quad \text{where } c{:}F \in \mathsf{CS}\ .$$

For instance, the deductive system $\mathsf{JD4_{CS}}$ consists of the axioms A1–A3, (jd), and (j4) and of the rules (MP) and (AN). There are $2^5 = 32$ combinations of the five axioms (jd), (jt), (j4), (j5), and (jb), but they yield only 24 series of logics $\mathsf{L_{CS}}$, because each instance of (jd) is also an instance of (jt).

For any justification logic $\mathsf{L_{CS}}$, we write $\mathsf{L_{CS}} \vdash A$ to mean that the *formula* A *is derivable in* $\mathsf{L_{CS}}$ and $\Delta \vdash_{\mathsf{L_{CS}}} A$ to mean that the *formula* A *is derivable in* $\mathsf{L_{CS}}$ *from the set of formulas* Δ. When the logic $\mathsf{L_{CS}}$ is clear from the context, the subscript $\mathsf{L_{CS}}$ is omitted. We write Δ, A for $\Delta \cup \{A\}$.

The *deduction theorem* is standard for justification logics. Therefore, we omit its proof here.

Theorem 2.1 (Deduction Theorem, [4]) *For any justification logic* $\mathsf{L_{CS}}$, *any* $\Delta \subseteq \mathcal{L}_j$, *and arbitrary* $A, B \in \mathcal{L}_j$, $\Delta, A \vdash_{\mathsf{L_{CS}}} B \iff \Delta \vdash_{\mathsf{L_{CS}}} A \to B$.

An important property of justification logics is their ability to internalize their own notion of proof, as stated in the following lemma, which can be easily proved by induction on the derivation.

Lemma 2.2 (Internalization for Variables, [4]) *Let* $\mathsf{L_{CS}}$ *be a justification logic with the axiomatically appropriate* CS. *For arbitrary formulas* $A, B_1, \ldots, B_n \in \mathcal{L}_j$, *if* $B_1, \ldots, B_n \vdash_{\mathsf{L_{CS}}} A$, *then there is a term* $t(x_1, \ldots, x_n) \in Tm$ *such that* $x_1{:}B_1, \ldots, x_n{:}B_n \vdash_{\mathsf{L_{CS}}} t(x_1, \ldots, x_n){:}A$ *for fresh variables* $x_1, \ldots, x_n$.

Corollary 2.3 (Constructive Necessitation, [4]) *Let* $\mathsf{L_{CS}}$ *be a justification logic with the axiomatically appropriate* CS. *For any formula* $A \in \mathcal{L}_j$, *if* $\mathsf{L_{CS}} \vdash A$, *then* $\mathsf{L_{CS}} \vdash t{:}A$ *for some ground term* $t \in Tm$.

Combining the previous results, we also obtain internalization for the case when the assumptions are justified by arbitrary terms.

Corollary 2.4 (Internalization for Arbitrary Terms) *Let* $\mathsf{L_{CS}}$ *be a justification logic with the axiomatically appropriate* CS. *For arbitrary formulas* $A, B_1, \ldots, B_n \in \mathcal{L}_j$ *and arbitrary terms* $s_1, \ldots, s_n \in Tm$, *if* $B_1, \ldots, B_n \vdash_{\mathsf{L_{CS}}} A$, *then there is a term* $t \in Tm$ *such that* $s_1{:}B_1, \ldots, s_n{:}B_n \vdash_{\mathsf{L_{CS}}} t{:}A$.

Proof. See Appendix A. □

Remark 2.5 There is another method for proving Corollary 2.4: use Internalization Lemma 2.2 to obtain $x_1{:}B_1, \ldots, x_n{:}B_n \vdash s'(x_1, \ldots, x_n){:}A$ and then replace $x_1, \ldots, x_n$ with $s_1, \ldots, s_n$ respectively. However, one has to be careful using this approach since substituting terms for variables need not preserve derivability. Consider $x{:}p \to p \in$ (jt) and let CS be an axiomatically appropriate constant specification for a logic with axioms $\mathsf{L} \ni$ (jt). Then, there is a constant c such that $\mathsf{L}_{\mathsf{CS}} \vdash c{:}(x{:}p \to p)$. Substituting a term t for x in $x{:}p \to p$ yields $t{:}p \to p \in$ (jt). Again, there is a constant d such that $\mathsf{L}_{\mathsf{CS}} \vdash d{:}(t{:}p \to p)$. However, without additional constraints on CS, there is no guarantee that $c = d$.

3 Basic Modular Models

Definition 3.1 *Let $X, Y \subseteq \mathcal{L}_j$ and $t \in Tm$. We define*

(i) $X \cdot Y := \{F \in \mathcal{L}_j \mid G \to F \in X$ *and* $G \in Y$ *for some formula* $G \in \mathcal{L}_j\}$;

(ii) $t{:}X := \{t{:}F \mid F \in X\}$.

Definition 3.2 (Basic evaluation) *A* basic evaluation for a logic L_{CS}, *or a basic L_{CS}-evaluation, is a function $*$ that maps atomic propositions to truth values 0, 1 and maps justification terms to sets of formulas, $*\colon Prop \to \{0,1\}$ and $*\colon Tm \to \mathcal{P}(\mathcal{L}_j)$, such that for arbitrary $s, t \in Tm$ and any $F \in \mathcal{L}_j$,*

(i) $s^* \cdot t^* \subseteq (s \cdot t)^*$;

(ii) $s^* \cup t^* \subseteq (s + t)^*$;

(iii) $F \in t^*$ *for any conclusion* $t{:}F$ *of* (AN) *or* (AN!), *whichever is a rule of* L_{CS};

(iv) $s{:}(s^*) \subseteq (!s)^*$ *for logics with* (j4) $\in \mathsf{L}$; *and*

(v) $F \notin t^*$ *implies* $\neg t{:}F \in (?t)^*$ *for logics with* (j5) $\in \mathsf{L}$.

Here p^* for $p \in Prop$ and t^* for $t \in Tm$ denote $*(p)$ and $*(t)$ respectively.

Definition 3.3 (Truth under a basic evaluation) *We define what it means for a formula to* hold under a basic evaluation $*$ *inductively as follows:*

$* \Vdash p$ *if and only if* $p^* = 1$ *for* $p \in Prop$;

$* \Vdash F \to G$ *if and only if* $* \nVdash F$ *or* $* \Vdash G$;

$* \Vdash \neg F$ *if and only if* $* \nVdash F$;

$* \Vdash t{:}F$ *if and only if* $F \in t^*$.

The definition does not depend on the logic for which $*$ is a basic evaluation. Thus, it is possible to talk about a basic evaluation without specifying its logic.

Definition 3.4 (Consistent, factive, and Brouwerian evaluation) *A basic L_{CS}-evaluation $*$ is called*

- consistent *if* $\bot \notin t^*$ *for any* $t \in Tm$;
- factive *if* $F \in t^*$ *implies* $* \Vdash F$ *for all* $t \in Tm$ *and* $F \in \mathcal{L}_j$;
- Brouwerian *if* $* \nVdash F$ *implies* $\neg t{:}F \in (\bar{?}t)^*$ *for all* $t \in Tm$ *and* $F \in \mathcal{L}_j$.

It is immediate from the last two definitions that a factive basic evaluation is always consistent.

Definition 3.5 (Basic modular model) *A* basic modular model for a logic L_{CS}, *or a basic modular L_{CS}-model, is a basic L_{CS}-evaluation $*$ such that*

$*$ *is*

(i) *consistent if* (jd) $\in$ L*;*

(ii) *factive if* (jt) $\in$ L*;*

(iii) *Brouwerian if* (jb) $\in$ L.

Theorem 3.6 (Soundness) *Let* L_{CS} *be a justification logic and* $F \in \mathcal{L}_j$.

$$\mathsf{L}_{\mathsf{CS}} \vdash F \quad\Longrightarrow\quad * \Vdash F \text{ for all basic modular } \mathsf{L}_{\mathsf{CS}}\text{-models } *.$$

Proof. See Appendix B. □

Completeness is established by a maximal consistent set construction.

Definition 3.7 (L_{CS}-consistent set) *Let* L_{CS} *be a justification logic. A set* $\Phi \subseteq \mathcal{L}_j$ *is called* L_{CS}-consistent *if there is* $F \in \mathcal{L}_j$ *such that* $\Phi \nvdash_{\mathsf{L}_{\mathsf{CS}}} F$. Φ *is called* maximal L_{CS}-consistent *if it is* L_{CS}*-consistent and has no* L_{CS}*-consistent proper extensions.*

As usual, a maximal L_{CS}-consistent set contains all instances of axioms from L and is closed under the inference rules of L_{CS}.

Theorem 3.8 (Completeness) *Let* L_{CS} *be a justification logic and* $F \in \mathcal{L}_j$.

$$* \Vdash F \text{ for all basic modular } \mathsf{L}_{\mathsf{CS}}\text{-models } * \quad\Longrightarrow\quad \mathsf{L}_{\mathsf{CS}} \vdash F\ .$$

Proof. See Appendix C. □

Remark 3.9 Basic modular models, introduced in [6], are closely related to M-models, introduced by Mkrtychev in [25] for the Logic of Proofs $\mathsf{LP}_{\mathsf{CS}}$, the logic with axioms $\mathsf{J} \cup \{(\text{jt}), (\text{j4})\}$, denoted $\mathcal{LP}_{AS}$ in [25]. Strictly speaking, Mkrtychev created two semantics, which he called *models* and *pre-models* and which he showed to be equivalent. M-models, which have been adapted to several other logics from the justification family in [20,21,26,27,28] and used to prove decidability of justification logics ([25]), to determine their complexity ([19,21,22]), and to study self-referentiality of modal logics ([23]), are the pre-models of [25]. However, the other semantics from [25], that of the models, is, in fact, exactly the semantics of basic modular models for $\mathsf{LP}_{\varnothing}$, i.e., of factive basic evaluations for $\mathsf{LP}_{\varnothing}$. The models of [25] are defined as *interpretations* with an additional condition, which is identical to our condition of factivity, while interpretations themselves are isomorphic modulo notation and terminology to basic evaluations for $\mathsf{LP}_{\varnothing}$. The machinery for handling the rule (AN) in [25] and in this paper is also essentially the same.

The difference between pre-models and models lies in the conditions under which $t{:}F$ holds. For models, it is the same as for basic evaluations, i.e., $F \in t^*$ is sufficient, whereas for pre-models, additionally $* \Vdash F$ is required. Clearly, this additional requirement replaces the requirement of factivity. Therefore, it should only be used when (jt) $\in \mathsf{L}$. Indeed, the M-models introduced in [21] for the logics with axioms J, $\mathsf{J} \cup \{(\text{jd})\}$, $\mathsf{J} \cup \{(\text{j4})\}$, or $\mathsf{J} \cup \{(\text{jd}), (\text{j4})\}$ are isomorphic to the basic modular models for these logics. By the same token, some of the basic modular models presented in this paper can and should be considered M-models.

Definition 3.10 (M-models for logics with (j5) but without (jt)) *An* M-model *for a logic* L_{CS} *with the axioms* $\mathsf{J} \cup \{(\mathrm{j5})\}$, $\mathsf{J} \cup \{(\mathrm{jd}), (\mathrm{j5})\}$, $\mathsf{J} \cup \{(\mathrm{j4}), (\mathrm{j5})\}$, *or* $\mathsf{J} \cup \{(\mathrm{jd}), (\mathrm{j4}), (\mathrm{j5})\}$ *is a basic modular model for* L_{CS}.

4 Epistemic Models and Modularity

Justification logics also have an epistemic semantics, developed by Fitting in [14]. An *F-model* for a justification logic is a quadruple $\mathcal{M} = (W, R, \mathcal{E}, \mathcal{V})$, where $(W, R, \mathcal{V})$ is a Kripke model and where the admissible evidence function $\mathcal{E} \colon W \times \mathrm{Tm} \to \mathcal{P}(\mathcal{L}_j)$ plays the role of a basic evaluation $*$ for each world. The crucial feature of F-models is that

$$\mathcal{M}, w \Vdash t{:}F \qquad \Longleftrightarrow \qquad F \in \mathcal{E}(w,t) \text{ and } \big(R(w,v) \Rightarrow \mathcal{M}, v \Vdash F\big) \ .$$

These epistemic models provide a way of comparing justification logics and their corresponding modal logics within the same semantics, as well as provide semantics for combinations of the two, which are sometimes called *logics of justifications and belief/knowledge*. However, F-models violate an important property of basic modular models: namely, the ontological separation of justification from truth. This separation is also violated in M-models for logics with the axiom (jt) (see Remark 3.9). Indeed, in such an M-model and in any F-model, to check whether a term t justifies a formula F, it must be observed whether F holds, in the M-model or in all accessible worlds of the F-model.

This prompted Artemov in [6] to introduce *modular models* for J_{CS} with a clear distinction between the truth and the justification of formulas. We now extend modular models to all the justification logics we are considering. Like F-models, these modular models can also be used for logics of justifications and belief/knowledge.

Definition 4.1 (Quasimodel) *A* quasimodel for L_{CS}, *or an* L_{CS}*-quasimodel, is a triple* $\mathcal{M} = (W, R, *)$, *where* $W \neq \varnothing$, $R \subseteq W \times W$, *and the* evaluation $*$ *maps each world* $w \in W$ *to a basic* L_{CS}*-evaluation* $*_w$. *We will write* p^*_w *instead of* $*_w(p)$ *and* t^*_w *instead of* $*_w(t)$.

Definition 4.2 (Truth in quasimodels) *We define what it means for a formula to* hold at a world $w \in W$ of a quasimodel $\mathcal{M} = (W, R, *)$ *inductively as follows:*

$\mathcal{M}, w \Vdash p$ *if and only if* $p^*_w = 1$ *for* $p \in Prop$;

$\mathcal{M}, w \Vdash F \to G$ *if and only if* $\mathcal{M}, w \nVdash F$ *or* $\mathcal{M}, w \Vdash G$;

$\mathcal{M}, w \Vdash \neg F$ *if and only if* $\mathcal{M}, w \nVdash F$;

$\mathcal{M}, w \Vdash t{:}F$ *if and only if* $F \in t^*_w$.

As in the case of basic evaluations, this definition does not depend on the logic for which $\mathcal{M}$ *is a quasimodel. We write* $\mathcal{M} \Vdash F$ *if* $\mathcal{M}, w \Vdash F$ *for all* $w \in W$.

For a given quasimodel $\mathcal{M} = (W, R, *)$ and a world $w \in W$, we define

$$\Box_w := \{F \in \mathcal{L}_j \mid \mathcal{M}, v \Vdash F \text{ whenever } R(w,v)\} \ . \tag{2}$$

By analogy with basic modular models, we define the following notions:

Definition 4.3 (Consistent, factive, and Brouwerian quasimodel) *An* L_{CS}*-quasimodel* $\mathcal{M} = (W, R, *)$ *is called*

- consistent *if* $\bot \notin t^*_w$ *for any* $w \in W$ *and any* $t \in Tm$*;*
- factive *if* $F \in t^*_w$ *implies* $\mathcal{M}, w \Vdash F$ *for all* $w \in W$*,* $t \in Tm$*, and* $F \in \mathcal{L}_j$*;*
- Brouwerian *if* $\mathcal{M}, w \nVdash F$ *implies* $\neg t{:}F \in (\bar{?}t)^*_w$ *for all* $w \in W$*,* $t \in Tm$*, and* $F \in \mathcal{L}_j$*.*

Definition 4.4 (Modular model) *A* modular model $\mathcal{M} = (W, R, *)$ for L_{CS}*, or a modular* L_{CS}*-model, is an* L_{CS}*-quasimodel that meets the following conditions:*

(i) $t^*_w \subseteq \Box_w$ *for all* $t \in Tm$ *and* $w \in W$*;* *(JYB)*

(ii) R *is serial if* (jd) $\in \mathsf{L}$*;*

(iii) R *is reflexive if* (jt) $\in \mathsf{L}$*;*

(iv) R *is transitive if* (j4) $\in \mathsf{L}$*;*

(v) R *is Euclidean if* (j5) $\in \mathsf{L}$*;*

(vi) R *is symmetric if* (jb) $\in \mathsf{L}$*;*

(vii) $\mathcal{M}$ *is Brouwerian if* (jb) $\in \mathsf{L}$.

Conditions (i)–(vi) may seem superfluous since R plays no role in determining the truth of formulas. But Conditions (ii)–(vi) are well known for the corresponding modal axioms in modal logic and, hence, are needed, so to say, for backward compatibility: they ensure that the same semantics can be used for justification logics, logics of justifications and belief/knowledge, and modal logics. Condition (i) plays, in this respect, the role of a catalyzer allowing for a transition between these three formalisms. This condition essentially says that *justification yields belief*, abbreviated JYB. Indeed, whenever $F \in t^*_w$, we have $\mathcal{M}, w \Vdash t{:}F$ so that F has a justification at w. The requirement that F belong to $\Box_w$ says that F must be believed at w in the sense of Kripke models, i.e., hold at all worlds considered possible at w.

Note that, unlike for the case of basic modular models, we do not require that modular models for logics with (jd) be consistent or those for logics with (jt) be factive. Instead, these properties are derived from JYB and the corresponding restrictions on R.

Lemma 4.5 (Reflexive modular models are factive) *Let* (jt) $\in \mathsf{L}$ *and let* $\mathcal{M} = (W, R, *)$ *be a modular* L_{CS}*-model. Then* $\mathcal{M}$ *is factive.*

Proof. Suppose $F \in t^*_w$. Then $F \in \Box_w$ by JYB. Since $R(w,w)$ by reflexivity of R, we obtain $\mathcal{M}, w \Vdash F$ from (2). □

Lemma 4.6 (Serial modular models are consistent) *Let* (jd) $\in$ L *and let* $\mathcal{M} = (W, R, *)$ *be a modular* L_{CS}*-model. Then* $\mathcal{M}$ *is consistent.*

Proof. Assume towards a contradiction that $\bot \in t^*_w$. Then $\bot \in \Box_w$ by JYB. By seriality of R, there is $v \in W$ such that $R(w, v)$ and, by (2), we would have $\mathcal{M}, v \Vdash \bot$, which is impossible. □

There is an additional property that follows from JYB but is peculiar to the possible-worlds scenario.

Lemma 4.7 (Monotonicity) *Let* (j4) $\in$ L *and* $\mathcal{M} = (W, R, *)$ *be a modular* L_{CS}*-model. Then for any* $t \in Tm$ *and for arbitrary* $a, b \in W$, $R(a, b)$ *implies* $t^*_a \subseteq t^*_b$.

Proof. Assume $R(a, b)$ and $F \in t^*_a$. Then $t{:}F \in (!t)^*_a$ because $*_a$ is a basic evaluation for L_{CS}. So $t{:}F \in \Box_a$ by JYB and $\mathcal{M}, b \Vdash t{:}F$ by (2), which means that $F \in t^*_b$. □

The soundness and completeness of justification logics with respect to modular models are almost obvious:

Theorem 4.8 (Soundness and Completeness, Modular Models I) *Let* L_{CS} *be a justification logic such that either* (jt) $\in$ L *or* (jd) $\notin$ L *and let* $F \in \mathcal{L}_j$.

$$\mathsf{L}_{\mathsf{CS}} \vdash F \qquad \Longleftrightarrow \qquad \mathcal{M} \Vdash F \text{ for all modular } \mathsf{L}_{\mathsf{CS}}\text{-models } \mathcal{M} \ . \qquad (3)$$

Proof. See Appendix D. □

Remark 4.9 Unfortunately, single-world modular models are insufficient for proving the completeness of logics that are consistent but not factive. Indeed, any serial single-world model is automatically reflexive. Thus, JYB for such a model would yield factivity by Lemma 4.5, making it impossible to distinguish between consistency and factivity. In this case, the simplest completeness proof is via the canonical model construction akin to that from the proof of Theorem 3.8.

Theorem 4.10 (Soundness and Completeness, Modular Models II) *Let* L_{CS} *be a logic with* (jt) $\notin$ L *and* (jd) $\in$ L. *Then* L_{CS} *is sound with respect to modular* L_{CS}*-models* ($\Longrightarrow$*-direction of* (3)). *If* CS *is axiomatically appropriate for* L, *then* L_{CS} *is also complete* ($\Longleftarrow$*-direction of* (3)).

Proof. See Appendix E. □

Remark 4.11 In fact, the completeness proof in Theorem 4.10 can be easily applied to all the logics covered by Theorems 4.8 and 4.10. The only addition would be the necessity to show R is reflexive if (jt) $\in$ L.

The relationship between F-models, mentioned at the beginning of this section, and modular models is rather straightforward. While an independent definition of F-models for all logics L_{CS}, except for those with (jb) $\in$ L, can be found in [4], we can describe them via modular models:

Definition 4.12 (F-models) *The definition of an* F-model $\mathcal{M}_{\mathsf{F}} = (W, R, *)$ for a logic L_{CS} *is identical to that of a modular* L_{CS}*-model (see Def. 4.4) with a new condition: if* (j4) $\in \mathsf{L}$*, then for all* $w, v \in W$ *and all* $t \in Tm$,

$$R(w,v) \quad \Longrightarrow \quad t^*_w \subseteq t^*_v \ ; \tag{4}$$

and with a restricted JYB analog: if (j5) $\in \mathsf{L}$ *or* (jb) $\in \mathsf{L}$*, then for all* $t \in Tm$*,* $F \in \mathcal{L}_j$*, and* $w \in W$*,*

$$F \in t^*_w \quad \Longrightarrow \quad F \in \Box^{\mathsf{F}}_w \ , \tag{5}$$

where

$$\Box^{\mathsf{F}}_w := \{G \in \mathcal{L}_j \mid \mathcal{M}_{\mathsf{F}}, v \Vdash_{\mathsf{F}} G \textit{ whenever } R(w,v)\} \ . \tag{6}$$

Condition (4) *for F-models is traditionally called* monotonicity, *whereas the most common equivalent form of* (5) *is called the* strong evidence property.

The absence of the JYB requirement in the case when neither (j5) $\in \mathsf{L}$ nor (jb) $\in \mathsf{L}$ means that in an F-model, $F \in t^*_w$ says that t is *admissible* as evidence for F at the world w, but it need not be *decisive* as it is in modular models. Accordingly, the notion of truth $\Vdash_{\mathsf{F}}$ in F-models differs from $\Vdash$ in quasimodels with respect to formulas of the form $t{:}F$.

Definition 4.13 (Truth in F-models) *The definition of* $\mathcal{M}_{\mathsf{F}}, w \Vdash_{\mathsf{F}} A$ *is identical to that of* $\mathcal{M}_{\mathsf{F}}, w \Vdash A$ *for the* L_{CS}*-quasimodel* $\mathcal{M}_{\mathsf{F}}$ *(see Def. 4.2) except that the last clause is replaced by* $\mathcal{M}_{\mathsf{F}}, w \Vdash_{\mathsf{F}} t{:}F$ *if and only if* $F \in t^*_w$ *and* $F \in \Box^{\mathsf{F}}_w$*. Note that* $\Vdash_{\mathsf{F}}$ *can be applied to any quasimodel irrespective of its logic.*

Lemma 4.14 (From Modular to F-Models) *Every modular* L_{CS}*-model* $\mathcal{M} = (W, R, *)$ *is also an F-model for* L_{CS} *such that for all* $w \in W$ *and* $F \in \mathcal{L}_j$*,*

$$\mathcal{M}, w \Vdash_{\mathsf{F}} F \quad \Longleftrightarrow \quad \mathcal{M}, w \Vdash F \ . \tag{7}$$

Proof. See Appendix F. □

Remark 4.15 It is clear from the proof of Lemma 4.14 that (5) holds for any F-model based on a modular L_{CS}-model, even if neither (j5) $\in \mathsf{L}$ nor (jb) $\in \mathsf{L}$.

The converse direction is more interesting. An F-model need not be a modular model itself, but it always induces an equivalent modular model.

Lemma 4.16 (From F- to Modular Models) *Let* $\mathcal{M}_{\mathsf{F}} = (W, R, *_{\mathsf{F}})$ *be an F-model for a logic* L_{CS}*. Then* $\mathcal{M} := (W, R, *)$ *with*

$$t^*_w := \{F \in \mathcal{L}_j \mid F \in t^{*_{\mathsf{F}}}_w \textit{ and } F \in \Box^{\mathsf{F}}_w\} = \{F \in \mathcal{L}_j \mid \mathcal{M}_{\mathsf{F}}, w \Vdash_{\mathsf{F}} t{:}F\} \tag{8}$$

is a modular L_{CS}*-model such that for all* $w \in W$ *and all* $F \in \mathcal{L}_j$*,*

$$\mathcal{M}, w \Vdash F \quad \Longleftrightarrow \quad \mathcal{M}_{\mathsf{F}}, w \Vdash_{\mathsf{F}} F \ . \tag{9}$$

Proof. See Appendix G. □

Remark 4.17 Soundness and completeness with respect to F-models for L_{CS} follow from Lemmas 4.14 and 4.16 and soundness and completeness with respect to modular L_{CS}-models, with the same requirement of axiomatic appropriateness for CS when (jd) $\in \mathsf{L}$. Thus, we have created F-models for the justification logics with (jb) $\in \mathsf{L}$.

5 Justifications and Belief

It should not be surprising that modular models can also be used for the joint language of justifications and belief. And while the condition JYB does not look out of place in justification logics, its real origins are, of course, modal, which is clearly seen in the following soundness proof. Many notions and conventions introduced in Sect. 2 are now generalized to the extended *language* $\mathcal{L}_\Box$ defined by the grammar:

$$F ::= p_i \mid \neg F \mid (F \to F) \mid t{:}F \mid \Box F \ .$$

For each set of axioms L considered earlier, we define the *set of axioms* $\mathsf{L}^\Box$ to consist of

- all the axioms of L in the extended language $\mathcal{L}_\Box$;
- axiom $\Box(A \to B) \to (\Box A \to \Box B)$;
- axiom $\Box\bot \to \bot$ if (jd) $\in \mathsf{L}$;
- axiom $\Box F \to F$ if (jt) $\in \mathsf{L}$;
- axiom $\Box F \to \Box\Box F$ if (j4) $\in \mathsf{L}$;
- axiom $\neg\Box F \to \Box\neg\Box F$ if (j5) $\in \mathsf{L}$;
- axiom $\neg F \to \Box\neg\Box F$ if (jb) $\in \mathsf{L}$; and
- axiom $t{:}F \to \Box F$.

The axiom $t{:}F \to \Box F$ is called the *connection axiom.* It formally states that justification yields belief. *Constant specifications for* $\mathsf{L}^\Box$ are defined in the obvious way. Given a constant specification CS for $\mathsf{L}^\Box$, the *deductive system* $\mathsf{L}^\Box_{\mathsf{CS}}$ is the Hilbert system given by the axioms $\mathsf{L}^\Box$ and by the rules (MP) and either (AN) or (AN!), as in L_{CS}, as well as by the usual necessitation rule from modal logic:

$$\frac{F}{\Box F}\,(\Box) \ .$$

A *basic evaluation for a logic* $\mathsf{L}^\Box_{\mathsf{CS}}$ and many other notions are defined in the same way as for its corresponding justification logic L_{CS} except that each instance of the language $\mathcal{L}_j$ should be replaced with $\mathcal{L}_\Box$ and those of the logic L_{CS} with $\mathsf{L}^\Box_{\mathsf{CS}}$. In particular, the definition of a *modular model for* $\mathsf{L}^\Box_{\mathsf{CS}}$ repeats that for L_{CS} with one extra clause for the truth of formulas $\Box F$:

$$\mathcal{M}, w \Vdash \Box F \qquad \Longleftrightarrow \qquad F \in \Box_w \ , \tag{10}$$

which is a standard definition recast in our notation.

We will not repeat definitions and proofs, unless there is a significant change due to the addition of modalities.

Theorem 5.1 (Soundness of Modular Models for $\mathsf{L}^{\Box}_{\mathsf{CS}}$) *Let* $\mathsf{L}^{\Box}_{\mathsf{CS}}$ *be a logic of justifications and belief and let* $F \in \mathcal{L}_{\Box}$.

$$\mathsf{L}^{\Box}_{\mathsf{CS}} \vdash F \quad \Longrightarrow \quad \mathcal{M} \Vdash F \text{ for all modular } \mathsf{L}^{\Box}_{\mathsf{CS}}\text{-models } \mathcal{M} \ .$$

Proof. Most of the proof repeats that of Theorem 3.6 or the standard argument for the modal axioms. The connection axiom $t{:}F \to \Box F$ is valid because of JYB and (10). □

Theorem 5.2 (Completeness of Modular Models for $\mathsf{L}^{\Box}_{\mathsf{CS}}$) *Let* $\mathsf{L}^{\Box}_{\mathsf{CS}}$ *be a logic of justifications and belief and let* $F \in \mathcal{L}_{\Box}$.

$$\mathcal{M} \Vdash F \text{ for all modular } \mathsf{L}^{\Box}_{\mathsf{CS}}\text{-models } \mathcal{M} \quad \Longrightarrow \quad \mathsf{L}^{\Box}_{\mathsf{CS}} \vdash F \ .$$

Proof. See Appendix H. □

6 Conservative Extensions of Modal Logics

Justification language and the closely related modal language provide a plethora of conservativity statements to be considered. Within justification logic itself, different sets of operations on justifications can be compared prompting the question whether larger sets are conservative over smaller ones, which is thoroughly studied in [15,24]. One can also ask whether logics of justifications and belief are conservative over the respective justification logics and/or the respective modal logics. While there are a few results of the latter type sprinkled through the literature so widely that listing all the papers here is impractical, the results of the former type are relatively rare. Artemov and Nogina in [7] introduced the logic $\mathsf{S4LP}$, which is $\mathsf{L}^{\Box}_{\mathsf{CS}}$ with $\mathsf{L} = \mathsf{J} \cup \{(\text{jt}), (\text{j4})\}$ in our notation. They also conjectured this logic to be conservative over L_{CS} (modulo a careful treatment of CS). In the introduction to [16], Fitting mentioned this conjecture as a fact without any proof. Most probably, Fitting's argument was semantic based on his semantics of F-models, which works both for these $\mathsf{L}^{\Box}_{\mathsf{CS}}$ and L_{CS} (see [13,14]). Finally, Ghari recently published a syntactic proof of the same fact in [17]. In this section, we use a semantic argument based on modular models to extend this conservativity result to other pairs of corresponding logics.

Theorem 6.1 (Conservativity)

(i) *Let* L_{CS} *be a justification logic with* (jd) $\notin \mathsf{L}$ *and let* $F \in \mathcal{L}_j$. *Then* $\mathsf{L}_{\mathsf{CS}} \vdash F \iff \mathsf{L}^{\Box}_{\mathsf{CS}} \vdash F$.

(ii) *Let* L_{CS} *be a justification logic with* (jd) $\in \mathsf{L}$, *let* CS *be an axiomatically appropriate constant specification for* L, *and let* $F \in \mathcal{L}_j$. *Then we have* $\mathsf{L}_{\mathsf{CS}} \vdash F \iff \mathsf{L}^{\Box}_{\mathsf{CS}} \vdash F$.

Proof. The statements in both cases follow from the fact that by Theorems 4.8, 4.10, 5.1, and 5.2, both logics are sound and complete with respect to the same class of modular models. □

The restriction on CS in the second part of this theorem originates from Theorem 4.10, whereas Theorem 5.2 does not feature any such restriction. The following example shows that this restriction is, in fact, essential for conservativity rather than being an artifact of the particular proof method.

Example 6.2 Consider $\mathsf{L}_\varnothing$ with $\mathsf{L} = \mathsf{J} \cup \{(\text{jd})\}$. We show that $\mathsf{L}_\varnothing^\Box$ is not conservative over $\mathsf{L}_\varnothing$. Indeed, $\neg y{:}x{:}\bot$ can be derived in $\mathsf{L}_\varnothing^\Box$ by taking the instance $x{:}\bot \to \bot$ of the axiom (jd) $\in \mathsf{L}^\Box$, applying normal modal reasoning to get $\Box x{:}\bot \to \Box\bot$, syllogizing with the modal seriality axiom $\Box\bot \to \bot \in \mathsf{L}^\Box$ to obtain $\Box x{:}\bot \to \bot$, and syllogizing once again with the instance $y{:}x{:}\bot \to \Box x{:}\bot$ of the connection principle $\in \mathsf{L}^\Box$. The final result is $\mathsf{L}_\varnothing^\Box \vdash y{:}x{:}\bot \to \bot$, or, equivalently, $\mathsf{L}_\varnothing^\Box \vdash \neg y{:}x{:}\bot$.

However, $y{:}x{:}\bot$ is shown to be satisfiable in an M-model for $\mathsf{L}_\varnothing$ in [22, Ex. 3.3.23]. Hence, $\mathsf{L}_\varnothing \nvdash \neg y{:}x{:}\bot$ due to the soundness of $\mathsf{L}_\varnothing$ with respect to its M-models or, equivalently, according to Remark 3.9, with respect to basic modular $\mathsf{L}_\varnothing$-models. Thus, $\mathsf{L}_\varnothing^\Box$ is not conservative over $\mathsf{L}_\varnothing$.

7 Fully Explanatory Models

Since in justification logics the modality of $\Box F$ is read existentially, i.e., as the existence of a justification for F, it is reasonable to ask whether the semantics we have presented supports this reading.

Definition 7.1 (Fully explanatory modular models) *A modular L_{CS}-model $\mathcal{M} = (W, R, *)$ is* fully explanatory *if for any $w \in W$ and any $F \in \mathcal{L}_j$, $F \in \Box_w$ implies $F \in t^*_w$ for some $t \in Tm$.*

This notion can be seen as the converse of JYB and, taking the latter into account, can be reformulated as $\Box_w = \bigcup_{t \in \mathrm{Tm}} t^*_w$. A similar notion was originally proposed by Fitting in [14] for F-models.

Theorem 7.2 *Let L_{CS} be a justification logic with the axiomatically appropriate* CS. *Then L_{CS} is sound and complete with respect to fully explanatory modular L_{CS}-models.*

Proof. Given Theorems 4.8 and 4.10 and Remark 4.11, it is sufficient to show that the canonical model $\mathcal{M}_c = (W, R, *)$ for L_{CS} constructed in the proof of Theorem 4.10 is fully explanatory. See Appendix I for details. □

8 Conclusion

Modular models provide an epistemic semantics for justification logics with a clear ontological separation of justification and truth. We have introduced modular models for the extensions of the basic justification logic J with any combination of the axioms (jd), (jt), (j4), (j5), and (jb).

One of the main properties of modular models is that justification yields belief, which has enabled us to study the relationship of modular models to more traditional epistemic semantics of F-models for justification logics and Kripke models for modal logics. We have also compared single-world variants of modular models to the existing symbolic semantics for justification logics. These comparisons have yielded several new semantics, including symbolic models for logics of belief with negative introspection (j5) and epistemic models for logics with the axiom (jb).

We have also extended the semantics of modular models to justification logics with an additional modal knowledge/belief operator and have exploited the common semantical framework to demonstrate that such extensions are typically conservative. All these conservativity results, with the exception of the conservativity of $\mathsf{S4LP}_{\mathsf{CS}}$ over $\mathsf{LP}_{\mathsf{CS}}$, are new.

Acknowledgments

We would like to thank Sergei Artemov for catalyzing the appearance of this paper. We thank the anonymous referees for useful comments. We are also indebted to Galina Savukova for helping us with the linguistic aspects of the paper.

References

[1] Artemov, S. N., *Operational modal logic*, Technical Report MSI 95–29, Cornell University (1995).

[2] Artemov, S. N., *Explicit provability and constructive semantics*, Bulletin of Symbolic Logic **7** (2001), pp. 1–36.

[3] Artemov, S. N., *Justified common knowledge*, Theoretical Computer Science **357** (2006), pp. 4–22.

[4] Artemov, S. N., *The logic of justification*, The Review of Symbolic Logic **1** (2008), pp. 477–513.

[5] Artemov, S. N., *Why do we need Justification Logic?*, in: J. van Benthem, A. Gupta and E. Pacuit, editors, *Games, Norms and Reasons: Logic at the Crossroads*, Synthese Library **353**, Springer, 2011 pp. 23–38.

[6] Artemov, S. N., *The ontology of justifications in the logical setting*, Studia Logica **100** (2012), pp. 17–30.

[7] Artemov, S. N. and E. Nogina, *Introducing justification into epistemic logic*, Journal of Logic and Computation **15** (2005), pp. 1059–1073.

[8] Brünnler, K., R. Goetschi and R. Kuznets, *A syntactic realization theorem for justification logics*, in: L. D. Beklemishev, V. Goranko and V. Shehtman, editors, *Advances in Modal Logic, Volume 8*, College Publications, 2010 pp. 39–58.

[9] Bucheli, S., "Justification Logics with Common Knowledge," Ph.D. thesis, Universität Bern (2012).

[10] Bucheli, S., R. Kuznets, B. Renne, J. Sack and T. Studer, *Justified belief change*, in: X. Arrazola and M. Ponte, editors, *LogKCA-10, Proceedings of the Second ILCLI International Workshop on Logic and Philosophy of Knowledge, Communication and Action*, University of the Basque Country Press, 2010 pp. 135–155.

[11] Bucheli, S., R. Kuznets and T. Studer, *Justifications for common knowledge*, Journal of Applied Non-Classical Logics **21** (2011), pp. 35–60.

[12] Bucheli, S., R. Kuznets and T. Studer, *Partial realization in dynamic justification logic*, in: L. D. Beklemishev and R. de Queiroz, editors, *Logic, Language, Information and Computation, 18th International Workshop, WoLLIC 2011, Philadelphia, PA, USA, May 18–20, 2011, Proceedings*, Lecture Notes in Artificial Intelligence **6642**, Springer, 2011 pp. 35–51.

[13] Fitting, M., *Semantics and tableaus for LPS4*, Technical Report TR–2004016, CUNY Ph.D. Program in Computer Science (2004).

[14] Fitting, M., *The logic of proofs, semantically*, Annals of Pure and Applied Logic **132** (2005), pp. 1–25.

[15] Fitting, M., *Justification logics, logics of knowledge, and conservativity*, Annals of Mathematics and Artificial Intelligence **53** (2008), pp. 153–167.

[16] Fitting, M., *S4LP and local realizability*, in: E. A. Hirsch, A. A. Razborov, A. Semenov and A. Slissenko, editors, *Computer Science — Theory and Applications, Third International Computer Science Symposium in Russia, CSR 2008, Moscow, Russia, June 7–12, 2008, Proceedings*, Lecture Notes in Computer Science **5010**, Springer, 2008 pp. 168–179.

[17] Ghari, M., *Cut elimination and realization for epistemic logics with justification*, Journal of Logic and Computation **Advance Access** (2011).

[18] Ghari, M., *Labelled sequent calculus for justification logics* (2012), unpublished manuscript.

[19] Krupski, N. V., *On the complexity of the reflected logic of proofs*, Theoretical Computer Science **357** (2006), pp. 136–142.

[20] Krupski, V. N., *On symbolic models for Single-Conclusion Logic of Proofs*, Sbornik: Mathematics **202** (2011), pp. 683–695, originally published in Russian.

[21] Kuznets, R., *On the complexity of explicit modal logics*, in: P. G. Clote and H. Schwichtenberg, editors, *Computer Science Logic, 14th International Workshop, CSL 2000, Annual Conference of the EACSL, Fischbachau, Germany, August 21–26, 2000, Proceedings*, Lecture Notes in Computer Science **1862**, Springer, 2000 pp. 371–383.

[22] Kuznets, R., "Complexity Issues in Justification Logic," Ph.D. thesis, CUNY Graduate Center (2008).

[23] Kuznets, R., *Self-referential justifications in epistemic logic*, Theory of Computing Systems **46** (2010), pp. 636–661.

[24] Milnikel, R. S., *Conservativity for logics of justified belief: Two approaches*, Annals of Pure and Applied Logic **163** (2012).

[25] Mkrtychev, A., *Models for the logic of proofs*, in: S. Adian and A. Nerode, editors, *Logical Foundations of Computer Science, 4th International Symposium, LFCS'97, Yaroslavl, Russia, July 6–12, 1997, Proceedings*, Lecture Notes in Computer Science **1234**, Springer, 1997 pp. 266–275.

[26] Rubtsova, N. M., *Logic of Proofs with substitution*, Mathematical Notes **82** (2007), pp. 816–826, originally published in Russian.

[27] Studer, T., *Decidability for some justification logics with negative introspection* (2011), preprint.
URL `http://www.iam.unibe.ch/~tstuder/papers/J5_Decidability.pdf`

[28] Yavorskaya (Sidon), T., *Interacting explicit evidence systems*, Theory of Computing Systems **43** (2008), pp. 272–293.

Appendix

A Proof of Corollary 2.4

Corollary 2.4 Let L_{CS} be a justification logic with the axiomatically appropriate CS. For arbitrary formulas $A, B_1, \ldots, B_n \in \mathcal{L}_j$ and arbitrary terms $s_1, \ldots, s_n \in \mathrm{Tm}$, if $B_1, \ldots, B_n \vdash_{\mathsf{L}_{\mathsf{CS}}} A$, then there is a term $t \in \mathrm{Tm}$ such that $s_1{:}B_1, \ldots, s_n{:}B_n \vdash_{\mathsf{L}_{\mathsf{CS}}} t{:}A$.

Proof. Assume $B_1, \ldots, B_n \vdash_{\mathsf{L}_{\mathsf{CS}}} A$. By Deduction Theorem 2.1,

$$\mathsf{L}_{\mathsf{CS}} \vdash B_1 \to (B_2 \to \cdots \to (B_n \to A) \cdots) \ .$$

By Constructive Necessitation 2.3, there is a ground term s' such that

$$\mathsf{L}_{\mathsf{CS}} \vdash s'{:}\Big(B_1 \to (B_2 \to \cdots \to (B_n \to A) \cdots)\Big) \ .$$

By repeated applications of A2 and modus ponens, for $t := s' \cdot s_1 \cdots s_n$,

$$s_1{:}B_1, \ldots, s_n{:}B_n \vdash_{\mathsf{L_{CS}}} t{:}A \ .$$

□

B Proof of Theorem 3.6

Theorem 3.6 Let $\mathsf{L_{CS}}$ be a justification logic and $F \in \mathcal{L}_j$. If $\mathsf{L_{CS}} \vdash F$, then $* \Vdash F$ for all basic modular $\mathsf{L_{CS}}$-models $*$.

Proof. As usual, the proof is by induction on the length of the derivation of F. Let $*$ be a basic modular $\mathsf{L_{CS}}$-model. It is obvious that all instances of propositional axioms hold under $*$ and the rule (MP) is respected by the semantics. Soundness of the axioms (A2), (A3), (j4), and (j5), as well as that of the rules (AN) and (AN!), immediately follows from the definition of a basic evaluation.

It is also easy to see that all instances of (jd) hold under all consistent basic evaluations, all instances of (jt) hold under all factive basic evaluations, and all instances of (jb) hold under all Brouwerian basic evaluations. The argument for (jt) is as follows: if $* \Vdash t{:}F$, then $F \in t^*$, so $* \Vdash F$ by factivity of $*$. □

C Proof of Theorem 3.8

Theorem 3.8 Let $\mathsf{L_{CS}}$ be a justification logic and $F \in \mathcal{L}_j$. If $* \Vdash F$ for all basic modular $\mathsf{L_{CS}}$-models $*$, then $\mathsf{L_{CS}} \vdash F$.

Proof. Assume that $\mathsf{L_{CS}} \nvdash F$. Then $\{\neg F\}$ is $\mathsf{L_{CS}}$-consistent and, hence, is contained in some maximal $\mathsf{L_{CS}}$-consistent set Φ. For this Φ, any $p \in \mathrm{Prop}$, and any $t \in \mathrm{Tm}$, we define

$$p^* := \begin{cases} 1 & p \in \Phi \\ 0 & p \notin \Phi \end{cases} \qquad \text{and} \qquad t^* := \{F \in \mathcal{L}_j \mid t{:}F \in \Phi\} \ . \tag{C.1}$$

It is easy to show that $*$ is a basic $\mathsf{L_{CS}}$-evaluation. By way of example, we will show Conditions (i) and (v); the rest is similar.

Condition (i) of Def. 3.2. Suppose $A \in s^* \cdot t^*$. Then there is $B \in \mathcal{L}_j$ such that $B \to A \in s^*$ and $B \in t^*$. By (C.1), $s{:}(B \to A) \in \Phi$ and $t{:}B \in \Phi$. By the maximal $\mathsf{L_{CS}}$-consistency of Φ, also $(s \cdot t){:}A \in \Phi$. Thus, $A \in (s \cdot t)^*$ by (C.1).

Condition (v) of Def. 3.2. Suppose (j5) $\in \mathsf{L}$ and $F \notin t^*$. By (C.1), we have $t{:}F \notin \Phi$. By the maximal $\mathsf{L_{CS}}$-consistency of Φ, then $\neg t{:}F \in \Phi$. Further, $?t{:}\neg t{:}F \in \Phi$ because (j5) $\in \mathsf{L}$. So $\neg t{:}F \in (?t)^*$ by (C.1).

We now show the so-called Truth Lemma: for all $D \in \mathcal{L}_j$,

$$D \in \Phi \qquad \Longleftrightarrow \qquad * \Vdash D \ . \tag{C.2}$$

We establish (C.2) by induction on the structure of D:

(i) $D = p \in \mathrm{Prop}$. Then $p \in \Phi \quad \Leftrightarrow \quad p^* = 1 \quad \Leftrightarrow \quad * \Vdash p$.

(ii) The cases when $D = \neg A$ and $D = A \to B$ are standard.

(iii) $D = t{:}A$. Then $t{:}A \in \Phi \quad \Leftrightarrow \quad A \in t^* \quad \Leftrightarrow \quad * \Vdash t{:}A$.

To show that $*$ is a basic modular L_{CS}-model, we need to check Conditions (i)–(iii) of Def. 3.5:

(i) Assume towards a contradiction that $\bot \in t^*$. Then $t{:}\bot \in \Phi$ by (C.1). Since (jd) $\in \mathsf{L}$, we would have $\bot \in \Phi$, which contradicts the consistency of Φ.

(ii) Suppose $F \in t^*$. Then $t{:}F \in \Phi$ by (C.1). Since (jt) $\in \mathsf{L}$, we have $F \in \Phi$. Now $* \Vdash F$ follows by (C.2).

(iii) Suppose $* \nVdash F$. Then $F \notin \Phi$ by (C.2). By the maximal L_{CS}-consistency of Φ, we have $\neg F \in \Phi$. Since (jb) $\in \mathsf{L}$, also $\bar{?}t{:}\neg t{:}F \in \Phi$, and $\neg t{:}F \in (\bar{?}t)^*$ follows by (C.1).

Since $\neg F \in \Phi$ by the construction of Φ, we find $* \Vdash \neg F$ by (C.2). Thus, $* \nVdash F$ for the constructed basic modular L_{CS}-model $*$. Completeness of L_{CS} follows by contraposition. □

D Proof of Theorem 4.8

Theorem 4.8 Let L_{CS} be a justification logic such that either (jt) $\in \mathsf{L}$ or (jd) $\notin \mathsf{L}$ and let $F \in \mathcal{L}_j$. Then $\mathsf{L}_{\mathsf{CS}} \vdash F$ if and only if $\mathcal{M} \Vdash F$ for all modular L_{CS}-models $\mathcal{M}$.

Proof. It is sufficient to prove that any formula refutable by a basic modular model can be refuted at a world in a modular model and vice versa.
Soundness. Since R plays no role in the definition of truth in modular models, for any modular model $\mathcal{M} = (W, R, *)$ and for any world $w \in W$, the basic L_{CS}-evaluation $*_w$ satisfies exactly the same formulas as the world w of $\mathcal{M}$ does, i.e.,

$$\mathcal{M}, w \Vdash F \qquad \Longleftrightarrow \qquad *_w \Vdash F \ . \tag{D.1}$$

In particular, $*_w$ is factive if $\mathcal{M}$ is and Brouwerian if $\mathcal{M}$ is. Thus, it follows from Lemma 4.5 and from Condition (vii) for modular models that $*_w$ is a basic modular L_{CS}-model, which refutes all formulas refuted at the world w of $\mathcal{M}$.
Completeness. For the opposite direction, let $*$ be a basic modular L_{CS}-model. We define an L_{CS}-quasimodel $\mathcal{M} := (\{1\}, R, \star)$ with $\star(1, t) := *(t)$. Since $* = \star_1$, by (D.1) we have $* \Vdash F$ iff $\mathcal{M}, 1 \Vdash F$. Thus, $\mathcal{M}$ is Brouwerian if $*$ is. To show that $\mathcal{M}$ is a modular L_{CS}-model, which refutes all formulas refuted by $*$, it remains to make sure all the restrictions on R and the condition JYB are met. The choice of R depends on the logic.

If (jt) $\in \mathsf{L}$, set $R := \{(1, 1)\}$. It is reflexive, symmetric, Euclidean, and transitive, so all restrictions on R are met. Further, if $F \in t_1^\star$, then $F \in t^*$ by the definition of $\star$. Since $*$ is factive, $* \Vdash F$. Thus, $\mathcal{M}, 1 \Vdash F$ by (D.1) and, consequently, $F \in \Box_1$. Thus, $\mathcal{M}$ meets JYB and is a modular L_{CS}-model.

If, on the contrary, (jt) $\notin \mathsf{L}$ and (jd) $\notin \mathsf{L}$, set $R := \varnothing$. It is symmetric, Euclidean, and transitive, and JYB in this case is met trivially, since $\Box_1 = \mathcal{L}_j$. □

E Proof of Theorem 4.10

Theorem 4.10 Let L_{CS} be a logic with (jt) $\notin \mathsf{L}$ and (jd) $\in \mathsf{L}$. Then L_{CS} is sound with respect to modular L_{CS}-models. If CS is axiomatically appropriate for L, then L_{CS} is also complete.

Proof. *Soundness.* Since $*_w$ from the soundness proof in Theorem 4.8 is consistent if $\mathcal{M}$ is and $\mathcal{M}$ is consistent by Lemma 4.6, the same soundness argument applies.
Completeness. We reuse the construction of a basic modular model $*_\Phi$ based on an L_{CS}-consistent set Φ from the proof of Theorem 3.8. From all such $*_\Phi$ we create the canonical modular L_{CS}-model and use (D.1) to transfer the properties of $*_\Phi$ proved for Theorem 3.8. Thus, we define $\mathcal{M}_c := (W, R, *)$, where $W := \{\Phi \subseteq \mathcal{L}_j \mid \Phi \text{ is a maximal } \mathsf{L}_{\mathsf{CS}}\text{-consistent set}\}$ and $*_\Phi$ for each $\Phi \in W$ is defined by (C.1). Finally, we set $R(\Phi, \Psi)$ iff $\Phi^\sharp \subseteq \Psi$, where $\Phi^\sharp := \{F \in \mathcal{L}_j \mid t{:}F \in \Phi\}$. If (jb) $\in \mathsf{L}$, then $\mathcal{M}_c$ is Brouwerian by (D.1) because each $*_\Phi$ is. To show that $\mathcal{M}_c$ is a modular L_{CS}-model, it remains to establish the appropriate properties of R and the condition JYB.

We start with the latter. Let $F \in t^*_\Phi$. Then $t{:}F \in \Phi$ by the definition of $*_\Phi$ and $F \in \Psi$ whenever $R(\Phi, \Psi)$ by the definition of R. By (C.2), $*_\Psi \Vdash F$, and $\mathcal{M}_c, \Psi \Vdash F$ by (D.1). Since Ψ is chosen arbitrarily, $F \in \Box_\Phi$.

In order to prove seriality of R, we have to use the axiomatical appropriateness of CS. It is sufficient to show that $\Phi^\sharp$ is consistent for any $\Phi \in W$. Then $\Phi^\sharp$ can be extended to a maximal consistent $\Psi \supseteq \Phi^\sharp$, which is accessible from Φ by the definition of R. Assume towards a contradiction that $\Phi^\sharp$ were not consistent. Then there would be $s_1{:}F_1, \dots s_n{:}F_n \in \Phi$ such that $F_1, \dots, F_n \vdash_{\mathsf{L}_{\mathsf{CS}}} \bot$. Since CS is axiomatically appropriate, by Corollary 2.4, there would be a term t such that $s_1{:}F_1, \dots s_n{:}F_n \vdash_{\mathsf{L}_{\mathsf{CS}}} t{:}\bot$. Hence, by (jd) and (MP), $s_1{:}F_1, \dots s_n{:}F_n \vdash_{\mathsf{L}_{\mathsf{CS}}} \bot$, which contradicts the consistency of Φ.

The argument for the other properties of R follows the same pattern. We only show the symmetry case. Let (jb) $\in \mathsf{L}$ and $R(\Phi, \Psi)$. To show that $R(\Psi, \Phi)$, assume towards a contradiction that $t{:}F \in \Psi$ but $F \notin \Phi$. Then $\neg F \in \Phi$ by the maximal L_{CS}-consistency of Φ and $\bar{?}t{:}\neg t{:}F \in \Phi$ for the same reason. Hence, $\neg t{:}F \in \Psi$ by the definition of R, which contradicts the consistency of Ψ. □

F Proof of Lemma 4.14

Lemma 4.14 Every modular L_{CS}-model $\mathcal{M} = (W, R, *)$ is also an F-model for L_{CS} such that for all $w \in W$ and $F \in \mathcal{L}_j$, we have $\mathcal{M}, w \Vdash_{\mathsf{F}} F$ if and only if $\mathcal{M}, w \Vdash F$.

Proof. We prove (7) for all $w \in W$ by induction on the structure of $F \in \mathcal{L}_j$. The only non-trivial case is when $F = t{:}G$, and the only non-trivial direction of this case is from right to left. If $\mathcal{M}, w \Vdash t{:}G$, then $G \in t^*_w$. Hence, $G \in \Box_w$ by JYB. By induction hypothesis, $G \in \Box^{\mathsf{F}}_w$. Thus, $\mathcal{M}, w \Vdash_{\mathsf{F}} t{:}G$.

It remains to show (4) for logics with (j4) $\in \mathsf{L}$ and (5) for those with (j5) $\in \mathsf{L}$ or (jb) $\in \mathsf{L}$. Since the former is exactly the statement of Lemma 4.7, we prove the latter. If $F \in t^*_w$, then $F \in \Box_w$ by JYB. By (7), $\Box_w = \Box^{\mathsf{F}}_w$ so that $F \in \Box^{\mathsf{F}}_w$.□

G Proof of Lemma 4.16

Lemma 4.16 Let $\mathcal{M}_{\mathsf{F}} = (W, R, *_{\mathsf{F}})$ be an F-model for a logic L_{CS}. Then $\mathcal{M} := (W, R, *)$ with

$$t^*_w := \{F \in \mathcal{L}_j \mid F \in t^{*_{\mathsf{F}}}_w \text{ and } F \in \Box^{\mathsf{F}}_w\} = \{F \in \mathcal{L}_j \mid \mathcal{M}_{\mathsf{F}}, w \Vdash_{\mathsf{F}} t{:}F\}$$

is a modular L_{CS}-model such that for all $w \in W$ and all $F \in \mathcal{L}_j$, we have $\mathcal{M}, w \Vdash F$ if and only if $\mathcal{M}_{\mathsf{F}}, w \Vdash_{\mathsf{F}} F$.

Proof. Although it is not yet proved that $\mathcal{M}$ is a quasimodel, we can still apply $\Vdash$ to it. Thus, we start by proving (9) for all $w \in W$ by induction on the structure of F. Again, the only non-trivial case is when $F = t{:}G$, and the statement for it follows immediately from (8).

We now use (9) to show that $\mathcal{M}$ is a modular L_{CS}-model. The conditions on R for F-models and modular models are identical, so we need to verify that $*_w$ is a basic L_{CS}-evaluation for each $w \in W$, that $\mathcal{M}$ is Brouwerian if (jb) $\in \mathsf{L}$, and that JYB holds. We check JYB first. Suppose $F \in t^*_w$. Then $F \in \Box^{\mathsf{F}}_w$ by (8). Hence, $F \in \Box_w$ by (9).

Suppose (jb) $\in \mathsf{L}$ and $\mathcal{M}, w \nVdash F$. Then $\mathcal{M}, w \nVdash_{\mathsf{F}} F$ by (9) so that $\neg t{:}F \in (\bar{?}t)^{*_{\mathsf{F}}}_w$ for any $t \in \mathrm{Tm}$ since $\mathcal{M}_{\mathsf{F}}$ is Brouwerian. Since (jb) $\in \mathsf{L}$, we can use JYB for $\mathcal{M}_{\mathsf{F}}$, which yields $\neg t{:}F \in \Box^{\mathsf{F}}_w$. Thus, $\neg t{:}F \in (\bar{?}t)^*_w$ by (8).

It remains to check that $*_w$ is a basic L_{CS}-evaluation.

(i) Suppose $F \in s^*_w \cdot t^*_w$. Then there must exist a formula $G \in \mathcal{L}_j$ such that $G \to F \in s^*_w$ and $G \in t^*_w$. By (8), $G \to F \in s^{*_{\mathsf{F}}}_w$ and $G \in t^{*_{\mathsf{F}}}_w$. Thus, (a) $F \in s^{*_{\mathsf{F}}}_w \cdot t^{*_{\mathsf{F}}}_w \subseteq (s \cdot t)^{*_{\mathsf{F}}}_w$ since $*_{\mathsf{F}}(w)$ is a basic L_{CS}-evaluation. Also by (8), $G \to F \in \Box^{\mathsf{F}}_w$ and $G \in \Box^{\mathsf{F}}_w$. In other words, $\mathcal{M}_{\mathsf{F}}, v \Vdash_{\mathsf{F}} G \to F$ and $\mathcal{M}_{\mathsf{F}}, v \Vdash_{\mathsf{F}} G$ whenever $R(w, v)$. Clearly, $\mathcal{M}_{\mathsf{F}}, v \Vdash_{\mathsf{F}} F$ whenever $R(w, v)$ so that (b) $F \in \Box^{\mathsf{F}}_w$. From (a) and (b), $F \in (s \cdot t)^*_w$ follows by (8).

(ii) The proof that $s^*_w \cup t^*_w \subseteq (s + t)^*_w$ is similar.

(iii) Suppose $t{:}F$ is a conclusion of the (AN) or the (AN!) rule, whichever is present in L_{CS}. Then $F \in t^{*_{\mathsf{F}}}_w$ because $*_{\mathsf{F}}(w)$ is a basic L_{CS}-evaluation. It is now sufficient to show that $F \in \Box^{\mathsf{F}}_w$, which follows from the soundness of F-models. In [4], the soundness is established for most of the logics except for those with (jb) $\in \mathsf{L}$, for which F-models have not been defined. Thus, we need to show the soundness of (jb) in F-models for such L_{CS}. Suppose (jb) $\in \mathsf{L}$ and $\mathcal{M}'_{\mathsf{F}}, w' \nVdash_{\mathsf{F}} G$ for an arbitrary F-model $\mathcal{M}'_{\mathsf{F}} = (W', R', *_{\mathsf{F}}')$ for L_{CS}. Then $\neg s{:}G \in (\bar{?}s)^{*_{\mathsf{F}}'}_{w'}$ for any $s \in \mathrm{Tm}$ because $\mathcal{M}'_{\mathsf{F}}$ is Brouwerian. Also, $\neg s{:}G \in \Box^{\mathsf{F}}_{w'}$ by JYB. Thus, $\mathcal{M}'_{\mathsf{F}}, w' \Vdash_{\mathsf{F}} \bar{?}s{:}\neg s{:}G$.

(iv) Suppose (j4) $\in \mathsf{L}$ and $s{:}F \in s{:}(s^*_w)$, i.e., $F \in s^*_w$. By (8), this implies $\mathcal{M}_{\mathsf{F}}, w \Vdash_{\mathsf{F}} s{:}F$. By soundness of (j4) in monotone F-models that satisfy (iv), we get $\mathcal{M}_{\mathsf{F}}, w \Vdash_{\mathsf{F}} !s{:}s{:}F$. Now $s{:}F \in (!s)^*_w$ follows from (8).

(v) Suppose (j5) $\in \mathsf{L}$ and $F \notin t^*_w$. By (8), this implies $\mathcal{M}_{\mathsf{F}}, w \nVdash_{\mathsf{F}} t{:}F$. By soundness of (j5) in F-models that satisfy JYB and (v), we get that $\mathcal{M}_{\mathsf{F}}, w \Vdash_{\mathsf{F}} ?t{:}\neg t{:}F$. Now $\neg t{:}F \in (?t)^*_w$ follows from (8).

$\Box$

H Proof of Theorem 5.2

Theorem 5.2 Let $\mathsf{L}_{\mathsf{CS}}^{\Box}$ be a logic of justifications and belief. For all formulas $F \in \mathcal{L}_{\Box}$, if $\mathcal{M} \Vdash F$ for all modular $\mathsf{L}_{\mathsf{CS}}^{\Box}$-models $\mathcal{M}$, then $\mathsf{L}_{\mathsf{CS}}^{\Box} \vdash F$.

Proof. We define the canonical model $\mathcal{M}_c^{\Box} = (W, R, *)$ as follows:

$$W := \{\Phi \subseteq \mathcal{L}_{\Box} \mid \Phi \text{ is a maximal } \mathsf{L}_{\mathsf{CS}}^{\Box}\text{-consistent set}\}$$

and $*_{\Phi}$ for each $\Phi \in W$ is defined by (C.1) except that t_{Φ}^* consists of $\mathcal{L}_{\Box}$-formulas instead of $\mathcal{L}_j$-formulas. Finally, we set $R(\Phi, \Psi)$ iff $\Phi^{\Box} \subseteq \Psi$, where $\Phi^{\Box} := \{F \in \mathcal{L}_{\Box} \mid \Box F \in \Phi\}$.

As usual, the Truth Lemma is established by induction on D: for all formulas $D \in \mathcal{L}_{\Box}$ and all maximal $\mathsf{L}_{\mathsf{CS}}^{\Box}$-consistent sets Φ,

$$D \in \Phi \qquad \Longleftrightarrow \qquad \mathcal{M}_c^{\Box}, \Phi \Vdash D \ . \tag{H.1}$$

The cases for propositions and Boolean connectives are straightforward. The case for $D = t{:}F$ does not involve R and is essentially the same as in the proof of Theorem 3.8. The case for $D = \Box F$ is proved by the standard modal argument because R is defined as in the modal canonical model rather than as in Theorem 3.8.

It remains to show that $\mathcal{M}_c^{\Box}$ is a modular $\mathsf{L}_{\mathsf{CS}}^{\Box}$-model. The proof that $*_{\Phi}$ is a basic $\mathsf{L}_{\mathsf{CS}}^{\Box}$-evaluation is almost literally the same as in Theorem 3.8. The conditions on R are established by the standard modal argument. If (jb) $\in \mathsf{L}$, the proof that $\mathcal{M}_c^{\Box}$ is Brouwerian follows the relevant part of the proof of Theorem 3.8, only referring to the Truth Lemma (H.1) instead of (C.2).

Thus, the proof of JYB is the only thing that needs to be redone due to the change in the definition of R, compared to Theorem 4.10. Suppose $F \in t_{\Phi}^*$. Then $t{:}F \in \Phi$ by (C.1). Using the axiom instance $t{:}F \to \Box F$ and the maximal $\mathsf{L}_{\mathsf{CS}}^{\Box}$-consistency of Φ, we conclude that $\Box F \in \Phi$. Hence, $F \in \Psi$ whenever $R(\Phi, \Psi)$ by the definition of R. Thus, $\mathcal{M}, \Psi \Vdash F$ whenever $R(\Phi, \Psi)$ by (H.1), i.e., $F \in \Box_{\Phi}$. □

I Proof of Theorem 7.2

Theorem 7.2 Let L_{CS} be a justification logic with the axiomatically appropriate CS. Then L_{CS} is sound and complete with respect to fully explanatory modular L_{CS}-models.

Proof. Given Theorems 4.8 and 4.10 and Remark 4.11, it is sufficient to show that the canonical model $\mathcal{M}_c = (W, R, *)$ forL_{CS} constructed in the proof of Theorem 4.10 is fully explanatory.

Assume towards a contradiction that $F \in \Box_{\Phi}$ for some $F \in \mathcal{L}_j$ and $\Phi \in W$ but $F \notin t_{\Phi}^*$ for any $t \in \mathrm{Tm}$. Then $\Phi^{\sharp} \cup \{\neg F\}$ would be L_{CS}-consistent.

Indeed, if $\Phi^{\sharp} \cup \{\neg F\}$ were L_{CS}-inconsistent, then $G_1, \dots, G_n, \neg F \vdash_{\mathsf{L}_{\mathsf{CS}}} \bot$ for some $G_1, \dots, G_n \in \Phi^{\sharp}$. Equivalently, there would be terms $s_1, \dots, s_n$ such that $s_i{:}G_i \in \Phi$ for $i = 1, \dots, n$ and $G_1, \dots, G_n \vdash_{\mathsf{L}_{\mathsf{CS}}} F$. By Corollary 2.4, given the axiomatic appropriateness of CS, there would be a term t such that $s_1{:}G_1, \ldots, s_n{:}G_n \vdash_{\mathsf{L_{CS}}} t{:}F$. By Deduction Theorem 2.1,

$$\mathsf{L_{CS}} \vdash s_1{:}G_1 \to (s_2{:}G_2 \to \cdots \to (s_n{:}G_n \to t{:}F)\cdots)$$

so that $t{:}F \in \Phi$ by the maximal $\mathsf{L_{CS}}$-consistency of Φ and $F \in t^*_\Phi$ by (C.1), contradicting our assumption.

Hence, the set $\Phi^\sharp \cup \{\neg F\}$ would be $\mathsf{L_{CS}}$-consistent and could be extended to a maximal $\mathsf{L_{CS}}$-consistent set Ψ. Clearly, $R(\Phi, \Psi)$ by the definition of R and $\mathcal{M}_c, \Psi \Vdash \neg F$ by (C.2) and (D.1). Thus, $\mathcal{M}_c, \Psi \nVdash F$, which contradicts our assumption that $F \in \Box_\Phi$. □

Interpolation and Beth Definability over the Minimal Logic

Larisa Maksimova [1]

Sobolev Institute of Mathematics
Siberian Branch of Russian Academy of Sciences
Novosibirsk 630090, Russia

Abstract

Extensions of the Johansson minimal logic J are investigated. It is proved that the weak interpolation property WIP is decidable over J. Well-composed logics with the Graig interpolation property CIP, restricted interpolation property IPR and projective Beth property PBP are fully described. It is proved that there are only finitely many well-composed logics with CIP, IPR or PBP; for any well-composed logic PBP is equivalent to IPR, and all the properties CIP, IPR and PBP are decidable on the class of well-composed logics..

Keywords: Interpolation, Beth property, minimal logic.

1 Superintuitionistic logics and J-logics

In this paper we consider extensions of the Johansson minimal logic J; this family extends the class of superintuitionistic (s.i.) logics. The main variants of the interpolation property are studied. In [4] we have proved that the weak interpolation property is decidable over J. There are only finitely many superintuitionistic logics with CIP, IPR or PBP, all of them are fully described [1,3], and CIP, IPR and PBP are decidable on the class of s.i. logics. Here we extend these results to the class of well-composed J-logics.

The language of J contains $\&, \vee, \to, \bot$ as primitive; negation is defined by $\neg A = A \to \bot$. The logic J can be given by the calculus, which has the same axiom schemes as the positive intuitionistic calculus Int^+, and the only rule of inference is modus ponens. By a J-*logic* we mean an arbitrary set of formulas containing all the axioms of J and closed under modus ponens and substitution rules. We denote

$\mathrm{Int} = \mathrm{J} + (\bot \to A)$, $\mathrm{Neg} = \mathrm{J} + \bot$, $\mathrm{Gl} = \mathrm{J} + (A \vee \neg A)$,
$\mathrm{Cl} = \mathrm{Int} + (A \vee \neg A)$, $\mathrm{JX} = \mathrm{J} + (\bot \to A) \vee (A \to \bot)$.

[1] Email: LMAKSI@math.nsc.ru
Supported by Russian Foundation for Basic Research, grant 12-01-00168a.

A J-logic is *superintuitionistic* if it contains the intuitionistic logic Int, and *negative* if contains Neg. A J-logic is *well-composed* if it contains JX. For a J-logic L, the family of J-logics containing L is denoted by $E(L)$.

If $\mathbf{p}$ is a list of variables, let $A(\mathbf{p})$ denote a formula whose all variables are in $\mathbf{p}$, and $\mathcal{F}(\mathbf{p})$ the set of all such formulas.

Let L be a logic. *The Craig interpolation property CIP, the restricted interpolation property IPR* and *the weak interpolation property WIP* are defined as follows (where the lists $\mathbf{p}, \mathbf{q}, \mathbf{r}$ are disjoint):

CIP. If $\vdash_L A(\mathbf{p},\mathbf{q}) \rightarrow B(\mathbf{p},\mathbf{r})$, then there is a formula $C(\mathbf{p})$ such that $\vdash_L A(\mathbf{p},\mathbf{q}) \rightarrow C(\mathbf{p})$ and $\vdash_L C\mathbf{p}) \rightarrow B(\mathbf{p},\mathbf{r})$.

IPR. If $A(\mathbf{p},\mathbf{q}), B(\mathbf{p},\mathbf{r}) \vdash_L C(\mathbf{p})$, then there exists a formula $A'(\mathbf{p})$ such that $A(\mathbf{p},\mathbf{q}) \vdash_L A'(\mathbf{p})$ and $A'(\mathbf{p}), B(\mathbf{p},\mathbf{r}) \vdash_L C(\mathbf{p})$.

WIP. If $A(\mathbf{p},\mathbf{q}), B(\mathbf{p},\mathbf{r}) \vdash_L \bot$, then there exists a formula $A'(\mathbf{p})$ such that $A(\mathbf{p},\mathbf{q}) \vdash_L A'(\mathbf{p})$ and $A'(\mathbf{p}), B(\mathbf{p},\mathbf{r}) \vdash_L \bot$.

Suppose that $\mathbf{p}$, $\mathbf{q}$, $\mathbf{q}'$ are disjoint lists of variables that do not contain x and y, $\mathbf{q}$ and $\mathbf{q}'$ are of the same length, and $A(\mathbf{p},\mathbf{q},x)$ is a formula. We define *the projective Beth property*:

PBP. If $A(\mathbf{p},\mathbf{q},x), A(\mathbf{p},\mathbf{q}',y) \vdash_L x \leftrightarrow y$, then $A(\mathbf{p},\mathbf{q},x) \vdash_L x \leftrightarrow B(\mathbf{p})$ for some $B(\mathbf{p})$.

The weaker *Beth property BP* arises from PBP by omitting $\mathbf{q}$ and $\mathbf{q}'$.

All J-logics satisfy BP, and for these logics the following hold:

- CIP $\Rightarrow$ PBP $\Rightarrow$ IPR $\Rightarrow$ WIP, PBP $\not\Rightarrow$ CIP, WIP $\not\Rightarrow$ IPR.

It is proved in [4] that WIP is decidable over J, i.e. there is an algorithm which, given a finite set Ax of axiom schemes, decides if the logic J+Ax has WIP. The families of J-logics with WIP and of J-logics without WIP have the continuum cardinality.

The logics J, Int, Neg, Gl, Cl and JX possess CIP and hence all other above-mentioned properties. It is known [3] that

- IPR $\Leftrightarrow$ PBP over Int and Neg.

It is known that there are only finitely many s.i. and negative logics with CIP, IPR and PBP [1,3]. Here we extend this result to all well-composed logics. Also we prove that IPR is equivalent to PBP in any well-composed logic, and CIP, IPR and PBP are decidable over JX.

2 Interpolation and amalgamation

The considered properties have natural algebraic equivalents. There is a duality between J-logics and varieties of J-algebras [6].

Algebraic semantics for J-logics is built via J-*algebras*, i.e. algebras $\mathbf{A} = < A; \&, \vee, \rightarrow, \bot, \top >$ such that A is a lattice w.r.t. $\&, \vee$ with the greatest element $\top$, $\bot$ is an arbitrary element of A, and

$z \leq x \rightarrow y \iff z\&x \leq y.$

A J-algebra $\mathbf{A}$ is a *Heyting algebra* if $\bot$ is the least element of A, and *a negative algebra* if $\bot$ is the greatest element of A; the algebra is *well-composed* if every its element is comparable with $\bot$. For any well-composed J-algebra $\mathbf{A}$, the set $\mathbf{A}^l = \{x|\ x \leq \bot\}$ forms a negative algebra, and the set $\mathbf{A}^l = \{x|\ x \geq \bot\}$ forms a Heyting algebra. If $\mathbf{B}$ is a negative algebra and $\mathbf{C}$ is a Heyting algebra, we denote by $\mathbf{B} \uparrow \mathbf{C}$ a well-composed algebra $\mathbf{A}$ such that $\mathbf{A}^l$ is isomorphic to $\mathbf{B}$ and $\mathbf{A}^u$ to $\mathbf{C}$. For a negative algebra $\mathbf{B}$, we denote by $\mathbf{B}^\Lambda$ a J-algebra arisen from $\mathbf{B}$ by adding a new greatest element $\top$.

A J-algebra $\mathbf{A}$ is *finitely indecomposable* if for all $x, y \in \mathbf{A}$:
$x \vee y = \top \Leftrightarrow (x = \top \text{ or } y = \top)$.

If A is a formula, $\mathbf{A}$ a J-algebra, then A is *valid in* $\mathbf{A}$ (in symbols, $\mathbf{A} \models A$) if the identity $A = \top$ is valid in $\mathbf{A}$. We write $\mathbf{A} \models L$ instead of $(\forall A \in L)(\mathbf{A} \models A)$. Let $V(L) = \{\mathbf{A}|\mathbf{A} \models L\}$. Each J-logic L is characterized by the variety $V(L)$.

We recall the definitions. A class V has *Amalgamation Property* if it satisfies

AP: For each $\mathbf{A}, \mathbf{B}, \mathbf{C} \in V$ such that $\mathbf{A}$ is a common subalgebra of $\mathbf{B}$ and $\mathbf{C}$, there exist an algebra $\mathbf{D}$ in V and monomorphisms $\delta : \mathbf{B} \to \mathbf{D}$ and $\epsilon : \mathbf{C} \to \mathbf{D}$ such that $\delta(x) = \epsilon(x)$ for all $x \in \mathbf{A}$.

Super-Amalgamation Property (SAP) is AP with extra conditions:

$$\delta(x) \leq \epsilon(y) \Leftrightarrow (\exists z \in \mathbf{A})(x \leq z \text{ and } z \leq y),$$

$$\delta(x) \geq \epsilon(y) \Leftrightarrow (\exists z \in \mathbf{A})(x \geq z \text{ and } z \geq y).$$

Restricted Amalgamation Property (RAP) and *Weak Amalgamation Property (WAPJ)* are defined as follows:

RAP: for any $\mathbf{A}, \mathbf{B}, \mathbf{C} \in V$ such that $\mathbf{A}$ is a common subalgebra of $\mathbf{B}$ and $\mathbf{C}$, there exist an algebra $\mathbf{D}$ in V and homomorphisms $g : \mathbf{B} \to \mathbf{D}$ and $h : \mathbf{C} \to \mathbf{D}$ such that $g(x) = h(x)$ for all $x \in \mathbf{A}$ and the restriction of g onto $\mathbf{A}$ is a monomorphism.

WAPJ: For each $\mathbf{A}, \mathbf{B}, \mathbf{C} \in V$ such that $\mathbf{A}$ is a common subalgebra of $\mathbf{B}$ and $\mathbf{C}$, there exist an algebra $\mathbf{D}$ in V and homomorphisms $\delta : \mathbf{B} \to \mathbf{D}$ and $\epsilon : \mathbf{C} \to \mathbf{D}$ such that $\delta(x) = \epsilon(x)$ for all $x \in \mathbf{A}$, and $\bot \neq \top$ in $\mathbf{D}$ whenever $\bot \neq \top$ in $\mathbf{A}$.

A class V has *Strong Epimorphisms Surjectivity* if it satisfies

SES: For each $\mathbf{A}, \mathbf{B}$ in V, for every monomorphism $\alpha : \mathbf{A} \to \mathbf{B}$ and for every $x \in \mathbf{B} - \alpha(\mathbf{A})$ there exist $\mathbf{C} \in V$ and homomorphisms $\beta : \mathbf{B} \to \mathbf{C}$, $\gamma : \mathbf{B} \to \mathbf{C}$ such that $\beta\alpha = \gamma\alpha$ and $\beta(x) \neq \gamma(x)$.

Theorem 2.1 ([2]) *For any* J*-logic* L*:*
(1) L *has CIP iff* $V(L)$ *has SAP iff* $V(L)$ *has AP,*
(2) L *has IPR iff* $V(L)$ *has RAP, (3)* L *has WIP iff* $V(L)$ *has WAPJ,*
(4) L *has PBP iff* $V(L)$ *has SES.*

In varieties of J-algebras: SAP $\Longleftrightarrow$ AP $\Rightarrow$ SES $\Rightarrow$ RAP $\Rightarrow$ WAPJ.

3 Weak interpolation and negative equivalence

For $L_1 \in E(\text{Neg})$, $L_2 \in E(\text{Int})$ we denote by $L_1 \uparrow L_2$ a logic characterized by all algebras of the form $\mathbf{A} \uparrow \mathbf{B}$, where $\mathbf{A} \models L_1$, $\mathbf{B} \models L_2$; a logic characterized by all algebras $\mathbf{A} \uparrow \mathbf{B}$, where $\mathbf{A}$ is a finitely decomposable algebra in $V(L_1)$

and $\mathbf{B} \in V(L_2)$, is denoted by $L_1 \Uparrow L_2$. Say that a J-logic is *primary* if it is of the form $L_1 \uparrow L_2$ or $L_1 \Uparrow L_2$.

In [2] an axiomatization was found for logics $L_1 \uparrow L_2$ and $L_1 \Uparrow L_2$, where L_1 is a negative and L_2 an s.i. logic.

All s.i. and negative logics have WIP. On the contrary, there are only finitely many s.i. and negative logics with CIP, IPR and PBP [1,2,3]. We give the list of all negative logics with CIP:

$$\text{Neg},\ \text{NC} = \text{Neg} + (p \to q) \vee (q \to p),\ \text{NE} = \text{Neg} + p \vee (p \to q),\ \text{For} = \text{Neg} + p.$$

It is proved in [4] that WIP is decidable over J, i.e. there is an algorithm which, given a finite set Ax of axiom schemes, decides if the logic J+Ax has WIP. A crucial role in the description of J-logics with WIP [4] belongs to the following list SL of eight *etalon logics*:

$$\{\text{For}, \text{Cl}, (\text{NE} \uparrow \text{Cl}), (\text{NC} \uparrow \text{Cl}), (\text{Neg} \uparrow \text{Cl}), (\text{NE} \Uparrow \text{Cl}), (\text{NC} \Uparrow \text{Cl}), (\text{Neg} \Uparrow \text{Cl})\}.$$

We say that a J-algebra is *central* if $\bot \neq \top$ and $x \leq \bot$ for any $x \neq \top$. For a J-logic L define the *center* $\Lambda(L)$ as the class of all central algebras validating L. Let a *central companion* L_{cn} of L be a logic generated by $\Lambda(L)$.

All etalon logics are generated by their centers, finitely axiomatizable, and finitely approximable [4]. A center of an etalon logic is said to be an *etalon center*.

Proposition 3.1 *For each etalon logic L_0 there is an algorithm which, given a finite set Ax of axiom schemes, decides if the logic* J + Ax *is equal to L_0.*

Theorem 3.2 ([4]) *For any* J*-logic L the following are equivalent:*

(i) *L has WIP,*

(ii) *$\Lambda(L)$ has the amalgamation property.*

(iii) *L has an etalon center.*

Two J-logics L and L' are *negatively equivalent* [6] if for any formula A

$$L \vdash \neg A \iff L' \vdash \neg A.$$

Theorem 3.3 *Two* J*-logics are negatively equivalent iff they have the same center.*

Theorem 3.4 *A J-logic has WIP iff it is negatively equivalent to one of the etalon logics.*

Theorem 3.5 ([4]) *WIP is decidable over* J.

4 Interpolation in well-composed J-logics

For any J-logic L define the *negative* and *intutionistic companions*:

$$L_{neg} = L + \bot,\ L_{int} = L + (\bot \to A).$$

The following theorem describes all well-composed logics with CIP.

Theorem 4.1 ([5]) *Let L be a well-composed logic. Then L has CIP if and only if L_{neg} and L_{int} have CIP, and L is representable as $L = L_{neg} \cap L_1$, where L_1 is a primary logic with an etalon center.*

The following theorem gives a full description of well-composed logics with IPR and PBP.

Theorem 4.2 ([5]) *For any well-composed logic L the following are equivalent:*

(i) *L has IPR,*

(ii) *L has PBP,*

(iii) *the companions L_{neg} and L_{int} have IPR, the central companion L_{cn} is an etalon logic, and L is representable as*

$$L = L_{neg} \cap L_{cn} \cap L_1,$$

where L_1 is a primary logic with an etalon center.

Corollary 4.3 *There are only finitely many well-composed logics with IPR; all of them are finitely axiomatizable and finitely approximable.*

Theorem 4.4 ([5]) *CIP, IPR and PBP are decidable on the class of well-composed logics.*

The following problems are still open.

Problem 1. How many J-logics have CIP, IPR or PBP?
Problem 2. Are IPR and PBP equivalent over J?
Problem 3. Are CIP, IPR and/or PBP decidable over J?

References

[1] Gabbay, D.M. and L. Maksimova. *Interpolation and Definability: Modal and Intuitionistic Logics*, Oxford University Press, Oxford, 2005.

[2] Maksimova, L.L.. *Interpolation and definability in extensions of the minimal logic*, Algebra and Logic, 44 (2005), pp. 726-750.

[3] Maksimova, L. *Problem of restricted interpolation in superintuitionistic and some modal logics*, Logic Journal of IGPL, 18 (2010), pp. 367-380.

[4] L.L.Maksimova, L.L. *Decidability of the weak interpolation property over the minimal logic*, Algebra and Logic, 50, no. 2 (2011), pp. 152-188.

[5] Maksimova, L.L. *The projective Beth property in well-composed logics*, Algebra and Logic, *to appear.*

[6] Odintsov, S P. *Constructive negation and paraconsistency*, Dordrecht, Springer-Verlag. 2008.

Finite Satisfiability of Modal Logic over Horn Definable Classes of Frames

Jakub Michaliszyn and Emanuel Kieroński [1]

Institute of Computer Science
University of Wrocław
Wrocław, Poland

Abstract

Modal logic plays an important role in various areas of computer science, including verification and knowledge representation. In many practical applications it is natural to consider some restrictions of classes of admissible frames. Traditionally classes of frames are defined by modal axioms. However, many important classes of frames, e.g. the class of transitive frames or the class of Euclidean frames, can be defined in a more natural way by first-order formulas. In a recent paper it was proved that the satisfiability problem for modal logic over the class of frames defined by a universally quantified, first-order Horn formula is decidable. In this paper we show that also the finite satisfiability problem for modal logic over such classes is decidable.

Keywords: modal logic, decidability, finite satisfiability

1 Introduction

Modal logic was introduced by philosophers to study modes of truth. The idea was to extend propositional logic by some new constructions, of which two most important were $\Diamond\varphi$ and $\Box\varphi$, originally read as *φ is possible* and *φ is necessary*, respectively. A typical question was, given a set of axioms $\mathcal{A}$, corresponding usually to some intuitively acceptable aspects of truth, what is the logic defined by $\mathcal{A}$, i.e. which formulas are provable from $\mathcal{A}$ in a Hilbert-style system.

One of the most important steps in the history of modal logic was the invention of a formal semantics based on the notion of the so-called Kripke structures. Basically, a Kripke structure is a directed graph, called a *frame*, together with a valuation of propositional variables. Vertices of this graph are called *worlds*. For each world truth values of all propositional variables can be defined independently. In this semantics, $\Diamond\varphi$ means *the current world is connected to some world in which φ is true*; and $\Box\varphi$, equivalent to $\neg\Diamond\neg\varphi$, means *φ is true in all worlds to which the current world is connected.*

[1] This work was supported by Polish Ministry of Science and Higher Education research grant N N206 371339.

It appeared that there is a beautiful connection between syntactic and semantic approaches to modal logic [21]: logics defined by axioms can be often equivalently defined by restricting classes of frames. E.g., the axiom $\Diamond\Diamond P \Rightarrow \Diamond P$ (*if it is possible that P is possible, then P is possible*), is valid precisely in the class of transitive frames; the axiom $P \Rightarrow \Diamond P$ (*if P is true, then P is possible*) – in the class of reflexive frames, $P \Rightarrow \Box\Diamond P$ (*if P is true, then it is necessary that P is possible*) – in the class of symmetric frames, and the axiom $\Diamond P \Rightarrow \Box\Diamond P$ (*if P is possible, then it is necessary that P is possible*) – in the class of Euclidean frames.

Many important classes of frames, in particular all the classes we mentioned above, can be defined by simple first-order formulas. For a given first-order sentence Φ over the signature consisting of a single binary symbol R we define $\mathcal{K}_\Phi$ to be the set of those frames which satisfy Φ.

Decidability over various classes of frames can be shown by employing the so-called standard translation of modal logic to first-order logic. Indeed, the satisfiability of a modal formula φ in $\mathcal{K}_\Phi$ is equivalent to satisfiability of $st(\varphi) \wedge \Phi$, where $st(\varphi)$ is the standard translation of φ. In this way, we can show that (multi)modal logic is decidable over any class defined by two-variable logic [17], even extended with a linear order [18], counting quantifiers [19,5,20], one transitive relation [22], two equivalence relations [12,14], or equivalence closures of two distinguished binary relations [11]. The same holds for formulas of the guarded fragment [4], even if we allow for some restricted application of transitive relations [23,13], fixed-points [6,1] and transitive closures [15]. In many cases the decidability results hold also when only finite frames are considered. The complexity bounds obtained this way, however, are high — usually between ExpTime and 2NExpTime.

Clearly, some modal logics defined by a first-order formula are undecidable. A stronger result was presented in [7] — it was shown that there exists a universal first-order formula with equality such that the global satisfiability problem over the class of frames that satisfy this formula is undecidable. In [9], this result was improved — it was shown that equality is not necessary. The proof from [9] works also for local satisfiability. Finally, in [10] it was shown that even a very simple formula with three variables without equality may lead to undecidability.

The classical classes of frames we mentioned earlier, i.e. transitive, reflexive, symmetric and Euclidean are decidable. They can be defined by first-order sentences even if we further restrict the language to universal Horn formulas without equality, UHF. Universal Horn formulas were considered in [8], where a dichotomy result was proved, that the satisfiability problem for modal logic over the class of frames defined by a UHF formula (with an arbitrary number of variables) is either in NP or PSpace-hard. The authors of [8] conjectured that the problem is decidable in PSpace for all universal Horn formulas. This conjecture was confirmed in [16].

In case of some UHF formulas Φ, decidability of corresponding modal logics is shown in [16] by demonstrating the finite model property with respect to $\mathcal{K}_\Phi$, i.e. by proving that every modal formula satisfiable over K_Φ has also a finite model in K_Φ. However, it is not always possible, as it is not hard to construct a UHF formula Φ, such that some modal formulas have only infinite models over K_Φ. Assume e.g that Φ enforces irreflexivity and transitivity, and consider the following modal formula: $\Diamond p \wedge \Box\Diamond p$.

This naturally leads to the question, whether for any UHF formula Φ the finite satisfiability problem for modal logic over K_Φ is decidable. This question is particularly important, if one considers practical applications, in which the structures (corresponding e.g. to knowledge bases or descriptions of programs) are usually required to be finite.

Decision procedures for the finite satisfiability problem for modal and related logics are very often more complex than procedures for general satisfiability. As argued in [25], the model theoretic reason for the good behavior of modal logics is the tree model property. A standard technique is to unravel an arbitrary model into a (usually infinite) tree. In [16] we also apply this idea (at least as a starting point of our constructions, as the obtained unravellings have to be sometimes modified to meet the requirements of the UHF formula defining the class of frames). Clearly such an approach, is not sufficient if we are interested only in finite models.

In this paper we are however able to positively answer the given question:

Theorem 1.1 *Let Φ be a universal Horn formula. Then the finite local and the finite global satisfiability problems for modal logic over $\mathcal{K}_\Phi$ are decidable.*

The precise statement of the results, containing also some complexity bounds, is given in Table 1.

Plan of the paper In Section 2 we define all important notions related to modal logic and Horn formulas, and then recall some definitions and results from [16]. The remaining part of the paper is divided into two sections each of which deals with one subclass of universal Horn formulas.

2 Preliminaries

2.1 Modal logic

As we work with both first-order logic and modal logic we help the reader to distinguish them in our notation: we denote first-order formulas with Greek capital letters, and modal formulas with Greek small letters.

We assume that the reader is familiar with first-order logic and propositional logic. Modal logic extends propositional logic with the $\Diamond$ operator and its dual $\Box$. Formulas of modal logic are interpreted in Kripke structures, which are triples of the form $\langle W, R, \pi \rangle$, where W is a set of elements, called *worlds*, $\langle W, R \rangle$ is a directed graph called a *frame*, and π is a function that assigns to each world a set of propositional variables which are true at this world. We say that a structure $\langle W, R, \pi \rangle$ is *based* on the frame $\langle W, R \rangle$. For a given class of frames $\mathcal{K}$, we say that a structure is $\mathcal{K}$-based if it is based on some frame from $\mathcal{K}$. We will use calligraphic letters $\mathcal{M}, \mathcal{N}$ to denote frames and Fraktur letters $\mathfrak{M}, \mathfrak{N}$ to denote structures.

For a frame $\langle W, R\rangle$ and a subset $W' \subseteq W$, we define $R_{\restriction W'} = R \cap (W' \times W')$. Similarly, for a labeling function π, we define $\pi_{\restriction W'}$ to be such that $\pi_{\restriction W'}(w) = \pi(w)$ for all $w \in W'$. We define the restriction of a frame $\langle W, R\rangle_{\restriction W'}$ to $W' \subseteq W$ as $\langle W', R_{\restriction W'}\rangle$.

The semantics of modal logic is defined recursively. A modal formula φ is (locally) *satisfied* in a world w of a model $\mathfrak{M} = \langle W, R, \pi\rangle$, denoted as $\mathfrak{M}, w \models \varphi$, if

(i) $\varphi = p$ where p is a variable and $\varphi \in \pi(w)$,

(ii) $\varphi = \neg\varphi'$ and this is not the case that $\mathfrak{M}, w \models \varphi'$,

(iii) $\varphi = \varphi_1 \vee \varphi_2$ and $\mathfrak{M}, w \models \varphi_1$ or $\mathfrak{M}, w \models \varphi_2$,

(iv) $\varphi = \Diamond\varphi'$ and there exists a world $v \in W$ such that $(w, v) \in R$ and $\mathfrak{M}, v \models \varphi'$,

The $\Box$ operator is dual to $\Diamond$: $\Box\varphi \equiv \neg\Diamond\neg\varphi$. Other logical connectives are defined in a standard way.

We say that a formula φ is *globally* satisfied in $\mathfrak{M}$, denoted $\mathfrak{M} \models \varphi$, if for all worlds w of $\mathfrak{M}$, we have $\mathfrak{M}, w \models \varphi$.

We consider satisfiability of modal formulas in restricted classes of frames. For a given class of frames $\mathcal{K}$ we say that a modal formula φ is *locally (globally) satisfiable over* $\mathcal{K}$ if there exists a $\mathcal{K}$-based structure $\mathfrak{M}$ and a world w of $\mathfrak{M}$ such that $\mathfrak{M}, w \models \varphi$ (resp. $\mathfrak{M} \models \varphi$). We are particularly interested in finite models. We say that a modal formula φ is *finitely locally (globally) satisfiable over* $\mathcal{K}$ if there exists a *finite* $\mathcal{K}$-based structure $\mathfrak{M}$ and a world w of $\mathfrak{M}$ such that $\mathfrak{M}, w \models \varphi$ (resp. $\mathfrak{M} \models \varphi$).

We define *local, global, finite local,* and *finite global satisfiability problem* for modal logic *over* $\mathcal{K}$ ($\mathcal{K}$-SAT, global-$\mathcal{K}$-SAT, $\mathcal{K}$-FINSAT, global-$\mathcal{K}$-FINSAT) as the question whether a given modal formula φ is locally, globally, finitely locally, resp. finitely globally satisfiable over $\mathcal{K}$.

We say that modal logic has the *finite model property* (resp. *finite global model property*), FMP, *with respect to a class of frames* $\mathcal{K}$, if any formula that is locally (resp. globally) satisfiable over $\mathcal{K}$ is also finitely locally (resp. globally) satisfiable over $\mathcal{K}$.

In our constructions we use the following terminology. A world w is *k-followed* (*k-preceded*) in a frame $\mathcal{M}$, if there exists a directed path $(w, u_1, u_2, \ldots, u_k)$ (resp. $(u_1, u_2, \ldots, u_k, w)$) in $\mathcal{M}$. Note that we do not require this path to consist of distinct elements. We say that a world w is *k-inner* in $\mathcal{M}$ if it is k-preceded and k-followed. We use also naturally defined notions of ∞-preceded, ∞-followed, and ∞-inner worlds. In particular, a world on a cycle is ∞-inner.

We employ a standard notion of a type. For a given formula φ, a Kripke structure $\mathfrak{M}$, and a world $w \in W$ we define the *type* of w (with respect to φ) in $\mathfrak{M}$ as $tp^{\varphi}_{\mathfrak{M}}(w) = \{\psi : \mathfrak{M}, w \models \psi$ and ψ is subformula of $\varphi\}$. We write $tp_{\mathfrak{M}}(w)$ if the formula is clear from context. Note that $|tp^{\varphi}_{\mathfrak{M}}(w)| \leq |\varphi|$, where $|\varphi|$ denotes the length of φ.

2.2 Universal Horn formulas

We use universal Horn formulas to define classes of frames. A Horn *clause* is a disjunctions of literals of which at most one is positive. The set of *universal Horn formulas*, UHF, is defined as the set of those Φ over the language $\{R\}$ (without equality) which are of the form $\forall \boldsymbol{x}.\Phi_1 \wedge \Phi_2 \wedge ... \wedge \Phi_n$, where each Φ_i is a Horn clause. We usually present Horn clauses as implications and skip the quantifiers. For example, the UHF formula $(xRy \wedge yRz \Rightarrow xRz) \wedge (xRx \Rightarrow \bot)$ defines the set of transitive and irreflexive frames. We assume without loss of generality that each Horn clause consists of variables x, y, ..., $z_1, z_2, \ldots$, and is of the form $\Psi \Rightarrow \bot$, $\Psi \Rightarrow xRx$, or $\Psi \Rightarrow xRy$. We define $\Psi(v_x, v_y, v_1, \ldots, v_k)$ as the instantiation of Ψ with $x = v_x$, $y = v_y$, $z_1 = v_1$, $z_2 = v_2, \ldots$, e.g. $(xRz_1 \wedge z_1Rz_2 \wedge z_2Ry \Rightarrow xRy)(a, b, c, d) = aRc \wedge cRd \wedge dRb \Rightarrow aRb$.

For a given $\Phi \in$ UHF, we write $\mathcal{K}_\Phi$ for the class of frames satisfying Φ. When considering $\mathcal{K}_\Phi$-SAT, global $\mathcal{K}_\Phi$-SAT, and their finite versions, the formula Φ is fixed and is not a part of the input. However, the complexity depends on this formula. To hide unnecessary details, we use a function $\mathfrak{g}$ to bound the size of models or complexity in the size of Φ. Please keep in mind that once Φ is fixed, $\mathfrak{g}(|\Phi|)$ can be treated as a constant, and therefore the precise value of $\mathfrak{g}$ is not important (see also [16]).

2.3 Consequences, closures and morphisms

In this and in the next subsection we recall some notions and results from [16]. Observations related to the content of this subsection appear also in [3].

We say that an edge (w, w') is a *consequence* of Φ in $\mathcal{M} = \langle W, R \rangle$, if for some worlds $v_1, \ldots, v_k \in W$ and a clause $\Psi_1 \Rightarrow \Psi_2$ of Φ we have $\mathcal{M} \models \Psi_1(w, w', v_1, \ldots, v_k)$, and $\Psi_2(w, w', v_1, \ldots, v_k) = wRw'$. We denote the set of all consequences of Φ in $\mathcal{M}$ by $\mathrm{C}^{\Phi}_{\rightsquigarrow}(\mathcal{M})$. We define the *consequence operator* as follows.

$$\textsc{Cons}_{\Phi,W}(R) = R \cup \mathrm{C}^{\Phi}_{\rightsquigarrow}(\langle W, R \rangle).$$

Now, the *closure operator* can be defined as the least fixed-point of *Cons*:

$$\textsc{Closure}_{\Phi,W}(R) = \bigcup_{i>0} \textsc{Cons}^{i}_{\Phi,W}(R).$$

For a given frame $\mathcal{M} = \langle W, R \rangle$ we denote by $\mathfrak{C}_\Phi(\mathcal{M})$ the frame $\langle W, \textsc{Closure}_{\Phi,W}(R) \rangle$. The following lemma says that when considering satisfiability over arbitrary models from $\mathcal{K}_\Phi$ we can restrict our attention to models which are closures of trees (note however that those trees are usually infinite).

Lemma 2.1 *Let φ be a modal formula and let $\Phi \in$ UHF. If φ is $\mathcal{K}_\Phi$-satisfiable, then there exists a tree $\mathcal{T}$ with the degree bounded by $|\varphi|$, and a labeling π, such that*

(i) $\langle \mathcal{T}, \pi_{\mathcal{T}} \rangle$ is a model of φ;

(ii) $\langle \mathfrak{C}_\Phi(\mathcal{T}), \pi_{\mathcal{T}} \rangle$ is a model of φ that satisfies Φ.

The result holds for local satisfiability and for global satisfiability.

In this paper we make a high use of morphisms. We say that a function f from a frame $\mathcal{M}_1$ into a frame $\mathcal{M}_2$ is a *morphism* iff for all worlds w, w' if $\mathcal{M}_1 \models wRw'$, then $\mathcal{M}_2 \models f(w)Rf(w')$. The following observation is straightforward.

Observation 2.2 *Let $\mathcal{M}_1, \mathcal{M}_2$ be frames, let $\Phi \in \mathsf{UHF}$ and let f be a function from $\mathcal{M}_1$ into $\mathcal{M}_2$. If f is a morphism from $\mathcal{M}_1$ into $\mathcal{M}_2$, then f is a morphism from $\mathfrak{C}_\Phi(\mathcal{M}_1)$ into $\mathfrak{C}_\Phi(\mathcal{M}_2)$.*

2.4 Properties of formulas, frames and models

In our analysis an important role is played by two simple frames. The linear frame $\mathcal{L}_\mathbb{Z}$, which is defined as $\langle\{\underline{i} : i \in \mathbb{Z}\}, \{(\underline{i}, \underline{i+1}) | i \in \mathbb{Z}\}\rangle$, and the infinite binary tree $\mathcal{T}_\infty$, defined as $\langle\{\underline{s} | s \in \{0,1\}^*\}, \{(\underline{s}, \underline{si}) | s \in \{0,1\}^* \wedge i \in \{0,1\}\}\rangle$. For each $s \in \mathbb{N}$, we define $\mathcal{I}_s = \mathcal{L}_{\mathbb{Z}\restriction W_s}$, where $W_s = \{\underline{i} | 0 \leq i < s\}$. Note that the frame $\mathcal{L}_\mathbb{Z}$ is based on the integers, so for any k, the shift function $sh_k(\underline{i}) = \underline{i+k}$ is an automorphisms of $\mathcal{L}_\mathbb{Z}$.

For an arbitrary tree $\mathcal{T}$, we define a morphism $h_\mathcal{T} : \mathcal{T} \to \mathcal{L}_\mathbb{Z}$ in such a way that for each v at the ith level of $\mathcal{T}$, $h_\mathcal{T}(v) = \underline{i}$.

We call a formula $\Phi \in \mathsf{UHF}$ *bounded* if $\mathfrak{C}_\Phi(\mathcal{L}_\mathbb{Z})$ is not a model of Φ, and *unbounded* otherwise.

We say that a formula $\Phi \in \mathsf{UHF}$ *forks at level* i if for all $\underline{s} \in \mathcal{T}_\infty$ with $|s| = i$ and $t, t' \in \{0,1\}^*$ there is no edge between $\underline{s0t}$ and $\underline{s1t'}$ in $\mathfrak{C}_\Phi(\mathcal{T}_\infty)$.

We say that an edge $(\underline{i}, \underline{j})$ is *forward* if $i < j$, *backward* if $i > j$, *short* if $|i - j| < 2$, and *long* if $|i - j| \geq 2$. We say that Φ *forces* long (resp. backward) edges if there is a long (resp. backward) edge in $\mathfrak{C}_\Phi(\mathcal{L}_\mathbb{Z})$ and that Φ *forces only long forward edges* if it forces long edges but it does not force backward edges.

We say that $\Phi \in \mathsf{UHF}$ satisfies

S1 if Φ does not force long edges,

S2 if Φ forces only long forward edges and there exist $l, a_1, a_2, \ldots, a_l$ bounded by $\mathfrak{g}(|\Phi|)$ such that for all worlds $\underline{i}$, $\underline{i+b}$, there is an edge from $\underline{i}$ to $\underline{i+b}$ in $\mathfrak{C}_\Phi(\mathcal{L}_\mathbb{Z})$ if and only if $b \geq 0$ and $b - 1$ is in the additive closure of $\{a_1, a_2, \ldots, a_l\}$.

S3 if Φ forces long and backward edges and there exists m bounded by $\mathfrak{g}(|\Phi|)$ such that for all worlds $\underline{i}$, $\underline{i+b}$, there is an edge from $\underline{i}$ to $\underline{i+b}$ in $\mathfrak{C}_\Phi(\mathcal{L}_\mathbb{Z})$ if and only if m divides $|b - 1|$.

It appears that the above subcases cover all possibilities.

Lemma 2.3 *Each $\Phi \in \mathsf{UHF}$ satisfies S1, S2, or S3.*

Lemma 2.4 *If $\Phi \in \mathsf{UHF}$ is bounded, then each formula satisfiable over $\mathcal{K}_\Phi$ has a model over $\mathcal{K}_\Phi$ with polynomially many worlds. The result holds for local satisfiability and for global satisfiability.*

Note that the above lemma implies that global $\mathcal{K}_\Phi$-FINSAT and global $\mathcal{K}_\Phi$-SAT are NP-complete for every consistent and bouned Φ. A similar bound on the size of models was also shown for all formulas satisfying S3:

Properties of Φ	global-$\mathcal{K}_\Phi$-FINSAT	$\mathcal{K}_\Phi$-FINSAT
inconsistent	P (TRIVIAL)	
consistent and bounded	FMP, NP-c (2.4)	
unbounded, satisfies S2	NEXPTIME (4.2)	
unbounded, satisfies S3	FMP, NP-c [16]	
is unbounded, satisfies S1 and ...		
...forks at all levels and merges at some level	Lack of FMP (3.3), PSPACE-c(3.8, 3.9)	FMP (3.1), PSPACE-c [16]
...forks at all levels and does not merge at any level	FMP (3.10), EXPTIME-c [16]	FMP (3.1), PSPACE-c [16]
...does not fork at some level	FMP (3.11), PSPACE-c [16]	FMP (3.1), NP-c [16]

Table 1
A summary of results for finite satisfiability of modal logic over classes of frames defined by Horn formulas.

Lemma 2.5 *If $\Phi \in$ UHF satisfies S3 then each formula satisfiable over $\mathcal{K}_\Phi$ has a model over $\mathcal{K}_\Phi$ with polynomially many worlds. The result holds for local satisfiability and for global satisfiability.*

Thus, in this paper it remains to investigate unbounded formulas satisfying S1 (Section 3) or S2 (Section 4). Two further technical lemmas will be helpful.

Lemma 2.6 *Let $\Phi \in$ UHF, $\mathcal{T}$ be a tree and v_i, v_j be $\mathfrak{g}(|\Phi|)$-inner worlds at the same path. Then there is an edge from v_i to v_j in $\mathfrak{C}_\Phi(\mathcal{T})$ if and only if there is an edge from $h_\mathcal{T}(v_i)$ to $h_\mathcal{T}(v_j)$ in $\mathfrak{C}_\Phi(\mathcal{L}_\mathbb{Z})$.*

We say that two worlds w, w' of a frame $\mathcal{M}$ are *equivalent* if for each world u we have uRw iff uRw'.

Lemma 2.7 *Let $\Phi \in$ UHF be a formula that does not fork, $\mathcal{T}$ be a tree with a bounded degree and w be a world at level $\mathfrak{g}(|\Phi|)$ in $\mathfrak{C}_\Phi(\mathcal{T})$. Then for $n = 2\mathfrak{g}(|\Phi|) + 1$ and all i, all the n-followed descendants of w at level $n + i$ are equivalent in the frame $\mathfrak{C}_\Phi(\mathcal{T})$.*

3 Formulas that do not force long edges

In this section, we consider unbounded formulas $\Phi \in$ UHF that satisfy S1. In the case of local satisfiability we show the finite model property, essentially by an application of a standard selection argument. The case of global satisfiability is much more complicated. In particular for some formulas Φ we have to deal with infinite models.

3.1 Local satisfiability

Proposition 3.1 *Let Φ be an unbounded UHF formula that does not force long edges. Then modal logic has the finite model property with respect to $\mathcal{K}_\Phi$.*

Proof. Assume that φ is locally $\mathcal{K}_\Phi$-satisfiable. Let $\mathcal{T}$ be a tree guaranteed by Lemma 2.1. Thus, there exists a model $\mathfrak{M}$ based on the frame $\mathfrak{C}_\Phi(\mathcal{T}) \in \mathcal{K}_\Phi$, such that $\mathfrak{M}, w \models \varphi$, where w is the root of $\mathcal{T}$. Recall the morphism $h_\mathcal{T} : \mathcal{T} \to \mathcal{L}_\mathbb{Z}$. By observation 2.2, $h_\mathcal{T}$ is also a morphism from $\mathfrak{C}_\Phi(\mathcal{T})$ to $\mathfrak{C}_\Phi(\mathcal{L}_\mathbb{Z})$. Since Φ does not force long edges, this implies that $\mathfrak{C}_\Phi(\mathcal{T})$ can only contain edges between nodes on the same level or on two consecutive levels.

In order to obtain a finite model, we simply remove from $\mathfrak{M}$ all worlds from levels greater than $|\varphi|$. Since the truth of φ depends only on the worlds that are reachable from the root w by a path whose length is bounded by $|\varphi|$ (more precisely: by the modal depth of φ), the resulting model is a finite model of φ and, of course, it satisfies Φ since Φ is universal. □

We showed that φ has a $\mathcal{K}_\Phi$-based model if and only if it has a finite $\mathcal{K}_\Phi$-based model, so $\mathcal{K}_\Phi$-FINSAT coincides with $\mathcal{K}_\Phi$-SAT, which was proved in [16] to be PSPACE-complete.

3.2 Global satisfiability

In the case of general satisfiability [16], it was enough to consider the behavior of a first order formula on $\mathcal{T}_\infty$ and $\mathcal{L}_\mathbb{Z}$. In the case of finite satisfiability, we need one more frame, which we call $\mathcal{X}$, that contains a world with in-degree 2.

Formally, we define the frame $\mathcal{X}$ as $\langle W_X, R_X \rangle$, where $W_X = \{\underline{i} | i \in \mathbb{Z}\} \cup \{\overline{i} | i \in \mathbb{Z} \setminus \{0\}\}$ and $R_X = \{(\underline{i}, \underline{i+1}) | i \in \mathbb{Z}\} \cup \{(\overline{i}, \overline{i+1}) | i \in \mathbb{Z} \setminus \{-1, 0\}\} \cup \{(\overline{-1}, \underline{0}), (\underline{0}, \overline{1})\}$. Fig. 1 shows a fragment of $\mathcal{X}$.

We say that a formula Φ *merges* at a level $k < 0$ if in $\mathfrak{C}_\Phi(\mathcal{X})$ there is an edge from $\underline{k-1}$ to $\overline{k}$. For example, the formula $\Phi = xRz \wedge zRv \wedge yRv \Rightarrow xRy$ merges. Note that $\mathcal{T}_\infty$ and $\mathcal{L}_\mathbb{Z}$ satisfy Φ.

We consider three cases. For each formula Φ of UHF such that Φ does not force long edges, merges at some level and forks at all levels, we show that modal logic does not have the finite global model property with respect to $\mathcal{K}_\Phi$ (Proposition 3.3), and that global $\mathcal{K}_\Phi$-FINSAT is PSPACE–complete (Propositions 3.8 and 3.9). For the cases of formulas that do not force long edges, do not merge at any level and fork at all levels, and formulas that do not force long edges and do not fork at all levels, the decidability follows from the finite model property (Propositions 3.10 and 3.11).

Formulas that merge. The following lemma shows an important regularity in models of formulas that merge.

Lemma 3.2 *Let Φ be an unbounded* UHF *formula that does not force long edges, and merges at a level k, $\mathfrak{M}$ be a model of Φ, $v_1, v_2, \ldots, v_l$ be a walk (i.e. a path, but not necessarily simple) in $\mathfrak{M}$ such that all v_i are ∞-inner.*

(i) If $v_l R v_{l-c}$ for some $c > 0$, then for all $i > c$, $v_i R v_{i-c}$.

(ii) If $v_{l-c} R v_l$ for some $c > 0$, then for all $i > c$, $v_{i-c} R v_i$.

Proof. Let $\ldots, v_{-2}, v_{-1}, v_0, v_1$ and $v_l, v_{l+1}, \ldots$ be infinite walks in $\mathfrak{M}$. Such walks exist since v_1 and v_l are ∞-inner.

We prove (i) by induction. Assume that for some $i > 0$, for all $j > i$ we

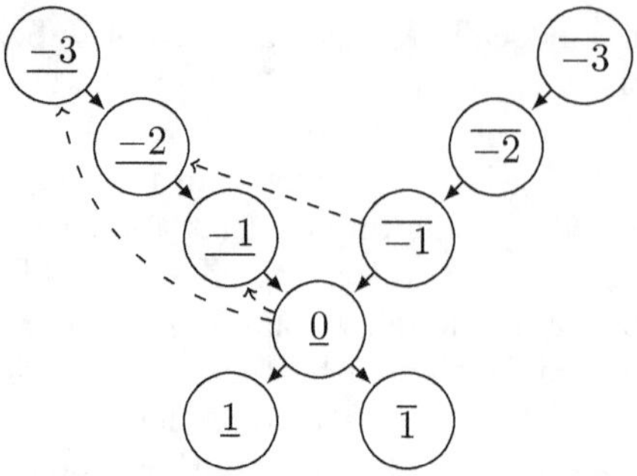

Fig. 1. A fragment of the frame $\mathcal{X}$ (circles and solid arrows). Consider a formula $\Phi = yRw \wedge wRv \wedge xRv \Rightarrow xRy$ that forces edge $(\overline{-1}, \underline{-2})$. When applied to $x = w = \underline{0}$, $y = \underline{-1}$ and $v = \underline{1}$, it forces edge $(\underline{0}, \underline{-1})$. Then, applied to $x = \underline{0}$, $v = \underline{-1}$, $w = \underline{-2}$ and $y = \underline{-3}$ it forces long edge $(\underline{0}, \underline{-3})$.

have $v_j R v_{j-c}$. We define a morphism h from $\mathcal{X}$ into $\mathfrak{M}$ as follows

$$h(w) = \begin{cases} v_{i+s+1} & \text{if } w = \underline{k+s} \text{ for some } s \leq 0 \\ v_{i-c+s} & \text{if } w = \underline{k+s} \text{ for some } s > 0 \\ v_{i-c+s} & \text{if } w = \overline{k+s} \text{ for some } s \in \mathbb{Z} \end{cases}$$

A quick check shows that h is indeed a morphism and since $\mathfrak{C}_\Phi(\mathcal{X})$ contains an edge from $\underline{k-1}$ to $\overline{k}$, $\mathfrak{M}$ has to contain edge from v_i to v_{i-c}.

The proof of (ii) is similar and thus omitted. □

Now we use the above lemma to show the lack of the finite model property.

Proposition 3.3 *Let Φ be an unbounded* UHF *formula that does not force long edges, merges at a level $k < 0$ and forks at all levels. Then modal logic lacks the finite global model property with respect to $\mathcal{K}_\Phi$.*

Proof. Let $\lambda = \bigwedge_{i \in \{0,1,2,3,4\}} \lambda_i$, where:

$$\lambda_0 = \bigvee_{i \in \{1,2,3,4\}} p_i \wedge \bigwedge_{i,j \in \{1,2,3,4\}, i \neq j} \neg(p_i \wedge p_j)$$

$$\lambda_1 = p_1 \Rightarrow (\Diamond p_2 \wedge \Box p_2)$$

$$\lambda_2 = p_2 \Rightarrow (\Diamond p_3 \wedge \Box p_3)$$

$$\lambda_3 = p_3 \Rightarrow (\Diamond p_2 \wedge \Diamond p_4 \wedge \Box(p_2 \vee p_4))$$

$$\lambda_4 = p_4 \Rightarrow (\Diamond p_1 \wedge \Box p_1)$$

An infinite model of λ is presented in Fig. 2. It is not hard to see that its frame belongs to $\mathcal{K}_\Phi$ for any Φ meeting the assumptions.

Assume that $\mathfrak{M}$ is a finite $\mathcal{K}_\Phi$-based model of λ and let w be a world that satisfies p_1 in $\mathfrak{M}$. Quick check shows that such a world must exist. Let $w, w_1, w_2, \ldots$ be an infinite path in $\mathfrak{M}$ such that for odd i, w_i satisfies p_2, and for even i, w_i satisfies p_3. Such a path is guaranteed by λ_1, λ_2 and λ_3.

Since $\mathfrak{M}$ is finite, it must be the case that for some $0 < r < l$ we have $\mathfrak{M} \models w_l R w_r$. Clearly, w_l and w_k are ∞-inner. It follows from Lemma 3.2 that

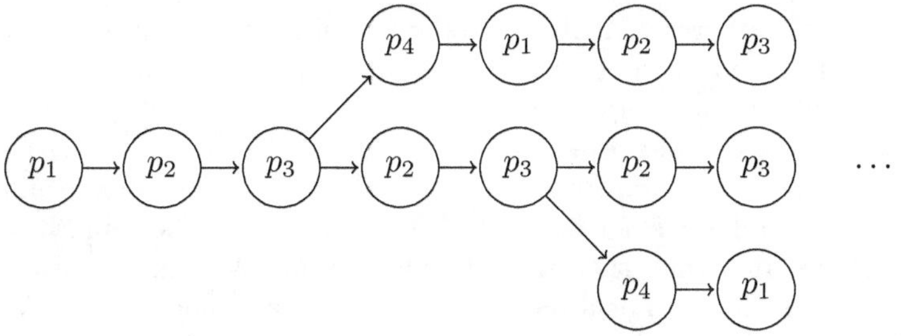

Fig. 2. An infinite model of τ.

$\mathfrak{M} \models w_{l-r}Rw$. But w_{l-r} satisfies p_2 or p_3, w satisfies p_0, and thus λ_2, λ_3 forbid this connection. Therefore there is no finite model of λ based on a frame from $\mathcal{K}_\Phi$. □

Before we show the algorithm that solves the satisfiability problem, we present a simple property of the finite global satisfiability problem, namely, that every satisfiable formula has a model which is strongly connected.

Lemma 3.4 *Let $\Phi \in \mathsf{UHF}$, φ be a modal formula and $\mathfrak{M}$ be a finite $\mathcal{K}_\Phi$-based model such that $\mathfrak{M} \models \varphi$. Then there is a $\mathcal{K}_\Phi$-based submodel $\mathfrak{N}$ of $\mathfrak{M}$ such that $\mathfrak{N} \models \varphi$ and the frame of $\mathfrak{N}$ is strongly connected.*

Proof. Consider a relation on the set of strongly connected components of $\mathfrak{M}$, defined in such a way that $\mathfrak{N} \leq \mathfrak{N}'$ iff there is a path from an element of $\mathfrak{N}'$ to an element of $\mathfrak{N}$, or if $\mathfrak{N} = \mathfrak{N}'$. It is not hard to see that $\leq$ is a partial order. Since $\mathfrak{M}$ is finite, there must be a component $\mathfrak{N}_{min}$ which is minimal with respect to $\leq$. As Φ is universal, $\mathfrak{N}_{min}$ satisfies Φ. Moreover, since each world from $\mathfrak{N}_{min}$ has all its successors in $\mathfrak{N}$ (there is no path to worlds in other connected components), $\mathfrak{N}_{min}$ satisfies φ. □

We say that a frame $\mathcal{M}$ is *k-periodic* if its universe can be divided into pairwise disjoint, non-empty sets of worlds $W_1, W_2, \ldots, W_k$ such that for each v, w from $\mathcal{M}$ there is an edge from v to w if and only if for some $i \leq k$, $v \in W_i$ and $w \in W_{(i \bmod k)+1}$. Notice that a 1-periodic frame is a clique. For each $k \in \mathbb{N}$ we define the cycle $\mathcal{C}_k$ as $\mathcal{I}_k$ with one additional edge, namely $(\underline{k-1}, \underline{0})$. Clearly, each $\mathcal{C}_k$ is k-periodic.

We are going to prove decidability by showing that each satisfiable formula has a model that is k-periodic for some k. In order to do so, we introduce two technical lemmas.

Lemma 3.5 *Let $\Phi \in \mathsf{UHF}$.*

(a) If Φ has a k-periodic model $\mathcal{M}$, then $\mathcal{C}_k$ is a model of Φ.

(b) If $\mathcal{C}_k$ is a model of Φ, then any k-periodic frame is a model of Φ.

(c) If $\mathcal{L}_\mathbb{Z}$ is a model of Φ, then for all $c > |\Phi|$, $\mathcal{C}_c$ is a model of Φ.

(d) If for some $k > |\Phi|$ the frame $\mathcal{C}_k$ is a model of Φ, then $\mathcal{L}_\mathbb{Z}$ is a model of Φ.

Proof. For (a), observe that if a periodic model $\mathcal{M}$ that consists of sets W_1, $W_2, \ldots, W_k$ is a model of Φ, then $\mathcal{C}_k$ is isomorphic with an induced substructure of $\mathfrak{M}$ that contains one world from every W_i.

We say that a morphism $h : \mathcal{M} \to \mathcal{M}'$ is *complete* if for all v, v' we have $h(v)Rh(v')$ if and only if vRv'. Note that if there is a complete morphism $h : \mathcal{M} \to \mathcal{M}'$ and Φ does not hold in $\mathcal{M}$, then it does not hold in $\mathcal{M}'$.

For (b), assume that there is a periodic frame $\mathcal{M}$ that consists of sets $W_1, W_2, \ldots, W_k$ and is not a model of Φ, but $\mathcal{C}_k$ is a model of Φ. We define a complete morphism $f : \mathcal{M} \to \mathcal{C}_k$ as $f(v) = \underline{i}$ for $v \in W_i$. Since Φ does not hold in $\mathcal{M}$ and f is a complete morphism, Φ does not hold in $\mathcal{C}_k$ — a contradiction.

We prove (c) as follows. Let $c > |\Phi|$. Assume that there is a clause Ψ satisfied in $\mathcal{L}_\mathbb{Z}$ but not in $\mathcal{C}_c$, and let $v_1, v_2, \ldots, v_n$ be worlds of $\mathcal{C}_c$ such that $\Psi(v_1, \ldots, v_n)$ is false. Let k be such that no world among $v_1, \ldots, v_n$ is equal $\underline{k}$. Consider the function $f : \mathcal{C}_{c\restriction\{v_1,\ldots,v_n\}} \to \mathcal{L}_\mathbb{Z}$ defined as

$$f(\underline{s}) = \begin{cases} \underline{s} & \text{for } s > k \\ \underline{c+s} & \text{for } s < k \end{cases}$$

A quick check shows that the function f is a complete morphism. Since $\Psi(v_1, \ldots, v_n)$ does not hold in $\mathcal{C}_c$, it follows that $\Psi(f(v_1), \ldots, f(v_n))$ does not hold in $\mathcal{L}_\mathbb{Z}$. But $\mathcal{L}_\mathbb{Z} \models \Psi$, a contradiction.

For the proof of (d), let $k > |\Phi|$, $\Psi \Rightarrow \Psi'$ be satisfied in $\mathcal{C}_k$ but not in $\mathcal{L}_\mathbb{Z}$. Let $v_1 = \underline{s}, v_1 = \underline{t}, v_3 \ldots, v_n$ be worlds of $\mathcal{L}_\mathbb{Z}$ such that $\Psi(v_1, \ldots, v_n)$ is true, $\Psi'(v_1, \ldots, v_n)$ is not, and $|s-t|$ is minimal. Let $f(\underline{i}) = \underline{i \mod k}$ be a morphism from $\mathcal{L}_\mathbb{Z}$ onto $\mathcal{C}_k$. If $t - s \mod k \neq 1$, then $\Psi \Rightarrow \Psi'(f(v_1), \ldots, f(v_n))$ does not hold and we have a contradiction. Otherwise, $|s-t| \geq k-1$ so there is a world $\underline{l}$ such that l is between s and t and $\underline{l}$ is different from all of $\underline{s}, \underline{t}, v_3, \ldots, v_n$. But then, morphism $g : \mathcal{L}_{\mathbb{Z}\restriction\{v_1,\ldots,v_n\}} \to \mathcal{L}_\mathbb{Z}$ defined as $g(\underline{s}) = \underline{s}$ for $s < l$ and $g(\underline{s}) = \underline{s-1}$ leads to the contradiction with the minimality of $|s-t|$. □

Lemma 3.6 *Let Φ be an unbounded* UHF *formula that does not force long edges and such that in $\mathfrak{C}_\Phi(\mathcal{X})$ for some $i, j < 0$ we have $\bar{i}R\underline{j}$ or $\underline{i}R\bar{j}$. Then $j - i = 1$ and $\mathfrak{C}_\Phi(\mathcal{L}_\mathbb{Z}) = \mathcal{L}_\mathbb{Z}$.*

Proof. As $\mathcal{X}$ is symmetric, $\mathfrak{C}_\Phi(\mathcal{X}) \models \underline{i}R\bar{j}$ implies $\mathfrak{C}_\Phi(\mathcal{X}) \models \bar{i}R\underline{j}$. So we assume that $\mathfrak{C}_\Phi(\mathcal{X}) \models \underline{i}R\bar{j}$.

Let us consider a morphism f from $\mathcal{X}$ into $\mathcal{L}_\mathbb{Z}$ defined as

$$f(\underline{k}) = f(\bar{k}) = \underline{k}$$

If $|j-i| > 1$, then there is a long edge in $\mathfrak{C}_\Phi(\mathcal{L}_\mathbb{Z})$ and it contradicts the assumption that $\mathcal{X}$ does not force long edges.

If $j - i = -1$, then the morphism f implies that there is an edge $(\underline{j}, \underline{j-1})$ in $\mathfrak{C}_\Phi(\mathcal{L}_\mathbb{Z})$ and, since $\mathfrak{C}_\Phi(\mathcal{L}_\mathbb{Z})$ is uniform, for all k there are edges $(\underline{k}, \underline{k-1})$ in $\mathfrak{C}_\Phi(\mathcal{L}_\mathbb{Z})$. We define another morphism g to show that then $\mathfrak{C}_\Phi(\mathcal{L}_\mathbb{Z})$ contains a long edge. Let g be a morphism from $\mathcal{X}$ into $\mathfrak{C}_\Phi(\mathcal{L}_\mathbb{Z})$ defined as

$$g(w) = \begin{cases} \underline{|k|} & \text{if } w = \underline{k} \text{ for some } k \\ \underline{-|k|} & \text{if } w = \bar{k} \text{ for some } k \end{cases}$$

It is not hard to see that g is indeed a morphism and therefore that $\mathfrak{C}_\Phi(\mathcal{L}_\mathbb{Z})$ contains a long edge $(\underline{|i|}, \overline{-|j|})$. An example is presented in Fig. 1.

If $j = i$, then the morphism f implies that there is a reflexive world in $\mathfrak{C}_\Phi(\mathcal{L}_\mathbb{Z})$, and therefore all worlds are reflexive. Consider a morphism h from $\mathcal{X}$ into $\mathfrak{C}_\Phi(\mathcal{L}_\mathbb{Z})$ defined as

$$h(w) = \begin{cases} \underline{1} & \text{if } w = \overline{k} \text{ for some } k \leq i \\ \underline{0} & \text{otherwise} \end{cases}$$

Since all worlds in $\mathfrak{C}_\Phi(\mathcal{L}_\mathbb{Z})$ are reflexive, h is indeed a morphism, so in $\mathfrak{C}_\Phi(\mathcal{L}_\mathbb{Z})$ there is edge $(\underline{1}, \underline{0})$ and, as in the previous case, all edges in $\mathfrak{C}_\Phi(\mathcal{L}_\mathbb{Z})$ are symmetric and therefore $\mathfrak{C}_\Phi(\mathcal{L}_\mathbb{Z})$ contains a long edge.

For the proof of $\mathfrak{C}_\Phi(\mathcal{L}_\mathbb{Z}) = \mathcal{L}_\mathbb{Z}$, recall that if $\mathfrak{C}_\Phi(\mathcal{L}_\mathbb{Z})$ contains a symmetric or reflexive edge, then it contains long edges. But Φ does not force long edges, and therefore $\mathfrak{C}_\Phi(\mathcal{L}_\mathbb{Z}) = \mathcal{L}_\mathbb{Z}$. □

In the proof of our next proposition we use the following simple fact, whose proof follows easily by an application of the Euclidean algorithm.

Fact 3.7 *Let X be a set of positive numbers. Then, there exists a finite subset X' of X such that $gcd(X) = gcd(X')$. Moreover, if X is closed under addition then for each $x > lcm(X')$, $gcd(X')$ divides x iff $x \in X$.*

For a given model $\mathfrak{M}$, we define a *characteristic cycle* of $\mathfrak{M}$ as a walk $v_0, v_1, \ldots, v_{l-1}$ that contains all worlds from $\mathfrak{M}$ and, moreover, in $\mathfrak{M}$ there is an edge from v_{l-1} to v_0. Note that for all strongly connected models containing at least two worlds such a characteristic cycle exists.

Proposition 3.8 *Let Φ be an unbounded* UHF *formula that does not force long edges, merges at a level $k < 0$ and forks at all levels. Then global $\mathcal{K}_\Phi$-FINSAT is in* PSPACE.

Proof. Let φ be a modal formula and $\mathfrak{M}$ be a strongly connected model of φ from $\mathcal{K}_\Phi$. Such a model exists due to Lemma 3.4. Assume that $\mathfrak{M}$ contains at least two worlds and let $v_0, v_1, \ldots, v_{l-1}$ be a characteristic cycle of $\mathfrak{M}$. For better readability, below we omit " mod l" in subscripts of vs.

Our aim is to show that $\mathfrak{M}$ is s-periodic for some s.

Let $\mathcal{X}_\mathfrak{M} \subseteq \mathrm{N}$ be such that $k \in \mathcal{X}_\mathfrak{M}$ if and only if there is v_i such that $\mathfrak{M} \models v_i R v_{i+k+1}$. Lemma 3.2 implies that for all v_i and $k \in \mathcal{X}_\mathfrak{M}$, $\mathfrak{M} \models v_i R v_{i+k+1}$.

We show that $\mathcal{X}_\mathfrak{M}$ is additively closed. Assume that $x, y \in \mathcal{X}_\mathfrak{M}$. It means that $\mathfrak{M}$ contains edges (v_{x+y+1}, v_{x+y+2}) , (v_{x+1}, v_{x+y+2}) and (v_0, v_{x+1}). We define a morphism h from $\mathcal{X}$ to $\mathcal{M}$, the frame of $\mathfrak{M}$, as

$$h(w) = \begin{cases} v_s & \text{if } w = \underline{k-1+s} \text{ for all } s \leq 0 \\ v_{x+1} & \text{if } w = \underline{k} \\ v_{x+y+1+s} & \text{if } w = \underline{k+s} \text{ for all } s > 0 \\ v_{x+y+1+s} & \text{if } w = \overline{k+s} \text{ for all } s \in \mathbb{Z} \end{cases}$$

We see that $h(\underline{k-1}) = v_0$ and $h(\overline{k}) = v_{x+y+1}$, and since in $\mathfrak{M}$ there is an edge from $\underline{k-1}$ to $\overline{k}$, $x+y \in \mathcal{X}_{\mathfrak{M}}$.

Let $\mathcal{X}^l_{\mathfrak{M}} = \{i \mod l | i \in \mathcal{X}_{\mathfrak{M}}\}$. By Fact 3.7, $\mathcal{X}^l_{\mathfrak{M}}$ can be represented as $\{i \cdot gcd(\mathcal{X}_{\mathfrak{M}}) \mod l | i \in \mathbb{N}\}$. Define $W_i = \{v_{i+j \cdot gcd(\mathcal{X}_{\mathfrak{M}})} | j \in \mathbb{N}\}$. It follows that all elements of W_i have all successors in W_{i+1}, and therefore $\mathfrak{M}$ is $gcd(\mathcal{X}_{\mathfrak{M}})$-periodic.

Now we show how to compress sets W_i. For each i and each subformula ψ of φ, if there is a world in W_i that satisfies ψ, we mark one such world. Then we remove unmarked worlds. It is easy to see that the types of worlds remain the same.

We have proved that all models of φ are s-periodic and that their sets can be compressed to a size bounded by $|\varphi|$, but the value of s can be arbitrary large. Now we show that there is an NPSPACE (=PSPACE) procedure that checks, for a given modal formula φ, if φ has a Φ-based finite global periodic model.

Our NPSPACE algorithm works as follows. First, it checks if there is a single world or a single clique (1-periodic set) with size bounded by $|\varphi|$, that satisfies both φ and Φ. If it is the case the algorithm returns "Yes". Otherwise, it guesses a set W_1 with size bounded by $|\varphi|$ and then, recursively, guesses the successive sets with size similarly bounded, checking if guessed worlds are consistent with their predecessor, and returns "no" otherwise. The algorithm stops after $\binom{2^{|\varphi|}}{|\varphi|} + 1$ steps and returns "yes".

If there is a model of φ, then the algorithm returns "yes". Indeed, we showed that φ has a single world model or an s-periodic model with size of sets bounded by $|\varphi|$, and the algorithm can simply guess this world or successively guess consecutive sets of this model.

If the algorithm returns "yes", then it visited two sets satisfying the same subformulas, so there is a sequence of sets $V_1, V_2, \ldots, V_k, V_1$ with $k \leq 2^{|\varphi|}$ such that each set contains all witnesses needed by its predecessors. We build an s-periodic model that contains sets $V_1, \ldots, V_k$ repeated $\lceil |\Phi|/k \rceil + 1$ times. Clearly, the obtained model satisfies φ. By Lemma 3.6, $\mathcal{L}_{\mathbb{Z}} = \mathfrak{C}_{\Phi}(\mathcal{L}_{\mathbb{Z}})$, and by Lemma 3.5 it is also a model of Φ. □

The corresponding lower bound can be shown by an encoding of a version of the corridor-tiling problem. A *tiling system* is a tuple $\mathcal{D} = (D, H_{\mathcal{D}}, V_{\mathcal{D}}, n)$, where D is a set of tiles, $H_{\mathcal{D}}, V_{\mathcal{D}} \subseteq D \times D$ are binary relations specifying admissible horizontal and vertical adjacencies, and n is a unary encoded natural number. For a given tiling system we ask if there exists a tiling of an infinite corridor of width n, respecting $H_{\mathcal{D}}$, $V_{\mathcal{D}}$ constraints. Formally, we ask if there exists a tiling $t : \{0, 1, \ldots, n-1\} \times \mathbb{N} \to D$, such that for all $0 \leq k < n$ and $l \in \mathbb{N}$ we have $(t(k,l), t(k,l+1)) \in V_{\mathcal{D}}$ and for all $0 \leq k < n-1$, $l \in \mathbb{N}$ we have $(t(k,l), t(k+1,l)) \in H_{\mathcal{D}}$. This problem is known to be PSPACE-complete [24].

Proposition 3.9 *Let Φ be an unbounded* UHF *formula that does not force long edges, merges at a level $k < 0$ and forks at all levels. Then global $\mathcal{K}_{\Phi}$-FINSAT* *is* PSPACE*-hard.*

Proof. Let $\mathcal{D} = (D, H_\mathcal{D}, V_\mathcal{D}, n)$ be an instance of the corridor-tiling problem. We will construct a modal formula η which is globally, finitely $\mathcal{K}_\Phi$-satisfiable iff $\mathcal{D}$ has a solution. In our intended model a single world represents a whole row of a solution.

We employ propositional variables p_i^d for $i < n$ and $d \in \mathcal{D}$. The intended meaning of p_i^d is that the point in column i is tiled by d.

We put $\eta = \eta^l \wedge \eta^h \wedge \eta^v$, where η^l that guarantees that each point is tiled by exactly one element of $\mathcal{D}$, η^h ensures that the tilling respects $H_\mathcal{D}$, and η^v ensures that each world has a successor that describes the row which is consistent with the current one with respect to the relation $V_\mathcal{D}$.

$$\eta^l = \bigwedge_{i<n} (\bigvee_{d\in\mathcal{D}} p_i^d \wedge \bigwedge_{d,d'\in\mathcal{D}, d\neq d'} \neg(p_i^d \wedge p_i^{d'}))$$

$$\eta^h = \bigwedge_{i<n-1} \bigvee_{(d,d')\in H_\mathcal{D}} (p_i^d \wedge p_{i+1}^{d'})$$

$$\eta^v = \Diamond\top \wedge \bigwedge_{i<n} \bigvee_{(d,d')\in V_\mathcal{D}} (p_i^d \wedge \Box p_i^{d'})$$

Assume that $\langle\mathcal{D}, V_\mathcal{D}, H_\mathcal{D}, n\rangle$ has a solution that consists of rows $r_1, r_2, \ldots$. Then among first n^n+1 of them some rows r_i, r_j with $i < j$ are tiled identically. Let $l = c(j-i)$, for some $c > |\Phi|$. We encode the solution on $\mathcal{C}_l$ in such a way that $\underline{s}$ represents row $i + (s \mod l)$. Note that by Lemma 3.6 and Lemma 3.5(c) it follows that $\mathcal{C}_l$ belongs to $\mathcal{K}_\Phi$. Conversely, if η has a model $\mathfrak{M}$ then we can construct a solution by starting from an arbitrary world of $\mathfrak{M}$, translating it to the initial row of a solution in a natural way, and recursively building successive rows as translations of the worlds guaranteed by η^v. □

Formulas that do not merge and fork at all levels. Now we prove that in the case of formulas Φ that do not force long edges, fork at all levels and do not merge at any level, modal logic has the finite model property with respect to $\mathcal{K}_\Phi$. In the proof, we start from an infinite tree–based model $\mathfrak{M}$, and construct a very large structure that locally looks like a part of $\mathfrak{M}$, but is finite. We need to do it carefully in order not to violate the first–order formula Φ.

Proposition 3.10 *Let Φ be an unbounded* UHF *formula that does not force long edges, forks at all levels and does not merge at any level $k < 0$. Then modal logic has the finite global model property with respect to $\mathcal{K}_\Phi$.*

Proof. Let $\mathfrak{M}^b$ be a tree-based model of φ and Φ, based on a tree $\mathcal{T}^b$, as guaranteed by Lemma 2.1. Let $n = |\varphi|$ and $N = |\Phi|$. If there is a world in $\mathfrak{M}^b$ without a proper successor, then the structure that contains only this world is a model of φ and Φ. Otherwise, all worlds are ∞-followed. We assume that every world has degree n – if a world has a smaller degree, then we can replicate any of its subtrees.

Let w be any $\mathfrak{g}(|\Phi|)$-inner world in $\mathcal{T}^b$, $\mathcal{T}$ be a subtree of $\mathcal{T}^b$ rooted at w, and $\mathfrak{M}$ be a substructure of $\mathfrak{M}^b$ that consists of the worlds from $\mathcal{T}$. Clearly, $\mathfrak{M}$ satisfies Φ and φ.

Let M be the universe of $\mathfrak{M}$. For each $w \in M$, we define a tree $\mathcal{S}'_w$ to be the subtree of $\mathcal{M}$ rooted in w, $\mathcal{S}_w$ to be the frame that contains first $2N$ levels of $\mathcal{S}'_w$, and $\mathfrak{S}_w$ to be the substructure of $\mathfrak{M}$ that contains the worlds from $\mathcal{S}_w$. Let $tp(\mathfrak{M})$ be a set of all types realized in $\mathfrak{M}$. For each type $t \in tp(\mathfrak{M})$, we pick one world w_t of this type and define $\mathfrak{S}_t = \mathfrak{S}_w$ and $\mathcal{S}_t = \mathcal{S}_w$

For each $\mathcal{S}_t$, we label leaves in $\mathcal{S}_t$ in a consecutive way, e.g. from left to right, such that leaves labeled with 1, 2, ..., n have the same parent and so on.

For each $s \in \{0,1\}$, $p \in \{1,\ldots,n\}$ and $t \in tp(\mathfrak{M})$, we define $\mathfrak{T}^s_{t,p}$ as a copy of $\mathfrak{S}_t$. We define the finite structure $\mathfrak{M}_s$ as a disjoint union of all possible $\mathfrak{T}^s_{t,p}$. We say that a world w is *at a level* k in $\mathfrak{T}^s_{t,p}$ if it is a copy of a world that is at a level k in $\mathcal{S}_t$ and that it is *at a level* k in $\mathfrak{M}_s$ if it is at a level k in some tree of $\mathfrak{M}_s$. We say that a world v is a *parent* of v' in $\mathfrak{M}_k$ if wRv, v is at a level k and v' is at a level $k+1$ for some k. For any two worlds v, v' that are in the same tree, we define $lca(v,v')$ as the lowest common ancestor of v and v' (w.r.t. the relation parent). We define $llca(v,v')$ as the level of $lca(v,v')$ if such world exists and $llca(v,v') = -1$ otherwise.

We define a structure $\mathfrak{M}'$ as a disjoint union of $\mathfrak{M}_0$ and $\mathfrak{M}_1$ with additional edges defined as follows. Consider tree $\mathfrak{T}^0_{t,p}$ and its leaf v labeled by p. Let w be a world in $\mathfrak{M}$ with the same type and $t_1,\ldots,t_k$ be types of successors of w in $\mathcal{T}$. For each $j \leq k$ we add an edge from v to the root of $\mathfrak{T}^1_{t_j,p}$ and, if some connection between w and its successors is symmetric, we make this edge symmetric as well. We do the same for the leaves from $\mathfrak{M}_1$, but we connect them with the roots from $\mathfrak{M}_0$.

It is not hard to see that all worlds in $\mathfrak{M}'$ satisfy φ. We prove that $\mathfrak{M}'$ satisfies Φ. Assume to the contrary that this is not the case. Let $\Psi \Rightarrow \Psi'$ be a formula which is not satisfied in $\mathfrak{M}'$. Then there are worlds $v_1,\ldots,v_n$ such that $\Psi(v_1,\ldots,v_n)$ holds but $\Psi'(v_1,\ldots,v_n)$ does not.

We define a function $\nu_k : \mathfrak{M}' \to \{0,\ldots,4N-1\}$ as

$$\nu_k(v) = \begin{cases} s-k & \text{for each } v \text{ at a level } s \geq k \text{ in } M_0 \\ s+2N-k & \text{for each } v \text{ at a level } s \text{ in } M_1 \\ s+4N-k & \text{for each } v \text{ at a level } s < k \text{ in } M_0 \end{cases}$$

Let k be such that no world among $v_1,\ldots,v_n$ is at level k in $\mathfrak{M}_0$ and $\mathfrak{M}_1$. A function $f : \mathfrak{M}'\restriction_{\{v_1,\ldots,v_n\}} \to \mathfrak{C}_\Phi(\mathcal{L}_\mathbb{Z})$ defined as

$$f(v) = \underline{\nu_k(v)}$$

is a morphism.

It is not possible that $\Psi' = \bot$, because then Φ would not be satisfied in $\mathfrak{C}_\Phi(\mathcal{L}_\mathbb{Z})$ and since Φ is unbounded, $\mathfrak{C}_\Phi(\mathcal{L}_\mathbb{Z})$ is a model of Φ. Similarly, if $\Psi' = xRx$, then some world in $\mathfrak{C}_\Phi(\mathcal{L}_\mathbb{Z})$ would be reflexive and, since all worlds in $\mathfrak{M}$ are $\mathfrak{g}(|\Phi|)$-inner in $\mathfrak{M}^b$, $\Psi'(v_1,\ldots,v_n)$ would be satisfied.

The only remaining case is $\Psi' = xRy$. Let v_1 be at a level l_1 in $\mathfrak{M}_{s_1}$ and v_2 be at a level l_2 is $\mathfrak{M}_{s_2}$. There are two cases: either $s_1 = s_2$ and $|l_1 - l_2| \leq 1$, or $s_1 \neq s_2$ and one of v_1, v_2 is a root and the other one is a leaf. Otherwise, Φ would force long edges.

Assume that $s_1 < s_2$ and let k be such that no world among $v_1, \ldots, v_n$ is at a level k in $\mathfrak{M}_0$. Consider a morphism $g : \mathfrak{M}'\restriction_{\{v_1,\ldots,v_n\}} \to \mathfrak{M}'$ defined as

$$g(v) = \begin{cases} v' & \text{if } v \text{ is at a level } i \geq k \text{ in } \mathfrak{M}_0 \text{ and } v' \text{ is a parent of } v \\ v & \text{otherwise} \end{cases}$$

It implies that Φ requires also an edge from some world that is not a leaf to some root, and so by the morphism f we can show that Φ forces long edges. The case when $s_1 > s_2$ is symmetric.

Assume that $s_1 = s_2 = 0$. If $v_1 = v_2$, then, by the morphism f, all worlds of $\mathfrak{C}_\Phi(\mathcal{L}_\mathbb{Z})$ are reflexive and Ψ' would be satisfied, as before. If v_2 is a parent of v_1, then, by the morphism f, all edges in $\mathfrak{C}_\Phi(\mathcal{L}_\mathbb{Z})$ are symmetric and Ψ' would be satisfied. So we can assume that v_1 and v_2 are not on the same path in $\mathfrak{M}_0$.

Assume that $l_1 \leq N$ and $l_2 \leq N$ and let $k > N$ be such that no world among $v_3, \ldots, v_N$ is at level k in $\mathfrak{M}_0$. We define a morphism $h_1 : \mathfrak{M}'\restriction_{\{v_1,\ldots,v_n\}} \to \mathcal{T}_\infty$ as follows.

$$h_1(v) = \begin{cases} \underline{0^{\nu_k(v)}} & \text{if } \nu_k(v) < 4N - k \\ \underline{0^{4N-k+llca(v,v_1)}1^{s-llca(v,v_1)}} & \text{if } v \text{ at level } s \text{ and } \nu_k(v) \geq 4N - k \end{cases}$$

Let $m = llca(v_1, v_2)$. Since v_1 and v_2 are not on the same path, $m < min(l_1, l_2)$. Since $h_1(v_1) = \underline{0^{4N-k+l_1}}$ and $h_1(v_2) = \underline{0^{4N-k+m}1^{l_2-m}}$ and h_1 is a morphism, it implies that Φ does not fork a the level $\underline{0^{4N-k+m}}$ — a contradiction.

Now consider the case when $l_1 \geq N$ and $l_2 \geq N$. Let $k < N$ be such that no world among $v_3, \ldots, v_N$ is at the level k in $\mathfrak{M}_0$.

If $llca(v_1, v_2) \leq k$, then Φ merges at some level. We prove it using the following morphism $h_2 : \mathfrak{M}'\restriction_{\{v_1,\ldots,v_n\}} \to \mathcal{X}$. Let $\mathfrak{T}^0_{t,p}$ be the tree that contains v_1.

$$h_2(v) = \begin{cases} \underline{s - 2N} & \text{if } v \text{ at a level } s \geq k \text{ in } \mathfrak{M}_0 \text{ and } llca(v_1, v) > k \\ \overline{s - 2N} & \text{if } v \text{ at a level } s \geq k \text{ in } \mathfrak{M}_0 \text{ and } llca(v_1, v) \leq k \\ \underline{s} & \text{if } v \text{ at a level } s \text{ in } \mathfrak{M}_1 \\ \underline{2N + s} & \text{if } v \text{ at a level } s \text{ in } \mathfrak{M}_0 \end{cases}$$

It is readily checkable that h_2 is a morphism and it implies that Φ merges at some level.

Let $llca(v_1, v_2) > k$. We prove that Φ does not fork at some level. To this end, let k' be such that no world among $v_3, \ldots, v_N$ is at the level k' in $\mathfrak{M}_1$. We define $V_1 = V_{M0} \cup V_{M1}$ as follows. Set $v \in V_{M0}$ if and only if v is at a level $s > k$ in $\mathfrak{M}_0$ and $lcm(v_1, v) \in \{v_1, v\}$ (in other worlds, v is an ancestor or descendant of v_1 in $\mathfrak{M}_0$). Finally, for each leaf w from V_{M0} labeled by m and each $t \in tp(\mathfrak{M})$, V_{M1} contains all worlds from levels less than k' in $\mathfrak{T}^1_{t,m}$.

Let $t = llca(v_1, v_2) - k$, We define a morphism $h_3 : \mathfrak{M}'\restriction_{\{v_1,\ldots,v_n\}} \to \mathcal{T}_\infty$.

$$h_3(v) = \begin{cases} \underline{0^{\nu_k(v)}} & \text{if } v \in V_1 \text{ or } \nu_k(v) < t \\ \underline{0^t 1^{\nu_k(v)-t}} & \text{otherwise} \end{cases}$$

It is readily checkable that h_3 is a morphism and it implies that Φ does not fork at the level t.

The case when $s_1 = s_2 = 1$ is symmetric. □

Formulas that do not merge and do not forks at some level. In the case of formulas that do not force long edges and do not fork at some level, the finite model property follows from the fact that each satisfiable formula has a k-periodic model for some k.

Proposition 3.11 *Let Φ be an unbounded* UHF *formula that does not force long edges and does not fork at some level $k > 0$. Then modal logic has the finite global model property with respect to $\mathcal{K}_\Phi$.*

Proof. Let $\mathfrak{M}$ be defined as in the proof of Proposition 3.10. First, observe that $\mathfrak{C}_\Phi(\mathcal{L}_\mathbb{Z}) = \mathcal{L}_\mathbb{Z}$ and, since Φ is unbounded, $\mathcal{L}_\mathbb{Z}$ is a model of Φ. Let v be a world at level $\mathfrak{g}(|\Phi|)$ and let $\mathfrak{M}'$ be the model that consists of all descendants of v from levels greater than $2\mathfrak{g}(|\Phi|)$. By Lemma 2.7, all worlds in $\mathfrak{M}'$ at the same level are equivalent. Since the number of types is bounded, there exist two levels k, l in $\mathfrak{M}'$ such that $k - l > |\Phi| + 1$ and the sets of types realized at levels k and l are equal. We create model $\mathfrak{M}''$ by removing all worlds from levels greater than or equal to k, and connecting all worlds from level $k-1$ to worlds from level l. Finally, we define $\mathfrak{M}'''$ by taking for each level one world of each type realized at this level. A quick check shows that models $\mathfrak{M}'$, $\mathfrak{M}''$, and $\mathfrak{M}'''$ satisfy φ and that $\mathfrak{M}'''$ is finite.

Now we justify that $\mathfrak{M}'''$ is a model of Φ. Since $\mathcal{L}_\mathbb{Z}$ is a model of Φ, Lemma 3.5 shows that C_{k-l} is a model of Φ, and the same lemma shows that therefore any k–l-periodic model is a model of Φ. Model $\mathfrak{M}'''$ is obviously k–l-periodic. □

4 Formulas that force long edges

As mentioned earlier, for formulas that satisfy S3, the polynomial model property follows from [16]. The rest of this section is devoted to formulas $\Phi \in$ UHF that satisfy S2.

First, observe that in this case modal logic may lack the finite model property (local and global) with respect to $\mathcal{K}_\Phi$. Consider, for example, $(xRz_1 \wedge z_1Ry \Rightarrow xRy) \wedge (xRx \Rightarrow \bot)$ and a modal formula $\Diamond\top \wedge \Box\Diamond\top$. A quick check shows that all models of these formulas are infinite (in local and global cases). On the other hand, modal logic has the finite model property with respect to the class defined by $xRx \wedge (xRz_1 \wedge z_1Ry \Rightarrow xRy)$.

To show decidability we prove that if a formula φ has a finite model (in local or global case), then it has a model of size bounded by $|\varphi|^{O(|\varphi|)}$. Clearly, it leads to a NExpTime algorithm that simply guesses such a model and verifies it.

Consider a modal formula φ and its $\mathcal{K}_\Phi$-based model $\mathfrak{M}$ with universe M. We say that a world w is *redundant* for φ and $\mathfrak{M}$ if $\mathfrak{M}_{\restriction M \setminus \{w\}}$ is a model of φ. We prove the following lemma by showing that a model that is large enough has to contain a redundant world.

Lemma 4.1 *Let Φ be an unbounded* UHF *formula that forces long edges. If φ has a finite $\mathcal{K}_\Phi$-based model, then it has a $\mathcal{K}_\Phi$-based model of size bounded by $|\varphi|^{O(|\varphi|)}$.*

Proof. Let Φ be an unbounded UHF formula that satisfies S2 for some l and $a_1, \ldots, a_l$, and φ be a modal formula with a $\mathcal{K}_\Phi$-based model $\mathfrak{M}$.

Let $c = a_1$. Observe that for all $i \in \mathbb{Z}$ and $k \geq 0$ we have $\mathfrak{C}_\Phi(\mathcal{L}_\mathbb{Z}) \models \underline{iRi + kc + 1}$.

We start from bounding the number of worlds that are not $\mathfrak{g}(|\Phi|)$-preceded. We use the standard selection technique [2] — we start from an arbitrary world that satisfies φ, and then recursively for each world added in the previous stage we pick at most $|\varphi|$ witnesses. Let $\mathfrak{M}'$ be a model obtained this way. We define *the royal part* of $\mathfrak{M}'$ as the set of worlds that contain all worlds that are not $\mathfrak{g}(|\Phi|)$-preceded and *the court* as the set of $\mathfrak{g}(|\Phi|)$-preceded worlds that were added as witnesses for some worlds from the royal part. Clearly, the total size of the royal part and the court can be bounded by $|\varphi|^{\mathfrak{g}(|\Phi|)+1}$.

Let w be a $\mathfrak{g}(|\Phi|)$-inner world not from the court such that for each subformula $\Diamond\psi$ of φ such that ψ is satisfied in w there exists a $\mathfrak{g}(|\Phi|)$-inner world $w_\psi \neq w$ that satisfies ψ and that there is a path from w to w_ψ with the length cj for some j. We show that w is redundant.

Consider any predecessor w' of w. If w' is not $\mathfrak{g}(|\Phi|)$-preceded, then it has all the required witnesses in the court and the royal part. Otherwise, let ψ be a subformula of φ such that w satisfies ψ. We show that there is an edge from w' to w_ψ. To this end, consider a path $v_1, v_2, \ldots, v_{\mathfrak{g}(|\Phi|)}, w', w, v'_1, v'_2, \ldots, v'_{cj}, w_\psi, v''_1, v''_2, \ldots, v''_{\mathfrak{g}(|\Phi|)}$. Such a path exists since w' is $\mathfrak{g}(|\Phi|)$-preceded and w_ψ is $\mathfrak{g}(|\Phi|)$-inner, and there is a straightforward morphism from $\mathcal{I}_{2\mathfrak{g}(|\Phi|)+2+cj}$ into this path. So it is enough to show that there is an edge from $\mathfrak{g}(|\Phi|)+1$ to $\mathfrak{g}(|\Phi|)+1+cj+1$ in $\mathfrak{C}(\mathcal{I}_{2\mathfrak{g}(|\Phi|)+2+j})$. By earlier observations, $\mathfrak{C}_\Phi(\mathcal{L}_\mathbb{Z})$ contains an edge from $\underline{\mathfrak{g}(|\Phi|)+1}$ to $\underline{\mathfrak{g}(|\Phi|)+1+cj+1}$, and Lemma 2.6 implies that there is an edge from $\underline{\mathfrak{g}(|\Phi|)+1}$ to $\underline{\mathfrak{g}(|\Phi|)+1+cj+1}$ in $\mathfrak{C}(\mathcal{I}_{2\mathfrak{g}(|\Phi|)+2+cj})$.

By iterating the above argument we can remove all $\mathfrak{g}(|\Phi|)$—inner worlds except for at most $|\varphi|^{c\cdot|\varphi|}$ worlds. Finally, we again use the selection technique to bound the number of worlds that are not $\mathfrak{g}(|\Phi|)$-followed by $|\varphi|^{c\cdot|\varphi|}\cdot|\varphi|^{\mathfrak{g}(|\Phi|)}$. Since Φ is not a part of an instance, we reduced the number of worlds to $|\varphi|^{O(|\varphi|)}$. □

The above lemma leads to the following result.

Proposition 4.2 *If Φ is an unbounded* UHF *formula that forces long edges, then $\mathcal{K}_\Phi$-FINSAT and global $\mathcal{K}_\Phi$-FINSAT are in* NExpTime.

Establishing better complexity bounds in the case of formulas satisfying S2 is left as an open problem.

References

[1] Bárány, V. and M. Bojanczyk, *Finite satisfiability for guarded fixpoint logic*, Information Processing Letters **112** (2012), pp. 371–375.

[2] Blackburn, P., M. de Rijke and Y. Venema, "Modal Logic," Cambridge Tracts in Theoretical Comp. Sc. **53**, Cambridge University Press, Cambridge, 2001.

[3] Gabbay, D. M. and V. B. Shehtman, *Products of modal logics, part 1*, Logic Journal of the IGPL **6** (1998), pp. 6–1.

[4] Grädel, E., *On the restraining power of guards*, Journal of Symbolic Logic **64** (1999), pp. 1719–1742.

[5] Grädel, E., M. Otto and E. Rosen, *Two-variable logic with counting is decidable*, in: *Twelfth Annual IEEE Symposium on Logic in Computer Science*, 1997, pp. 306–317.

[6] Grädel, E. and I. Walukiewicz, *Guarded fixed point logic*, in: *Fourteenth Annual IEEE Symposium on Logic in Computer Science*, 1999, pp. 45–54.

[7] Hemaspaandra, E., *The price of universality*, Notre Dame Journal of Formal Logic **37** (1996), pp. 174–203.

[8] Hemaspaandra, E. and H. Schnoor, *On the complexity of elementary modal logics*, in: *STACS*, LIPIcs **1** (2008), pp. 349–360.

[9] Hemaspaandra, E. and H. Schnoor, *A universally defined undecidable unimodal logic*, in: *MFCS*, Lecture Notes in Computer Science **6907** (2011), pp. 364–375.

[10] Kieroński, E., J. Michaliszyn and J. Otop, *Modal logics definable by universal three-variable formulas*, in: *FSTTCS*, LIPIcs **13** (2011), pp. 264–275.

[11] Kieroński, E., J. Michaliszyn, I. Pratt-Hartmann and L. Tendera, *Extending two-variable first-order logic with equivalence closure*, in: *LICS '12: Proceedings of the 29th IEEE symposium on Logic in Computer Science*, 2012.

[12] Kieroński, E. and M. Otto, *Small substructures and decidability issues for first-order logic with two variables.*, in: *LICS*, 2005, pp. 448–457.

[13] Kieroński, E. and L. Tendera, *On finite satisfiability of the guarded fragment with equivalence or transitive guards*, LPAR **4790** (2007), pp. 318–332.

[14] Kieroński, E. and L. Tendera, *On finite satisfiability of two-variable first-order logic with equivalence relations*, LICS (2009), pp. 123–132.

[15] Michaliszyn, J., *Decidability of the guarded fragment with the transitive closure*, in: *ICALP (2)*, Lecture Notes in Computer Science **5556** (2009), pp. 261–272.

[16] Michaliszyn, J. and J. Otop, *Decidable elementary modal logics*, in: *LICS '12: Proceedings of the 29th IEEE symposium on Logic in Computer Science*, 2012.

[17] Mortimer, M., *On languages with two variables*, Mathematical Logic Quarterly **21** (1975), pp. 135–140.

[18] Otto, M., *Two variable first-order logic over ordered domains*, Journal of Symbolic Logic **66** (1998), pp. 685–702.

[19] Pacholski, L., W. Szwast and L. Tendera, *Complexity of two-variable logic with counting*, in: *Twelfth Annual IEEE Symposium on Logic in Computer Science*, 1997, pp. 318–327.

[20] Pratt-Hartmann, I., *Complexity of the two-variable fragment with (binary-coded) counting quantifiers*, Journal of Logic, Language and Information (2005).

[21] Sahlqvist, H., *Completeness and correspondence in the first and second order semantics for modal logic*, Proceedings of the Third Scandinavian Logic Symposium (1973).

[22] Szwast, W. and L. Tendera, *FO^2 with one transitive relation is decidable*, unpublished.

[23] Szwast, W. and L. Tendera, *The guarded fragment with transitive guards*, Annals of Pure and Applied Logic **128** (2004), pp. 227–276.

[24] van Emde Boas, P., *The convenience of tilings*, in: *Complexity, Logic, and Recursion Theory* (1997), pp. 331–363.

[25] Vardi, M. Y., *Why is modal logic so robustly decidable?*, DIMACS Series in Discrete Mathematics and Theoretical Computer Science, **31** (1997), pp. 149–184.

A Remark on a Peculiarity in the Functor Semantics for Superintuitionistic Predicate Logics with (or without) Equality

Dmitrij Skvortsov [1]

All-Russian Institute of Scientific and Technical Information
Usievicha 20, 125190, Moscow, RUSSIA /
Molodogvardejskaja 22, korp.3, kv.29, 121351, Moscow, RUSSIA

Abstract

We notice the following slightly curious (and perhaps, slightly unexpected) logical property of the functor semantics for superintuitionistic predicate logics, contrasting with a well-known property of the usual Kripke semantics. Namely, for a category $\mathcal{C}$ its logic (i.e., the logic of all $\mathcal{C}$-sets with the given, fixed $\mathcal{C}$) in general is not reducible to cones (i.e., restrictions of $\mathcal{C}$ to upward closed rooted subsets of its frame representation $W = Ob(\mathcal{C})$). Related notions and observations are discussed as well.

Keywords: superintuitionistic predicate logics, Kripke semantics, functor semantics, categories.

Recall that a *cone* (a point-generated subframe) in an intuitionistic propositional Kripke frame (i.e., a pre-ordered, or in particular, a partially ordered set) W is $W^u = \{v \in W \mid u \leq v\}$ (for $u \in W$), ordered by the restriction of $\leq$ from W to W^u. It is well known that the (propositional) validity in Kripke frames is reducible to that in their cones, i.e.,

$$\mathbf{PL}(W) = \bigcap_{u \in W} \mathbf{PL}(W^u), \tag{0}$$

where $\mathbf{PL}(W)$ is the propositional logic of W (i.e., the set of formulas valid in W). The similar reducibility to cones holds for predicate logics as well, i.e.,

$$\mathbf{KL}(W) = \bigcap_{u \in W} \mathbf{KL}(W^u) \quad \text{for a pre-ordered set } W, \tag{1}$$

$\mathbf{KL}(W)$ being the set of predicate formulas valid in all predicate Kripke frames $F = (W, \overline{D})$ (with systems of expanding non-empty domains $\overline{D} = (D_u : u \in W)$) based on W; this claim holds for logics both without and with equality (in

---

[1] e-mail: skvortsovd@yandex.ru

the latter case we will write $\mathbf{KL}^{=}(W)$). This statement is well known and usually is considered as obvious, however it is not so straightforward. Namely, an immediate predicate counterpart of (0) looks as follows:

$$\mathbf{L}(F) = \bigcap_{u \in W} \mathbf{L}(F^u) \quad \text{for a predicate Kripke frame } F \text{ based on } W, \qquad (1)'$$

where the cone F^u (for $u \in W$) is the restriction of F to W^u. Now, to deduce (1), the following simple claim needs to be used:

$$\begin{array}{l} \text{every frame } F_0 \text{ over } W^u \text{ can be extended} \\ \qquad\qquad \text{to a frame } F \text{ over } W \text{ such that } F^u = F_0. \end{array} \qquad (1)''$$

This auxiliary claim for the ordinary predicate Kripke semantics is rather obvious. Nevertheless, for more general Kripke-style semantics it becomes not so trivial. Here we consider the functor semantics [2] for superintuitionistic predicate logics and describe a natural counterpart of $(1)''$ that fails for this semantics; moreover, we show that the corresponding counterpart of (1) fails for predicate logics with equality. We believe that it would fail for logics without equality as well, although we do not have such an example. This observation confirms that the behavior of rather general predicate Kripke-style semantics often differs from that of the propositional Kripke semantics.

1 Preliminary notions: The functor semantics

We consider *predicate formulas* (without or with equality, and in any case without function symbols) built as usual, by using the connectives $\&, \vee, \rightarrow$, the propositional constant $\bot$ ('falsity'), and the quantifiers $\forall, \exists$. We use the standard abbreviations: $(A \leftrightarrow B) = (A \rightarrow B)\&(B \rightarrow A)$ and $\neg A = (A \rightarrow \bot)$.

We regard *superintuitionistic predicate logics* (without and with equality) in the usual way, i.e., as sets of predicate formulas containing all axioms of intuitionistic (Heyting) predicate logic **QH** (or $\mathbf{QH}^{=}$ for the case with equality), and closed under modus ponens, generalization, and substitution of arbitrary formulas for atomic ones.

To begin with, let us recall necessary notions related to the functor semantics, see [2] (also cf. [1, Sect. 5.6: Def. 5.6.3 etc.] or e.g. [6, Sect.4.1]).

Let $\mathcal{C}$ be a category with a frame representation W. This means that $W = Ob(\mathcal{C})$ is the set of objects of $\mathcal{C}$ pre-ordered by the following relation:

$u \leq v$ iff $\mathcal{C}(u, v) \neq \varnothing$, i.e., iff in $\mathcal{C}$ there exists a morphism from u to v.

A *$\mathcal{C}$-set* (a SET-valued functor or a presheaf over $\mathcal{C}$, inhabited, i.e., with non-emptiness assumption) is a triple $\mathbb{F} = (W, \overline{D}, \overline{E})$ in which $\overline{D} = (D_u : u \in W)$ is a family of disjoint non-empty domains and $\overline{E} = (E_\mu : \mu \in Mor(\mathcal{C}))$ is a family of functions with $E_\mu : D_u \rightarrow D_v$ whenever $\mu \in \mathcal{C}(u, v)$ (i.e., μ is a morphism from u to v). As usual, it is required that $E_{\mu \circ \mu'} = E_{\mu'} \circ E_\mu$ for $\mu \in \mathcal{C}(u, v)$, $\mu' \in \mathcal{C}(v, w)$ (i.e., $E_{\mu \circ \mu'}(a) = E_{\mu'}(E_\mu(a))$ for any $a \in D_u$), and $E_{1_u} = 1_{D_u}$ (the identity function on D_u corresponds to the identity morphism $1_u \in \mathcal{C}(u, u)$, $u \in W$). Sometimes we can admit $\mathcal{C}$-sets with non-disjoint domains; in this case we regard them as $\mathcal{C}$-sets with disjoint domains $D'_u = \{u\} \times D_u = \{\langle u, a \rangle \mid a \in D_u\}$.

Let $D_n = \bigcup_{u \in W} (D_u)^n$ for $n > 0$ and $D_0 = W$. We define a pre-order $\leq_n$ on D_n (for $n > 0$) by:

$$[(a_1, \ldots, a_n) \leq_n (b_1, \ldots, b_n)] \text{ iff } \exists \mu \in Mor(\mathcal{C}) [\bigwedge_{i=1}^{n} (E_\mu(a_i) = b_i)];$$

sometimes we write $(E_\mu(\mathbf{a}) = \mathbf{b})$ for $\bigwedge_{i=1}^{n} (E_\mu(a_i) = b_i)$, where $\mathbf{a} = (a_1, \ldots, a_n)$ and $\mathbf{b} = (b_1, \ldots, b_n)$.[2] We identify $\leq_0$ with the original pre-order $\leq$ on $D_0 = W$.

Finally let: $[(a_1, \ldots, a_n) <_n (b_1, \ldots, b_n)]$ iff
$[(a_1, \ldots, a_n) \leq_n (b_1, \ldots, b_n)] \,\&\, \neg[(b_1, \ldots, b_n) \leq_n (a_1, \ldots, a_n)]$.

A *valuation* in a $\mathcal{C}$-set $\mathbb{F}$ is a function ξ sending every n-place predicate symbol P to an upward closed (by $\leq_n$) subset $\xi(P)$ of D_n.

A valuation ξ in $\mathbb{F}$ gives rise to the forcing relation $u \vDash_\xi A(\mathbf{a})$ between points $u \in W$ and formulas $A(\mathbf{a})$ (with parameters replaced by elements of D_u); here $\mathbf{a} = (a_1, \ldots, a_n) \in (D_u)^n$. Namely, $\vDash$ is defined inductively:[3]

$u \vDash P(\mathbf{a}) \Leftrightarrow (\mathbf{a} \in \xi(P))$ for a predicate symbol P; $\quad u \nvDash \bot$;
$u \vDash (B \& C)(\mathbf{a}) \Leftrightarrow (u \vDash B(\mathbf{a})) \,\&\, (u \vDash C(\mathbf{a}))$;
$u \vDash (B \vee C)(\mathbf{a}) \Leftrightarrow (u \vDash B(\mathbf{a})) \vee (u \vDash C(\mathbf{a}))$;
$u \vDash (B \to C)(\mathbf{a}) \Leftrightarrow \forall v \geq u \, \forall \mathbf{b} \in (D_v)^n \, [(\mathbf{a} \leq_n \mathbf{b}) \& (v \vDash B(\mathbf{b})) \Rightarrow (v \vDash C(\mathbf{b}))]$;
$u \vDash \forall x B(\mathbf{a}, x) \Leftrightarrow \forall v \geq u \, \forall \mathbf{b} \in (D_v)^n \, \forall c \in D_v \, [(\mathbf{a} \leq_n \mathbf{b}) \Rightarrow (v \vDash B(\mathbf{b}, c))]$;
$u \vDash \exists x B(\mathbf{a}, x) \Leftrightarrow \exists c \in D_u \, [u \vDash B(\mathbf{a}, c)]$, and $u \vDash (a = b) \Leftrightarrow (a = b)$.

It is easily shown (by induction on A) that the forcing is preserved upwards:

$$(\mathbf{a} \leq_n \mathbf{b}) \Rightarrow (u \vDash A(\mathbf{a}) \Rightarrow v \vDash A(\mathbf{b}))$$

for any formula A and $\mathbf{a} \in (D_u)^n$, $\mathbf{b} \in (D_v)^n$, $u \leq v$.

A predicate formula A with parameters $\mathbf{x} = (x_1, \ldots, x_n)$ is *true* w.r.t. ξ if $u \vDash A(\mathbf{a})$ for every $u \in W$ and $\mathbf{a} \in (D_u)^n$. A formula A is *valid* in a $\mathcal{C}$-set $\mathbb{F}$ (notation: $\mathbb{F} \vDash A$) if it is true w.r.t. all valuations in $\mathbb{F}$. The *predicate logic* (without or with equality) *of a $\mathcal{C}$-set* $\mathbb{F}$ is the set

$\mathbf{L}^{(=)}(\mathbb{F}) = \{A \mid$ all substitution instances of A are valid in $\mathbb{F}\}$;

again we write $\mathbf{L}(\mathbb{F})$ or $\mathbf{L}^{=}(\mathbb{F})$ for the logics without or with equality (respectively), and write $\mathbf{L}^{(=)}(\mathbb{F})$, when we mean both of these logics. It is known (cf. e.g. [1, Proposition 5.6.27] that $\mathbf{L}^{(=)}(\mathbb{F})$ is indeed a superintuitionistic logic.[4]

A formula is called *valid* in a category $\mathcal{C}$ (or $\mathcal{C}$-*valid*, for short) if it is valid in all $\mathcal{C}$-sets.

[2] Here $(D_n, \leq_n)$ are well-defined, since domains D_u are disjoint. This is perhaps the only reason, why the disjointness of D_u is convenient and useful.

[3] Usually we omit subscript ξ (for readability) and write $\vDash$ for $\vDash_\xi$, $\vDash'$ for $\vDash_{\xi'}$, etc. Sometimes we may call $\vDash$ a valuation in $\mathbb{F}$ as well.

[4] By the way, there exists another, slightly more explicit description of the logic $\mathbf{L}^{(=)}(\mathbb{F})$; we present (and use) this description in Appendix (see Section 5).

Respectively, for a category $\mathcal{C}$ its *predicate logic* (without or with equality)

$$\mathbf{L}^{(=)}(\mathcal{C}) = \bigcap(\,\mathbf{L}^{(=)}(\mathbb{F}) : \mathbb{F} \text{ is a } \mathcal{C}\text{-set }\,)$$

is the set of formulas, all substitution instances of which are $\mathcal{C}$-valid.

Note that a usual predicate Kripke frame $F = (W, \overline{D})$ can be presented as a $\mathcal{C}$-set (based on W) with expanding domains (i.e., $D_u \subseteq D_v$ for $u \leq v$) and with inclusion mappings E_μ for $\mu \in \mathcal{C}(u, v)$.

2 The logics of categories: Non-reducibility to cones

Let $\mathcal{C}$ be a category based on a pre-ordered set W, and let $\mathbb{F}$ be a $\mathcal{C}$-set. Their *cones* $\mathcal{C}^u$ and $\mathbb{F}^u$ (for $u \in W$) are defined in a natural way, as the restrictions to the cone $W^u = \{v \in W \mid u \leq v\}$ of W; clearly, $\mathbb{F}^u$ is a $\mathcal{C}^u$-set. The restriction of a valuation in a $\mathcal{C}$-set $\mathbb{F}$ to $\mathbb{F}^u$ is a valuation in $\mathbb{F}^u$ (for atomic formulas), and the corresponding forcing relation $\models$ in $\mathbb{F}^u$ (for all formulas) is obtained as the restriction of $\models$ from $\mathbb{F}$. On the other hand, any valuation in $\mathbb{F}^u$ can be easily extended to a valuation in $\mathbb{F}$. Hence we conclude that:

$$\mathbf{L}^{(=)}(\mathbb{F}) \;=\; \bigcap_{u \in W} \mathbf{L}^{(=)}(\mathbb{F}^u) \text{ for any } \mathcal{C}\text{-set } \mathbb{F} \text{ based on } W, \qquad (2)'$$

i.e., the natural counterpart of $(1)'$ for the functor semantics holds. On the other hand, a similar property for categories fails:

Theorem *There exists a category $\mathcal{C}_0$ over a three-element chain $W_0 = \{v_0, v_1, v_2\}$ (where $v_0 < v_1 < v_2$) such that*

$$\mathbf{L}^{=}(\mathcal{C}) \not\subseteq \mathbf{L}^{=}(\mathcal{C}^u) \qquad (\overline{2})$$

for $\mathcal{C} = \mathcal{C}_0$ and $u = v_1$.

Namely, define the category $\mathcal{C}_0$ (based on W_0) with the following morphisms: $\mathcal{C}_0(v, v) = \{1_v\}$ for all $v \in W_0$, $\mathcal{C}_0(v_0, v_1) = \{\mu_0\}$, $\mathcal{C}_0(v_1, v_2) = \{\mu_1, \mu_2\}$, $\mathcal{C}_0(v_0, v_2) = \{\mu^*\}$, and with the composition $\mu_0 \circ \mu_i = \mu^*$ for $i = 1, 2$.

Put $\Phi_0 = (\Phi_1 \to \Phi_2)$, where $\Phi_1 = \forall x \forall y\,[(x = y) \vee P \vee \neg P]$, $\Phi_2 = \neg P \vee \neg\neg P$.

Define the $\mathcal{C}_0^{v_1}$-set $\mathbb{F}_0 = (W_0^{v_1}, \overline{D}, \overline{E})$, where $D_{v_1} = \{a_0\}$, $D_{v_2} = \{a_1, a_2\}$, and $E_{\mu_i}(a_0) = a_i$ for $i = 1, 2$. Our theorem follows from the subsequent claim:

Lemma 1 *(1) $\Phi_0 \notin \mathbf{L}^{=}(\mathbb{F}_0)$, and so $\Phi_0 \notin \mathbf{L}^{=}(\mathcal{C}_0^{v_1})$; (2) $\Phi_0 \in \mathbf{L}^{=}(\mathcal{C}_0)$.*

Proof. (1) The substitution instance $\Phi_0^1(z) = (\Phi_1^1(z) \to \Phi_2^1(z))$, where $\Phi_1^1(z) = \forall x \forall y\,[(x = y) \vee P_1(z) \vee \neg P_1(z)]$, $\Phi_2^1(z) = \neg P_1(z) \vee \neg\neg P_1(z)$, is not valid in $\mathbb{F}_0$. Indeed, consider a valuation

$$v_j \models P_1(a_i) \;\Leftrightarrow\; (i = j = 2)$$

in $\mathbb{F}_0$. Then $v_1 \models \Phi_1^1(a_0)$ (since $v_2 \models P_1(a_i) \vee \neg P_1(a_i)$ for $i = 1, 2$, and D_{v_1} is one-element) and $v_1 \not\models \Phi_2^1(a_0)$ (since $v_2 \models P_1(a_2)$ and $v_2 \models \neg P_1(a_1)$).

(2) Let a substitution instance of Φ_0 be given: $\Phi_0^A(\mathbf{z}) = (\Phi_1^A(\mathbf{z}) \to \Phi_2^A(\mathbf{z}))$, where $\Phi_1^A(\mathbf{z}) = \forall x \forall y\,[(x{=}y) \vee A(\mathbf{z}) \vee \neg A(\mathbf{z})]$, $\Phi_2^A(\mathbf{z}) = \neg A(\mathbf{z}) \vee \neg\neg A(\mathbf{z})$, $\mathbf{z} = (z_1, \ldots, z_n)$ being the list of parameters of A (all z_i are distinct from x, y). Let $\mathbb{F} = (W_0, \overline{D}, \overline{E})$ be a $\mathcal{C}_0$-set. Suppose that $u \vDash \Phi_1^A(\mathbf{d})$ and $u \nvDash \Phi_2^A(\mathbf{d})$ for some $u \in W_0$ and $\mathbf{d} \in (D_u)^n$. Then $v_2 \vDash A(\mathbf{d}')$ and $v_2 \vDash \neg A(\mathbf{d}'')$ for some $\mathbf{d}', \mathbf{d}'' \in (D_{v_2})^n$ such that $\mathbf{d} <_n \mathbf{d}'$ and $\mathbf{d} <_n \mathbf{d}''$ (and so $u < v_2$). Then clearly, $u \nvDash A(\mathbf{d})$ and $u \nvDash \neg A(\mathbf{d})$, hence the domain D_u is one-element (since $u \vDash (a_1{=}a_2)$ for any $a_1, a_2 \in D_u$).

First, if $u{=}v_0$, then $\mathbf{d}' = E_{\mu^*}(\mathbf{d}) = \mathbf{d}''$, and this leads to a contradiction.

Second, let $u{=}v_1$, $D_u = \{a_1\}$. Then $\mathbf{d} = (a_1, \ldots, a_1) = E_{\mu_0}(\mathbf{d}^*)$, where $\mathbf{d}^* = (a_0, \ldots, a_0)$ for an arbitrary $a_0 \in D_{v_0}$. Now, $\mathbf{d}' = E_{\mu_i}(\mathbf{d})$ for some $i \in \{1, 2\}$, hence $\mathbf{d}' = E_{\mu_i}(E_{\mu_0}(\mathbf{d}^*)) = E_{\mu^*}(\mathbf{d}^*)$, and similarly $\mathbf{d}'' = E_{\mu^*}(\mathbf{d}^*)$. Thus $\mathbf{d}' = \mathbf{d}''$ again. □

The key idea. For our category $\mathcal{C} = \mathcal{C}_0$ *and* $u{=}v_1$ we have:

$$\text{there exists a } \mathcal{C}^u\text{-set } \mathbb{F}_0 \text{ that cannot be extended to a } \mathcal{C}\text{-set } \mathbb{F} \text{ such that } \mathbb{F}^u = \mathbb{F}_0. \qquad (\overline{2})''$$

Clearly, this property $(\overline{2})''$ holds for any category $\mathcal{C}$ (based on W) and $u \in W$ satisfying $(\overline{2})$.

We conjecture that the peculiarity $(\overline{2})$ actually transfers to logics without equality as well, i.e., we hope that there exists a category $\mathcal{C}$ (based on W) such that

$$\mathbf{L}(\mathcal{C}) \not\subseteq \mathbf{L}(\mathcal{C}^u) \text{ for some } u \in W. \qquad (\overline{2})$$

However we could not construct a formula without equality that 'reflects' and exploits the property of our category $\mathcal{C}_0$, which does not allow to extend the $\mathcal{C}_0^{v_1}$-set $\mathbb{F}_0$ to a $\mathcal{C}_0$-set.

On the other hand, the following claim is obvious:

Lemma 2 *Let $\mathcal{C}$ be a category based on W. A formula A is $\mathcal{C}$-valid if it is $\mathcal{C}^u$-valid for all $u \in W$.*

Indeed, if a $\mathcal{C}$-set $\mathbb{F} \nvDash A$ is given, then $u \nvDash A$ for some $u \in W$ and a valuation $\vDash$ in $\mathbb{F}$. Hence $u \nvDash^u A$ for the restriction $\vDash^u$ of $\vDash$ to $\mathbb{F}^u$, and so $\mathbb{F}^u \nvDash A$. □

Hence we obtain

Corollary 3 *For any category $\mathcal{C}$ based on W:*

$$\bigcap_{u \in W} \mathbf{L}^{(=)}(\mathcal{C}^u) \subseteq \mathbf{L}^{(=)}(\mathcal{C}). \qquad (2)$$

Clearly, this inclusion is proper iff $\mathcal{C}$ satisfies $(\overline{2})$ (for the logics without or with equality, respectively).

3 Reducibility to cones for ∀-positive formulas

Now we describe a class of formulas, for which $\mathcal{C}$-validity is reducible to $\mathcal{C}^u$-validity (for all categories $\mathcal{C}$).

Recall that an occurrence of a subformula or a quantifier etc. in a formula A is called *positive* (or *negative*) if it occurs in an even (respectively, odd) number of premises of implications. We call a formula A (without or with equality) *∀-positive* if all occurrences of ∀ in A are positive and all occurrences of ∃ in A are negative. Similarly, a formula A is *∃-positive* if all occurrences of ∃ in A are positive and all occurrences of ∀ in A are negative. The following simple lemma gives an inductive description of these notions:

Lemma 4

(1) *Any atomic formula is both ∀-positive and ∃-positive.*

(2) *Formulas $A_1 \& A_2$, $A_1 \vee A_2$ are ∀-positive (or ∃-positive) iff both A_1 and A_2 are ∀-positive (respectively, ∃-positive).*

(3) *A formula $(A_1 \to A_2)$ is ∀-positive iff A_1 is ∃-positive and A_2 is ∀-positive.*
A formula $(A_1 \to A_2)$ is ∃-positive iff A_1 is ∀-positive and A_2 is ∃-positive.

(4) *A formula $\forall x\, A$ is ∀-positive iff A is ∀-positive.*
Also $\forall x\, A$ is not ∃-positive (for any A).

(5) *A formula $\exists x\, A$ is ∃-positive iff A is ∃-positive.*
Also $\exists x\, A$ is not ∀-positive (for any A).

We present a proof of the subsequent claim in Appendix (see Section 5):

Proposition 1 *Let $\mathcal{C}$ be a category based on W, and let A be a ∀-positive formula (without or with equality). Then:*

$$A \in \mathbf{L}^{(=)}(\mathcal{C}) \Rightarrow A \in \mathbf{L}^{(=)}(\mathcal{C}^u) \text{ for all } u \in W.$$

Therefore,

$$(A \in \mathbf{L}^{(=)}(\mathcal{C})) \text{ iff } (A \in \mathbf{L}^{(=)}(\mathcal{C}^u) \text{ for all } u \in W),$$

for any ∀-positive formula A (since the converse implication readily follows from Corollary 3).

Let us call all negative occurrences of ∀ and all positive occurrences of ∃ in a formula A its *critical* occurrences. Clearly, a formula is ∀-positive iff it has no critical occurrences of quantifiers.

The formula Φ_0 constructed in Section 2 (for our Theorem) is intuitionistically equivalent to the formula

$$\Phi_0' = (\Phi_1' \to \Phi_2), \qquad \text{where } \Phi_1' = \exists x \forall y\, [(x{=}y) \vee P \vee \neg P],$$

since $\mathbf{QH}^{=} \vdash (\Phi_1 \leftrightarrow \Phi_1')$. The formula Φ_0' has only one critical occurrence $\forall y$ (and one non-critical $\exists x$). Also Φ_0' is equivalent to $\forall x\, \Phi_0''$, where

$$\Phi_0'' = (\forall y\,[(x{=}y) \vee P \vee \neg P] \rightarrow \neg P \vee \neg\neg P\,).$$

This $\exists$-free formula has one critical occurrence $\forall y$ (and one non-critical $\forall x$). Moreover, $\forall x\, \Phi_0''$ can be replaced with the non-closed $\exists$-free formula $\Phi_0''(x)$ with only one occurrence of $\forall$ (clearly, critical). On the other hand, we do not know, whether Proposition 1 can be transferred to $\forall$-free formulas (with critical, i.e., positive occurrences of $\exists$).

4 The logics of pre-ordered sets: Reducibility to cones

In Section 2 we presented a natural counterpart of the claim (1) that fails for the functor semantics (at least, in the case with equality). Nevertheless, a more literal, straightforward counterpart of (1) holds.

Namely, let us define the *predicate logic* (without or with equality) *of* a pre-ordered set W *in the functor semantics* as follows:

$$\mathbf{FL}^{(=)}(W) = \bigcap (\,\mathbf{L}^{(=)}(\mathcal{C}) : \mathcal{C} \text{ is a category based on } W\,).$$

The following claim holds (the proof is given in Appendix, see Section 5):

Proposition 2 $\mathbf{FL}^{(=)}(W) \subseteq \mathbf{FL}^{(=)}(W^u)$ *for any pre-ordered* W *and* $u{\in}W$.

Therefore,

$$\mathbf{FL}^{(=)}(W) = \bigcap_{u\in W} \mathbf{FL}^{(=)}(W^u) \quad \text{for any pre-ordered set } W; \tag{3}$$

indeed, the converse inclusion readily follows from Corollary 3.

However, for logics without equality this claim is not interesting, because a strictly stronger statement actually holds (its proof requires extra preparation, techniques, and accuracy, and will be given in the continuation [5] of this paper):

Proposition 2* $\mathbf{FL}(W) = \mathbf{QH}$ *(intuitionistic predicate logic) for a one-element partially ordered set* W*, and hence for any pre-ordered* W.

In other words, the logic without equality $\mathbf{FL}(W)$ does not depend on W.

So to say, the functor semantics is too powerful to be considered at the level of propositional Kripke bases (i.e., pre-ordered sets), because at this level it becomes degenerated (at least, for logics without equality). That is why we assume that the level of categories (i.e., the logics $\mathbf{L}^{(=)}(\mathcal{C})$ introduced in Section 1) is more adequate. In particular, (2) seems to be a more natural counterpart of (1) for the functor semantics than (3), and the peculiarity stated in Section 2 shows the limits of this correspondence between (1) and (2).

On the other hand, Proposition 2* is not transferred to logics with equality $\mathbf{FL}^{=}(W)$. Namely, all these logics have the intuitionistic equality-free fragment, however they are distinct for different W. For example, it is easily seen that the formula

$$\Psi = [\,\forall x\, \forall y\, (x = y) \rightarrow P \vee \neg P\,]$$

belongs to $\mathbf{FL}^{=}(W)$ for a one-element W (since any $\mathcal{C}$-set over one-element partially ordered set with one-element domain is a classical one-element model), but does not belong, e.g., to $\mathbf{FL}^{=}(W)$ for a two-element chain. Actually,

$$\Psi \in \mathbf{FL}^{=}(W) \quad \text{iff} \quad (\ \leq \text{ is an equivalence relation on } W\ )$$

(i.e., iff the skeleton of W is an antichain).

Moreover, for a propositional formula A, let us denote

$$\Psi^A = [\,\forall x\, \forall y\, (x = y) \to A\,].$$

Proposition 3 *For a propositional formula A and a pre-ordered set W we have:*

$\Psi^A \in \mathbf{FL}^{=}(W)$ iff $A \in \mathbf{PL}(W)$ *(the superintuitionistic propositional logic of W).*

Indeed, a $\mathcal{C}$-set with one-element domains is essentially a usual predicate Kripke frame with one-element constant domain; it is sufficient to identify its individual domains, i.e., we suppose that $D_u = D = \{a_0\}$ for all $u \in W$, and then any E_μ (for $\mu \in \mathcal{C}(u,v)$, $u \leq v$ in W) becomes the identical mapping on D.[5] □

This statement means that the propositional logic $\mathbf{PL}(W)$ is embeddable into the corresponding logic with equality $\mathbf{FL}^{=}(W)$. Therefore we conclude that:

Corollary 5 *There exists a continuum of different logics with equality of the form $\mathbf{FL}^{=}(W)$ (for different W).*

Indeed, there exists a continuum of Kripke-complete superintuitionistic propositional logics (cf. [3]). □

Open problem 1 Try to describe the logics with equality $\mathbf{FL}^{=}(W)$ for natural and simple partially ordered (or pre-ordered) sets W; e.g., for a one-element W.

We do not know, whether these logics are recursively (or finitely) axiomatizable; clearly, this question arises only for sets W, whose propositional logics $\mathbf{PL}(W)$ are recursively axiomatizable (e.g., for finite W). We do not know, whether $\mathbf{FL}^{=}(W)$ for a pre-ordered set W equals the logic of its partially ordered skeleton. And more questions remain open.

We say that a pre-ordered set W is $\mathbf{QH}^{=}$*-complete in the functor semantics* if $\mathbf{FL}^{=}(W) = \mathbf{QH}^{=}$. Proposition 3 implies the following simple consequence:

Corollary 6 *A pre-ordered set W is $\mathbf{QH}^{=}$-complete in the functor semantics only if $\mathbf{PL}(W) = \mathbf{H}$ (intuitionistic propositional logic).*

---

5 By the way, note that here we mention only propositional formulas A, because it is easily seen that in a Kripke frame with one-element constant domain any predicate formula A is reducible to a propositional formula $\alpha(A)$, cf. [4, Sect. 5 (and 6)].

Indeed, definitely $\mathbf{QH}^= \not\vdash \Psi^A$ for an intuitionistically unprovable propositional formula A. □

Now, let us say that W is $\mathbf{QH}^=$*-complete in the Kripke semantics* if $\mathbf{KL}^=(W) = \mathbf{QH}^=$. The most familiar examples are the infinite tree ω^* of all finite sequences of natural numbers or the binary tree $\{0,1\}^*$ of all finite $\{0,1\}$-sequences, etc.

Open problem 2 Does there exist a pre-ordered (or a partially ordered) set W that is $\mathbf{QH}^=$-complete in the functor semantics but not in the Kripke semantics?

Let us mention two possible candidates.

First, let W_{fin} be the disjoint union of all finite trees (or e.g., all Jaskowski's trees). Then $\mathbf{PL}(W_{fin}) = \mathbf{H}$, whereas $\mathbf{KL}^=(W) \neq \mathbf{QH}^=$, because e.g. the well-known Kuroda's formula is Kripke-valid in W_{fin}:

$$K = \neg\neg\forall x(P(x) \vee \neg P(x)).$$

On the other hand, $K \notin \mathbf{FL}^=(W_{fin})$, since it is easily shown that $K \notin \mathbf{FL}^=(W)$ for a one-element W (and then apply Proposition 2).

Second, let $W = \overline{\omega^*}$ be the tree obtained by adding maximal points above all points of ω^* (or above all branches in ω^*, etc.), or a similar binary tree $W = \overline{\{0,1\}^*}$. Then $\mathbf{KL}^=(W) = (\mathbf{QH}^= + K)$. By the way, note that it is easily shown that $\mathbf{FL}^=(\overline{\omega^*}) \subseteq \mathbf{FL}^=(W_{fin})$, because there exist p-morphisms of $\overline{\omega^*}$ onto all finite trees. So $\overline{\omega^*}$ would be definitely $\mathbf{QH}^=$-complete if W_{fin} were.

5 Appendix: The proofs of Propositions 1 and 2

To establish Proposition 1, we use the following notion.

Let $\mathcal{C}$ be a category based on W, and let $\mathbb{F}', \mathbb{F}''$ be $\mathcal{C}$-sets. Say that $\mathbb{F}'$ is a $\mathcal{C}$-*subset* of $\mathbb{F}''$ (and $\mathbb{F}''$ is a $\mathcal{C}$-*extension* of $\mathbb{F}'$) if $D'_u \subseteq D''_u$ for every $u \in W$ and E'_μ is the restriction of E''_μ to D'_u for $\mu \in \mathcal{C}(u,v)$, $u \leq v$; naturally, we suppose that the functions E'_μ are well-defined, i.e., $E''_\mu(D'_u) \subseteq D'_v$ for any $\mu \in \mathcal{C}(u,v)$. Now, a valuation ξ'' in $\mathbb{F}''$ gives rise to the valuation ξ' in $\mathbb{F}'$ defined as the restriction of ξ'' to $\mathbb{F}'$ (for atomic formulas); note that the corresponding forcing relation $\models'$ in $\mathbb{F}'$ in general is not the restriction of $\models''$ from $\mathbb{F}''$ to $\mathbb{F}'$ (for non-atomic formulas).

Lemma 7 *Let $\mathcal{C}$ be a category based on W, let $\mathbb{F}'$ be a $\mathcal{C}$-subset of $\mathbb{F}''$, and let ξ' be the restriction (to $\mathbb{F}'$) of a valuation ξ'' in $\mathbb{F}''$. Let $u \in W$, and let $A(\mathbf{a})$ be a formula (with parameters replaced by elements of D'_u). Then:*

(1) *if A is $\forall$-positive, then: $(u \models'' A(\mathbf{a})$ in $\mathbb{F}'') \Rightarrow (u \models' A(\mathbf{a})$ in $\mathbb{F}')$;*

(2) *if A is $\exists$-positive, then: $(u \models' A(\mathbf{a})$ in $\mathbb{F}') \Rightarrow (u \models'' A(\mathbf{a})$ in $\mathbb{F}'')$.*

Proof is obtained by a straightforward induction on A (use Lemma 4 !). Let us mention the three principal cases (other cases are obvious).

(I) $A = (A_1 \rightarrow A_2)$.

(1) Let A be $\forall$-positive, so A_1 is $\exists$-positive and A_2 is $\forall$-positive. Assume that $u \not\models' A(\mathbf{a})$, i.e., $v \models' A_1(E'_\mu(\mathbf{a}))$ and $v \not\models' A_2(E'_\mu(\mathbf{a}))$ for some $v \geq u$, $\mu \in \mathcal{C}(u,v)$. Then $v \models'' A_1(E''_\mu(\mathbf{a}))$ and $v \not\models'' A_2(E''_\mu(\mathbf{a}))$ by inductive hypothesis (note that here $E''_\mu(\mathbf{a}) = E'_\mu(\mathbf{a})$), and so $u \not\models'' A(\mathbf{a})$. (2) is shown similarly.

(II) $A = \exists x\, A_0$.

(2) Let A (and so A_0) be $\exists$-positive. Assume that $u \models' A(\mathbf{a})$, i.e., $u \models' A_0(b, \mathbf{a})$ for some $b \in D'_v \subseteq D''_v$. Then $u \models'' A_0(b, \mathbf{a})$, and so $u \models'' A(\mathbf{a})$.

Also, (1) is vacuous, since A definitely is not $\forall$-positive.

(III) $A = \forall x\, A_0$.

(1) Let A (and so A_0) be $\forall$-positive. Now we have to assume that $u \not\models' A(\mathbf{a})$, i.e., $v \not\models' A_0(b, E'_\mu(\mathbf{a}))$ for some $v \geq u$, $\mu \in \mathcal{C}(u,v)$, $b \in D'_v \subseteq D''_v$. Then we conclude that $v \not\models'' A_0(b, E''_\mu(\mathbf{a}))$, and so $u \not\models'' A(\mathbf{a})$.

And (2) is vacuous again. □

Any $\mathcal{C}$-set $\mathbb{F}$ gives rise to its *simple extension* $\mathbb{F}^+$, where $D^+_u = D_u \cup \{e_u\}$ (here $e_u \notin D_u$) for all $u \in W$, and $E^+_\mu \mid_{D_u} = E_\mu$, $E^+_\mu(e_u) = e_v$ for $\mu \in \mathcal{C}(u,v)$.

Lemma 8 *Let $\mathcal{C}$ be a category based on W, let $u_0 \in W$, and let A be a $\forall$-positive formula. Then the $\mathcal{C}$-validity of A implies its $\mathcal{C}^{u_0}$-validity.*

Proof. Let a $\forall$-positive A be not $\mathcal{C}^{u_0}$-valid., i.e., $w \not\models_0 A(\mathbf{a})$ for some valuation ξ_0 in a $\mathcal{C}^{u_0}$-set $\mathbb{F}_0$, some $w \in W^{u_0}$, and $\mathbf{a} \in (D_w)^n$. Consider a $\mathcal{C}^{u_0}$-set $\mathbb{F}$ such that $\mathbb{F}^{u_0} = (\mathbb{F}_0)^+$, $D_u = \{e_u\}$ for $u \notin W^{u_0}$, and $E^+_\mu(e_u) = e_v$ for $\mu \in \mathcal{C}(u,v)$ (for all $u, v \in W$, $u \leq v$). Extend ξ_0 to a valuation ξ in $\mathbb{F}$; e.g. put a valuation $\xi(P) = \xi_0(P)$ for all predicate symbols P (so atoms are true only at points from W^{u_0}). Then $w \not\models A(\mathbf{a})$ by Lemma 7(1), and so A is not $\mathcal{C}$-valid. □

However, this claim does not immediately imply Proposition 1, since a substitution instance of a $\forall$-positive formula in general is not $\forall$-positive. So we apply the following, slightly more explicit description of the logic $\mathbf{L}^{(=)}(\mathbb{F})$ of a $\mathcal{C}$-set.

Let A be a predicate formula, and let k_i-ary symbols P_i (for $i = 1, \ldots, m$) be all predicate symbols occurring in A (besides equality!). Now, for $n \geq 0$, let $\mathbf{y} = (y_1, \ldots, y_n)$ be a list of different variables not occurring in A, and let Q_i be different $(k_i + n)$-ary predicate symbols. Then *n-shift* A^n of A is the substitution instance of A obtained by simultaneously replacing of all atomic subformulas $P_i(\mathbf{x})$ in A with $Q_i(\mathbf{x}, \mathbf{y})$ $(i = 1, \ldots, m)$.

It is known that $A \in \mathbf{L}^{(=)}(\mathbb{F})$ iff all A^n (for $n \geq 0$) are valid in $\mathbb{F}$ (see [1, Proposition 5.6.26(2)] or cf. e.g. [6, Sect.4.1]). So to say, to check, whether $A \in \mathbf{L}^{(=)}(\mathbb{F})$, it is sufficient to check the validity of A 'with an arbitrary number of additional parameters' (and then the validity for all substitution instances of A readily follows). Therefore, for a category $\mathcal{C}$ we conclude that

$$A \in \mathbf{L}^{(=)}(\mathcal{C}) \text{ iff all } A^n \text{ (for } n \geq 0\text{) are } \mathcal{C}\text{-valid.}$$

Clearly, for a $\forall$-positive A, all A^n are $\forall$-positive as well, and therefore now Proposition 1 immediately follows from Lemma 8.

Next, Proposition 2 follows from (2)′ (see Section 2) and the subsequent:

Lemma 9 *Let W be a pre-ordered set and $u_0 \in W$. Then every category $\mathcal{C}$ based on W^{u_0} and a $\mathcal{C}$-set $\mathbb{F}$ can be extended to a category $\mathcal{C}^*$ based on W and a $\mathcal{C}^*$-set $\mathbb{F}^*$ such that $(\mathcal{C}^*)^{u_0} = \mathcal{C}$ and $(\mathbb{F}^*)^{u_0} = \mathbb{F}$.*

Proof. Put $\mathcal{C}^*(u,v) = \mathcal{C}(u,v)$ for $u, v \in W^{u_0}$, $\mathcal{C}^*(u,v) = \{1_{u,v}\}$ for $u, v \notin W^{u_0}$, $\mathcal{C}^*(u,v) = \{\langle \mu, u \rangle \mid \mu \in \mathcal{C}(u_0, v)\}$ for $u \notin W^{u_0}$, $v \in W^{u_0}$ (here $\langle \mu, u \rangle$ is a 'copy' of μ at $u \notin W^{u_0}$). Let us describe $\mu \circ \mu'$ for $\mu \in \mathcal{C}^*(u,v)$, $\mu' \in \mathcal{C}^*(v,w)$. First, $\mu \circ \mu'$ is taken from $\mathcal{C}$ for $u \in W^{u_0}$. Second, $1_{u,v} \circ 1_{v,w} = 1_{u,w}$ for $w \notin W^{u_0}$ and $1_{u,v} \circ \langle \mu, v \rangle = \langle \mu, u \rangle$ for $v \notin W^{u_0}$, $w \in W^{u_0}$ (here $\mu \in \mathcal{C}(u_0, w)$). Finally, $\langle \mu, u \rangle \circ \mu' = \langle \mu \circ \mu', u \rangle$ for $u \notin W^{u_0}$, $v \in W^{u_0}$ (here $\mu \in \mathcal{C}(u_0, v)$, $\mu' \in \mathcal{C}(v, w)$, so $\mu \circ \mu' \in \mathcal{C}(u_0, w)$). The associativity for the composition in $\mathcal{C}^*$ is easily checked.

Now we extend $\mathbb{F}$ to $\mathbb{F}^*$. Namely, put $D^*_u = D_u$ for $u \in W^{u_0}$ and $D^*_u = D_{u_0}$ for $u \notin W^{u_0}$. Also define E^*_μ for $\mu \in \mathcal{C}^*(u,v)$ as follows: $E^*_\mu = E_\mu$ for $u, v \in W^{u_0}$,
$E^*_{1_{u,v}} = 1_{D_{u_0}}$ (the identity function on $D_{u_0} = D^*_u = D^*_v$) for $u, v \notin W^{u_0}$,
$E^*_{\langle \mu, u \rangle} = E_\mu$ for $u \notin W^{u_0}$, $v \in W^{u_0}$ (here $\mu \in \mathcal{C}(u_0, v)$ and $D^*_u = D_{u_0}$).
The key property $E^*_{\mu \circ \mu'} = E^*_{\mu'} \circ E^*_\mu$ for $\mu \in \mathcal{C}^*(u,v)$, $\mu' \in \mathcal{C}^*(v,w)$ (cf. Section 1) obviously holds. □

Acknowledgements.

The author would like to thank V. Shehtman, E. Zolin, and the anonymous referees for useful comments, which helped improve the exposition of the paper.

References

[1] Gabbay, D., V. Shehtman and D. Skvortsov, "Quantification in nonclassical logic, Vol. 1, Section 5.6, Studies in Logic and the Foundations of Mathematics 153," Elsevier, 2009.

[2] Ghilardi, S., *Presheaf semantics and independence results for some non-classical first-order logics*, Archive for Math. Logic **29** (1989), pp. 125–136.

[3] Jankov, V., *Constructing a sequence of strongly independent superintuitionistic propositional calculi*, Soviet Mathematics, Doklady **9** (1968), pp. 806–807.

[4] Ono, H., *A study of intermediate predicate logics*, Publications of RIMS, Kyoto Univ. **8** (1972-1973), pp. 619–649.

[5] Skvortsov, D., *On a peculiarity in the functor semantics for superintuitionistic predicate logics, II*, In preparation.

[6] Skvortsov, D., *On intermediate predicate logics of some finite Kripke frames, I. Levelwise uniform trees*, Studia Logica **77** (2004), pp. 295–323.

Morphisms on Bi-approximation Semantics

Tomoyuki Suzuki [1]

Department of Computer Science, University of Leicester
Leicester, LE1 7RH, United Kingdom

Abstract

In the present paper, we introduce bounded morphisms on bi-approximation semantics, show the so-called p-morphism lemma on bi-approximation semantics, and investigate the dual representation of the morphisms. In addition, we study three properties, namely B-embedding, B-separating and B-reflecting, to preserve validity of sequents on frames. These bounded morphisms do not look like embedding, surjective and isomorphic p-morphisms on Kripke semantics in modal logic. Nevertheless, with help of auxiliary relations or properties held via the dual representation, we notice that the notion of our bounded morphisms on bi-approximation semantics is a natural generalisation of the one on Kripke semantics in modal logic.

Keywords: Bi-approximation semantics, substructural logic, Stone-style representation, Ghilardi and Meloni's canonicity methodology.

1 Introduction

What is the right notion of relational semantics of non-distributive, i.e. not necessarily distributive, lattice-based logics, such as for example substructural logic? The main problem to deal with non-distributive lattice-based logics on relational semantics is how to avoid validating the distributive law, i.e. $\phi \wedge (\psi \vee \chi)$ implies $(\phi \wedge \psi) \vee (\phi \wedge \chi)$. That is, on a Kripke-style semantics, if we interpret conjunction $\wedge$ as "and" and disjunction $\vee$ as "or" as in the case of modal logic, each state naturally satisfies the distributive law. In the literature, we can find some relational-type semantics for non-distributive lattice-based logics by introducing non-standard interpretations of disjunction $\vee$ to reject the distributivity, see e.g. [10,13,12,8], also [15].

Bi-approximation semantics has been introduced in [17] to investigate another potential relational-type semantics for non-distributive lattice-based logics. Unlike what happens in normal relational semantics, we reason about logical consequences (sequents), instead of formulae, on bi-approximation semantics. Note that, recently, a relational-type semantics, named *residuated frames* [6], was introduced as a complete analog of the proof system of substructural logic, which can be useful to the decidability property. On the other hand,

[1] Email: *tomoyuki.suzuki@mcs.le.ac.uk*

bi-approximation semantics was introduced as a canonicity-friendly framework to characterise *Ghilardi and Meloni's canonicity methodology* [9], which is applicative to other lattice-based logics. Furthermore, since bi-approximation semantics is canonicity-friendly, we can also discuss a Sahlqvist theorem on bi-approximation semantics [16].

In [17], we have investigated the idea of reasoning about sequents on bi-approximation semantics, and shown the object-level Stone-style representation theorem between p-frames and FL-algebras. However, we have not studied how morphisms are defined on bi-approximation semantics, and if they exist, how they are related to bounded morphisms in Kripke-style semantics. The main purpose of the current paper is to give a possible answer for those questions. In this paper, we shall introduce bounded morphisms to preserve truth values of formulae, and discuss the so-called *p-morphism lemma*, e.g. [11,2,3]. Based on these bounded morphisms, we shall think about invariance of validity of sequents on p-frames via specific morphisms, which are not exactly the same as the validity preserving p-morphisms in Kripke semantics, called *embedding*, *surjective* and *isomorphic* p-morphisms. On the other hand, we shall analyse how our bounded morphisms are related to p-morphisms on Kripke-style semantics. Besides, we shall also prove the dual representation of morphisms between the category of lattice expansions and strict homomorphisms and the category of p-frames and bounded morphisms. In the end, we observe that our bounded morphism is a natural generalisation of p-morphisms on Kripke-style semantics.

We outline the structure of the current paper. We briefly recall the basic terminology of substructural logic in Section 2. In Section 3, bi-approximation semantics are summarised and the fundamental results of bi-approximation semantics are reviewed. Bounded morphisms for bi-approximation semantics are introduced in Section 4. Further, we shall prove the preservation of truth value on bi-approximation models, the so-called p-morphism lemma, for the three different satisfaction relations. Notice that there are three satisfaction relations: one is for assumptions (formulae), the other is for conclusions (formulae), and the last one is for truth value of sequents. In addition, we shall introduce the notion of the validity preserving bounded morphisms on bi-approximation semantics, and show the invariance of validity of sequents. In Section 5, the dual representation of morphisms between lattice expansions and bi-approximation semantics will be argued. As a result, we shall notice that our notion of bounded morphisms also satisfies the same properties as the dual representation of morphisms between Boolean algebras and Kripke semantics via the Stone representation. Finally, in Section 6, we shall give some conclusive remarks. We also note that, because of the strict page restriction, many proofs are in the appendix.

Acknowledgements: The author would like to thank the anonymous reviewers for their valuable comments to improve not just this paper but also his subsequent works.

2 Substructural logic

In this paper, we denote propositional variables by $p, q, r, p_1, \ldots$, the set of all propositional variables by Φ, and $\mathbf{t}$ and $\mathbf{f}$ are logical constants representing *true* and *false*, respectively. As logical connectives, we use *disjunction* $\vee$, *conjunction* $\wedge$, *fusion* (*multiplication*) $\circ$, *implications* (*residuals*) $\rightarrow$ and $\leftarrow$. Formulae of substructural logic are denoted by ϕ, ψ, χ, $\phi_1, \ldots$, and $\psi_1, \ldots$, and the set of all formulae is denoted by Λ. The following BNF generates formulae of substructural logic.

$$\phi ::= p \mid \mathbf{t} \mid \mathbf{f} \mid \phi \vee \phi \mid \phi \wedge \phi \mid \phi \circ \phi \mid \phi \rightarrow \phi \mid \phi \leftarrow \phi$$

Γ, Δ, Σ, Π are (possibly empty) finite lists of formulae, and θ is a list of at most one formula. Then, we call $\Gamma \Rightarrow \theta$ a *sequent*, see e.g. [14].

For sequents, we consider a sequent calculus for substructural logic, called *FL* as in Fig. 1. In the sequent calculus FL, a formula ϕ is *provable in FL* if

Initial sequents.

$$\phi \Rightarrow \phi \qquad\qquad \Rightarrow \mathbf{t} \qquad\qquad \mathbf{f} \Rightarrow$$

Cut rule.

$$\frac{\Gamma \Rightarrow \phi \qquad \Sigma, \phi, \Pi \Rightarrow \theta}{\Sigma, \Gamma, \Pi \Rightarrow \theta}\ (\mathrm{cut})$$

Rules for logical constants.

$$\frac{\Gamma, \Delta \Rightarrow \theta}{\Gamma, \mathbf{t}, \Delta \Rightarrow \theta}\ (\mathbf{t}\mathrm{w}) \qquad\qquad \frac{\Gamma \Rightarrow}{\Gamma \Rightarrow \mathbf{f}}\ (\mathbf{f}\mathrm{w})$$

Rules for logical connectives.

$$\frac{\Gamma, \phi, \Delta \Rightarrow \theta \qquad \Gamma, \psi, \Delta \Rightarrow \theta}{\Gamma, \phi \vee \psi, \Delta \Rightarrow \theta}\ (\vee \Rightarrow)$$

$$\frac{\Gamma \Rightarrow \phi}{\Gamma \Rightarrow \phi \vee \psi}\ (\Rightarrow \vee_1) \qquad\qquad \frac{\Gamma \Rightarrow \psi}{\Gamma \Rightarrow \phi \vee \psi}\ (\Rightarrow \vee_2)$$

$$\frac{\Gamma, \phi, \Delta \Rightarrow \theta}{\Gamma, \phi \wedge \psi, \Delta \Rightarrow \theta}\ (\wedge_1 \Rightarrow) \qquad\qquad \frac{\Gamma, \psi, \Delta \Rightarrow \theta}{\Gamma, \phi \wedge \psi, \Delta \Rightarrow \theta}\ (\wedge_2 \Rightarrow)$$

$$\frac{\Gamma \Rightarrow \phi \qquad \Gamma \Rightarrow \psi}{\Gamma \Rightarrow \phi \wedge \psi}\ (\Rightarrow \wedge)$$

$$\frac{\Gamma, \phi, \psi, \Delta \Rightarrow \theta}{\Gamma, \phi \circ \psi, \Delta \Rightarrow \theta}\ (\circ \Rightarrow) \qquad\qquad \frac{\Gamma \Rightarrow \phi \qquad \Sigma \Rightarrow \psi}{\Gamma, \Sigma \Rightarrow \phi \circ \psi}\ (\Rightarrow \circ)$$

$$\frac{\Gamma \Rightarrow \phi \qquad \Sigma, \psi, \Pi \Rightarrow \theta}{\Sigma, \Gamma, \phi \rightarrow \psi, \Pi \Rightarrow \theta}\ (\rightarrow\Rightarrow) \qquad\qquad \frac{\phi, \Gamma \Rightarrow \psi}{\Gamma \Rightarrow \phi \rightarrow \psi}\ (\Rightarrow\rightarrow)$$

$$\frac{\Gamma \Rightarrow \phi \qquad \Sigma, \psi, \Pi \Rightarrow \theta}{\Sigma, \psi \leftarrow \phi, \Gamma, \Pi \Rightarrow \theta}\ (\leftarrow\Rightarrow) \qquad\qquad \frac{\Gamma, \phi \Rightarrow \psi}{\Gamma \Rightarrow \psi \leftarrow \phi}\ (\Rightarrow\leftarrow)$$

Fig. 1. The sequent calculus FL

the sequent $\Rightarrow \phi$ is derivable in FL. The substructural logic **FL** is the set of all provable formulae in FL.

Proposition 2.1 *For all formulae ϕ and ψ, we have*

(i) *ϕ is provable if and only if $\mathbf{t} \Rightarrow \phi$ is derivable,*

(ii) *$\phi \Rightarrow \psi$ is derivable if and only if $\phi \to \psi$ is provable in FL if and only if $\psi \leftarrow \phi$ is provable in FL,*

(iii) *$\phi_1, \ldots, \phi_n \Rightarrow \theta$ is derivable in FL if and only if $\phi_1 \circ \cdots \circ \phi_n \Rightarrow \theta$ is derivable in FL.*

By Proposition 2.1, we sometimes state that the substructural logic **FL** is the set of all sequents derivable in FL. Besides, we also think about a sequent $\phi_1, \ldots, \phi_n \Rightarrow \theta$ as a pair of two formulae $\phi_1 \circ \cdots \circ \phi_n$ and θ. In particular, if the left-hand side is empty, we let the first formula **t**, and if the right-hand side is empty, we let the second formula **f**. Hence, hereinafter, we may state that a sequent $\phi \Rightarrow \psi$ is a pair of two formulae.

The algebraic counterparts of the substructural logic **FL** are known as FL-algebras [7].

Definition 2.2 (FL-algebra) *An 8-tuple $\mathbb{A} = \langle A, \vee, \wedge, *, \backslash, /, 1, 0\rangle$ is an* FL-algebra, *if $\langle A, \vee, \wedge\rangle$ forms a lattice, $\langle A, *, 1\rangle$ is a monoid, 0 is a constant in A, and for all $a, b, c \in A$, we have*

$$a * b \leq c \iff b \leq a\backslash c \iff a \leq c/b.$$

On FL-algebras, each formula ϕ is interpreted to the corresponding term t as usual, i.e. $\circ$, $\to$, $\leftarrow$, **t** and **f** are interpreted to $*$, $\backslash$, $/$, 1 and 0. The corresponding term of a formula ϕ is denoted by s_ϕ or by t_ϕ. Furthermore, each sequent $\phi_1, \ldots, \phi_n \Rightarrow \theta$ is interpreted as an inequality $s_{\phi_1} * \cdots * s_{\phi_n} \leq t_\theta$. We sometimes abuse the notation for a list of formulae, i.e. s_Γ is the term $s_{\phi_1} * \cdots * s_{\phi_n}$, when Γ is the finite list of formulae $\phi_1, \ldots, \phi_n$. Note that if the left hand side of a sequent is empty then the corresponding term is 1, and if the right hand side of a sequent is empty then the corresponding term is 0.

3 Bi-approximation semantics

Bi-approximation semantics is designed as a canonicity-friendly relational semantics for substructural logic [17]. A novelty of this semantics is to evaluate not only formulae but also sequents based on polarities. A *polarity* is a triple $\langle X, Y, B\rangle$ where X and Y are not-necessarily disjoint and non-empty sets and B is a binary relation on $X \times Y$, see e.g. [1,4,8] for polarities. Note that, for each polarity $\langle X, Y, B\rangle$, the binary relation B is naturally extended to a preorder order $\leq$ on $X \cup Y$, see [8,17]. For $x_1, x_2 \in X$, we let $x_1 \leq x_2 \iff \forall y \in Y. [x_2 B y \Longrightarrow x_1 B y]$ and for $y_1, y_2 \in Y$, we let $y_1 \leq y_2 \iff \forall x \in X. [x B y_1 \Longrightarrow x B y_2]$. Hence, we may call the triple $\langle X, Y, \leq\rangle$ with the extended preorder order $\leq$ a *polarity*. Unlike the standard relational semantics like Kripke semantics or Routley-Meyer semantics, bi-approximation semantics reasons about sequents on a polarity $\langle X, Y, \leq \rangle$ as follows: we evaluate *premises* on X, *conclusions* on Y and sequents (logical consequences) as the binary relation $\leq$ between premises and conclusions.

Definition 3.1 *A* polarity frame for substructural logic, p-frame *for short, is an 8-tuple* $\mathbb{F} = \langle X, Y, \leq, R, O_X, O_Y, N_X, N_Y \rangle$, *where* $\langle X, Y, \leq \rangle$ *is a polarity,* R *is a ternary relation on* $X \times X \times Y$, O_X *is a non-empty subset of* X, N_X *is a subset of* X, O_Y *and* N_Y *are subsets of* Y, *and* $\mathbb{F}$ *satisfies*

R-order: *for all* $x, x' \in X$,

$$x' \leq x \text{ if and only if } \exists o \in O_X.\big[R^\circ(x, o, x') \text{ or } R^\circ(o, x, x')\big];$$

R-identity: *for each* $x \in X$,

$$\exists o_2 \in O_X.\big[R^\circ(x, o_2, x)\big] \text{ and } \exists o_1 \in O_X.\big[R^\circ(o_1, x, x)\big];$$

R-transitivity: *for all* $x_1, x_1', x_2, x_2' \in X$ *and* $y, y' \in Y$,

$$x_1' \leq x_1, x_2' \leq x_2, y \leq y' \text{ and } R(x_1, x_2, y) \Longrightarrow R(x_1', x_2', y');$$

R-associativity: *for all* $x_1, x_2, x_3, x \in X$,

$$\exists x' \in X.\big[R^\circ(x_1, x', x) \text{ and } R^\circ(x_2, x_3, x')\big]$$
$$\iff \exists x'' \in X.\big[R^\circ(x_1, x_2, x'') \text{ and } R^\circ(x'', x_3, x)\big];$$

O: $O_X = \{x \in X \mid \forall y \in O_Y. x \leq y\}$ *and* $O_Y = \{y \in Y \mid \forall x \in O_X. x \leq y\}$;

N: $N_X = \{x \in X \mid \forall y \in N_Y. x \leq y\}$ *and* $N_Y = \{y \in Y \mid \forall x \in N_X. x \leq y\}$;

$\circ$***-tightness:*** *for all* $x_1, x_2 \in X$ *and* $y \in Y$,

$$\forall x \in X.\big[R^\circ(x_1, x_2, x) \Longrightarrow x \leq y\big] \Longrightarrow R(x_1, x_2, y);$$

$\rightarrow$***-tightness:*** *for all* $x_1, x_2 \in X$ *and* $y \in Y$,

$$\forall y_2 \in Y.\big[R^\rightarrow(x_1, y_2, y) \Longrightarrow x_2 \leq y_2\big] \Longrightarrow R(x_1, x_2, y);$$

$\leftarrow$***-tightness:*** *for all* $x_1, x_2 \in X$ *and* $y \in Y$,

$$\forall y_1 \in Y.\big[R^\leftarrow(y_1, x_2, y) \Longrightarrow x_1 \leq y_1\big] \Longrightarrow R(x_1, x_2, y);$$

where $R^\circ(x_1, x_2, x)$, $R^\rightarrow(x_1, y_2, y)$ *and* $R^\leftarrow(y_1, x_2, y)$ *are auxiliary relations of* R *defined as follows:*

R°: $R^\circ(x_1, x_2, x) \iff \forall y \in Y.\big[R(x_1, x_2, y) \Longrightarrow x \leq y\big]$;

$R^\rightarrow$: $R^\rightarrow(x_1, y_2, y) \iff \forall x_2 \in X.\big[R(x_1, x_2, y) \Longrightarrow x_2 \leq y_2\big]$;

$R^\leftarrow$: $R^\leftarrow(y_1, x_2, y) \iff \forall x_1 \in X.\big[R(x_1, x_2, y) \Longrightarrow x_1 \leq y_1\big]$.

Remark 3.2 In [17], the tightness conditions are given by more restricted forms. However, we can weaken those tightness conditions as above to show all results in [17], and they are actually even more direct. But the most important reason to introduce the above tightness conditions is that the slight difference affects the current author's subsequent works.

Remark 3.3 The conditions of p-frames are not completely independent (e.g. the R-transitivity follows from the other conditions). However, we keep the above definition to discuss similarities to Kripke-type semantics for distributive substructural logics.

On p-frames, we define doppelgänger valuations to give compatible truth value to atomic sequents, namely we want to make $p \mapsto p$ valid for each propositional variable $p \in \Phi$.

Definition 3.4 (Doppelgänger valuation) *Given a p-frame $\mathbb{F}$, a pair $V = (V^{\downarrow}, V_{\uparrow})$ of two functions $V^{\downarrow}\colon \Phi \to \wp(X)$ and $V_{\uparrow}\colon \Phi \to \wp(Y)$, where $\wp(X)$ and $\wp(Y)$ are the powersets of X and Y, is a* doppelgänger valuation *on $\mathbb{F}$, if $V^{\downarrow}(p) = \{x \in X \mid \forall y \in V_{\uparrow}(p).\, x \le y\}$ and $V_{\uparrow}(p) = \{y \in Y \mid \forall x \in V^{\downarrow}(p).\, x \le y\}$ for each propositional variable $p \in \Phi$.*

Given a p-frame $\mathbb{F}$ and a doppelgänger valuation V on $\mathbb{F}$, we call the pair $\mathbb{M} = \langle \mathbb{F}, V\rangle$ a *bi-approximation model.* On a bi-approximation model $\mathbb{M}$, we inductively define a satisfaction relation $\models$ as follows: for all $x \in X$ and $y \in Y$, we let

X-1 $\mathbb{M} \models^{x} p \iff x \in V^{\downarrow}(p)$ for each propositional variable $p \in \Phi$,

X-2 $\mathbb{M} \models^{x} \mathbf{t} \iff x \in O_X$,

X-3 $\mathbb{M} \models^{x} \mathbf{f} \iff x \in N_X$,

X-4 $\mathbb{M} \models^{x} \phi \vee \psi \iff \forall y \in Y.\big[\mathbb{M} \models_{y} \phi \vee \psi \Longrightarrow x \le y\big]$,

X-5 $\mathbb{M} \models^{x} \phi \wedge \psi \iff \mathbb{M} \models^{x} \phi$ and $\mathbb{M} \models^{x} \psi$,

X-6 $\mathbb{M} \models^{x} \phi \circ \psi \iff \forall y \in Y.\big[\mathbb{M} \models_{y} \phi \circ \psi \Longrightarrow x \le y\big]$,

X-7 $\mathbb{M} \models^{x} \phi \to \psi \iff \forall x' \in X, y \in Y.\big[\mathbb{M} \models^{x'} \phi \text{ and } \mathbb{M} \models_{y} \psi \Longrightarrow R(x', x, y)\big]$,

X-8 $\mathbb{M} \models^{x} \psi \leftarrow \phi \iff \forall x' \in X, y \in Y.\big[\mathbb{M} \models^{x'} \phi \text{ and } \mathbb{M} \models_{y} \psi \Longrightarrow R(x, x', y)\big]$,

Y-1 $\mathbb{M} \models_{y} p \iff y \in V_{\uparrow}(p)$ for each propositional variable $p \in \Phi$,

Y-2 $\mathbb{M} \models_{y} \mathbf{t} \iff y \in O_Y$,

Y-3 $\mathbb{M} \models_{y} \mathbf{f} \iff y \in N_Y$,

Y-4 $\mathbb{M} \models_{y} \phi \vee \psi \iff \mathbb{M} \models_{y} \phi$ and $\mathbb{M} \models_{y} \psi$,

Y-5 $\mathbb{M} \models_{y} \phi \wedge \psi \iff \forall x \in X.\big[\mathbb{M} \models^{x} \phi \wedge \psi \Longrightarrow x \le y\big]$,

Y-6 $\mathbb{M} \models_{y} \phi \circ \psi \iff \forall x_1, x_2 \in X.\big[\mathbb{M} \models^{x_1} \phi \text{ and } \mathbb{M} \models^{x_2} \psi \Longrightarrow R(x_1, x_2, y)\big]$,

Y-7 $\mathbb{M} \models_{y} \phi \to \psi \iff \forall x \in X.\big[\mathbb{M} \models^{x} \phi \to \psi \Longrightarrow x \le y\big]$,

Y-8 $\mathbb{M} \models_{y} \psi \leftarrow \phi \iff \forall x \in X.\big[\mathbb{M} \models^{x} \psi \leftarrow \phi \Longrightarrow x \le y\big]$,

S-1 $\mathbb{M} \models^{x}_{y} \phi \mapsto \psi \iff$ if $\mathbb{M} \models^{x} \phi$ and $\mathbb{M} \models_{y} \psi$ then $x \le y$,

S-2 $\mathbb{M} \models \phi \mapsto \psi \iff \forall x \in X, y \in Y.\big[\mathbb{M} \models^{x}_{y} \phi \mapsto \psi\big]$,

S-3 $\mathbb{F} \models \phi \mapsto \psi \iff \langle \mathbb{F}, V \rangle \models \phi \mapsto \psi$ for every doppelgänger valuation V.

Definition 3.5 (Truth value) *The satisfaction relation* $\models$ *is interpreted as follows:*

(i) $\mathbb{M} \models^{x} \phi$: *a formula* ϕ *is* assumed *at* x *in* $\mathbb{M}$,

(ii) $\mathbb{M} \models_{y} \psi$: *a formula* ψ *is* concluded *at* y *in* $\mathbb{M}$,

(iii) $\mathbb{M} \models^{x}_{y} \phi \mapsto \psi$: *a sequent* $\phi \mapsto \psi$ *is* true *between* x *and* y *in* $\mathbb{M}$,

(iv) $\mathbb{M} \models \phi \mapsto \psi$: *a sequent* $\phi \mapsto \psi$ *is* universally true *on* $\mathbb{M}$,

(v) $\mathbb{F} \models \phi \mapsto \psi$: *a sequent* $\phi \mapsto \psi$ *is* valid *on* $\mathbb{F}$.

Preceding results on bi-approximation semantics. Thanks to the tightness conditions in Definition 3.1, we obtain the following lemma for auxiliary relations R°, $R^{\rightarrow}$ and $R^{\leftarrow}$.

Lemma 3.6 (Redefinition of R) *For each p-frame* $\mathbb{F}$, *we have*

(i) $R(x_1, x_2, y) \iff \forall x \in X.\big[R^{\circ}(x_1, x_2, x) \Longrightarrow x \leq y\big]$,

(ii) $R(x_1, x_2, y) \iff \forall y_2 \in Y.\big[R^{\rightarrow}(x_1, y_2, y) \Longrightarrow x_2 \leq y_2\big]$,

(iii) $R(x_1, x_2, y) \iff \forall y_1 \in Y.\big[R^{\leftarrow}(y_1, x_2, y) \Longrightarrow x_1 \leq y_1\big]$.

The so-called *Hereditary property* on bi-approximation semantics is given as follows.

Proposition 3.7 (Hereditary) *Let* $\mathbb{M}$ *be a bi-approximation model. For all* $x, x' \in X$ *and* $y, y' \in Y$, *we have*

(i) *if* $x' \leq x$ *and* $\mathbb{M} \models^{x} \phi$ *then* $\mathbb{M} \models^{x'} \phi$,

(ii) *if* $y \leq y'$ *and* $\mathbb{M} \models_{y} \psi$ *then* $\mathbb{M} \models_{y'} \psi$.

The following proposition states that every doppelgänger valuation is naturally extended from propositional variables Φ to all formulae Λ: see also [17, Corollary 3.11].

Proposition 3.8 *Let* $\mathbb{M}$ *be a bi-approximation model. For each* $x \in X$, *each* $y \in Y$ *and all formulae* $\phi, \psi \in \Lambda$, *we have*

(i) $\mathbb{M} \models^{x} \phi \iff \forall y \in Y.\big[\mathbb{M} \models_{y} \phi \Longrightarrow x \leq y\big]$,

(ii) $\mathbb{M} \models_{y} \psi \iff \forall x \in X.\big[\mathbb{M} \models^{x} \psi \Longrightarrow x \leq y\big]$.

Furthermore, we can also prove the soundness and Sahlqvist completeness theorem on bi-approximation semantics.

Theorem 3.9 (Soundness [17]) *Let* $\phi \mapsto \psi$ *be a sequent. If* $\phi \mapsto \psi$ *is derivable in the sequent system FL then it is valid on any p-frame.*

Theorem 3.10 (Sahlqvist completeness [18,16]) *Let* Ω *be a set of sequents which have consistent variable occurrence: see [18] for the definition of consistent variable occurrence. The substructural logic* $\mathbf{FL} \oplus \Omega$, *which is* $\mathbf{FL}$

extended by Ω, is elementary and canonical, hence complete with respect to a class of first-order definable p-frames.

4 Bounded morphisms on bi-approximation semantics

In this section, we introduce *bounded morphisms* by focusing on invariance of the satisfaction relation $\models$: see Lemmata 4.5 and 4.6, and Theorem 4.10.

Definition 4.1 (Bounded morphisms) *Given two p-frames $\mathbb{F} = \langle X_1, Y_1, \leq_1, R_1, O_{X1}, O_{Y1}, N_{X1}, N_{Y1}\rangle$ and $\mathbb{G} = \langle X_2, Y_2, \leq_2, R_2, O_{X2}, O_{Y2}, N_{X2}, N_{Y2}\rangle$, a pair $\langle\sigma|\tau\rangle$ of two functions $\sigma\colon X_1 \to X_2$ and $\tau\colon Y_1 \to Y_2$ is a* bounded morphism *from $\mathbb{F}$ to $\mathbb{G}$, denoted by $\langle\sigma|\tau\rangle\colon \mathbb{F} \to \mathbb{G}$, if $\langle\sigma|\tau\rangle$ satisfies*

(i) *for all $x \in X_1$ and $y \in Y_1$, $\sigma(x) \leq_2 \tau(y) \Longrightarrow x \leq_1 y$;*

(ii) *for all $x \in X_1$ and $y' \in Y_2$,*

$$\forall y \in Y_1.\big[y' \leq_2 \tau(y) \Longrightarrow x \leq_1 y\big] \Longrightarrow \sigma(x) \leq_2 y';$$

(iii) *for all $x' \in X_2$ and $y \in Y_1$,*

$$\forall x \in X_1.\big[\sigma(x) \leq_2 x' \Longrightarrow x \leq_1 y\big] \Longrightarrow x' \leq_2 \tau(y);$$

(iv) *for all $x_1, x_2 \in X_1$ and $y \in Y_1$, $R_2(\sigma(x_1), \sigma(x_2), \tau(y)) \Longrightarrow R_1(x_1, x_2, y)$;*

(v) *for all $x'_1, x'_2 \in X_2$ and $y \in Y_1$,*

$$\forall x_1, x_2 \in X_1.\big[\sigma(x_1) \leq_2 x'_1 \text{ and } \sigma(x_2) \leq_2 x'_2 \Longrightarrow R_1(x_1, x_2, y)\big] \Longrightarrow R_2(x'_1, x'_2, \tau(y));$$

(vi) *for all $x'_1 \in X_2$, $x_2 \in X_1$ and $y' \in Y_2$,*

$$\forall x_1 \in X_1, y \in Y_1.\big[\sigma(x_1) \leq_2 x'_1 \text{ and } y' \leq_2 \tau(y) \Longrightarrow R_1(x_1, x_2, y)\big] \Longrightarrow R_2(x'_1, \sigma(x_2), y');$$

(vii) *for all $x_1 \in X_1$, $x'_2 \in X_2$ and $y' \in Y_2$,*

$$\forall x_2 \in X_1, y \in Y_1.\big[\sigma(x_2) \leq_2 x'_2 \text{ and } y' \leq_2 \tau(y) \Longrightarrow R_1(x_1, x_2, y)\big] \Longrightarrow R_2(\sigma(x_1), x'_2, y');$$

(viii) *for each $x \in X_1$,*

$$\big[x \in O_{X1} \iff \sigma(x) \in O_{X2}\big] \text{ and } \big[x \in N_{X1} \iff \sigma(x) \in N_{X2}\big];$$

(ix) *for each $y \in Y_1$,*

$$\big[y \in O_{Y1} \iff \tau(y) \in O_{Y2}\big] \text{ and } \big[y \in N_{Y1} \iff \tau(y) \in N_{Y2}\big].$$

Moreover, a bounded morphisms $\langle\sigma|\tau\rangle\colon \mathbb{F} \to \mathbb{G}$ is a bounded morphism on bi-approximation models from $\langle\mathbb{F}, U\rangle$ to $\langle\mathbb{G}, V\rangle$, if $\langle\sigma|\tau\rangle$ additionally satisfies

(x) for each $x \in X_1$, $x \in U^{\downarrow}(p) \iff \sigma(x) \in V^{\downarrow}(p)$,

(xi) for each $y \in Y_1$, $y \in U_{\uparrow}(p) \iff \tau(y) \in V_{\uparrow}(p)$,

for each propositional variable $p \in \Phi$.

Remark 4.2 Bounded morphisms for polarities are given by the first three conditions.

Remark 4.3 In [8], morphisms on generalized Kripke frames are induced by the complete homomorphisms on the dual algebras. On the setting, we cannot define morphisms on two-sorted frames in our setting unless they are surjective. On the other hand, our setting allows us to have complete homomorphisms on the dual morphisms which cannot be represented by our bounded morphisms, and this is not the case of modal logic.

At first, one may feel that the conditions of bounded morphisms for bi-approximation semantics are far from those for Kripke semantics. For example, a p-morphism $\rho\colon \langle W, R\rangle \to \langle W', R'\rangle$ for Kripke frames satisfies the following condition: for all $w_1, w_2 \in W_1$, if $R(w_1, w_2)$ then $R'(\rho(w_1), \rho(w_2))$. This condition looks completely opposite of (i) in Definition 4.1. However, with help of the definition of $\leq$ and the auxiliary relations R°, $R^\rightarrow$ and $R^\leftarrow$ (see Definition 3.1), we can find some similarities.

Proposition 4.4 *Let $\mathbb{F}$ and $\mathbb{G}$ be p-frames, and $\langle\sigma|\tau\rangle\colon \mathbb{F} \to \mathbb{G}$ a bounded morphism. Then, we have*

(i) *for all $x_1, x_2 \in X_1$, $x_1 \leq_1 x_2 \Longrightarrow \sigma(x_1) \leq_2 \sigma(x_2)$,*[2]

(ii) *for all $y_1, y_2 \in Y_1$, $y_1 \leq_1 y_2 \Longrightarrow \tau(y_1) \leq_2 \tau(y_1)$,*

(iii) *for all $x_1, x_2, x \in X_1$, $R_1^\circ(x_1, x_2, x) \Longrightarrow R_2^\circ(\sigma(x_1), \sigma(x_2), \sigma(x))$,*

(iv) *for all $x_1 \in X_1$, $y_2, y \in Y_1$, $R_1^\rightarrow(x_1, y_2, y) \Longrightarrow R_2^\rightarrow(\sigma(x_1), \tau(y_2), \tau(y))$,*

(v) *for all $x_2 \in X_2$, $y_1, y \in Y_1$, $R_1^\leftarrow(y_1, x_2, y) \Longrightarrow R_2^\leftarrow(\tau(y_1), \sigma(x_2), \tau(y))$.*

Now we shall show the so-called p-morphism lemma for bi-approximation models. However, unlike the p-morphism lemma for Kripke models, in our case, there are three satisfaction relations which we should respect, i.e. two types of $\models$ for formulae and one $\models$ for sequents.

Lemma 4.5 (for formulae) *Let $\mathbb{M}_1$ and $\mathbb{M}_2$ be bi-approximation models, and $\langle\sigma|\tau\rangle\colon \mathbb{M}_1 \to \mathbb{M}_2$ a bounded morphism. For all formulae $\phi, \psi \in \Lambda$, each $x \in X_1$ and each $y \in Y_1$, we have*

(i) $\mathbb{M}_1 \overset{x}{\models} \phi \iff \mathbb{M}_2 \overset{\sigma(x)}{\models} \phi$,

(ii) $\mathbb{M}_1 \underset{y}{\models} \psi \iff \mathbb{M}_2 \underset{\tau(y)}{\models} \psi$.

Next, we show the p-morphism lemma for sequents.

Lemma 4.6 (for sequents) *Let $\mathbb{M}_1$ and $\mathbb{M}_2$ be bi-approximation models, and $\langle\sigma|\tau\rangle\colon \mathbb{M}_1 \to \mathbb{M}_2$ a bounded morphism. For every sequent $\phi \mapsto \psi$, we have*

$$\mathbb{M}_1 \models \phi \mapsto \psi \iff \forall x \in X_1, y \in Y_1.\big[\mathbb{M}_2 \overset{\sigma(x)}{\underset{\tau(y)}{\models}} \phi \mapsto \psi\big].$$

[2] This follows from (iii) and R-order on the current setting. However, this holds on the setting of bounded morphisms for polarities as well.

Proof. ($\Rightarrow$). For arbitrary $x \in X_1$ and $y \in Y_1$, suppose that $\mathbb{M}_2 \stackrel{\sigma(x)}{\Vdash} \phi$ and $\mathbb{M}_2 \underset{\tau(y)}{\Vdash} \psi$. To show that $\sigma(x) \leq_2 \tau(y)$, we shall use (ii) of Definition 4.1. That is, it suffices to show that, for each $y' \in Y_1$, if $\tau(y) \leq_2 \tau(y')$ then $x \leq_1 y'$. By the Hereditary condition, if $\tau(y) \leq_2 \tau(y')$, we have $\mathbb{M}_2 \underset{\tau(y')}{\Vdash} \psi$. Thanks to Lemma 4.5, we obtain that $\mathbb{M}_1 \stackrel{x}{\Vdash} \phi$ and $\mathbb{M}_1 \underset{y'}{\Vdash} \psi$. Since $\mathbb{M}_1 \models \phi \mapsto \psi$, we get $x \leq_1 y'$. Therefore, $\sigma(x) \leq_2 \tau(y)$.

($\Leftarrow$). For arbitrary $x \in X_1$ and $y \in Y_1$, if $\mathbb{M}_1 \stackrel{x}{\Vdash} \phi$ and $\mathbb{M}_1 \underset{y}{\Vdash} \psi$, by Lemma 4.5, we have that $\mathbb{M}_2 \stackrel{\sigma(x)}{\Vdash} \phi$ and $\mathbb{M}_2 \underset{\tau(y)}{\Vdash} \psi$. By our assumption $\mathbb{M}_2 \underset{\tau(y)}{\stackrel{\sigma(x)}{\Vdash}} \phi \mapsto \psi$, we obtain that $\sigma(x) \leq_2 \tau(y)$. By (i) of Definition 4.1, we conclude $x \leq_1 y$. □

Remark 4.7 Unlike what happens in the setting of modal logic, our bounded morphisms are not strong enough to show the local correspondence property, i.e. $\mathbb{M}_1 \underset{y}{\stackrel{x}{\Vdash}} \phi \mapsto \psi \iff \mathbb{M}_2 \underset{\tau(y)}{\stackrel{\sigma(x)}{\Vdash}} \phi \mapsto \psi$, in general. To prove it, we need an extra condition: see Lemma 4.9.

Next, our discussion is heading toward invariance of validity of sequents, namely p-morphism lemmata for p-frames. To do so, we first introduce the following special bounded morphisms. Let $\mathbb{F}$ and $\mathbb{G}$ be p-frames, and $\langle \sigma | \tau \rangle \colon \mathbb{F} \to \mathbb{G}$ a bounded morphism.

B-embedding: for all $x \in X_1$ and $y \in Y_1$, $x \leq_1 y \Longrightarrow \sigma(x) \leq_2 \tau(y)$;

B-separating: for all $x' \in X_2$ and $y' \in Y_2$,

$$\forall x \in X_1, y \in Y_1. \big[\sigma(x) \leq_2 x' \text{ and } y' \leq_2 \tau(y) \Longrightarrow x \leq_1 y\big] \Longrightarrow x' \leq_2 y';$$

B-reflecting: both B-embedding and B-separating.

Intuitively, B-embedding and B-separating are sort of the order-embedding and the surjectivity for bi-approximation semantics, respectively: see also Theorem 4.10. Therefore, B-reflecting is sort of the isomorphism. However, on bi-approximation semantics, since truth value of sequents is *approximated*, we may have states which do not affect to evaluate sequents. In other words, the surjectivity is not vital to argue invariance of validity of sequents. B-reflecting is designed to capture the essence of the approximation. So, B-reflecting is not always isomorphic, but it is perfectly describing the approximation of sequents on $\mathbb{F}$ as the approximation of sequents on $\mathbb{G}$, and vise versa. For B-embedding bounded morphisms, we can obtain the following.

Proposition 4.8 *Let $\mathbb{F}$ and $\mathbb{G}$ be p-frames, and $\langle \sigma | \tau \rangle \colon \mathbb{F} \to \mathbb{G}$ a B-embedding bounded morphism. Then we have*

(i) *for all $x_1, x_2 \in X_1$, $\sigma(x_1) \leq_2 \sigma(x_2) \Longrightarrow x_1 \leq_1 x_2$,*

(ii) *for all $y_1, y_2 \in Y_1$, $\tau(y_1) \leq_2 \tau(y_2) \Longrightarrow y_1 \leq_1 y_2$.*

Proof. Here, we show (i) only, but (ii) can be analogously proved. For arbitrary $x_1, x_2 \in X_1$, assume $\sigma(x_1) \leq_2 \sigma(x_2)$. For any $y \in Y_1$, if $x_2 \leq_1 y$ then, as

$\langle\sigma|\tau\rangle$ is B-embedding, $\sigma(x_2) \leq_2 \tau(y)$. By transitivity, we have $\sigma(x_1) \leq \tau(y)$. By (i) of Definition 4.1, we obtain $x_1 \leq_1 y$, which concludes $x_1 \leq_1 x_2$. □

Lemma 4.9 (Local p-morphism lemma for sequents) *Let $\mathbb{M}_1$ and $\mathbb{M}_2$ be bi-approximation models, and $\langle\sigma|\tau\rangle\colon \mathbb{M}_1 \to \mathbb{M}_2$ a B-embedding bounded morphism. For each sequent $\phi \mapsto \psi$, each $x \in X_1$ and each $y \in Y_1$, we have*

$$\mathbb{M}_1 \mathrel{\|\!\overset{x}{\underset{y}{=}}} \phi \mapsto \psi \iff \mathbb{M}_2 \mathrel{\|\!\overset{\sigma(x)}{\underset{\tau(y)}{=\!=}}} \phi \mapsto \psi.$$

Proof. ($\Rightarrow$). Suppose that $\mathbb{M}_2 \mathrel{\|\!\overset{\sigma(x)}{=\!=}} \phi$ and $\mathbb{M}_2 \mathrel{\|\!\underset{\tau(y)}{=\!=}} \psi$. Thanks to Lemma 4.5, we have that $\mathbb{M}_1 \mathrel{\|\!\overset{x}{=}} \phi$ and $\mathbb{M}_1 \mathrel{\|\!\underset{y}{=}} \psi$. By our assumption, we obtain $x \leq_1 y$. Here, as $\langle\sigma|\tau\rangle$ is B-embedding, $\sigma(x) \leq_2 \tau(y)$ holds.

($\Leftarrow$). Assume that $\mathbb{M}_1 \mathrel{\|\!\overset{x}{=}} \phi$ and $\mathbb{M}_1 \mathrel{\|\!\underset{y}{=}} \psi$. Thanks to Lemma 4.5, we have that $\mathbb{M}_2 \mathrel{\|\!\overset{\sigma(x)}{=\!=}} \phi$ and $\mathbb{M}_2 \mathrel{\|\!\underset{\tau(y)}{=\!=}} \psi$. As $\mathbb{M}_2 \mathrel{\|\!\overset{\sigma(x)}{\underset{\tau(y)}{=\!=}}} \phi \mapsto \psi$, we get $\sigma(x) \leq_2 \tau(y)$. By (i) of Definition 4.1, $x \leq_1 y$, which concludes $\mathbb{M}_1 \mathrel{\|\!\overset{x}{\underset{y}{=}}} \phi \mapsto \psi$. □

Now, we show the following invariance of validity of sequents on bi-approximation semantics.

Theorem 4.10 *Let $\mathbb{F}$ and $\mathbb{G}$ be p-frames, and $\langle\sigma|\tau\rangle\colon \mathbb{F} \to \mathbb{G}$ a bounded morphism. For each sequent $\phi \mapsto \psi$, we have*

(i) *if $\langle\sigma|\tau\rangle$ is B-embedding then $\mathbb{G} \models \phi \mapsto \psi \Longrightarrow \mathbb{F} \models \phi \mapsto \psi$,*

(ii) *if $\langle\sigma|\tau\rangle$ is B-separating then $\mathbb{F} \models \phi \mapsto \psi \Longrightarrow \mathbb{G} \models \phi \mapsto \psi$,*

(iii) *if $\langle\sigma|\tau\rangle$ is B-reflecting then $\mathbb{F} \models \phi \mapsto \psi \iff \mathbb{G} \models \phi \mapsto \psi$.*

Proof. Here, we prove only (i), by contraposition. Suppose $\mathbb{F} \not\models \phi \mapsto \psi$. Then, there exists a doppelgänger valuation U on $\mathbb{F}$, $x \in X_1$ and $y \in Y_1$ such that $\mathbb{F}, U \mathrel{\not\models\!\overset{x}{\underset{y}{}}} \phi \mapsto \psi$. Firstly, we claim that there exists a doppelgänger valuation V on $\mathbb{G}$ which makes $\langle\sigma|\tau\rangle$ a bounded morphism from $\langle\mathbb{F}, U\rangle$ to $\langle\mathbb{G}, V\rangle$.

For the doppelgänger valuation U on $\mathbb{F}$, we let

(i) $V^{\downarrow}(p) := \{x' \in X_2 \mid \forall y \in U_{\uparrow}(p).\, x' \leq_2 \tau(y)\}$,

(ii) $V_{\uparrow}(p) := \{y' \in Y_2 \mid \forall x' \in V^{\downarrow}(p).\, x' \leq_2 y'\}$,

for each propositional variable $p \in \Phi$. Now, we show that V is a doppelgänger valuation on $\mathbb{G}$. It suffices to show

$$V^{\downarrow}(p) = \{x' \in X_2 \mid \forall y' \in V_{\uparrow}(p).\, x' \leq_2 y'\}.$$

($\subseteq$). For arbitrary $x' \in V^{\downarrow}(p)$ and $y' \in V_{\uparrow}(p)$, by the definition of $V_{\uparrow}(p)$, we have $x' \leq_2 y'$. ($\supseteq$). We prove this by contraposition. Suppose $x' \notin V^{\downarrow}(p)$. Then, there exists $y \in U_{\uparrow}(p)$ such that $x' \not\leq_2 \tau(y)$. By definition, we have $\tau[U_{\uparrow}(p)] \subseteq V_{\uparrow}(p)$. Hence, $\tau(y) \in V_{\uparrow}(p)$, which shows the statement. Therefore, V is a doppelgänger valuation on $\mathbb{G}$.

Next, we show $\langle\sigma|\tau\rangle$ is a bounded morphism from $\langle\mathbb{F}, U\rangle$ to $\langle\mathbb{G}, V\rangle$. That is, $x \in U^{\downarrow}(p) \iff \sigma(x) \in V^{\downarrow}(p)$ and $y \in U_{\uparrow}(p) \iff \tau(y) \in V_{\uparrow}(p)$.

$x \in U^{\downarrow}(p) \iff \sigma(x) \in V^{\downarrow}(p)$. ($\Rightarrow$). For arbitrary $x \in U^{\downarrow}(p)$ and $y \in U_{\uparrow}(p)$, we have $x \leq_1 y$. Since $\langle \sigma | \tau \rangle$ is B-embedding, we obtain $\sigma(x) \leq_2 \tau(y)$, so $\sigma(x) \in V^{\downarrow}(p)$. ($\Leftarrow$). For arbitrary $\sigma(x) \in V^{\downarrow}(p)$ and $y \in U_{\uparrow}(p)$, by the definition of $V^{\downarrow}(p)$, we obtain $\sigma(x) \leq_2 \tau(y)$. By (i) of Definition 4.1, $x \leq_1 y$, hence $x \in U^{\downarrow}(p)$.

$y \in U_{\uparrow}(p) \iff \tau(y) \in V_{\uparrow}(p)$. ($\Rightarrow$). For arbitrary $y \in U_{\uparrow}(p)$ and $x' \in V^{\downarrow}(p)$, by definition, we have $x' \leq_2 \tau(y)$, hence $\tau(y) \in V_{\uparrow}(p)$. ($\Leftarrow$). For arbitrary $\tau(y) \in V_{\uparrow}(p)$ and $x \in U^{\downarrow}(p)$, because $x \in U^{\downarrow}(p) \iff \sigma(x) \in V^{\downarrow}(p)$ (as we saw above), we have $\sigma(x) \leq_2 \tau(y)$. By (i) of Definition 4.1, we obtain $x \leq_1 y$, hence $y \in U_{\uparrow}(p)$. Therefore, $\langle \sigma | \tau \rangle$ is a bounded morphism from $\langle \mathbb{F}, U \rangle$ to $\langle \mathbb{G}, V \rangle$.

Now, by Lemma 4.9 and our assumption, i.e. $\mathbb{F}, U \not\models^{x}_{y} \phi \mapsto \psi$, we obtain that $\mathbb{G}, V \not\models^{\sigma(x)}_{\tau(y)} \phi \mapsto \psi$, which derives $\mathbb{G} \not\models \phi \mapsto \psi$. □

Remark 4.11 To prove (i) of Theorem 4.10, we construct a doppelgänger valuation V on $\mathbb{G}$ from a doppelgänger valuation U on $\mathbb{F}$. There are two natural way to do this. One is in the proof of Theorem 4.10. The other is the following: for each proposition variable $p \in \Phi$,

(i) $V^{\downarrow}(p) := \{x' \in X_2 \mid \forall y' \in V_{\uparrow}(p).\, x' \leq_2 y'\}$,

(ii) $V_{\uparrow}(p) := \{y' \in Y_2 \mid \forall x \in U^{\downarrow}(p).\, \sigma(x) \leq_2 y'\}$.

In general, these two doppelgänger valuations do not coincide. However, when $\langle \sigma | \tau \rangle$ satisfies B-separating as well, i.e. B-reflecting, they are always identical.

5 The dual representation

Here, we state that a homomorphism $h \colon \mathbb{A} \to \mathbb{B}$ is *strict*, if for each element $b \in B$ there exist $\underline{a}, \overline{a} \in A$ such that $h(\underline{a}) \leq_B b$ and $b \leq_B h(\overline{a})$. Note that, on bounded lattices (bounded lattice expansions), every homomorphism is strict, because it preserves the constants top $\top$ and bottom $\bot$.

To discuss the dual representation of bi-approximation semantics, we introduce the following categories.

- $\mathcal{FL}$: the category of FL-algebras and strict homomorphisms.
- $\mathcal{BFL}$: the category of bounded FL-algebras and homomorphisms.
- $\mathcal{POL}$: the category of p-frames and bounded morphisms.

We mention that $\mathcal{POL}$ is named after *polarity frames*.

Proposition 5.1 *$\mathcal{BFL}$ is a full and faithful subcategory of $\mathcal{FL}$.*

Below we briefly recall the *object-level* representation of bi-approximation semantics in [17].

Dual algebras of p-frames. For each p-frame $\mathbb{F}$, we construct two isomorphic FL-algebras *in parallel* based on a Galois connection between $\wp(X)$ and $\wp(Y)^{\partial}$, where $\wp(X)$ is the poset of the powerset of X and the set-inclusion $\subseteq$ and $\wp(Y)^{\partial}$ is the poset of the powerset of Y and the set-reverse-inclusion $\supseteq$. Note that the superscript $\_^{\partial}$ indicates that the order is the reverse of the standard inclusion $\subseteq$. We introduce the following two functions $\lambda\colon \wp(X) \to \wp(Y)^\partial$ and $\upsilon\colon \wp(Y)^\partial \to \wp(X)$: for each $\mathfrak{X} \in \wp(X)$ and each $\mathfrak{Y} \in \wp(Y)^\partial$, we let

(i) $\lambda(\mathfrak{X}) := \{y \in Y \mid \forall x \in \mathfrak{X}. x \leq y\}$,

(ii) $\upsilon(\mathfrak{Y}) := \{x \in X \mid \forall y \in \mathfrak{Y}. x \leq y\}$.

Since λ and υ form a Galois connection, the images are isomorphic as posets, i.e. $\upsilon[\wp(Y)^\partial] \cong \lambda[\wp(X)]$. Hereafter, we denote the images $\upsilon[\wp(Y)^\partial]$ by $\mathbb{D}$ and $\lambda[\wp(X)]$ by $\mathbb{U}$. Note that each element in $\mathbb{D}$ is a downward closed subset of X and every element in $\mathbb{U}$ is an upward closed subset of Y. To extend $\mathbb{D}$ and $\mathbb{U}$ as two isomorphic FL-algebras, we define the operations $\vee$, $\wedge$, $*$, $\backslash$ and $/$, on top of $\mathbb{D}$ and $\mathbb{U}$ as follows: for all $\alpha^\downarrow, \beta^\downarrow \in \mathbb{D}$ and $\alpha_\uparrow, \beta_\uparrow \in \mathbb{U}$ (note that $\alpha^\downarrow$ and $\alpha_\uparrow$, and $\beta^\downarrow$ and $\beta_\uparrow$ are corresponding elements in $\mathbb{D}$ and $\mathbb{U}$)

(i) $\alpha^\downarrow \vee \beta^\downarrow := \upsilon(\alpha_\uparrow \vee \beta_\uparrow)$; $\quad \alpha_\uparrow \vee \beta_\uparrow := \alpha_\uparrow \cap \beta_\uparrow$;

(ii) $\alpha^\downarrow \wedge \beta^\downarrow := \alpha^\downarrow \cap \beta^\downarrow$; $\quad \alpha_\uparrow \wedge \beta_\uparrow := \lambda(\alpha^\downarrow \wedge \beta^\downarrow)$;

(iii) $\alpha^\downarrow * \beta^\downarrow := \upsilon(\alpha_\uparrow * \beta_\uparrow)$; $\alpha_\uparrow * \beta_\uparrow := \{y \in Y \mid \forall x_1 \in \alpha^\downarrow, x_2 \in \beta^\downarrow. R(x_1, x_2, y)\}$;

(iv) $\alpha^\downarrow \backslash \beta^\downarrow := \{x_2 \in X \mid \forall x_1 \in \alpha^\downarrow, y \in \beta_\uparrow. R(x_1, x_2, y)\}$; $\alpha_\uparrow \backslash \beta_\uparrow := \lambda(\alpha^\downarrow \backslash \beta^\downarrow)$;

(v) $\beta_\uparrow / \alpha^\downarrow := \{x_1 \in X \mid \forall x_2 \in \alpha^\downarrow, y \in \beta_\uparrow. R(x_1, x_2, y)\}$; $\beta_\uparrow / \alpha_\uparrow := \lambda(\beta^\downarrow / \alpha^\downarrow)$.

Theorem 5.2 ([17]) *$\langle \mathbb{D}, \vee, \wedge, *, \backslash, /, O_X, N_X\rangle$ and $\langle \mathbb{U}, \vee, \wedge, *, \backslash, /, O_Y, N_Y\rangle$ are FL-algebras. Furthermore, they are isomorphic.*

Definition 5.3 (Dual algebra) *Let $\mathbb{F}$ be a p-frame.* The dual algebra of $\mathbb{F}$ *is an abstract FL-algebra $\mathbb{F}^+ = \langle A, \vee, \wedge, *, \backslash, /, 1, 0\rangle$ which is isomorphic to $\langle \mathbb{D}, \vee, \wedge, *, \backslash, /, O_X, N_X\rangle$ and $\langle \mathbb{U}, \vee, \wedge, *, \backslash, /, O_Y, N_Y\rangle$.*[3]

Theorem 5.4 ([17]) *Let $\mathbb{F}$ be a p-frame. For each sequent $\phi \Rightarrow \psi$, we have*

$$\mathbb{F} \models \phi \Rightarrow \psi \iff \mathbb{F}^+ \models s_\phi \leq t_\psi.$$

Recall that s_ϕ and t_ψ are the algebraic terms for ϕ and ψ: see Section 2.

Dual frames of FL-algebras. Here we show the construction of the dual frames of FL-algebras. We mention that the dual frames correspond to *the intermediate level* in [9] but see also [5,18].

Let $\mathbb{A} = \langle A, \vee, \wedge, *, \backslash, /, 1, 0\rangle$ be an FL-algebra. On $\mathbb{A}$, we introduce the following polarity $\langle \mathcal{F}, \mathcal{I}, \sqsubseteq\rangle$, where $\mathcal{F}$ is the set of all filters of $\mathbb{A}$, $\mathcal{I}$ is the set of all ideals of $\mathbb{A}$, and $F \sqsubseteq I \iff F \cap I \neq \emptyset$ for $F \in \mathcal{F}$ and $I \in \mathcal{I}$. Note that we have $F_1 \sqsubseteq F_2 \iff F_2 \subseteq F_1$ and $I_1 \sqsubseteq I_2 \iff I_1 \subseteq I_2$ by definition. Now, on top of this polarity, we put extra structures: R, $O_\mathcal{F}$, $O_\mathcal{I}$, $N_\mathcal{F}$ and $N_\mathcal{I}$ as follows: for all $F_1, F_2 \in \mathcal{F}$ and $I \in \mathcal{I}$, we let $R(F_1, F_2, I) \iff F_1 * F_2 \sqsubseteq I$, where $F_1 * F_2 := \{a \in A \mid \exists f_1 \in F_1, f_2 \in F_2. f_1 * f_2 \leq a\}$. And, we let $O_\mathcal{F}$, $O_\mathcal{I}$, $N_\mathcal{F}$ and $N_\mathcal{I}$ be the set of all filters containing 1, the set of all ideals containing 1, the set of all filters containing 0 and the set of all ideals containing 0. Then, we call the 8-tuple $\mathbb{A}_+ = \langle \mathcal{F}, \mathcal{I}, \sqsubseteq, R, O_\mathcal{F}, O_\mathcal{I}, N_\mathcal{F}, N_\mathcal{I}\rangle$ *the dual frame of* $\mathbb{A}$.

---

[3] Yet another concrete construction of the dual algebra is suggested by a reviewer. But, we keep this definition at least in the current paper.

Theorem 5.5 ([17]) *For any FL-algebra $\mathbb{A}$, the dual frame $\mathbb{A}_+$ is a p-frame.*

Also, we mention that, on the dual frame $\mathbb{A}_+$, we have that

(i) $F_1 * F_2 := \{a \in A \mid \exists f_1 \in F_1, f_2 \in F_2.\, f_1 * f_2 \leq a\}$ is a filter,

(ii) $F \backslash I := \{a \in A \mid \exists f \in F, i \in I.\, a \leq f \backslash i\}$ is an ideal,

(iii) $I/F := \{a \in A \mid \exists f \in F, i \in I.\, a \leq i/f\}$ is an ideal,

(iv) $R^\circ(F_1, F_2, F) \iff F \sqsubseteq F_1 * F_2$,

(v) $R^\rightarrow(F_1, I_2, I) \iff F_1 \backslash I \sqsubseteq I_2$,

(vi) $R^\leftarrow(I_1, F_2, I) \iff I/F_2 \sqsubseteq I_1$.

Theorem 5.6 ([17]) *Let $\mathbb{A}$ be an FL-algebra. For each sequent $\phi \mapsto \psi$,*

$$\mathbb{A} \models s_\phi \leq t_\psi \iff \mathbb{A}_+ \Vdash \phi \mapsto \psi,$$

where s_ϕ and t_ψ are the algebraic terms corresponding to ϕ and ψ.

The dual representation of morphisms. Now, we consider the dual representation of morphisms based on the object-level dual representation.

Let $\mathbb{A}$ and $\mathbb{B}$ be FL-algebras, $\mathcal{F}_A$ the set of all filters of $\mathbb{A}$, $\mathcal{F}_B$ the set of all filters of $\mathbb{B}$, $\mathcal{I}_A$ the set of all ideals of $\mathbb{A}$ and $\mathcal{I}_B$ the set of all ideals of $\mathbb{B}$. For every strict homomorphism $h\colon \mathbb{A} \to \mathbb{B}$, we define a pair of maps $h_+ \colon \mathcal{F}_B \to \mathcal{F}_A$ and $h_- \colon \mathcal{I}_B \to \mathcal{I}_A$ as follows:

(i) for each $F \in \mathcal{F}_B$, we let $h_+(F) := \{a \in A \mid h(a) \in F\}$,

(ii) for each $I \in \mathcal{I}_B$, we let $h_-(I) := \{a \in A \mid h(a) \in I\}$.

It is straightforward to show that h_+ and h_- are well-defined. But, note that the strictness is mandatory to prove the non-emptiness of $h_+(F)$ and $h_-(I)$.

Theorem 5.7 *Let $\mathbb{A}$ and $\mathbb{B}$ be FL-algebras, and $h\colon \mathbb{A} \to \mathbb{B}$ a strict homomorphism. The pair of maps h_+ and h_- forms a bounded morphism from $\mathbb{B}_+$ to $\mathbb{A}_+$, i.e. $\langle h_+ | h_- \rangle \colon \mathbb{B}_+ \to \mathbb{A}_+$.*

Conversely, for p-frames $\mathbb{F}$ and $\mathbb{G}$ and a bounded morphism $\langle \sigma | \tau \rangle \colon \mathbb{F} \to \mathbb{G}$, we introduce two maps $\sigma^+ \colon \mathbb{D}_2 \to \mathbb{U}_1$ and $\tau^- \colon \mathbb{U}_2 \to \mathbb{D}_1$, where $\mathbb{F}^+ \cong \mathbb{D}_1 \cong \mathbb{U}_1$ and $\mathbb{G}^+ \cong \mathbb{D}_2 \cong \mathbb{U}_2$: for each $\alpha \in \mathbb{G}^+$ which is $\alpha^\downarrow \in \mathbb{D}$ and $\alpha_\uparrow \in \mathbb{U}$, we let

(i) $\sigma^+(\alpha^\downarrow) := \{y \in Y_1 \mid \forall x \in \sigma^{-1}[\alpha^\downarrow].\, x \leq_1 y\}$,

(ii) $\tau^-(\alpha_\uparrow) := \{x \in X_1 \mid \forall y \in \tau^{-1}[\alpha_\uparrow].\, x \leq_1 y\}$,

where σ^{-1} and τ^{-1} are the inverse images of σ and τ. For these maps, we can prove the following important facts.

Proposition 5.8 (Coherence) *Let $\mathbb{F}$ and $\mathbb{G}$ be p-frames, and $\langle \sigma | \tau \rangle \colon \mathbb{F} \to \mathbb{G}$ a bounded morphism. For all $\alpha^\downarrow \in \mathbb{D}_2$ and $\alpha_\uparrow \in \mathbb{U}_2$, if they are corresponding points, i.e. $\lambda(\alpha^\downarrow) = \alpha_\uparrow$ and $\upsilon(\alpha_\uparrow) = \alpha^\downarrow$, we have*

(i) $\sigma^+(\alpha^\downarrow) = \tau^{-1}[\alpha_\uparrow]$,

(ii) $\tau^-(\alpha_\uparrow) = \sigma^{-1}[\alpha^\downarrow]$.

Proof. Here, we show (ii) only, but (i) can be analogously proved. ($\subseteq$). For every $x \in \tau^{-}(\alpha_{\uparrow})$, it suffices to show that $\sigma(x) \leq_2 y'$ for each $y' \in \alpha_{\uparrow}$. Let y' be any element in $\alpha_{\uparrow}$. We use (ii) of Definition 4.1. For every $y \in Y_1$, if $y' \leq_2 \tau(y)$, since $\alpha_{\uparrow}$ is upward closed, $\tau(y) \in \alpha_{\uparrow}$, hence $y \in \tau^{-1}[\alpha_{\uparrow}]$. Now, by the definition of τ^{-}, we obtain $x \leq_1 y$. Therefore, $\sigma(x) \leq_2 y'$. ($\supseteq$). Let $x \in \sigma^{-1}[\alpha^{\downarrow}]$, i.e. $\sigma(x) \in \alpha^{\downarrow}$. For each $y \in \tau^{-1}[\alpha_{\uparrow}]$, since $\tau(y) \in \alpha_{\uparrow}$, we have $\sigma(x) \leq_2 \tau(y)$. By (i) of Definition 4.1, $x \leq_1 y$, hence $x \in \tau^{-}(\alpha_{\uparrow})$. □

Proposition 5.9 (Interdefinability) *Let $\mathbb{F}$ and $\mathbb{G}$ be p-frames, and $\langle\sigma|\tau\rangle \colon \mathbb{F} \to \mathbb{G}$ a bounded morphism. σ^{+} and τ^{-} coincide on the dual algebras. That is, for each $\alpha \in \mathbb{G}^{+}$, i.e. $\alpha^{\downarrow} \in \mathbb{D}_2$ and $\alpha_{\uparrow} \in \mathbb{U}_2$, we have $\sigma^{+}(\alpha^{\downarrow}) := \{y \in Y_1 \mid \forall x \in \tau^{-}(\alpha_{\uparrow}).\, x \leq_1 y\}$ and $\tau^{-}(\alpha_{\uparrow}) := \{x \in X_1 \mid \forall y \in \sigma^{+}(\alpha^{\downarrow}).\, x \leq_1 y\}$.*

With respect to the coherence of σ^{+} and τ^{-} in Proposition 5.8 and the interdefinability of σ^{+} and τ^{-} in Proposition 5.9, hereafter, we treat the two maps σ^{+} and τ^{-} as a map, denoted by $\langle\sigma^{+}|\tau^{-}\rangle$. We sum up the maps in Fig. 2.

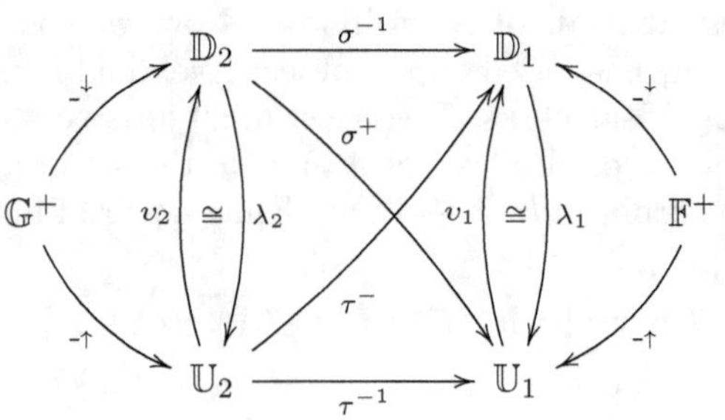

Fig. 2. The relationships of σ^{+}, τ^{-}, σ^{-1} and τ^{-1}

Theorem 5.10 *Let $\mathbb{F}, \mathbb{G}$ be p-frames and $\langle\sigma|\tau\rangle \colon \mathbb{F} \to \mathbb{G}$ a bounded morphism. The map $\langle\sigma^{+}|\tau^{-}\rangle \colon \mathbb{G}^{+} \to \mathbb{F}^{+}$ is a strict homomorphism from $\mathbb{G}^{+}$ to $\mathbb{F}^{+}$.*

Theorem 5.11 *Let $\mathbb{A}$ and $\mathbb{B}$ be FL-algebras, $h \colon \mathbb{A} \to \mathbb{B}$ a strict homomorphism, $\mathbb{F}$ and $\mathbb{G}$ p-frames, and $\langle\sigma|\tau\rangle \colon \mathbb{F} \to \mathbb{G}$ a bounded morphism.*

(i) *If $h \colon \mathbb{A} \to \mathbb{B}$ is injective, $\langle h_{+}|h_{-}\rangle \colon \mathbb{B}_{+} \to \mathbb{A}_{+}$ is B-separating.*

(ii) *If $h \colon \mathbb{A} \to \mathbb{B}$ is surjective, $\langle h_{+}|h_{-}\rangle \colon \mathbb{B}_{+} \to \mathbb{A}_{+}$ is B-embedding.*

(iii) *If $\langle\sigma|\tau\rangle \colon \mathbb{F} \to \mathbb{G}$ is B-separating, $\langle\sigma^{+}|\tau^{-}\rangle \colon \mathbb{G}^{+} \to \mathbb{F}^{+}$ is injective.*

(iv) *If $\langle\sigma|\tau\rangle \colon \mathbb{F} \to \mathbb{G}$ is B-embedding, $\langle\sigma^{+}|\tau^{-}\rangle \colon \mathbb{G}^{+} \to \mathbb{F}^{+}$ is surjective.*

6 Conclusion

In the current paper, we have introduced the notion of bounded morphisms on bi-approximation semantics by focusing on invariance of the satisfaction relation on bi-approximation models. Based on the notion of bounded morphisms, we have investigated the so-called p-morphism lemma on bi-approximation semantics. Apart from Kripke models, on bi-approximation semantics, we evaluate not only formulae (assumptions and conclusions) but also sequents (logical consequences). Nevertheless, we have shown that the bounded morphism can preserve all three satisfaction relations on bi-approximation models. Also, we have shown the similarity to p-morphisms in Kripke semantics. In addition, we have discussed invariance of validity of sequents on p-frames via B-embedding, B-separating and B-reflecting bounded morphisms as well. As we have seen in Section 4, the concepts of those bounded morphisms are not exactly the same as those on Kripke semantics. However, the dual representation of morphisms between lattice expansions and bi-approximation semantics satisfies the same properties as the dual representation of morphisms between modal algebras and Kripke semantics in modal logic, e.g. the dual morphisms coincide with the inverse maps. Therefore, the bounded morphisms can be seen as a natural generalisation of p-morphisms on Kripke semantics. Further results on bi-approximation semantics have already discussed by means of bounded morphisms, which will appear in the current author's future work.

Appendix

Proof. [Proposition 4.4] (i). For arbitrary $x_1, x_2 \in X_1$, let $x_1 \leq_1 x_2$. By the definition of $\sigma(x_1) \leq_2 \sigma(x_2)$, we want to show that, for each $y' \in Y_2$, if $\sigma(x_2) \leq_2 y'$ then $\sigma(x_1) \leq_2 y'$. To show that, we use (ii) of Definition 4.1. Assume that $\sigma(x_2) \leq_2 y'$. For each $y \in Y_1$, if $y' \leq_2 \tau(y)$, by transitivity, $\sigma(x_2) \leq_2 \tau(y)$. By (i) of Definition 4.1, we have $x_2 \leq_1 y$. Again, by transitivity, we obtain that $x_1 \leq_1 y$. So, we have $\sigma(x_1) \leq_2 y'$, which concludes $\sigma(x_1) \leq_2 \sigma(x_2)$.

(ii). For arbitrary $y_1, y_2 \in Y_1$, suppose that $y_1 \leq_1 y_2$. Assume that $x' \leq_2 \tau(y_1)$. For any $x \in X_1$, if $\sigma(x) \leq_2 x'$, by transitivity, $\sigma(x) \leq_2 \tau(y_1)$. By (i) of Definition 4.1, we have $x \leq_1 y_1$, hence $x \leq_1 y_2$ by transitivity. Therefore, we obtain $x' \leq_2 \tau(y_2)$, which means $\tau(y_1) \leq_2 \tau(y_2)$.

(iii). For arbitrary $x_1, x_2, x \in X_1$, suppose that $R_1^{\circ}(x_1, x_2, x)$. We want to use (ii) of Definition 4.1 to show $R_2^{\circ}(\sigma(x_1), \sigma(x_2), \sigma(x))$. For any $y \in Y_1$ and $y' \in Y_2$, if $y' \leq_2 \tau(y)$ and $R_2(\sigma(x_1), \sigma(x_2), y')$ then, by R_2-transitivity, $R_2(\sigma(x_1), \sigma(x_2), \tau(y))$. By (iv) of Definition 4.1, we obtain $R_1(x_1, x_2, y)$. By the definition of R_1°, we get $x \leq_1 y$, hence $\sigma(x) \leq_2 y'$ by (ii) of Definition 4.1, which concludes $R_2^{\circ}(\sigma(x_1), \sigma(x_2), \sigma(x))$.

(v). For arbitrary $x_2 \in X_1$, $y_1, y \in Y_1$, assume $R_1^{\leftarrow}(y_1, x_2, y)$. For any $x_1' \in X_2$ satisfying $R_2(x_1', \sigma(x_2), \tau(y))$, and each $x_1 \in X_1$, if $\sigma(x_1) \leq_2 x_1'$, by R_2-transitivity, $R_2^{\leftarrow}(\sigma(x_1), \sigma(x_2), \tau(y))$, so $R_1(x_1, x_2, y)$. By the definition of $R_1^{\leftarrow}$, we obtain $x_1 \leq_1 y_1$. By (iii) of Definition 4.1, we get $x_1' \leq_2 \tau(y_1)$, which means $R_2^{\leftarrow}(\tau(y_1), \sigma(x_2), \tau(y))$. □

Proof. [Lemma 4.5] Parallel induction. Base cases hold by definition.

Inductive steps: for each $x \in X_1$ and each $y \in Y_1$,

$\vee$: (ii). $\mathbb{M}_1 \Vdash\!\!=_{y} \phi \vee \psi$ is, by definition, $\mathbb{M}_1 \Vdash\!\!=_{y} \phi$ and $\mathbb{M}_1 \Vdash\!\!=_{y} \psi$. By induction hypothesis, they are equivalent to $\mathbb{M}_2 \Vdash\!\!=_{\tau(y)} \phi$ and $\mathbb{M}_2 \Vdash\!\!=_{\tau(y)} \psi$, which means $\mathbb{M}_2 \Vdash\!\!=_{\tau(y)} \phi \vee \psi$.

(i). ($\Rightarrow$). Assume $\mathbb{M}_1 \Vdash^{x} \phi \vee \psi$. For any $y' \in Y_2$, suppose $\mathbb{M}_2 \Vdash_{y'} \phi \vee \psi$. We use (ii) of Definition 4.1. For each $y \in Y_1$, if $y' \leq_2 \tau(y)$, by the Hereditary, we have $\mathbb{M}_2 \Vdash_{\tau(y)} \phi \vee \psi$. By (ii), we have $\mathbb{M}_1 \Vdash_{y} \phi \vee \psi$, hence $x \leq_1 y$ by our assumption. So, we get $\sigma(x) \leq_2 y'$, which concludes $\mathbb{M}_2 \Vdash^{\sigma(x)} \phi \vee \psi$. ($\Leftarrow$). Assume $\mathbb{M}_2 \Vdash^{\sigma(x)} \phi \vee \psi$. For any $y \in Y_1$, if $\mathbb{M}_1 \Vdash_{y} \phi \vee \psi$, by (ii), we have $\mathbb{M}_2 \Vdash_{\tau(y)} \phi \vee \psi$, hence $\sigma(x) \leq_2 \tau(y)$ by our assumption. By (i) of Definition 4.1, we obtain $x \leq_1 y$, which derives $\mathbb{M}_1 \Vdash^{x} \phi \vee \psi$.

$\wedge$: (i). $\mathbb{M}_1 \Vdash^{x} \phi \wedge \psi$ is, by definition, $\mathbb{M}_1 \Vdash^{x} \phi$ and $\mathbb{M}_1 \Vdash^{x} \psi$. By induction hypothesis, they are equivalent to $\mathbb{M}_2 \Vdash^{\sigma(x)} \phi$ and $\mathbb{M}_2 \Vdash^{\sigma(x)} \psi$, which concludes $\mathbb{M}_2 \Vdash^{\sigma(x)} \phi \wedge \psi$.

(ii). ($\Rightarrow$). Assume $\mathbb{M}_1 \Vdash_{y} \phi \wedge \psi$. For any $x' \in X_2$, suppose $\mathbb{M}_2 \Vdash^{x'} \phi \wedge \psi$. We use (iii) of Definition 4.1. For every $x \in X_1$, if $\sigma(x) \leq_2 x'$, by the Hereditary, we have $\mathbb{M}_2 \Vdash^{\sigma(x)} \phi \wedge \psi$. By our assumption, we obtain $\sigma(x) \leq_2 \tau(y)$, hence $x \leq_1 y$. So, $x' \leq_2 \tau(y)$, which concludes $\mathbb{M}_2 \Vdash_{\tau(y)} \phi \wedge \psi$. ($\Leftarrow$). Assume $\mathbb{M}_2 \Vdash_{\tau(y)} \phi \wedge \psi$. For any $x \in X_1$, if $\mathbb{M}_1 \Vdash^{x} \phi \wedge \psi$, by (i), we have $\mathbb{M}_2 \Vdash^{\sigma(x)} \phi \wedge \psi$, hence $\sigma(x) \leq_2 \tau(y)$, by our assumption. So, we obtain $x \leq_1 y$, which derives $\mathbb{M}_1 \Vdash_{y} \phi \wedge \psi$.

$\circ$: (ii). ($\Rightarrow$). Assume $\mathbb{M}_1 \Vdash_{y} \phi \circ \psi$. For arbitrary $x'_1, x'_2 \in X_2$, suppose that $\mathbb{M}_2 \Vdash^{x'_1} \phi$ and $\mathbb{M}_2 \Vdash^{x'_2} \psi$. We use (v) of Definition 4.1. For all $x_1, x_2 \in X_1$, if $\sigma(x_1) \leq_2 x'_1$ and $\sigma(x_2) \leq_2 x'_2$, by the Hereditary, we have that $\mathbb{M}_2 \Vdash^{\sigma(x_1)} \phi$ and $\mathbb{M}_2 \Vdash^{\sigma(x_2)} \psi$. By induction hypothesis, we have that $\mathbb{M}_1 \Vdash^{x_1} \phi$ and $\mathbb{M}_1 \Vdash^{x_2} \psi$. By our assumption, we obtain $R_1(x_1, x_2, y)$. Therefore, we get $R_2(x'_1, x'_2, \tau(y))$, which concludes $\mathbb{M}_2 \Vdash_{\tau(y)} \phi \circ \psi$. ($\Leftarrow$). Assume that $\mathbb{M}_2 \Vdash_{\tau(y)} \phi \circ \psi$. For arbitrary $x_1, x_2 \in X_1$, if $\mathbb{M}_1 \Vdash^{x_1} \phi$ and $\mathbb{M}_1 \Vdash^{x_2} \psi$, by induction hypothesis, we have that $\mathbb{M}_2 \Vdash^{\sigma(x_1)} \phi$ and $\mathbb{M}_2 \Vdash^{\sigma(x_2)} \psi$, hence $R_1(x_1, x_2, y)$, which derives $\mathbb{M}_1 \Vdash_{y} \phi \circ \psi$.

(ii). This is the same as the case of $\mathbb{M}_1 \Vdash^{x} \phi \vee \psi \iff \mathbb{M}_2 \Vdash^{\sigma(x)} \phi \vee \psi$.

$\rightarrow$: (i). ($\Rightarrow$). Assume $\mathbb{M}_1 \Vdash^{x} \phi \rightarrow \psi$. For arbitrary $x'_1 \in X_2$ and $y' \in Y_2$, suppose $\mathbb{M}_2 \Vdash^{x'_1} \phi$ and $\mathbb{M}_2 \Vdash_{y'} \psi$. We use (vi) of Definition 4.1. For all $x_1 \in X_1$ and $y \in Y_1$, if $\sigma(x_1) \leq_2 x'_1$ and $y' \leq_2 \tau(y)$, by the Hereditary, we have that $\mathbb{M}_2 \Vdash^{\sigma(x_1)} \phi$ and $\mathbb{M}_2 \Vdash_{\tau(y)} \psi$. By induction hypothesis, we obtain that $\mathbb{M}_1 \Vdash^{x_1} \phi$ and $\mathbb{M}_1 \Vdash_{y} \psi$. Since $\mathbb{M}_1 \Vdash^{x} \phi \rightarrow \psi$, we get $R_1(x_1, x, y)$. Therefore, $R_2(x'_1, \sigma(x), y')$, which means $\mathbb{M}_2 \Vdash^{\sigma(x)} \phi \rightarrow \psi$. ($\Leftarrow$). Assume $\mathbb{M}_2 \Vdash^{\sigma(x)} \phi \rightarrow \psi$. For arbitrary $x_1 \in X_1$ and $y \in Y_1$, if $\mathbb{M}_1 \Vdash^{x_1} \phi$ and

$\mathbb{M}_1 \underset{y}{\models} \psi$, by induction hypothesis, we have that $\mathbb{M}_2 \overset{\sigma(x_1)}{\models} \phi$ and $\mathbb{M}_2 \underset{\tau(y)}{\models} \psi$. Because of $\mathbb{M}_2 \overset{\sigma(x)}{\models} \phi \to \psi$, we obtain $R_2(\sigma(x_1), \sigma(x), \tau(y))$. Hence, $R_1(x_1, x, y)$, which concludes $\mathbb{M}_1 \overset{x}{\models} \phi \to \psi$.

(ii). This is analogous to (ii) of $\leftarrow$.

$\leftarrow$: (i). This is analogous to (i) of $\to$.

(ii). ($\Rightarrow$). Assume $\mathbb{M}_1 \underset{y}{\models} \psi \leftarrow \phi$. For any $x' \in X_2$, suppose $\mathbb{M}_2 \overset{x'}{\models} \psi \leftarrow \phi$. We use (iii) of Definition 4.1. For each $x \in X_1$, if $\sigma(x) \leq_2 x'$, by the Hereditary, we have $\mathbb{M}_2 \overset{\sigma(x)}{\models} \psi \leftarrow \phi$. By (i), we obtain $\mathbb{M}_1 \overset{x}{\models} \psi \leftarrow \psi$. By our assumption, we get $x \leq_1 y$. So, $x' \leq_2 \tau(y)$, which means $\mathbb{M}_2 \underset{\tau(y)}{\models} \psi \leftarrow \phi$. ($\Leftarrow$). Suppose $\mathbb{M}_2 \underset{\tau(y)}{\models} \psi \leftarrow \phi$. For any $x \in X_1$, if $\mathbb{M}_1 \overset{x}{\models} \psi \leftarrow \phi$, by (i), $\mathbb{M}_2 \overset{\sigma(x)}{\models} \psi \leftarrow \phi$. Since $\mathbb{M}_2 \underset{\tau(y)}{\models} \psi \leftarrow \phi$, we obtain $\sigma(x) \leq_2 \tau(y)$, hence $x \leq_1 y$, which concludes $\mathbb{M}_1 \underset{y}{\models} \psi \leftarrow \phi$.

□

Proof. [(ii) and (iii) of Theorem 4.10] (ii). We prove by contraposition. Suppose $\mathbb{G} \not\models \phi \Mapsto \psi$. Then, there exists a doppelgänger valuation V on $\mathbb{G}$ such that $\mathbb{G}, V \not\models \phi \Mapsto \psi$. For the doppelgänger valuation V, we induce a doppelgänger valuation U on $\mathbb{F}$, which makes $\langle \sigma | \tau \rangle$ is a bounded morphism from $\langle \mathbb{F}, U \rangle$ to $\langle \mathbb{G}, V \rangle$. For any propositional variable $p \in \Phi$, we let

(1) $U^{\downarrow}(p) := \{x \in X_1 \mid \forall y' \in V_{\uparrow}(p). \sigma(x) \leq_2 y'\}$,

(2) $U_{\uparrow}(p) := \{y \in Y_1 \mid \forall x' \in V^{\downarrow}(p). x' \leq_2 \tau(y)\}$.

Now, we show that $\langle \sigma | \tau \rangle$ is a bounded morphism from $\langle \mathbb{F}, U \rangle$ to $\langle \mathbb{G}, V \rangle$, i.e. $x \in U^{\downarrow}(p) \iff \sigma(x) \in V^{\downarrow}(p)$ and $y \in U_{\uparrow}(p) \iff \tau(y) \in V_{\uparrow}(p)$.

($x \in U^{\downarrow}(p) \iff \sigma(x) \in V^{\downarrow}(p)$). ($\Rightarrow$). For arbitrary $x \in U^{\downarrow}(p)$ and $y' \in V_{\uparrow}(p)$, we use (ii) of Definition 4.1. For any $y \in Y_1$, if $y' \leq_2 \tau(y)$, as $V_{\uparrow}(p)$ is upward closed, we have $\tau(y) \in V_{\uparrow}(p)$. By the definition of $U^{\downarrow}(p)$, we get $\sigma(x) \leq_2 \tau(y)$. By (i) of Definition 4.1, $x \leq_1 y$. So, $\sigma(x) \leq_2 y'$, hence $\sigma(x) \in V^{\downarrow}(p)$ ($\Leftarrow$). This is trivial by definition.

($y \in U_{\uparrow}(p) \iff \tau(y) \in V_{\uparrow}(p)$). ($\Rightarrow$). For arbitrary $y \in U_{\uparrow}(p)$ and $x' \in V^{\downarrow}(p)$, we use (iii) of Definition 4.1. For any $x \in X_1$, if $\sigma(x) \leq_2 x'$, as $V^{\downarrow}(p)$ is downward closed, we have $\sigma(x) \in V^{\downarrow}(p)$. By the definition of $U_{\uparrow}(p)$, we obtain that $\sigma(x) \leq_2 \tau(y)$. By (i) of Definition 4.1, $x \leq_1 y$. So, $x' \leq_2 \tau(y)$, hence $\tau(y) \in V_{\uparrow}(p)$. ($\Leftarrow$). This is trivial by definition.

Next, we prove that U is a doppelgänger valuation on $\mathbb{F}$. That is,

(1) $U^{\downarrow}(p) = \{x \in X_1 \mid \forall y \in U_{\uparrow}(p). x \leq_1 y\}$

($\subseteq$). For arbitrary $x \in U^{\downarrow}(p)$ and $y \in U_{\uparrow}(p)$, since $\sigma(x) \in V^{\downarrow}(p)$ and $\tau(y) \in V_{\uparrow}(p)$ (as we saw above) and V is a doppelgänger valuation on $\mathbb{G}$, we obtain $\sigma(x) \leq_2 \tau(y)$, hence $x \leq_1 y$ by (i) of Definition 4.1. ($\supseteq$). Contraposition. Suppose $x \notin U^{\downarrow}(p)$. There exists $y' \in V_{\uparrow}(p)$ such that $\sigma(x) \not\leq_2 y'$. By (ii) of Definition 4.1, there exists $y \in Y_1$ such that $y' \leq_2 \tau(y)$ but $x \not\leq_1 y$. As $V_{\uparrow}(p)$ is upward closed, $\tau(y) \in V_{\uparrow}(p)$, hence $y \in U_{\uparrow}(p)$.

Therefore, there exists $y \in U_{\uparrow}(p)$ such that $x \not\lesssim_1 y$.

(2) $U_{\uparrow}(p) = \{y \in Y_1 \mid \forall x \in U^{\downarrow}(p). x \leq_1 y\}$

($\subseteq$). For arbitrary $y \in U_{\uparrow}(p)$ and $x \in U^{\downarrow}(p)$, since $\tau(y) \in V_{\uparrow}(p)$ and $\sigma(x) \in V^{\downarrow}(p)$ (as we saw above) and V is a doppelgänger valuation on $\mathbb{G}$, we obtain $\sigma(x) \leq_2 \tau(y)$, hence $x \leq_1 y$. ($\supseteq$). Contraposition. Suppose $y \notin U_{\uparrow}(p)$. There exists $x' \in V^{\downarrow}(p)$ such that $x' \not\lesssim_2 \tau(y)$. By (iii) of Definition 4.1, there exists $x \in X_1$ such that $\sigma(x) \leq_2 x'$ but $x \not\lesssim_1 y$. Since $V^{\downarrow}(p)$ is downward closed, $\sigma(x) \in V^{\downarrow}(p)$, hence $x \in U^{\downarrow}(p)$. So, there exists $x \in U^{\downarrow}(p)$ such that $x \not\lesssim_1 y$.

Therefore, U is a doppelgänger valuation on $\mathbb{F}$.

By our assumption, i.e. $\mathbb{G}, V \not\Vdash \phi \Rrightarrow \psi$, there exists $x' \in X_2$ and $y' \in Y_2$ such that $\mathbb{G}, V \not\Vdash^{x'}_{y'} \phi \Rrightarrow \psi$, which means $\mathbb{G}, V \Vdash^{x'} \phi$, $\mathbb{G}, V \Vdash_{y'} \psi$ but $x' \not\lesssim_2 y'$. Now, as $\langle \sigma | \tau \rangle$ is B-separating, there exist $x \in X_1$ and $y \in Y_1$ such that $\sigma(x) \leq_2 x'$, $y' \leq_2 \tau(y)$ and $x \leq_1 y$. By the Hereditary, we obtain that $\mathbb{G}, V \Vdash^{\sigma(x)} \phi$ and $\mathbb{G}, V \Vdash_{\tau(y)} \psi$. So, by Lemma 4.5, we also have that $\mathbb{F}, U \Vdash^{x} \phi$ and $\mathbb{F}, U \Vdash_{y} \psi$. However, since $x \not\lesssim_1 y$, we conclude $\mathbb{F}, U \not\Vdash^{x}_{y} \phi \Rrightarrow \psi$, hence $\mathbb{F} \not\Vdash \phi \Rrightarrow \psi$.

(iii). This follows directly from (i) and (ii). □

Proof. [Theorem 5.7] By definition, h_+ is a function from $\mathcal{F}_B$ to $\mathcal{F}_A$ and h_- is a function from $\mathcal{I}_B$ to $\mathcal{I}_A$. All we need to show here is to check the conditions in Definition 4.1.

(i). For arbitrary $F \in \mathcal{F}_B$ and $I \in \mathcal{I}_B$, if $h_+(F) \sqsubseteq_A h_-(I)$, there exists $a \in A$ such that $a \in h_+(F) \cap h_-(I)$. By definition, $h(a) \in F$ and $h(a) \in I$, hence $h(a) \in F \cap I$, which concludes $F \sqsubseteq_B I$.

(iii). Contraposition. For arbitrary $G \in \mathcal{F}_A$ and $I \in \mathcal{I}_B$, suppose $G \not\sqsubseteq_A h_-(I)$. Let F be the generated filter by the image $h[G]$, i.e. $F := \uparrow h[G]$. By definition, we have $h[G] \subseteq_B F$, hence $h_+(F) \sqsubseteq_A G$. Plus, by our assumption $G \not\sqsubseteq_A h_-(I)$, we obtain $F \not\sqsubseteq_B I$ (otherwise, it contradicts to $G \not\sqsubseteq_A h_-(I)$). Therefore, there exists $F \in \mathcal{F}_B$ such that $h_+(F) \sqsubseteq_B G$ but $F \not\sqsubseteq_B I$.

(iv). For all $F_1, F_2 \in \mathcal{F}_B$ and $I \in \mathcal{I}_B$, assume $R_A(h_+(F_1), h_+(F_2), h_-(I))$, namely $h_+(F_1) *_A h_+(F_2) \sqsubseteq_A h_-(I)$. Then, there exist $a_1 \in h_+(F_1)$ and $a_2 \in h_+(F_2)$ such that $a_1 *_A a_2 \in h_-(I)$, which means $h(a_1 *_A a_2) \in I$. Since h is homomorphic, we have $h(a_1) *_B h(a_2) \in I$. Moreover, as $h(a_1) \in F_1$ and $h(a_2) \in F_2$, we conclude $F_1 *_B F_2 \sqsubseteq_B I$, i.e. $R_B(F_1, F_2, I)$.

(v). Contraposition. For arbitrary $G_1, G_2 \in \mathcal{F}_A$ and $I \in \mathcal{I}_B$, suppose that $R_A(G_1, G_2, h_-(I))$ does not hold, i.e. $G_1 *_A G_2 \not\sqsubseteq_A h_-(I)$. Let F_1 and F_2 be the generated filters by the images $h[G_1]$ and $h[G_2]$, that is, $F_1 := \uparrow h[G_1]$ and $F_2 := \uparrow h[G_2]$. By definition, we obtain that $h[G_1] \subseteq_B F_1$ and $h[G_2] \subseteq_B F_2$, hence $h_+(F_1) \sqsubseteq_A G_1$ and $h_+(F_2) \sqsubseteq_A G_2$. In addition, for any $b \in F_1 *_B F_2$, there exist $a_1 \in G_1$ and $a_2 \in G_2$ such that $h(a_1) *_B h(a_2) \leq_B b$. As h is homomorphic, we have $h(a_1 *_A a_2) \leq_B b$. Now, if $F_1 *_B F_2 \sqsubseteq_B I$ then $h(a_1 *_A a_2) \in I$, which contradicts to $G_1 *_A G_2 \not\sqsubseteq_A h_-(I)$. Therefore, $F_1 *_B F_2 \not\sqsubseteq_B I$, i.e. $R_B(F_1, F_2, I)$ does not hold.

(vi). Contraposition. For arbitrary $G_1 \in \mathcal{F}_A$, $F_2 \in \mathcal{F}_B$ and $J \in \mathcal{I}_A$, suppose that $R_A(G_1, h_+(F_2), J)$ does not hold, i.e. $G_1 *_A h_+(F_2) \not\sqsubseteq_A J$. Let F_1 be the generated filter by the image $h[G_1]$, i.e. $F_1 := \uparrow h[G_1]$, and I the generated ideal by the image $h[J]$, i.e. $I := \downarrow h[J]$. By definition, we have that $h[G_1] \subseteq_B F_1$ and $h[J] \subseteq_B I$, hence $h_+(F_1) \sqsubseteq_A G_1$ and $J \sqsubseteq_A h_-(I)$. Furthermore, if $F_1 *_B F_2 \sqsubseteq_B I$, there exist $a_a \in G_1$ and $a \in I$ such that $h(a_1)\backslash_B h(a) = h(a_1 \backslash_A a) \in F_2$, which contradicts to $G_1 *_A h_+(F_2) \not\sqsubseteq_A J$. Therefore, $F_1 *_B F_2 \not\sqsubseteq_B I$, i.e. $R_B(F_1, F_2, I)$ does not hold.

(viii). For any $F \in O_{\mathcal{F}_B}$, by definition $1_B \in F$. Because h is homomorphic, $h(1_A) = 1_B \in F$, which derives $1_1 \in h_+(F)$. So, $h_+(F) \in O_{\mathcal{F}_A}$. Conversely, if $h_+(F) \in O_{\mathcal{F}_A}$ then $1_A \in h_+(F)$, so $h(1_A) = 1_B \in F$. Therefore, $F \in O_{\mathbb{F}_B}$. The other case is analogous. □

Proof. [Proposition 5.9] By the definition of σ^+ and τ^-, and Proposition 5.8.□

Proof. [Theorem 5.10] It suffices to show that $\langle \sigma^+ | \tau^- \rangle \colon \mathbb{G}^+ \to \mathbb{F}^+$ is homomorphic. Note that the strictness follows from the preservability of top $\top$ and bottom $\bot$, because $\mathbb{F}^+$ and $\mathbb{G}^+$ are bounded.

($\vee$). To prove $\langle \sigma^+ | \tau^- \rangle(\alpha \vee \beta) = \langle \sigma^+ | \tau^- \rangle(\alpha) \vee \langle \sigma^+ | \tau^- \rangle(\beta)$, it suffices to show that $\sigma^+(\alpha^\downarrow \vee \beta^\downarrow) = \sigma^+(\alpha^\downarrow) \vee \sigma^+(\beta^\downarrow)$, i.e. $\tau^{-1}[\alpha_\uparrow \cap \beta_\uparrow] = \tau^{-1}[\alpha_\uparrow] \cap \tau^{-1}[\beta_\uparrow]$ by Proposition 5.8. But, this is straightforward.

($\wedge$). To prove $\langle \sigma^+ | \tau^- \rangle(\alpha \wedge \beta) = \langle \sigma^+ | \tau^- \rangle(\alpha) \wedge \langle \sigma^+ | \tau^- \rangle(\beta)$, it suffices to show that $\tau^-(\alpha_\uparrow \wedge \beta_\uparrow) = \tau^-(\alpha_\uparrow) \wedge \tau^-(\beta_\uparrow)$, i.e. $\sigma^{-1}[\alpha^\downarrow \cap \beta^\downarrow] = \sigma^{-1}[\alpha^\downarrow] \cap \sigma^{-1}[\beta^\downarrow]$ by Proposition 5.8. But, this is straightforward.

($*$). To prove $\langle \sigma^+ | \tau^- \rangle(\alpha * \beta) = \langle \sigma^+ | \tau^- \rangle(\alpha) * \langle \sigma^+ | \tau^- \rangle(\beta)$, it suffices to show $\sigma^+(\alpha^\downarrow * \beta^\downarrow) = \sigma^+(\alpha^\downarrow) * \sigma^+(\beta^\downarrow)$, i.e. $\tau^{-1}[\alpha_\uparrow * \beta_\uparrow] = \tau^{-1}[\alpha_\uparrow] * \tau^{-1}[\beta_\uparrow]$ by Proposition 5.8. ($\subseteq$). For each $y \in \tau^{-1}[\alpha_\uparrow * \beta_\uparrow]$, to prove $y \in \tau^{-1}[\alpha_\uparrow] * \tau^{-1}[\beta_\uparrow]$, we need to show that $R_1(x_1, x_2, y)$ for arbitrary $x_1 \in \upsilon_1(\tau^{-1}[\alpha_\uparrow])$ and $x_2 \in \upsilon_1(\tau^{-1}[\beta_\uparrow])$. Let y, x_1 and x_2 be arbitrary elements in $\tau^{-1}[\alpha_\uparrow * \beta_\uparrow]$, $\upsilon_1(\tau^{-1}[\alpha_\uparrow])$ and $\upsilon_1(\tau^{-1}[\beta_\uparrow])$. By Proposition 5.9 and Proposition 5.8, $x_1 \in \sigma^{-1}[\alpha^\downarrow]$ and $x_2 \in \sigma^{-1}[\beta^\downarrow]$. By definition, we have that $\tau(y) \in \alpha_\uparrow * \beta_\uparrow$, $\sigma(x_1) \in \alpha^\downarrow$ and $\sigma(x_2) \in \beta^\downarrow$. Further, by the definition of $*$ on $\mathbb{U}_2$, we get $R_2(\sigma(x_1), \sigma(x_2), \tau(y))$. By (iv) of Definition 4.1, we obtain $R_1(x_1, x_2, y)$. ($\supseteq$). For each $y \in \tau^{-1}[\alpha_\uparrow] * \tau^{-1}[\beta_\uparrow]$, we want to show that $R_2(x_1', x_2', \tau(y))$ for arbitrary $x_1' \in \alpha^\downarrow$ and $x_2' \in \beta^\downarrow$. Let y, x_1' and x_2' be arbitrary elements in $\tau^{-1}[\alpha_\uparrow * \beta_\uparrow]$, $\alpha^\downarrow$ and $\beta^\downarrow$. We use (v) of Definition 4.1. For all $x_1, x_2 \in X_1$, if $\sigma(x_1) \leq_2 x_1'$ and $\sigma(x_2) \leq_2 x_2'$, since $\alpha^\downarrow$ and $\beta^\downarrow$ are downward closed, $\sigma(x_1) \in \alpha^\downarrow$ and $\sigma(x_2) \in \beta^\downarrow$, hence $x_1 \in \sigma^{-1}[\alpha^\downarrow]$ and $x_2 \in \sigma^{-1}[\beta^\downarrow]$. By Proposition 5.8 and Proposition 5.9, we have that $x_1 \in \upsilon_1(\tau^{-1}[\alpha_\uparrow]])$ and $x_2 \in \upsilon_1(\tau^{-1}[\beta_\uparrow])$. As $y \in \tau^{-1}[\alpha_\uparrow] * \tau^{-1}[\beta_\uparrow]$, we obtain $R_1(x_1, x_2, y)$, which concludes $R_2(x_1', x_2', \tau(y))$.

($\backslash$). To prove $\langle \sigma^+ | \tau^- \rangle(\alpha \backslash \beta) = \langle \sigma^+ | \tau^- \rangle(\alpha) \backslash \langle \sigma^+ | \tau^- \rangle(\beta)$, it suffices to show $\tau^-(\alpha \backslash \beta) = \tau^-(\alpha) \backslash \tau^-(\beta)$, that is, $\sigma^{-1}(\alpha^\downarrow \backslash \beta^\downarrow) = \sigma^{-1}[\alpha^\downarrow] \backslash \sigma^{-1}[\beta^\downarrow]$ by Proposition 5.8. ($\subseteq$). Let $x_2 \in \sigma^{-1}[\alpha^\downarrow \backslash \beta^\downarrow]$. To show $x_2 \in \sigma^{-1}[\alpha^\downarrow] \backslash \sigma^{-1}[\beta^\downarrow]$, it suffices to show that $R_1(x_1, x_2, y)$ holds for arbitrary $x_1 \in \sigma^{-1}[\alpha^\downarrow]$ and $y \in \lambda_1(\sigma^{-1}[\beta^\downarrow])$. For all $x_1 \in \sigma^{-1}[\alpha^\downarrow]$ and $y \in \lambda_1(\sigma^{-1}[\beta^\downarrow])$, by Proposition 5.8 and Proposition 5.9, we have $y \in \tau^{-1}[\beta_\uparrow]$, hence $\sigma(x_1) \in \alpha^\downarrow$ and $\tau(y) \in \beta_\uparrow$. By our assumption

$x_2 \in \sigma^{-1}[\alpha^{\downarrow}\backslash\beta^{\downarrow}]$, we obtain $R_2(\sigma(x_1), \sigma(x_2), \tau(y))$. By (iv) of Definition 4.1, we conclude $R_1(x_1, x_2, y)$. ($\supseteq$). Let x_2 be any element in $\sigma^{-1}[\alpha^{\downarrow}]\backslash\sigma^{-1}[\beta^{\downarrow}]$. To show $x_2 \in \sigma^{-1}[\alpha^{\downarrow}\backslash\beta^{\downarrow}]$, we need to prove that $\sigma(x_2) \in \alpha^{\downarrow}\backslash\beta^{\downarrow}$. For arbitrary $x'_1 \in \alpha^{\downarrow}$ and $y' \in \beta_{\uparrow}$, we use (vi) of Definition 4.1. For all $x_1 \in X_1$ and $y \in Y_1$, if $\sigma(x_1) \leq_2 x'_1$ and $y' \leq_2 \tau(y)$, as $\alpha^{\downarrow}$ is downward closed and $\beta_{\uparrow}$ is upward closed, $\sigma(x) \in \alpha^{\downarrow}$ and $\tau(y) \in \beta_{\uparrow}$. Since $x_2 \in \sigma^{-1}[\alpha^{\downarrow}]\backslash\sigma^{-1}[\beta^{\downarrow}]$, we obtain $R_1(x_1, x_2, y)$, hence $x_2 \in \sigma^{-1}[\alpha^{\downarrow}\backslash\beta^{\downarrow}]$.

(/). This is analogous to the case of (\).

(1). To prove $\langle\sigma^{+}|\tau^{-}\rangle(1) = 1$, it suffices to show that $\tau^{-}(1) = 1$, i.e. $\sigma^{-1}[O_{X2}] = O_{X1}$ by Proposition 5.8. But, this is straightforward.

(0), ($\top$) and ($\bot$) are analogous to the case of (1). □

Proof. [Theorem 5.11] (i). We prove by contraposition. For arbitrary $G \in \mathcal{F}_A$ and $J \in \mathcal{I}_A$, suppose $G \not\sqsubseteq_A J$, i.e. $G \cap J = \emptyset$. Let F be the generated filter by the image $h[G]$, i.e. $F := \uparrow h[G]$, and I the generated ideal by the image $h[J]$, i.e. $I := \downarrow h[J]$. By definition, we have that $h[G] \subseteq_B F$ and $h[J] \subseteq_B I$. So, we have that $h_{+}(F) \sqsubseteq_A G$ and $J \sqsubseteq_A h_{-}(I)$. Now, if $F \sqsubseteq_B I$, i.e. $F \cap I \neq \emptyset$, there exist $a_1 \in G$ and $a_2 \in J$ such that $h(a_1) \leq_B h(a_2)$, that is $h(a_1) \vee_B h(a_2) = h(a_1 \vee_A a_2) = h(a_2)$. Since h is injective, we obtain $a_1 \leq_A a_2$. Hence, $G \sqsubseteq_A J$, which contradicts to $G \not\sqsubseteq_A J$. Therefore, $F \not\sqsubseteq_B I$.

(ii). For arbitrary $F \in \mathcal{F}_B$ and $I \in \mathcal{I}_B$, if $F \sqsubseteq_B I$, there exists $b \in F \cap I$. As h is surjective, there exists $a \in A$ such that $h(a) = b \in F \cap I$. Then, $a \in h_{+}(F)$ and $a \in h_{-}(I)$, hence $h_{+}(F) \sqsubseteq_A h_{-}(I)$.

(iii). Since every lattice is anti-symmetric, it suffices to show that $\langle\sigma^{+}|\tau^{-}\rangle$ is order-embedding. That is, we show that, for all $\alpha, \beta \in \mathbb{G}^{+}$, if $\langle\sigma^{+}|\tau^{-}\rangle(\alpha) \leq_1 \langle\sigma^{+}|\tau^{-}\rangle(\beta)$ then $\alpha \leq_2 \beta$. We prove it by contraposition. Suppose $\alpha \not\leq_2 \beta$. Then, there exists $x' \in \alpha^{\downarrow}$ and $y' \in \beta_{\uparrow}$ such that $x' \not\leq_2 y'$. Since $\langle\sigma|\tau\rangle$ is B-separating, there exist $x \in X_1$ and $y \in Y_1$ such that $\sigma(x) \leq_2 x'$, $y' \leq_2 \tau(y)$ and $x \not\leq_1 y$. By (i) of Definition 4.1, $\sigma(x) \not\leq_2 \tau(y)$, which derives $\sigma(x) \notin \beta^{\downarrow}$, i.e. $x \notin \sigma^{-1}[\beta^{\downarrow}]$. Moreover, as $\alpha^{\downarrow}$ is downward closed and $\beta_{\uparrow}$ is upward closed, we have that $\sigma(x) \in \alpha^{\downarrow}$ and $\tau(y) \in \beta_{\uparrow}$, hence $x \in \sigma^{-1}[\alpha^{\downarrow}]$. That is, there exists $x \in X_1$ such that $x \in \sigma^{-1}[\alpha^{\downarrow}]$ but $x \notin \sigma^{-1}[\beta^{\downarrow}]$. By Proposition 5.8, we conclude $\tau^{-}(\alpha_{\uparrow}) \not\subseteq_1 \tau^{-}(\beta_{\uparrow})$, i.e. $\langle\sigma^{+}|\tau^{-}\rangle(\alpha) \not\leq_1 \langle\sigma^{+}|\tau^{-}\rangle(\beta)$.

(iv). For each element $\beta \in \mathbb{F}^{+}$, we let an element $\alpha \in \mathbb{G}^{+}$ as follows:

(i) $\alpha^{\downarrow} := \{x' \in X_2 \mid \forall y \in \beta_{\uparrow}. x' \leq_2 \tau(y)\}$,

(ii) $\alpha_{\uparrow} := \{y' \in Y_2 \mid \forall x' \in \alpha^{\downarrow}. x' \leq_2 y'\}$.

Note that there is another natural way to introduce α as we saw in Remark 4.11. Now, we check that they coincide, namely $\alpha^{\downarrow} = \upsilon_2(\alpha_{\uparrow})$ and $\alpha_{\uparrow} = \lambda_2(\alpha^{\downarrow})$. But, the latter is trivial by definition.

($\alpha^{\downarrow} = \upsilon_2(\alpha_{\uparrow})$). ($\subseteq$). For arbitrary $x' \in \alpha^{\downarrow}$ and $y' \in \alpha_{\uparrow}$, by the definition of $\alpha_{\uparrow}$, we trivially have $x' \leq_2 y'$. ($\supseteq$). Contraposition. Suppose $x' \notin \alpha^{\downarrow}$. There exists $y \in \beta_{\uparrow}$ such that $x' \not\leq_2 \tau(y)$. By definition, we have $\tau[\beta_{\uparrow}] \subseteq \alpha_{\uparrow}$, so $\tau(y) \in \alpha_{\uparrow}$. But, $x' \not\leq_2 \tau(y)$, which concludes $x' \notin \upsilon_2(\alpha_{\uparrow})$.

Next we show $\tau^{-}(\alpha_{\uparrow}) = \beta^{\downarrow}$ and $\sigma^{+}(\alpha^{\downarrow}) = \beta_{\uparrow}$.

($\tau^{-}(\alpha_{\uparrow}) = \beta^{\downarrow}$). By Proposition 5.8, $\tau^{-}(\alpha_{\uparrow}) = \sigma^{-1}[\alpha^{\downarrow}]$. ($\subseteq$). For arbitrary

$x \in \sigma^{-1}[\alpha^{\downarrow}]$ and $y \in \beta_{\uparrow}$, we have $\sigma(x) \in \alpha^{\downarrow}$. By the definition of $\alpha^{\downarrow}$, we obtain $\sigma(x) \leq_2 \tau(y)$. By (i) of Definition 4.1, $x \leq_1 y$, hence $x \in \beta^{\downarrow}$. ($\supseteq$). For arbitrary $x \in \beta^{\downarrow}$ and $y \in \beta_{\uparrow}$, we have $x \leq_1 y$. As $\langle \sigma | \tau \rangle$ is B-embedding, we obtain $\sigma(x) \leq_2 \tau(y)$, hence $\sigma(x) \in \alpha^{\downarrow}$, which means $x \in \sigma^{-1}[\alpha^{\downarrow}]$.

($\sigma^{+}(\alpha^{\downarrow}) = \beta_{\uparrow}$). By Proposition 5.8, $\sigma^{+}(\alpha^{\downarrow}) = \tau^{-1}[\alpha_{\uparrow}]$. ($\subseteq$). For arbitrary $y \in \tau^{-1}[\alpha_{\uparrow}]$ and $x \in \beta^{\downarrow}$, we have $\tau(y) \in \alpha_{\uparrow}$. Further, as we saw above, $\sigma^{-1}[\alpha^{\downarrow}] = \beta^{\downarrow}$, hence $\sigma(x) \in \alpha^{\downarrow}$. By the definition of $\alpha_{\uparrow}$, we obtain $\sigma(x) \leq_2 \tau(y)$. By (i) of Definition 4.1, $x \leq_1 y$. Therefore, $y \in \beta_{\uparrow}$. ($\supseteq$). For arbitrary $y \in \beta_{\uparrow}$ and $x' \in \alpha^{\downarrow}$, by the definition of $\alpha^{\downarrow}$, we have $x' \leq_2 \tau(y)$. Therefore, $\tau(y) \in \alpha_{\uparrow}$, i.e. $y \in \tau^{-1}[\alpha_{\uparrow}]$.

As a conclusion, for each $\beta \in \mathbb{F}^{+}$, there exists $\alpha \in \mathbb{G}^{+}$ such that $\langle \sigma^{+} | \tau^{-} \rangle(\alpha) = \beta$, hence $\langle \sigma^{+} | \tau^{-} \rangle$ is surjective. □

References

[1] Birkhoff, G., "Lattice Theory," American Mathematical Society Colloquium Publications **XXV**, American Mathematical Society, Providence, 1973, third edition.

[2] Blackburn, P., M. de Rijke and Y. Venema, "Modal logic," Cambridge Tracts in Theoretical Computer Science **53**, Cambridge University Press, Cambridge, 2002.

[3] Chagrov, A. and M. Zakharyaschev, "Modal logic," Oxford Logic Guides **35**, Oxford Science Publications, New York, 1997.

[4] Davey, B. and H. Priestley, "Introduction to Lattices and Order," Cambridge University Press, Cambridge, 2002, 2nd edition.

[5] Dunn, M., M. Gehrke and A. Palmigiano, *Canonical extensions and relational completeness of some substructural logics*, The Journal of Symbolic Logic **70** (2005), pp. 713–740.

[6] Galatos, N. and P. Jipsen, *Residuated frames with applications to decidability* Accepted in the Transactions of the AMS.

[7] Galatos, N., P. Jipsen, T. Kowalski and H. Ono, "Residuated lattices: an algebraic glimpse at substructural logics," Studies in Logics and the Foundation of Mathematics **151**, Elsevier, Amsterdam, 2007.

[8] Gehrke, M., *Generalized Kripke frames*, Studia Logica **84** (2006), pp. 241–275.

[9] Ghilardi, S. and G. Meloni, *Constructive canonicity in non-classical logics*, Annals of Pure and Applied Logic **86** (1997), pp. 1–32.

[10] Goldblatt, R., *Semantic analysis of orthologic*, Journal of Philosophical Logic **3** (1974), pp. 19–35.

[11] Goldblatt, R., "Logics of time and computation," CSLI Lecture Notes **7**, CSLI Publications, Stanford, 1992, revised and expanded edition.

[12] Hartonas, C., *Duality for lattice-ordered algebras and normal algebraizable logics*, Studia Logica **58** (1997), pp. 403–450.

[13] Hartonas, C. and J. M. Dunn, *Stone duality for lattices*, Algebra Universalis **37** (1997), pp. 391–401.

[14] Ono, H., *Substructural logics and residuated lattices - an introduction*, in: V. F. Hendricks and J. Malinowski, editors, *50 Years of Studia Logica: Trends in Logic*, Kluwer Academic Publishers, Dordrecht, 2003 pp. 193–228.

[15] Restall, G., "An Introduction to Substructural Logics," Routledge, London, 2000.

[16] Suzuki, T., *A Sahlqvist theorem for substructural logic*, The Review of Symbolic Logic Forthcoming.

[17] Suzuki, T., *Bi-approximation semantics for substructural logic at work*, Advances in Modal Logic **8** (2010), pp. 411–433.

[18] Suzuki, T., *Canonicity results of substructural and lattice-based logics*, The Review of Symbolic Logic **4** (2011), pp. 1–42.

Grammar Logics in Nested Sequent Calculus: Proof Theory and Decision Procedures

Alwen Tiu

Research School of Computer Science, The Australian National University

Egor Ianovski

Department of Computer Science, University of Auckland

Rajeev Goré

Research School of Computer Science, The Australian National University

Abstract

A grammar logic refers to an extension of the multi-modal logic K in which the modal axioms are generated from a formal grammar. We consider a proof theory, in nested sequent calculus, of grammar logics with converse, i.e., every modal operator $[a]$ comes with a converse $[\bar{a}]$. Extending previous works on nested sequent systems for tense logics, we show all grammar logics (with or without converse) can be formalised in nested sequent calculi, where the axioms are internalised in the calculi as structural rules. Syntactic cut-elimination for these calculi is proved using a procedure similar to that for display logics. If the grammar is context-free, then one can get rid of all structural rules, in favor of deep inference and additional propagation rules. We give a novel semi-decision procedure for context-free grammar logics, using nested sequent calculus with deep inference, and show that, in the case where the given context-free grammar is regular, this procedure terminates. Unlike all other existing decision procedures for regular grammar logics in the literature, our procedure does not assume that a finite state automaton encoding the axioms is given.

Keywords: Nested sequent calculus, display calculus, modal logics, deep inference.

1 Introduction

A grammar logic refers to an extension of the multi-modal logic K in which the modal axioms are generated from a formal grammar. Thus given a set Σ of indices, and a grammar production rule as shown below left, where each a_i and b_j are in Σ, we extend K with the multi-modal axiom shown below right:

$$a_1 a_2 \cdots a_l \to b_1 b_2 \cdots b_r \qquad\qquad [a_1][a_2]\cdots[a_l]A \supset [b_1][b_2]\cdots[b_r]A$$

The logic is a context-free grammar logic if $l = 1$ and furthermore, is a right linear grammar logic if the production rules also define a right linear grammar.

The logic is a regular grammar logic if the set of words generated from each $a \in \Sigma$ using the grammar production rules is a regular language. A right linear grammar logic is also a regular grammar logic since a right linear grammar can be converted to a finite automaton in polynomial time. Adding "converse" gives us alphabet symbols like $\bar{a}$ which correspond to the converse modality $[\bar{a}]$ and lead to multi-modal extensions of tense logic Kt where each modality $[a]$ and its converse $[\bar{a}]$ obey the interaction axioms $A \supset [a]\langle\bar{a}\rangle A$ and $A \supset [\bar{a}]\langle a\rangle A$.

Display calculi [2] can handle grammar logics with converse since they all fall into the primitive fragment identified by Kracht [21]. Display calculi all enjoy Belnap's general cut-elimination theorem, but it is well-known that they are not suitable for proof-search. Our work is motivated by the problem of automating proof search for display calculus. As in our previous work [11,12,13], we have chosen to work not directly in display calculus, but in a slightly different calculus based on nested sequents [19,4], which we call shallow nested sequent calculi. The syntactic constructs of nested sequents are closer to traditional sequent calculus, so as to allow us to use familiar notions in sequent calculus proof search procedures, such as the notions of saturation and loop checking, to automate proof search. A common feature of shallow nested sequent calculus and display calculus is the use of display postulates and other complex structural rules. These structural rules are the main obstacle to effective proof search, and our (proof theoretic) methodology for designing proof search calculi is guided by the problem of eliminating these structural rules entirely. We show here how our methodology can be used to derive proof search calculi for context-free grammar logics.

The general satisfiability problem for a grammar logic is to decide the satisfiability of a given formula when given a set of production rules for the underlying grammar. Nguyen and Szałas [23] give an excellent summary of what is known about this problem, as outlined next. Grammar logics were introduced by Fariñas del Cerro and Penttonen [8]. Baldoni et al [1] used prefixed tableaux to show that this problem is decidable for right linear logics but is undecidable for context free grammar logics. Demri [6] used an embedding into propositional dynamic logic with converse to prove this problem is EXPTIME-complete for right linear logics. Demri and de Nivelle [7] gave an embedding of the satisfiability problem for regular grammar logics into the two-variable guarded fragment of first-order logic and showed that satisfiability of regular grammar logics with converse is also EXPTIME-complete. Seen as description logics with inverse roles and complex role inclusions, decision procedures for regular grammar logics have also been studied extensively by Horrocks, et. al., see, e.g., [18,17,20]. Goré and Nguyen [10] gave an EXPTIME tableau decision procedure for the satisfiability of regular grammar logics using formulae labelled with automata states. Finally, Nguyen and Szałas [22,23] gave an extension of this method to handle converse by using the cut rule. In an unpublished manuscript, Nguyen has shown how to use the techniques of Goré and Widmann [15] to avoid the use of the cut rule. But as far as we know, there is no comprehensive sequent-style proof theory for grammar logics with converse which enjoys a syntactic cut-elimination theorem and which is amenable to proof-search.

We consider a proof theory, in nested sequent calculus, of grammar logics with converse, i.e., every modal operator $[a]$ comes with a converse $[\bar{a}]$. Extending previous works on nested sequent systems for (bi-)modal logics [11,13], we show, in Section 3, that all grammar logics (with or without converse) can be formalised in (shallow) nested sequent calculi, where the axioms are internalised in the calculi as structural rules. Syntactic cut-elimination for these calculi is proved using a procedure similar to that for display logics. We then show, in Section 4, that if the grammar is context-free, then one can get rid of all structural rules, in favor of *deep inference* [16,4] and propagation rules.

We then recast the problem of deciding grammar logics for the specific cases where the grammars are regular, using nested sequent calculus with deep inference. We first give, in Section 6.1, a decision procedure in the case where the regular grammar is given in the form of a FSA. This procedure is similar to existing tableaux-based decision procedures [17,22,23], where the states and transitions of the FSA are incorporated into proof rules for propagation of diamond-formulae. This procedure serves as a stepping stone to defining the more general decision procedure which does not depend on an explicit representation of axioms as a FSA in Section 6.2. The procedure in Section 6.2 is actually a semi-decision procedure that works on any finite set of context-free grammar axioms. However, we show that, in the case where the given grammar is regular, this procedure terminates. The procedure avoids the requirement to provide a FSA for the given axioms. This is significantly different from existing decision procedures for regular grammar logics [7,10,23,22], where it is assumed that a FSA encoding the axioms of the logics is given.

In this work, we follow Demri and de Nivelle's presentation of grammar axioms as a semi-Thue system [7]. The problem of deciding whether a context-free semi-Thue system is regular or not appears to be still open; see [20] for a discussion on this matter. Termination of our generic procedure for regular grammar logics of course does not imply solvability of this problem as it is dependent on the assumption that the given grammar is regular (see Theorem 6.11).

There are several proof systems to be presented next; we briefly describe the naming scheme used to differentiate these various sytems. We use Km to denote the Hilbert system for the basic multi-modal logic. Given a semi-Thue system S, we denote with Km(S) the extension of Km with axioms generated from S (see Section 2). There are two style of nested sequent calculus we consider here: the shallow-inference calculi (Section 3) and the deep-inference calculi (Section 4). Names of proof systems for the former are prefixed with an 'S', while names of proof systems for the latter are prefixed with a 'D'. For example, the corresponding shallow nested sequent calculus for Km (resp., Km(S)) is called SKm (resp., SKm(S)). In Section 5, we also consider shallow and deep nested sequent calculi parameterised by an automaton $\mathcal{A}$; these will be denoted by SKm($\mathcal{A}$) and DKm($\mathcal{A}$), respectively.

2 Grammar logics

The language of a multi-modal logic is defined w.r.t. to an alphabet Σ, used to index the modal operators. We use a, b and c, possibly with subscripts, for elements of Σ and use u and v, for elements of Σ^*, the set of finite strings over Σ. We use ϵ for the empty string. We define an operation $\bar{\cdot}$ (converse) on alphabets to capture converse modalities following Demri [7]. The converse operation satisfies $\bar{\bar{a}} = a$. We assume that Σ can be partitioned into two distinct sets Σ^+ and Σ^- such that $a \in \Sigma^+$ iff $\bar{a} \in \Sigma^-$. The converse operation is extended to strings in Σ^* as follows: if $u = a_1 a_2 \dots a_n$, then $\bar{u} = \bar{a}_n \bar{a}_{n-1} \dots \bar{a}_2 \bar{a}_1$, where $n \geq 0$. Note that if $u = \epsilon$ then $\bar{u} = \epsilon$.

We assume a given denumerable set of atomic formulae, ranged over by p, q, and r. The language of formulae is given by the following, where $a \in \Sigma$:

$$A ::= p \mid \neg A \mid A \vee A \mid A \wedge A \mid [a]A \mid \langle a \rangle A$$

Given a formula A, we write $A^\perp$ for the negation normal form (nnf) of $\neg A$. Implication $A \supset B$ is defined as $\neg A \vee B$.

Definition 2.1 *A* Σ-frame *is a pair* $\langle W, R \rangle$ *of a non-empty set* W *of worlds and a set* R *of binary relations* $\{R_a\}_{a \in \Sigma}$ *over* W *satisfying, for every* $a \in \Sigma$, $R_a = \{(x, y) \mid R_{\bar{a}}(y, x)\}$. *A* valuation V *is a mapping from propositional variables to sets of worlds. A* model $\mathfrak{M}$ *is a triple* $\langle W, R, V \rangle$ *where* $\langle W, R \rangle$ *is a frame and* V *is a valuation. The relation* $\models$ *is defined inductively as follows:*

- $\mathfrak{M}, x \models p$ *iff* $x \in V(p)$.
- $\mathfrak{M}, x \models \neg A$ *iff* $\mathfrak{M}, x \not\models A$.
- $\mathfrak{M}, x \models A \wedge B$ *iff* $\mathfrak{M}, x \models A$ *and* $\mathfrak{M}, x \models B$.
- $\mathfrak{M}, x \models A \vee B$ *iff* $\mathfrak{M}, x \models A$ *or* $\mathfrak{M}, x \models B$.
- *For every* $a \in \Sigma$, $\mathfrak{M}, x \models [a]A$ *iff for every* y *such that* $R_a(x, y)$, $\mathfrak{M}, y \models A$.
- *For every* $a \in \Sigma$, $\mathfrak{M}, x \models \langle a \rangle A$ *iff there exists* y *such that* $R_a(x, y)$, $\mathfrak{M}, y \models A$.

A formula A *is* satisfiable *iff there exists a* Σ*-model* $\mathfrak{M} = \langle W, R, V \rangle$ *and a world* $x \in W$ *such that* $\mathfrak{M}, x \models A$.

We now define a class of multi-modal logics, given Σ, that is induced by *production rules* for strings from Σ^*. We follow the framework in [7], using semi-Thue systems to define the logics. A production rule is a binary relation over strings in Σ^*, interpreted as a rewrite rule on strings. We use the notation $u \to v$ to denote a production rule which rewrites u to v. A *semi-Thue* system is a set S of production rules. It is *closed* if $u \to v \in S$ implies $\bar{u} \to \bar{v} \in S$.

Given a Σ-frame $\langle W, R \rangle$, we define another family of accessibility relations indexed by Σ^* as follows: $R_\epsilon = \{(x, x) \mid x \in W\}$ and for every $u \in \Sigma^*$ and for every $a \in \Sigma$, $R_{ua} = \{(x, y) \mid (x, z) \in R_u, (z, y) \in R_a, \text{ for some } z \in W\}$.

Definition 2.2 *Let* $u \to v$ *be a production rule and let* $\mathcal{F} = \langle W, R \rangle$ *be a* Σ*-frame.* $\mathcal{F}$ *is said to satisfy* $u \to v$ *if* $R_v \subseteq R_u$. $\mathcal{F}$ *satisfies a semi-Thue system* S *if it satisfies every production rule in* S.

Definition 2.3 *Let S be a semi-Thue system. A formula A is said to be S-*satisfiable *iff there is a model $\mathfrak{M} = \langle W, R, V\rangle$ such that $\langle W, R\rangle$ satisfies S and $\mathfrak{M}, x \models A$ for some $x \in W$. A is said to be S-*valid *if for every Σ-model $\mathfrak{M} = \langle W, R, V\rangle$ that satisfies S, we have $\mathfrak{M}, x \models A$ for every $x \in W$.*

Given a string $u = a_1 a_2 \ldots a_n$ and a formula A, we write $\langle u\rangle A$ for the formula $\langle a_1\rangle\langle a_2\rangle \cdots \langle a_n\rangle A$. The notation $[u]A$ is defined analogously. If $u = \epsilon$ then $\langle u\rangle A = [u]A = A$.

Definition 2.4 *Let S be a closed semi-Thue system over an alphabet Σ. The system* Km(S) *is an extension of the standard Hilbert system for multi-modal* Km *(see, e.g., [3]) with the following axioms:*

- *for each $a \in \Sigma$, a* residuation axiom: $A \supset [a]\langle \bar{a}\rangle A$
- *and for each $u \to v \in S$, an axiom $[u]A \supset [v]A$.*

Because S is closed, each axiom $[u]A \supset [v]A$ has an *inverted version* $[\bar{u}]A \supset [\bar{v}]A$. The following theorem can be proved following a similar soundness and completeness proof for Hilbert systems for modal logics (see, e.g., [3]).

Theorem 2.5 *A formula F is S-valid iff F is provable in* Km(S).

3 Nested sequent calculi with shallow inference

We now give a sequent calculus SKm for Km(S), by using the framework of nested sequent calculus [19,4,11,13]. We follow the notation used in [19,13], extended to the multi-modal case. From this section onwards, we shall be concerned only with formulae in nnf, so we can restrict to one-sided sequents.

A nested sequent is a multiset of the form shown below at left where each A_i is a formula and each Δ_i is a nested sequent:

$$A_1, \ldots, A_m, (a_1)\{\Delta_1\}, \ldots, (a_n)\{\Delta_n\} \qquad A_1 \vee \cdots \vee A_m \vee [a_1]B_1 \vee \cdots \vee [a_n]B_n$$

We shall often drop the outermost braces when writing nested sequents, e.g., writing Γ, Δ instead of $\{\Gamma, \Delta\}$. The structural connective $(a)\{.\}$ is a proxy for the modality $[a]$, so this nested sequent can be interpreted as the formula shown above right (modulo associativity and commutativity of $\vee$), where each B_i is the interpretation of Δ_i. We shall write $(u)\{\Delta\}$, where $u = a_1 \cdots a_n \in \Sigma^*$, to denote the structure:

$$(a_1)\{(a_2)\{\cdots (a_n)\{\Delta\}\} \cdots\}.$$

A *context* is a nested sequent with a 'hole' [ ] in place of a formula: this notation should not be confused with the modality $[a]$. We use $\Gamma[\;]$, $\Delta[\;]$, etc. for contexts. Given a context $\Gamma[\;]$ and a nested sequent Δ, we write $\Gamma[\Delta]$ to denote the nested sequent obtained by replacing the hole in $\Gamma[\;]$ with Δ.

The core inference rules for multi-modal SKm (without axioms) are given in Figure 1. The rule r is called a *residuation rule* (or display postulate) and corresponds to the residuation axioms.

$$\frac{}{\Gamma, p, \neg p}\ id \qquad \frac{\Gamma, A \quad \Delta, A^{\perp}}{\Gamma, \Delta}\ cut \qquad \frac{\Gamma, \Delta, \Delta}{\Gamma, \Delta}\ ctr \qquad \frac{\Gamma}{\Gamma, \Delta}\ wk \qquad \frac{\Gamma, (a)\{\Delta\}}{(\bar{a})\{\Gamma\}, \Delta}\ r$$

$$\frac{\Gamma, A \quad \Gamma, B}{\Gamma, A \wedge B}\ \wedge \qquad \frac{\Gamma, A, B}{\Gamma, A \vee B}\ \vee \qquad \frac{\Gamma, (a)\{A\}}{\Gamma, [a]A}\ [a] \qquad \frac{\Gamma, (a)\{\Delta, A\}}{\Gamma, (a)\{\Delta\}, \langle a \rangle A}\ \langle a \rangle$$

Fig. 1. The inference rules of the shallow nested sequent calculus SKm

To capture Km(S), we need to convert each axiom generated from S to an inference rule. Each production rule $u \to v$ gives rise to the axiom $[u]A \supset [v]A$, or equivalently, $\langle \bar{v} \rangle A \supset \langle \bar{u} \rangle A$. The latter is an instance of Kracht's *primitive axioms* [21] (generalised to the multimodal case). Thus, we can convert the axiom into a structural rule following Kracht's rule scheme for primitive axioms:

$$\frac{(u)\{\Delta\}, \Gamma}{(v)\{\Delta\}, \Gamma}$$

Let $\rho(S)$ be the set of structural rules induced by the semi-Thue system S.

Definition 3.1 *Let S be a closed semi-Thue system S over an alphabet Σ.* SKm(S) *is the proof system obtained by extending* SKm *with* $\rho(S)$.

We say that two proof systems are equivalent if and only if they prove the same set of formulae.

Theorem 3.2 *The system* SKm(S) *and* Km(S) *are equivalent.*

Proof. *(Outline).* In one direction, from SKm(S) to Km(S), we show that, for each inference rule of SKm(S), if the formula interpretation of the premise(s) is valid then the formula interpretation of the conclusion is also valid. For the converse, it is enough to show that all axioms of Km(S) are derivable in SKm(S). It can be shown that both the residuation axioms and the axioms generated from S can be derived using the structural rules r and $\rho(S)$. □

The cut-elimination proof for SKm(S) follows a similar generic procedure for display calculi [2,21], which has been adapted to nested sequent in [13]. The key to cut-elimination is to show that SKm(S) has the *display property.*

Lemma 3.3 *Let $\Gamma[\Delta]$ be a nested sequent. Then there exists a nested sequent Γ' such that $\Gamma[\Delta]$ is derivable from the nested sequent Γ', Δ, and vice versa, using only the residuation rule r. Intuitively, the nested Δ in $\Gamma[\Delta]$ is displayed at the top level in Γ', Δ.*

Theorem 3.4 *Cut elimination holds for* SKm(S).

Proof. This is a straightforward adaptation of the cut-elimination proof in [13] for tense logic. □

4 Deep inference calculi

Although the shallow system SKm(S) enjoys cut-elimination, proof search in its cut-free fragment is difficult to automate, due to the presence of structural

$$\frac{}{\Gamma[p, \neg p]}\ id_d \qquad \frac{\Gamma[A \wedge B, A] \quad \Gamma[A \wedge B, B]}{\Gamma[A \wedge B]}\ \wedge_d \qquad \frac{\Gamma[A \vee B, A, B]}{\Gamma[A \vee B]}\ \vee_d$$

$$\frac{\Gamma[[a]A, (a)\{A\}]}{\Gamma[[a]A]}\ [a]_d \qquad \frac{\Gamma[(a)\{\Delta, A\}, \langle a \rangle A]}{\Gamma[(a)\{\Delta\}, \langle a \rangle A]}\ \langle a\!\uparrow\rangle \qquad \frac{\Gamma[(a)\{\Delta, \langle \bar{a} \rangle A\}, A]}{\Gamma[(a)\{\Delta, \langle \bar{a} \rangle A\}]}\ \langle a\!\downarrow\rangle$$

Fig. 2. The inference rules of DKm

rules. To reduce the non-determinism caused by structural rules, we consider next a proof system in which all structural rules (including those induced by grammar axioms) can be absorbed into logical rules. As the display property in Lemma 3.3 suggests, the residuation rule allows one to essentially apply an inference rule to a particular subsequent nested inside a nested sequent, by displaying that subsequent at the top level and undisplaying it back to its original position in the nested sequent. It is therefore quite intuitive that one way to get rid of the residuation rule is to allow *deep inference* rules, that apply deeply within any arbitrary context in a nested sequent.

The deep inference system DKm, which corresponds to SKm, is given in Figure 2. As can be readily seen, the residuation rule is absent and contraction and weakening are absorbed into logical rules.

To fully absorb the residuation rule and other structural rules induced by semi-Thue systems, we need to introduce additional introduction rules for the diamond operator $\langle a \rangle$, which we call *propagation rules.* We shall show next how these propagation rules are generated from a semi-Thue system.

Let S be a closed semi-Thue system over alphabet Σ. We write $u \Rightarrow_S v$ to mean that the string v can be reached from u by applying the production rules (as rewrite rules) in S successively to u. Define $L_a(S) = \{u \mid a \Rightarrow_S u\}$. Then $L_a(S)$ defines a language generated from S with the start symbol a.

A nested sequent can be seen as a tree whose nodes are multisets of formulae, and whose edges are labeled with elements of Σ. We assume that each node in a nested sequent can be identified uniquely, i.e., we can consider each node as labeled with a unique position identifier. An internal node of a nested sequent is a node which is not a leaf node. We write $\Gamma[\ ]_i$ to denote a context in which the hole is located in the node at position i in the tree representing $\Gamma[\ ]$. This generalises to multi-contexts, so $\Gamma[\ ]_i[\ ]_j$ denotes a two-hole context, one hole located at i and the other at j (they can be the same location). From now on, we shall often identify a nested sequent with its tree representation, so when we speak of a node in Γ, we mean a node in the tree of Γ. If i and j are nodes in Γ, we write $i \succ^a j$ when j is a child node of i and the edge from i to j is labeled with a. If i is a node in the tree of Γ, we write $\Gamma|i$ to denote the multiset of formula occuring in the node i. Let Δ and Γ be nested sequents. Suppose i is a node in Γ. Then we write $\Gamma(i \ll \Delta)$ for the nested sequent obtained from Γ by adding Δ to node i in Γ. Note that for such an addition to preserve the uniqueness of the position identifiers of the resulting tree, we need to rename the identifiers in Δ to avoid clashes. We shall assume implicitly that such a renaming is carried out when we perform this addition.

Definition 4.1 (Propagation automaton.) *A propagation automaton is a finite state automaton $\mathcal{P} = (\Sigma, Q, I, F, \delta)$ where Q is a finite set of states, $I = \{s\}$ is a singleton set of an initial state and $F = \{t\}$ is a singleton set of a final state with $s, t \in Q$, and for every $i, j \in Q$, if $i \xrightarrow{a} j \in \delta$ then $j \xrightarrow{\bar{a}} i \in \delta$.*

In other words, a propagation automaton is just a finite state automaton (FSA) where each transition has a dual transition.

Definition 4.2 *Let $\mathcal{A} = (\Sigma, Q, I, F, \delta)$ be a FSA. Let $\boldsymbol{i} = i_1, \dots, i_n$ and $\boldsymbol{j} = j_1, \dots, j_n$ be two sequences of states in Q. Let $[i_1 := j_1, \dots, i_n := j_n]$ (we shall abbreviate this as $[\boldsymbol{i} := \boldsymbol{j}]$) be a (postfix) mapping from Q to Q that maps i_m to j_m, where $1 \leq m \leq n$, and is the identity map otherwise. This mapping is extended to a (postfix) mapping between sets of states as follows: given $Q' \subseteq Q$, $Q'[\boldsymbol{i} := \boldsymbol{j}] = \{k[\boldsymbol{i} := \boldsymbol{j}] \mid k \in Q'\}$. The automaton $\mathcal{A}[\boldsymbol{i} := \boldsymbol{j}]$ is the tuple $(\Sigma, Q[\boldsymbol{i} := \boldsymbol{j}], I[\boldsymbol{i} := \boldsymbol{j}], F[\boldsymbol{i} := \boldsymbol{j}], \delta')$ where*

$$\delta' = \{k[\boldsymbol{i} := \boldsymbol{j}] \xrightarrow{a} l[\boldsymbol{i} := \boldsymbol{j}] \mid k \xrightarrow{a} l \in \delta\}.$$

To each nested sequent Γ, and nodes i and j in Γ, we associate a propagation automaton $\mathcal{R}(\Gamma, i, j)$ as follows:

(i) the states of $\mathcal{R}(\Gamma, i, j)$ are the nodes of (the tree of) Γ;

(ii) i is the initial state of $\mathcal{R}(\Gamma, i, j)$ and j is its final state;

(iii) each edge $x \succ^a y$ in Γ corresponds to two transitions in $\mathcal{R}(\Gamma, i, j)$: namely, $x \xrightarrow{a} y$ and $y \xrightarrow{\bar{a}} x$.

Note that although propagation automata are defined for nested sequents, they can be similarly defined for (multi-)contexts as well, as contexts are just sequents containing a special symbol [ ] denoting a hole. So in the following, we shall often treat a context as though it is a nested sequent.

A semi-Thue system S over alphabet Σ is *context-free* if its production rules are all of the form $a \to u$ for some $a \in \Sigma$.

In the following, to simplify presentation, we shall use the same notation to refer to an automaton $\mathcal{A}$ and the regular language it accepts. Given a context-free closed semi-Thue system S, the *propagation rules for S* are all the rules of the following form where i and j are two (not necessarily distinct) nodes of Γ:

$$\frac{\Gamma[\langle a\rangle A]_i[A]_j}{\Gamma[\langle a\rangle A]_i[\emptyset]_j}\ p_S, \text{ provided } \mathcal{R}(\Gamma[\ ]_i[\ ]_j, i, j) \cap L_a(S) \neq \emptyset.$$

Note that the intersection of a regular language and a context-free language is a context-free language (see, e.g., Chapter 3 in [9] for a construction of the intersection), and since the problem of checking emptiness for context-free languages is decidable [9], the rule p_S can be effectively mechanised.

Definition 4.3 *Given a context-free closed semi-Thue system S over an alphabet Σ, the proof system* DKm(S) *is obtained by extending* DKm *with* p_S.

We now show that DKm(S) is equivalent to SKm(S). The proof relies on a series of lemmas showing admissibility of all structural rules of SKm(S) in DKm(S). The proof follows the same outline as in the case for tense logic [13]. The adaptation of the proof in [13] is quite straightforward, so we shall not go into detailed proofs but instead just outline the required lemmas. Some of their proofs are outlined in the appendix. In the following lemmas, we shall assume that S is a closed context-free semi-Thue system over some Σ.

Given a derivation Π, we denote with $|\Pi|$ the *height* of Π, which is simply the length (i.e., the number of edges) of the longest branch in Π. A single premise rule ρ is said to be *admissible* in DKm(S) if provability of its premise in DKm(S) implies provability of its conclusion in DKm(S). It is *height-preserving admissible* if whenever the premise has a derivation then the conclusion has a derivation of the same height, in DKm(S).

Admissibility of the weakening rule is a consequence of the following lemma.

Lemma 4.4 *Let Π be a derivation of $\Gamma[\emptyset]$ in* DKm(S). *Then, for all nested sequents Δ, there exists a derivation Π' of $\Gamma[\Delta]$ in* DKm(S) *such that $|\Pi| = |\Pi'|$.*

The admissibility proofs of the remaining structural rules all follow the same pattern: the most important property to prove is that, if a propagation path for a diamond formula exists between two nodes in the premise, then there exists a propagation path for the same formula, between the same nodes, in the conclusion of the rule.

Lemma 4.5 *The rule r is height-preserving admissible in* DKm(S).

Admissibility of contraction is proved indirectly by showing that it can be replaced by a formula contraction rule and a distributivity rule:

$$\frac{\Gamma[A, A]}{\Gamma[A]}\ actr \qquad \frac{\Gamma[(a)\{\Delta_1\}, (a)\{\Delta_2\}]}{\Gamma[(a)\{\Delta_1, \Delta_2\}]}\ m$$

The rule m is also called a *medial* rule and is typically used to show admissibility of contraction in deep inference [5].

Lemma 4.6 *The rule ctr is admissible in* DKm(S) *plus actr and m.*

Lemma 4.7 *The rules actr and m are height-preserving admissible in* DKm(S).

Admissibility of contraction then follows immediately.

Lemma 4.8 *The contraction rule ctr is admissible in* DKm(S).

Lemma 4.9 *The structural rules $\rho(S)$ of* SKm(S) *are height-preserving admissible in* DKm(S).

Theorem 4.10 *For every context-free closed semi-Thue system S, the proof systems* SKm(S) *and* DKm(S) *are equivalent.*

5 Regular grammar logics

A context free semi-Thue system S over Σ is regular if for every $a \in \Sigma$, the language $L_a(S)$ is a regular language. In this section, we consider logics generated by regular closed semi-Thue systems. We assume in this case that the union of the regular languages $\{L_a(S) \mid a \in \Sigma\}$ is represented explicitly as an FSA $\mathcal{A}$ with no silent transitions. Thus $\mathcal{A} = (\Sigma, Q, I, F, \delta)$ where Q is a finite set of states, $I \subseteq Q$ is the set of initial states, $F \subseteq Q$ is the set of final states, and δ is the transition relation. Given $\mathcal{A}$ as above, we write $s \xrightarrow{a}_{\mathcal{A}} t$ to mean $s \xrightarrow{a} t \in \delta$. We assume that each $a \in \Sigma$ has a unique initial state $init_a \in I$.

We shall now define an alternative deep inference system given this explicit representation of the grammar axioms as an FSA. Following similar tableaux systems in the literature that utilise such an automaton representation [17,22,23], we use the states of the FSA to index formulae in a nested sequent to record stages of propagation. For this, we first introduce a form of labeled formula, written $s : A$, where $s \in Q$. The propagation rules corresponding to $\mathcal{A}$ are:

$$\frac{\Gamma[\langle a\rangle A, init_a : A]}{\Gamma[\langle a\rangle A]}\; i \qquad \frac{\Gamma[s : A, (a)\{s' : A, \Delta\}]}{\Gamma[s : A, (a)\{\Delta\}]}\; t\!\uparrow,\ \text{if } s \xrightarrow{a}_{\mathcal{A}} s'$$

$$\frac{\Gamma[s : A, A]}{\Gamma[s : A]}\; f,\ \text{if } s \in F \qquad \frac{\Gamma[(a)\{s : A, \Delta\}, s' : A]}{\Gamma[(a)\{s : A, \Delta\}]}\; t\!\downarrow,\ \text{if } s \xrightarrow{\bar{a}}_{\mathcal{A}} s'.$$

Definition 5.1 *Let S be a regular closed semi-Thue system over Σ and let $\mathcal{A}$ be an FSA representing the regular language generated by S and Σ.* DKm($\mathcal{A}$) *is the proof system* DKm *extended with the rules $\{i, f, t\!\downarrow, t\!\uparrow\}$ for $\mathcal{A}$.*

It is intuitively clear that DKm($\mathcal{A}$) and DKm(S) are equivalent, when $\mathcal{A}$ defines the same language as $L(S)$. Essentially, a propagation rule in DKm(S) can be simulated by DKm($\mathcal{A}$) using one or more propagations of labeled formulae. The other direction follows from the fact that when a diamond formula $\langle a\rangle A$ is propagated, via the use of labeled formulae, to a labeled formula $s : A$ where s is a final state, then there must be a chain of transitions between labeled formulae for A whose string forms an element of $\mathcal{A}$, hence also in $L_a(S)$. One can then propagate directly $\langle a\rangle A$ in DKm(S).

Theorem 5.2 *Let S be a regular closed semi-Thue system over Σ and let $\mathcal{A}$ be a FSA representing the regular language generated by S and Σ. Then* DKm(S) *and* DKm($\mathcal{A}$) *are equivalent.*

6 Decision procedures

We now show how the proof systems DKm($\mathcal{A}$) and DKm(S) can be turned into decision procedures for regular grammar logics. Our aim is to derive the decision procedure for DKm(S) directly without the need to convert S explicitly to an automaton; the decision procedure DKm($\mathcal{A}$) will serve as a stepping stone towards this aim. The decision procedure for DKm(S) is a departure from all existing decision procedures for regular grammar logics (with or without converse) [17,7,10,22,23] that assume that an FSA representing S is given.

$Prove_1(\mathcal{A}, \Gamma)$

(i) If $\Gamma = \Gamma'[p, \neg p]$, return $\top$.

(ii) If Γ is $\mathcal{A}$-stable, return $\bot$.

(iii) If Γ is not saturated:
  (a) If $A \vee B \in \Gamma|i$ but $A \notin \Gamma|i$ or $B \notin \Gamma|i$, then let $\Gamma' := \Gamma(i \ll \{A, B\})$ and return $Prove_1(\mathcal{A}, \Gamma')$.
  (b) Suppose $A_1 \wedge A_2 \in \Gamma|i$ but neither $A_1 \in \Gamma|i$ nor $A_2 \in \Gamma|i$. Let $\Gamma_1 = \Gamma(i \ll \{A_1\})$ and $\Gamma_2 = \Gamma(i \ll \{A_2\})$. Then return $\bot$ if $Prove_1(\mathcal{A}, \Gamma_j) = \bot$ for some $j \in \{1, 2\}$. Otherwise return $\top$.

(iv) If Γ is not $\mathcal{A}$-propagated: then there is a node i s.t. one of the following applies:
  (a) $\langle a \rangle A \in \Gamma|i$ but $init_a : A \notin \Gamma|i$. Then let $\Gamma' := \Gamma(i \ll \{init_a : A\})$.
  (b) $s : A \in \Gamma|i$ and $s \in F$, but $A \notin \Gamma|i$. Then let $\Gamma' := \Gamma(i \ll \{A\})$.
  (c) $s : A \in \Gamma|i$, there is j s.t. $i \succ^a j$ and $s \xrightarrow{a}_{\mathcal{A}} t$, but $t : A \notin \Gamma|j$. Then let $\Gamma' := \Gamma(j \ll \{t : A\})$.
  (d) $s : A \in \Gamma|i$, there is j s.t. $j \succ^a i$ and $s \xrightarrow{\bar{a}}_{\mathcal{A}} t$, but $t : A \notin \Gamma|j$. Then let $\Gamma' := \Gamma(j \ll \{t : A\})$.

  Return $Prove_1(\mathcal{A}, \Gamma')$.

(v) If there is an internal node i in Γ that is not realised: Then there is $[a]A \in \Gamma|i$ such that $A \notin \Gamma|j$ for every j s.t. $i \succ^a j$. Let $\Gamma' := \Gamma(i \ll (a)\{A\})$. Return $Prove_1(\mathcal{A}, \Gamma')$.

(vi) If there is a leaf node i that is not realised and is not a loop node: Then there is $[a]A \in \Gamma|i$. Let $\Gamma' := \Gamma(i \ll (a)\{A\})$. Return $Prove_1(\mathcal{A}, \Gamma')$.

Fig. 3. An automata-based prove procedure.

6.1 An automata-based procedure

The decision procedure for $\mathrm{DKm}(\mathcal{A})$ is basically just backward proof search, where one tries to saturate each sequent in the tree of sequents until either the id_d rule is applicable, or a certain stable state is reached. When the latter is reached, we show that a counter model to the original nested sequent can be constructed. Although we obtain this procedure via a different route, the end result is very similar to the tableaux-based decision procedure in [17]. In particular, our notion of a stable state (see Definition 6.3) used to block proof search is the same as the blocking condition in tableaux systems [17,7,10,23,22], which takes advantange of the labeling of formulae with the states of the automaton.

Recall the notation $\Gamma|i$ refers to the multiset of formulae at node i in Γ.

Definition 6.1 (Saturation and realisation) *A node i in Γ is* saturated *if the following hold:*

(i) *If $A \in \Gamma|i$ then $A^\perp \notin \Gamma|i$.*

(ii) *If $A \vee B \in \Gamma|i$ then $A \in \Gamma|i$ and $B \in \Gamma|i$.*

(iii) *If $A \wedge B \in \Gamma|i$ then $A \in \Gamma|i$ or $B \in \Gamma|i$.*

$\Gamma|i$ *is* realised *if* $[a]A \in \Gamma|i$ *implies that there exists* j *such that* $i \succ^a j$ *and* $A \in \Gamma|j$.

Definition 6.2 ($\mathcal{A}$-propagation) *Let* $\mathcal{A} = (\Sigma, Q, I, F, \delta)$. *A nested sequent* Γ *is said to be* $\mathcal{A}$*-propagated if for every node* i *in* Γ*, the following hold:*

(i) *If* $\langle a \rangle A \in \Gamma|i$ *then* $init_a : A \in \Gamma|i$ *for any* $a \in \Sigma$.

(ii) *If* $s : A \in \Gamma|i$ *and* $s \in F$*, then* $A \in \Gamma|i$.

(iii) *For all* j, a, s *and* t*, such that* $i \succ^a j$ *and* $s \xrightarrow{a}_{\mathcal{A}} t$*, if* $s : A \in \Gamma|i$ *then* $t : A \in \Gamma|j$.

(iv) *For all* j, a, s *and* t*, such that* $j \succ^a i$ *and* $s \xrightarrow{\bar{a}}_{\mathcal{A}} t$*, if* $s : A \in \Gamma|i$ *then* $t : A \in \Gamma|j$.

Definition 6.3 ($\mathcal{A}$-stability) *A nested sequent* Γ *is* $\mathcal{A}$-stable *if*

(i) *Every node is saturated.*

(ii) Γ *is* $\mathcal{A}$*-propagated.*

(iii) *Every internal node is realised.*

(iv) *For every leaf node* i*, one of the following holds:*
   (a) *There is an ancestor node* j *of* i *such that* $\Gamma|i = \Gamma|j$. *We call the node* i *a* loop node.
   (b) $\Gamma|i$ *is realised (i.e., it cannot have a member of the form* $[a]A$*).*

The prove procedure for DKm($\mathcal{A}$) is given in Figure 3. We show that the procedure is sound and complete with respect to DKm($\mathcal{A}$). The proofs of the following theorems can be found in the appendix.

Theorem 6.4 *If* $Prove_1(\mathcal{A}, \{F\})$ *returns* $\top$ *then* F *is provable in* DKm($\mathcal{A}$). *If* $Prove_1(\mathcal{A}, \{F\})$ *returns* $\bot$ *then* F *is not provable in* DKm($\mathcal{A}$).

Theorem 6.5 *For every formula* A, $Prove_1(\mathcal{A}, \{A\})$ *terminates.*

Corollary 6.6 *The proof system* DKm($\mathcal{A}$) *is decidable.*

6.2 A grammar-based procedure

The grammar-based procedure differs from the automaton-based procedure in the notion of propagation and that of a stable nested sequent.

In the following, given a function θ from labels to labels, and a list $\boldsymbol{i} = i_1, \ldots, i_n$ of labels, we write $\theta(\boldsymbol{i})$ to denote the list $\theta(i_1), \ldots, \theta(i_n)$. We write $[\boldsymbol{i} := \theta(\boldsymbol{i})]$ to mean the mapping $[i_1 := \theta(i_1), \ldots, i_n := \theta(i_n)]$.

In the following definitions, S is assumed to be a context-free semi-Thue system over some alphabet Σ.

Definition 6.7 (S-propagation) *Let* Γ *be a nested sequent. Let* $\mathcal{P} = (\Sigma, Q, \{i\}, \{j\}, \delta)$ *be a propagation automata, where* Q *is a subset of the nodes in* Γ. *We say that* Γ *is* $(S, \mathcal{P})$-propagated *if the following holds:* $\langle a \rangle A \in \Gamma|i$ *and* $\mathcal{P} \cap L_a(S) \neq \emptyset$ *imply* $A \in \Gamma|j$. Γ *is* S-propagated *if it is* $(S, \mathcal{R}(\Gamma, i, j))$*-propagated for every node* i *and* j *in* Γ.

Definition 6.8 (S-stability) *A nested sequent* Γ *is* S-stable *if*

$Prove_2(S, \Gamma, k)$

(i) If $\Gamma = \Gamma'[p, \neg p]$, return $\top$.

(ii) If Γ is S-stable, return $\bot$.

(iii) If Γ is not saturated:
- If $A \vee B \in \Gamma|i$ but $A \notin \Gamma|i$ or $B \notin \Gamma|i$, then let $\Gamma' := \Gamma(i \ll \{A, B\})$ and return $Prove_2(S, \Gamma', k)$.
- Suppose $A_1 \wedge A_2 \in \Gamma|i$ but neither $A_1 \in \Gamma|i$ nor $A_2 \in \Gamma|i$. Let $\Gamma_1 = \Gamma(i \ll \{A_1\})$ and $\Gamma_2 = \Gamma(i \ll \{A_2\})$. If $Prove_2(S, \Gamma_j, k) = \bot$, for some $j \in \{1, 2\}$, then return $\bot$. Otherwise, if $Prove_2(S, \Gamma_j, k) = \star$ for some $j \in \{1, 2\}$, then return $\star$. If $Prove_2(S, \Gamma_j, k) = \top$, for all $j \in \{1, 2\}$, then return $\top$.

(iv) If Γ is not S-propagated: then there must be nodes i and j such that $\langle a \rangle A \in \Gamma|i$ and $\mathcal{R}(\Gamma, i, j) \cap L_a(S) \neq \emptyset$, but $A \notin \Gamma|j$. Let $\Gamma' := \Gamma(j \ll \{A\})$. Return $Prove_2(S, \Gamma', k)$.

(v) If there is an internal node i in Γ that is not realised: Then there is $[a]A \in \Gamma|i$ such that $A \notin \Gamma|j$ for every j s.t. $i \succ^a j$. Let $\Gamma' := \Gamma(i \ll (a)\{A\})$. Return $Prove_2(S, \Gamma', k)$.

(vi) Non-deterministically choose a leaf node i that is not realised and is at height equal to or lower than k in Γ: Then there is $[a]A \in \Gamma|i$. Let $\Gamma' := \Gamma(i \ll (a)\{A\})$. Return $Prove_2(S, \Gamma', k)$.

(vii) Return $\star$.

$Prove(S, \Gamma)$

(i) $k := 0$.

(ii) If $Prove_2(S, \Gamma, k) = \top$ or $Prove_2(S, \Gamma, k) = \bot$, return $\top$ or $\bot$ respectively.

(iii) $k := k + 1$. Go to step (ii).

Fig. 4. A grammar-based prove procedure.

(i) *Every node is saturated.*

(ii) Γ *is* S*-propagated.*

(iii) *Every internal node is realised.*

(iv) *Let* $\boldsymbol{x} = x_1, \ldots, x_n$ *be the list of all unrealised leaf nodes. There is a function* λ *assigning each unrealised leaf node* x_m *to an ancestor* $\lambda(x_m)$ *of* x_m *such that* $\Gamma|x_m = \Gamma|\lambda(x_m)$ *and for every node* y *and* z, Γ *is* $(S, \mathcal{P})$*-propagated, where* $\mathcal{P} = \mathcal{R}(\Gamma, y, z)[\boldsymbol{x} := \lambda(\boldsymbol{x})]$.

Now we define a non-deterministic prove procedure $Prove_2(S, \Gamma, k)$ as in Figure 4, where k is an integer and S is a context-free closed semi-Thue system. Given a nested sequent Γ, and a node i in Γ, the *height* of i in Γ is the length of the branch from the root of Γ to node i. The procedure $Prove_2(S, \Gamma, k)$ tries to construct a derivation of Γ, but is limited to exploring only those nested sequents derived from Γ that has height at most k. The procedure *Prove* given below is essentially an iterative deepening procedure that calls $Prove_2$ repeatedly with increasing values of k. If an input sequent is not valid, the procedure will try to guess the smallest S-stable sequent that refutes the input sequent, i.e., it essentially tries to construct a finite countermodel. The procedure *Prove* gives a semi-decision procedure for context-free grammar logics. This uses the following lemma about S-stable sequents, which shows how to extract a countermodel from an S-stable sequent.

Lemma 6.9 *Let S be a context-free closed semi-Thue system. If Γ is an S-stable nested sequent, then there exists a model $\mathfrak{M}$ such that for every node x in Γ, there exists a world w in $\mathfrak{M}$ such that for every $A \in \Gamma|x$, we have $\mathfrak{M}, w \not\models A$.*

Theorem 6.10 *Let S be a context-free closed semi-Thue system. For every formula F, $Prove(S, \{F\})$ returns $\top$ if and only if F is provable in* $\mathrm{DKm}(S)$.

We next show that $Prove(S, \Gamma)$ terminates when S is regular. The key is to bound the size of S-stable sequents, hence the non-deterministic iterative deepening will eventually find an S-stable sequent, when Γ is not provable.

Theorem 6.11 *Let S be a regular closed semi-Thue system over an alphabet Σ. Then for every formula F, the procedure $Prove(S, \{F\})$ terminates.*

The proof relies on the fact that there exists a minimal FSA $\mathcal{A}$ encoding S, so one can simulate steps of $Prove_1(\mathcal{A}, \{F\})$ in $Prove(S, \{F\})$. It is not difficult to show that if a run of $Prove_1(\mathcal{A}, \{F\})$ reaches an $\mathcal{A}$-stable nested sequent Γ', then one can find a k such that a run of $Prove_2(S, \{F\}, k)$ reaches a saturated and S-propagated nested sequent Δ, such that Γ' and Δ are identical except for the labeled formulae in Γ'. The interesting part is in showing that Δ is S-stable. The details are in the appendix.

The following is then a corollary of Theorem 6.10 and Theorem 6.11.

Corollary 6.12 *Let S be a regular closed semi-Thue system over an alphabet Σ. Then the procedure Prove is a decision procedure for* $\mathrm{DKm}(S)$.

7 Conclusion and future work

Nested sequent calculus is closely related to display calculi, allowing us to benefit from well-studied proof theoretic techniques in display calculi, such as Belnap's generic cut-elimination procedure, to prove cut-elimination for $\mathrm{SKm}(S)$. At the more practical end, we have established via proof theoretic means that nested sequent calculi for regular grammar logics can be effectively mechanised. This work and our previous work [11,13] suggests that nested sequent calculus could potentially be a good intermediate framework to study both proof theory and decision procedures, at least for modal and substructural logics.

Nested sequent calculus can be seen as a special case of labelled sequent calculus, as the tree structure in a nested sequent can be encoded using labels and accessibility relations among these labels in labelled calculi. The relation between the two has recently been established in [24], where the author shows that, if one gets rid of the frame rules in labelled calculi and structural rules in nested sequent calculi, there is a direct mapping between derivations of formulae between the two frameworks. However, it seems that the key to this connection, i.e., admissibility of the frame rules, has already been established in Simpson's thesis [25], where he shows admissibility of a class of frame rules in favour of propagation rules obtained by applying a closure operation on these frame rules. The latter is similar to our notion of propagation rules. Thus it seems that structural rules in (shallow) nested sequent calculus play a role similar to the frame rules in labelled calculi. We plan to investigate this connection further, e.g., under what conditions are the structural rules admissible in deep inference calculi, and whether those conditions translate into any meaningful characterisations in terms of (first-order) properties of frames.

The two decision procedures for regular grammar logics we have presented are not optimal. As can be seen from the termination proofs, their complexity is at least EXPSPACE. We plan to refine the procedures further to achieve optimal EXPTIME complexity, e.g, by extending our deep nested sequent calculi with "global caching" techniques from tableaux systems [14].

Acknowledgment The authors would like to thank Paola Bruscoli and anonymous referrees of previous drafts of this paper for their detailed and useful comments. The first author is supported by the Australian Research Council Discovery Grant DP110103173.

Appendix

A Proofs

Lemma 4.5. The rule r is height-preserving admissible in DKm(S).

Proof. Suppose Π is a derivation of $\Gamma, (a)\{\Delta\}$. We show by induction on $|\Pi|$ that there exists a derivation Π' of $(\bar{a})\{\Gamma\}, \Delta$ such that $|\Pi| = |\Pi'|$. This is mostly straightforward, except for the case where Π ends with a propagation rule. In this case, it is enough to show that the propagation automata for $\Gamma, (a)\{\Delta\}$ is in fact exactly the same as the propagation automata of $(\bar{a})\{\Gamma\}, \Delta$. □

Lemma 4.7. The rules *actr* and m are height-preserving admissible.

Proof. Admissibility of *actr* is trivial. To show admissibility of m, the non-trivial case is when we need to permute m over p_S. Suppose Π is a derivation of $\Gamma[(a)\{\Delta_1\}, (a)\{\Delta_2\}]$ ending with a propagation rule. Suppose i is the node where Δ_1 is located and j is the node where Δ_2 is located. If $\mathcal{P}$ is a propagation automata between nodes k and l in $\Gamma[(a)\{\Delta_1\}, (a)\{\Delta_2\}]$, then $\mathcal{P}[j := i]$ is a propagation automata between nodes $k[j := i]$ and $l[j := i]$ in $\Gamma[(a)\{\Delta_1, \Delta_2\}]$. So all potential propagations of diamond formulae are preserved in the conclusion of m. So m can be permuted up over the propagation rule and by the induction hypothesis it can be eventually eliminated. □

Lemma 4.9. The structural rules $\rho(S)$ of SKm(S) are height-preserving admissible in DKm(S).

Proof. Suppose Π is a derivation of $\Gamma[(a)\{\Delta\}]$. We show that there is a derivation Π' of $\Gamma[(u)\{\Delta\}]$, where $u = a_1 \cdots a_n$ such that $a \to u \in S$. This is mostly straightforward except when Π ends with a propagation rule. Suppose the hole in $\Gamma[\ ]$ is located at node k and Δ is located at node l, with $k \succ^a l$. In this case we need to show that if a diamond formula $\langle b\rangle A$ can be propagated from a node i to node j in $\Gamma[(a)\{\Delta\}]$ then there is also a propagation path between i and j in $\Gamma[(u)\{\Delta\}]$ for the same formula. Suppose $\mathcal{P}_1$ is the propagation automata $\mathcal{R}(\Gamma[(a)\{\Delta\}], i, j)$. Then the propagation automata $\mathcal{P}_2 = \mathcal{R}(\Gamma[(u)\{\Delta\}], i, j)$ is obtained from $\mathcal{P}_1$ by adding $n-1$ new states $k_1, \ldots, k_{n-1}$ between k and l, and the following transitions: $k \xrightarrow{a_1} k_1$, $k_1 \xrightarrow{a_{m+1}} k_{m+1}$, for $2 \leq m < n$ and $k_{n-1} \xrightarrow{a_n} l$, and their dual transitions.

Suppose $i \xrightarrow{v} j$ is a propagation path in $\Gamma[(a)\{\Delta\}]$. If v does not go through the edge $k \succ^a l$ (in either direction, up or down) then the same path also exists in $\Gamma[(u)\{\Delta\}]$. If it does pass through $k \succ^a l$, then the path must contain one or more transitions of the form $k \xrightarrow{a} l$ or $l \xrightarrow{\bar{a}} k$. Then one can simulate the path $i \xrightarrow{v} j$ with a path $i \xrightarrow{v'} j$ in $\mathcal{P}_2$, where v' is obtained from v by replacing each $k \xrightarrow{a} l$ with $k \xrightarrow{u} l$ and each $l \xrightarrow{\bar{a}} k$ with $l \xrightarrow{\bar{u}} k$. It remains to show that $v' \in \mathcal{P}_2 \cap L_b(S)$. But this follows from the fact that $a \to u \in S$ and $\bar{a} \to \bar{u} \in S$ (because S is a closed), so $v \Rightarrow_S v' \in L_b(S)$. □

Theorem 4.10. For every context-free closed semi-Thue system S, the proof systems SKm(S) and DKm(S) are equivalent.

Proof. One direction, from SKm(S) to DKm(S) follows from the admissibility of structural rules of SKm(S) in DKm(S). To show the other direction, given a derivation Π in DKm(S), we show, by induction on the number of occurrences of p_S, with a subinduction on the height of Π, that Π can be transformed into a derivation in SKm(S). As rules other than p_S can be derived directly in SKm(S), the only interesting case to consider is when Π ends with p_S:

$$\frac{\Gamma[\langle a\rangle A]_i[A]_j}{\Gamma[\langle a\rangle A]_i[\emptyset]_j}\ p_S,\ \text{where } \mathcal{R}(\Gamma[\ ]_i[\ ]_j, i, j) \cap L_a(S) \neq \emptyset$$

and $\Gamma[\langle a\rangle A]_i[A]_j$ is derivable via a derivation Π' in DKm(S). Choose some $u \in \mathcal{R}(\Gamma[\ ]_i[\ ]_j, i, j) \cap L_a(S)$. Then we can derive the implication $\langle u\rangle A \supset \langle a\rangle A$ in SKm(S). Using this implication, the display property and the cut rule, it can be shown that the following rule is derivable in SKm(S).

$$\frac{\Gamma[\langle a\rangle A, \langle u\rangle A]}{\Gamma[\langle a\rangle A]}\ d$$

Then it is not difficult to show that the rule p_S can be simulated by the derived rule d above, with chains of r and $\langle a\rangle$-rules in SKm(S), and utilising the weakening lemma (Lemma 4.4). □

Theorem 5.2. Let S be a regular closed semi-Thue system over Σ and let $\mathcal{A}$ be a FSA representing the regular language generated by S and Σ. Then DKm(S) and DKm($\mathcal{A}$) are equivalent.

Proof. *(Outline).* In one direction, i.e., showing that a derivation of a formula B in DKm(S) can be translated into a derivation, we do induction on the height of derivations in DKm(S). As all non-propagation rules between the two systems are identical, it is enough to show that the propagation rules p_S of DKm(S) is admissible in DKm($\mathcal{A}$), which we show next.

Suppose we have a derivation Π of $\Gamma[\langle a\rangle A]_i[A]_j$ in DKm($\mathcal{A}$), and suppose that $\mathcal{R}(\Gamma[]_i[]_j, i, j) \cap L_a(S) \neq \emptyset$. We show that $\Gamma[\langle a\rangle A]_i[\emptyset]_j$ is derivable in DKm($\mathcal{A}$). In other words, this says that p_S is admissible in DKm($\mathcal{A}$).

Since $\mathcal{R}(\Gamma[]_i[]_j, i, j) \cap L_a(S) \neq \emptyset$, there must exist a sequence of transitions in $\mathcal{R}(\Gamma[\ ]_i[\ ]_j, i, j)$: $i \xrightarrow{a_1} i_1 \xrightarrow{a_2} \cdots \xrightarrow{a_{n-1}} i_{n-1} \xrightarrow{a_n} j$, where $a_1 \cdots a_n \in L_a(S)$ and where each i_k, for $1 \leq k \leq n-1$, is a node in $\Gamma[\ ]_i[\ ]_j$ and

- either $i \succ^{a_1} i_1$ or $i_1 \succ^{\bar{a}_1} i$,
- either $i_{k-1} \succ^{a_k} i_k$ or $i_k \succ^{\bar{a}_k} i_{k-1}$, for $2 \leq k < n-1$
- and either $i_{n-1} \succ^{a_n} j$ or $j \succ^{\bar{a}_n} i_{n-1}$.

Since $\mathcal{A}$ accepts $L_a(S)$, there must exist a sequence of transitions in $\mathcal{A}$ such that: $init_a \xrightarrow{a_1} s_1 \xrightarrow{a_2} \cdots \xrightarrow{a_{n-1}} s_{n-1} \xrightarrow{a_n} f$, where f is a final state in $\mathcal{A}$. The propagation path $a_1 \cdots a_n$ can then be simulated in DKm($\mathcal{A}$) as follows. First, define a sequence of nested sequents as follows:

- $\Gamma_0 := \Gamma[\langle a\rangle A]_i[\emptyset]_j$, $\Gamma_1 := \Gamma[\langle a\rangle A, init_a : A]_i[\emptyset]_j$.
- $\Gamma_{k+1} := \Gamma_k(i_k \ll \{s_k : A\})$, for $1 \leq k \leq n-1$.
- $\Gamma_{n+1} := \Gamma_n(j \ll \{f : A\})$ and $\Gamma_{n+2} := \Gamma_{n+1}(j \ll \{A\})$.

Then Γ_0 can be obtained from Γ_{n+2} by a series of applications of propagation rules of DKm($\mathcal{A}$). That is, Γ_0 is obtained from Γ_1 by applying the rule i; Γ_k is obtained from Γ_{k+1} by applying either the rule $t\!\downarrow$ or $t\!\uparrow$, for $1 \leq k \leq n-1$, at node i_k and Γ_n is obtained from Γ_{n+1} by applying the rule $t\!\downarrow$ or $t\!\uparrow$ at node j, and Γ_{n+1} is obtained from Γ_{n+2} by applying the rule f at node j. Note that Γ_{n+2} is a weakening of $\Gamma[\langle a\rangle A]_i[A]_j$ with labeled formulae spread in some nodes between i and j. It remains to show that Γ_{n+2} is derivable. This is obtained simply by applying weakening to Π (this weakening lemma for DKm($\mathcal{A}$) can be proved similarly as in the proof of Lemma 4.4).

For the other direction, assume we have a DKm($\mathcal{A}$)-derivation Ψ of B. We show how to construct a derivation Ψ' of B in DKm(S). The derivation Ψ' is constructed as follows: First, remove all labelled formulae from Ψ; then remove the rules $t\!\uparrow$, $t\!\downarrow$ and i, and finally, replace the rule f with p_S. The rules $t\!\uparrow$, $t\!\downarrow$ and i from Ψ simply disappear in Ψ' because with labelled formulae removed, the premise and the conclusion of any of the rules in Ψ map to the same sequent in Ψ'. Instances of the other rules in Ψ map to the same rules in Ψ'. We need to show that Ψ' is indeed a derivation in DKm(S). The only non-trivial case is to show that the mapping from the rule f to the rule p_S is correct, i.e., the resulting instances of p_S in Ψ' are indeed valid instances. The proof is rather involved, but essentially it shows that if a labelled formula $s : A$ is present in a node in a nested sequent Γ constructed during proof search of B, then it must be a product of a sequence of propagation of labelled formulae starting from some $\langle a \rangle A$ in a node in Γ. The complete proof is omitted here, but it is available in the extended version of this paper. □

Theorem 6.4. If $Prove_1(\mathcal{A}, \{F\}) = \top$ then F is provable in DKm($\mathcal{A}$). If $Prove_1(\mathcal{A}, \{F\}) = \bot$ then F is not provable in DKm($\mathcal{A}$).

Proof. The proof of the first statement is straightforward, since the steps of $Prove_1$ are just backward applications of rules of DKm($\mathcal{A}$). To prove the second statement, we show that if $Prove_1(\mathcal{A}, \{F\}) = \bot$ then there exists a model $\mathfrak{M} = (W, R, V)$, where $R = \{R_a\}_{a \in \Sigma}$, such that $\mathfrak{M} \not\models F$. By the completeness of DKm($\mathcal{A}$), it will follow that F is not provable in DKm($\mathcal{A}$).

Since $Prove_1(\mathcal{A}, \{F\}) = \bot$ the procedure must generate an $\mathcal{A}$-stable Δ, with F in the root node of Δ. Let W be the set of all the realised nodes of Δ. For every pair $i, j \in W$, construct an automaton $\mathcal{P}(i, j)$ by modifying the propagation automaton $\mathcal{R}(\Delta, i, j)$ by identifying every unrealised node k' with its closest ancestor k such that $\Delta|k = \Delta|k'$. That is, replace every transition of the form $s \xrightarrow{a} k'$ with $s \xrightarrow{a} k$ and $k' \xrightarrow{a} s$ with $k \xrightarrow{a} s$. Then define $R_a(x, y)$ iff $\mathcal{P}(x, y) \cap L(\mathcal{A}_a) \neq \emptyset$, where $\mathcal{A}_a$ is $\mathcal{A}$ with only $init_a$ as the initial state. Suppose S is a closed semi-Thue system that corresponds to $\mathcal{A}$. Then it can be shown that the frame $\langle W, R \rangle$ satisfies S.

To complete the model, let $x \in V(p)$ iff $\neg p \in \Delta|x$. We claim that for every $x \in W$ and every $A \in \Delta|x$, we have $\mathfrak{M}, x \not\models A$. We shall prove this by induction on the size of A. Note that we ignore the labelled formulae in Δ; they are just a bookeeping mechanism. As F is in the root node of Δ, this will also prove $\mathfrak{M} \not\models F$. We show here the interesting case involving the diamond operators.

Suppose $\langle a \rangle A \in \Delta|x$. Assume for a contradiction that $\mathfrak{M}, x \models \langle a \rangle A$. That is, $R_a(x, y)$ and $\mathfrak{M}, y \models A$. If $R_a(x, y)$ then there is a accepting path $p_a(x, y)$ in $\mathcal{P}(x, y)$ of the form: $x_0 \xrightarrow{a_1} x_1 \xrightarrow{a_2} x_2 \cdots x_{n-1} \xrightarrow{a_n} x_n$, where $x_0 = x$ and $x_n = y$ such that $u = a_1 \ldots a_n \in L(\mathcal{A}_a)$. Then because $u \in L(\mathcal{A}_a)$, there must be a sequence of states $s_0, s_1, \ldots, s_n$ of $\mathcal{A}$ such that $s_0 = init_a \in I$ and $s_n \in F$ and the transitions between states: $s_0 \xrightarrow{a_1} s_1 \xrightarrow{a_2} s_2 \cdots s_{n-1} \xrightarrow{a_n} s_n$. We show by induction on the length of transtions that that $s_i : A \in \Delta|x_i$ for $0 \leq i \leq n$. In the base case, because $\langle a \rangle A \in \Delta|x$, by $\mathcal{A}$-propagation, we have $s_0 : A \in \Delta|x_0$. For the inductive cases, suppose $s_i : A \in \Delta|x_i$, for $n > i \geq 0$. There are two cases to consider. Suppose the transition $x_i \xrightarrow{a_{i+1}}_{\mathcal{P}(x,y)} x_{i+1}$ is present in $\mathcal{R}(\Delta, x, y)$. Then either $x_i \succ^{a_{i+1}} x_{i+1}$ or $x_{i+1} \succ^{\bar{a}_{i+1}} x_i$. In either case, by $\mathcal{A}$-propagation of Δ, we must have $s_{i+1} : A \in \Delta|x_{i+1}$.

If $x_i \xrightarrow{a_{i+1}}_{\mathcal{P}(x,y)} x_{i+1}$ is not a transition in $\mathcal{R}(\Delta, x, y)$, then this transition must have resulted from a use of a loop node. There are two subcases: either x_i or x_{i+1} is the closest ancestor of a loop node x' with $\Delta|x_i = \Delta|x'$ or, respectively, $\Delta|x_{i+1} = \Delta|x'$. Suppose x_i is the closest ancestor of x' with $\Delta|x_i = \Delta|x'$. By the definition of $\mathcal{P}(x, y)$, this means we have $x' \xrightarrow{a_{i+1}} x_{i+1}$ in $\mathcal{R}(\Delta, x, y)$. Because $\Delta|x_i = \Delta|x'$ and $s_i : A \in \Delta|x_i$, we have $s_i : A \in \Delta|x'$. Then by $\mathcal{A}$-propagation, it must be the case that $s_{i+1} : A \in \Delta|x_{i+1}$. Suppose x_{i+1} is the closest ancestor of x' with $\Delta|x' = \Delta|x_{i+1}$. Then $x_i \xrightarrow{a_{i+1}} x'$ is a transition in $\mathcal{R}(\Delta, x, y)$. By $\mathcal{A}$-propagation, it must be the case that $s_{i+1} : A \in \Delta|x'$, and therefore also $s_{i+1} : A \in \Delta|x_{i+1}$.

So we have $s_n : A \in \Delta|y$. But, again by $\mathcal{A}$-propagation, this means $A \in \Delta|y$ (because s_n is a final state). Then by the induction hypothesis, we have $\mathfrak{M}, y \not\models A$, contradicting the assumption. □

Theorem 6.5. For every formula A, $Prove_1(\mathcal{A}, \{A\})$ terminates.

Proof. *(Outline)* We say that a nested sequent Γ is a *set-based nested sequent* if in every node of Γ, every (labelled) formula occurs at most once (a formula C and its labelled versions are considered distinct). By inspection of the procedure $Prove_1$, it is clear that all the intermediate sequents created during proof search for $Prove_1(\mathcal{A}, \{A\})$ are set-based sequents.

The only possible cause of non-termination is steps (v) and (vi), where the input nested sequent is extended with new nodes. The blocking condition in (vi) ensures that the height of any nested sequent generated during proof search is bounded. Let m be the number of states in $\mathcal{A}$ and let n be the number of subformulae of A. Then the total number of different sets of formulae and labeled formulae (with labels from $\mathcal{A}$) is bounded by $2^{(m+1)n}$. Therefore, any set-based nested sequent generated during proof search will have height at most $2^{(m+1)n}$, as the loop checking ensures they are never expanded beyond this height. Given a nested sequent of a fixed height, the expansion in step (v) only adds to the width of the nested sequent. This expansion is limited by the number of 'boxed' subformulae of A. So the size of the nested sequents generated during proof search is bounded, and therefore each branch of the search has a finite length, thus the procedure must terminate. □

Lemma 6.9. Let S be a context-free closed semi-Thue system. If Γ is an S-stable nested sequent, then there exists a model $\mathfrak{M}$ such that for every node x in Γ there exists a world w in $\mathfrak{M}$ such that for every $A \in \Gamma|x$, we have $\mathfrak{M}, w \not\models A$.

Proof. Let $\boldsymbol{x} = x_1, \ldots, x_n$ be the list of (pairwise distinct) unrealised leaf nodes in Γ. Because Γ is S-stable, we have a function λ assigning each unrealised leaf node x_i to an ancestor node $\lambda(x_i)$ such that $\Gamma|x_i = \Gamma|\lambda(x_i)$, and for every node y and z in Γ, we have that Γ is $(S, \mathcal{P}(y,z))$-propagated, where $\mathcal{P}(y,z) = \mathcal{R}(\Gamma, y, z)[\boldsymbol{x} := \lambda(\boldsymbol{x})]$. Then define $\mathfrak{M} = \langle W, \{R_a \mid a \in \Sigma\}, V\rangle$ where

- W is the set of nodes of Γ minus the nodes $\boldsymbol{x}$,
- for every $x, y \in W$, $R_a(x,y)$ iff $\mathcal{P}(x,y) \cap L_a(S) \neq \emptyset$, and
- $V(p) = \{x \in W \mid \neg p \in \Gamma|x\}$.

We now show that for every node v in Γ, there exists a $w \in W$ such that if $A \in \Gamma|v$ then $\mathfrak{M}, w \not\models A$, where the world w is determined by v as follows: if v is in $\boldsymbol{x}$, then $w = \lambda(v)$; otherwise, $w = v$. We prove this by induction on the size of A. The only interesting cases are those where $A = \langle a\rangle C$ or $A = [a]C$ for some a and C.

- Suppose $A = \langle a \rangle C$. Suppose, for a contradiction, that $\mathfrak{M}, w \models \langle a \rangle C$. That means there exists a w' such that $R_a(w, w')$ and $\mathfrak{M}, w' \models C$. By the definition of R_a, we have that $\mathcal{P}(w, w') \cap L_a(S) \neq \emptyset$. Because Γ is S-stable, by Definition 6.8(iv), it is $(S, \mathcal{P}(w, w'))$-propagated. This means that $C \in \Gamma|w'$. Then by the induction hypothesis, $\mathfrak{M}, w' \not\models C$, which contradicts our assumption.
- Suppose $A = [a]C$. To show $\mathfrak{M}, w \not\models [a]C$, it is enough to show there exists w' such that $R_a(w, w')$ and $\mathfrak{M}, w' \not\models C$.

  Note that w must be an internal node in Γ, so by the S-stability of Γ, node w in Γ must be realised. Therefore there exists a node z such that $w \succ^a z$ in Γ and $C \in \Gamma|z$. If $z \notin \boldsymbol{x}$, then let $w' = z$; otherwise, let $w' = \lambda(z)$. In either case, $\Gamma|z = \Gamma|w'$, so in particular, $C \in \Gamma|w'$. Also, in either case, the propagation automata $\mathcal{P}(w, w')$ contains a transition $w \xrightarrow{a}_{\mathcal{P}(w,w')} w'$ (in the case where $z \in \boldsymbol{x}$, this is because $\lambda(z)$ is identified with z). Obviously, $a \in L_a(S)$, so $L_a(S) \cap \mathcal{P}(w, w') \neq \emptyset$, so by the definition of R_a, we have $R_a(w, w')$. Since $C \in \Gamma|w'$, by the induction hypothesis, $\mathfrak{M}, w' \not\models C$. So we have $R_a(w, w')$, and $\mathfrak{M}, w' \not\models C$, therefore $\mathfrak{M}, w \not\models [a]C$.

□

Theorem 6.10. Let S be a context-free closed semi-Thue system. For every formula F, *Prove*$(S, \{F\})$ returns $\top$ if and only if F is provable in DKm(S).

Proof. *(Outline)* One direction, i.e., *Prove*$(S, \{F\}) = \top$ implies that F is provable in DKm(S), follows from the fact that steps of *Prove* are simply backward applications of rules of DKm(S). To prove the other direction, we note that if F has a derivation in DKm(S), it has a derivation of a minimal length, say Π. In particular, in such an derivation, there are no two identical nested sequents in any branch of the derivation. Because in DKm(S) each backward application of a rule retains the principal formula of the rule, every application of a rule in Π will eventually be covered by one of the steps of *Prove*. Since there are only finitely many rule applications in Π, eventually these will all be covered by *Prove* and therefore it will terminate. For example, if Π ends with a diamond (propagation) rule applied to a non-saturated sequent, the *Prove* procedure will choose to first saturate the sequent before applying the propagation rule. Since all rules are invertible, we do not lose any provability of the original sequent, but the *Prove* procedure may end up doing more steps. We need to show, additionally, that every sequent arising from the execution of *Prove*$(S, \{F\})$ is not S-stable. Suppose otherwise, i.e., the procedure produces an S-stable sequent Δ. Now it must be the case that F is in the root node of Δ. By Lemma 6.9, this means there exists a countermodel that falsifies F, contrary to the validity of F. □

Theorem 6.11. Let S be a regular closed semi-Thue system. Then for every formula F, the procedure *Prove*$(S, \{F\})$ terminates.

Proof. Since S is regular, there exists a minimal deterministic FSA $\mathcal{A}$ corresponding to S such that *Prove*$_1(\mathcal{A}, \{F\})$ terminates.

Suppose *Prove*$_1(\mathcal{A}, \{F\}) = \top$. Then F must be derivable in DKm$(\mathcal{A})$

by Theorem 6.4. Since DKm($\mathcal{A}$) and DKm(S) are equivalent (Theorem 5.2), there must also be a derivation of F in DKm(S). Then by Theorem 6.10, $Prove(S, \{F\})$ must terminate and return $\top$.

Suppose $Prove_1(\mathcal{A}, \Gamma) = \bot$. Then there exists an $\mathcal{A}$-stable Γ' that can be constructed from Γ in the execution of $Prove_1(\mathcal{A}, \Gamma)$. It can be shown that a Δ that is identical to Γ' without any labelled formulae can be constructed in the execution of $Prove_2(S, \Gamma, d)$ for some d. We claim that Δ is S-stable. Saturation, propagation and the realisation of internal nodes follow immediately from the construction, it remains to find a function λ as in Definition 6.8. We claim that such a function is given by $\lambda(x) = y$ where y is the closest ancestor of x in Γ' such that $\Gamma'|x = \Gamma'|y$. That is, we identify each unrealised leaf with the same node it would have been identified with in $Prove_1(\mathcal{A}, \Gamma)$.

Let $\boldsymbol{i} = i_1, \ldots, i_l$ be the list of all unrealised leaf nodes in Δ and let $\mathcal{P}(x, y) = \mathcal{R}(\Delta, x, y)[\boldsymbol{i} := \lambda(\boldsymbol{i})]$. (Note that as the tree structures of Γ' and Δ are identical, we also have $\mathcal{P}(x, y) = \mathcal{R}(\Gamma', x, y)[\boldsymbol{i} := \lambda(\boldsymbol{i})]$.) For a contradiction, suppose there exists j and k such that Δ is not $(S, \mathcal{P}(j, k))$-propagated, i.e., there exist $\langle a \rangle A \in \Delta|j$, such that $A \notin \Delta|k$ but $\mathcal{P}(j,k) \cap L_a(S) \neq \emptyset$. In other words, there is a word $b_1 \ldots b_n \in \mathcal{P}(j,k) \cap L_a(S)$, and a sequence of states $x_0, \ldots, x_n$ in $\mathcal{P}(j,k)$ such that $x_0 = j, x_n = k, x_{m-1} \xrightarrow{b_m}_{\mathcal{P}(j,k)} x_m$, where $1 \leq m < n$. We will show that there exists a function St assigning states of $\mathcal{A}$ to nodes of Γ' satisfying: $St(x_0) \in I$, $St(x_{m-1}) \xrightarrow{b_m}_{\mathcal{A}} St(x_m)$, $St(x_n) \in F$, and $St(x_m) : A \in \Gamma'|x_m$. This will establish that $St(x_n) : A \in \Gamma'|x_n$ where $St(x_n) \in F$. Then by $\mathcal{A}$-propagation, it will follow that $A \in \Gamma'|k$, and therefore $A \in \Delta|k$, contradicting our assumption that $A \notin \Delta|k$.

Let $s_0, \ldots, s_n$ be the run of $\mathcal{A}_a$ associated with input $b_1 \ldots b_n$. Let $St(x_m) = s_m$. As $L(\mathcal{A}_a) = L_a(S)$, we know that $s_0, \ldots, s_n$ is an accepting run. This gives us $St(x_0) \in I, St(x_{m-1}) \xrightarrow{b_m}_{\mathcal{A}} St(x_m)$ and $St(x_n) \in F$. It remains to show that $St(x_m) : A \in \Gamma'|x_m$. We will do so by induction on m.

Base case: As $\langle a \rangle A \in \Gamma'|x_0$, by $\mathcal{A}$-propagation we obtain $s_0 : A \in \Gamma'|x_0$.

Inductive case: Suppose $x_m \xrightarrow{b_{m+1}}_{\mathcal{P}(j,k)} x_{m+1}$. By the inductive hypothesis, $s_m : A \in \Gamma'|x_m$. There are two cases to consider:

- The transition $x_m \xrightarrow{b_{m+1}}_{\mathcal{P}(j,k)} x_{m+1}$ also exists in $\mathcal{R}(\Gamma', j, k)$. In this case, by $\mathcal{A}$-propagation, we have $s_{m+1} : A \in \Gamma'|x_{m+1}$.

- The transition $x_m \xrightarrow{b_{m+1}}_{\mathcal{P}(j,k)} x_{m+1}$ is obtained from $\mathcal{R}(\Gamma', j, k)$ through the identification of unrealised leaf nodes with their closest ancestors. There are two subcases:
  - $x_m = \lambda(y)$ for some unrealised leaf node y such that $\Gamma'|x_m = \Gamma'|y$, and $y \xrightarrow{b_{m+1}}_{\mathcal{R}(\Gamma',j,k)} x_{m+1}$. Since $\Gamma'|x_m = \Gamma'|y$, we have that $s_m : A \in \Gamma'|y$ and it follows by $\mathcal{A}$-propagation that $s_{m+1} : A \in \Gamma'|x_{m+1}$.
  - $x_{m+1} = \lambda(y)$ for some unrealised leaf node y such that that $\Gamma'|x_{m+1} = \Gamma'|y$, and $x_m \xrightarrow{b_{m+1}}_{\mathcal{R}(\Gamma',j,k)} y$. By $\mathcal{A}$-propagation, $s_{m+1} : A \in \Gamma'|y = \Gamma'|x_{m+1}$.

Thus when $Prove(S, \Gamma)$ calls $Prove_2(S, \Gamma, d)$, it will construct an S-stable sequent and terminate. □

References

[1] Baldoni, M., L. Giordano and A. Martelli, *A tableau for multimodal logics and some (un)decidability results*, in: *TABLEAUX*, LNCS **1397** (1998), pp. 44–59.

[2] Belnap, N., *Display logic*, Journal of Philosophical Logic **11** (1982), pp. 375–417.

[3] Blackburn, P., J. van Benthem and F. Wolter, "Handbook of Modal Logic," Studies in Logic and Practical Reasoning, Elsevier, 2007.

[4] Brünnler, K., *Deep sequent systems for modal logic*, Archive for Mathematical Logic **48** (2009), pp. 551–577.

[5] Brünnler, K. and A. Tiu, *A local system for classical logic*, in: *LPAR*, LNCS **2250** (2001), pp. 347–361.

[6] Demri, S., *The complexity of regularity in grammar logics and related modal logics*, J. Log. Comput. **11** (2001), pp. 933–960.

[7] Demri, S. and H. de Nivelle, *Deciding regular grammar logics with converse through first-order logic*, Journal of Logic, Language and Information **14** (2005), pp. 289–329.

[8] Fariñas del Cerro, L. and M. Penttonen, *Grammar logics*, Logique et Analyse **121-122** (1998), pp. 123–134.

[9] Ginsburg, S., "The Mathematical Theory of Context-Free Languages," McGraw-Hill, Inc., New York, NY, USA, 1966.

[10] Goré, R. and L. A. Nguyen, *A tableau calculus with automaton-labelled formulae for regular grammar logics*, in: *TABLEAUX*, LNCS **3702** (2005), pp. 138–152.

[11] Goré, R., L. Postniece and A. Tiu, *Taming displayed tense logics using nested sequents with deep inference*, in: *TABLEAUX*, LNCS **5607** (2009), pp. 189–204.

[12] Goré, R., L. Postniece and A. Tiu, *Cut-elimination and proof search for bi-intuitionistic tense logic*, in: *Advances in Modal Logic* (2010), pp. 156–177.

[13] Goré, R., L. Postniece and A. Tiu, *On the correspondence between display postulates and deep inference in nested sequent calculi for tense logics*, Logical Methods in Computer Science **7** (2011).

[14] Goré, R. and F. Widmann, *Sound global state caching for $\mathcal{ALC}$ with inverse roles*, in: *TABLEAUX*, Lecture Notes in Computer Science **5607**, 2009, pp. 205–219.

[15] Goré, R. and F. Widmann, *Optimal and cut-free tableaux for propositional dynamic logic with converse*, in: *IJCAR*, LNCS **6173** (2010), pp. 225–239.

[16] Guglielmi, A., *A system of interaction and structure*, ACM Trans. Comput. Log. **8** (2007).

[17] Horrocks, I., O. Kutz and U. Sattler, *The even more irresistible $\mathcal{SROIQ}$*, in: *KR* (2006), pp. 57–67.

[18] Horrocks, I. and U. Sattler, *Decidability of $\mathcal{SHIQ}$ with complex role inclusion axioms*, Artif. Intell. **160** (2004), pp. 79–104.

[19] Kashima, R., *Cut-free sequent calculi for some tense logics*, Studia Logica **53** (1994), pp. 119–135.

[20] Kazakov, Y., *$\mathcal{RIQ}$ and $\mathcal{SROIQ}$ are harder than $\mathcal{SHOIQ}$*, in: *KR* (2008), pp. 274–284.

[21] Kracht, M., *Power and weakness of the modal display calculus*, in: *Proof theory of modal logic (Hamburg, 1993)*, Applied Logic Series **2**, Kluwer Acad. Publ., 1996 pp. 93–121.

[22] Nguyen, L. A. and A. Szałas, *A tableau calculus for regular grammar logics with converse*, in: *CADE*, LNCS **5663**, 2009, pp. 421–436.

[23] Nguyen, L. A. and A. Szałas, *Exptime tableau decision procedures for regular grammar logics with converse*, Studia Logica **98** (2011), pp. 387–428.

[24] Ramanayake, D. R. S., "Cut-elimination for provability logics and some results in display logic," Ph.D. thesis, The Australian National University (2011).

[25] Simpson, A. K., "The Proof Theory and Semantics of Intuitionistic Modal Logics," Ph.D. thesis, University of Edinburgh (1994).

Dynamic Mereotopology II: Axiomatizing some Whiteheadean Type Space-time Logics

Dimiter Vakarelov

Sofia University "St. Kliment Ohridski"
5 James Bourchier blvd 1164 Sofia, Bulgaria,
e-mail dvak@fmi.uni-sofia.bg

Abstract

In this paper we present an Whiteheadean style point-free theory of space and time. Here "point-free" means that neither space points, nor time moments are assumed as primitives. The algebraic formulation of the theory, called dynamic contact algebra (DCA), is a Boolean algebra whose elements symbolize dynamic regions changing in time. It has three spatio-temporal relations between dynamic regions: *space contact*, *time contact* and *preceding*. We prove a representation theorem for DCA-s of topological type, reflecting the dynamic nature of regions, which is a reason to call DCA-s *dynamic mereotopoly*. We also present several complete quantifier-free logics based on the language of DCA-s.

Keywords: Dynamic mereotopology, point-free theory of space and time, representation theorem, quantifier-free spatio-temporal logic.

Introduction

Alfred Whitehead is well-known as the co-author with Bertrand Russell of the famous book "Principia Mathematica". They intended to write a special part of the book related to the foundation of geometry, but due to some disagreements between them this part has not been written. Later on Whitehead formulated his own program for a new theory of space and time. The best articulation of this program is in this quote from [24], page 195 (boldface is ours):

> "...Our space concepts are concepts of relations between material things in space. Thus there is no such entity as a self-subsistent point. A point is merely the name for some peculiarity of the relations between matter which is, in common language, said to be in space.
>
> **It follows from relativity theory that a point should be definable in terms of the relations between material things.** So far as I am aware, this outcome of the theory has escaped the notice of mathematicians, who have invariably assumed the point as the ultimate starting ground of their reasoning. ... **Similar explanations apply to time.** Before the theories of space and time have been carried to a satisfactory conclusion on the relational basis, a long and careful scrutiny of the definitions of points of space and instants of time will have

to be undertaken, and many ways of effecting these definitions will have to be tried and compared. **This is an unwritten chapter of mathematics**..."

According to the above program Whitehead should be considered as the initiator of point-free approach to the theory of space, known now as *Region Based Theory of Space (RBTS)*. In his famous book "Process and Reality" [26] Whitehead presented a more detailed program of how to build mathematical formalizations of some versions of RBTS. His main primitive is the notion of *region* as a formal analog of physical body, and some relations between regions like *part-of*, *overlap*, and *contact*. Whitehead shows how to define points, lines and planes by means of certain classes of regions, and in general how to rebuild the whole geometry on this new basis in an equivalent way. Other sources of RBTS are, for instance, de Laguna [5] and Tarski [21], who showed how to rebuild Euclidean geometry on the base of mereology and the primitive notion of *ball*. Survey papers about RBTS are [3,15,22].

Part-of and overlap are studied in *mereology* considered as a philosophical theory of parts and wholes [19], and according to Tarski (see [19]) mereology is equivalent in certain sense to Boolean algebra. Extensions of mereology with contact or some contact-like relations, is now called *mereotopology*, which can be considered as a theoretical tool for RBTS. Motivation for such a terminology is that the main point models of mereotopologies are topological. Recent forms of mereotopology are various notions of contact algebras (see [9,20,7]), which are Boolean algebras enriched with an additional relation called contact, and the simplest one for which we reserve the name "contact algebra", was introduced in [7]. Standard point models of contact algebras are the Boolean algebras of regular closed sets in topological spaces and two regular closed sets are in a contact if they have a nonempty intersection. The paper [7] contains point-free characterizations of contact algebras of regular closed sets of various classes of topological spaces. Let us mention that the pioneering work in this area was given by H. de Vries [6] but mainly oriented to applications in topology, which was one of the reasons his work to be unknown for a long time among the community of researchers interested in RBTS. Let us note also that Whitehead's ideas for RBTS were in a sense reinvented in computer science and Artificial Intelligence as a more suitable formalism for representing spatial information. This fact generated a very intensive study of spatial formalisms related to RBTS with various applications (see for this [4]).

Whitehead's theory of time was developed mainly in [25] and [26] and was called *Epochal Theory of Time (ETT)*. Whitehead claims that the theory of time should not be separated from the theory of space and their integrated theory has to be extracted from the existing things in reality and some of their spatio-temporal relations. This integrated theory should be point-free in a double sense: that both space points and time points (moments of time) should be definable by the other primitives of the theory. Unfortunately, unlike his program how to build mathematical theory of space, given in [26], Whitehead did not describe analogous program for his integrated theory of space and time. He presented his ideas for ETT quite informally and in a pure philosophical manner, which makes extremely difficult to extract from his texts clear mathematical theory corresponding to ETT. So, in this paper we will follow mainly Whitehead's ideas described in the quote cited above and will try to present an extension of RBTS, considered as an integrated theory of space and time, containing neither points, nor moments of time as primitive notions.

The present paper is a second one in the series of papers started by [23] with the aim to present some integrated theory of space and time in a point-free Whiteheadean style. The main idea in [23] was to obtain an analog of contact algebra, called *dynamic contact algebra* (DCA), which formalizes changing regions. The standard point model of DCA having explicit set T of time points (moments of time) is defined as follows. Suppose we are observing some area of regions. If they are not changing then their spatial structure forms a contact algebra. If the regions are changing in time then the information which we need for a given dynamic region a is to know its instance a_m at each moment of time m. So, a has to be considered as a vector, or a function defined in T, with coordinates a_m, $m \in T$. It is natural to consider a_m as an element of a contact algebra $(\mathbf{B}_m, C_m) = (B_m, 0_m, 1_m, \leq_m, ., +, *, C_m)$, (called coordinate contact algebra corresponding to the time moment m), assuming in this way that $\mathbf{B}_m$ as a snapshot of the whole state of affairs at the moment m. We may assume further that dynamic regions form a Boolean algebra with operations defined coordinatewise. Hence this Boolean algebra is a sublagebra of the Cartesian product of all coordinate algebras. Just to make things not very complicated we assumed in [23] that the set T has no internal structure of the intended time ordering and considered as primitives only two very simple spatio-temporal relations between dynamic regions: *stable contact* denoted by $C^{\forall}$, and *unstable contact* denoted by $C^{\exists}$, with the following definitions in the standard model:

$aC^{\forall}b$ iff $(\forall m \in T)a_m C_m b_m$,

i.e. a is in a *stable contact* with b if a and b are in a *contact* at each moment of time m. Analogously

$aC^{\exists}b$ iff $(\exists m \in T)a_m C_m b_m$,

i.e. a is in an *unstable contact* with b if a and b are in a *contact* at some moment of time m.

Note that these relations do not depend on any time ordering. In order to make all this free of the explicit use of time we axiomatized this structure in an abstract algebraic form obtaining an abstract definition of DCA. The main result in [23] was a representation theorem of DCA into standard dynamic contact algebras.

In the present paper we extend the point-free approach from [23] considering standard dynamic models in which the set of time moments T is supplied with an intended time ordering denoted by $\prec$ and satisfying some reasonable conditions. So, in order to introduce it we need a new relation between dynamic regions depending on time order. There are many such relations and the main problem is to find a suitable one which guarantees the expected representation theorem with some natural properties of the time order. In this paper we use one, denoted by $\mathcal{B}$, called *precedence relation*, with a very simple formal properties in the expected axiomatization. The intuitive meaning of $a\mathcal{B}b$ is "a exists at some moment of time and b exists at a later moment". In the standard model this definition sounds as follows:

- $a\mathcal{B}b$ iff $(\exists m, n \in T)(m \prec n$, and $a_m \neq 0_m$ and $b_n \neq 0_n)$.

Here $a_m \neq 0_m$ just says that a exists at the moment m, and the same for $b_n \neq 0_n$.

We consider also two other relations:

- $aC^s b$ iff $(\exists m \in T)a_m C_m b_m$, *space contact*,
- $aC^t b$ iff $(\exists m \in T)a_m \neq 0_m$ and $b_m \neq 0_m$, *time contact.*

Space contact coincides with the unstable contact $C^{\exists}$ and we renamed it, because it is the natural contact relation between dynamic regions which ensures that the coordinate algebras $\mathbf{B}_m$ are contact algebras and consequently it is responsible for the definition of space points. While space contact means having a common space point, time contact indeed means having a common time point and it is also responsible for the definition of time points. Note that it is a kind of *simultaneity* or *contemporaneity* relation mentioned in Whitehead texts. Let us note that Whitehead did not use something like our precedence relation $a\mathcal{B}b$. We take it by two reasons: first it is responsible in the definition of time ordering, and second, although its simplicity in the axiomatization, it together with the time contact is able to characterize point-free many natural properties of time ordering: left and right seriality, reflexivity, transitivity, linearity, density, up-directness and down-directness, different subsets of which define different space-time theories.

The rest of the paper is organized as follows. Section 1 lists some facts about contact and precontact algebras. In Section 2 we introduce dynamic models of space with explicit moments of time and time ordering. Section 3 is devoted to the main notion of the paper - dynamic contact algebra (DCA). In Section 4 we developed representation theory of DCA-s. Section 5 contains some quantifier-free constraint logics based on the language of DCA. In Section 6 we discuss relations with other works, some open problems, and plans for future research.

1 Preliminaries about contact and precontact algebra

Definition 1.1 ([7]) *Let $(B, 0, 1, ., +, ^*)$ be a non-degenerate Boolean algebra with * denoting its Boolean complement. A binary relation C in B is called a contact relation if it satisfies the following conditions:*

(C1) *If aCb, then $a, b \neq 0$,*
(C2) *$aC(b+c)$ iff aCb or aCc,*
(C3) *If aCb, then bCa,*
(C4) *If $a.b \neq 0$, then aCb.*

If B is a Boolean algebra and C a contact relation *in B then the pair (B, C) is called a* contact algebra *over B. The elements of B are called* regions. *The negation of C is denoted by $\overline{C}$. The element 0 is called zero region and is considered as a region which does not exist (here "exists" is considered as a predicate). Then $a \neq 0$ means that a exists. If $a \leq b$, where $\leq$ is the Boolean* *ordering, then this will be read as a is a part of b, so $\leq$ is the mereological relation part-of. The region 1 is called the unit region, the region which has as its parts all other regions. The relation aOb iff $a.b \neq 0$ is the mereological relation* overlap *in B.*

The following lemma lists some easy consequences of axioms.

Lemma 1.2 *(i) aCb and $a \leq a'$ and $b \leq b'$ implies $a'Cb'$,*
(ii) $a \neq 0$ iff aCa

Let us note that each Boolean algebra has at least two contact relations: the overlap O, which by axiom (C4) is the *smallest contact* in B, and $aC^{max}b \leftrightarrow_{def} a \neq 0$ and $b \neq 0$, which by axiom C1 is the *maximal contact* in B.

Contact algebras of regular closed sets in a topological space. Let X be a non-empty topological space with the closure and interior operations denoted respectively by $Cl(a)$ and $Int(a)$. A subset a of X is *regular closed* if $a = Cl(Int(a))$. The set of all regular closed subsets of X is denoted by $RC(X)$. It is a well-known fact that regular closed sets with the operations $a+b = a \cup b$, $a.b = Cl(Int(a \cap b))$, $a^* = Cl(X \setminus a)$, $0 = \varnothing$ and $1 = X$ form a Boolean algebra. If we define the contact by $a\, C_X\, b$ iff $a \cap b \neq \varnothing$ then C_X satisfies the axioms $(C1)$–$(C4)$. Such a contact is called standard contact for regular closed sets and the contact algebra of this example and any its subalgebra is said to be *standard contact algebra of regular closed sets*, or simply *topological contact algebra.* Topological space with a closed base of regular closed sets is called *semiregular* (for other topological notions related to contact algebras see [7] or [22]).

The following representation theorem for contact algebras is proved in [7]:

Theorem 1.3 (Topological representation thm. for contact algebras) *For every contact algebra (B, C) there exists a semi-regular and compact T_0 space X and an embedding h into the contact algebra $RC(X)$.*

Theorem 1.4 ([23] Joint embedding theorem for contact algebras) *Let (B_t, C_t), $t \in T$ be a nonempty family of contact algebras. Then there exist a semiregular and compact T_0 space X and a family of embeddings g_t of (B_t, C_t) into $RC(X)$, $t \in T$.*

Abstract points of contact algebras. The abstract points which are used in the representation theory of contact algebras developed in [7] are called clans (see [7] for the origin of this name). The definition is the following:

Definition 1.5 ([7]) *Let (B, C) be a contact algebra. A subset $\Gamma \subseteq B$ is called a clan if the following conditions are satisfied:*
(i) $1 \in \Gamma$ and $0 \notin \Gamma$,
(ii) If $a \in \Gamma$ and $a \leq b$ then $b \in \Gamma$,
(iii) If $a + b \in \Gamma$ then $a \in \Gamma$ or $b \in \Gamma$,
(iv) If $a, b \in \Gamma$ then aCb.

The set of all clans is denoted by $CLANS(B)$. For $a \in B$ we denote by $h(a) = \{\Gamma \in CLANS(B) : a \in \Gamma\}$. It is shown in [7] that the set $CLANS(B)$

with the set $\{h(a) : a \in B\}$ considered as a closed base of a topology in CLANS(B), defines a semiregular and compact T_0 topology and h is the embedding of (B, C) into $RC(CLANS(B))$ considered in Theorem 1.3. The following lemma, used in the proof of Theorem 1.3, characterizes contact relation on terms of clans.

Lemma 1.6 aCb *iff* $(\exists\Gamma \in CLANS(B))(a, b \in \Gamma)$ *iff* $h(a) \cap h(b) \neq \varnothing$ [7].

Precontact algebras. A slight generalization of contact algebra is the notion of *precontact algebra*, studied in [8] under the name *proximity algebra.* The definition sounds as follows.

Definition 1.7 (Precontact algebras) *Let $\underline{B}$ be a Boolean algebra and C is a binary relation on B. C is called a* precontact *relation on B if it satisfies the following conditions:*

(P1) $aCb \to a \neq 0$ *and* $b \neq 0$,
(P2a) $aC(b + c) \leftrightarrow aCb$ *or* aCc,
(P2b) $(a + b)Cc \leftrightarrow aCc$ *or* bCc.
The pair (B, C) is called a precontact algebra.

Obviously every contact relation is a precontact relation.
The following property is an easy consequence from the axioms:
(Mono C) $aCb \wedge a \leq a' \wedge b \leq b' \to a'Cb'$.

2 Dynamic models of space with explicit moments of time and time ordering

Choosing the right definition. In this section we will present a dynamic model of time with explicitly given time moments and precedence relation between them. The construction is based on the following intuition. Suppose we have a domain with changing regions in time and a camera making snapshots for each moment of time. Then each snapshot describes the picture of the state of affairs in the corresponding moment of time m. We assume that the regions at each moment m form a contact algebra - $\underline{B}_m$, which describes their spatial interrelations. In this way to each m from a given set T of time moments we associate a contact algebra $\underline{B}_m$. Each changing region has a trajectory, which can be considered as a vector with coordinates indexed by the elements of T, and each coordinate a_m being from the contact algebra $\underline{B}_m$. We identify changing regions by their trajectories and assume that they form a Boolean algebra with Boolean operations defined coordinate-wise. In this way they form a Boolean subalgebra of the cartesian product $\prod_{m \in T} \underline{B}_m$ of the family of contact algebras $\{\underline{B}_m : m \in B_m\}$. We may assume also that T is supplied with some natural ordering relation $\prec$.

The above informal reasoning suggests the following formal definition.

Definition 2.1 (Dynamic models of space with explicit moments of time and time ordering) *Let T be a non-empty set, which elements are called time moments and $\prec$ be a binary relation on T, called* time ordering, *or* before-after *relation ($m \prec n$ is read: m is before n, or n is after m).*

Then the system $\underline{T} = (T, \prec)$, $T \neq \varnothing$ *is called a* time structure. *Let for each* $m \in T$, $(\underline{B}_m, C_m) = (\underline{B}_m, 0_m, 1_m, \leq_m, ., +, *, C_m)$ *be a contact algebra. Let* $B(\underline{T}) = \prod_{m \in T} B_m$ *denote the Cartesian product of the family* $\{\underline{B}_m : m \in T\}$ *considering its members as Boolean algebras. Since contact algebras are non-degenerated Boolean algebras, then their product* $B(\underline{T})$ *is also a non-degenerated Boolean algebra (note that this fact does not depend on the Axiom of Choice). The algebras* $(\underline{B}_m, C_m)$, $m \in T$, *are called* coordinate contact algebras *of* $B(\underline{T})$. *The elements of* $B(\underline{T})$ *are vectors with coordinates in the coordinate contact algebras, so they can be considered as possible trajectories of changing regions. So we name the elements of* $B(\underline{T})$ *dynamic regions. The product* $B(\underline{T})$ *contains all possible trajectories of changing regions, which is an extreme case, so it is natural to consider subalgebras of the product. Any such subalgebra is called a* dynamic model of space with explicit moments of time and time ordering, *shortly* dynamic model of space.

Let $\mathbf{B}$ *be a dynamic model of space.* $\mathbf{B}$ *is called* full *if it coincides with the product* $B(\underline{T})$; $\mathbf{B}$ *is called* rich *if it contains all vectors in which there are no coordinates different from zero and one. Obviously full models are rich.*

In dynamic model of space time moments and the precedence relation are given explicitly by the time structure $(T, \prec)$. By the topological representation theory of contact algebras each coordinate contact algebra $(\underline{B}_m, C_m)$ determines its topological space X_m, called *coordinate space.* By the joint embedding theorem 1.4 all coordinate algebras can be embedded into the contact algebra of regular closed sets $RC(X)$ of a single space X, which may be considered as the set of points of the dynamic model of space.

We will use the following notations for dynamic regions. If $a \in B(\underline{T})$ and $m \in T$, then a_m denotes the m-th coordinate of a and a_m is *the region* a *at the moment* m. For instance the expression $a_m \neq 0_m$ means that a *exists at* m, and the expression $a_m C_m b_m$ means that a *and* b *are in a contact at the moment* m.

Dynamic model of space is a quite rich spatio-temporal structure in which one can give explicit definitions of various spatio-temporal relations between dynamic regions. In this paper we shall study the following three relations between dynamic regions:

- $aC^s b$ iff $(\exists m \in T)(a_m C_m b_m)$, called *space contact,*
- $aC^t b$ iff $(\exists m \in T)(a_m \neq 0_m$ and $b_m \neq 0_m)$, called *time contact,*
- $a\mathcal{B}b$ iff $(\exists m, n \in T)(m \prec n$ and $a_m \neq 0_m$ and $b_n \neq 0_n)$, called *precedence.*

While space contact means having a common space point, the time contact indeed means having a common time point. It also is a kind of *simultaneity relation* or *contemporaneity* relation used in Whitehead's works.

Definition 2.2 *Dynamic model of space supplied with the three relations* C^s, C^t, *and* $\mathcal{B}$ *is called a* standard dynamic contact algebra. *It is called* rich *if the dynamic model is rich.*

Our aim is to give abstract point-free characterization of standard dynamic contact algebras. First we will study some formal properties of the three relations, which in the abstract setting will be taken as axioms.

Lemma 2.3 *(i)* C^s *is a contact relation,*
(ii) C^t *is a contact relation satisfying the additional condition*
$(C^s \to C^t)$ $aC^sb \to aC^tb$.
(iii) $\mathcal{B}$ *is a* precontact relation.

Proof. Direct verification. □

Correspondence results for time structures. Below we list some correspondences between conditions on time ordering $\prec$ and conditions on dynamic regions in standard dynamic contact algebras formulated in terms of C^t and $\mathcal{B}$.
(RS) *Right seriality* $(\forall m)(\exists n)(m \prec n) \Longleftrightarrow$ **(rs)** $aC^tb \to a\mathcal{B}p \vee b\mathcal{B}p^*$,
(LS)*Left seriality* $(\forall m)(\exists n)(n \prec m) \Longleftrightarrow$ **(ls)** $aC^tb \to p\mathcal{B}a \vee p^*\mathcal{B}b$,
(Up Dir) *Updirectness* $(\forall i, j)(\exists k)(i \prec k$ and $j \prec k) \Longleftrightarrow$
(up dir) $aC^tb \wedge cC^td \wedge a' + b' + c' + d' = 1 \to a\mathcal{B}a' \vee b\mathcal{B}b' \vee c\mathcal{B}c' \vee d\mathcal{B}d'$,
(Down Dir) *Downdirectness* $(\forall i, j)(\exists k)(k \prec i$ and $k \prec j) \Longleftrightarrow$
(down dir) $aC^tb \wedge cC^td \wedge a' + b' + c' + d' = 1 \to a'\mathcal{B}a \vee b'\mathcal{B}b \vee c'\mathcal{B}c \vee d'\mathcal{B}d$,
(Dens) *Density* $i \prec j \to (\exists k)(i \prec k \wedge k \prec j) \Longleftrightarrow$ **(dens)** $a\mathcal{B}b \to a\mathcal{B}p \vee p^*\mathcal{B}b$,
(Ref) *Reflexivity* $(\forall m)(m \prec m) \Longleftrightarrow$ **(ref)** $aC^tb \to a\mathcal{B}b$,
(Lin) *Linearity* $(\forall m, n)(m \prec n \vee n \prec m) \Longleftrightarrow$ **(lin)** $aC^tb \wedge cC^td \to a\mathcal{B}c \vee d\mathcal{B}b$,
(Tr) *Transitivity* $i \prec j$ *and* $j \prec k \to i \prec k \Longleftrightarrow$ **(tr)** $a\overline{\mathcal{B}}b \to (\exists c)(a\overline{\mathcal{B}}c \wedge c^*\overline{\mathcal{B}}b)$.

Note 1 The correspondence ... $\Longleftrightarrow$ ... means that the condition from the left side is universally true in the time structure $(T, \prec)$ iff the condition from the right side is universally true in **B**. Note that the above listed conditions for time ordering are not independent, for instance (Ref) implies (RS), (LS) and (Dens); (Tr) implies (Up Dir) and (Down Dir). Taking some meaningful subsets of these conditions we obtain various different notions of time order. For instance the subsets $\{(Ref), (Tr), (Lin)\}$ and $\{(RS), (LS), (Tr), (Lin), (Dens)\}$ are typical for classical (reflexive or non-reflexive) time, while the subsets $\{(Ref), (Tr), (UpDir), (DownDir)\}$ or $\{(RS), (LS), (Tr), (UpDir), (DownDir)\}$ are used to characterize some types of relativistic time (for the later see, for instance, [12,18,17,16]). Note that irreflexivity of $\prec$ and linearity for irreflexive $\prec$ in the form $m = n \vee m \prec n \vee n \prec m$ are also good properties of time ordering, but we did not find suitable correspondence conditions for them.

Lemma 2.4 (Correspondence Lemma) *Let* **B** *be a standard dynamic contact algebra with time structure* $(T, \prec)$*, and consider the above listed correspondences except transitivity. They are true under the following conditions: (1) for the implication from left to the right* **B** *is arbitrary, (2) for the implication from right to the left* **B** *is supposed to be rich. (3) for the case of transitivity in both directions* **B** *is supposed to be rich.*

Proof. See Appendix A. □

3 Dynamic contact algebras

Definition 3.1 *By a dynamic contact algebra (DCA for short) we mean any system* $\underline{B} = (B, C^s, C^t, \mathcal{B}) = (B, 0, 1, ., +, ^*, C^s, C^t, \mathcal{B})$ *where* $(B, 0, 1, ., +, ^*)$ *is a non-degenerate Boolean algebra, satisfying the following conditions:*

(i) C^s *is a contact relation on* B *called* space contact,

(ii) C^t *is a contact relation on* B, *called* time contact *satisfying the following additional axiom*

$(C^s \to C^t)$ $aC^sb \to aC^tb$.

(iii) $\mathcal{B}$ *is a precontact relation.*

The elements of B *are called dynamic regions. We will consider DCA-s satisfying some of the eight conditions (rs), (ls),(up dir), (down dir), (ref), (lin), (dens) and (tr) listed in Section 2. Since these axioms determine the properties of the time ordering between moments of time (which will be defined in the next section), they are called shortly* "time axioms". *Since DCA-s are algebraic systems, we adopt for them the standard definitions of subalgebra, homomorphism, isomorphism, isomorphic embedding, etc.*

Note 2 In [23] the name "dynamic contact algebra" (DCA) was used for another, but similar notion. We consider DCA as an integral name for a wider class of algebras formalizing some aspects of an integrated point-free theory of space and time. In this paper, however, it will be used as given by Definition 3.1.

Typical examples of DCA-s are standard dynamic contact algebras introduced in Section 2. In the next section we will show that each DCA is isomorphic to an algebra of such a kind.

Ultrafilter characterizations of the relations C^s, C^t and $\mathcal{B}$. Let $\underline{B}$ be a DCA. We denote by $UF(\underline{B})$ the set of ultrafilters in $\underline{B}$. We define three relations $R^s, R^t, \prec$ between ultrafilters as follows: UR^sV iff $U \times V \subseteq C^s$, UR^tV iff $U \times V \subseteq C^t$, $U \prec V$ iff $U \times V \subseteq \mathcal{B}$. Note that these three definitions are meaningful not only for ultrafilters but also for arbitrary subsets of B.

Lemma 3.2 *Let* C *denote any of the three relations* $C^s, C^t, \mathcal{B}$ *and let* R *be the above defined corresponding relation between filters. Then:*

(i) If F, G *are filters and* FRG, *then there are ultrafilters* U, V *such that* $F \subseteq U$, $G \subseteq V$ *and* URV.

(ii) aCb *iff there exist ultrafilters* U, V *such that* URV, $a \in U$ *and* $b \in V$.

(iii) R^s *and* R^t *are reflexive and symmetric,*

(iv) $R^s \subseteq R^t$

Proof. Conditions (i), (ii) and (iii) are from [8] and (iv) follows directly from axiom $(C^s \to C^t)$. □

Now we list some correspondences between the eight conditions (rs)– (tr) listed in Section 2 and corresponding conditions between ultrafilter relations R^t and $\prec$. They are similar to the corresponding relations (RS)–(Tr). We give the new relations the same short notations but in square brackets - [RS] – [Tr].

[RS] $U_1 R^t U_2 \to (\exists V)(U_1 \prec V \wedge U_2 \prec V) \iff$ **(rs)** $aC^t b \to a\mathcal{B}p \vee b\mathcal{B}p^*$,
[LS] $U_1 R^t U_2 \to (\exists V)(V \prec U_1 \wedge V \prec U_2) \iff$ **(ls)** $aC^t b \to p\mathcal{B}a \vee p^*\mathcal{B}b$,
[Up Dir] $U_1 R^t U_2 \wedge U_3 R^t U_4 \to (\exists V)(U_1 \prec V \wedge U_2 \prec V \wedge U_3 \prec V \wedge U_4 \prec V)$
$\iff$ **(up dir)** $aC^t b \wedge cC^t d \wedge a' + b' + c' + d' = 1 \to a\mathcal{B}a' \vee b\mathcal{B}b' \vee c\mathcal{B}c' \vee d\mathcal{B}d'$,
[Down Dir] $U_1 R^t U_2 \wedge U_3 R^t U_4 \to (\exists V)(V \prec U_1 \wedge V \prec U_2 \wedge V \prec U_3 \wedge V \prec U_4)$
$\iff$ **(down dir)** $aC^t b \wedge cC^t d \wedge a' + b' + c' + d' = 1 \to a'\mathcal{B}a \vee b'\mathcal{B}b \vee c'\mathcal{B}c \vee d'\mathcal{B}d$,
[Dens] $U \prec V \to (\exists W)(U \prec W \wedge W \prec V) \iff$ **(dens)** $a\mathcal{B}b \to a\mathcal{B}p \vee p^*\mathcal{B}b$,
[Ref] $UR^t V \to U \prec V \iff$ **(ref)** $aC^t b \to a\mathcal{B}b$,
[Lin] $U_1 R^t U_2 \wedge V_1 R^t V_2 \to (U_1 \prec V_1 \vee V_2 \prec U_2) \iff$
(lin) $aC^t b \wedge cC^t d \to a\mathcal{B}c \vee d\mathcal{B}b$,
[Tr] $U \prec V \wedge V \prec W \to U \prec W \iff$ **(tr)** $a\overline{\mathcal{B}}b \to (\exists c)(a\overline{\mathcal{B}}c \wedge c^*\overline{\mathcal{B}}b)$.

Lemma 3.3 (Ultrafilter correspondences) *Let $\underline{B} = (B, C^s, C^t, \mathcal{B})$ be a DCA. Then all eight correspondences listed above are true in the following sense: for a given correspondence ... $\iff$ ..., the left side is universally true in the set $UF(\underline{B})$) iff the right side is universally true in $\underline{B}$.*

Proof. See Appendix B. □

4 Representation theory of dynamic contact algebras

In this section we will show that each DCA, with possible extension with some additional time axioms, can be represented as a dynamic model of space with explicit moments of time and time ordering, satisfying some reasonable properties. The representation theorem will be based on a canonical construction of the dynamic model of space associated to the given DCA $\underline{B} = (B, C^s, C^t, \mathcal{B})$. Our strategy is the following. First we define the set $T = T(\underline{B})$ of moments of time and before-after relation $\prec$ in T. The time moments are defined as some sets of ultrafilters in T with a canonical definition of $\prec$ with properties determined by the corresponding time axioms. The next step is to associate to each time moment the corresponding coordinate contact algebra and to define the canonical dynamic model of space and the canonical embedding isomorphism. The final step is to show that it indeed embeds the algebra into the obtained dynamic model of space.

Defining the time structure.

Definition 4.1 *Let $\underline{B} = (B, C^s, C^t, \mathcal{B})$ be a DCA. A set α of ultrafilters in $\underline{B}$ is called a* time moment *if it satisfies the following conditions:*

(tm1) *α is non-empty,*

(tm2) *If $U, V \in \alpha$, then $UR^t V$,*

(tm3) *If $\underline{B}$ satisfies some of the axioms (rs),(ls), (up dir), (down dir) and (dens) then α has at most two different ultrafilters.*

The set of time moments of $\underline{B}$ is denoted by $T(\underline{B})$ or simply by T. Obviously T is nonempty, because, for instance, each singleton set $\{U\}$, where U is an ultrafilter, is a time point (condition **(tm1)** *is obviously satisfied and* **(tm2)** *is satisfied because of the reflexivity of R^t).*

The definition of time ordering $\prec$ *is this: for $\alpha, \beta \in T$*

(to) $\alpha \prec \beta$ *iff* $(\forall U, V \in UF(\underline{B}))(U \in \alpha \wedge V \in \beta \rightarrow U \prec V)$.

The pair $\underline{T}(\underline{B}) = (T(\underline{B}), \prec)$ *(denoted sometimes shortly* $(T, \prec)$*) is called the time structure of* $\underline{B}$.

Note that condition **(tm3)** is taken only by technical reasons for the proofs of the properties of time ordering, and depends on the fact that the correspondents of the axioms mentioned in the condition, namely [RS], [LS], [Up Dir], [Down Dir] and [Dens], are not universal sentences, while the correspondents of the other time axioms are universal sentences.

Lemma 4.2 (Properties of time ordering) *Let* $\underline{B} = (B, C^s, C^t, \mathcal{B})$ *be a dynamic contact algebra and* $(T, \prec)$ *be its time structure. Consider the correspondences* $\ldots \Longleftrightarrow \ldots$ *from Section 2. Then each correspondence is true in the following sense: the condition from the left side of* $\ldots \Longleftrightarrow \ldots$ *is universally true in the time structure* $(T, \prec)$*) iff the right side is universally true in the algebra* $\underline{B}$.

Proof. See Appendix C. □

Defining coordinate contact algebras, the canonical dynamic model of space and the isomorphic embedding. To define coordinate contact algebras we will use the following construction of *factor contact algebra* from sets of ultrafilters in a given contact algebra.

Let α be a non-empty set of ultrafilters in a contact algebra $\underline{B} = (B, C)$. Define the following equivalence relation on B depending on α:

$a \equiv_\alpha b$ iff $(\forall U \in \alpha)(a \in U \leftrightarrow b \in U)$.

It is easy to see that $\equiv_\alpha$ is a congruence on $\underline{B}$ and let $\underline{B}/\alpha$ denote the factor algebra $\underline{B}/\equiv_\alpha$. Denote the Boolean ordering on $\underline{B}/\alpha$ by $\leq_\alpha$. Define the relation $|a|_\alpha C_\alpha |b|_\alpha$ iff there exist ultrafilters $U, V \in \alpha$ such that URV and $a \in U$ and $b \in V$, ($URV \leftrightarrow_{def} U \times V \subseteq C$, see Lemma 3.2).

Lemma 4.3 *(i)* B_α *is a non-degenerated Boolean algebra.*

(ii) $|a|_\alpha \neq |0|_\alpha$ *iff there exists an ultrafilter* $U \in \alpha$ *such that* $a \in U$.

(iii) The relation C_α *is well defined and the pair* (B_α, C_α) *is a contact algebra.*

Proof. Direct verification. □

Definition 4.4 (Coordinate contact algebras, the canonical dynamic model of space and the embedding) *Let* $\underline{B}$ *be a DCA and* α *be a moment of time in* $\underline{B}$. *Define the factor Boolean algebra* $\underline{B}_\alpha$ *by the construction described above and the contact relation*

$|a|_\alpha C_\alpha |b|_\alpha$ *iff* $(\exists U, V \in \alpha)(UR^sV,\ a \in U$ *and* $b \in V)$.

Then by Lemma 4.3 $(\underline{B}_\alpha, C_\alpha)$ *is a contact algebra, called the coordinate contact algebra associated to* α. *The canonical dynamic model of space, denoted here* $\mathbf{B}^{can}$*, and the relations* C^s, C^t *and* $\mathcal{B}$ *in it are defined by the just defined coordinate contact algebras as in Section 2. The embedding* h *is defined coordinatewise as follows: for each* $a \in B$,

$h(a)_\alpha = |a|_\alpha$.

Note 3 The definition of the coordinate contact algebra $\underline{B}_\alpha$ as a factor algebra with respect to the set of ultrafilters of the time moment α is based on the following intuition. If we look at dynamic regions as trajectories of changing regions, then for different a and b we may have that $a_\alpha = b_\alpha$, which is an equivalence relation determined by α. The formal definition of this equivalence is just the relation $\equiv_\alpha$, which determines the coordinate contact algebra $\underline{B}_\alpha$. So the elements of $\underline{B}_\alpha$ are the equivalence classes $|a|_\alpha$.

Lemma 4.5 (Embedding Lemma) *Let $\underline{B}$ be a DCA. Then:*

(i) $aC^s b$ iff there exists $\alpha \in T$ such that $|a|_\alpha C_\alpha |b|_\alpha$ iff $h(a)C^s h(b)$,

(ii) $aC^t b$ iff there exists $\alpha \in T$ such that $|a|_\alpha \neq |0|_\alpha$ and $|b|_\alpha \neq |0|_\alpha$ iff $h(a)C^t h(b)$,

(iii) $a\mathcal{B}b$ iff there exist $\alpha, \beta \in T$ such that $\alpha \prec \beta$ and $|a|_\alpha \neq |0|_\alpha$ and $|b|_\beta \neq |0|_\beta$ iff $h(a)\mathcal{B}h(b)$,

(iv) $a \leq b$ iff for all $\alpha \in T$ $|a|_\alpha \leq_\alpha |b|_\alpha$ iff $h(a) \leq h(b)$,

(v) h preserves Boolean operations.

Proof. See Appendix D. □

Lemma 4.6 *Let $\underline{B}$ be a DCA and $\mathbf{B}^{can}$ be its canonical dynamic model of space. Then for each time axiom Ax from the list $\langle (rs), (ls), \ldots, (tr) \rangle$ the following equivalence is true: Ax holds in $\underline{B}$ iff Ax holds in $\mathbf{B}^{can}$.*

Proof. The proof for all cases for Ax is the same, so we will illustrate it for $Ax = (rs)$. Namely we have the following chain of equivalencies:
The condition (rs) holds in $\underline{B} \Longleftrightarrow$ (by Lemma 4.2) the condition (RS) holds in the time structure $(T, \prec)$ of $\underline{B} \Longleftrightarrow$ (by Lemma 2.4) the condition (rs) holds in $\mathbf{B}^{can}$. □

Theorem 4.7 (Representation Theorem for DCA-s) *Let $\underline{B}$ be a DCA. Then there exists a dynamic model of space $\mathbf{B}$ and an isomorphic embedding h of $\underline{B}$ into $\mathbf{B}$. Moreover, $\underline{B}$ satisfies some of the time axioms iff the same axioms are satisfied in $\mathbf{B}$.*

Proof. The proof is a direct corollary of Lemma 4.5 and Lemma 4.6. □

5 Quantifier-free logics based on dynamic contact algebras

In this section we will present a complete axiomatization of some quantifier-free logics based on the language of dynamic contact algebras. The motivation to consider quantifier-free logics is to obtain decidable fragments. The language contains a denumerable set of variables, called Boolean variables, two symbols for the Boolean constants 0 and 1, symbols $+, ., ^*$ for the Boolean operations, equality $=$ and three two place predicate symbols $C^s, C^t, \mathcal{B}$. Terms of this language (called Boolean terms) are build in the standard way from Boolean variables and constants. Atomic formulas are of the form $a = b$, $aC^s b$, $aC^t b$ and $a\mathcal{B}b$, where a, b are Boolean terms. Formulas are build from the atomic formulas by means of the propositional operations $\neg, \wedge, \vee, \Rightarrow, \Leftrightarrow$ in a standard way.

The intended semantics of the introduced language is in various classes of DCA-s. Let $\underline{B}$ be a DCA, a valuation v in $\underline{B}$ is a mapping from the set of Boolean variables and constants to B extended in a standard way to the set of all terms. The pair (B, v) is called a model. The truth of a formula α in the model (B, v), denoted by $(B, v) \models \alpha$ is defined inductively as follows:

$(B, v) \models a = b$ iff $v(a) = v(b)$,

$(B, v) \models aC^s b$ iff $v(a)C^s v(b)$, and similarly for $aC^t b$ and $a\mathcal{B}b$.

$(B, v) \models \neg\alpha$ iff $(B, v) \not\models \alpha$,

$(B, v) \models \alpha \wedge \beta$ iff $(B, v) \models \alpha$ and $(B, v) \models \beta$,

and similarly for the other propositional connectives.

We say that α is true in the algebra $\underline{B}$ if it is true in all models over $\underline{B}$; α is true in a class Σ of DCA-s if it is true in all algebras from Σ. A formula α has a model in a given class of DCA-s if there is a model in this class in which α is true. A set A of formulas has a model in a given class of DCA-s if there is a model in that class in which all members of A are true.

We denote by $\mathbb{L}_{all}$ the logic corresponding to the class of all DCA-s. Since all axioms of DCA are universal formulas, axiomatization of $\mathbb{L}_{all}$ can be done quantifier-free on the base of propositional logic as follows:

Axiom system for $\mathbb{L}_{all}$.

(I) Axiom schemes for classical propositional logic

Here one can use any Hilbert-style axiomatization of classical propositional logic with axiom schemes with metavariables ranging in the set of formulas.

(II) Axiom schemes for Boolean algebra with the axioms of equality.

Since Boolean algebras can be axiomatized by universal formulas in a first-order logic with equality, one can take any such set of axiom schemes plus the axioms of equality, all written in a quantifier-free form.

(III) Specific axioms for DCA

Since the axioms of DCA are all universal sentences, we rewrite them as formulas of our language

Rules of inference: The only rule is Modus Ponens (MP) $\frac{\alpha, \alpha \Rightarrow \beta}{\beta}$

Since all time axioms except the axiom (tr) are also universal formulas, extensions of $\mathbb{L}_{all}$ with such axioms can be done just by adding them as axiom schemes to the above axiom system. Since the axiom of transitivity (tr) is not a universal formula, we can not add it to the axiom system of $\mathbb{L}_{all}$, but instead we can add the following additional rule of inference, which in a sense imitates the corresponding axiom:

The rule of transitivity TR: $\dfrac{\alpha \Rightarrow (a\mathcal{B}p \vee p^*\mathcal{B}b)}{\alpha \Rightarrow a\mathcal{B}b}$, where p is a Boolean variable that does not occur in a, b, and α.

If Ax is a set of time axioms then $\mathbb{L}_{Ax}$ will denote the extensions of $\mathbb{L}_{all}$ with the axiom schemes from Ax, where for the case of transitivity axiom (tr) we consider the rule of transitivity **TR**.

Let us note that the rule of transitivity **TR** was studied in [1] under the name of **rule of normality (NOR)**.

The following theorem is the main result in this section:

Theorem 5.1 (Completeness theorem for $\mathbb{L}_{Ax}$) *The logic $\mathbb{L}_{Ax}$ is strongly sound and complete in the class of all dynamic contact algebras satisfying the axioms Ax.*

Proof. See appendix E. □

Let us note that due to the representation Theorem 4.7 for dynamic contact algebras we obtain as a corollary from Theorem 5.1 also completeness of $\mathbb{L}_{Ax}$ with respect to the corresponding class of standard dynamic contact algebras.

Theorem 5.2 (Decidability of some $\mathbb{L}_{Ax}$) *Let Ax be a subset of time axioms not containing the axiom (tr). Then the logic $\mathbb{L}_{Ax}$ is decidable.*

Proof. We shall show that the logic has finite model property. Let α be a formula which is not a theorem. Then by the completeness theorem there is a model (B, v) based on some dynamic contact algebra $\underline{B}$ from the class of algebras in which $\mathbb{L}_{Ax}$ is complete, such that $(B, v) \not\models \alpha$. Let $\underline{B}_{fin}$ be the finite Boolean subalgebra of $\underline{B}$ generated by the set $\{v(b_1), \ldots, v(b_n)\}$, where $b_1, \ldots, b_n$ are all variables of α. Since all axioms from Ax are universal formulas, then they are satisfied in $\underline{B}_{fin}$ and hence it is in the class of algebras in which the logic is complete. So α is falsified in a finite algebra with the size $\leq 2^{2^n}$ where n is the number of the variables of α. □

Let us note that we do not know if decidability is preserved by adding to the logic the rule of transitivity. We formulate also as an open problem the complexity of the decision problems related to discussed logics.

6 Concluding remarks

Related works. Some other works quite similar to our approach are [9,10]. They are point-free with respect to space points but not with respect to time points in the sense that the set of time points is explicitly given in the axiomatization. A point-free version of *dynamic mereology* based on some natural stable and unstable mereological relations is [14]. The survey paper [13] is devoted to various combinations of spatial and temporal logics, concerning mainly expressivity and complexity of formalisms, but not point-free representations (see also [2,11]). Modal logics for Minkowski space-time, based on different ideas, are considered in [12,18,17,16].

Discussion and open problems. We want to note here that this "second attempt" to present an integrated theory of space and time in an Whiteheadian manner shows that the structure of time can also be characterized in a point-free way considering as primitives neither time points, nor their intended internal ordering structure. The three primitive spatio-temporal relations between dynamic regions – space contact C^s, time contact C^t and precedence $\mathcal{B}$ studied in the present paper form in a sense the minimal set of primitives which guarantee the definitions of space points and the coordinate contact algebras and the set time points and the corresponding time structure $(T, \prec)$. There are, however, many other spatio-temporal relations which are not considered in this paper. For instance we do not consider stable contact $C^{\forall}$, considered in [23]. The reason is that canonical constructions used in the representation theory in [23] are not compatible with the canonical constructions used in the present paper and it is an open problem to characterize $C^{\forall}$ in the context of the primitives C^t and $\mathcal{B}$. So we formulate as an open problem the extension of the language of dynamic contact algebras with other sensible spatio-temporal relations between dynamic regions depending on time order. For instance we may define in the standard model the following one-place predicate – *"a has only one period of life"* with the following formal definition: there exist two moments of time $m \neq n$ and $m \prec n$ such that for all k in the set $[m, n] =_{def} \{k \in T : k = m \vee k = n \vee m \prec k \prec n\}$, a exists at k (formally $a_k \neq 0_k$) and for all $k \notin [m, n]$, a does not exist". For instance living organisms have only one period of life in the above sense. Another open problem is to ensure coordinate contact algebras to determine some good classes of topological spaces: connected spaces, regular spaces, Hausdorff spaces, Euclidean spaces.

Let us mention that the time contact C^t corresponds to the natural notion of *simultaneity* or *contemporaneity*. It can be generalized for any finite set A of regions: $CON(A)$ (the members of A are contemporaries) iff there exists $m \in T$ such that for all $a \in A$, $a_m \neq 0_m$ (a_m exists at the moment m). Such a polyadic version is considered also by Whitehead and we plan to axiomatize this generalization in an extended version of this paper. Let us note, however, that our notion of contemporaneity, which is based on its standard use in the ordinary language, differs from the meaning used by Whitehead, who considered it as a kind of *"causal independency"* (introduced rather informally). Causal independency in Whitehead is related to some other relations named *"causal future"* and *"causal past"*, which are influenced by relativity theory (see [24], part IV). The exact definitions of similar causal relations, considered as relations between points in Minkowski space, are studied in some modal logics of space-time [12,18,17,16]. It will be nice to have good formal analogs of such relations considered as relations between dynamic regions.

We plan to apply a translation of the quantifier-free logics studied in this paper in suitable modal logics with universal modality in order to see if the rule of transitivity **TR** can be eliminated and to use this fact for the study of decidability and complexity problems related to these logics.

Thanks. The paper is sponsored by the contract with the Bulgarian National Science Fund, project name: "Theories of Space and Time: algebraic, topological and logical approaches". Thanks are due to the anonymous referees for their useful suggestions and also to Philippe Balbiani and Tinko Tinchev for their fruitful comments and discussion on an earlier version of this paper. I am indebted also to Veselin Petrov, a Bulgarian philosopher and mathematician, who introduced me to Whitehead's philosophy.

References

[1] Balbiani, P., T. Tinchev and D. Vakarelov, *Modal logics for region-based theory of space*, Fundamenta Informaticae **81** (2007), pp. 29–82.

[2] Bennett, B., A.Cohn, F. Wolter and M. Zakharyaschev, *Multidimensional modal logic as a framework for spatio-temporal reasoning*, Applied Intelligence **17** (2002), pp. 239–251.

[3] Bennett, B. and I. Düntsch, *Axioms, Algebras and Topology*, in: M. Aielo, I. Pratt-Hartmann and J. van Benthem, editors, *Handbook of Spatial Logics*, Springer, 2007 pp. 99–160.

[4] Cohn, A. and J. Renz, *Qualitative spatial representation and reasoning*, in: F. van Hermelen, V. Lifschitz and B. Porter, editors, *Handbook of Knowledge Representation*, Elsevier, 2008 pp. 551–596.

[5] de Laguna, T., *Point, line and surface as sets of solids*, The Journal of Philosophy (1922), pp. 449–461.

[6] de Vries, H., "Compact Spaces and Compactifications," Van Gorcum, 1962.

[7] Dimov, G. and D. Vakarelov, *Contact algebras and region-based theory of space. A proximity approach. I and II*, Fundamenta Informaticae **74** (2006), pp. 209–249, 251–282.

[8] Düntsch, I. and D. Vakarelov, *Region-based theory of discrete spaces*, Annals of Mathematics and Artificial Intelligence **49** (2007), pp. 5–14.

[9] Düntsch, I. and M. Winter, *A representation theorem for Boolean contact algebras*, Theoretical Computer Science (B) **347** (2005), pp. 498–512.

[10] Düntsch, I. and M. Winter, *Moving spaces*, in: S. Demri and C. S. Jensen, editors, *Proceedings of 15 International Symposium on Temporal Representation and Reasoning (TIME-2008)*, IEEE Computer Society, 2008, pp. 59–763.

[11] Gerevini, A. and B. Nebel, *Qualitative spatio-temporal reasoning with RCC-8 and Allen's Interval Calculus: Computational complexity*, in: *ECAI - 2002*, 2002, pp. 312–316.

[12] Goldblatt, R., *Diodorean Modality in Minkowski Space-time*, Studia Logica (1980), pp. 219–236.

[13] Kontchakov, R., A. Kurucz, F. Wolter and M. Zakharyaschev, *Spatial logic + temporal logic = ?*, in: M. Aielo, I. Pratt-Hartmann and J. van Benthem, editors, *Handbook of Spatial Logics*, Springer, 2007 pp. 497–564.

[14] Nenchev, V., *Logics for stable and unstable mereological relations*, Central European Journal of Mathematics **9** (2011), pp. 1354–1379.

[15] Pratt-Hartmann, I., *First-order region-based theory of space*, in: M. Aielo, I. Pratt-Hartmann and J. van Benthem, editors, *Handbook of Spatial Logics*, Springer, 2007 pp. 13–97.

[16] Shapirowski, I. and V. Shehtman, *Modal logics for regions and Minkowski Spacetime*, Journal of Logic and Computation **15**, pp. 559–574.

[17] Shapirowski, I. and V. Shehtman, *Chronological Modality in Minkowski Spacetime*, in: P. Balbiani, N.-Y. Suzuki, F. Wolter and M. Zakharyaschev, editors, *Advances in Modal Logic, vol. 4*, King's College Publications, 2003 pp. 437–459.

[18] Shehtman, V., *Modal Logics for Domains in the Real Plane*, Studia Logica **42** (1983), pp. 63–80.

[19] Simons, P., "Parts: A Study in Ontology," Oxford University Press, 1987.

[20] Stell, J. G., *Boolean Connection Algebras: A new Approach to the Boolean Connection Calculus*, Artificial Intelligence **122** (2000), pp. 111–136.

[21] Tarski, A., *Foundations of geometry of solids (translation of summary of address given by Tarski to the First Polish Mathematical Congress, 1927)*, in: *Logic, Semantics, Metamathematics*, Clarendon Press, 1956 pp. 24–29.

[22] Vakarelov, D., *Region-based theory of space: Algebras of regions, representation theory and logics*, in: D. Gabbay, M. Zakharyaschev and S. S. Goncharov, editors, *Mathematical Problems from Applied Logic II. Logics for the XXI-st Century*, Springer pp. 267–348.

[23] Vakarelov, D., *Dynamic mereotopology: A point-free theory of changing regions. I. Stable and unstable mereotopological relations*, Fundamenta Informaticae **100** (2010), pp. 159–180.

[24] Whitehead, A. N., "The Organization of Thought," London, William and Norgate, 1917.

[25] Whitehead, A. N., "Science and the Modern World," New York: MacMillan, 1925.
[26] Whitehead, A. N., "Process and Reality," New York: MacMillan, 1929.

Appendix A: proof of Lemma 2.4 (Correspondence Lemma)

Proof. We consider first the case of transitivity. Suppose that **B** is rich.

(**Tr**) $\Longrightarrow$ (**tr**) Suppose that $\prec$ is transitive and let $a\overline{\mathcal{B}}b$. We have to find c such that ($a\overline{\mathcal{B}}c$ and $c^*\overline{\mathcal{B}}b$). Define c coordinatewise as follows:

$$c_k = \begin{cases} 1_k, & \text{if } (\exists j)(k \prec j \wedge b_j \neq 0_j) \\ 0_k, & \text{if } (\exists i)(i \prec k \wedge a_i \neq 0_i) \end{cases}$$

The definition of c_k is correct, because if the two conditions are satisfied simultaneously then by transitivity we obtain $i \prec j$ which together with $a_i \neq 0_i$ and $b_j \neq 0_j$ imply $a\mathcal{B}b$ - a contradiction. Obviously by richness $c \in \mathbf{B}$.

To show $a\overline{\mathcal{B}}c$ suppose the contrary. Then there exist $i, k \in T$ such that $i \prec k$, $a_i \neq 0_i$ and $c_k \neq 0_k$. From here and by the definition of c_k we get $c_k = 1_k$. By the definition of c_k, this implies that there exists $j \in T$ such that $k \prec j$ and $b_j \neq 0_j$. From $i \prec k$ and $k \prec j$ we obtain (by transitivity) $i \prec j$, which together with $a_i \neq 0_i$ and $b_j \neq 0_j$ imply $a\mathcal{B}b$ - a contradiction with the assumption $a\overline{\mathcal{B}}b$.

To show $c^*\overline{\mathcal{B}}b$ suppose again the contrary. Then there exists $k, j \in T$ such that $k \prec j$, $c^*_k \neq 0_k$ and $b_j \neq 0_j$. From here we get $c^*_k = 1_k$ and $c_k = 0_k$. This implies that there exists $i \in T$ such that $i \prec k$ and $a_i \neq 0_i$. Conditions $i \prec k$ and $k \prec j$ imply $i \prec j$ which together with $a_i \neq 0_i$ and $b_j \neq 0_j$ imply $a\mathcal{B}b$ - again a contradiction with the assumption $a\overline{\mathcal{B}}b$.

(**Tr**) $\Longleftarrow$ (**tr**) Suppose that transitivity does not hold for $\prec$, then for some i', j', k' we have $i' \prec j'$, $j' \prec k'$ but $i' \not\prec k'$. Define a and b coordinatewise as follows:

$$a_i = \begin{cases} 1_i, & \text{if } i = i' \\ 0_i, & \text{if } i \neq i' \end{cases}, \quad b_k = \begin{cases} 1_k, & \text{if } k = k' \\ 0_k, & \text{if } k \neq k' \end{cases}.$$

By richness a and b belong to **B**. We will show that $a\overline{\mathcal{B}}b \rightarrow (\exists c)(a\overline{\mathcal{B}}c \wedge c^*\overline{\mathcal{B}}b)$ does not hold, i.e. $a\overline{\mathcal{B}}b$ and $\neg(\exists c)(a\overline{\mathcal{B}}c \wedge c^*\overline{\mathcal{B}}b)$.

First we show $a\overline{\mathcal{B}}b$. Suppose the contrary. Then for some $i, k \in T$ we have $i \prec k$, $a_i \neq 0_i$ and $b_k \neq 0_k$. This implies $i = i'$, $k = k'$ and hence $i' \prec k'$ - a contradiction with the assumption $i' \not\prec k'$.

Now we will show $\neg(\exists c)(a\overline{\mathcal{B}}c \wedge c^*\overline{\mathcal{B}}b)$. Suppose the contrary, i.e. that there exists c such that $a\overline{\mathcal{B}}c$ and $c^*\overline{\mathcal{B}}b$.

Case 1. $c_{j'} \neq 0_{j'}$. We have $i' \prec j'$, $a_{i'} = 1_{i'} \neq 0_{i'}$. This implies $a\mathcal{B}c$ - a contradiction with the assumption $a\overline{\mathcal{B}}c$.

Case 2. $c_{j'} = 0_{j'}$. Then $c^*_{j'} = 1_{j'} \neq 0_{j'}$. We have $j' \prec k'$ and $b_{k'} = 1_{k'} \neq 0_{k'}$. This implies $c^*\mathcal{B}b$ - a contradiction with the assumption $c^*\overline{\mathcal{B}}b$.

(**RS**) $\Longrightarrow$ (**rs**) Suppose that $\prec$ is right-serial and let aC^tb. Then for some $m \in T$ we have $a_m \neq 0_m$ and $b_m \neq 0_m$. By right-seriality there exists $n \in T$ such that $m \prec n$. Let p be an arbitrary region.

Case $p_n \neq 0_n$. This implies $a\mathcal{B}p$.

Case $p_n = 0_n$. This implies $p_n^* \neq 0_n$ and hence $b\mathcal{B}p^*$.

So, in both cases we obtain $aC^tb \to a\mathcal{B}p \vee b\mathcal{B}p^*$.

(RS) $\Longleftarrow$ **(rs)** In this case we will reason by contraposition. Suppose that $\prec$ is not right-serial so $(\exists m')(\forall n)(m' \not\prec n)$. Under this assumption we will proceed to show that $aC^tb \to a\mathcal{B}p \vee b\mathcal{B}p^*$ does not hold, so for some a, b, p aC^tb, $a\overline{\mathcal{B}}p$ and $b\overline{\mathcal{B}}p^*$. Let $p = 1$ and define a, b coordinatewise as follows:

$$a_m = b_m = \begin{cases} 1_m, & \text{if } m = m' \\ 0_m, & \text{if } m \neq m' \end{cases},$$

By richness a, b are in $\mathbf{B}$.

First we will show aC^tb. Observe that we have $a_{m'} = b_{m'} = 1_{m'} \neq 0_{m'}$ which implies aC^tb.

To show $a\overline{\mathcal{B}}p$ suppose the contrary, i.e. that $a\mathcal{B}p$, i.e. $a\mathcal{B}1$ holds. Then for some $m \prec n$ we have $a_m \neq 0_m$ and $1_n \neq 0_n$. By the definition of a we obtain $a_m = 1_m$ and hence $m = m'$, which implies $m' \prec n$. This contradicts the assumption $(\forall n)(m' \not\prec n)$.

Since $p = 1$, then $p^* = 0$ and hence the condition $b\overline{\mathcal{B}}p^*$ trivially holds.

The proofs for the other cases of the lemma are similar. □

Appendix B: proof of Lemma 3.3 (Ultrafilter correspondences)

Proof. For the proof of Lemma 3.3 we will need the following facts about filters in Boolean algebra and one lemma which we formulated without proofs:
Facts. *If F, G are filters then $F \oplus G =_{def} \{c : (\exists a \in F)(\exists b \in G)(a.b \leq c)\}$ is the smallest filter containing F and G. $F \oplus G$ is not a proper filter (i.e. $0 \in F \oplus G$) iff there exists p such that $p^* \in F$ and $p \in G$. The operation $\oplus$ is associative and commutative. Every proper filter can be extended into an ultrafilter.*

Lemma 6.1 *Let U, V be filters and let $\mathbf{F}_I(U) =_{def} \{b : (\exists a \in U)a\overline{C}b^*\}$ and $\mathbf{F}_{II}(V) =_{def} \{a : (\exists b \in V)a^*\overline{C}b\}$. Then:*

(i) $\mathbf{F}_I(U)$ and $\mathbf{F}_{II}(V)$ are filters,

(ii) If U is a filter and V is an ultrafilter then: $U \times V \leq C$ iff $\mathbf{F}_I(U) \subseteq V$,

(iii If U is an ultrafilter and V is a filter then: $U \times V \leq C$ iff $\mathbf{F}_{II}(V) \subseteq U$.

[RS] $\Rightarrow$ **(rs)**. Suppose **[RS]** holds and let aC^tb. Then by Lemma 3.2 There are ultrafilters U_1, U_2 such that $U_1R^tU_2$, $a \in U_1$ and $b \in U_2$. By [RS] there exists an ultrafilter V such that $U_1 \prec V$ and $U_2 \prec V$. Let p be an arbitrary region.

Case 1: $p \in V$. Then by $U_1 \prec V$ and $a \in U_1$ we conclude by Lemma 3.2 that $a\mathcal{B}p$.

Case 2: $p^* \in V$. As in Case 1, this implies $b\mathcal{B}p^*$.

[RS] $\Leftarrow$ **(rs)**. Suppose **(rs)** holds and let $U_1R^tU_2$. We shall show that there exists an ultrafilter V such that $U_1 \prec V$ and $U_2 \prec V$.

To prove this we shall show first that $\mathbf{F}_I(U_1) \oplus \mathbf{F}_{II}(U_2)$ is a proper filter (see Lemma 6.1 for notations). Suppose that this is not the case. Then there exists p such that $p^* \in \mathbf{F}_I(U_1)$ and $p \in \mathbf{F}_{II}(U_2)$. This implies that there exists $a \in U_1$ such that $a\overline{\mathcal{B}}p$ and that there exists $b \in U_2$ such that $b\overline{\mathcal{B}}p^*$. Since $U_1 R^t U_2$, this implies $aC^t b$, which by $a\overline{\mathcal{B}}p$ and $b\overline{\mathcal{B}}p^*$ shows that **(rs)** does not hold - a contradiction. Thus $\mathbf{F}_I(U_1) \oplus \mathbf{F}_{II}(U_2)$ is a proper filter. Then it can be extended into an ultrafilter V. Consequently $\mathbf{F}_I(U_1) \subseteq V$ (which implies $U_1 \prec V$) and $\mathbf{F}_{II}(U_2) \subseteq V$ (which implies $U_2 \prec V$), which ends the proof of this case.

The other cases of the lemma can be proved in a similar way making use of Lemma 6.1, Lemma 3.2 and and the above mentioned **Facts** many times. Since (tr) is not an universal sentence, as an illustration we will demonstrate the proof of the implication **[Tr]**$\Rightarrow$**(tr)**. Another proof of this implication can be found in [8].

[Tr]$\Rightarrow$**(tr)**. Suppose that **[Tr]** holds and let $a\overline{\mathcal{B}}b$. Suppose that there is no c such that $a\overline{\mathcal{B}}c$ and $c^*\overline{\mathcal{B}}b$. We will proceed to obtain a contradiction as follows. First we will show that there are ultrafiltres U, V, W, such that $a \in U$, $b \in W$, $U \prec V$ and $V \prec W$. Then by transitivity we get $U \prec W$. But $a \in U$, $b \in W$ and $U \prec W$ implies $a\mathcal{B}b$ - the desired contradiction.

Now to realize the above strategy define $[a) =_{def} \{a' : a \leq a'\}$, $[b) =_{def} \{b' : b \leq b'\}$. $[a)$ is a filter containing a and $[b)$ is a filter containing b. We shall show that the filter $\mathbf{F}_I([a)) \oplus \mathbf{F}_{II}([b))$ is a proper filter. Otherwise there are $p^* \in \mathbf{F}_I([a))$, $p \in \mathbf{F}_{II}([b))$, which implies (see Lemma 6.1) that there exists $a' \in [a)$ (so $a \leq a'$) such that $a'\overline{\mathcal{B}}p$ and that there exists $b' \in [b)$ (so $b \leq b'$) such that $p^*\overline{\mathcal{B}}b'$. By the monotonicity of $\mathcal{B}$ we obtain $a\overline{\mathcal{B}}p$ and $p^*\overline{\mathcal{B}}b$. This contradicts the assumption that there is no c such that $a\overline{\mathcal{B}}c$ and $c^*\overline{\mathcal{B}}b$ – simply take $c = p$. Consequently $\mathbf{F}_I([a)) \oplus \mathbf{F}_{II}([b))$ is a proper filter. Then it can be extended into an ultrafilter V. Hence we get $\mathbf{F}_I([a)) \subseteq V$ and $\mathbf{F}_{II}([b)) \subseteq V$. This, by 6.1 implies $[b) \times V \subseteq \mathcal{B}$ and $[a) \times V \subseteq \mathcal{B}$. Then applying Lemma 3.2 we can extend $[a)$ into an ultrafilter U such that $U \times V \subseteq \mathcal{B}$,(so $U \prec V$), and similarly to extend $[b)$ into an ultrafilter W such that $V \times W \subseteq \mathcal{B}$, (so $V \prec W$). Obviously $a \in U$ and $b \in W$. Thus we have obtained $U \prec V$, $V \prec W$, $a \in U$ and $b \in W$ - the strategy is fulfilled, which ends the proof of this case. □

Appendix C: proof of Lemma 4.2 (Properties of time ordering)

Proof. The proofs of all cases are similar, so we will demonstrate only two examples: $(Tr) \iff (tr)$ and $(RS) \iff (rs)$.

$(Tr) \implies (tr)$. Suppose the contrary, i.e. $\prec$ is a transitive relation on T and that (tr) does not hold in $\underline{B}$. Then by Lemma 3.3 there are ultrafilters U, V, W such that $U \prec V$, $V \prec W$ but $U \not\prec W$. Let $\alpha = \{U\}$, $\beta = \{V\}$ and $\gamma = \{W\}$. Since R^t is a reflexive relation, then α, β, γ are time points and hence are elements of T. By the definition of $\prec$ we get $\alpha \prec \beta$, $\beta \prec \gamma$, but not $\alpha \prec \gamma$, which contradicts the transitivity of $\prec$ in T.

$(Tr) \Longleftarrow (tr)$. Suppose that (tr) holds in $\underline{B}$. Then by Lemma 3.3 the relation $\prec$ defined in the set of ultrafilters of $\underline{B}$ is a transitive relation. We will show that $\prec$, as a relation between time moments, is a transitive relation on T. Suppose α, β, γ are from T, $\alpha \prec \beta$, $\beta \prec \gamma$ but $\alpha \not\prec \gamma$. Then there are ultrafilters $U \in \alpha$ and $W \in \gamma$ such that $U \not\prec W$. Since β is non-empty, let $V \in \beta$. Then we obtain $U \prec V$, $V \prec W$, which together with $U \not\prec W$ contradicts the transitivity of $\prec$ in the set of ultrafilters.
$(RS) \Longrightarrow (rs)$. Suppose that the condition **(RS)** holds in T. We will show that $\underline{B}$ satisfies **(rs)**. To this end suppose that aC^tb holds in $\underline{B}$ and proceed to show that ether $a\mathcal{B}p$ holds or $b\mathcal{B}p^*$ holds.

From aC^tb it follows by Lemma 3.2 that there are ultrafilters U, V such that UR^tV, $a \in U$ and $b \in V$. Let $\alpha = \{U, V\}$. Since UR^tV, we obtain that $\alpha \in T$. Then by the condition **(RS)** there exists $\beta \in T$ such that $\alpha \prec \beta$. Let $W \in \beta$. Then by the definition of $\prec$ in T we have $U \prec W$ and $V \prec W$. Let p be an arbitrary region in $\underline{B}$.
Case 1: $p \in W$. We have $a \in U$ and $U \prec W$. This by Lemma 3.2 implies $a\mathcal{B}p$.
Case 2: $p^* \in W$. We have $b \in V$ and $V \prec W$. This by Lemma 3.2 implies $b\mathcal{B}p^*$.
$(RS) \Longleftarrow (rs)$. Suppose that (rs) holds in $\underline{B}$, then by Lemma 3.3 [**RS**] holds in the set of ultrafilters in $\underline{B}$. We shall show that **(RS)** holds in T. Let $\alpha \in T$. Now, since we are working with the condition (rs) the condition **(tm3)** is fulfilled and hence α is in the form $\alpha = \{U_1, U_2\}$. Then by **(tm2)** $U_1R^tU_2$ and by [**RS**] there is an ultrafilter V such that $U_1 \prec V$ and $U_2 \prec V$. Define $\beta = \{V\}$. By the reflexivity of R^t and **(tm2)** we have $\beta \in T$. By the definition of $\prec$ in T we obtain $\alpha \prec \beta$, so **(RS)** holds in T. □

Appendix D: proof of Lemma 4.5 (Embedding Lemma)

Proof. (i)($\Rightarrow$). Suppose aC^sb. By Lemma 3.2 there exist ultrafilters U, V such that UC^sV, $a \in U$ and $b \in V$. Define $\alpha = \{U, V\}$. By Lemma 3.2 we have also UC^tV, so $\alpha \in T$ and hence $|a|_\alpha C_\alpha |b|_\alpha$, which is equivalent to $h(a)C^sh(b)$.

($\Leftarrow$). Suppose $h(a)C^sh(b)$. Then for some $\alpha \in T$ we have $|a|_\alpha C_\alpha |b|_\alpha$ and by the definition of C_α there are ultrafilters $U, V \in \alpha$ such that UR^sV, $a \in U$ and $b \in V$. Then by Lemma 3.2 we obtain aC^sb.

(ii)($\Rightarrow$). Suppose aC^tb. By Lemma 3.2 aC^tb implies that there exist ultrafilters U, V such that UC^tV, $a \in U$ and $b \in V$. Define $\alpha = \{U, V\}$. Since UC^tV we get $\alpha \in T$. By Lemma 4.3 $a \in U \in \alpha$ and $b \in V \in \alpha$ is equivalent to $|a|_\alpha \neq |0|_\alpha$ and $|b|_\alpha \neq |0|_\alpha$, which is equivalent to $h(a)C^th(b)$. Thus aC^tb implies $h(a)C^th(b)$.

($\Leftarrow$). Suppose $h(a)C^th(b)$. This implies that for some $\alpha \in T$, $|a|_\alpha \neq |0|_\alpha$ and $|b|_\alpha \neq |0|_\alpha$. Then by Lemma 4.3 there exist $U, V \in \alpha$ such that $a \in U$ and $b \in V$. Condition $U, V \in \alpha \in T$ implies by **(tm2)** that UR^tV, which together with $a \in U$ and $b \in V$ implies by Lemma 3.2 that aC^tb.

(iii) ($\Rightarrow$). Suppose $a\mathcal{B}b$. By Lemma 3.2 $a\mathcal{B}b$ implies that there exist ultrafilters U, V such that $U \prec V$, $a \in U$ and $b \in V$. Define $\alpha = \{U\}$ and $\beta = \{V\}$, and by reflexivity of R^t and **(tm2)** we obtain $\alpha, \beta \in T$. Since $U \prec V$ we get $\alpha \prec \beta$. By Lemma 4.3 $a \in U \in \alpha$ and $b \in V \in \beta$ is equivalent to $|a|_\alpha \neq |0|_\alpha$ and $|b|_\beta \neq |0|_\beta$, which together with $\alpha \prec \beta$ implies $h(a)\mathcal{B}h(b)$. Thus $a\mathcal{B}b$ implies $h(a)\mathcal{B}h(b)$.

($\Leftarrow$). Suppose $h(a)\mathcal{B}h(b)$. This implies that there are $\alpha, \beta inT$ such that $\alpha \prec \beta$, $|a|_\alpha \neq |0|_\alpha$ and $|b|_\beta \neq |0|_\beta$. This by Lemma 4.3 implies that there are ultrafilters $U \in \alpha$ and $V \in \beta$ such that $a \in U$ and $b \in V$. Since $\alpha \prec \beta$, this implies by the definition of $\prec$ in T that $U \prec V$. Conditions $a \in U$, $b \in V$ and $U \prec V$ imply by Lemma refultrafilter relations that $a\mathcal{B}b$. Thus $h(a)\mathcal{B}h(b)$ implies $a\mathcal{B}b$.

(iv). In this case we will reason by contraposition: $a \not\leq b$ iff $a.b^* \neq 0$ iff $(a.b^*)C^s(a.b^*)$ iff (by (i)) there exists $\alpha \in T$ such that $(|a|_\alpha.|b|^*_\alpha)C_\alpha(|a|_\alpha.|b|^*_\alpha)$ iff there exists $\alpha \in T$ such that $|a|_\alpha.|b|^*_\alpha \neq |0|_\alpha$ iff there exists $\alpha \in T$ such that $|a|_\alpha \not\leq_\alpha |b|_\alpha$ iff $h(a) \not\leq h(b)$.

(v) The condition follows from the fact that $a \mapsto |a|_\alpha$ is a homomorphism with respect to Boolean operations. □

Appendix E: Proof of Theorem 5.1

Proof. The soundness part of the theorem is easy. For the completeness part we have to show that each consistent set A of formulas has a model. For the proof we will use a kind of canonical model construction. This construction is a variant of the Henkin proof of the completeness theorem for the first-order logic adapted for the logics with additional rules like the rule of transitivity **TR**. This construction is described in [1] Sec. 7 (see also [22] Sec. 3.3), so we refer the reader to consult for the details the above references. The main idea is shortly the following.

Each consistent set A can be extended into a maximal consistent set Γ with some special properties depending on the rules of the logic:

(1) Γ contains all theorems of the logic and is closed under the rule modus ponens,

(2) If the conclusion $\alpha \Rightarrow a\mathcal{B}b$ of the rule **TR** does not belong to Γ then the premise $\alpha \Rightarrow (a\mathcal{B}p \vee p^*\mathcal{B}b)$ also does not belong to Γ for some variable p.

Then, using Γ, one can construct in a canonical way a dynamic contact algebra $\underline{B}$ as follows: define in the set of Boolean terms $a \equiv b$ iff $a = b \in \Gamma$. It can be proved that this is a congruence relation with respect to the Boolean operations which makes possible to define a Boolean algebra over the classes $|a|$ modulo this congruence. We define $|a|C^s|b|$ iff $aC^sb \in \Gamma$ and similarly for the other relations C^t and $\mathcal{B}$. The axioms of dynamic contact algebra guarantee that $\underline{B}$ is a dynamic contact algebra. Moreover the above properties of Γ and the additional axioms and the rule **TR** of the logic guarantee that the obtained dynamic contact algebra satisfies all axioms of the set Ax.

By means of Γ one can define a canonical valuation v in $\underline{B}$ as follows: $v(p) = |p|$ and to prove that $(B, v) \models \alpha$ iff $\alpha \in \Gamma$. Then this shows that (B, v) is a model of Γ and hence a model of A. □

Not All Those Who Wander Are Lost: Dynamic Epistemic Reasoning in Navigation

Yanjing Wang Yanjun Li

Department of Philosophy
Peking University
{y.wang, lyj2010}@pku.edu.cn

Abstract

In everyday life, people get lost even when they have the map: they simply may not know where they are in the map. However, when moving forward they may have new observations which can help to locate themselves by reasoning. In this paper, we propose and develop a semantic-driven dynamic epistemic framework to handle epistemic reasoning in such navigation scenarios. Our framework can be viewed as a careful blend of dynamic epistemic logic and epistemic temporal logic, thus enjoying features from both frameworks. We made an in-depth study on many model theoretical aspects of the proposed framework and provide a complete axiomatization.

Keywords: dynamic epistemic logic, epistemic temporal logic, navigation, planning

1 Introduction

1.1 Motivation

Have you ever been lost with a map? Almost everyone had such an experience as a tourist in an unfamiliar city: even when you have the map of the city, it is sometimes still hard to figure out where you are exactly and how to reach your destination. There are typical cases when you just cannot find the street name, or you are on a long long street with a lot of turns (welcome to Amsterdam!). In such scenarios, a little bit of wandering and reasoning may help:

> *This circular street along the canal is called Prinsengracht, but I am not sure whether I am at place A or B. Let me walk a bit further. Now I see that I can turn left but according to the map if I were at B I would not be able to turn left so soon. Thus I must have been A. Now I know my way to Leidseplein.*

In the above reasoning process, the important elements are: the map, the uncertainties about your location and your observations of the current available actions. We reason by matching the actual available moves with the moves at the possible current locations according to the map.

[1] The main title is taken from the poem *All that is gold does not glitter* by J.R.R. Tolkien.

Sometimes, it is more important to reach your destination than locating yourself exactly. In the *Mission Impossible*-like films, the secret agent sneaking in an enemy building is usually guided by his headquarters (often a geek sitting behind a laptop). However, the communication with the HQ will almost always be lost at some point for some reason. Finally the agent has to find his own way. Suppose the agent has the following floor plan with safety zones marked (though there are no special signs at those places), but does not know whether he is currently at s_2 or at s_3 (denoted by the dashed line):

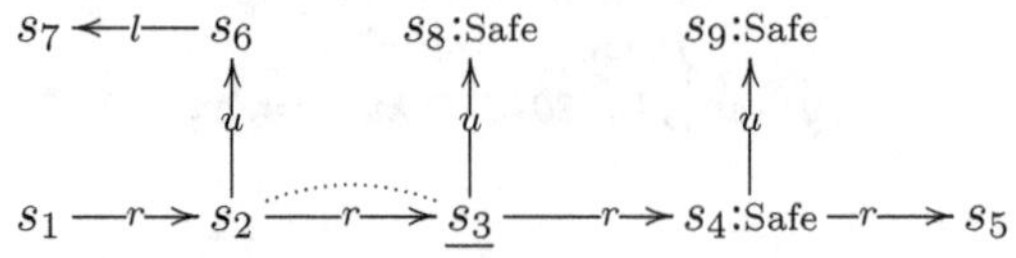

Now suppose that the agent is *actually* at s_3 (the underlined state) and he can *only* observe the available moves at his current location, e.g., at s_3 he only observes that he may move right (r) or move up (u). Let us consider the following scenarios:

- Knowing the actual location of the agent, the HQ may guide the agent to move right (do r) to a safe place (s_4). However, merely following the command, the agent may not know that he is safe after doing r, since if he were at s_2, doing r would get him to an unsafe place s_3, but s_3 and s_4 share exactly the same available moves (r and u), thus he cannot distinguish them.
- The HQ may alternatively guide the agent to move up (u) to s_8. This time the agent should know that he is safe: he sees that he cannot move any further, however, if he were at s_2 initially and thus at s_6 after moving u, then he would be able to move left which contradicts his current observations.
- Suppose the communication with the HQ is lost, the agent may make his own plan as follows: he knows that no matter where exactly he is right now, moving first r and then u will make sure that he is safe, although afterwards he will still not know where he is exactly.

In this paper, we formalize the epistemic reasoning behind such scenarios by proposing a semantic-driven dynamic epistemic logical framework with the following real life applications in mind as the long term goals:

- Global navigation: given the map with uncertainties and the actual location of a subject (human or robot), navigate it to guarantee certain (epistemic) goals, e.g., transport the prisoners to a safe place without letting them know where they are.
- Local navigation: given only the map with uncertainties of the current location of a subject, let the subject navigate itself to guarantee certain (epistemic) goals, e.g., the robot should plan its own way in an endangered nuclear power plant to make sure it "knows" that it will reach all the critical machines that need to be shut down.

1.2 Related work

Related Work *Dynamic epistemic logics* (DEL) are designed to handle knowledge updates caused by events (cf. e.g, [20,1,24]). Arguably the most general framework of DEL is the one using event models proposed in [2]. It is natural to apply the existing techniques of DEL with event models in the navigation setting which is also about knowledge updates after actions. However, as we will show in the later part of the paper, the standard event model approach (even extended with protocols as in [21,11]) is not suitable to handle epistemic reasoning in such scenarios. On the other hand, algebraic approaches inspired by DEL have been proposed to model the robot navigation in [17,18,10]. Despite the apparent differences in frameworks (algebra vs. logic), we depart from this series of works in the way of handling the map information and actions. In [17,18,10], the nodes of the map are encoded by basic propositions and thus the moves in the map are taken to be actions that change the truth value of basic propositions (encodings of the current position). In our semantic-driven approach, we simply take the maps with uncertainties as models and moving in a map does not change the truth values of any basic propositions but the current position and epistemic uncertainties. Instead of the *theorem proving* in the algebraic approach we can *model check* a rich class of desired properties expressed by a natural yet simple logic language, which can be fully automated.

Another usual framework for reasoning about knowledge and developments of a system is the *epistemic temporal logic* (ETL) proposed in [7,19]. Essentially, ETL and DEL are instances of two-dimensional modal logic, which is also used in other multi-agent systems (cf. e.g. [16,14]). Efforts have been made to merge the frameworks of ETL and DEL [21,11]. Our approach can also be viewed as a careful blend of ETL and DEL in the sense that the temporal development is explicitly encoded in the map as in ETL but the epistemic developments are computed in spirit of DEL.

The planning problem with uncertainties and non-deterministic actions (conformant planning) are well-studied in Artificial Intelligence (cf. e.g., [8]), since it was raised in [15]. Our models are similar to the belief spaces used in solving such planning problems (cf. e.g., [4]). The focus there, however, is on the algorithms and heuristics to the planning problem while we would like to present a semantics-driven logic for reasoning about knowledge, which also differs from the situation calculus based logical planning approaches such as [13]. We hope to encode various planning problems by model checking problems in the extensions of our framework, which we leave for further occasions. [2]

The technical contributions and the structure of the paper are summarized as follows:

- In Section 2, we propose a dynamic epistemic framework on maps with un-

---

[2] The connections to belief space planning was suggested to us by Prof. Bernhard Nebel, Dr. Christian Becker-Asano and Dr. Andreas Witzel, when the first author was visiting Isaac Newton Institute in 2012 for a project coordinated by Prof. Benedikt Löwe.

certainties. The semantics is non-standard in the sense that we only assign truth values to the formulas on certain states of the models (not all of them!).

- A substitution-closed axiomatization is provided in Section 3 to capture the validity of the logic and the completeness is proved by using a detour technique handling the interactions of the epistemic operator and the action operators.
- Section 4 discusses some model theoretical properties of the proposed logic: the structural invariance, the finite model property and notably a non-trivial normal form theorem which says that any formula is equivalent to an (exponentially longer) formula where K operator only appear outside the scopes of action operators.
- In Section 5, we compare our logic with ETL via an intuitive translation. We also show that, due to technical reasons, DEL with event models and protocols are not suitable for handling navigation tasks compared to our logic.

2 Preliminaries

2.1 Kripke model with uncertainties

Given a set $\mathbf{P}$ of basic propositions and a set $\mathbf{A}$ of basic actions, a *multimodal Kripke model* $\mathcal{N}$ w.r.t. $\mathbf{P}$ and $\mathbf{A}$ is a tuple: $\mathcal{N} = \langle S, \{R_a \mid a \in \mathbf{A}\}, V\rangle$ where S is a non-empty set of states (or locations), $R_a \subseteq S \times S$ is a binary relation, $V : \mathbf{P} \to \mathcal{P}(S)$ is a valuation function. To simplify notations, we write $s \xrightarrow{a} t$ for sR_at. Given a Kripke model $\mathcal{N}$, we denote its set of states, relations and valuation by $S_{\mathcal{N}}$, $\xrightarrow{a}_{\mathcal{N}}$ and $V_{\mathcal{N}}$. Given an $s \in S_{\mathcal{N}}$, let $e(s)$ be the set of available actions at s, i.e., $e(s) = \{a \mid \exists s' \in S_{\mathcal{N}}$ such that $s \xrightarrow{a} s'\}$. Such a Kripke model may be viewed as an abstract "map" with some basic facts decorating the states. Note that *non-deterministic* actions are allowed: executing a at the same state may result in different states.[3]

An *uncertainty map (UM)* is a Kripke model with a set of uncertainties about the current location of an agent. Formally, a UM model $\mathcal{M}$ is a tuple

$$\langle S, \{R_a \mid a \in \mathbf{A}\}, V, U\rangle$$

where $\langle S, \{R_a \mid a \in \mathbf{A}\}, V\rangle$ is a Kripke model and $U \subseteq S$ is a non-empty set such that for all $s, t \in U$: $e(s) = e(t)$. The requirement of U actually says that the uncertainties should comply with the observation about the available actions. We use $U_{\mathcal{M}}$ to denote the uncertainty set of $\mathcal{M}$. A *pointed UM model* $(\mathcal{M}, s)$ is a UM model $\mathcal{M}$ with a designated state $s \in U_{\mathcal{M}}$ representing the actual location of the agent. Given a model $\mathcal{M}$, let $E(s)$ be the set of states that share the same available actions as s, i.e., $E(s) = \{t \in S \mid e(s) = e(t)\}$.

The graph mentioned in the introduction can be viewed as an illustration

[3] The non-determinism can also be used to model uncertainties about actions, e.g., without a compass, moving east, west, south, north may look exactly the same to an agent at a cross road, thus in the map we may use the same action to stand for these four moves.

of a UM model w.r.t $\mathbf{P} = \{\text{Safe}\}$[4] and $\mathbf{A} = \{l, u, r\}$ with the uncertainty set $\{s_2, s_3\}$ (the states connected by the dotted line).

2.2 Language and semantics

To reason about knowledge and actions in the scenarios mentioned earlier, we use the following simplest modal language $\mathtt{EAL}_{\mathbf{P}}^{\mathbf{A}}$ (*Epistemic Action Language*) with knowledge and actions as modalities:

$$\phi ::= \top \mid p \mid \neg\phi \mid \phi \wedge \phi \mid [a]\phi \mid K\phi$$

where $p \in \mathbf{P}$, $a \in \mathbf{A}$. As usual, we use the following abbreviations: $\bot := \neg\top$, $\phi \vee \psi := \neg(\neg\phi \wedge \neg\psi), \phi \to \psi := \neg\phi \vee \psi, \langle a\rangle\phi := \neg[a]\neg\phi, \hat{K}\phi := \neg K\neg\phi$. Intuitively, $K\phi$ says that the agent knows that ϕ and $[a]\phi$ expresses that if the agent can move forward by a, then after doing a, ϕ holds (a may be non-deterministic).

Given any UM model $\mathcal{M} = \langle S, \{R_a \mid a \in \mathbf{A}\}, V, U\rangle$ and any point $s \in U$ the satisfaction relation is defined on pointed UM model $\mathcal{M}, s$ as:

$$\begin{array}{rcl}
\mathcal{M}, s \vDash \top & \iff & \text{always} \\
\mathcal{M}, s \vDash p & \iff & s \in V(p) \\
\mathcal{M}, s \vDash \neg\phi & \iff & \mathcal{M}, s \nvDash \phi \\
\mathcal{M}, s \vDash \phi \wedge \psi & \iff & \mathcal{M}, s \vDash \phi \text{ and } \mathcal{M}, s \vDash \phi \\
\mathcal{M}, s \vDash [a]\phi & \iff & \forall t \in S : s \xrightarrow{a} t \text{ implies } \mathcal{M}|_t^a, t \vDash \phi \\
\mathcal{M}, s \vDash K\phi & \iff & \forall u \in U : \mathcal{M}, u \vDash \phi
\end{array}$$

where $\mathcal{M}|_t^a = \langle S, \{R_a \mid a \in \mathbf{A}\}, V, U|_t^a\rangle$ and $U|_t^a = U|^a \cap E(t)$ with $U|^a = \{r' \mid \exists r \in U \text{ such that } r \xrightarrow{a} r'\}$.

It is easy to check that in the clause of $[a]\phi$, $U|_t^a \subseteq E(t)$ and $t \in U|_t^a$ thus $\mathcal{M}|_t^a, t$ is indeed a pointed UM model.

The semantics of $K\phi$ is rather intuitive as in epistemic logic. The intuition behind the semantics of $[a]\phi$ formulas is as follows: if you can move forward by a and then end up at t, your uncertainty set should be *carried forward* with you along the possible a moves, which explains the first set in the definition of $U|_t^a$; As for the second part, note that you may eliminate some uncertainties according to the actual observation about the available actions at t.

Here are a few points we have to highlight before moving further:

- We define semantics on pointed UM models and *only* the states in $U_{\mathcal{M}}$ can be taken as the designated points to evaluate formulas. This means that the truth values of $\mathtt{EAL}_{\mathbf{P}}^{\mathbf{A}}$ formulas are *not* defined on all the states in a model.
- In particular, your knowledge at a certain state in the model *only* become clear when you have moved there or one of its indistinguishable states, thus the knowledge is essentially path-dependent (see the example below).

[4] If there is no label *Safe* at a state, then it means the proposition *Safe* is not true there.

- Therefore, we say that a formula ϕ is *valid* ($\vDash \phi$) iff for any pointed UM model $\mathcal{M}, s$: $\mathcal{M}, s \vDash \phi$.

Let us consider the following example (it is a tweaked version of a common example used in [17,18,10]).

Example 2.1 The left and right graphs below depict the initial pointed model $\mathcal{M}, s_1$, and the pointed model after a b move ($\mathcal{M}|^b_{s_3}, s_3$) respectively.

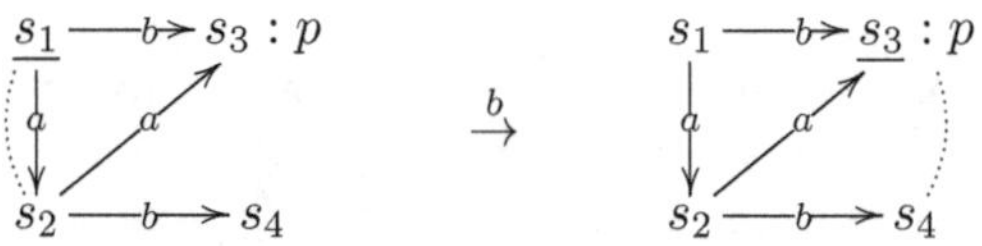

It is easy to verify that $\mathcal{M}, s_1 \vDash K\neg p \wedge \langle b\rangle \neg Kp$.

The left, middle, and right graphs below depict the pointed models $\mathcal{M}, s_1$, $\mathcal{M}|^a_{s_2}, s_2$ and $(\mathcal{M}|^a_{s_2})|^a_{s_3}, s_3$ respectively.

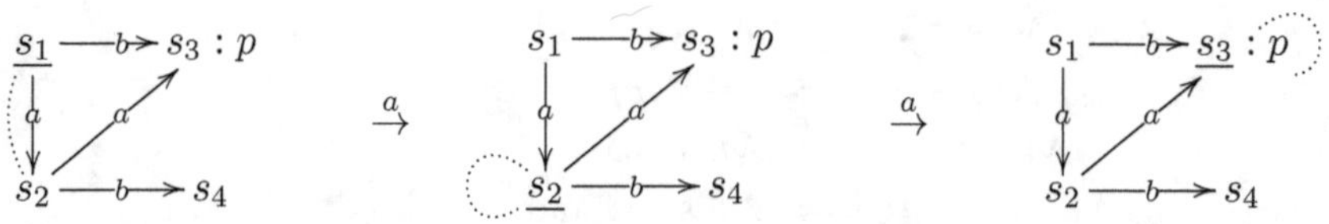

Now we see that $\mathcal{M}, s_1 \vDash K\neg p \wedge \langle a\rangle\langle a\rangle Kp$. Compare $(\mathcal{M}|^a_{s_2})|^a_{s_3}, s_3$ and $\mathcal{M}|^b_{s_3}, s_3$, it is clear that checking whether Kp is true at s_3 depends on how do you get to s_3. It does not mean much to evaluate the knowledge of an agent on the states that he thinks he cannot be currently. The agent may know more or stay ignorant after wandering around.

Going back to our "mission impossible" example in the introduction, we can now verify the claims about three scenarios w.r.t. the model (call it $\mathcal{M}_{\mathrm{MI}}$):

- $\mathcal{M}_{\mathrm{MI}}, s_3 \vDash \langle r\rangle(\mathrm{Safe} \wedge \neg K\mathrm{Safe})$ (HQ guides you safe but you do not know it)
- $\mathcal{M}_{\mathrm{MI}}, s_3 \vDash \langle u\rangle(\mathrm{Safe} \wedge K\mathrm{Safe})$ (HQ guides you safe and you know it)
- $\mathcal{M}_{\mathrm{MI}}, s_3 \vDash K(\langle r\rangle\langle u\rangle\mathrm{Safe} \wedge [r][u]\mathrm{Safe})$ (You know the plan will make you safe)

Given a UM model pointed $\mathcal{M}, s$, a goal expressed by a $\mathtt{EAL}^{\mathbf{A}}_{\mathbf{P}}$ formula ϕ and a plan as a sequence of actions $a_1 \cdots a_n$, we can verify whether the plan can possibly satisfy the goal by checking $\mathcal{M}, s \vDash \langle a_1\rangle \cdots \langle a_n\rangle\phi$.

3 Axiomatization

In this section, we provide a sound and complete axiomatization of $\mathtt{EAL}^{\mathbf{A}}_{\mathbf{P}}$ on UM models. Recall that a formula is valid if it holds on all the *pointed models*. In the sequel we assume that $\mathbf{A}$ is finite.

Given a UM model $\mathcal{M}$, let $\mathcal{M}^{\mathtt{ML}}$ be the Kripke "core" of $\mathcal{M}$ (by simply ignoring the uncertainty set $U_{\mathcal{M}}$); let $\mathcal{M}^{\mathtt{EL}}$ be the **S5** model $\langle U_{\mathcal{M}}, \sim, V'\rangle$ where $\sim\, = U_{\mathcal{M}} \times U_{\mathcal{M}}$ and $V' = V_{\mathcal{M}}|_{U_{\mathcal{M}}}$. Let $\vDash_{\mathtt{ML}}$ and $\vDash_{\mathtt{EL}}$ denote the standard semantics for multimodal logic and epistemic logic respectively (cf. e.g., [3]).

Two easy observations follow immediately from the semantics of $\mathtt{EAL}^{\mathbf{A}}_{\mathbf{P}}$:

Proposition 3.1 *For any K-free $EAL^{\mathbf{A}}_{\mathbf{P}}$-formula ϕ: $\mathcal{M}, s \vDash \phi$ iff $\mathcal{M}^{ML}, s \vDash_{ML} \phi$. For any $[\cdot]$-free $EAL^{\mathbf{A}}_{\mathbf{P}}$-formula ϕ: $\mathcal{M}, s \vDash \phi$ iff $\mathcal{M}^{EL}, s \vDash_{EL} \phi$.*

However, it is clear that $\mathtt{EAL}^{\mathbf{A}}_{\mathbf{P}}$ cannot be reduced, qua expressive power, to any of these two fragments of $\mathtt{EAL}^{\mathbf{A}}_{\mathbf{P}}$, due to the two dimensional nature (action and knowledge) of the UM models. [5] This means that the usual axiomatization of DEL-style logic (e.g., [20] and [1]) via reductions does not work here. In the axiomatization, we include the axioms of epistemic logic and multimodal logic with extra axioms capturing the dynamics in terms of the interaction between $\langle a \rangle$ and K. Inspired by [25], the Henkin-style completeness proof makes use of an auxiliary semantics which transforms dynamics of models into static relations in the canonical model.

3.1 Finite axiomatization $\mathbf{S}_{\mathtt{EAL}^{\mathbf{A}}_{\mathbf{P}}}$

System $\mathbf{S}_{\mathtt{EAL}^{\mathbf{A}}_{\mathbf{P}}}$

Axioms		**Rules**	
TAUT	all the axioms of propositional logic	MP	$\dfrac{\phi, \phi \to \psi}{\psi}$
DISTK	$K(p \to q) \to (Kp \to Kq)$	NECK	$\dfrac{\phi}{K\phi}$
DIST(a)	$[a](p \to q) \to ([a]p \to [a]q)$	NEC(a)	$\dfrac{\phi}{[a]\phi}$
OBS(a)	$K\langle a \rangle \top \vee K\neg\langle a \rangle \top$	SUB	$\dfrac{\phi(p)}{\phi(\psi)}$
T	$Kp \to p$		
4	$Kp \to KKp$		
5	$\neg Kp \to K\neg Kp$		
PR(a)	$\langle a \rangle \hat{K} p \to \hat{K}\langle a \rangle p$		
NL(a)	$\bigwedge_{\mathbf{B} \subseteq \mathbf{A}} (\hat{K}\langle a \rangle (p \wedge \psi_{\mathbf{B}}) \to [a](\psi_{\mathbf{B}} \to \hat{K} p))$		

where a ranges over $\mathbf{A}$, p, q range over $\mathbf{P}$ and in the last clause, $\psi_{\mathbf{B}} = (\bigwedge_{b \in \mathbf{B}} \langle b \rangle \top) \wedge (\bigwedge_{b \notin \mathbf{B}} \neg \langle b \rangle \top)$. Since $\mathbf{A}$ is finite, $\mathtt{EAL}^{\mathbf{A}}_{\mathbf{P}}$ is a finite axiomatic system. PR($\cdot$) and NL($\cdot$) denote the axioms of *perfect recall* and *no learning* respectively, following the convention in the literature (cf. [9]).

Based on the semantics of $\mathtt{EAL}^{\mathbf{A}}_{\mathbf{P}}$ and Proposition 3.1, it is easy to verify that the following axioms and rules are valid: DISTK, DIST($\cdot$), T, 4, 5, NECK, NEC($\cdot$). The validity of OBS($\cdot$) is due to the requirement on the uncertainty sets in UM models. Note that the uniform substitution SUB is also valid according to our semantics, which is different from the usual DEL-style logics (cf. [24]).

To prove the soundness of $\mathtt{EAL}^{\mathbf{A}}_{\mathbf{P}}$, we still need to show that PR($\cdot$) and NL($\cdot$)

[5] E.g., you cannot find an epistemic formula to tell two UM models apart if they share the same epistemic core but are quite different in the map part. Similar for the K-free formulas.

are valid. In the following, we verify the corresponding axiom schemas where p is replaced by an arbitrary ϕ.

Proposition 3.2 *For any* $a \in \mathbf{A}$: $\vDash \langle a\rangle\hat{K}\phi \to \hat{K}\langle a\rangle\phi$

Proof For any $\mathcal{M}, s$, if $\mathcal{M}, s \vDash \langle a\rangle\hat{K}\phi$, then there is a $t \in S$, such that $s \overset{a}{\to} t$ and $\mathcal{M}|_t^a, t \vDash \hat{K}\phi$, thus there is also a $v \in U_{\mathcal{M}}|_t^a$, $\mathcal{M}|_t^a, v \vDash \phi$. Because $v \in U_{\mathcal{M}}|_t^a = U_{\mathcal{M}}|^a \cap E(t)$, then there is a $u \in U_{\mathcal{M}}$, such that $u \overset{a}{\to} v$ and $E(v) = E(t)$. Thus $U_{\mathcal{M}}|_v^a = U_{\mathcal{M}}|_t^a$, then $\mathcal{M}|_t^a = \mathcal{M}|_v^a$, thus $\mathcal{M}|_v^a, v \vDash \phi$. Since $u \overset{a}{\to} v$, $\mathcal{M}, u \vDash \langle a\rangle\phi$, because $u \in U_{\mathcal{M}}$ then $\mathcal{M}, s \vDash \hat{K}\langle a\rangle\phi$. □

Proposition 3.3 *For any* $a \in \mathbf{A}$: $\vDash \bigwedge_{\mathbf{B}\subseteq\mathbf{A}}(\hat{K}\langle a\rangle(\phi \wedge \psi_{\mathbf{B}}) \to [a](\psi_{\mathbf{B}} \to \hat{K}\phi))$

Proof For any $\mathcal{M}, s$, we need to prove that for any $\mathbf{B} \subseteq \mathbf{A}$, $\mathcal{M}, s \vDash \hat{K}\langle a\rangle(\phi \wedge \psi_{\mathbf{B}}) \to [a](\psi_{\mathbf{B}} \to \hat{K}\phi)$. If $\mathcal{M}, s \vDash \hat{K}\langle a\rangle(\phi \wedge \psi_{\mathbf{B}})$, then there is a $u \in U_{\mathcal{M}}$, such that $\mathcal{M}, u \vDash \langle a\rangle(\phi \wedge \psi_{\mathbf{B}})$, thus there is also a $v \in S_{\mathcal{M}}$, such that $u \overset{a}{\to} v$ and $\mathcal{M}|_v^a, v \vDash \phi \wedge \psi_{\mathbf{B}}$. Then we need to prove that $\mathcal{M}, s \vDash [a](\psi_{\mathbf{B}} \to \hat{K}\phi)$. Namely, for any $t \in S_{\mathcal{M}}$, assuming $s \overset{a}{\to} t$ and $\mathcal{M}|_t^a, t \vDash \psi_{\mathbf{B}}$, we need to show that $\mathcal{M}|_t^a, t \vDash \hat{K}\phi$. Since $\mathcal{M}|_v^a, v \vDash \psi_{\mathbf{B}}$, then $E(t) = E(v)$, thus $U_{\mathcal{M}}|_t^a = U_{\mathcal{M}}|_v^a$ and $\mathcal{M}|_t^a = \mathcal{M}|_v^a$. Since $\mathcal{M}|_v^a, v \vDash \phi$, $\mathcal{M}|_t^a, v \vDash \phi$. Now since $v \in U_{\mathcal{M}}|_t^a$, $\mathcal{M}|_t^a, t \vDash \hat{K}\phi$. □

Since we include DIST$(\cdot)$, DISTK, NEC$(\cdot)$, NECK in the system, it is easy to verify the following propositions as standard exercises in *normal modal logic*.

Proposition 3.4 $\vdash [a](\phi \wedge \psi) \leftrightarrow ([a]\phi \wedge [a]\psi)$, $\vdash [a]\phi \vee [a]\psi \to [a](\phi \vee \psi)$, $\vdash K(\phi \wedge \psi) \leftrightarrow (K\phi \wedge K\psi)$.

Proposition 3.5 *If* $\vdash \phi \leftrightarrow \phi'$, $\vdash \psi \leftrightarrow \psi'$, *then* $\vdash \neg\phi \leftrightarrow \neg\phi'$, $\vdash \phi \wedge \psi \leftrightarrow \phi' \wedge \psi'$, $\vdash \langle a\rangle\phi \leftrightarrow \langle a\rangle\phi'$, *and* $\vdash K\phi \leftrightarrow K\phi'$.

Based on the above propositions, we can show the useful inference rule of *replacements of equivalents* is an admissible rule of the system $\mathbf{S}_{\mathtt{EAL}_{\mathbf{P}}^{\mathbf{A}}}$.

Proposition 3.6 *If* $\vdash \psi \leftrightarrow \psi'$, *and* ϕ' *is obtained by replacing some occurrences of* ψ *in* ϕ *with* ψ', *then* $\vdash \phi \leftrightarrow \phi'$.

3.2 Completeness

To prove the completeness, we will use an auxiliary semantics of $\mathtt{EAL}_{\mathbf{P}}^{\mathbf{A}}$ on *epistemic multimodal models (EM models)*. Formally, an EM model $\mathcal{N}$ is a tuple $\langle S, \{R_a \mid a \in \mathbf{A}\}, V, \sim\rangle$, where $\langle S, \{R_a \mid a \in \mathbf{A}\}, V\rangle$ is a multimodal Kripke model and $\sim$ is an equivalence relation over S such that $s \sim t$ implies $e(s) = e(t)$. Note that $\sim$ can also be viewed as a partition of S. Therefore the difference between a UM model $\mathcal{M}$ and an EM model $\mathcal{N}$ is that $U_{\mathcal{M}}$ denotes a single equivalence class while $\sim$ denotes a set of the equivalence classes which form a partition of $S_{\mathcal{N}}$. $\mathtt{EAL}_{\mathbf{P}}^{\mathbf{A}}$ formulas can be interpreted on EM models with the usual Kripke semantics (denoted as $\Vdash$):

$$\mathcal{N}, s \Vdash [a]\phi \iff \forall t : s \overset{a}{\to} t \text{ implies } \mathcal{N}, t \Vdash \phi$$

$$\mathcal{N}, s \Vdash K\phi \iff \forall t : s \sim t \text{ implies } \mathcal{N}, t \Vdash \phi$$

However, we cannot always transform a UM model $\mathcal{M}$ into an EM model $\mathcal{M}'$ by simply adding more equivalence classes such that for any $\mathtt{EAL}^{\mathbf{A}}_{\mathbf{P}}$-formula ϕ: $\mathcal{M}, s \vDash \phi \iff \mathcal{M}', s \Vdash \phi$. In Example 2.1, based on the initial UM model, it is impossible to assign an equivalence class including s_3 to make sure that $\langle b \rangle \neg Kp$ and $\langle a \rangle \langle a \rangle Kp$ both hold at s_1.

On the other hand, an EM model can also be viewed as a UM model with extra epistemic information. Given an EM model $\mathcal{N} = \langle S, \{R_a \mid a \in \mathbf{A}\}, V, \sim \rangle$ and $s \in S$, let $\mathcal{N}_s$ be the UM model $\langle S, \{R_a \mid a \in \mathbf{A}\}, V, U_s \rangle$ where $U_s = \{t \mid s \sim t \text{ in } \mathcal{N}\}$. We say $\vDash$ and $\Vdash$ *coincide* on an EM model $\mathcal{N}$ if for any $s \in S_{\mathcal{N}}$ and any $\mathtt{EAL}^{\mathbf{A}}_{\mathbf{P}}$-formula ϕ: $\mathcal{N}, s \Vdash \phi \iff \mathcal{N}_s, s \vDash \phi$. It is not hard to see that the two semantics do not coincide on arbitrary EM models in general, however, as we will show later, the two semantics do coincide on the canonical EM model which is essential in the proof of completeness.

Our proof strategy can be summarised as follows:

(i) Prove the Lindebaum-like lemma: every $\mathbf{S}_{\mathtt{EAL}^{\mathbf{A}}_{\mathbf{P}}}$-consistent set of formulas can be extended into a maximal consistent set ($\mathbf{S}_{\mathtt{EAL}^{\mathbf{A}}_{\mathbf{P}}}$-MCS).

(ii) Construct a *canonical EM model* $\mathcal{C}$ and prove the truth lemma w.r.t. the auxiliary semantics ($\Vdash$).

(iii) Show that $\vDash$ and $\Vdash$ coincide on the canonical model thus obtaining the truth lemma w.r.t. $\vDash$. Finally, given a $\mathbf{S}_{\mathtt{EAL}^{\mathbf{A}}_{\mathbf{P}}}$-MCS Γ, $(\mathcal{C}_\Gamma, \Gamma)$ is the pointed UM model which satisfies all the formulas in Γ.

The Lindebaum lemma is routine. We define a canonical EM model based on MCSs of $\mathbf{S}_{\mathtt{EAL}^{\mathbf{A}}_{\mathbf{P}}}$ as usual for normal modal logics (cf. e.g., [3]):

$$\mathcal{C} = < S^c, \{R^c_a \mid a \in \mathbf{A}\}, \sim^c, V^c >$$

where:

- S^c is the set of all $\mathbf{S}_{\mathtt{EAL}^{\mathbf{A}}_{\mathbf{P}}}$-MCSs;
- $sR^c_a t \iff$ for any $\phi \in t$ then $\langle a \rangle \phi \in s \iff$ for any $[a]\phi \in s$ then $\phi \in t$;
- $s \sim^c t \iff$ for any $\phi \in t$ then $\hat{K}\phi \in s \iff$ for any $K\phi \in s$ then $\phi \in t$;
- $V^c(p) = \{s \mid p \in s\}$.

According to the canonicity of axioms T, 4, and 5, we know that $\sim^c$ is indeed an equivalence relation on S^c. To verify that $\mathcal{C}$ is indeed an EM model, we need to verify that $s \sim^c t$ implies $e(s) = e(t)$.

Proposition 3.7 *In the canonical model $\mathcal{C}$, $a \in e(s) \iff \langle a \rangle \top \in s$.*

Proof $\Rightarrow$: If $a \in e(s)$, according to the definition of $e(s)$, there is a $t \in S^c, s \xrightarrow{a} t$,because $\top \in t$, then $\langle a \rangle \top \in s$.

$\Leftarrow$: Let $D = \{\phi \mid [a]\phi \in s\}$. Since $\vdash [a](\phi \wedge \psi) \leftrightarrow [a]\phi \wedge [a]\psi$, s is closed under finite conjunctions. If D is not consistent, then there is $\phi \in D, \vdash \phi \rightarrow \bot$. By the rule $\mathtt{NEC}(a)$, $\vdash [a](\phi \rightarrow \bot)$, thus by $\mathtt{DIST}(a)$, $\vdash [a]\phi \rightarrow [a]\bot$, namely $\vdash [a]\phi \rightarrow \neg\langle a \rangle \top$. Since $[a]\phi \in s$, $\neg\langle a \rangle \top \in s$ which is contradictory to $\langle a \rangle \top \in s$.

Therefore there is a maximal consistent set t such that $D \subseteq t$. According to the definition of R^c_a, we have $sR^c_a t$ thus $a \in e(s)$. □

Proposition 3.8 *In the canonical model $\mathcal{C}$, if $s \sim^c t$ then $e(s) = e(t)$.*

Proof For any $a \in \mathbf{A}$, if $a \in e(s)$, then according to Proposition 3.7, $\langle a\rangle\top \in s$. By axioms $\texttt{OBS}(a)$ and $\texttt{T}$, $K\langle a\rangle\top \in s$. Since $s \sim^c t$, $\langle a\rangle\top \in t$, by Proposition 3.7, $a \in e(t)$, namely $e(s) \subseteq e(t)$. It is symmetric to show $e(t) \subseteq e(s)$. □

In the rest of this section, we will show that the two semantics coincide on $\mathcal{C}$. To prove this, the key idea is to show that the equivalence classes of $\mathcal{C}$ capture all the possible dynamics of the uncertainty sets, e.g., if you move from a state s in an equivalence class U_s in $\mathcal{C}$ to a state t, then the updated uncertainty set $(U_s)|^a_t$ is exactly the equivalence class that t belongs to in $\mathcal{C}$. Formally, we have the following proposition (recall that $U_s = \{t \mid s \sim^c t \text{ in } \mathcal{C}\}$).

Proposition 3.9 *In the canonical model $\mathcal{C}$, if $s \xrightarrow{a} t$, then $(U_s)|^a \cap E(t) = U_t$. Namely $U_s|^a_t = U_t$ and thus $\mathcal{C}_s|^a_t = \mathcal{C}_t$.*

Proof $\subseteq$: Assuming $v \in (U_s)|^a \cap E(t)$, we need to prove that $v \in U_t$, namely $v \sim^c t$. Since $v \in (U_s)|^a$, there is a u, such that $u \sim^c s$ and $u \xrightarrow{a} v$. Let $\mathbf{B} = \{a \mid a \in \mathbf{A} \text{ and } \langle a\rangle\top \in v\}$, then $\psi_{\mathbf{B}} \in v$. For any $\phi \in v$, it is clear that $\phi \wedge \psi_{\mathbf{B}} \in v$. Since $u \xrightarrow{a} v$, $\langle a\rangle(\phi \wedge \psi_{\mathbf{B}}) \in u$. By $u \sim^c s$ we have $\hat{K}\langle a\rangle(\phi \wedge \psi_{\mathbf{B}}) \in s$. Now by axiom $\texttt{NL}(a)$ and rule $\texttt{SUB}$, $[a](\psi_{\mathbf{B}} \to \hat{K}\phi) \in s$. Since $s \xrightarrow{a} t$, $\psi_{\mathbf{B}} \to \hat{K}\phi \in t$. Because $v \in E(t)$, then $\psi_{\mathbf{B}} \in t$ thus $\hat{K}\phi \in t$. By the definition of $\sim^c$, we have $v \sim^c t$, namely $v \in U_t$.

$\supseteq$: If $v \in U_t$, by Proposition 3.8, $e(v) = e(t)$ then $v \in E(t)$. In order to prove $v \in (U_s)|^a \cap E(t)$, we only need to show that $v \in (U_s)|^a$. In the following we will construct an MCS u such that $s \sim u$ and $u \xrightarrow{a} v$. Let $D = \{\psi \mid K\psi \in s\} \cup \{\langle a\rangle\phi \mid \phi \in v\}$. It is easy to see that $\{\psi \mid K\psi \in s\}$ is closed under finite conjunctions. If D is not consistent, we must have $\vdash \psi \wedge \langle a\rangle\phi_1 \wedge \cdots \wedge \langle a\rangle\phi_n \to \bot$ for some $K\psi \in s$ and $\phi_1 \ldots \phi_n \in v$, then $\vdash \psi \to ([a]\neg\phi_1 \vee \cdots \vee [a]\neg\phi_n)$. Because $\vdash [a]\neg\phi_1 \vee \cdots \vee [a]\neg\phi_n \to [a](\neg\phi_1 \vee \cdots \vee \phi_n)$, then $\vdash \psi \to [a](\neg\phi_1 \vee \cdots \vee \neg\phi_n)$. By $\texttt{NECK}$ and $\texttt{DISTK}$, $\vdash K\psi \to K[a](\neg\phi_1 \vee \cdots \vee \neg\phi_n)$. By $\texttt{PR}(a)$ and $\texttt{SUB}$, $\vdash K[a](\neg\phi_1 \vee \cdots \vee \neg\phi_n) \to [a]K(\neg\phi_1 \vee \cdots \vee \neg\phi_n)$, then $\vdash K\psi \to [a]K(\neg\phi_1 \vee \cdots \vee \neg\phi_n)$. Since $K\psi \in s$, $[a]K(\neg\phi_1 \vee \cdots \vee \neg\phi_n) \in s$. Due to the fact that $s \xrightarrow{a} t$, $K(\neg\phi_1 \vee \cdots \vee \neg\phi_n) \in t$. Since $v \sim^c t$, then $\neg\phi_1 \vee \cdots \vee \neg\phi_n \in v$. This is contradictory to $\phi_1 \ldots \phi_n \in v$ and that v is consistent. Therefore D is consistent, then there is a maximal consistent set u, such that $D \subseteq u$. Clearly $u \sim^c s$ and $u \xrightarrow{a} v$, thus $v \in (U_s)|^a$. In sum, $v \in (U_s)|^a \cap E(t)$. □

To prove the truth lemma w.r.t. $\vDash$ we make use of the following truth lemma w.r.t. $\Vdash$ as a standard exercise for normal modal logic (cf. e.g., [3]).

Lemma 3.10 *For any $\mathbf{EAL}^{\mathbf{A}}_{\mathbf{P}}$ formula ϕ, any s in $\mathcal{C}$: $\mathcal{C}, s \Vdash \phi \iff \phi \in s$.*

All we need now is to show that two semantics coincide on $\mathcal{C}$.

Lemma 3.11 *For any* $\mathtt{EAL}_{\mathbf{P}}^{\mathbf{A}}$ *formula* ϕ*, any* s *in* $\mathcal{C}$*:* $\mathcal{C}, s \Vdash \phi \iff \mathcal{C}_s, s \vDash \phi$

Proof Do induction on the structure of ϕ. The cases for $\phi = p$, $\phi = \neg\psi$, and $\phi = \phi_1 \wedge \phi_2$ are immediate.

$\phi = [a]\psi$, if $\mathcal{C}, s \Vdash [a]\psi$, but $\mathcal{C}_s, s \nvDash [a]\psi$, then there is a $t \in S^c$, such that $s \xrightarrow{a} t$ and $\mathcal{C}_s|^a_t, t \nvDash \psi$, by Proposition 3.9, $\mathcal{C}_s|^a_t = \mathcal{C}_t$, then $\mathcal{C}_t, t \nvDash \phi$, by IH, $\mathcal{C}, t \nVdash \phi$. Since $s \xrightarrow{a} t$, $\mathcal{C}, s \Vdash [a]\psi$, contradiction. On the other hand, if $\mathcal{C}_s, s \vDash [a]\psi$, but $\mathcal{C}, s \nVdash [a]\psi$, then there is a $t \in S^c$, such that $s \xrightarrow{a} t$, and $\mathcal{C}, t \nVdash \psi$. According to IH, $\mathcal{C}_t, t \nvDash \psi$. Now from Proposition 3.9, we have $\mathcal{C}_s|^a_t = \mathcal{C}_t$, then $\mathcal{C}_s|^a_t, t \nvDash \psi$. However, $s \xrightarrow{a} t$ and $\mathcal{C}_s, s \vDash [a]\psi$, contradiction.

$\phi = K\psi$, if $\mathcal{C}, s \Vdash K\psi$, but $\mathcal{C}_s, s \nvDash K\psi$, then there is a $u \in U_s$ such that $\mathcal{C}_s, u \nvDash \psi$. Since $u \in U_s$, then $U_s = U_u$, thus $\mathcal{C}_s = \mathcal{C}_u$, therefore $\mathcal{C}_u, u \nvDash \psi$. By IH $\mathcal{C}, u \nVdash \psi$ which is contradictory to the facts $u \sim^c s$ and $\mathcal{C}, s \Vdash K\psi$. On the other hand, if $\mathcal{C}_s, s \vDash K\phi$, but $\mathcal{C}, s \nVdash K\psi$, then there is a u, such that $s \sim^c u$ and $\mathcal{C}, u \nVdash \psi$. By IH, $\mathcal{C}_u, u \nvDash \psi$. Since $s \sim^c u$, then $U_s = U_u$ thus $\mathcal{C}_s = \mathcal{C}_u$. Therefore $\mathcal{C}_s, u \nvDash \psi$ which is contradictory to $u \in U_s$ and $\mathcal{C}_s, s \vDash K\psi$. □

Based on Lemmata 3.10 and 3.11, every $\mathbf{S}_{\mathtt{EAL}_{\mathbf{P}}^{\mathbf{A}}}$-consistent set of formulas has a model $\mathcal{C}_s, s$, thus the completeness is immediate.

Theorem 3.12 $S_{\mathtt{EAL}_{\mathbf{P}}^{\mathbf{A}}}$ *is sound and complete on UM models.*

4 Modal theoretical properties of $\mathtt{EAL}_{\mathbf{P}}^{\mathbf{A}}$

In this section, we prove three results which can help us to understand $\mathtt{EAL}_{\mathbf{P}}^{\mathbf{A}}$ on UM models better. We first show that $\mathtt{EAL}_{\mathbf{P}}^{\mathbf{A}}$ is invariant under a special notion of bisimulation between UM models, which inspires a normal form theorem as our second result, and finally we prove the finite model property of $\mathtt{EAL}_{\mathbf{P}}^{\mathbf{A}}$ based on these insights.

4.1 Structural invariance for $\mathtt{EAL}_{\mathbf{P}}^{\mathbf{A}}$

Recall that given a UM model $\mathcal{M} = \langle S, \{R_a \mid a \in \mathbf{A}\}, U, V\rangle$, $\mathcal{M}^{\mathtt{ML}}$ is the multimodal model without U, i.e. $\mathcal{M}^{\mathtt{ML}} = \langle S, \{R_a \mid a \in \mathbf{A}\}, V\rangle$. Let $\mathtt{ML}_{\mathbf{P}}^{\mathbf{A}}$ be the K-free fragment of $\mathtt{EAL}_{\mathbf{P}}^{\mathbf{A}}$.

Next, we define a structural relation between UM models based on the notion of bisimilarity ($\underline{\leftrightarrow}$) on multimodal models w.r.t. $\mathbf{P}$ and $\mathbf{A}$ (cf. e.g., [3]).

Definition 4.1 *For any UM models* $\mathcal{M}$ *and* $\mathcal{N}$*, we say that* $\mathcal{M}$ *is U-bisimilar to* $\mathcal{N}$ *(notation:* $\mathcal{M} \rightleftarrows \mathcal{N}$*) iff:*

- *for any* $u \in U_{\mathcal{M}}$*, there is a* $u' \in U_{\mathcal{N}}$*, such that* $\mathcal{M}^{ML}, u \underline{\leftrightarrow} \mathcal{N}^{ML}, u'$*,*
- *for any* $u' \in U_{\mathcal{N}}$*, there is a* $u \in U_{\mathcal{M}}$*, such that* $\mathcal{M}^{ML}, u \underline{\leftrightarrow} \mathcal{N}^{ML}, u'$*.*

We say two pointed UM models are U-bisimilar ($\mathcal{M}, u \rightleftarrows \mathcal{N}, u'$*) iff* $\mathcal{M}^{ML}, u \underline{\leftrightarrow} \mathcal{N}^{ML}, u'$ *and* $\mathcal{M} \rightleftarrows \mathcal{N}$*.*

Now we prove that the moves in UM models preserve U-bisimilarity.

Proposition 4.2 *If* $\mathcal{M}, s \rightleftarrows \mathcal{N}, u$*,* $s \xrightarrow{a} t$ *in* $\mathcal{M}$*,* $u \xrightarrow{a} v$ *in* $\mathcal{N}$*, and* $\mathcal{M}^{ML}, t \underline{\leftrightarrow} \mathcal{N}^{ML}, v$*, then* $\mathcal{M}|^a_t, t \rightleftarrows \mathcal{N}|^a_v, v$*.*

Proof Since $(\mathcal{M}|_t^a)^{\texttt{ML}} = \mathcal{M}^{\texttt{ML}}, (\mathcal{N}|_v^a)^{\texttt{ML}} = \mathcal{N}^{\texttt{ML}}$ and $\mathcal{M}^{\texttt{ML}}, t \leftrightarrows \mathcal{N}^{\texttt{ML}}, v$, then by the definition of $\rightleftarrows$, we only need to prove that $\mathcal{M}|_t^a \rightleftarrows \mathcal{N}|_v^a$, namely for each x' in $U_{\mathcal{M}}|_t^a$ there is a y' in $U_{\mathcal{N}}|_v^a$ such that $(\mathcal{M}|_t^a)^{\texttt{ML}}, x' \leftrightarrows (\mathcal{N}|_v^a)^{\texttt{ML}}, y'$ (the reverse condition can be proved symmetrically).

For each $x' \in U_{\mathcal{M}}|_t^a = U_{\mathcal{M}}|^a \cap E(t)$, there is an $x \in U_{\mathcal{M}}$ and $x \overset{a}{\to} x'$ in $\mathcal{M}$. Since $\mathcal{M} \rightleftarrows \mathcal{N}$, then there is a $y \in U_{\mathcal{N}}$, such that $\mathcal{M}^{\texttt{ML}}, x \leftrightarrows \mathcal{N}^{\texttt{ML}}, y$. Since $x \overset{a}{\to} x'$ in $\mathcal{M}$, then according to the definition of bisimilarity, there is a $y' \in S_{\mathcal{N}}$, such that $y \overset{a}{\to} y'$ in $\mathcal{N}$ and $\mathcal{M}^{\texttt{ML}}, x' \leftrightarrows \mathcal{N}^{\texttt{ML}}, y'$ (thus $(\mathcal{M}|_t^a)^{\texttt{ML}}, x' \leftrightarrows (\mathcal{N}|_v^a)^{\texttt{ML}}, y'$.). Clearly $y' \in U_{\mathcal{N}}|^a$. We only need to show that $y' \in E(v)$ in order to prove that $y' \in U_{\mathcal{N}}|_v^a$. Since $\mathcal{M}^{\texttt{ML}}, x' \leftrightarrows \mathcal{N}^{\texttt{ML}}, y'$, $e(y') = e(x')$. Now due to the fact that $x' \in U_{\mathcal{M}}|_t^a$, we have $e(x') = e(t)$. Since $\mathcal{M}^{\texttt{ML}}, t \leftrightarrows \mathcal{N}^{\texttt{ML}}, v$, then it is easy to see that $e(t) = e(v)$, thus $e(y') = e(x') = e(t) = e(v)$, therefore $y' \in E(v)$, namely $y' \in U_{\mathcal{N}}|_v^a$. □

We say a UM model $\mathcal{M}$ is *image-finite* if $U_{\mathcal{M}}$ is finite, and for any $s \in S_{\mathcal{M}}$ and any $a \in \mathbf{A}$: $\{t \mid s \overset{a}{\to} t\}$ is finite. Let $\equiv_{\texttt{EAL}_\mathbf{P}^\mathbf{A}}$ be the logical equivalence relation between pointed models. As an easy exercise, we can show that U-bisimilarity indeed preserves the truth values of the $\texttt{EAL}_\mathbf{P}^\mathbf{A}$-formulas based on Proposition 4.2 (we omit the proof due to limited space).

Proposition 4.3 *For any pointed models $\mathcal{M}, u$, $\mathcal{N}, u'$: $\mathcal{M}, u \rightleftarrows \mathcal{N}, u'$ implies $\mathcal{M}, u \equiv_{\texttt{EAL}_\mathbf{P}^\mathbf{A}} \mathcal{N}, u'$. The converse holds when restricted to image-finite models.*

4.2 Normal form

Proposition 4.3 says that the distinguishing power of $\texttt{EAL}_\mathbf{P}^\mathbf{A}$ is bounded by the U-bisimilarity. A closer look reveals something more surprising: qua expressive power, the full language of $\texttt{EAL}_\mathbf{P}^\mathbf{A}$ is equivalent to its fragment where knowledge operator only appears outside the scopes of the action modalities. Formally, formulas ϕ in this fragment ($\texttt{EALK}_\mathbf{P}^\mathbf{A}$) can be generated by:

$$\begin{array}{rcl}\phi & ::= & \top \mid p \mid \psi \mid \neg\phi \mid \phi \wedge \phi \mid K\phi \\ \psi & ::= & \top \mid p \mid \neg\psi \mid \psi \wedge \psi \mid [a]\psi\end{array}$$

where $a \in \mathbf{A}$ and $p \in \mathbf{P}$.

In this subsection we will show that every $\texttt{EAL}_\mathbf{P}^\mathbf{A}$ formula is equivalent to an (exponentially longer) $\texttt{EALK}_\mathbf{P}^\mathbf{A}$ formula. Note that although Proposition 4.3 already suggests that $\texttt{EAL}_\mathbf{P}^\mathbf{A}$ and $\texttt{EALK}_\mathbf{P}^\mathbf{A}$ have the same distinguishing power, their expressive powers may still differ,[6] thus the result does not follow from Proposition 4.3.

Definition 4.4 *We define the K-degree of $\texttt{EAL}_\mathbf{P}^\mathbf{A}$ formulas ($kd(\phi)$) as follows:*

$$\begin{array}{ll} kd(\top) = 0 & kd(p) = 0 \\ kd(\neg\phi) = kd(\phi) & kd(\phi \wedge \psi) = max\{kd(\phi), kd(\psi)\} \\ kd([a]\phi) = 0 & kd(K\phi) = 1 + kd(\phi)\end{array}$$

[6] Here by *distinguishing power* we mean the power of a language to tell two models apart while expressive power measures the power of the language to define classes of models (properties of the models).

where $p \in \boldsymbol{P}$ *and* $a \in \boldsymbol{A}$.

Note that we treat the outmost $[\cdot]\phi$ (not in the scope of any other $[\cdot]$) as atomic formulas by setting $kd([\cdot]\phi) = 0$, e.g., $kd(K[a]K[b]p) = 1$.

Definition 4.5 *An* $EAL^{\mathbf{A}}_{\mathbf{P}}$ *formula* ϕ *is in* K*-conjunctive normal form (K-CNF) iff:*

- $\phi = \alpha_1 \wedge \cdots \wedge \alpha_n$ *such that* $\forall 1 \leq i \leq n : \alpha_i = \beta_{i_1} \vee \cdots \vee \beta_{i_m}$ *for some* $m \geq 1$,
- *each* β_{i_j} *is in the shape of* $p, \neg p, [\cdot]\psi, \neg[\cdot]\psi$, $K\chi$ *or* $\hat{K}\chi$ *where* $kd(\chi) = 0$.

Note that $\mathbf{S}_{\mathtt{EAL}^{\mathbf{A}}_{\mathbf{P}}}$ includes all the axioms and rules of **S5**. Thus by using the standard result for **S5** logic, we can turn each $\mathtt{EAL}^{\mathbf{A}}_{\mathbf{P}}$ formula into K-CNF.

Proposition 4.6 *For any* $EAL^{\mathbf{A}}_{\mathbf{P}}$*-formula* ϕ, *there is an* $EAL^{\mathbf{A}}_{\mathbf{P}}$ *formula* ϕ', *such that* $\vDash \phi \leftrightarrow \phi'$ *and* ϕ' *is in K-CNF. In particular, for each* $EALK^{\mathbf{A}}_{\mathbf{P}}$*-formula there is an equivalent* $EALK^{\mathbf{A}}_{\mathbf{P}}$*-formula in K-CNF.*

Proof We treat the outmost $[\cdot]\psi$ formulas as atomic formulas when massaging the original formula according to the standard normal form result for **S5** (cf. e.g., [12]), thus keeping the formulas in $\mathtt{EALK}^{\mathbf{A}}_{\mathbf{P}}$. □

Note that in an $\mathtt{EAL}^{\mathbf{A}}_{\mathbf{P}}$ formula of K-CNF, there may still be some occurrences of K modality inside the scope of $[\cdot]$ modalities. In the sequel, we will try to *push* the K operator out. Here are two crucial results proved using the spirit behind the validity of axioms $\mathtt{PR}(\cdot)$ and $\mathtt{NL}(\cdot)$. [7]

Proposition 4.7

$$(1) \vDash [a](K\phi \vee \chi) \leftrightarrow \bigwedge_{B \subseteq A} (\langle a \rangle(\psi_B \wedge \neg\chi) \to K[a](\psi_B \to \phi))$$

$$(2) \vDash [a](\hat{K}\phi \vee \chi) \leftrightarrow \bigwedge_{B \subseteq A} (\langle a \rangle(\psi_B \wedge \neg\chi) \to \hat{K}\langle a \rangle(\psi_B \wedge \phi))$$

Proof

(1) Left to right: If $\mathcal{M}, s \vDash [a](K\phi \vee \chi)$, we need to prove that for any $\mathbf{B} \subseteq \mathbf{A}$, $\mathcal{M}, s \vDash \langle a \rangle(\psi_{\mathbf{B}} \wedge \neg\chi) \to K[a](\psi_{\mathbf{B}} \to \phi)$. Now suppose $\mathcal{M}, s \vDash \langle a \rangle(\psi_{\mathbf{B}} \wedge \neg\chi)$, then there is a t, such that $s \xrightarrow{a} t$, and $\mathcal{M}|^a_t, t \vDash \psi_{\mathbf{B}} \wedge \neg\chi$. Because $\mathcal{M}, s \vDash [a](K\phi \vee \chi)$, then $\mathcal{M}|^a_t, t \vDash K\phi \vee \chi$, then $\mathcal{M}|^a_t, t \vDash \psi_{\mathbf{B}} \wedge K\phi$.

We need to prove that $\mathcal{M}, s \vDash K[a](\psi_{\mathbf{B}} \to \phi)$, namely for any $u \in U_{\mathcal{M}}$ we need to show $\mathcal{M}, u \vDash [a](\psi_{\mathbf{B}} \to \phi)$. That is, for any v such that $u \xrightarrow{a} v$, we need to show $\mathcal{M}|^a_v, v \vDash \psi_{\mathbf{B}} \to \phi$. Suppose $\mathcal{M}|^a_v, v \vDash \psi_{\mathbf{B}}$, we then have $E(v) = E(t)$ since $\mathcal{M}|^a_t, t \vDash \psi_{\mathbf{B}}$. Therefore $U_{\mathcal{M}}|^a_t = U_{\mathcal{M}}|^a \cap E(t) = U_{\mathcal{M}}|^a \cap E(v) = U_{\mathcal{M}}|^a_v$, thus $\mathcal{M}|^a_t = \mathcal{M}|^a_v$ and $v \in U_{\mathcal{M}}|^a_t$. Since $\mathcal{M}|^a_t, t \vDash K\phi$, we have $\mathcal{M}|^a_t, v \vDash \phi$, thus $\mathcal{M}|^a_v, v \vDash \phi$.

(1) Right to left: Suppose for any $\mathbf{B} \subseteq \mathbf{A}$, $\mathcal{M}, s \vDash \langle a \rangle(\psi_{\mathbf{B}} \wedge \neg\chi) \to K[a](\psi_{\mathbf{B}} \to \phi)$, we need to show that $\mathcal{M}, s \vDash [a](K\phi \vee \chi)$. Suppose not, then

[7] The equivalences can be proved in $\mathbf{S}_{\mathtt{EAL}^{\mathbf{A}}_{\mathbf{P}}}$ too.

there is a t, such that $s \xrightarrow{a} t$ and $\mathcal{M}|^a_t, t \vDash \neg K\phi \wedge \neg\chi \wedge \psi_{\mathbf{B}}$ for some $\mathbf{B} = e(t)$. Then $\mathcal{M}, s \vDash \langle a\rangle(\psi_{\mathbf{B}} \wedge \neg\chi)$. Since $\mathcal{M}, s \vDash \langle a\rangle(\psi_{\mathbf{B}} \wedge \neg\chi) \rightarrow K[a](\psi_{\mathbf{B}} \rightarrow \phi)$, then $\mathcal{M}, s \vDash K[a](\psi_{\mathbf{B}} \rightarrow \phi)$. Since $\mathcal{M}|^a_t, t \vDash \neg K\phi$, then there is $v \in U_{\mathcal{M}}|^a_t$, such that $\mathcal{M}|^a_t, v \vDash \neg\phi$ (*). Since $v \in U_{\mathcal{M}}|^a_t$, then $v \in E(t)$ and there is a $u \in U_{\mathcal{M}}$ such that $u \xrightarrow{a} v$. Since $\mathcal{M}, s \vDash K[a](\psi_{\mathbf{B}} \rightarrow \phi)$, then $\mathcal{M}, u \vDash [a](\psi_{\mathbf{B}} \rightarrow \phi)$. Since $u \xrightarrow{a} v$, $\mathcal{M}|^a_v, v \vDash \psi_{\mathbf{B}} \rightarrow \phi$. Since $v \in E(t)$ and $\mathbf{B} = e(t)$, we have $\mathcal{M}|^a_v, v \vDash \psi_{\mathbf{B}}$ thus $\mathcal{M}|^a_v, v \vDash \phi$. Again by the fact that $v \in E(t)$ we have $U_{\mathcal{M}}|^a_t = U_{\mathcal{M}}|^a_v$, thus $\mathcal{M}|^a_t = \mathcal{M}|^a_v$. Now it is easy to see that $\mathcal{M}|^a_t, v \vDash \phi$ which is contradictory to (*). Thus there is no such t that $s \xrightarrow{a} t$ and $\mathcal{M}|^a_t, t \vDash \neg K\phi \wedge \neg\chi$, therefore $\mathcal{M}, s \vDash [a](K\phi \vee \chi)$.

(2) Left to right: Assuming $\mathcal{M}, s \vDash [a](\hat{K}\phi \vee \chi)$, we need to prove that for any $\mathbf{B} \subseteq \mathbf{A}$: $\mathcal{M}, s \vDash \langle a\rangle(\psi_{\mathbf{B}} \wedge \neg\chi) \rightarrow \hat{K}\langle a\rangle(\psi_{\mathbf{B}} \wedge \phi)$.

Now suppose $\mathcal{M}, s \vDash \langle a\rangle(\psi_{\mathbf{B}} \wedge \neg\chi)$, then there is a t, such that $s \xrightarrow{a} t$ and $\mathcal{M}|^a_t, t \vDash \psi_{\mathbf{B}} \wedge \neg\chi$. Since $\mathcal{M}, s \vDash [a](\hat{K}\phi \vee \chi)$, then $\mathcal{M}|^a_t, t \vDash \hat{K}\phi \vee \chi$, then $\mathcal{M}|^a_t, t \vDash \psi_{\mathbf{B}} \wedge \hat{K}\phi$. Thus there is a $v \in U_{\mathcal{M}}|^a_t$, such that $\mathcal{M}|^a_t, v \vDash \phi$. Since $v \in U_{\mathcal{M}}|^a_t$, there is a $u \in U_{\mathcal{M}}$ such that $u \xrightarrow{a} v$ and $v \in E(t)$. $v \in E(t)$ implies that $\mathcal{M}|^a_t, v \vDash \psi_{\mathbf{B}}$. Then $\mathcal{M}|^a_t, v \vDash \psi_{\mathbf{B}} \wedge \phi$. Because $v \in E(t)$, then $U_{\mathcal{M}}|^a_t = U_{\mathcal{M}}|^a_v$, thus $\mathcal{M}|^a_t = \mathcal{M}|^a_v$. Therefore $\mathcal{M}|^a_v, v \vDash \psi_{\mathbf{B}} \wedge \phi$, and then we have $\mathcal{M}, u \vDash \langle a\rangle(\psi_{\mathbf{B}} \wedge \phi)$. Since $s, u \in U_{\mathcal{M}}$, $\mathcal{M}, s \vDash \hat{K}\langle a\rangle(\psi_{\mathbf{B}} \wedge \phi)$.

(2) Right to left: Suppose for any $\mathbf{B} \subseteq \mathbf{A}$, $\mathcal{M}, s \vDash \langle a\rangle(\psi_{\mathbf{B}} \wedge \neg\chi) \rightarrow \hat{K}\langle a\rangle(\psi_{\mathbf{B}} \wedge \phi)$ but $\mathcal{M}, s \nvDash [a](\hat{K}\phi \vee \chi)$, then there is a t, such that $s \xrightarrow{a} t$ and $\mathcal{M}|^a_t, t \vDash K\neg\phi \wedge \neg\chi \wedge \psi_{\mathbf{B}}$ for some $\mathbf{B} = e(t) \subseteq \mathbf{A}$. Therefore $\mathcal{M}, s \vDash \langle a\rangle(\psi_{\mathbf{B}} \wedge \neg\chi)$, thus $\mathcal{M}, s \vDash \hat{K}\langle a\rangle(\psi_{\mathbf{B}} \wedge \phi)$ due to the assumption that $\mathcal{M}, s \vDash \langle a\rangle(\psi_{\mathbf{B}} \wedge \neg\chi) \rightarrow \hat{K}\langle a\rangle(\psi_{\mathbf{B}} \wedge \phi)$. Now there is a $u \in U_{\mathcal{M}}$ such that $\mathcal{M}, u \vDash \langle a\rangle(\psi_{\mathbf{B}} \wedge \phi)$. Then there is a v, $u \xrightarrow{a} v$ and $\mathcal{M}|^a_v, v \vDash \psi_{\mathbf{B}} \wedge \phi$. Since $\mathcal{M}|^a_v, v \vDash \psi_{\mathbf{B}}$ and $e(t) = \mathbf{B}$, we have $v \in E(t)$, thus $U_{\mathcal{M}}|^a_t = U_{\mathcal{M}}|^a_v$. Therefore $\mathcal{M}|^a_t = \mathcal{M}|^a_v$, then $\mathcal{M}|^a_t, v \vDash \phi$ which contradicts to $\mathcal{M}|^a_t, t \vDash K\neg\phi$. Thus there is no such t that $s \xrightarrow{a} t$ and $\mathcal{M}|^a_t, t \vDash K\neg\phi \wedge \neg\chi$, therefore $\mathcal{M}, s \vDash [a](\hat{K}\phi \vee \chi)$. □

Now we are ready to prove the main theorem of this subsection.

Theorem 4.8 *For any* $\mathtt{EAL}^{\mathbf{A}}_{\mathbf{P}}$ *formula* ϕ*, there is an* $\mathtt{EALK}^{\mathbf{A}}_{\mathbf{P}}$ *formula* ϕ'*, such that* $\vDash \phi \leftrightarrow \phi'$*.*

Proof Before we start the proof, note that by Proposition 3.6 and Theorem 3.12, the replacements of the equals preserve validity. We will use it repeatedly. We prove the theorem by induction on the structure of ϕ:

The cases for $\phi = p, \neg\phi', \phi_1 \wedge \phi_2$, and $K\phi'$ can be easily proved by IH and the replacement of equals.

$\phi = [a]\psi$, by IH, there is an $\mathtt{EALK}^{\mathbf{A}}_{\mathbf{P}}$-formula ψ', such that $\vDash \psi \leftrightarrow \psi'$. By Proposition 4.6, there is an $\mathtt{EALK}^{\mathbf{A}}_{\mathbf{P}}$-formula χ in K-CNF, such that $\vDash \chi \leftrightarrow \psi'$. Since χ is in K-CNF, then we can assume that $\chi = \alpha_1 \wedge \cdots \wedge \alpha_n$, then $[a]\chi$ is clearly equivalent to $[a]\alpha_1 \wedge \cdots \wedge [a]\alpha_n$. We want to show that for each $1 \leq i \leq n$, $[a]\alpha_i$ is equivalent to an $\mathtt{EALK}^{\mathbf{A}}_{\mathbf{P}}$-formula since then $[a]\alpha_1 \wedge \cdots \wedge [a]\alpha_n$ is also equivalent to an $\mathtt{EALK}^{\mathbf{A}}_{\mathbf{P}}$-formula.

Since χ is an $\mathtt{EALK}^{\mathbf{A}}_{\mathbf{P}}$-formula, each α_i is also an $\mathtt{EALK}^{\mathbf{A}}_{\mathbf{P}}$-formula. By the definition of K-CNF, each α_i is in the shape of $\beta_1 \vee \cdots \vee \beta_m$, then $[a]\alpha_i = [a](\beta_1 \vee \cdots \vee \beta_m)$. It is clear that for any $1 \leq j \leq m$, β_j is an $\mathtt{EALK}_{\mathbf{P}}^{\mathbf{A}}$-formula. By definition of K-CNF, each β_j is in the shape of $p, \neg p, [\cdot]\psi, \neg[\cdot]\psi$, $K\chi$ or $\hat{K}\chi$ where $kd(\chi) = 0$. Note that for $\mathtt{EALK}_{\mathbf{P}}^{\mathbf{A}}$- formulas $\phi : kd(\phi) = 0 \iff \phi$ is K-free. Now since β_j is in $\mathtt{EALK}_{\mathbf{P}}^{\mathbf{A}}$, then it is not hard to see that β_j contains K operator iff $\beta_j = K\chi$ or $\beta_j = \hat{K}\chi$ where χ is K-free. Then we can sort all the β_j into two categories depending on whether it is K-free, and rearrange the disjuncts in α_i as $\beta_{i_1} \vee \cdots \vee \beta_{i_h} \vee \cdots \vee \beta_{i_m}$, such that $kd(\beta_{i_k}) = 1$ for $1 \leq k \leq h$ and $kd(\beta_{i_k}) = 0$ for $h < k \leq m$. We will prove the following claim ($\star$):

For any $h \geq 0$ and any $m > h$ there is an $\mathtt{EALK}_{\mathbf{P}}^{\mathbf{A}}$-formula γ, such that $\models \gamma \leftrightarrow [a](\beta_{i_1} \vee \cdots \vee \beta_{i_h} \vee \cdots \vee \beta_{i_m})$.

We prove it by induction on h. The case of $h = 0$ is trivial since all the β_{i_k} $(1 \leq k \leq m)$ are K-free.

Now suppose the claim holds when $h = n$ (for all $m > h$) we need to prove the case of $h = n + 1$. Let $\chi = \beta_{i_1} \vee \cdots \vee \beta_{i_n} \vee \beta_{i_{n+2}} \vee \cdots \vee \beta_{i_m}$, then $[a](\beta_{i_1} \vee \cdots \vee \beta_{i_m})$ is equivalent to $[a](\chi \vee \beta_{i_{n+1}})$. Since $kd(\beta_{i_{n+1}}) = 1$, thus $\beta_{i_{n+1}}$ contains K then $\beta_{i_{n+1}} = K\chi'$ or $\beta_{i_{n+1}} = \hat{K}\chi'$, where χ' is K-free.

(i) If $\beta_{i_{n+1}} = K\chi'$, then $[a](\chi \vee \beta_{i_{n+1}}) = [a](\chi \vee K\chi')$. By Proposition 4.7 (1), $[a](\chi \vee K\chi')$ is equivalent to

$$\bigwedge_{\mathbf{B} \subseteq \mathbf{A}} (\neg[a](\neg\psi_{\mathbf{B}} \vee \chi) \rightarrow K[a](\neg\psi_{\mathbf{B}} \vee \chi'))$$

Note that given an $\mathbf{B} \subseteq \mathbf{A}$, $\psi_{\mathbf{B}}$ is a K-free $\mathtt{EALK}_{\mathbf{P}}^{\mathbf{A}}$-formula in the shape of $\bigwedge_{a \in \mathbf{B}} \langle a \rangle \top \wedge \bigwedge_{b \notin \mathbf{B}} \neg \langle b \rangle \top$. Therefore $\neg\psi_{\mathbf{B}}$ is still an $\mathtt{EALK}_{\mathbf{P}}^{\mathbf{A}}$-formula which is equivalent to $\bigvee_{a \in \mathbf{B}} [a]\bot \vee \bigvee_{b \notin \mathbf{B}} \neg[b]\bot$. Therefore $\neg\psi_{\mathbf{B}} \vee \chi$ can be massaged into a right disjunctive form $\beta_{i_1} \vee \cdots \vee \beta_{i_n} \vee \cdots \vee \beta_{i_{m+|\mathbf{A}|}}$. Now by IH, there is an $\mathtt{EALK}_{\mathbf{P}}^{\mathbf{A}}$-formula $\gamma_{\mathbf{B}}$, such that $\gamma_{\mathbf{B}}$ is equivalent to $[a](\neg\psi_{\mathbf{B}} \vee \chi)$. Since χ' is K-free, then $K[a](\neg\psi_{\mathbf{B}} \vee \chi')$ is already an $\mathtt{EALK}_{\mathbf{P}}^{\mathbf{A}}$-formula. Now let $\theta_{\mathbf{B}} = \neg\gamma_{\mathbf{B}} \rightarrow K[a](\neg\psi_{\mathbf{B}} \vee \chi')$ we can see that $\theta_{\mathbf{B}}$ is an $\mathtt{EALK}_{\mathbf{P}}^{\mathbf{A}}$-formula equivalent to $\neg[a](\neg\psi_{\mathbf{B}} \vee \chi) \rightarrow K[a](\neg\psi_{\mathbf{B}} \vee \chi')$.

(ii) The case for $\beta_{i_{n+1}} = \hat{K}\chi'$ can be proved similarly by using Proposition 4.7 (2). Hereby we complete the proof for claim ($\star$).

In sum, for each i: $[a]\alpha_i$ is equivalent to an $\mathtt{EALK}_{\mathbf{P}}^{\mathbf{A}}$-formula thus $[a]\psi$ is equivalent to an $\mathtt{EALK}_{\mathbf{P}}^{\mathbf{A}}$-formula therefore completing the proof of the theorem. □

The above proof also suggests a naive algorithm to translate an $\mathtt{EAL}_{\mathbf{P}}^{\mathbf{A}}$-formula into an $\mathtt{EALK}_{\mathbf{P}}^{\mathbf{A}}$-formula, which works in the inside-out fashion:

(i) Find the minimal sub-formulas which are not in $\mathtt{EALK}_{\mathbf{P}}^{\mathbf{A}}$, massage them into K-CNF, and then use the method described in the above proof to translate them into equivalent $\mathtt{EALK}_{\mathbf{P}}^{\mathbf{A}}$-formulas by pushing K out.

(ii) Replacing those sub-formulas in the original formula by their $\mathtt{EALK}_{\mathbf{P}}^{\mathbf{A}}$ correspondents.

(iii) Repeat step (i) until all the subformulas are in the right shapes. Every step pushes the K operator one level out towards the outmost positions thus the procedure terminates eventually.

For example let $\mathbf{A} = \{a\}$, $[a]Kp$ can be translated to a K-CNF $\mathtt{EALK}^{\mathbf{A}}_{\mathbf{P}}$formula:

$$([a][a]\bot \vee K[a](\langle a\rangle\top \to p)) \wedge ([a]\langle a\rangle\top \vee K[a]([a]\bot \to p))$$

Let $\chi_1 = [a][a]\bot$, $\chi_2 = [a]\langle a\rangle\top$ and $\phi_1 = [a](\langle a\rangle\top \to p)$ and $\phi_2 = [a]([a]\bot \to p)$. Thus $[a][a]Kp$ is equivalent to: $[a](\chi_1 \vee K\phi_1) \wedge [a](\chi_2 \vee K\phi_2)$ and then to

$$\bigwedge_{i=1,2} (((\langle a\rangle(\langle a\rangle\top\wedge\neg\chi_i) \to K[a](\langle a\rangle\top \to \phi_i))\wedge(\langle a\rangle(\neg\langle a\rangle\top\wedge\neg\chi_i) \to K[a](\neg\langle a\rangle\top \to \phi_i)))$$

Clearly, the translated formula is at least exponentially longer. We leave the discussion on the succinctness of $\mathtt{EAL}^{\mathbf{A}}_{\mathbf{P}}$ compared to $\mathtt{EALK}^{\mathbf{A}}_{\mathbf{P}}$ for future work.

Remark 4.9 Is there a simpler translation? In many DEL-style logics, we can often define a simple recursive translation from the full language to its fragment by swapping the connectives and modalities, e.g., in public announcement logic, $[\psi]\neg\phi \iff \psi \to \neg[\psi]\phi$ (cf. e.g., [20]). However, such idea may not work here: it seems there is no general equivalence-preserving rule to swap $\langle a\rangle$ and $\neg$.

4.3 Finite model property

Theorem 4.8 also suggests that $\mathtt{EAL}^{\mathbf{A}}_{\mathbf{P}}$ has the *finite model property*: First of all, it is not hard to see that for any $\mathtt{EALK}^{\mathbf{A}}_{\mathbf{P}}$ formula ϕ, ϕ has a UM model iff ϕ has an EM model (w.r.t. $\Vdash$); Secondly, $\mathtt{EALK}^{\mathbf{A}}_{\mathbf{P}}$ on EM model has the finite model property (an easy exercise for normal modal logic); Thirdly any pointed EM model of an $\mathtt{EALK}^{\mathbf{A}}_{\mathbf{P}}$-formula can be viewed as an $\mathtt{EALK}^{\mathbf{A}}_{\mathbf{P}}$-equivalent UM model by ignoring the equivalence classes that do not contain the designated point.

In the rest of this section, we directly prove the finite model property of $\mathtt{EAL}^{\mathbf{A}}_{\mathbf{P}}$ on UM models by using finite approximations of U-bisimilarity.

Definition 4.10 *The modal degree of $\mathtt{EAL}^{\mathbf{A}}_{\mathbf{P}}$-formulas ($md(\phi)$) is defined as follows:*

$$md(\top) = 0 \qquad md(p) = 0$$
$$md(\neg\phi) = md(\phi) \qquad md(\phi \wedge \psi) = max\{md(\phi), md(\psi)\}$$
$$md(\langle a\rangle\phi) = 1 + md(\phi) \qquad md(K\phi) = md(\phi)$$

Note that here K does not count for modal degrees. Let $\mathtt{EAL}^{\mathbf{A}}_{\mathbf{P}\,n} = \{\phi \mid \phi \in \mathtt{EAL}^{\mathbf{A}}_{\mathbf{P}}$, and $md(\phi) \leq n\}$

Now we define the finite approximation of $\rightleftarrows$ based on n-bisimilarity w.r.t. **P** and **A** (cf. [3]).

Definition 4.11 *EM model $\mathcal{M}$ and $\mathcal{N}$ are n-U-bisimilar ($\mathcal{M} \rightleftarrows_n \mathcal{N}$) iff for any $u \in U_{\mathcal{M}}$, there is a $u' \in U_{\mathcal{N}}$ such that $\mathcal{M}^{ML}, u \underline{\leftrightarrow}_n \mathcal{N}^{ML}, u'$ and for any $u' \in U_{\mathcal{N}}$, there is a $u \in U_{\mathcal{M}}$ such that $\mathcal{M}^{ML}, u \underline{\leftrightarrow}_n \mathcal{N}^{ML}, u'$. For pointed models, $\mathcal{M}, s \rightleftarrows_n \mathcal{N}, u$ iff $\mathcal{M}^{ML}, s \underline{\leftrightarrow}_n \mathcal{N}^{ML}, u$ and $\mathcal{M} \rightleftarrows_n \mathcal{N}$.*

Proposition 4.12 *For $n > 0$: if $\mathcal{M}, s \rightleftarrows_{n+1} \mathcal{N}, u$, $s \xrightarrow{a} t$, $u \xrightarrow{a} v$, and $\mathcal{M}^{\mathtt{ML}}, t \leftrightarroweq_n \mathcal{N}^{\mathtt{ML}}, v$, then $\mathcal{M}|_t^a, t \rightleftarrows_n \mathcal{N}|_v^a, v$.*

Proof Since $(\mathcal{M}|_t^a)^{\mathtt{ML}} = \mathcal{M}^{\mathtt{ML}}$ and $(\mathcal{N}|_v^a)^{\mathtt{ML}} = \mathcal{N}^{\mathtt{ML}}$, $(\mathcal{M}|_t^a)^{\mathtt{ML}}, t \leftrightarroweq_n (\mathcal{N}|_v^a)^{\mathtt{ML}}, v$, thus we only need to prove $\mathcal{M}|_t^a \rightleftarrows_n \mathcal{N}|_v^a$.

For any $t' \in U_{\mathcal{M}}|_t^a = U_{\mathcal{M}}^a \cap E(t)$, there is an $s' \in U_{\mathcal{M}}$, such that $s' \xrightarrow{a} t'$. Since $\mathcal{M}, s \rightleftarrows_{n+1} \mathcal{N}, u$, then there is a $u' \in U_{\mathcal{N}}$, such that $\mathcal{M}^{\mathtt{ML}}, s' \leftrightarroweq_{n+1} \mathcal{N}^{\mathtt{ML}}, u'$. Therefore there is a v', such that $u' \xrightarrow{a} v'$ and $\mathcal{M}^{\mathtt{ML}}, t' \leftrightarroweq_n \mathcal{N}^{\mathtt{ML}}, v'$. Therefore $v' \in U_{\mathcal{N}}|^a$ and $e(v') = e(t')$ (due to the fact that $n > 0$ and $\mathcal{M}^{\mathtt{ML}}, t' \leftrightarroweq_n \mathcal{N}^{\mathtt{ML}}, v'$). Since $e(t') = e(t)$ and $e(t) = e(v)$, we have $e(v') = e(v)$, thus $v' \in E(v)$, therefore $v' \in U_{\mathcal{N}}|_v^a$. Now we have proved that for any $t' \in U_{\mathcal{M}}|_t^a$, there is a $v' \in U_{\mathcal{N}}|_v^a$, such that $\mathcal{M}^{\mathtt{ML}}, t' \leftrightarroweq_n \mathcal{N}^{\mathtt{ML}}, v'$, namely $(\mathcal{M}|_t^a)^{\mathtt{ML}}, t \leftrightarroweq_n (\mathcal{N}|_v^a)^{\mathtt{ML}}, v$. The other direction is totally symmetric. □

Proposition 4.13 $\mathcal{M}, s \rightleftarrows_{n+1} \mathcal{N}, u \implies \mathcal{M}, s \equiv_{EAL^A_{P\,n}} \mathcal{N}, u.$

Proof The proof is based on induction on n: For $n = 0$, we can easily check that all the $\mathtt{EAL}^{\mathbf{A}}_{\mathbf{P}\,0}$-formulas are preserved under $\rightleftarrows_1$. Now suppose $\mathcal{M}, s \rightleftarrows_{k+1} \mathcal{N}, u$ implies $\mathcal{M}, s \equiv_k \mathcal{N}, u$. We need to show $\mathcal{M}, s \rightleftarrows_{k+1+1} \mathcal{N}, u$ implies for all $\mathtt{EAL}^{\mathbf{A}}_{\mathbf{P}\,k+1}$ formula ϕ: $\mathcal{M}, s \vDash \phi \iff \mathcal{N}, u \vDash \phi$. Now suppose that $\mathcal{M}, s \rightleftarrows_{k+1+1} \mathcal{N}, u$, we proceed by induction on the structure ϕ:

For Boolean cases, it is obvious.

$\phi = [a]\psi$: If $\mathcal{N}, u \vDash [a]\psi$ but $\mathcal{M}, s \nvDash [a]\psi$, then there is a t, such that $s \xrightarrow{a} t$, and $\mathcal{M}|_t^a, t \nvDash \psi$. Since $\mathcal{M}, s \leftrightarroweq_{k+2} \mathcal{N}, u$, then there is a v, $u \xrightarrow{a} v$, and $\mathcal{M}^{\mathtt{ML}}, t \leftrightarroweq_{k+1} \mathcal{N}^{\mathtt{ML}}, v$. By Proposition 4.12, $\mathcal{M}|_t^a, t \rightleftarrows_{k+1} \mathcal{N}|_v^a, v$. Since $md(\psi) \leq k$, and by IH for k, $\mathcal{N}|_v^a, v \nvDash \psi$, this is contradictory to the fact that $\mathcal{N}, u \vDash [a]\psi$. The other direction is symmetric.

$\phi = K\psi$, if $\mathcal{N}, u \vDash K\psi$ but $\mathcal{M}, s \nvDash K\psi$ then there is a $s' \in U_{\mathcal{M}}$, $\mathcal{M}, s' \nvDash \psi$. Since $\mathcal{M}, s \rightleftarrows_{k+2} \mathcal{N}, u$ then there is a $u' \in U_{\mathcal{N}}$, $\mathcal{M}^{\mathtt{ML}}, s' \leftrightarroweq_{k+2} \mathcal{N}^{\mathtt{ML}}, u'$ then $\mathcal{M}, s' \rightleftarrows_{k+2} \mathcal{N}, u'$. By IH for simpler ϕ, $\mathcal{N}, u' \nvDash \psi$. Then $\mathcal{N}, u \nvDash K\psi$, contradiction. The other direction is symmetric. □

The reader may wonder about the mismatch between n and $n + 1$ in the above proposition. Actually it is not surprising since in the semantics we actually look one step forward to observe the available actions. The translation of the previous subsection also showed that the $\mathtt{EALK}^{\mathbf{A}}_{\mathbf{P}}$ equivalent translation may have larger modality depth than the original formula (check the example of $[a]Kp$ in the previous subsection).[8]

Theorem 4.14 *For each EAL^A_P-formula ϕ, if it has a UM model then it has a finite tree-like model with the depth of at most $n + 1$ where $n = md(\phi)$.*

Proof (Sketch:) Without loss of generality, we assume **P** and **A** are finite (since a formula is about at most finitely many symbols). Let $n = md(\phi)$. Suppose ϕ has a UM model $\mathcal{M}$, we first "contract" the set $U_{\mathcal{M}}$ according to

[8] Actually, a closer analysis would reveal that this can only happen when $md(\phi) = 1$. Proposition 4.13 can be strengthened to : for $n > 1$: $\mathcal{M}, s \rightleftarrows_n \mathcal{N}, u \iff \mathcal{M}, s \equiv_{\mathtt{EAL}^{\mathbf{A}}_{\mathbf{P}\,n}} \mathcal{N}, u$.

$\leftrightarrows_{n+1}$ (equivalently, according to $\equiv_{\mathtt{ML}_{n+1}}$, where $\mathtt{ML}_{n+1}$ is the K-free fragment of $\mathtt{EAL}^{\mathbf{A}}_{\mathbf{P}\,n+1}$). Note that since $\mathtt{ML}_{n+1}$ is essentially a finite language modulo logical equivalence, after the contraction there are only finitely many representatives which form a finite set U'. Then we finitely (up to $n+1$) unravel the pointed multimodal models based on these states and prune the branches to make a finite model. Finally we make a disjoint union of these unravellings with the uncertainty set U'. □

To squeeze the model even further we also develop a highly non-trivial filtration technique which we left for the full version of this paper. Based on such a finite model property and the fact that there is a finite complete axiomatization, we can conclude that $\mathtt{EAL}^{\mathbf{A}}_{\mathbf{P}}$ is decidable on UM models.

5 Comparisons

We claimed in the introduction that our $\mathtt{EAL}^{\mathbf{A}}_{\mathbf{P}}$ framework is a blend of ETL and DEL frameworks. In this section, we make it more precise by comparing $\mathtt{EAL}^{\mathbf{A}}_{\mathbf{P}}$ to ETL and DEL. The conclusions can be summarized as follows:

- Our UM models can be viewed as compact representations of ETL structures where the epistemic relations are computed based on the following: 1. the previous epistemic uncertainties, 2. the executed actions, and 3. the new observations, which is, in spirit, similar to the DEL-like epistemic updates.
- On the other hand, the DEL approach via product updates on epistemic models with protocols can also be viewed as a logic on particular ETL structures, however, satisfying a property which is violated in the navigation scenarios. Due to this and other difficulties, the standard DEL with event models is not suitable to handle the reasoning in the navigation scenarios.

To facilitate the comparisons, let us fix some notations first. Given a UM model $\mathcal{M}$ for $\mathtt{EAL}^{\mathbf{A}}_{\mathbf{P}}$, we say $\rho = s_0a_1s_1a_2s_2\cdots a_ns_n$ is a *path* in $\mathcal{M}$ if $s_0 \in U_{\mathcal{M}}$, $n \geq 0$ and for any $0 \leq i \leq n-1$: $s_i \overset{a_{i+1}}{\to} s_{i+1}$ in $\mathcal{M}$. Given a path $\rho = s_0a_1s_1a_2s_2\cdots a_ns_n$ let $len(\rho) = n$ be the *length* of ρ. Note that there are paths of length 0 consisting of a single state.

5.1 Comparison with ETL

Technically speaking, a single-agent ETL model is just a tree-like EM model. We can unravel a UM model into such an ETL model.

Definition 5.1 *Given a UM model $\mathcal{M} = \langle S, \{R_a \mid a \in \boldsymbol{A}\}, U, V\rangle$, we define $\mathcal{M}^{ETL}$ as $\langle S^\bullet, \{R^\bullet_a \mid a \in \boldsymbol{A}\}, \sim, V^\bullet\rangle$ where:*

(i) *$S^\bullet = \{\rho \mid \rho$ is a path in $\mathcal{M}$ starting with some $s \in U\}$*

(ii) *$\rho \overset{a}{\to} \rho'$ in $\mathcal{M}^{ETL}$ iff $\rho' = \rho a t$ for some $t \in S$ and $a \in \boldsymbol{A}$.*

(iii) *For any two paths $\rho = s_0a_1\cdots a_ns_n$ and $\rho' = t_0b_1\cdots b_mt_m$ in $S^\bullet$: $\rho \sim \rho'$ in $\mathcal{M}^\bullet$ iff ($n = m$, for all $i \leq n$: $a_i = b_i$ and $e(s_i) = e(t_i)$).*

(iv) *$V^\bullet(s_0a_1\cdots a_ns_n) = V(s_n)$*

It is easy to show that $\sim$ is indeed an equivalence relation. Moreover we can define $\sim$ more explicitly:

Proposition 5.2 *$\sim \subseteq S^\bullet \times S^\bullet$ is the minimal set of pairs satisfying the following conditions:*

(i) *$s \sim t$ for any $s, t \in U$*

(ii) *$\rho a s \sim \rho' a' s'$ if $\rho \sim \rho'$, $a = a'$ and $e(s) = e(s')$.*

Let us unravel the initial model in Example 2.1.

Example 5.3 Given the initial model as $\mathcal{M}$, $\mathcal{M}^{\mathrm{ETL}}$ can be depicted as follows (where dotted lines denote the $\sim$ relation while omitting the reflexive arrows):

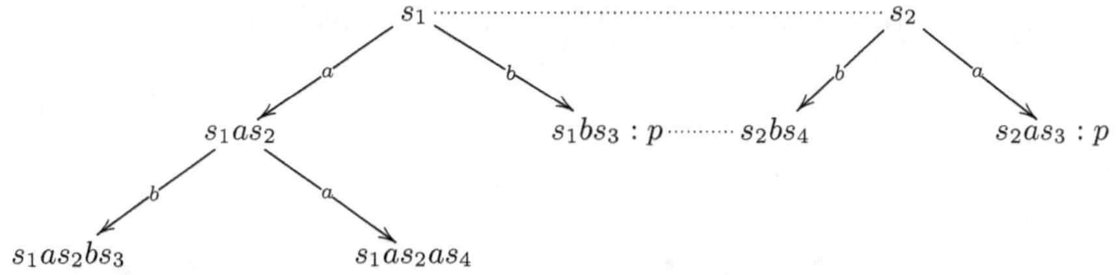

The following result is crucial to prove that $\mathcal{M}^{\mathrm{ETL}}$ is indeed a good transformation preserving the truth values of $\mathtt{EAL}^{\mathbf{A}}_{\mathbf{P}}$ formulas.

Proposition 5.4 *Let $\mathcal{M} = \langle S, \{R_a \mid a \in \boldsymbol{A}\}, U, V\rangle$ and $s \in U$. If there exists an $s' \in S$ such that $s \xrightarrow{a} s'$, then $\mathcal{M}^\bullet, sas' \leftrightarroweq (\mathcal{M}|^a_{s'})^\bullet, s'$ (here the bisimilarity is w.r.t. $\boldsymbol{P}, \boldsymbol{A}$ and also $\sim$).*

Proof We define a binary relation Z on $S_{\mathcal{M}^\bullet} \times S_{(\mathcal{M}|^a_{s'})^\bullet}$ as follows $Z = \{(\rho, \rho') \mid \rho \in S_{\mathcal{M}^\bullet}, \rho' \in S_{(\mathcal{M}|^a_{s'})^\bullet}$ and there exists an $u \in U_{\mathcal{M}}$ such that $\rho = ua\rho'\}$.

Clearly $sas'Zs'$, thus Z is non-empty. Now we prove that Z is a bisimulation. The propositional invariance condition and the back-and-forth conditions for $\xrightarrow{a}$ are obvious. We only need to check the back-and-forth conditions for $\sim$. In the sequel, suppose $\rho = ua\rho'$ for some path ρ' in $S_{(\mathcal{M}|^a_{s'})^\bullet}$, then it is clear that ρ' starts with some state $t \in S_{\mathcal{M}}$ such that $e(t) = e(s')$.

Now suppose $\rho \sim \xi$ then according to the definition of $\sim$, ξ must be in the shape of $va\xi'$ where $v \in U_{\mathcal{M}}$ and ξ' must start with a state $t' \in S_{\mathcal{M}}$ such that $e(t') = e(t) = e(s')$. Therefore $t' \in U_{\mathcal{M}|^a_{s'}}$, then it is not hard to see that $\rho' \sim \xi'$ in $\mathcal{M}|^a_{s'}$. For the other direction, suppose $\rho' \sim \xi'$ in $\mathcal{M}|^a_{s'}$ then it is easy to see that there is a $v \in U_{\mathcal{M}}$ such that $ua\rho' \sim va\xi'$ by definition of $\sim$ in $\mathcal{M}$. □

Recall that $\Vdash$ denotes the satisfaction relation of the auxiliary semantics of $\mathtt{EAL}^{\mathbf{A}}_{\mathbf{P}}$ language on EM models used in Section 3. The following preservation result can be proved based on Proposition 5.4 without much efforts.

Theorem 5.5 *For any $\mathtt{EAL}^{\mathbf{A}}_{\mathbf{P}}$ formula ϕ: $\mathcal{M}, s \vDash \phi \iff \mathcal{M}^\bullet, s \Vdash \phi$.*

Theorem 5.5 established the equivalence of our framework and ETL framework (restricted to the models with epistemic relations computed in a particular way). However, it is not reasonable to work with ETL models explicitly since the unravelling transforms a finite map into an infinite forest if there are loops.

5.2 Comparison with DEL

As we mentioned in the introduction, there are efforts trying to merge ETL and DEL frameworks. Most notably, [21] characterizes the DEL-generated ETL models (under protocols) by a few properties. In [6] the authors argue that *synchronicity* is not an inherent feature of DEL but is introduced by the specific translation used in [21]. However, it is usually agreed that the property of *(local) no miracles (LNM)* is inevitable for a DEL-generated ETL model. [9] Formally, LNM says that in the DEL-generated ETL model, for any states $s, s', t, t', w, w', v, v'$ and any action labels c, d:

$$((s \xrightarrow{c} s', t \xrightarrow{d} t', s' \sim t') \text{ and } (s \sim w \sim v, w \xrightarrow{c} w', v \xrightarrow{d} v')) \text{ implies } w' \sim v'.$$

In picture (whether s and t are indistinguishable is unknown):

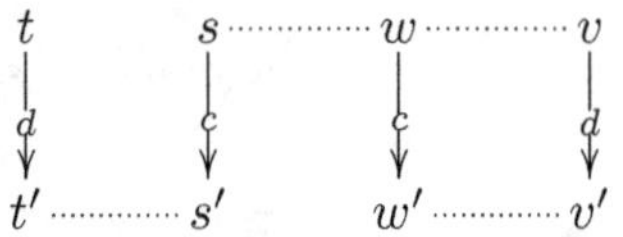

In words, LNM roughly says that whenever the results of executing two actions are indistinguishable, then executing these two actions on indistinguishable states must result in indistinguishable states again. Actually, this property is inherent in the definition of the updated epistemic relations according to the standard product update [1], where two updated states are indistinguishable after executing two actions on two old states respectively iff the two old states are indistinguishable and the two actions are indistinguishable too.

However, in our navigation scenarios, it is often the case that executing the same action (thus indistinguishable from itself) on previously indistinguishable states will result in distinguishable states, e.g., in Example 5.3, performing action a on indistinguishable states s_1 and s_2 will result in distinguishable states. To see that it is an example violating LNM, let $s = t = s_2, s' = t' = s_2as_3, c = d = a, w = s_1, v = s_2$ and $w' = s_1as_2, v' = s_2as_3$ and check the definition of LNM [10].

The feature of our epistemic update mechanism is reflected in Proposition 5.2: the three conditions in the inductive case actually say (1) the old uncertainties on states are respected; (2) we do not have uncertainties about actions with different names; (3) the observations at the new states may affect the new uncertainties. It is the feature (3) which makes us deviate from LNM.

Merely technically speaking, we may still try to mimic the $\mathtt{EAL}_{\mathbf{P}}^{\mathbf{A}}$ framework by the standard DEL via event models. The difficulties and the potential solutions are summarized below, interested readers may consult the algebraic DEL-approaches in [17,18,10].

[9] Although we take the notion of LNM as in [21], our counterexample also works for the other no miracle notions in [6].

[10] Actually LNM also prevents the application of standard DEL in security verification as observed in [5].

- Failure of LNM: try to split one action into different actions w.r.t. different conditions, e.g., in Example 5.3 the two a moves must be treated differently in the event model.
- The standard DEL-model does not contain procedural information: use state-dependent protocols to encode the moves in the map (cf. e.g., [21]).
- No location changes: try to capture the changes of current location by factual changing actions (cf. e.g., [23,18]).
- DEL-updates are functional: to model non-deterministic actions, multi-pointed event models should be used.

6 Conclusion and future work

In this paper, we lay out a logical framework for dynamic epistemic reasoning in navigation. The "philosophy" behind our work is summarized as follows:

- Keep the logical language and its model as simple and natural as possible, while put the complexity on the semantics.
- Combine the spirits from ETL and DEL by having the temporal possibilities encoded in the model while computing the epistemic developments step by step according to the update semantics.
- When proving theoretical results, try to reduce the dynamics of the models into static relations in a larger model.

We think this is just the opening of an interesting story. A few future directions are mentioned as follows. For the current framework, we have not discussed the computational issues such as the complexity of satisfiability and model checking problems and the succinctness compared to $\mathtt{EALK}^{\mathbf{A}}_{\mathbf{P}}$. To generalize the current framework, we may consider general observations instead of observations of the currently available actions only, for which we may use propositional variables to encode available observations as in [22]. Similar techniques for axiomatization should work in the more general case. As in [18], the converse action operator may be introduced to express "I know where I was", although we conjecture that it can be eliminated qua expressive power. Finally it is natural to ask whether our framework can be extended to multi-agent.

As we mentioned in the introduction, we are aiming at real-life applications of navigation and planning, for which an epistemic Propositional Dynamic Logic (EPDL) language is more attractive due to its program language. We can then reduce the planning problem into model checking problem of EPDL-formulas expressing sentences like "there is a plan that can make sure he knows ϕ". This extension may require new techniques and we leave it for future work.

Acknowledgement This work is supported by SSFC grant 11CZX054. The first author would like to thank Dr. Mehrnoosh Sadrzadeh for pointing out the problem of robot navigation during her visit to Amsterdam in 2010. The authors are also grateful to the anonymous referees of AiML2012 for their insightful comments on the early version of this paper.

References

[1] Baltag, A. and L. Moss, *Logics for epistemic programs*, Synthese **139** (2004), pp. 165–224.

[2] Baltag, A., L. Moss and S. Solecki, *The logic of public announcements, common knowledge, and private suspicions*, in: *Proceedings of TARK'98* (1998), pp. 43–56.

[3] Blackburn, P., M. de Rijke and Y. Venema, "Modal Logic," Cambridge University Press, 2002.

[4] Bonet, B. and H. Geffner, *Planning with incomplete information as heuristic search in belief space*, in: *Artificial Intelligence Planning Systems*, 2000, pp. 52–61.

[5] Dechesne, F. and Y. Wang, *To know or not to know: Epistemic approaches to security protocol verification*, Synthese **177** (2010), pp. 51–76.

[6] Dégremont, C., B. Löwe and A. Witzel, *The synchronicity of dynamic epistemic logic*, in: *Proceedings of TARK'11*, 2011, pp. 145–152.

[7] Fagin, R., J. Halpern, Y. Moses and M. Vardi, "Reasoning about knowledge," MIT Press, Cambridge, MA, USA, 1995.

[8] Goldman, R. and M. Boddy, *Expressive planning and explicit knowledge*, in: *Proceedings of AIPS'96)*, 1996, pp. 110–117.

[9] Halpern, J., R. van der Meyden and M. Vardi, *Complete axiomatizations for reasoning about knowledge and time*, SIAM Journal on Computing **33** (2004), pp. 674–703.

[10] Horn, A., *Dynamic epistemic algebra with post-conditions to reason about robot navigation*, in: L. Beklemishev and R. de Queiroz, editors, *Proceedings of WoLLIC'11*, LNCS **6642** (2011), pp. 161–175.

[11] Hoshi, T., "Epistemic Dynamics and Protocol Information." Ph.D. thesis, Stanford (2009).

[12] Hughes, G. E. and M. J. Cresswell, "A New Introduction to Modal Logic," Routledge, 1996.

[13] Levesque, H. J., R. Reiter, Y. Lespérance, F. Lin and R. B. Scherl, *GOLOG: A logic programming language for dynamic domains*, Journal of Logic Programming **31** (1997), pp. 1–3.

[14] Meyer, J.-J. C., W. van der Hoek and B. van Linder, *A logical approach to the dynamics of commitments*, Artifitial Intellegence **113** (1999), pp. 1–40.

[15] Michie, D., *Machine intelligence at edinburgh*, in: *On Machine Intelligence*, 1974 pp. 143–155.

[16] Moore, R., *A formal theory of knowledge and action*, in: J. Hobbs and R. Moore, editors, *Formal Theories of the Commonsense World* (1985), pp. 319–358.

[17] Panangaden, P., C. Phillips, D. Precup and M. Sadrzadeh, *An algebraic approach to dynamic epistemic logic*, in: *Proceeedings of Description Logics'10*, 2010.

[18] Panangaden, P. and M. Sadrzadeh, *Learning in a changing world, an algebraic modal logical approach*, in: *AMAST'10*, 2010, pp. 128–141.

[19] Parikh, R. and R. Ramanujam, *Distributed processes and the logic of knowledge*, in: *Proceedings of Conference on Logic of Programs* (1985), pp. 256–268.

[20] Plaza, J. A., *Logics of public communications*, in: M. L. Emrich, M. S. Pfeifer, M. Hadzikadic and Z. W. Ras, editors, *Proceedings of the 4th International Symposium on Methodologies for Intelligent Systems*, 1989, pp. 201–216.

[21] van Benthem, J., J. Gerbrandy, T. Hoshi and E. Pacuit, *Merging frameworks for interaction*, Journal of Philosophical Logic **38** (2009), pp. 491–526.

[22] van der Hoek, W., N. Troquard and M. Wooldridge, *Knowledge and control*, in: S. Tumer, Yolum and Stone, editors, *Proceedings of AAMAS'11*, 2011, pp. 719–726.

[23] van Ditmarsch, H., W. van der Hoek and B. Kooi, *Dynamic epistemic logic with assignment*, in: *Proceedings of AAMAS'05*, 2005, pp. 141–148.

[24] van Ditmarsch, H., W. van der Hoek and B. Kooi, "Dynamic Epistemic Logic," (Synthese Library), Springer, 2007, 1st edition.

[25] Wang, Y. and Q. Cao, *On axiomatizations of public announcement logic* (2012), manuscript under submission.

www.ingramcontent.com/pod-product-compliance
Lightning Source LLC
Chambersburg PA
CBHW060218230426
43664CB00011B/1474